AF574507

Exploration Methods for Sandstone Reservoirs

Exploration Methods for Sandstone Reservoirs

D. A. Busch and D. A. Link

OGCI Publications
Oil & Gas Consultants International Inc.
Tulsa

4554 South Harvard Avenue
Tulsa, Oklahoma 74135

Printed in the United States of America

Library of Congress Catalog Card Number: 84-062623
International Standard Book Number: 0-930972-07-4

CONTENTS

PREFACE

The purpose of this book is to present an integration of information from academia and industry that can be used in exploring for sandstone reservoirs in the subsurface. This book will be of interest to both advanced undergraduates and graduates in universities and to practicing earth scientists in petroleum geology, geophysics, and engineering.

In this book, we attempt to identify those critical aspects of sedimentation and stratigraphy that readily lend themselves to studies of subsurface sandstone reservoirs. To understand sandstones in the subsurface, a backgrond knowledge of modern sandstone depositional environments is essential. It is for this reason that selected modern analogues are discussed and illustrated for each of the major environments of clastic deposition. In addition, geometric aspects (shape, size, and orientation) of subsurface sandstone reservoirs are stressed over sedimentary textures and structures because the latter are difficult to determine using geophysical logs,—and cores are frequently not available.

Many petroleum explorationists search for a single diagnostic criterion to identify subsurface sandstone bodies, not realizing that multiple criteria are available and must be employed in their studies. We attempt to present these criteria along with concepts and techniques that will help in identifying and tracing sandstone reservoirs in the subsurface.

Many of the illustrations have been borrowed from numerous authors of technical papers and books, and their release for republication is gratefully acknowledged. Wherever feasible, the authors have incorporated their own ideas, based on their individual experience in the area of petroleum exploration. For these ideas, they assume sole responsibility for their validity.

The authors express their sincere gratitude to Tricia Duyfhuizen, who helped initiate the project, and Jeanie Allen, who supervised and implemented the myriad details relative to organization and editing of the manuscript. Verma Hughes' efforts in obtaining releases from numerous authors and unstinting efforts in preparation of the manuscript for the type setter also are gratefully acknowledged. Our thanks also are extended to Carol Gerald, Kathy Hart, Fran Kelsey, Linda Dillard, and Brenda Davis for typing the several drafts of the manuscript.

1 SEDIMENTARY ENVIRONMENTS

INTRODUCTION

The subject matter of this book is built around sedimentary depositional environments. By this means, the authors hope to demonstrate the importance they attach to this subject as a basis for identifying and tracing sandstone reservoirs in the subsurface. Most of our knowledge of sedimentary environments has been borrowed from the students of Holocene sediments, the majority of whom manifest little interest in the problems of identifying their ancient counterparts in the subsurface. In studies of Recent sediments, direct observations may be made of a wide assortment of variables, e.g., geomorphology, agents of weathering, erosion and deposition, subaqueous depth, temperature and salinity variations, current directions, fauna and flora, etc. A change in any one of these variables may be capable of producing changes in the others. Subaerial (continental) environments may be sites of either erosion or deposition, whereas subaqueous environments generally are areas of deposition. Combinations of the two are not unusual. No two similar environments are ever identical, and transitional boundaries may occur between one distinct environment and a laterally adjacent different environment.

A summary of the principal sedimentary environments of clastic sedimentation follows that serves as a basic outline for later chapters. The outline begins with the sediments deposited nearest the prime source area, progresses figuratively in a downstream direction to the coastline and, finally, into the deepest portion of a basin.

- Continental (subaerial or nonmarine)
    - Alluvial fan
    - Braided stream
    - Youthful valley
    - Mature valley (meander belt)
    - Eolian (occur in both continental and transitional environments)
    - Lacustrine
- Transitional
    - Deltaic
        - River dominated
        - Wave dominated
        - Tide dominated
- Coastal-interdeltaic
    - Chenier and chenier plain
    - Barrier bar (with tidal inlets, flood-tidal and ebb-tidal deltas, and back-barrier lagoon
- Marine
    - Transgressive and regressive marine
    - Submarine canyon
    - Submarine fan
    - Basin floor
    - Basin canyon

In treating the sandstone reservoirs deposited within the environments listed in this outline, model studies are presented that include more detailed information than generally is applicable to any one subsurface study. This is because the tools available to the explorationist generally are quite different from those of the student of Recent sediments. The principal tools of the explorationist consist of an assortment of geophysical logs, rotary (or cable tool) samples, and occasional cores. To use these tools, however, the explorationist must have an extensive background knowledge of reservoir geometry, lithology, shape, texture, sedimentary structure, paleocurrent patterns and, in some cases, paleontology.

GEOMETRY

The authors stress geometric aspects because these parameters generally can be ascertained from detailed correlations of geophysical logs, which involve length, width, thickness, cross-sectional shape, orientation, and relationship to enclosing strata. One aspect of reservoir geometry that is all too frequently neglected is fitting it into its proper paleogeomorphological setting. To do this the explorationist should initially reconstruct a major segment of the depobasin configuration. This is accomplished by constructing an isopach map of the genetic sequence of which the reservoir sandstone is a component part. The significance and method for doing this are discussed in the next chapter. This is particularly important when it comes to mapping a "shoestring" reservoir sandstone. All such sand bodies are elongate and lenticular, but similarities stop there: such a sand body could be a channel sand, delta-distributary sand, a chenier, a barrier bar, an offshore bar, a tidal-current ridge, a sand wave, etc.

Recent developments in reflection seismology enable identification of delta foresets, geographic extent of deep-water submarine fans, and sometimes even individual channel sandstones and barrier bars. Such seismic techniques are discussed and illustrated by Lyons and Dobrin (1972), Harms and Tackenberg (1972), and Sheriff (1976).

LITHOLOGY

The lithology of a reservoir can be easily determined by examination of either drill cuttings or cores. Such studies, however, are seldom diagnostic of the depositional environment. The lithology may reflect the types of rock exposed at the source area, mode of transportation, or the environment of accumulation—or all three. Thus, lithology alone is of limited value.

Clay minerals, as indicators of depositional environment, have been extensively studied, but they too are related to such variables as the composition of the rocks in the provenance area, the climate that weathered it, and diagenetic history. Weaver (1958) concluded on the basis of hundreds of analyses of clay minerals in ancient strata that no one clay mineral is diagnostic of any one depositional environment, not even marine versus nonmarine. There is an indication, however, that illite and montmorillonite are more characteristic of marine strata than nonmarine. Kaolinite, on the other hand, occurs more commonly in nonmarine strata, especially in fluviatile deposits.

The mineral glauconite is a complex alumimum silicate containing magnesium, iron, and potassium and occurs principally as an internal mold of "forams" and as a replacement of faecal material (pellets). This mineral, however, is unstable and is thought to form during the early stages of diagenesis of marine sediments. Prior to lithification, glauconite can be reworked and concentrated in shallow-water sands and even transported by currents into deeper sites of marine sand accumulation. As it is a chemically unstable mineral, it is extremely susceptible to destruction upon exposure to weathering processes. For this reason it has been rarely observed in association with nonmarine reservoir sands. Selley (1978) states, "the presence of glauconite in a sand is a useful criterion for a marine origin. The absence of glauconite is, of course, inconclusive and does not indicate a non-marine environment. . . ." It can be identified from cuttings in sandstones where there are no cores and that are paleontologically barren.

TEXTURE

Grain size, generally, is an indicator of the energy level of a depositional environment. We are reminded by Selley that:

> The coarser the grain size the higher the energy level of the depositing current and the better the sorting the more prolonged its action. These generalizations have long been applied to terrigenous sediments. . . . However, even these generalizations of the correlation of grain size and sorting to energy level must be interpreted with caution. No matter how strong a current it cannot deposit sediment coarser than that available in the source.

The literature abounds with statistical textural studies of both modern and ancient sandstones, but these have proven to have limited value in the matter of indexing environments of clastic sedimentary accumulations. Grain size studies of such parameters as sorting, skewness, kurtosis, etc., clearly reveal too much overlap to be diagnostic of a single environment of deposition, and such studies are now on the decline. This type of investigation is what the authors refer to as "sedimentological overkill" if it is conducted solely for the purpose of identifying depositional environments. Failure of statistical studies of modern sands of known depositional environment to serve as analogues for their ancient counterparts is due primarily to three factors. First, texture is related to not only the environment of deposition but also to its previous history. For example, if only a well-sorted sand is weathered and eroded in a source area then the recycled sediment (upon redeposition) will have the same texture regardless of environment of deposition. Second, interstitial clays may have been deposited either simultaneously (syngenetic) with the reservoir sand, or they may have been washed in at a later date (epigenetic). They could be even the result of diagenesis of chemically unstable sand grains. The third factor has to do with postdepositional solution and individual grain overgrowths. Thus, the overall texture of a sandstone might be so seriously altered as to render disaggregation impossible. In such instances thin section measurements of grain size have been employed, but direct comparisons with the results of sieve analyses of Recent sediments (of known depositonal environments) are less than satisfactory.

SHAPE

The shapes of sand grains also have been studied in considerable detail. Selley (1978) points out that:

> recent glacial sands tend to be more angular and of lower sphericity than water-borne grains, while dune sands are often very well rounded. Kuenen (1960) summarizes a series of papers describing the experimental abrasion of sand grains by various processes. His data confirm that wind action is a more effective rounding agent than running water. Grain shape however is not just a function of the last depositional process which affected it but also of its previous history and original shape. Polycyclic sands will tend to be well-rounded regardless of the various processes to which they have been subjected.

SEDIMENTARY STRUCTURES

In model studies of the various reservoir sandstones treated in this text, considerable emphasis is given to sedimentary structure, especially the modern analogues. This is due to these structures being generated in place and not generally modified by fluid invasion. One serious modification that does occur, however, is that brought about by burrowing organisms (bioturbation). The principal internal structures include horizontal bedding, cross-stratification, lamination, and microcross-lamination (ripple marks). Selley (1978) states that "cross-bedding is a sedimentary structure which is common, morphologically variable, and extensively documented. . . . Much can be learned about the process which formed cross-bedding, but, since few processes are restricted to any one environment, this knowledge is of limited value. . . . Large scale eolian cross-bedding may be an exception to this rule. . . ." The value of sedimentary structures to the exploration geologist is generally limited solely because of the relative paucity of available cores. When available, however, they constitute a meaningful supplementary tool for identifying certain reservoir types. This is especially true when working with point bars. Here one should expect to find a zone of giant "ripples" (festoon bedding) between the basal zone of very poorly bedded, variable textured sandstone, and the overlying zone of horizontally bedded sandstone, siltstone, and shale.

PALEOCURRENT PATTERNS

Paleocurrent patterns constitute another feature of sedimentary rocks, and there are numerous references on this subject. Their principal application, however, is to Recent sediments and outcropping ancient sediments. Current patterns usually cannot be determined from geophysical logs, cores, and rotary cuttings. They can be surmised in some cases, however, from the geometrical aspects of the sediments.

PALEONTOLOGY

Micropaleontological data seldom are available from drilling operations of the Paleozoic and Mesozoic sections. Their principal use in North America has been in subsurface studies of the Tertiary section of the Gulf Coastal area of the United States and Mexico. Extensive use of microfauna and flora has been made here as indicators of depositional environment. In fact, it is almost an indispensable tool in early recognition of water-depth conditions, temperature, salinity, and current turbulence. Sandstone depositional environments in coastal areas generally are not conducive to the preservation of the skeletal remains of Foraminifera due to the high energy (wave, longshore currents, and tides) conditions. They are much more readily preserved in the shales that are interbedded with such sandstones. For example, inner neritic benthonic Foraminifera taken from such shale "breaks" are indicative of a coastal environment where such reservoir types as deltas, barrier bars, and shallow-water offshore bars may occur. Shales associated with deeper water reservoir sandstones (turbidites) will contain benthonic Foraminifera indicative of water depth ranges from middle neritic to abyssal. In working with turbidites, however, inner, middle, and outer neritic forams intimately associated with bathyal and abyssal forams may be found. This is the result of sediment displacement by submarine currents and submarine slump blocks. In southeastern Mexico, the Chicontepec submarine canyon fill contains forams ranging in age from Late Jurassic to Cretaceous to Paleocene and Early Eocene. In any one sample set an abundance of forams of this age range may occur. This is positive evidence for recycled sediments, even though the submarine canyon was eroded in Paleocene time.

From the foregoing comments, it should be apparent that studies of geometry, lithology, texture, shape, sedimentary structures, paleocurrent patterns, and microfauna have very unequal application in determining depositional environments. They are all of great importance in studies of Recent sediments and outcrops but are of considerably more restricted value in subsurface studies. Of the five listed features of sedimentary units, geometry and microfauna have the most application in the determination of subsurface depositional environments. A good background knowledge of all of these features, however, is considered highly desirable for the geologist to be effective in the exploration of sandstone reservoirs.

UNCONFORMITIES

Unconformities occur abundantly in the subsurface and play a very significant roll in prospecting for stratigraphic traps. They are much more readily identified from detailed geophysical log correlations than by seismic means. The seismic technique, however, is quite applicable when dealing with angular unconformities where the strata subcropping the unconformity have an appreciable dip. Dipmeter logs also may be quite definitive in this situation. Many unconformities have a considerable paleotopographic expression that predetermines not only shoreline trends of the next transgressive sea but also the positions and trends of paleodrainage courses. In several chapters of this book, considerable attention is devoted to methods of reconstructing these old eroded surfaces, for the purpose of enabling the explorationist to fit the geometry of the reservoir facies into them. This is predicated on the idea that there frequently is a direct relationship between paleotopography and reservoir geometry.

GEOPHYSICAL LOGS

The spontaneous potential (SP) and gamma-ray (GR) logs have considerable value as grain-size profiles of individual sandstones. The amplitude of the SP curve varies with the cumulative effects of electrofiltration and electro-osmosis of the strata exposed in the borehole. Thus, this curve is an indirect measure (in millivolts) of the permeability rather than porosity. Selley (1978) points out that "shales are impermeable and that sands are (usually) permeable. Furthermore, in sands permeability tends to decrease with grain size (see, for example, Pryor, 1973). This is because the amount of clay matrix tends to increase with the decreasing grain size, and the matrix blocks the throat passages between pores. Thus, because the SP log essentially records permeability and because permeability is generally related to particle size, then the log may be used as a continuous vertical grain size profile." This applies only in sediments that have primary intergranular porosity. Thus, this curve may be used only in sandstones that have not been substantially altered by diagenesis. In many cases the GR curve also may be used as an indicator of grain size. Clays are much more radioactive than quartz sand. Thus, an increase in gamma radiation indicates an increase in the clay present in the sediment. An increase in clay content is accompanied by an apparent decrease in grain-size. The amplitude of gamma-ray curves may be affected by factors other than clay content. It has been noted by Weber (1971), for example, that the transitional zone at the base of a barrier bar may have a higher gamma radiation than normally is attributed to marine clay. In a study of a series of sidewall cores from this high radiation zone in one of the Nigerian fields he noted, "The source of the radiation appears to be the high percentage of silt size zircon. Zircon contains in general from 200–2,000 ppm Th and from 600–6,000 ppm U." Other minerals such as mica and glauconite, when occurring in abundance, also may cause a sandstone to appear fine-grained and shaly on the gamma log. Variations in the diameter of the borehole also affect the amplitude of the gamma-ray curve. The amplitude generally is lessened opposite zones of extensive caving. Such zones, however, are readily identified on the caliper log.

Many geologists draw a vertical line on the SP curve of electrical logs indicating what is called the "shale baseline." Then they draw a second vertical line, to the left of the shale base, to indicate reservoir quality sandstone. The position of this second line is arbitrary and may be at 60, 70, or 80 mv to the left of the shale baseline. When this is done with all of the logs of a given study area, isopach maps are then prepared to illustrate the thickness and distribution of reservoir quality sandstones. Figure 1-1 is an

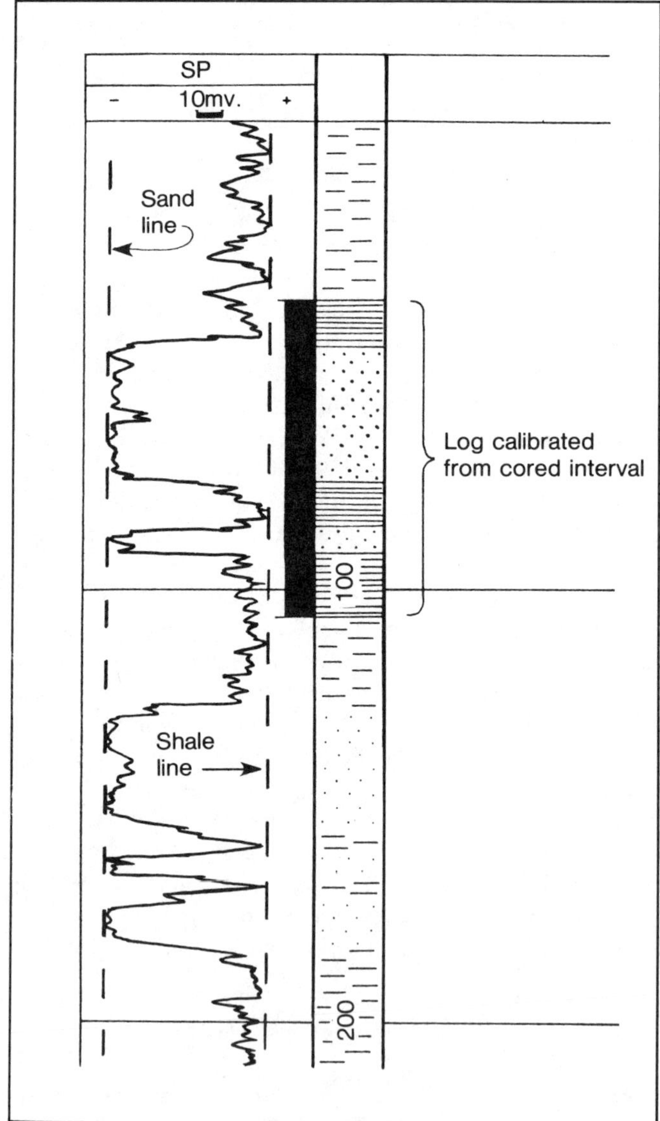

Fig. 1-1. Spontaneous potential (SP) curve of an electrical log, showing positions of the shale baseline and the line of maximum amplitude of the sandstone. The grain size of the sandstones can be calibrated with the SP curve when cores or cuttings are available. (© 1970, 1978 by Richard C. Shelly. Used by permission of the publisher, Cornell University Press).

illustration of a SP curve showing both the shale base line and the line of maximum amplitude of the sandstones.

For many years geologists have been attempting to use SP and GR "signatures" as a means of identifying environments of sand deposition, but only with limited success. Although characteristic "signatures" reflect grain-size profiles, there is no one diagnostic profile for a given environment. This is because similar "signatures" occur for the sandstones deposited within different environments. Although certain profiles may be suggestive of one or more environments, a wireline log pattern should never be interpreted as diagnostic of any one environment of deposition. It is only when the explorationist uses "signatures" in conjunction with other criteria (such as paleontological data, presence or absence of glauconite and carbonaceous material, core data, geometry, orientation relative to reconstructed depobasin shape, etc.) that SP-GR profiles assume importance.

In many areas a given sandstone type may occur repeatedly within a several hundred- (to thousand-) foot stratigraphic interval, reflecting a depositional environment that remained fairly static in a particular geographic area. For example, numerous fluvial sandstones may occur in the same borehole of an alluvial fan or within a braided stream environment. In other cases, several point bars and barrier bars may be "stacked." Barrier bars may have distributary channel sandstone cutting obliquely across or at right angles, to the long axis. Turbidites may have hundreds of sandstones (especially in deep-sea fans).

In other cases, dissimilar SP patterns related to one environment may occur vertically in a borehole. Deltas contain many subenvironments that may be superimposed vertically on one another as the delta switches back and forth, and the SP signature of the subenvironments may be different.

There are a number of basic SP-GR log shapes that are classified with descriptive terminology in Figure 1-2. The smooth shapes generally are more applicable to SP log shapes, whereas the serrated shapes are more characteristic of GR logs. This is because the two logs are reflecting entirely different properties of the rock exposed in a borehole. Not withstanding this fact, the two sets of logs frequently may be used interchangeably in many studies of depositional environment.

Many authors have attempted to catalogue environments of deposition on the basis of SP-GR log shapes. In all cases such attempts should be considered as suggestive and not diagnostic. It is only when they are used in conjunction with other criteria that they may assume significance. Their lack of diagnostic value is due to the fact that in a number of instances similar log shapes are known to occur in some sandstones laid down in entirely different environments. Furthermore, within a given depositional environment (i.e., barrier-bar and deltaic), there may be an assortment of log shapes that characterize the subenvironments which collectively make up the larger environment.

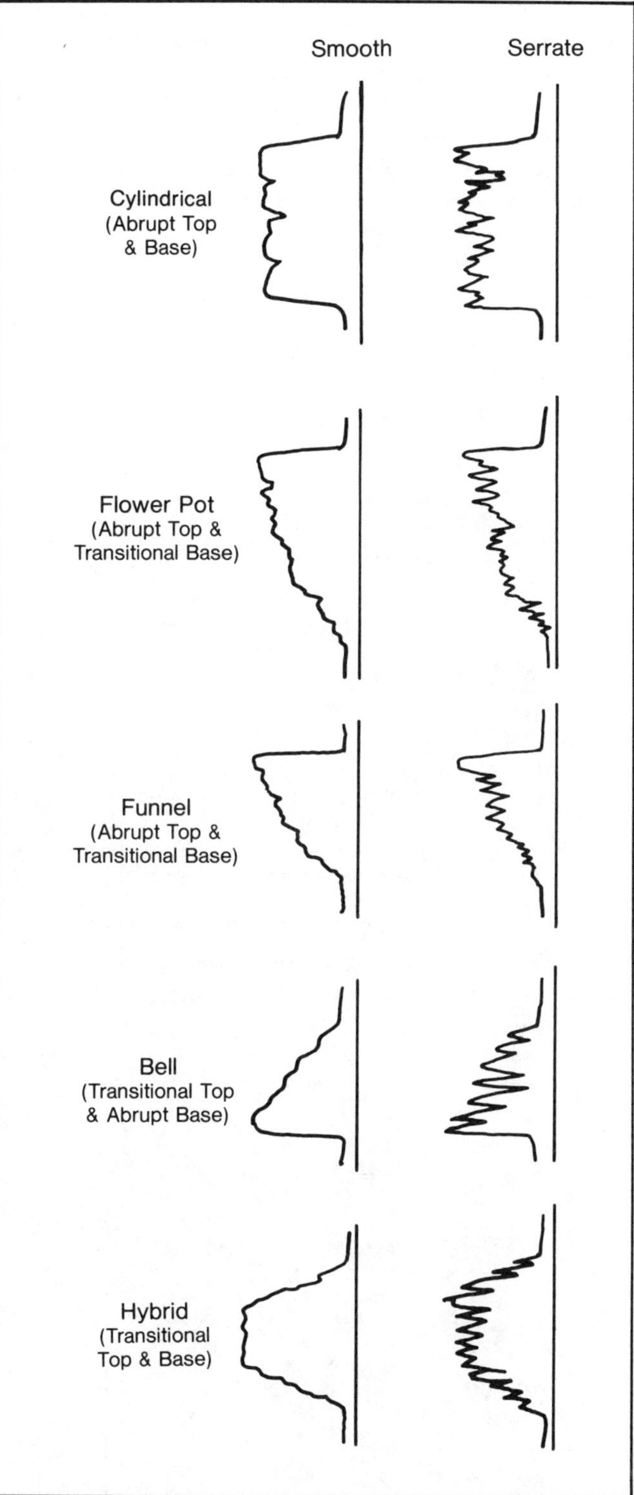

Fig. 1-2. Descriptive classification of basic SP and GR log shapes.

Coleman and Prior (1982) have presented a series of SP-

GR "signatures" for a variety of environments and have indexed them to the different local subenvironments. In each instance they also show the resistivity curves, which are partial mirror images of the SP-GR curves. Such indexing serves a very useful purpose in cautioning the explorationist that overgeneralization is a distinct hazard and must be avoided. Instructive examples of their work are shown in Figures 1-3 to 1-10.

The logs of braided stream deposits exhibit multiple stacked sandstones, many of which show fining upward in texture, as illustrated in Figure 1-3. Coleman and Prior state that, "Directional properties within each cycle often display narrow directional spread and are fairly representative of the long-axis orientation or downstream direction of the channel. . . . Log response often shows an overall blocky shape, with numerous sharp 'kickouts' representing local coarse sand-filled scours." Correlations of individual sandstones cannot be carried any great distance normal to the directional flow of the streams. Siltstone and clay "breaks" between the sandstones are numerous but not ex-

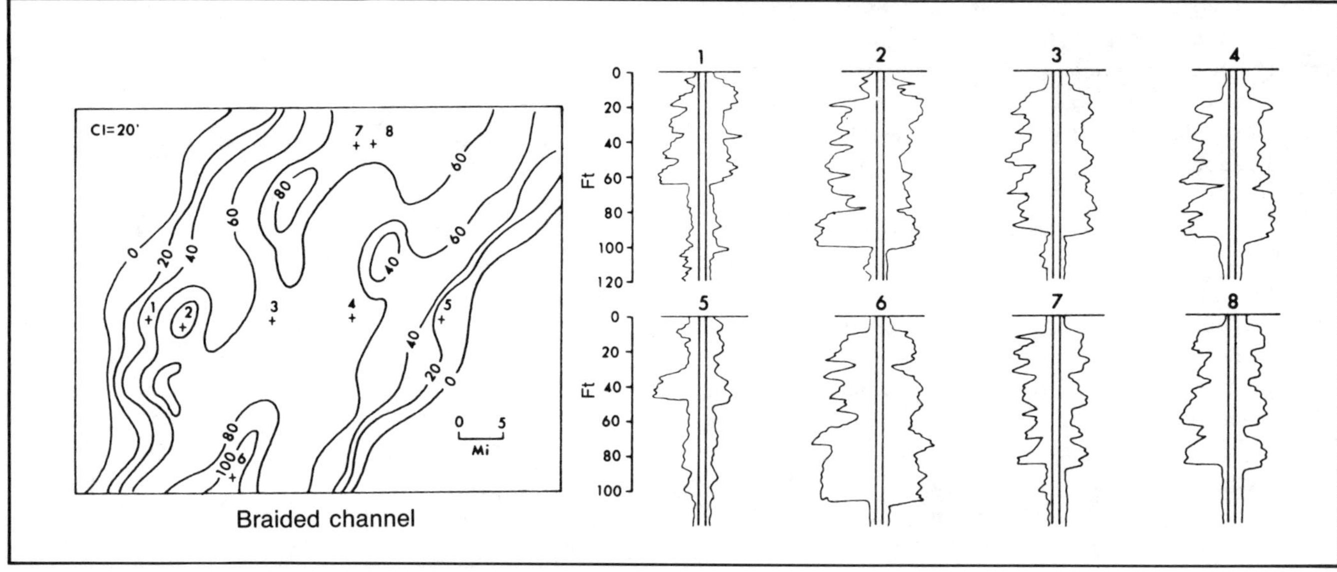

Fig. 1-3. Summary of log shapes showing the major characteristics of braided stream deposits and their respective locations within the braided stream complex. (Modified after Colemean and Prior, 1982; permission to publish by AAPG).

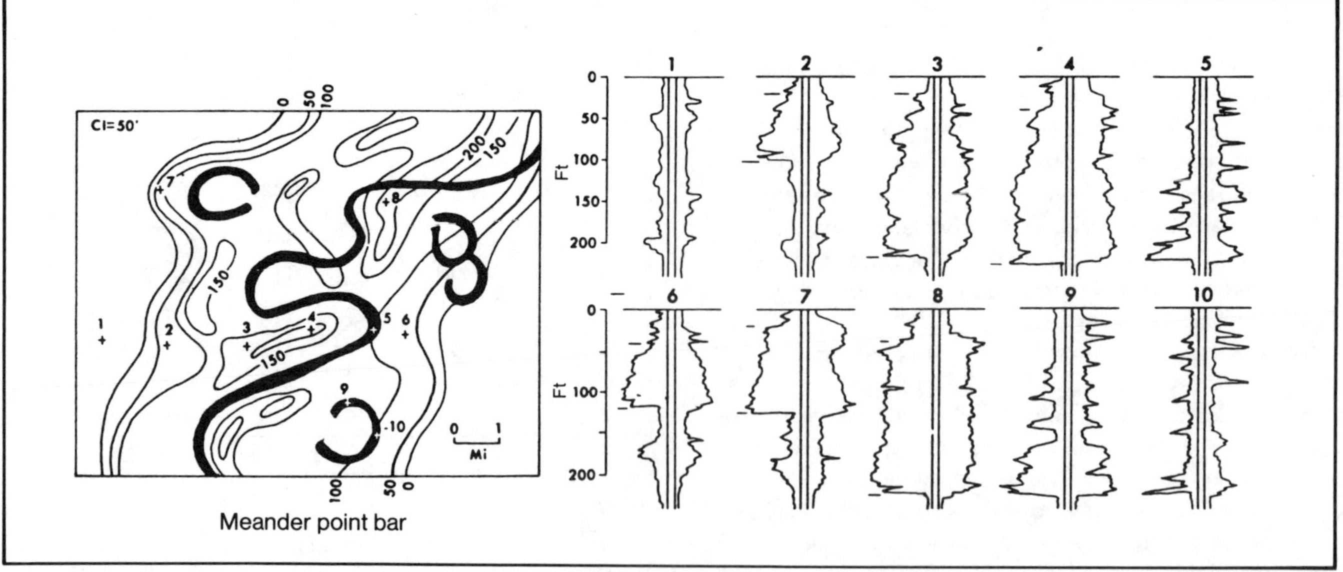

Fig. 1-4. Summary of log shapes showing the major characteristics of meandering point-bar deposits and their respective locations within the meander belt. (Modified after Coleman and Prior, 1982; permission to publish by AAPG).

tensive laterally. These are backswamp-marsh-floodplain deposits.

Most of the logs of wells drilled into the point-bar sandstones of a meander belt have a bell-shaped profile, indicating a fining-upward sequence. Profiles 5, 9, and 10 of Figure 1-4 are dominated by siltstone and shales, rather than sandstone, because all three drillsites are located in clay plugs where significant sand accumulation rarely occurs. Within a mature meander belt, it is not unusual to encounter all, or parts of, two or three "stacked" point bars in the same borehole.

Typical log shapes for a lacustrine delta fill are illustrated in Figure 1-5. Such sandstones as these are the result of a breakout of a delta distributary into a fresh-water lake, which previously had been a topographically isolated area within an interdistributary environment. Coleman and Prior note that "Interdistributary deltaic lakes range in size from only a few square kilometers to 80 to 100 sq km . . .

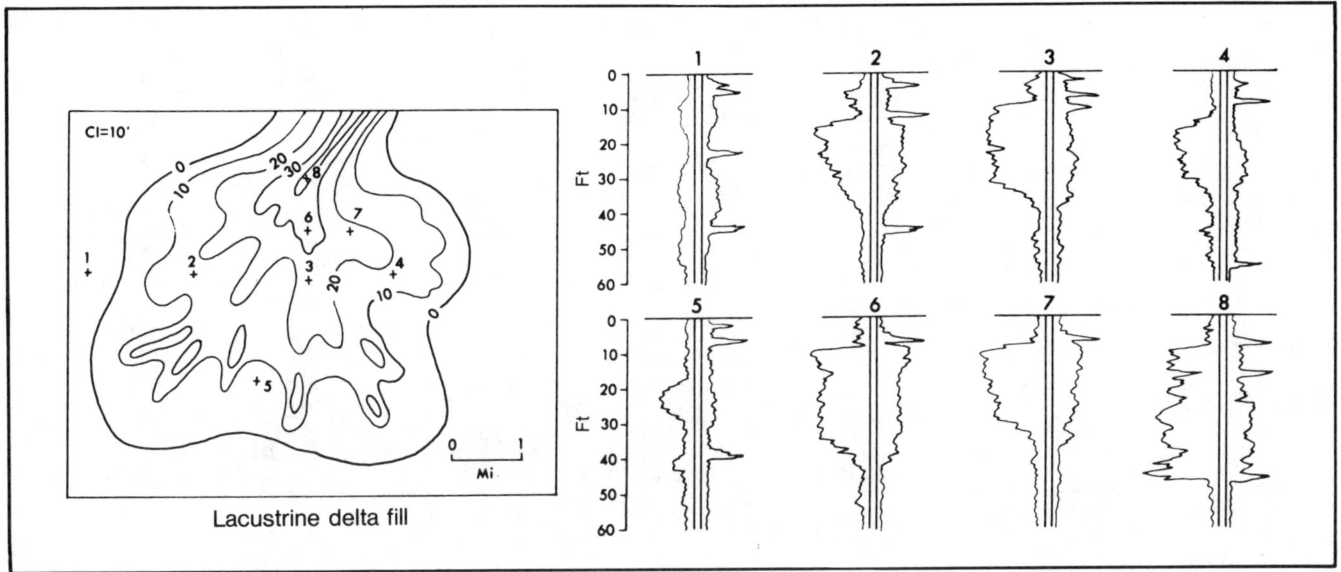

Fig. 1-5. Summary of log shapes showing the major characteristics of lacustrine delta fill deposits and their respective locations within the upper delta plain. (Modified after Colemen and Prior, 1982; permission to publish by AAPG).

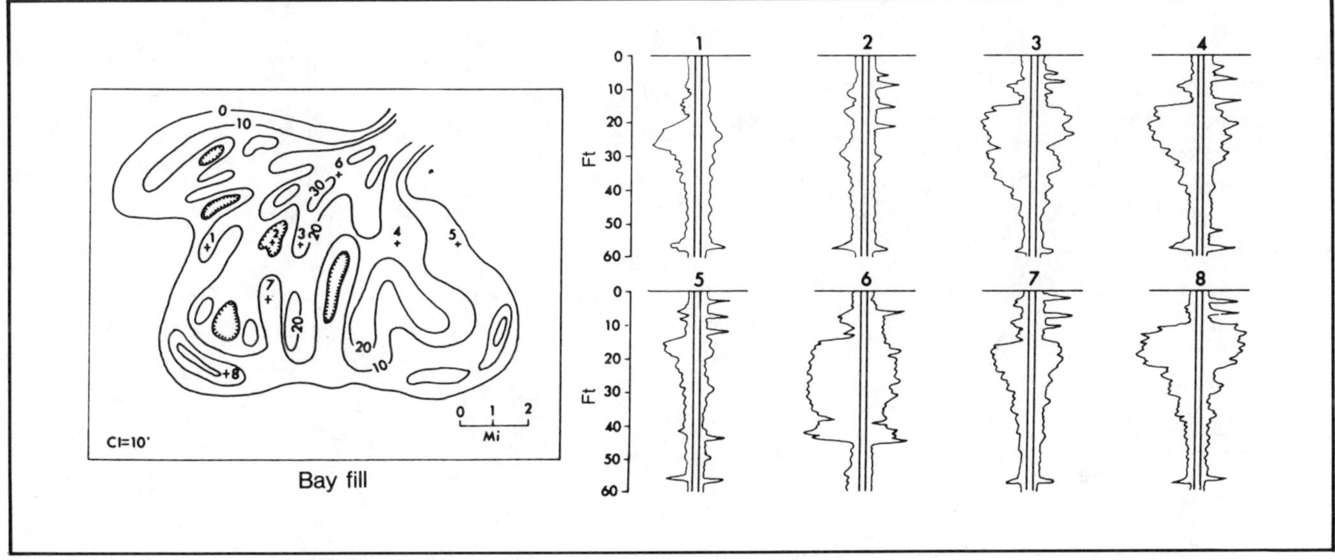

Fig. 1-6. Summary of log shapes showing the major characteristics of bay fill deposits and their respective locations within the lower delta plain. (Modified after Coleman and Prior, 1982; permission to publish by AAPG).

often resistivity kicks are extremely common within such a setting and are responses to lignite, coal, and iron-rich seams which form within these essentially red bed deposits. The sand body itself normally displays a graded base; however, in some areas, generally near breaks in the river bank, thick, sharp-based sands can often accumulate immediately within the region of the actual crevassing."

Figure 1-6 shows the various E-log shapes one might expect in various boreholes drilled into a bay fill and the accompanying isopach map, which is shown on the left. Such bay fills occur in the interdistributary areas within the lower delta plain. As such, they constitute crevasse splays that build out into an open marginal marine environment. Coleman and Prior point out that, "Often, sands in this vicinity display a sharp base scoured into the underlying interdistributary bay and marsh deposits. Away from the initial break, however, the typical coarsening upward sequence (or inverted bell-shaped logs), becomes the most common type of log response. Within the overall sand body there are areas where sands have not accumulated to any

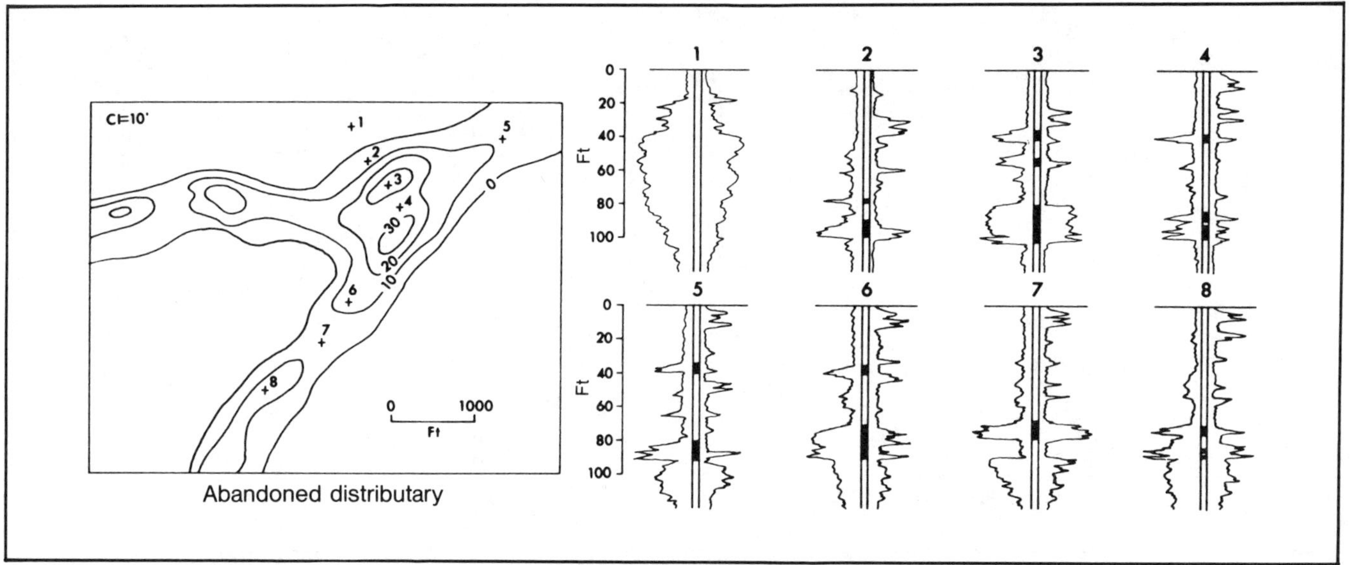

Fig. 1-7. Summary of log shapes showing the major characteristics of abandoned distributary deposits and their respective locations in the lower delta plain. (Modified after Coleman and Prior, 1982; permission to publish by AAPG).

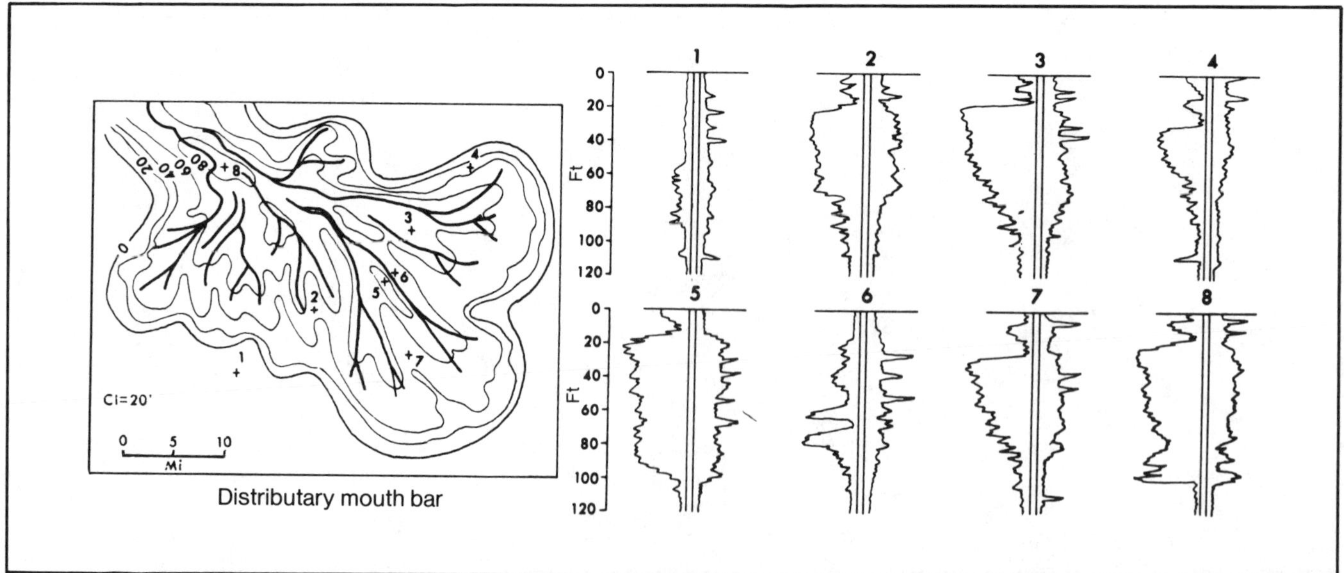

Fig. 1-8. Summary of log shapes showing the major characteristics of distributary-mouth bar deposits and their respective locations in the subaqueous delta plain. (Modified after Coleman and Prior, 1982; permission to publish by AAPG).

great thickness and therefore (borehole 2) in which virtually no sand can be found and the entire sequence consists of interdistributary-bay silts and clays, grading upwards to marsh deposits."

Figure 1-7 exhibits the extremely variable log responses that might be reasonably expected when drilling into an abandoned delta channel-fill sequence. The isopach map shows a maximum sand thickness both upstream and downstream from the area of bifurcation. Such abandoned channels are commonly eroded (then filled) through the thicker sands of their respective distributary mouth bars. It is for this reason that the log of borehole 1 of Figure 1-7 shows the maximum sand thickness outside of the abandoned distributary. Grain size of the channel sand may decrease upward or there may be virtually no change in grain size. Generally, log responses in an abandoned distributary are extremely erratic with the principal sand development being at or near the base of the channel.

Figure 1-8 shows a variety of log shapes to be expected within a distributary mouth bar complex. The isopach map

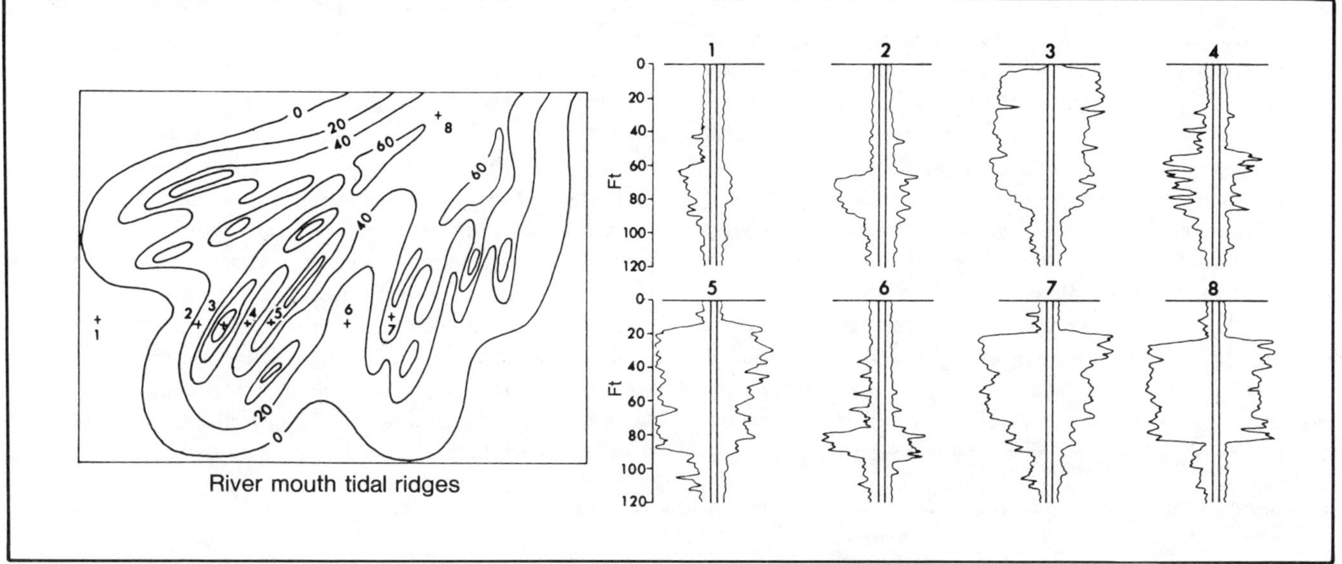

Fig. 1-9. Summary of log shapes showing the characteristics of river-mouth tidal ridge deposits and their respective locations in the subaqueous delta plain. (Modified after Coleman and Prior, 1982; permission to publish by AAPG).

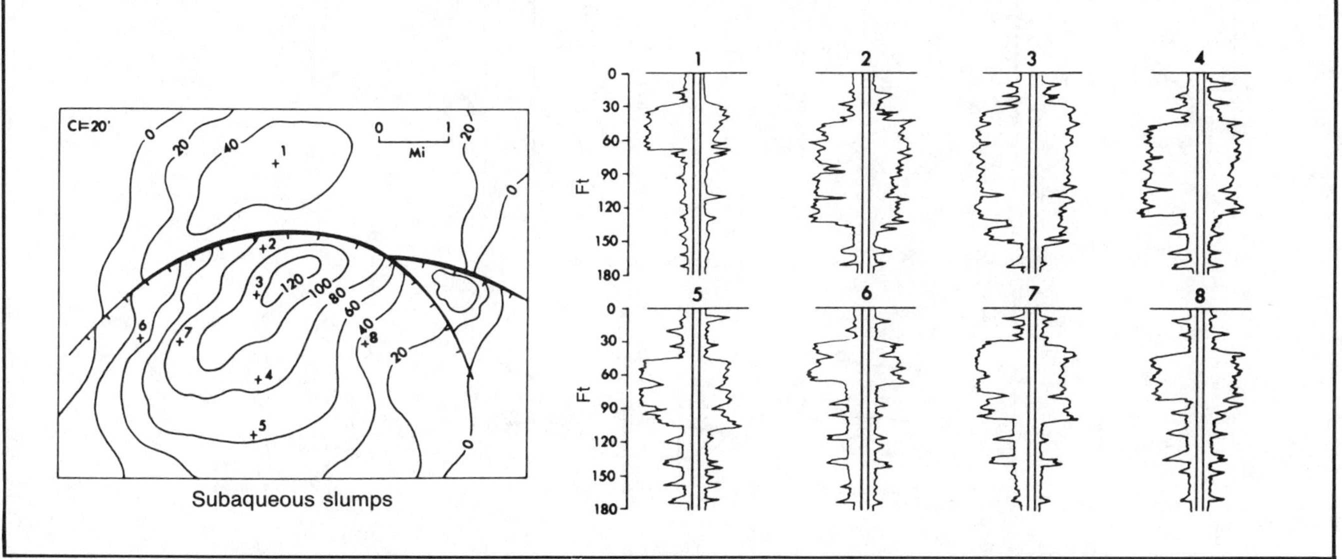

Fig. 1-10. Summary of log shapes showing the characteristics of slump deposits and their respective locations in the subaqueous delta plain. (Modified after Coleman and Prior, 1982; permission to publish by AAPG).

is that of a distributary mouth bar system in which individual distributary mouth bars have laterally coalesced to form a delta-front type of sand body. Most of the log responses show a coarsening upward sequence with considerable variation in sand thickness, depending on the location of the borehole. Coleman and Prior note, "In general, the nearer the boring to the axis of the distributary, the sharper the base of the sand body, and gradational contacts become less well defined. Distally (away from the distributary channel axis), the sequence displays a much greater tendency toward a large transition from distal bar to distributary mouth bar."

In a tidal-dominated river mouth environment, a series of subparallel tidal ridges will occur, such as those shown by isopach contours in Figure 1-9. Data relative to log shapes of such sand bodies in the subsurface are sparse. Although the log shapes shown on the right half of Figure 1-9 are a bit speculative, they are, nevertheless, consistent with observations made of modern analogues. Sand thicknesses vary considerably. The log shapes of core holes 3, 5, and 7 have flowerpot profiles with a general increase in grain size upward. The abrupt base of the cylindrical profile of core hole 8 is attributed to the sand resting directly on a scoured surface at this drillsite. Sand thicknesses vary considerably depending on whether the borehole was drilled into a swell (thick) or swale (thin) portion of the sand body.

Figure 1-10 illustrates log responses in boreholes scattered principally over the downdropped block of a faulted structure. Coleman and Prior note that "one of the most characteristic features of sand bodies deposited by slumping processes is the extremely blocky character of the electric log response. Sands generally tend to be sharp based, producing rather uniform log response. Correlation of individual kicks on the sand body is extremely tentative because of the nature of the slumping process." In areas such as the Niger delta, growth fault slump blocks occur in a prograding fashion directly opposite the distal ends of the individual distributaries in quite shallow water. Where this is the case, log responses are similar to those shown in Figure 1-10, and well-by-well correlations are easy within the individual downdropped depocenters.

Figure 1-11 illustrates a variety of log shapes observed in various depositional environments. The bell shape of log 1 is suggestive of a subaerial channel sand fill and illustrates a fining upward sequence. The hybrid profile of log 2 also may represent a subaerial channel fill with a slight gradient effect at both the top and the base. Neither profile, however, should be considered as diagnostic. For example, the profile of log 1 also could represent the sand section on the updrift side of the tidal inlet of a barrier bar, as discussed in Chapter 9. Delta distributaries also are known to be filled with sand that may exhibit either a bell

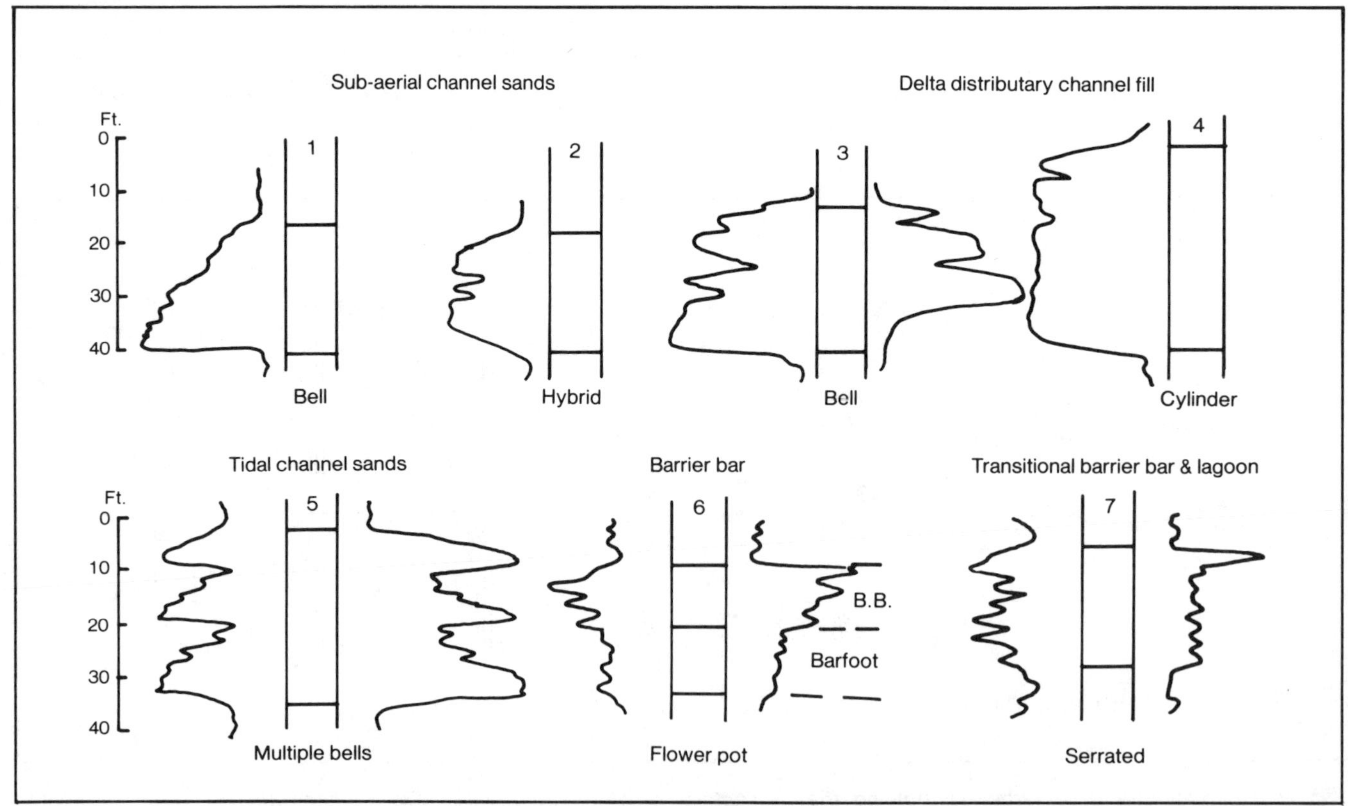

Fig. 1-11. Assortment of log shapes suggestive of depositional environments.

shape or a cylinder shape, as shown in profiles 3 and 4, respectively. Generally speaking, delta distributary sandstones frequently exhibit a cylindrical profile with an abupt base and a fairly abrupt top. Tidal channel sandstones frequently occur in multiples with each of the member sandstones exhibiting a bell shape, as illustrated log 5 of Figure 1-11.

No one log response will suffice to illustrate the barrier-bar complex. Log 6 of Figure 1-11 is the one most frequently considered representative. The bar foot (Weber, 1971) in this profile is all transitional indicating a fining-downward grain size. This is the zone representative of an oscillating wave base. The transition zone between a barrier bar and the lagoon, which occurs on the landward side, is generally quite abrupt and narrow. Wells are seldom drilled into this narrow zone, but when they are, the SP-GR response is that of a serrated profile, as illustrated in No. 7 borehole of Figure 1-11. Note that there is no bar foot in this portion of the barrier bar. Other equally significant log shapes occur within the barrier-bar complex, and these are discussed and illustrated in Chapter 9.

Turbidite sediments vary in log response, depending on whether the well was drilled in a canyon, deep-water fan, or a basin floor. Log sections frequently involve many hundreds (even thousands) of feet of section.

This discussion of geophysical log shapes is admittedly incomplete but should serve as a basis for demonstrating that SR-GR "signatures" are suggestive (but not diagnostic) of various depositional environments. Similar log shapes are known to occur in sandstones of differing environments of deposition. They should always be used in conjunction with other available internal and external data and not as a single definitive criterion.

SELECTED BIBLIOGRAPHY

Beerbower, J. R., 1964, Cyclothems and cyclic depositional mechanisms in alluvial plain sedimentation, *in* D. F. Merriam, ed., Symposium on cyclic sedimentation: Kans. Geol. Surv. Bull. 169, p. 31-42.

Bjerkli, R., and J. S. Ostmo-Saeter, 1973, Formation of glauconite in foraminiferal shells on the continental shelf off Norway: Mar. Geol., v. 14, p. 169-178.

Campbell, R. L., 1968, Stratigraphic applications of dipmeter data in mid-continent: AAPG Bull., v. 52, no. 9, p. 1700-1719.

Coleman, J. M., and D. B. Prior, 1982, Deltaic environments, *in* P. A. Scholle and D. Spearing, eds., Sandstone depositional environments: AAPG Mem. 31, p. 139-178.

Conybeare, C. E. B., 1976, Geomorphology of oil and gas fields in sandstone bodies: Amsterdam, Elsevier, 341 p.

Crosby, E. J., 1972, Classification of sedimentary environments, *in* J. K. Rigby and W. K. Hamblin, eds., Recognition of ancient sedimentary environments: SEPM Sp. Pub., no. 16, p. 1-11.

Folk, R. L., 1967, A review of grain size parameters: Sedimentology, v. 6, no. 2, p. 73-94.

Galloway, W. E., 1968, Depositional systems of the Lower Wilcox group, North Central Gulf Coast basin: Gulf Coast Assoc. Geol. Socs. Trans., v. 18, p. 275-289.

Grim, R. E., 1958, Concept of diageneses in argillallous sediments: AAPG Bull., v. 42, no. 2, p. 246-253.

Harms, J. C., and P. Tackenberg, 1972, Seismic signatures of sedimentation models: Geophysics, v. 37, p. 45-58.

Heckel, P. H., 1972, Recognition of ancient shallow marine environments, *in* J. K. Rigby and W. K. Hamblin, eds., SEPM Sp. Pub, no. 16, p. 226-286.

Jageler, A. H., and D. R. Matuszak, 1972, Use of well logs and dipmeter in stratigraphic trap exploration, *in* R. E. King, ed., Stratigraphic oil and gas fields: AAPG Sp. Pub. no. 16, p. 107–135.

Krumbein, W. C., and L. L. Sloss, 1958, Principles of stratigraphy and sedimentation: San Francisco, W. H. Freeman & Co., 497 p.

Kuenen, P. H., 1960, Experimental abrasion of sand grains: Repto. Int. Geol. Cong. 21 Session, pt. 10, Submarine Geology.

Le Blanc, R. J., 1972, Geometry of sandstone reservoir bodies, *in* T. O. Cook, ed., Underground waste management and environmental implications: AAPG Mem. 18, p. 133-189.

Lyons, P. L., and M. B. Dobrin, 1972, Seismic exploration for stratigraphic traps, *in* R. E. King, ed., Stratigraphic oil and gas fields: AAPG Sp. Pub. no. 16, p. 225-243.

Martin, R., 1966, Paleogeomorphology and its application to exploration for oil and gas (with examples from western Canada): AAPG Bull., v. 50, no. 10, p. 2277-2311.

Moiola, R. J., and D. Weiser, 1968, Textural parameters: An evaluation: Jour. Sed. Petrol., v. 38, no. 1, p. 45-53.

Peterson, J. A., and J. C. Osmond, eds., 1961, Geometry of sandstone bodies: Tulsa, AAPG, 240 p.

Pettijohn, F. J., and P. E. Potter, 1964, Atlas and glossary of primary sedimentary structures: New York, Springer-Verlag, 370 p.

Potter, P. E., 1967, Sand bodies and sedimentary environments: a review: AAPG Bull., v. 51, no. 3, p. 337-365.

Pryor, W. A., 1973, Permeability–porosity patterns and variations in some Holocene sand bodies: AAPG Bull., v. 57, no. 1, p. 162-189.

Selley, R. C., 1978, Ancient sedimentary environments: Ithaca, N.Y., Cornell Univ. Press. 287 p.

Shelton, J. W., 1967, Stratigraphic models and general criteria for recognition of alluvial, barrier-bar, and turbidity current sand deposits; AAPG Bull., v. 51, no. 12, p. 2441-2460.

Sheriff, R. E., 1976, Inferring stratigraphy from seismic data: AAPG Bull., v. 60, no. 4, p. 528-576.

Weaver, C. E., 1958, Geologic interpretation of argillaceous sediments., Pt. 1, origin and significance of clay minerals in sedimentary rocks: AAPG Bull., v. 42, no. 2, p. 254-271.

Weber, K. J., 1971, Sedimentological aspects of oil fields in the Niger delta: Geologie en Mijnbouw, v. 50, no. 3, p. 559-576.

2 GENETIC UNITS OF STRATIGRAPHY

INTRODUCTION

Time stratigraphic classification and terminology and the designation of stratigraphic units have received considerable discussion in the literature (Hedberg, 1948, 1958; Wheeler and Mallory, 1953, 1956, Forgotson, 1957; Mitchum, Vail, and Thompson, 1977). Much of the emphasis, however, deals with fossil age-dating of sediments, facies changes, terminology, etc., with insufficient consideration given to practical genetic units that can be identified and utilized by the exploration geologist in subsurface studies. Forgotson was among the first to recognize a lack of suitable terminology in subsurface regional correlations and the mapping of rock units on a regional scale. For this reason, he proposed the term *format* "to bridge the missing link in the three-dimensional classification of stratigraphic units." His classification of lithogic units is as follows:

1. Three dimensional rock units as a whole—the lithosome of Wheeler and Mallory.
2. Laterally segregated rock units—lithofacies, the expression of lateral variation in lithology of a stratigraphic interval.
3. Vertically segregated units:
   - A. Formal units—groups, formations, and members of classical lithostratigraphy where the principal application is in surface mapping.
   - B. Informal units—divided into two general categories:
     - i. Attribute-defined units—mineral zones, insoluble residue zones, trace element, porosity, velocity, resistivity, radioactive, temperature zones, etc.
     - ii. Marker-defined units—involving strata between marker beds and horizons where boundaries may involve "unconformities, corrosion surfaces, reflecting and refracting surfaces, certain bentonite horizons and coal, etc. The markers are identified by characteristics recognized on logs of one kind or another."

Forgotson stressed that "correlations and regional relationships of laterally continuous rock units should not be hindered by lack of suitable terminoloy." Lithic units occurring between well-defined marker beds or horizons generally are not recognized by our formal system of stratigraphic nomenclature. These units may be mapped beyond "the limits of recognition of their original defining markers by interval correlation and (vertical) position with respect to additional markers which appear in the section." He further suggests this type of lithologic unit does not fit the definition of a formation, a lithosome, or a facies. An isopach map of such a unit constitutes an extremely useful tool in reconstructing a portion of a depositional basin.

In proposing the term *format*, Forgotson points out that it is a marker-defined operational unit that consists of "segregations of strata sandwiched between markers which can be traced through facies changes affecting the enclosed strata." Occasionally such units already have a formation name, but this is purely coincidental. Although the authors fully subscribe to the format concept, it is their conviction that it is not the ultimate in subsurface stratigraphic analysis. It is for this reason that Busch (1971) proposed an expansion of the format concept, which was later developed in more detail (1974).

Geophysical tools employed by the petroleum geologist consist of combinations of electrical and radioactivity logs and are well known to the petroleum industry. Most of the knowledge that the petroleum geologist has of cores and cuttings is gained on the job as a well-site geologist. Sample data supplement his use of combinations of electrical and radioactivity logs, which are essential in calculating S_W and porosity. Geophysical logs are perhaps the petroleum geologist's most powerful tool. He can best learn their use through special courses, extensive reading, and considerable practice. There are over 30 types of geophysical logs, and the geologist must be able to use most of them.

MARKER BEDS AND HORIZONS

The presence of marker beds (or horizons) is critical to the recognition and mapping of genetic units. A marker bed generally is thin, widespread, and was deposited essentially parallel with sea level. It represents essentially contemporaneity of deposition and serves as an excellent datum of reference in (1) reconstructing ancient erosion surfaces, (2) in determining rates and direction of down-to-basin sedimentary divergence, and (3) the establishment of paleodepositional strike. All of this antedates and, therefore, ignores any effects resulting from postdepositional tectonics.

The most frequently used marker beds (sometimes horizons) are as follows:

1. Thin limestone
2. Bentonite
3. Coal
4. Evaporites
5. Siltstone
6. Thin marine shale
7. Radioactive "hot streaks"
8. Index fossils, faunal and floral zones

None of these time marker beds are infallible, as there are certain exceptions in each instance. Potential marker beds are subparallel in a paleodepositional strike direction and generally diverge at a uniform rate in a down-to-basin direction. It is only when two or more of them exhibit these characteristics that either of them may be used as a reference datum in the construction of a stratigraphic profile or in the reconstruction of paleodepositional slopes. Thin limestones are the most frequently used marker beds. Bentonites are perhaps the most widespread and represent the shortest interval of geological time. Coal beds require judicious use as they may be quite lenticular or may possibly transgress or regress time lines if traced over too great a distance. Evaporite beds (i.e., gypsum or anhydrite), being chemical precipitates, make excellent marker beds and are readily correlated on mechanical logs. Unfortunately, they do not occur as abundantly within the stratigraphic column as the other marker beds. Siltstones frequently are seaward extensions of sandstones that were deposited in the shallower portions of the neritic environment. As such, they may have a linear trend of many miles in a paleodepositional strike direction, but have a much more limited width (10–15 miles; 16–22 km).

The use of thin shale beds as time markers requires considerable judgement. Perhaps their best use as reference data is in a paralic sequence, which may occur on the downthrown side of a growth fault in a deltaic environment. These marine shales immediately overlie and underlie a paralic sequence and represent fairly rapid regression prior to the deposition of either a barrier bar, a point bar, or a distributary channel fill.

There are many basin areas where thick marine shales dominate the stratigraphic section, and the first six marker beds, listed previously, may have little or no value. By utilizing gamma-ray logs of thick marine shale sections, however, one can almost invariably find "hot streaks" that have time-stratigraphic significance. These are manifestations of thin intervals of relatively high gamma-ray radiation. Some, but not all, of them are fireclays such as occur subjacent to coal beds.

The use of index fossils and faunal zones requires precise vertical positioning of the fauna in a wellbore. Rotary cuttings all too often do not afford such precision. In thick sequences of shale, where there may be no other indication of time markers, they have proven to be invaluable, especially in Tertiary sequences.

GENETIC INCREMENT OF STRATA (GIS)

Genetically associated sediment types in the marginal marine environment may be recognized in either of two ways, the *Genetic Increment of Strata* (GIS) and the *Genetic Sequence of Strata* (GSS). Selected isopach maps of the GIS and GSS intervals can serve to solve many of the problems encountered in exploration for hydrocarbon-bearing reservoirs in the Appalachian basin, the Eastern Interior basin, the Mid-Continent basins and the Rocky Mountain basins of the United States. There also is considerable application in the Alberta Geosyncline of western Canada. However, applications in the Gulf Coast Tertiary and the California intermountain basins are extremely limited. This is due primarily to the paucity of traceable marker beds and numerous contemporaneous and postdepositional faults.

By preparing isopach maps of genetic units between marker beds, the geologist may reconstruct segments of basin configuration and, thus, permits the placement of individual reservoirs (generally sandstone) into their paleodepositional environments. The basic concept and application of the GIS and GSS is not restricted to sandstone and shale; they also may be applied to carbonate-evaporite-shale sequences.

A GIS consists of an interval of strata deposited during one cycle of sedimentation in which each lithologic component is genetically related to all the others. It is essential that the upper boundary be a lithologic-time marker, and the lower boundary may be either a lithologic-time marker, an unconformity, or a facies change from marine to nonmarine. Examples of each of these conditions are illustrated in Figure 2-1. In each case the GIS consists of all the sediments deposited during one stage of either cyclic subsidence or cyclic emergence of the depositional surface relative to sea level. Abrupt changes in sea level, under conditions of a stable depositional surface, would produce the same effects. Figures 2-1 A-D diagrammatically illustrate four examples of GIS, all of which are the result of cyclic marine transgression. Figure 2-1E is an example of cyclic regression. Thus, cyclic transgression may be the result of either cyclic subsidence of the depositional surface or cyclic emergence of sea level. Conversely, cyclic regression may be due to either cyclic emergence of the depositional surface or cyclic lowering of sea level. The GIS shown in Figure 2-1A is defined at the top by a thin limestone and at the base by an unconformity. The orientation of this profile is parallel with the paleodepositional strike of the basin, as indicated by like interval thicknesses at both ends. The long axis of the thicker channel sandstone is oriented essentially normal to the paleodepositional strike. This is consistent with the basic tenet that the stream that eroded the channel flowed essentially perpendicular to the regional strike and parallel with the direction of regional tilt. Figure 2-2 is a hypothetical isopach map of a GIS, showing the location of the cross sections of Figures 2-1A and B. Note that Figure 2-1B is perpendicular to the trend of the basin contours and, therefore, diverges southward.

Multiple criteria are employed as a basis for identifying any one sandstone type in the subsurface. These criteria are summarized at the end of several of the discussions of the different sandstone reservoir types. An almost diagnostic criterion for a channel sandstone is downward thickening at the expense of the underlying beds (Fig. 2-1A). Its orientation relative to basin shape, as shown in Figure 2-2, is also very significant. The practical value of the GIS isopach map of Figure 2-2 is that it reveals a systematic thickening basinward. Most channel sandstones trend in the same general direction as the thickening of the GIS. The paleogradient of the channel may be determined by dividing the increase in interval thickness of the GIS (from the marker bed, above, to the unconformity at the base) along the channel axis by the distance in miles. It may be noted that the barrier bar of Figure 2-1B is flat on the base, thickens upward, is asymmetrical in cross section, and trends parallel with the basin contours of Figure 2-2. These are just several of the multiple criteria one may glean from the geometrical and internal aspects of these near shore reservoir types. For more detailed discussions of criteria for these two reservoir types, see the discussions of channel and barrier-bar sandstones, respectively.

In Figure 2-1B the lower limit of the GIS is a thin limestone bed, and the upper limit is a bed of bentonite. An isopach map of such a GIS will show systematic divergence between the two marker beds in a basinward direction. Either an anomalous "terracing" effect or local thickening takes place parallel with the contours. This thickening is due to the relative noncompactability of the sandstone of the barrier bar as compared with the enclosing shale. The position and trend of this barrier bar could be determined by mapping variations in thickness of the lagoonal and marine shales alone. Mapping of the reciprocal thinning of the shale section above and laterally adjacent to the sandstone (down to the horizon of its base) also is a valid method for tracing the sandstone body. The mapping of this shale interval is recommended in areas where most of the producing wells were stopped short of the oil- or gas-water contact, and, as a result, too few of the control points penetrate the entire thickness of the barrier-bar sandstone.

Of the five cross sections shown in Figure 2-1 only profile B illustrates the Forgotson concept of a format. It is an interval defined at the top and bottom by marker beds and consists predominantly of shale locally interrupted by a lenticular sandstone (in this instance a barrier bar). With the accompanying cross sections, it can be seen that the GIS concept not only embraces the format but goes considerably beyond its implications.

Figure 2-1C is a thick wedge of shale with a transgressive sheet of sandstone at the base. The upper surface of the sandstone is "shingled," or en échelon, and each wedge of the sandstone is the result of a stillstand stage of the shoreline under conditions of cyclic marine transgression. An isopach map of this GIS shows the approximate paleotopography of the unconformity at the base. Such a map is quite useful in tracing paleodepositional trends of beach sandstones, especially in instances where the en échelon sandstones are separated by distinct shale "breaks."

In Figure 2-1D there is an abrupt facies change at the base of the sandstone, where it rests principally on nonmarine shale. Although the sandstone blankets the nonmarine sediments, there are two locally thick areas that are the result of stillstands of the shoreline under conditions of cyclic marine transgression. These stillstands favor ideal conditions for wave energy to winnow out the fines (silt and clay) and leave strand zones of excellent potential reservoir sandstone. The areas separating the two thicker sandstone accumulations consist of poorly sorted sand, silt and shale, and are devoid of reservoir properties. An isopach map of this GIS serves as an excellent basis for tracing and extrapolating reservoir trends for many miles in a paleodepositional strike direction. An excellent example of this type of sandstone is the Point Lookout sandstone (Hollenshead and Pritchard, 1961).

Figure 2-1E is a GIS resulting from cyclic regression. Such a profile is the result of either cyclic lowering of sea level or cyclic emergence of the bordering land area. The marine

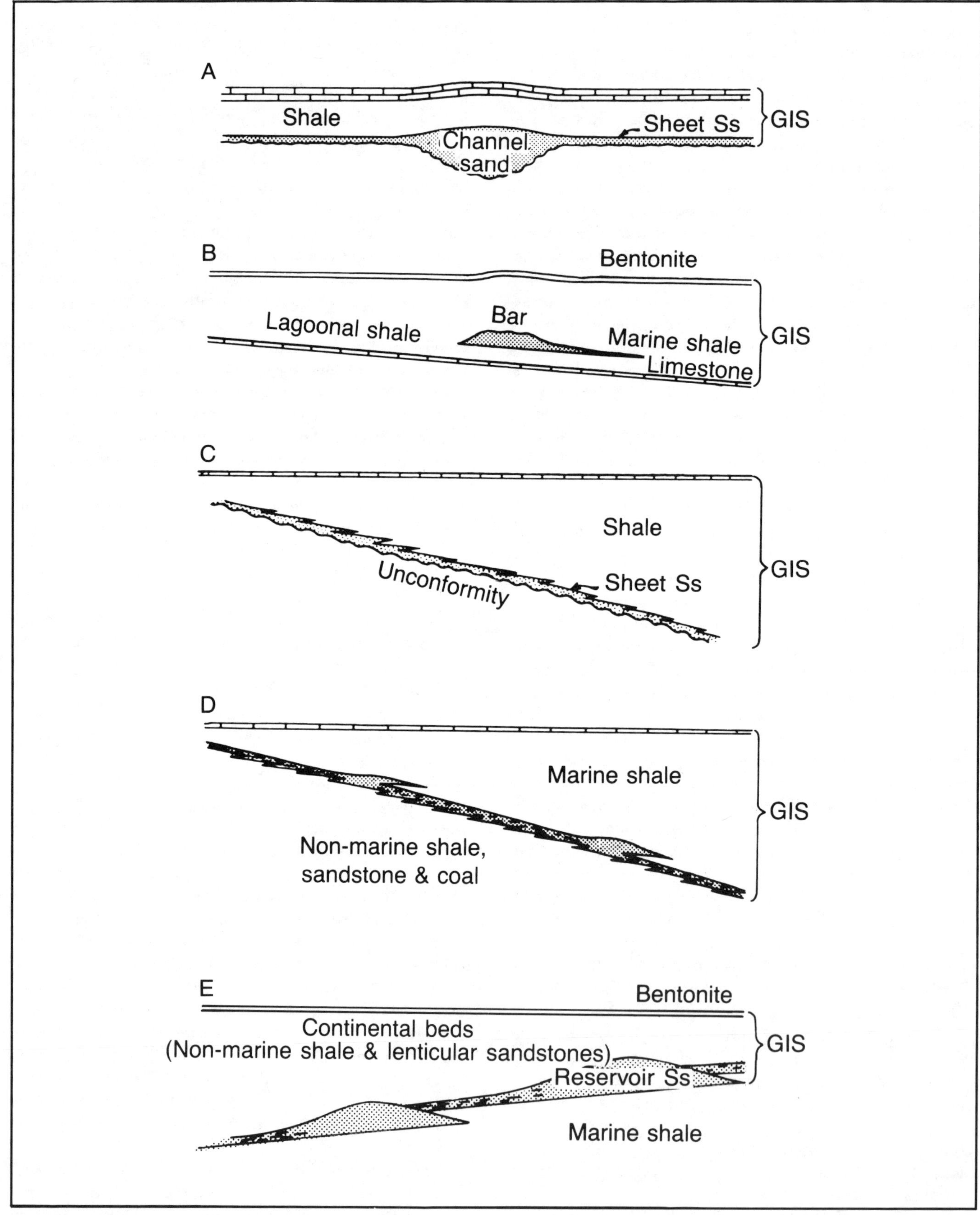
A
Shale
Sheet Ss
GIS
Channel sand
B
Bentonite
Bar
Lagoonal shale
Marine shale
GIS
Limestone
C
Shale
GIS
Unconformity
Sheet Ss
D
Marine shale
GIS
Non-marine shale, sandstone & coal
E
Bentonite
Continental beds (Non-marine shale & lenticular sandstones)
GIS
Reservoir Ss
Marine shale

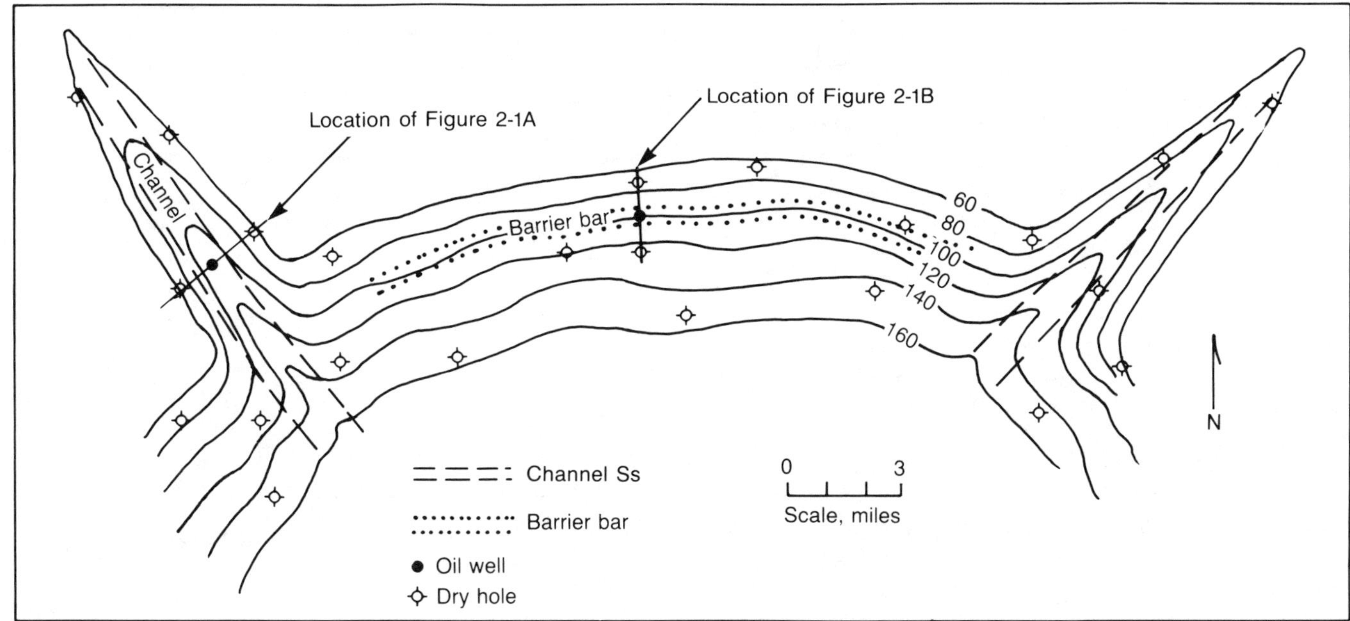

Fig. 2-2. Hypothetical isopach map of a GIS showing locations of two channel sandstones, both oriented normal to basin contours, and a barrier bar oriented parallel with the basin contours.

shale is "blanketed" by a sheet of sandstone that exhibits two locally thick accumulations. Each of the thicker areas is the result of a stillstand of the shoreline that favored the winnowing out of silt and clay by wave energy. The result is excellent reservoir sand conditions where the thick pods occur and poor reservoir conditions along where the lithology consists of poorly sorted sand, silt, and shale. In this example the regression of the shoreline was too rapid to allow wave energy to winnow out the fines. It may be noted that the thicker areas of sandstone development interfinger basinward with the *underlying* marine shale, positive proof of the cyclic regressive nature of the shoreline.

GENETIC SEQUENCE OF STRATA (GSS)

A GSS consists of two or more contiguous GIS's and is the result of more or less continuous sedimentation. Disconformities of limited geographic extent occasionally may be present, but there must not be any angular unconformities within the selected stratigraphic interval.

Figure 2-3 illustrates the relationship of a GIS to a GSS. It includes three different types of GIS's that collectively constitute the GSS. The uppermost GIS is defined by thin limestone marker beds at both the top and bottom and contains a wedge of shale between. The middle GIS is de-

---

Fig. 2-1. Diagrammatic illustrations of Genetic Increments of Strata (GIS).

Fig. 2-1A. Channel sandstone grading laterally into a sheet sandstone with overlying shale and limestone; upper limit of GIS is defined by lithologic-time marker (limestone), base defined by an unconformity.

Fig. 2-1B. Barrier-bar sandstone bordered by lagoonal shale on one side and marine shale on the other; upper and lower limits of GIS are defined by lithologic-time marker beds (bentonite and limestone).

Fig. 2-1C. Sheet sandstone, resulting from either cyclic subsidence of land or cyclic emergence of sea level, overlain by basinward thickening of marine shale; upper limit of GIS is defined by lithologic-time marker bed (limestone), base by an unconformity.

Fig. 2-1D. Sheet sandstone resulting from either cyclic subsidence of land or cyclic emergence of sea level, showing two areas of locally thicker (and better-sorted) sandstone at former stillstand positions of shoreline; upper boundary is defined by lithologic-time marker bed (limestone), base by facies change from continental to marine beds.

Fig. 2-1E. Sheet sandstone resulting from either cyclic emergence of land or cyclic subsidence of sea level, showing two areas of locally thicker (and better-sorted) sandstone at former stillstand positions of shoreline; upper boundary is defined by lithologic-time marker bed (bentonite), base by abrupt facies change from marginal-marine sandstone to marine shale.

In C and D the en échelon sandstone wedges interfinger with the overlying marine shale, whereas in E, they interfinger with the underlying marine shale. (Modified and supplemented after Busch, 1971).

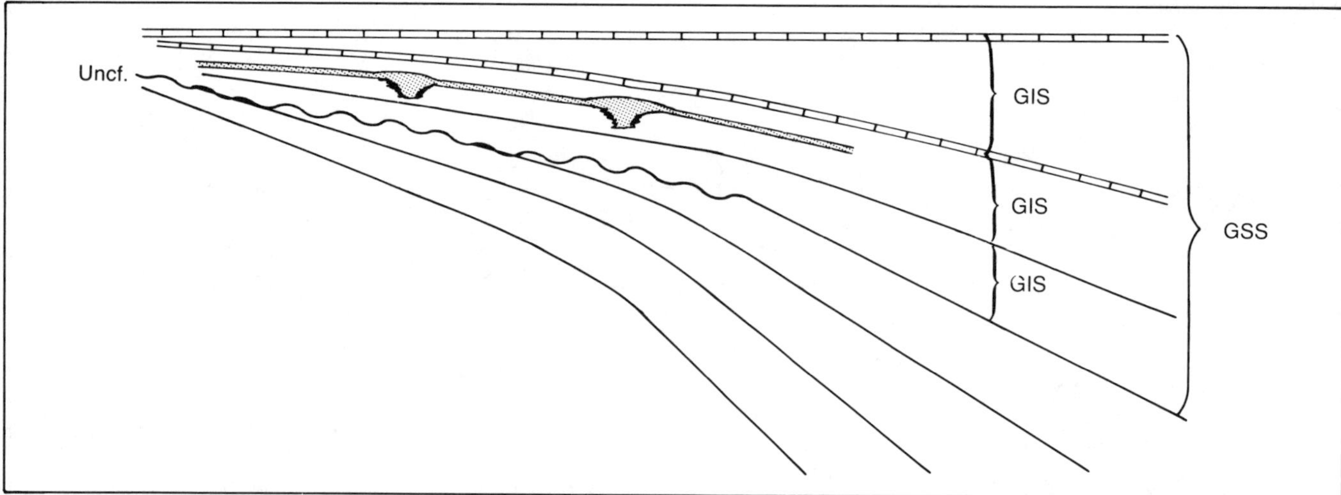

Fig. 2-3. Relationship of Genetic Sequence of Strata (GSS) to several Genetic Increments of Strata (GIS). Upper limit is defined by a thin limestone base by an unconformity that disappears basinward and grades into a marker horizon. (Modified after Busch, 1971; permission to publish by AAPG).

fined by a thin limestone at the top and a marker horizon at the base, which may consist of a readily correlated resistivity "signature," a gamma-ray log "hot streak," a thin siltstone, etc. A fairly extensive thin sheet sandstone occurs in the middle of a predominantly shale section. Locally, the sandstone thickens downward into channel sandstone fills. The lowermost GIS is defined at the top by a marker horizon and at the base by a marker horizon (to the right) that grades into an unconformity (to the left).

The three lowest stratigraphic units shown in Figure 2-3 do not constitute either a GIS or a GSS where the unconformity occurs because an unconformity should *not* be used to define the top of either a GIS or a GSS. Basinward from the unconformity, however, there are three GIS's and one GSS.

One might surmise that a GSS is nothing more than an expanded GIS, since they both consist of sedimentary wedges that generally diverge basinward. The principal uses of the GSS concept, however, involve the following considerations:

1. Generally applicable to areas of sparse, widespread well control.
2. Serves to define basin size and shape (paleodepositional configuration).
3. Serves as excellent clue to paleodepositional shoreline trends.
4. Yields early information regarding principal directions of sedimentary transport.
5. Distinguishes shelf areas from the less-stable basin areas (which generally are thickest and dominated by shale accumulation).
6. Establishes necessary (early) framework for mapping a system(s) of deltaic distributaries.
7. Is completely independent of (and antedates) the present structural configuration.

Figure 2-4 is a hypothetical contour map of a GSS illustrating a large segment of a depobasin. It is based on widely scattered well control (may be all dry holes) and clearly serves to distinguish the shelf area from the adjacent less-stable basin area. The shelf area is characterized by widely spread and irregular contour lines; whereas the less stable basin area is recognized by closely spaced, subparallel contour lines.

In exploring a sparsely drilled area of little or no established production, the shelf area should receive most of the initial attention from an exploration standpoint because marine transgressions and regressions were most extensive in this area. The rate of slope of the shelf usually was much less than what one might surmise from a GSS isopach map. This is because the GSS consists of multiple GIS's, the thicknesses of which are additive. The rate of basinward thickening of GIS, however, is a much better indicator of rate of shelf slope than can be gleaned from a GSS.

Marine regressions on a shelf area frequently involve many miles of shoreline retreat. With consequent exposure of previously deposited marine sediments, stream erosion extends basinward many miles, with the trends of the channels essentially normal to the shoreline. Stillstands of the shoreline favor the deposition of barrier bars in shallow water and are oriented parallel with the shoreline. Practically all regressions are followed by marine transgressions, which may be either more extensive or less extensive than the preceding regression. For example, in the Gulf Coast Ter-

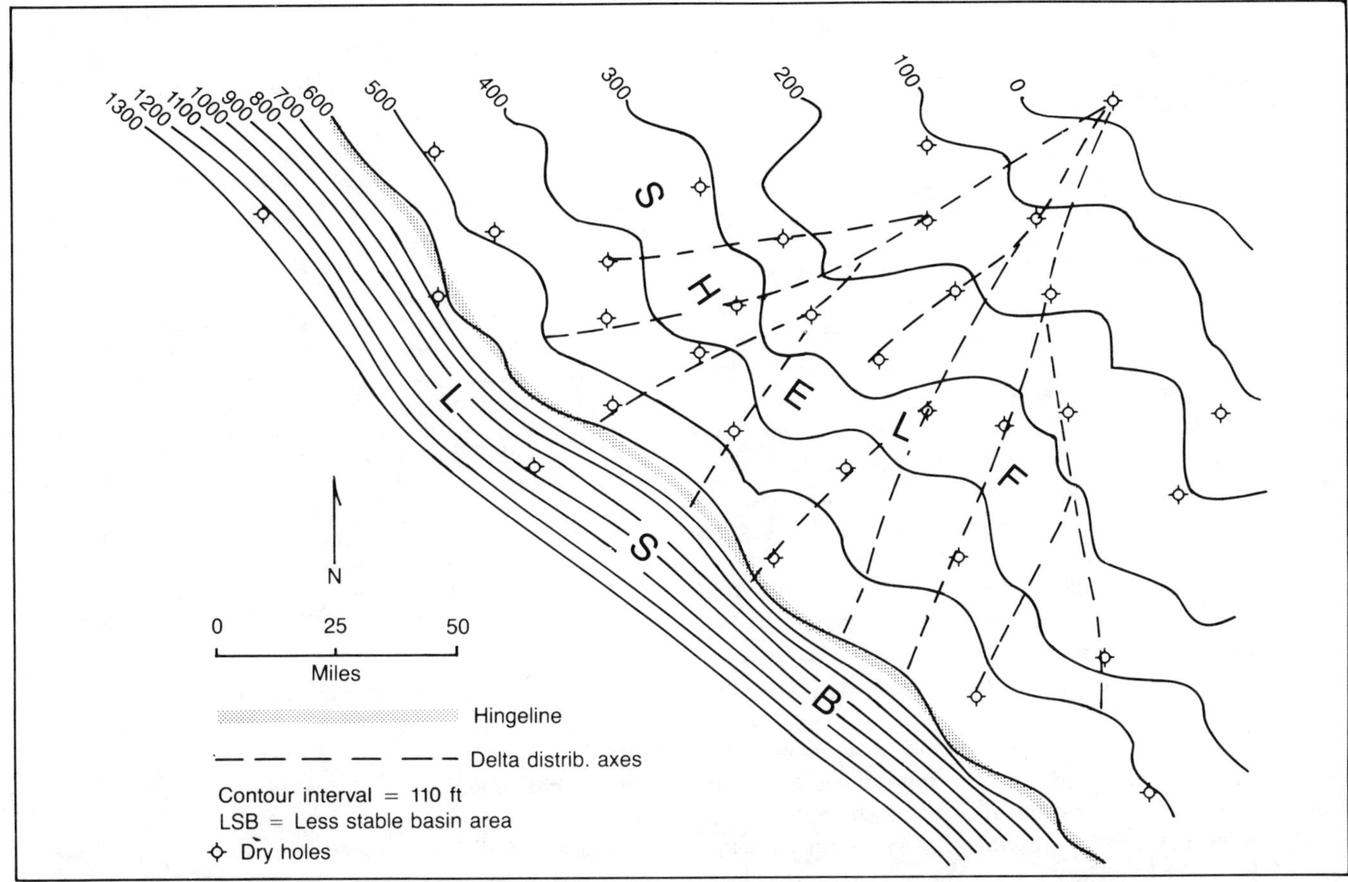

Fig. 2-4. Hypothetical isopach map of a GSS showing area of shelf (contour lines widely spaced and irregular) and less stable basin area (contour lines, closely spaced and nearly parallel). Dashed lines are axial trends of a system of bifurcating distributary channels of a large delta. This is a generalized reconstruction of a large segment of a basin and is completely independent of present-day structure.

tiary and Quaternary the entire sedimentary history is one of alternating regression and transgression of the shoreline, with regression being dominant. During the transgressive stage, multiple offshore bars frequently were deposited. In addition, the previously eroded stream channels became clogged with clastics as the base level of deposition of each stream moved landward. The result was channel sand development which may be locally shaly.

Each transgression of the shoreline may be accompanied by the deposition of an extensive, sheet-like thin limestone basinward from the strand zone. Such limestones serve as excellent marker beds for the delineation of GIS's. The Pennsylvanian section of the Mid-Continent area of the United States and the eastern shelf of the Midland Basin offer excellent examples of thin limestones that have time-stratigraphic significance.

The diverging dashed lines of Figure 2-4 are axial trends of a system of delta distributaries. All of these trends are either very close to or pass through wells in which channel sandstones occur at the same stratigraphic position within the GSS. The prevailing direction of divergence is essentially basinward. The same arrangement of control points could be used as a basis for "fanning" the delta fingers in any direction of the compass. To spread those axes, however, in any other direction than that shown on Figure 2-4 is to completely ignore basin shape and the direction of subaerial fluvial transport. The GSS map is an absolute prerequisite for determining the prevailing directional flow of the distributaries.

In the discussion of deltas, it is pointed out that barrier bars frequently are deposited along the seaward margins of deltas parallel with the coastline. In Figure 2-4 the place to look for a barrier bar is along the hingeline or barely basinward of it. An excellent example of an ancient subsurface delta with an associated barrier bar is the Booch (Pennsylvanian) delta of east-central Oklahoma (Busch 1953, 1959, 1971).

The GIS generally is too thin a stratigraphic unit to be identified and isopached from seismic data. However, a GSS frequently can be identified and mapped from variable density seismic profile data, but should be integrated with available well log data. When integrating the seismic and electric log data, the seismic data should be corrected for variations in interval velocity to fit the thickness values picked from electric logs.

DEPOSITIONAL SEQUENCE

John L. Rich (1950, 1951) was among the first to recognize the significance of the fact that marine strata do not always increase in thickness out to the center of an epeiric sea. To describe the situation involving basinward thinning of strata, he coined the phrase "impoverished basin," or "hungry basin." His conclusions have been confirmed by seismic profiling of sediments of the shelf, bathyal, and abyssal environments. These sediments and environments are illustrated in Figure 2-5. The undaform of Rich (1951) corresponds to the shelf, the clinoform to the bathyal, and the fondoform to the abyssal environments, respectively. The sediments deposited in these several environments are the undathem, the clinothem, and the fondothem. These sediments may be seen to thin as they "cascade" down the clinoform.

In "impoverished" basin areas clinothem sediments thin in the direction of the fondothem, in spite of the fact that these sediments constitute a genetic unit. Such genetic units should not be called either a GIS or a GSS inasmuch as a part of the upper boundary is not defined by a marker bed everywhere. This is well illustrated by Mitchum et al. (1977) and is shown in Figure 2-6. In Figure 2-6A they illustrate what they refer to as a "depositional sequence." The numbered units from 6–14, inclusive, make up this sequence. It should be noted that an unconformity truncates units 6–8 and the upper portion of unit 9. Thus, this diachronous boundary represents a hiatus of variable duration due to both erosion and nondeposition. This unconformity, however, has time-stratigraphic significance in that it generally produces a seismic reflection related to a velocity-density contrast in the sediments above and below. Disconformities and bedding planes also are capable of producing similar seismic events. It is such stratigraphic intervals as those embracing all of units 6 - 14 in Figure 2-6 that Mitchum et al. have identified as a depositional sequence, which they define as:

> a stratigraphic unit composed of a relatively conformable succession of genetically related strata and bounded at its top and base by unconformities or their correlative conformities. A depositional sequence is chronostratigraphically significant because it was deposited during a given interval of geologic time limited by the ages of the sequence boundaries where they are conformities, although the age range of the strata within the sequence may differ from place to place where the boundaries are unconformities. Two types of chronostratigraphic surfaces are related to sequences: (1) unconformities and their correlative conformities forming sequence boundaries, and (2) stratal (bedding) surfaces within sequences.

Depositional sequences may be identified just as readily from geophysical logs as from seismic data. In the case of electrical logs, thin limestones and siltstones generally serve to identify the boundaries of the inclined units of the clinothem.

Figure 2-7 (Rascoe, 1982) is a vertically exaggerated cross section of some Upper Pennsylvanian (Virgilian) sediments of the northeastern shelf of the Anadarko Basin of Beaver County, Oklahoma. The Laverty-Hoover depositional sequence diverges systematically from east to west to the B-3 well and then thins basinward to the west. Practically all of the section between wells H-1 and B-7 consists of clinothem. The shelf edge of the undathem is shown in well H-1. The edge of the fondothem is shown only in well

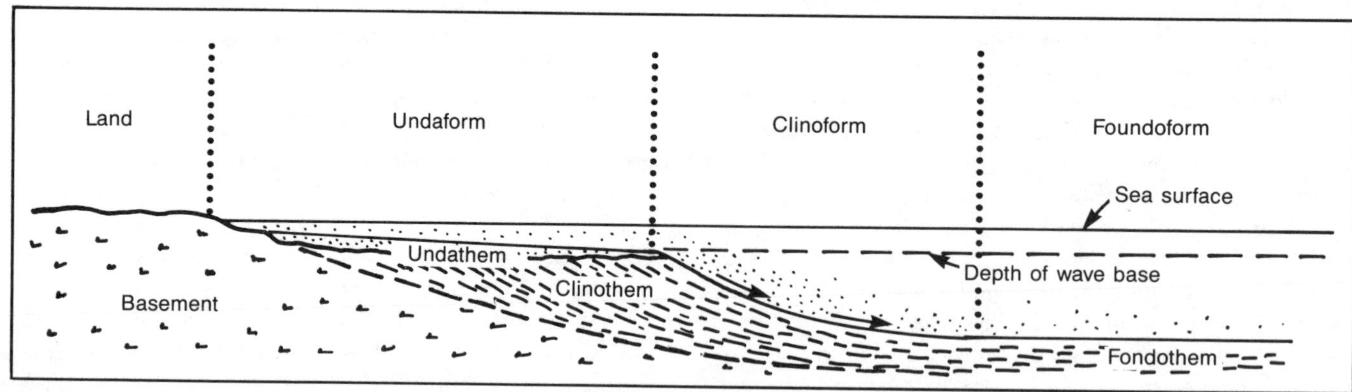

Fig. 2-5. Sediments and their respective environments of deposition along a continental margin. Similar conditions occur locally within an epi-continental seaway. Sediments include the undathem, clinothem, and the fondothem. Their respective environments of deposition are the undaform, clinoform, and the fondoform. Muddy water after a storm is shown by stippling and density currents by arrows. The vertical scale is greatly exaggerated. (Modified after John L. Rich, 1951; permission to publish by GSA).

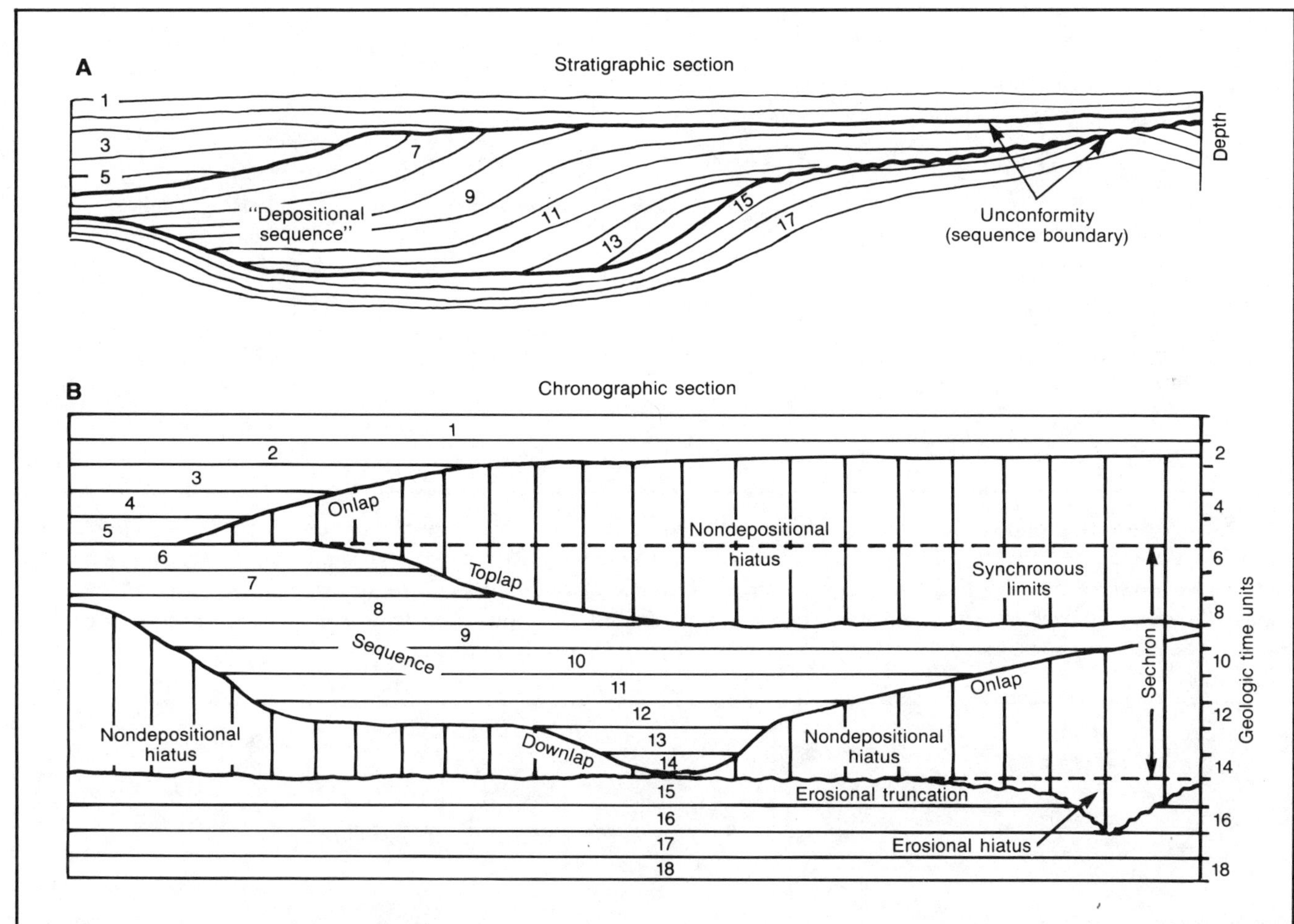

Fig. 2-6. Basic concepts of a Depositional Sequence. A Depositional Sequence is a stratigraphic unit composed of relatively conformable successions of genetically related strata and bounded at its top and base by unconformities or their correlative conformities.

A. *Generalized stratigraphic section of a sequence. Sequence boundaries (heavy lines) pass laterally from unconformities to correlative conformities. Individual units of strata 1 through 18 are traced by following stratification surfaces and assumed conformable where successive strata are present. Where units of strata are missing, hiatuses are evident.*

B. *Generalized chronostratigraphic section of a sequence. Stratigraphic relations shown in A are replotted here in chronostratigraphic section (geologic time is the ordinate). Geologic-time ranges of all individual units of strata are given as equal. Geologic-time range of sequence varies from place to place, but variation is confined within synchronous limits. These limits are determined by those parts of sequence boundaries that are conformities. Here, limits occur at beginning of unit 14 and end of unit 6. A sechron is defined as maximum geologic-time range of a sequence.*

(Modified after Mitchum, Vail, and Thompson, 1977; permission to publish by AAPG).

B-8. The Laverty-Hoover sandstone lens occurs entirely within this clinothem and affords ideal conditions for a stratigraphic trap. Khaiwka (1968) documented the Hoover pool of this area as occurring under exactly these conditions. The much smaller, isolated lens of Hoover sandstone between wells B-6 and B-7 occurs a similar distance below the top of the Brownsville limestone (Fig. 2-7) as the much larger "Laverty" Hoover sandstone to the east, but it is a distinctly younger sandstone lens.

The eastern shelf and slope deposits of the Midland Basin of north-central Texas present a classic example for a study of genetic units of stratigraphy. This was first pointed out by Brown (1969). He states that the sediments include "10 to 15 repetitive sequences including open shelf, deltaic, fluviatile and interdeltaic depositional systems."

> Elongate sandstones are generally arranged parallel to paleoslope in vertically offset patterns controlled by differential compaction of fluvial and deltaic sands and interdistributary

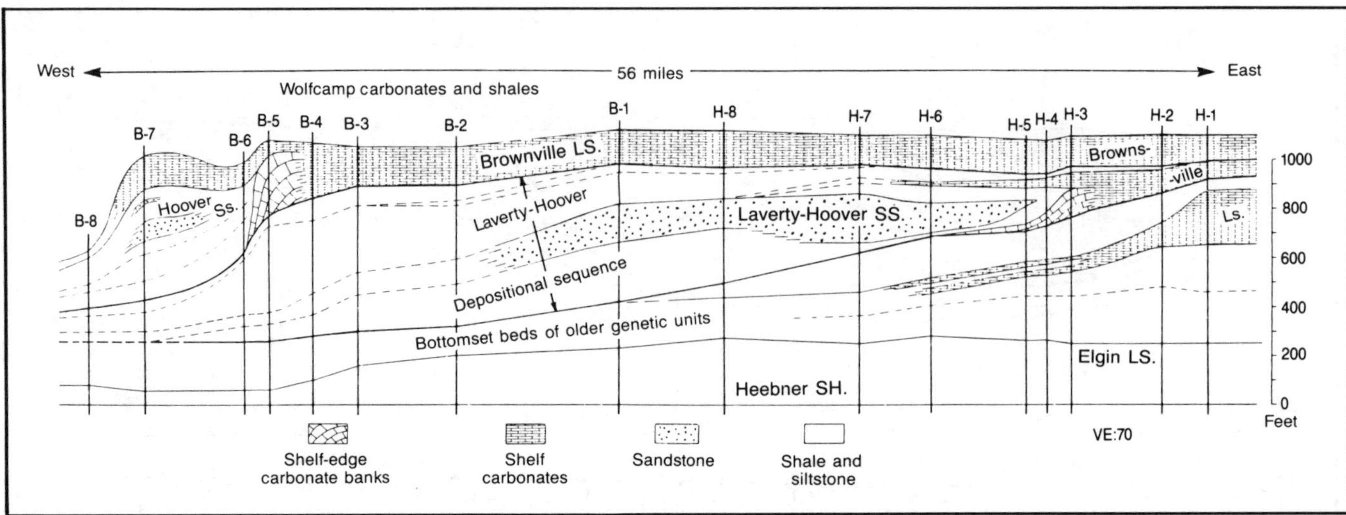

Fig. 2-7. West-east stratigraphic profile illustrating the lens shape of the Laverty-Hoover depositional sequence on the northeast flank of the Anadarko Basin in Beaver and Harper counties, Oklahoma. The small isolated Hoover sandstone lens between wells B-6 and B-7 occurs a similar vertical distance below the top of the Brownsville limestone as the much larger "Laverty" Hoover sandstone to the east but it is a distinctly younger sandstone lens. (Modified after Rascoe, 1982; permission to publish by AAPG).

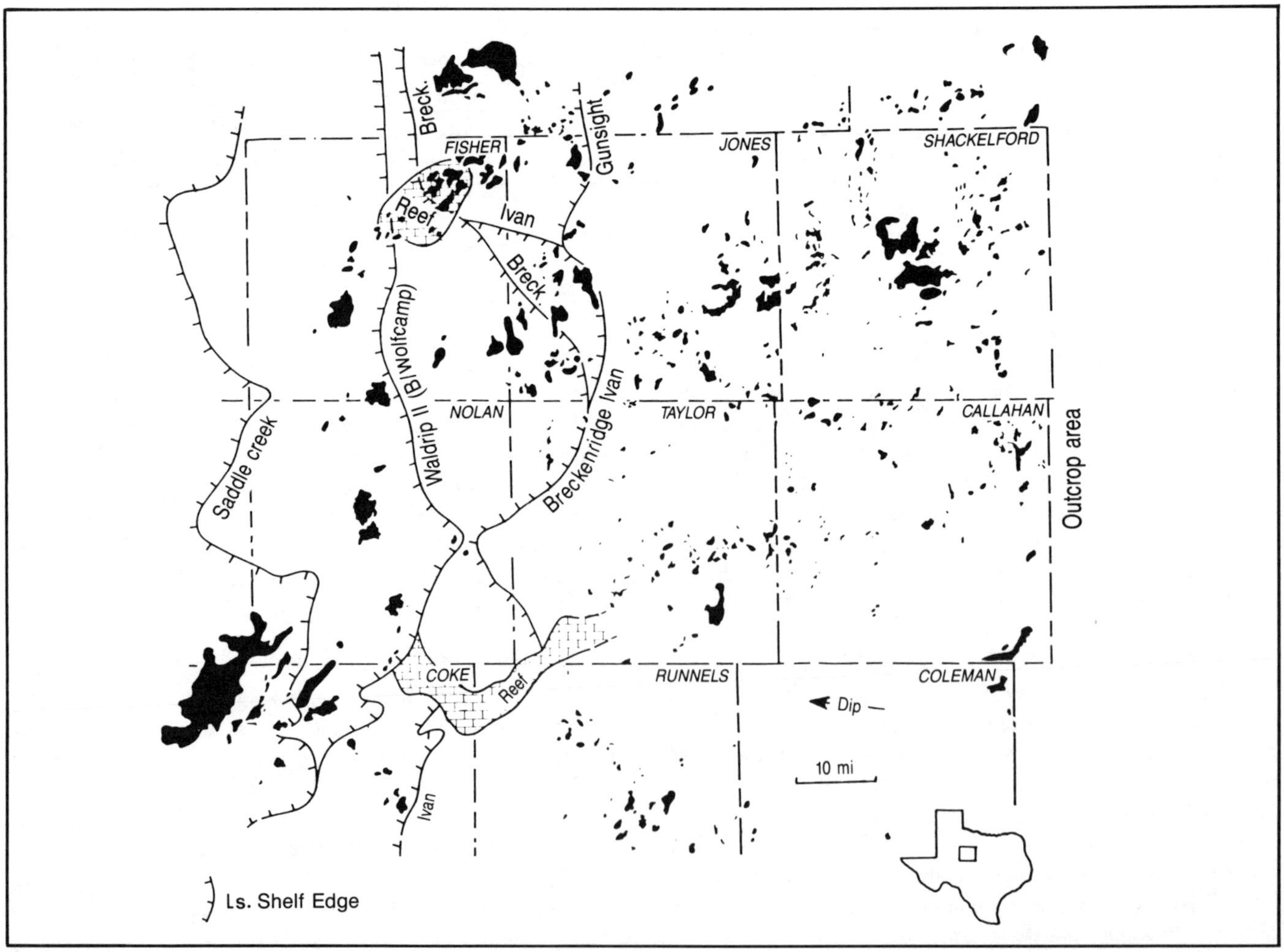

> muds . . . Initial Cisco deltas followed a paleosurface grain controlled by underlying bank limestones; this orientation was maintained during previous systems and, in turn, provided control for the next deltaic episode . . . Superposed sequences of limestone-bounded dominantly terrigenous clastic rocks are extensively distributed throughout approximately 25 counties on the Eastern Shelf. Bounding limestone units are not necessarily time-stratigraphic but represent the best time-markers in the section. Limestone-bound stratigraphic units or sequences are persistent, mappable rock units at the outcrop and in the subsurface and have genetic significance important in understanding the origin of the repetitive shelf facies. In this report, these subdivisions are called *formats* (Forgotson, 1957). The marker-defined operational units are informal stratigraphic subdivisions designated by the names of bounding limestones (e.g., Home Creek Bunger format). Virgil and Wolf Creek formats are commonly 100 feet thick in upslope areas, but several superposed sequences may be combined into thicker formats useful in specific problems of stratigraphic analysis.

The Virgil and Wolfcamp sediments on the eastern shelf of the Midland Basin cover approximately 25 counties in an area 75 mi (120 km) wide (east-west). Brown (1969) states the Virgil and Wolfcamp strata of this shelf consist "of thin, persistent limestones which are interstratified with thicker mudstone or shale units containing sheet sandstones, elongate sandstones, and less common coal or bituminous shales. Ten to 15 mudstone and sandstone sequences (or "cycles") are separated by regionally persistent limestone beds within the 1,200 ft (360 m) section." A portion of this shelf area is illustrated in Figure 2-8 and includes all of the oil pools producing within this area. As the shelf sediments were deposited, they prograded progressively further to the west. The hinge lines of successively higher stratigraphic units are indicated in this figure. Multiple GIS's occur on the shelf area (where there is progressive divergence westward). Each of these genetic units is defined at the top by a limestone marker bed.

There is considerable oil production west of the several hinge lines shown in Figures 2-8 and 2-9. All of the reservoirs west of these several hinge lines consist of lenses of sandstone that make up a portion of the clinothem. In a classic unpublished paper, presented as a Distinguished Lecture for AAPG, Bloomer (1981) presented a detailed analysis of several of the pools in the eastern portion of the Midland Basin. Figure 2-10 is one of his cross sections that clearly illustrates several westward prograding shelves and the positions of oil-bearing sandstones. The stratigraphic interval between the Flippen and Waldrip II limestones constitutes a depositional sequence. It occurs in the clinoform and thins conspicuously to the east to well 13, which is on the edge of the Waldrip II hinge line. There is

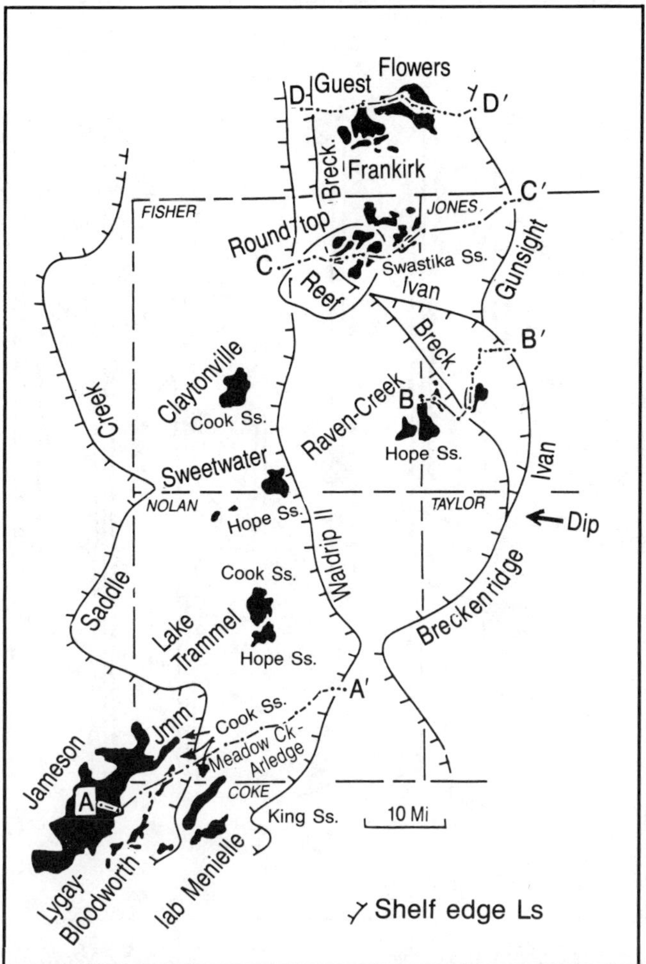

Fig. 2-9. Slope sandstone reservoirs on the east side of the Midland Basin of north-central Texas. Westward, prograding shelf edges (hinge lines) are indicated by hachured lines. The location of cross section A-A′ (Fig. 10) is shown in the southwestern portion of the map. The Jameson pool in this area is the largest Cook sandstone reservoir discovered to date and occurs midway down the slope. (Modified after Bloomer, 1981; permission to publish by AAPG).

more gradual thinning to the west-southwest as the fondoform is approached. This profile points to the fallacy in matters of correlation in trying, for example, to maintain a uniform interval distance below the Saddle Creek limestone while following the Cook sandstone, (Group 4000), which occurs just below the Flippen limestone. Figure 2-11 is an isopach map of the Cook sandstone in a part of the Jameson field. This sand consists of a slope fan delta sit-

Fig. 2-8. Eastern portion of Midland Basin, north-central Texas, showing Virgillian and Wolfcamp U. Penna and L. Perm. sandstone reservoirs on the shelf (east one-half) and slopes (west one-half). Irregular westward prograding shelf edges are indicated by hachured lines. (Modified after Bloomer, 1981; permission to publish by AAPG).

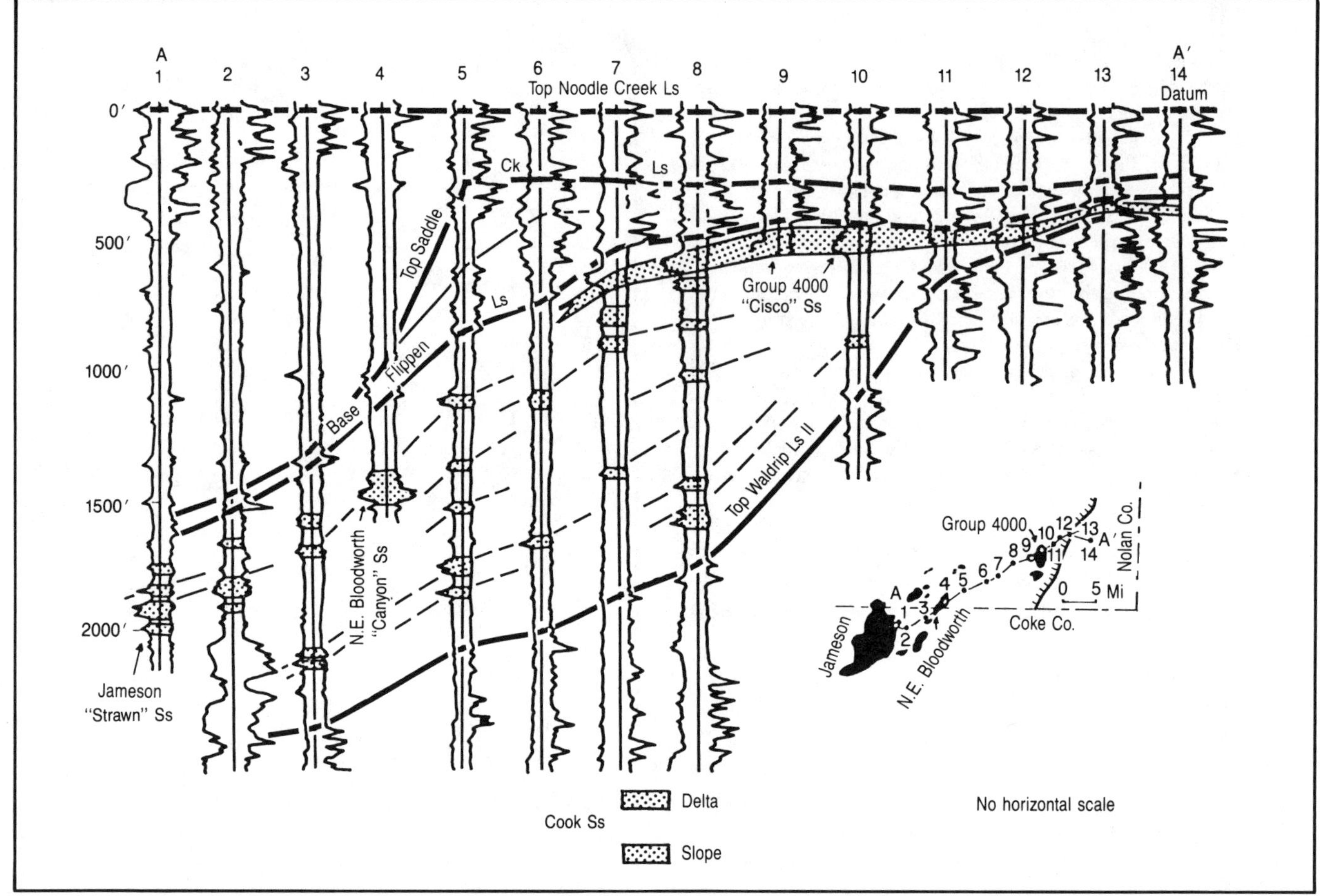

Fig. 2-10. Stratigraphic profile A-A′ of the Lower Wolfcamp series on the east side of the Midland Basin of north-central Texas, illustrating an excellent example of a westward dipping depositional sequence. This sequence occurs between the Flippen and Waldrip II limestones. It occurs in the clinoform environment, thins abruptly at the shelf edge (well 13) and more gradually to the west-southwest. The Cook sandstone is the principal producing reservoir within this depositional sequence. The location of the profile is shown on Figure 2-9. (Modified after Bloomer, 1981; permission to publish by AAPG).

uated in the middle portion of the clinoform, well below the shelf edge. Axial divergent trends of maximum sandstone thickness are indicated by the heaviest solid lines.

PARALIC SEQUENCE

The term *paralic* was first proposed by Krumbein and Sloss (1958) and pertains to environments of the marine borders, such as lagoonal, littoral, shallow neritic, etc.; however, it remained for Weber (1971) to apply this term to the phrase *Paralic Sequence*. His studies of vertical sequences of sediments in shallow boreholes in Recent and deep boreholes in the Tertiary of the Niger delta serve as a basis for recognizing the cyclic nature of the sedimentation within the numerous depocenters of this large delta. A complete cycle may range from 45 to 330 ft (15 to 100 m) in thickness but they are seldom more than 180 ft (60 m). Weber states that "a complete cycle generally consists of a thin fossiliferous transgressive marine sand followed by an offlap sequence which commences with a marine shale and continues with laminated fluviomarine sediments. Barrier-bar and/or fluviatile sediments may follow before another transgression terminates the cycle. This type of cycle is also known from other deltas described for instance by Kruit (1955) and Oomkens (1967)."

Figure 2-12 is an example of a series of offlap paralic sequences in the Uzere West Field of the Niger delta. Seven cycles of sedimentation are shown. Weber states, "Most cycles begin with the erosion of the underlying sand unit and the deposition of a thin fossiliferous transgressive marine sand. These sands can be recognized by their relatively high resistivity because their pores are partly filled with carbonate cement. Often the gamma radiation emitted by the transgressive sands is also high (e.g., in the topmost cycle in Fig. 2-12) due to a high percentage of the potassium-rich glauconite . . . This type of transgressive sand

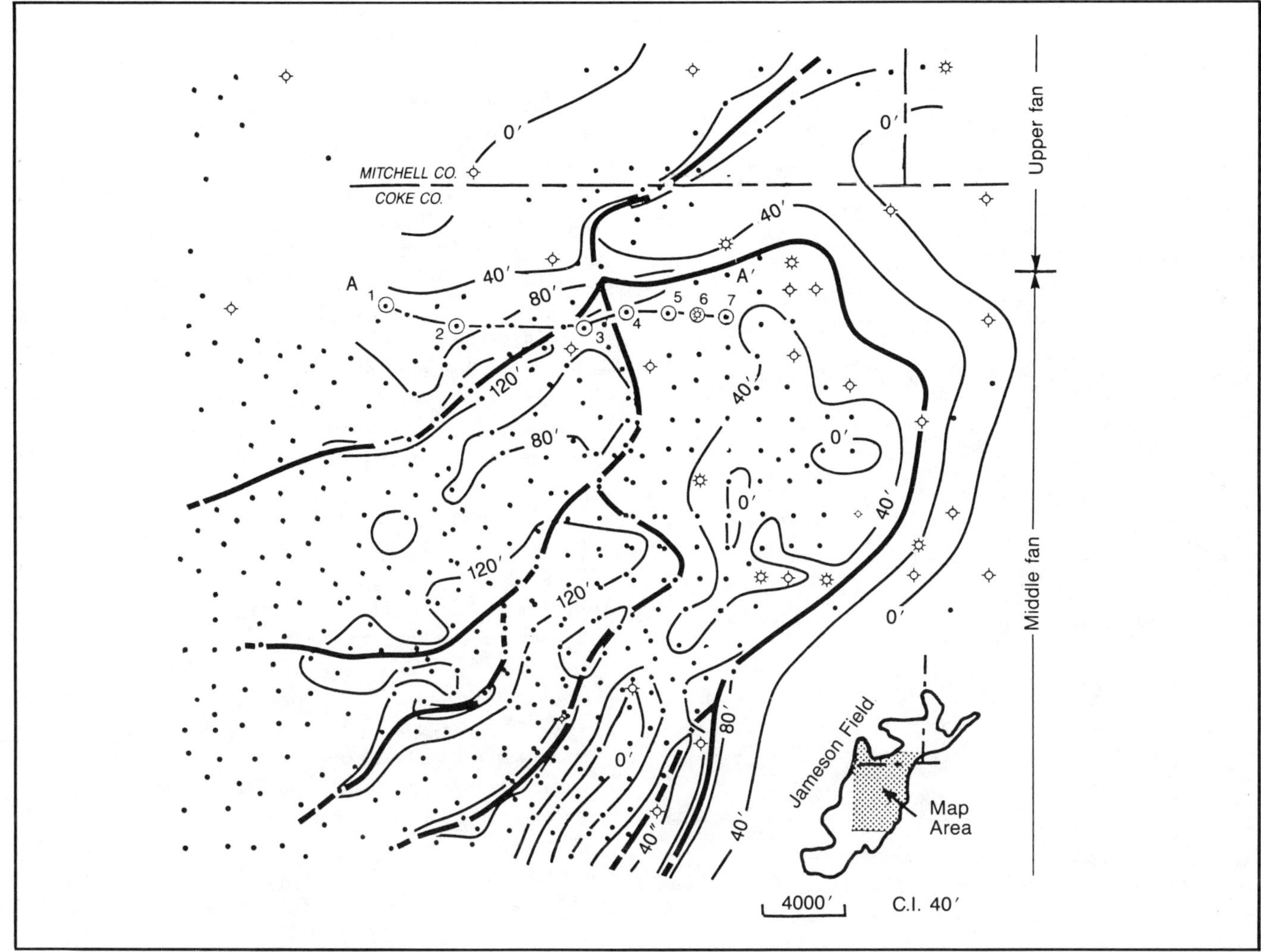

Fig. 2-11. Isopach map of the Cook sandstone in a part of the Jameson pool of north-central Texas, showing a series of southwestward diverging distributary axes of a slope fan delta (heavy, sinuous, solid lines). (Modified after Bloomer, 1981; permission to publish by AAPG).

is probably associated with a regional transgression which pushed the shoreline back over a considerable distance." Such transgressions are directly the result of a vertical pulse of movement along a growth fault on the landward side of a local depocenter. These very thin, high-resistivity sands are derived mainly from reworking and winnowing of the subjacent eroded beds. Thus, there is a minor unconformity at the base.

Marine clays (shaded on Fig. 2-12) overlie the thin transgressive sands and usually are quite silty and sandy. Weber states, "A gradual decrease of faunal remains and a diminishing diversity in species indicates an upwardly increasing sedimentation rate. Burrowing is common and the original layering is usually destroyed; . . . clay and plant remains can be abundant. The faunal interpretation, combined with the occasional abundance of plant remains, indicate that the clays were deposited in the inner to middle neritic zones . . . The marine clays are of great importance with respect to the hydrocarbon accumulations because they form the seal over the top of the reservoirs."

Fluvial-marine sandstones almost invariably overlie the marine clays. These sandstones might consist of barrier bar(s), point bar(s), distributary channel sand, tidal channel sands, or combinations of all four. If the sandstone above the marine clay is a barrier bar, the two lithologies are separated by a transitional zone called the bar foot. This bar foot consists of alternating fine-grain sandstone, siltstone, and shale and owes its presence to an oscillating wave base. Gamma radiation in this "barrier foot" is frequently higher than that of the marine clay. This is attributed to the presence of silt size zircon with its relatively high thorium and uranium content.

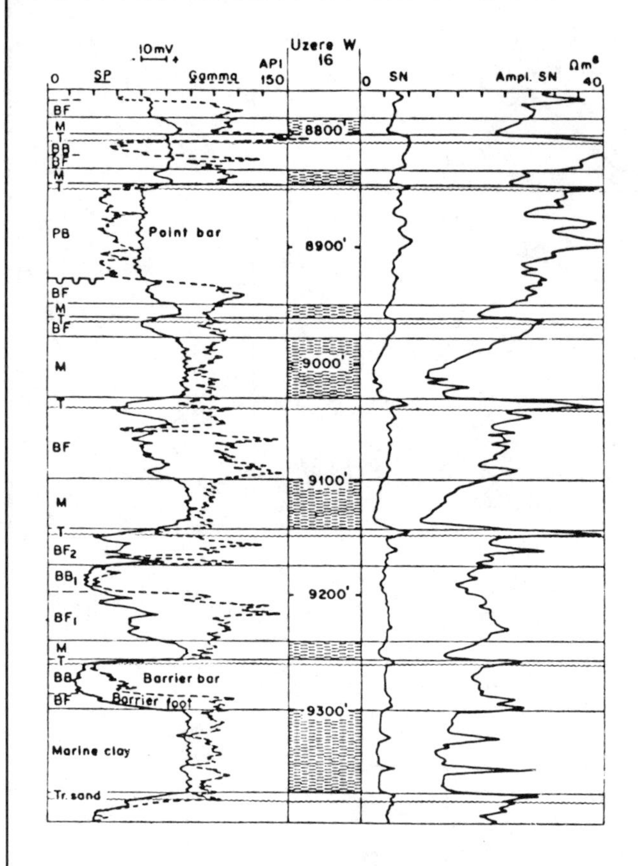

Fig. 2-12. Example of a series of paralic sequences in the Uzere West Field, Niger delta, Nigeria. (Modified after Weber, 1971; permission to publish by Geologie en Mijnbouw).

SELECTED BIBLIOGRAPHY

Bloomer, R. R., 1973, Fluvial, deltaic, and slope reservoir sandstone in west-central Texas (abs.): AAPG Bull., v. 57, no. 2, p. 420.

———, 1981, Depositional environments and reservoir morphologies of channel sandstones (abs.): Distinguished Lecture Tour abstracts 1981-1982, AAPG Bull., v. 65, no. 11, p. 2465.

Brown, Jr., L. F., 1969, Geometry and distribution of fluvial and deltaic sandstones (Pennsylvanian and Permian), north-central Texas: Bur. Econ. Geol., Geol. Circ. 69-4; also reprinted from Gulf Coast Assoc. Geol. Socs. Trans., v. 19, p. 23-47.

Busch, D. A., 1953, The significance of deltas in subsurface exploration: Tulsa Geol. Soc. Digest, v. 21, p. 71-80.

———, 1959, Prospecting for stratigraphic traps: AAPG Bull., v. 43, no. 12, p. 2829-2943.

———, 1971, Genetic units in delta prospecting: AAPG Bull., v. 55, no. 8, p. 1137-1154.

———, 1974, Stratigraphic traps in sandstones-exploration techniques: AAPG Mem. 21, 174 p.

Forgotson, J. M., 1957, Nature, usage, and definition of marker-defined vertically segregated rock units: AAPG Bull., v. 41, no. 9, p. 2108-2133.

Hedberg, H. D., 1948, Time stratigraphic classification of sedimentary rocks: GSA Bull., v. 59, no. 5, p. 447-462.

———, 1958, Time stratigraphic classification and terminology: AAPG Bull., v. 42, no. 8, p. 1881-1896.

Hollenshead, C. T. and R. L. Pritchard, 1961, Geometry of producing Mesa Verde sandstones, San Juan basin, *in* J. A. Peterson and J. C. Osmond, eds., Geometry of Sandstone Bodies: Tulsa, AAPG, p. 98-118.

Khaiwka, M. H., 1968, Geometry and depositional environments of Pennsylvanian reservoir sandstones, northwestern Oklahoma, Ph.D. Diss., Univ. of Oklahoma, 126 p.

Kruit, C., 1955, Sediments of the Rhone delta, I. Grain size and microfauna: Verh. Kon. Ned. Geol. Mijnb. Gen. Geol. Ser. 15, p. 357-514.

Krumbein, W. C., and L. L. Sloss, 1958, Principles of stratigraphy and sedimentation: San Francisco, W. H. Freeman and Co., 497 p.

Mitchum, R. M., Jr., P. R. Vail, and S. Thompson, III, 1977, The depositional sequence as a basic unit for stratigraphic analysis; part 2, seismic stratigraphy and global changes of sea level, *in* Seismic stratigraphy application in hydrocarbon exploration: AAPG Mem. 26, p. 53-62.

Oliver, T. A., and N. W. Cowper, 1965, Depositional environments of Ireton formation, central Alberta: AAPG Bull., v. 49, no. 9, p. 1140-1425.

Oomkens, E., 1967, Depositional sequences and sand distribution in a deltaic complex, a sedimentological investigation of the post-glacial Rhone delta complex: Geologie en Mijnbouw, 46, p. 265-278.

Rascoe, B., Jr., 1962, Regional stratigraphic analysis of Pennsylvanian and Permian rocks in western mid-continent, Colorado, Kansas, Oklahoma, Texas: AAPG Bull., v. 46, no. 8, p. 1345-1370.

———, 1982, Genetic units in petroleum exploration, Unpub. report, 10 p.

Rich, J. L., 1950, Flow markings, groovings, and intrastratal crumplings as criteria for recognition of slope deposits, with illustrations from Silurian rocks of Wales: AAPG Bull., v. 34, no. 4, p. 717-741.

———, 1951, Three critical environments of deposition and criteria for recognition of rocks deposited in each of them: GSA Bull., v. 62, no. 1, p. 1-20.

Shepard, F. P., 1932, Sediments of continental shelves: GSA Bull., v. 43, no. 12, p. 1017-1039.

——— and G. V. Cohee, 1936, Continental shelf sediments of the Mid-Atlantic States, GSA Bull., v. 47, no. 3, p. 441-458.

Van Sicklen, De W. C., 1958, Depositional topography—examples and theory, AAPG Bull., v. 42, no. 8, p. 1897-1913.

Wanless, H. R., and J. M. Weller, 1932, Correlation and extent of Pennsylvanian cyclothems: GSA Bull., v. 43, no. 4, p. 1003-16.

Weber, K. J., 1971, Sedimentological aspects of the Niger delta: Geologie en Mijnbouw, v. 50, no. 3, p. 559-576.

Weirick, T. E., 1953, Shelf principle of oil origin, migration, and accumulation: AAPG Bull., v. 37, no. 10, p. 2027-2045.

Wheeler, H. E., and V. S. Mallory, 1953, Designation of stratigraphic units: AAPG Bull., v. 37, no. 8, p. 2407-21.

——— and ———, 1956, Factors in lithostratigraphy: AAPG Bull., v. 40, no. 11, p. 2711-23.

3 ALLUVIAL FANS

INTRODUCTION

An alluvial fan forms where a stream emerges from a mountain front or highland and deposits a sedimentary unit with a surface resembling the segment of a cone (Bull, 1972). Fan deposits are radial from the source and have concave-up longitudinal profiles and convex-up transverse profiles (Fig. 3-1). Fan slopes, ranging from 1° to 25° (Denny, 1965; Bull, 1977), vary directly with sediment grain size (Hooke, 1968), which varies from boulders to mud. Fans may be solitary bodies or coalesce to form bajadas, and they occur in all climates from arctic to tropical and arid to humid. Progradation of a fan into a standing body of water produces a fan delta, e.g., the Yallahs fan delta in southeastern Jamaica (Wescott and Ethridge, 1980).

ALLUVIAL FAN DIVISIONS

A fan may be divided into three parts (Fig. 3-2A; McGowen and Groat, 1971). Inner, middle, and outer fan divisions are similar to those found on subaqueous fans (Chapter 12). Near the fanhead or apex is the inner fan. A straight, entrenched, relatively broad and deep channel occurs here that conducts the source stream and its sediments out onto the fan. Channel deposits are very coarse grained. Channel margin, levee, and interchannel deposits are finer grained and may be interspersed with landslide and debris flow deposits derived from the divides between principal drainages (Fig. 3-2A; Nilsen, 1982).

The middle fan is a broad area of many shallow, radiating channels, only a few of which are active at any one time. Channels migrate rapidly and are sites of rapid sedimentation. Channel type depends on sediment character and climate and may be straight, braided, or meandering (Nilsen, 1982).

Below the middle fan is the slightly channeled to unchanneled outer fan. Sediments are fine-grained and laterally extensive compared with those higher on the fan. Deposition results from streamflows emerging from middle fan channels and spreading laterally. Outer fan sediments may be truncated by or interfinger with deposits from systems active along the basin axis, e.g., eolian, lacustrine, alluvial valley (Fig. 3-2B; Nilsen, 1982).

DEPOSITIONAL PROCESSES

Running water and debris flow are the two main depositional processes on an alluvial fan. Fan deposition may be dominated by one process or the other, or any combination of the two. Depositional process also changes downfan with debris flows being more common nearer the fanhead and water-laid deposits more common distally (Hooke, 1967; Bull, 1972; Nilsen, 1982).

Water-laid Deposits

Three categories of water-laid deposits occur on alluvial fans, streamflow, sheetflood, and sieve deposits (Bull, 1972, Gloppen and Steel, 1981; Nilsen, 1982). Rudaceous water-laid deposits may be interbedded with laminated sandstones and are distinguished from comparable debris flow rudites by being better sorted, better rounded, and better stratified (Gloppen and Steel, 1981).

Streamflow deposits. These are channel fill, channel margin, levee, and overbank interchannel deposits (Nilsen, 1982), which are generally poorly sorted, poorly bedded, and very coarse- to fine-grained (Bull, 1972). Deposits are long, narrow, overlapping (Bull, 1972; Fig. 3-1, transverse section C-D), and occur from the inner to the outer fan. Sedimentary structures include: erosional bases; large- to

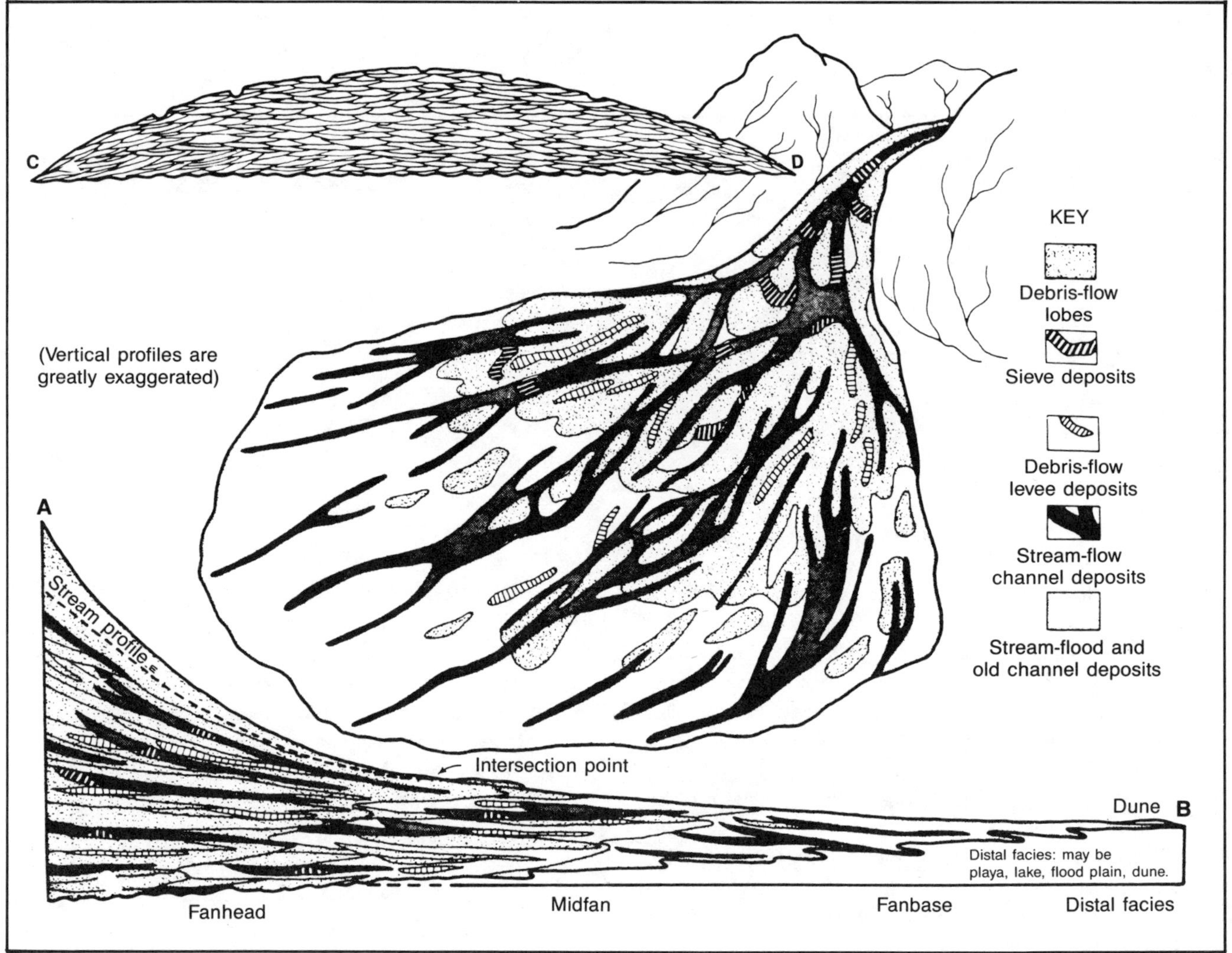

Fig. 3-1. Schematic representation of fan surface, longitudinal profile (AB), transverse profile (CD), and distribution of sedimentary facies. (From Spearing, 1974; permission to publish by GSA).

small-scale cross-bedding related to longitudinal and transverse bars, ripples, and channel fillings; reactivation surfaces; and parallel stratification from deposition in the upper flow regime. Conglomerates and pebbly sandstones over one meter thick become finer-grained and thinner downfan. Grain size and bed thickness of overbank deposits also decrease laterally away from channels (Gloppen and Steel, 1981; Nilsen, 1982).

Sheetflood deposits. Sediment-laden streams emerging from channels deposit their loads due to loss of competence as the flow spreads laterally. Deposits of granule and small pebble conglomerate and sandstone are relatively well-sorted and interbedded with sheets of finer-grained sediment. Beds are thin and laterally extensive, with massive, cross-bedded, and laminated internal stratification (Bull, 1972; Nilsen, 1982). Gloppen and Steel (1981), however, indicate limited lateral extent of beds with individual unit widths of 20 m or less on Devonian fans in Norway. Sheetflood deposits are restricted to the lower parts of fans, and waning flood stages frequently cut shallow channels into the deposits (Gloppen and Steel, 1981; Nilsen, 1982).

Sieve deposits. If the surface fan deposits are sufficiently coarse and permeable such that flood waters completely infiltrate the fan prior to reaching the toe, a lobe of coarse-grained debris may be formed. The lobe will act as a sieve to subsequent flows and strain the sediment from water flowing through the lobe. Successive sieve deposits reduce the gradient and finer-grained materials accumulate behind the coarse lobe (Fig. 3-3). When infilling behind the coarse material is complete, more coarse sediment is brought to the lobe front and deposited immediately upfan from the previous lobe (Hooke, 1967), and the process is repeated.

Characteristics of sieve deposits on modern, arid region fans are good sorting, subangular clasts, and massive bedding with poorly defined contacts (Bull, 1972). Later infiltration of the open framework by fine-grained sediment

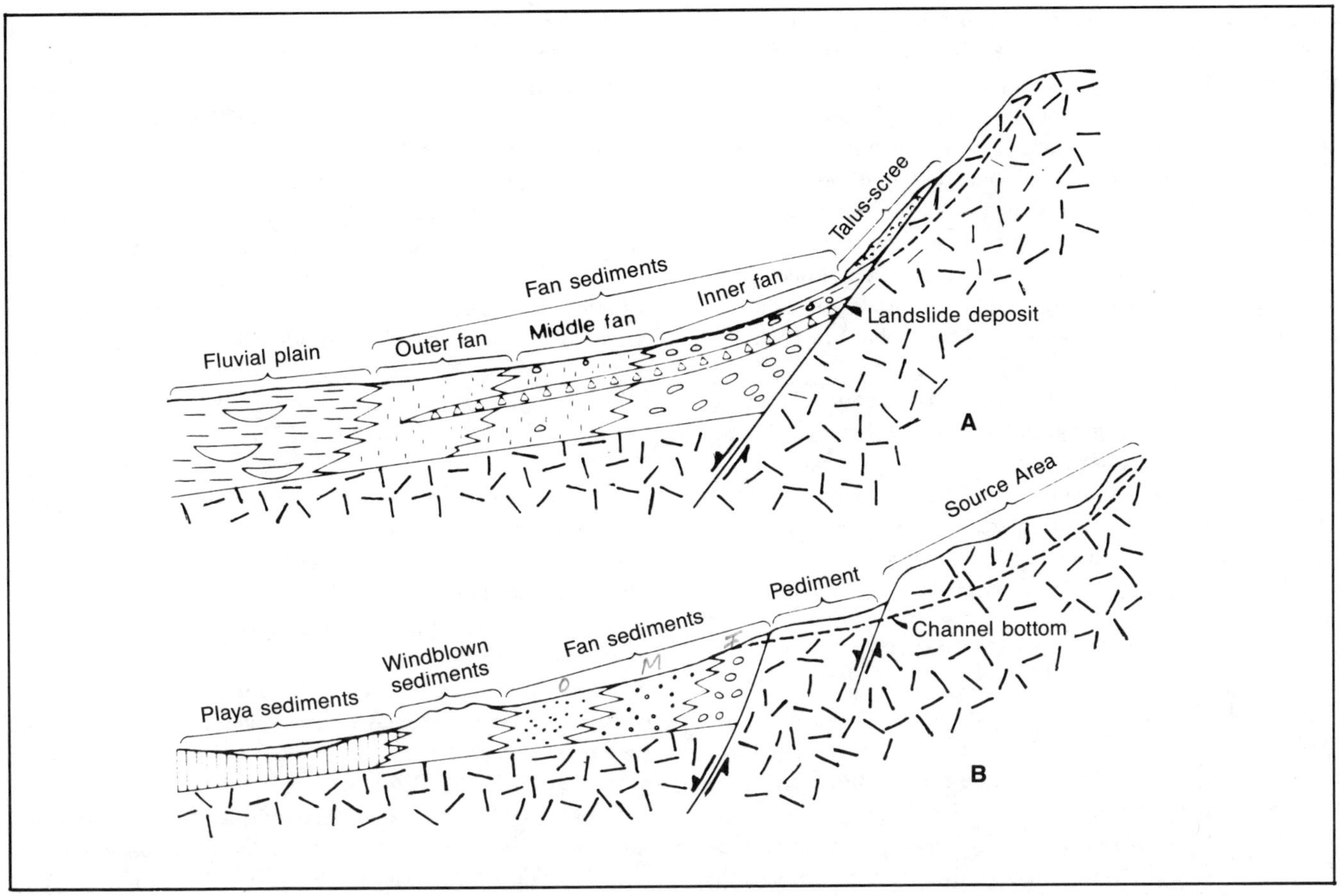

Fig. 3-2. Cross sections showing fan divisions and relationships to adjacent facies. (From Nilsen, 1982; permission to publish by AAPG).

makes the deposit bimodal (Reid, 1974; Gloppen and Steel, 1981).

Debris-Flow Deposits

A debris flow is a mixture of granule and larger clasts in a matrix of silt, clay, and water. Fine sediment concentration in the matrix is sufficiently high that it behaves like a plastic and has a finite yield strength. Matrix viscosity and density reach 1000 poises and 2.0 to 2.4 gm/cc, respectively (Sharp and Nobles, 1953). Equivalent values for water are 0.01 poise and l gm/cc. Density contrast between the clasts being transported and the matrix is so low that

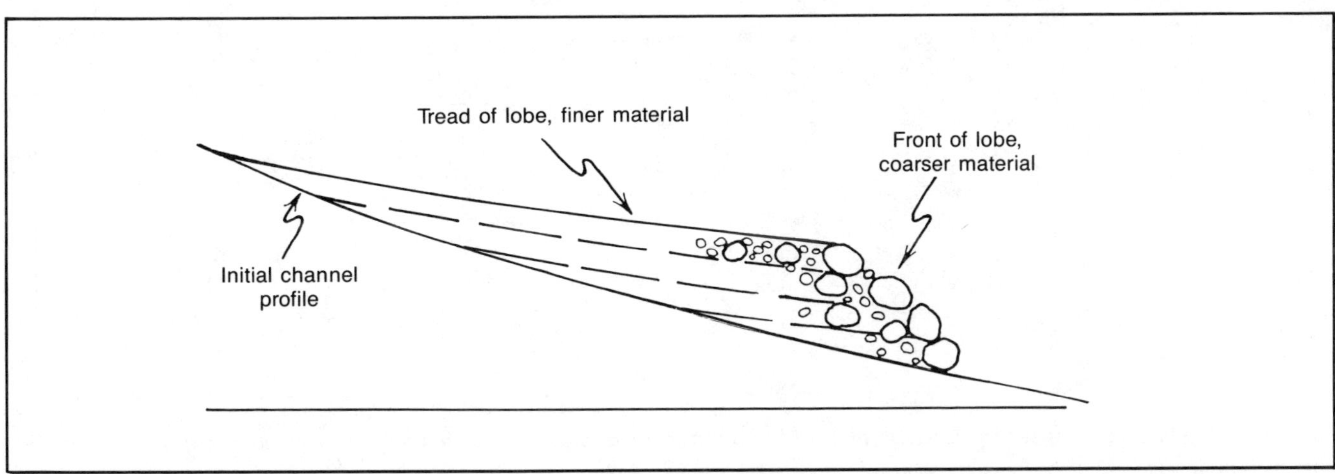

Fig. 3-3. Sketch of growth of sieve deposit. (From Hooke, 1967; permission to publish by Journal of Geology, University of Chicago).

the settling velocity of the clasts is greatly reduced and very large clasts may be transported (Hooke, 1967). Debris-flow deposits are lobate tongues of unstratified, poorly sorted, generally matrix-supported conglomerate. Basal contacts are sharp and margins are well defined (Hooke, 1967; Bull, 1972). Beds are ungraded to slightly reverse graded at the base. Grain size and bed thickness decrease downfan. Overlying an ancient flow may be a thinner, better-sorted, fine, massive conglomerate or laminated granule sandstone deposited by late-stage waning floods and winnowing of the top of the debris flow (Jahns, 1947; Johnson, 1970; Gloppen and Steel, 1981).

Several conditions must be met in a fan source area for a debris flow to occur. Included are production of fine-grained sediment by weathering of bedrock, steep slopes with little vegetation to inhibit erosion, and large quantities of water in short periods of time at irregular intervals (Bull, 1972). Intense storms over high-relief desert mountain basins with fine-grained bedrock meet the conditions.

ALLUVIAL FAN GEOMETRY

Bull (1972) recognizes three longitudinal section morphologies for alluvial fans. Uplift along a mountain front produces a fan that is thickest adjacent to the front (Fig. 3-4). High sediment yield from the elevated source area allows the fan to prograde over distal basin deposits producing a wedge-shaped fan body. Different radial section morphology occurs when uplift is continuous with fan deposition (Fig. 3-5). A lenticular fan body forms with the thickest accumulation occurring away from the mountain front. When uplift ceases and erosion of the highland continues the mountain front retreats forming a pediment (Fig. 3-6). A wedge-shaped body that thickens away from the mountain front is produced and any part of the fan above the pediment is eroded and deposited downfan.

If conditions are proper for the formation of one alluvial fan in an area, chances are other fans derived from other drainages also will occur. Individual fans and nonfan environments of deposition adjacent to an ancient fan in the subsurface may be differentiated with logs (Figs. 3-4 and 3-5).

FAN DELTAS

An alluvial fan prograding from a highland into a standing body of water forms a fan delta. Aside from normal alluvial fan deposits, a fan delta has two additional facies. One is a transitional facies of reworked fan deposits at the basin margin, and the other is a subaqueous fan facies basinward of the transitional facies. Characteristics of the two facies depend on the nature of the shelf over which the fan delta is prograding. Facies accumulating on a narrow, steep shelf or slope are different from those accumulating on wide, more gentle shelves (Wescott and Ethridge, 1980).

Depositional models and descriptions of transitional and subaqueous facies are based on studies of modern and ancient fan deltas. Wescott and Ethridge (1980) use the Yallahs fan delta as an example of deposition on a relatively steep slope. Located in southeastern Jamaica, the 10.4 sq

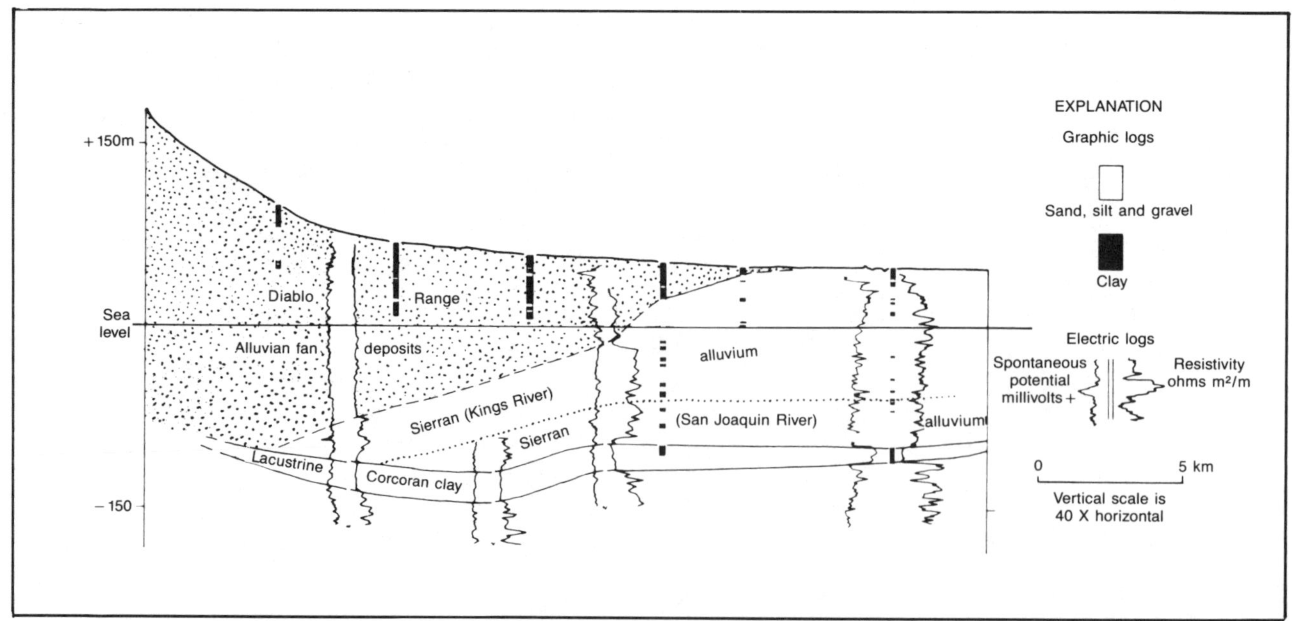

Fig. 3-4. Longitudinal cross section through alluvial fan deposits with thickest deposits adjacent to mountain front. (Modified after Magleby and Klein, 1965, by Bull, 1972; permission to publish by SEPM).

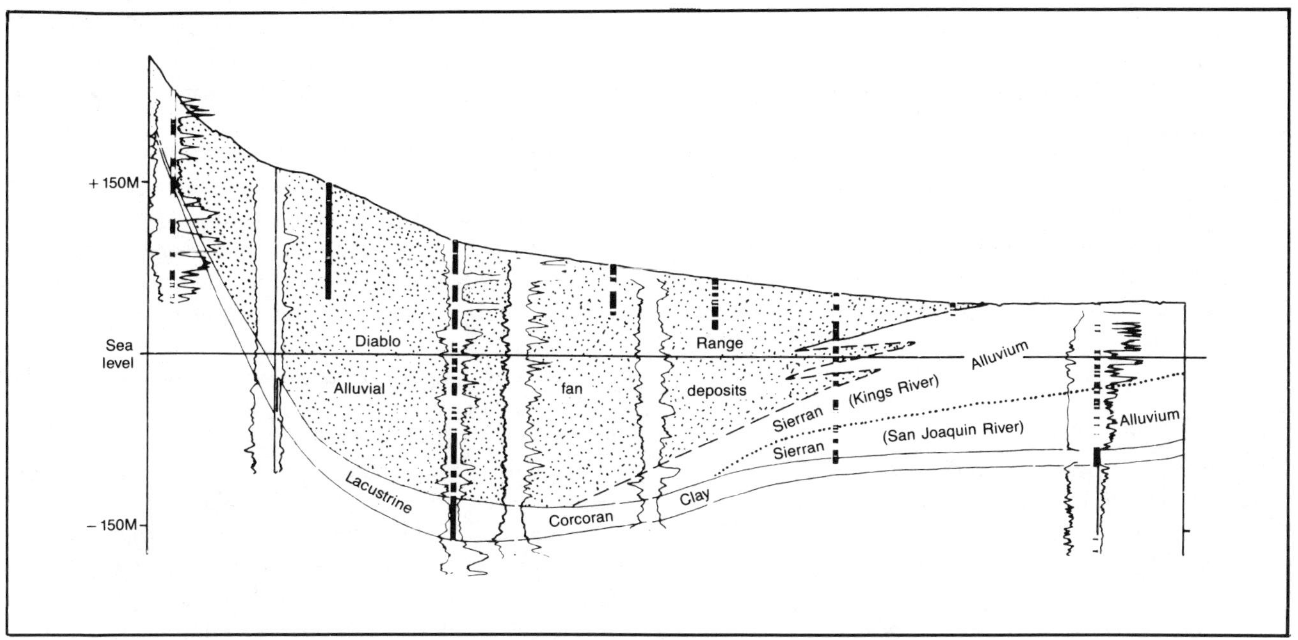

Fig. 3-5. Longitudinal cross section through lenticular alluvial fan deposits. (Modified after Magleby and Klein, 1965, by Bull, 1972; permission to publish by SEPM).

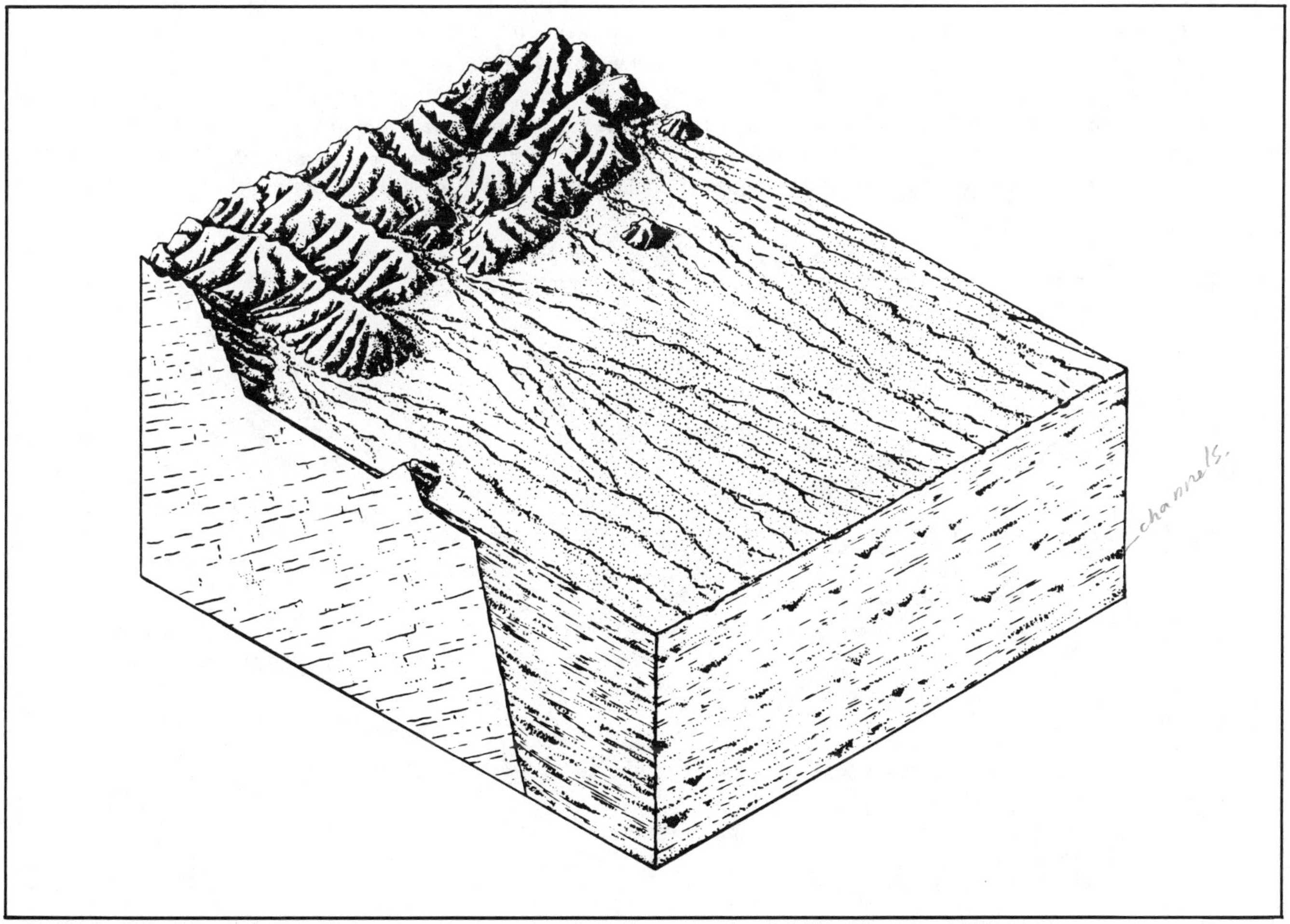

Fig. 3-6. Block diagram of retreating mountain front, pediment, and fan deposits that thicken away from the mountains. (From Bull, 1972; permission to publish by SEPM).

km delta is prograding a slope cut by three submarine canyons (Fig. 3-7). For a model of deposition on a shelf, Wescott and Ethridge (1980) use descriptions of proglacial fans along the southeast coast of Alaska published by Boothroyd and Ashley (1975), Boothroyd (1976), Galloway (1976), and Boothroyd and Nummedal (1978). Transitional and subaqueous fan delta deposits that prograded into lakes in the Devonian age Hornelen basin, Norway, are described by Steel et al. (1977), and Gloppen and Steel (1981).

Transitional Zone Facies

Deposits in the transitional zone are alluvial fan sediments that have prograded into a basin and have been modified by marine or lacustrine currents, waves, or tides. Wescott and Ethridge's model for a fan delta prograding onto a slope is the Yallahs alluvial fan where deposits are reworked in the beach environment. Two types of beaches, depositional and erosional, occur in the transitional zone and their sedimentary characteristics differ significantly. Depositional beaches are broad and sandy with a wide backshore, well-developed berm, and gently sloping foreshore. Backshore sands are flat or slope slightly landward and foreshore sands dip slightly seaward. Grain size increases from the backshore, across the berm, and down the foreshore to the plunge point. Wave-reworked conglomerate horizons interbedded with the beach sands are less lenticular and better bedded than alluvial fan gravel zones and are more common in the lower foreshore and storm berms.

Erosional beaches occur at the bases of scarps (Fig. 3-8). There is little or no backshore and the foreshore is steep and narrow. Sediments are lag deposits from marine erosion of the scarps and consist of boulders to very coarse sand from which the fine-grained material has been winnowed. Conglomerates are clast supported and interstitial materials are poorly sorted sands. Eolian deposits of limited extent also occur in the Yallahs transition zone. They are well-sorted and massive to poorly stratified (Wescott and Ethridge, 1980).

Rather than prograding onto a slope, coastal Alaskan fan deltas prograde across a wide, relatively gently sloping shelf. Sand beaches, offshore bars, or barriers separated from the fan by lagoons form the transitional zone. Intertidal mud flats and marsh muds may occur landward of the beaches (Wescott and Ethridge, 1980).

Devonian alluvial fans that prograded into lakes in the Hornelen basin in Norway differ from the Yallahs and Alaskan alluvial fans by being dominated by debris flows rather than water-laid deposits (Fig. 3-9). Debris flow deposits were modified by relatively low energy waves along the lacustrine shoreline. Transitional zone deposits are irregularly distributed conglomerates which are occasionally bimodal, better sorted, better rounded, and display a more closely packed clast configuration than other debris flow deposits. They pass laterally into and are overlain by lam-

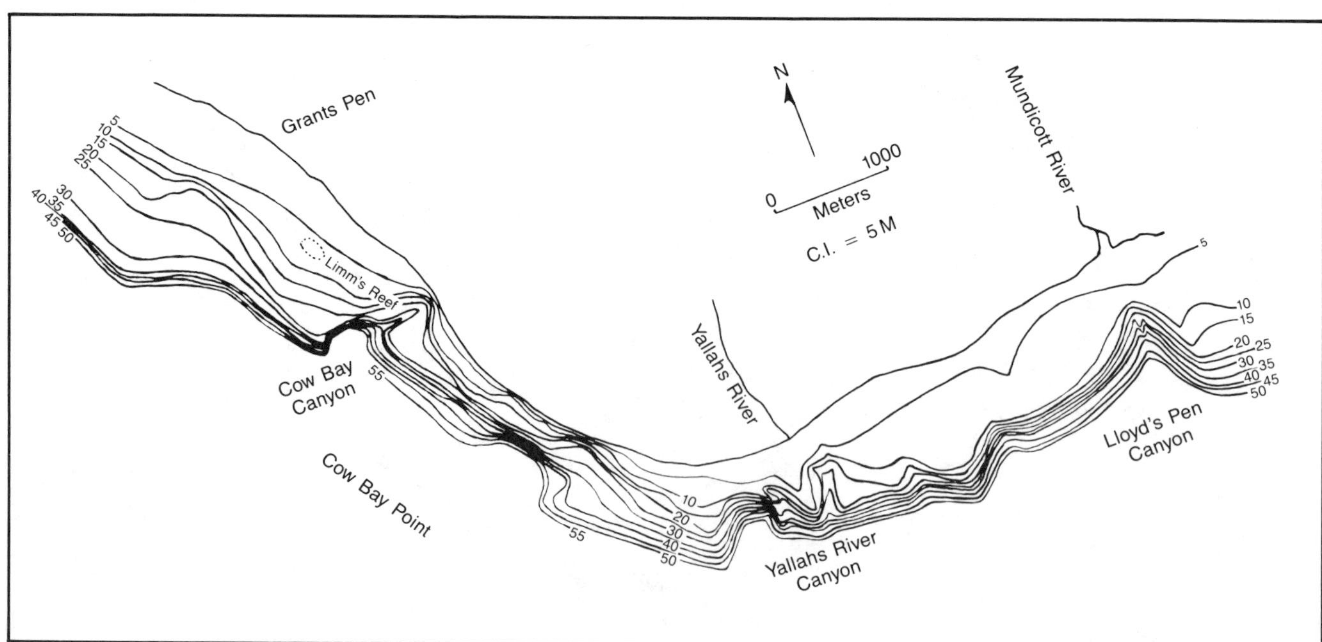

Fig. 3-7. Bathymetry of southeast Jamaica around the Yallahs River. (From Wescott and Ethridge, 1980; permission to publish by AAPG).

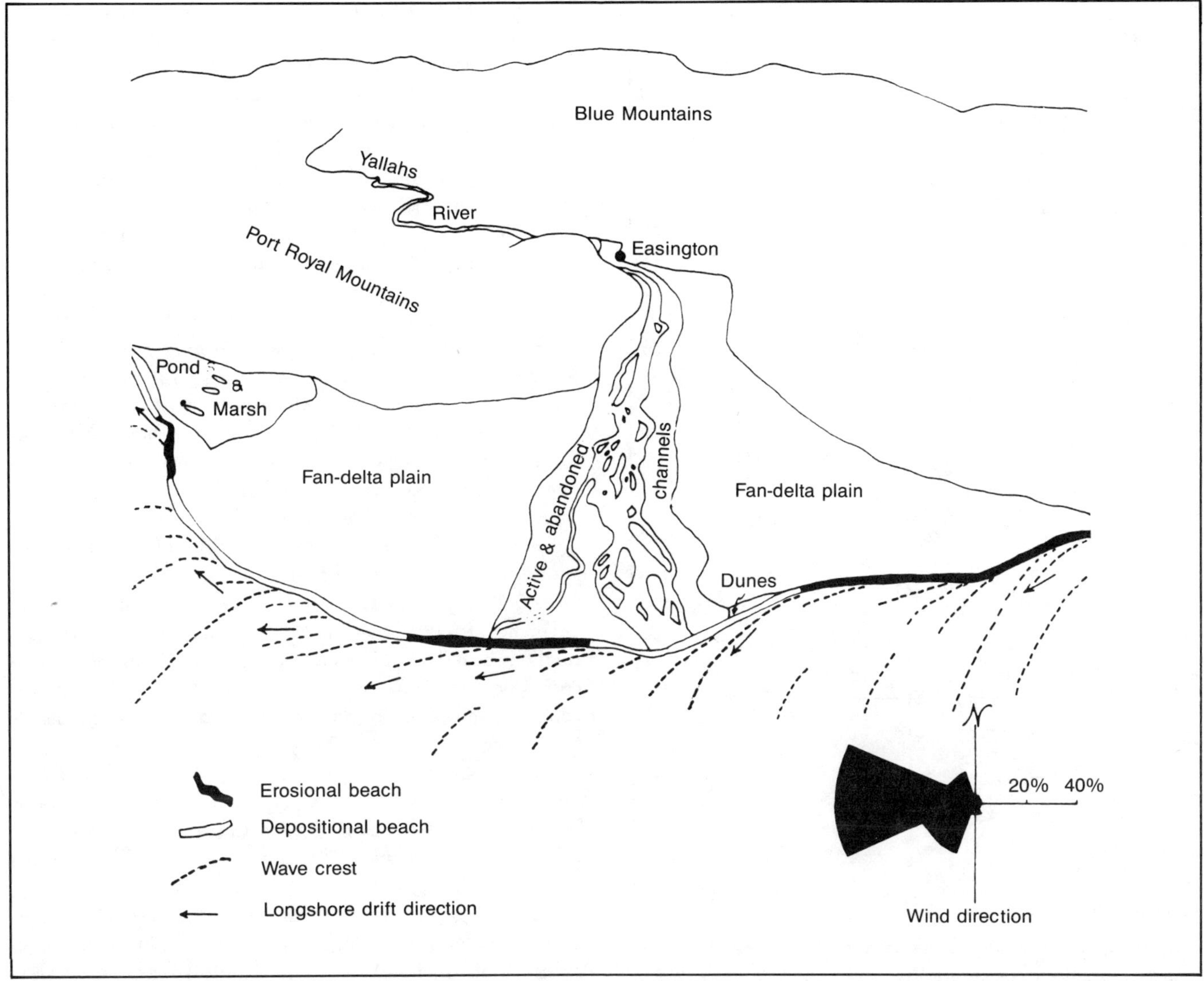

Fig. 3-8. Sketch map of the Yallahs fan delta and principal environments. (From Wescott and Ethridge, 1980; permission to publish by AAPG).

inated granule sandstones produced by the waves reworking the fans (Gloppen and Steel, 1981).

Subaqueous Fan Delta Facies

The alluvial fan of the Yallahs system is prograding onto a narrow shelf and a steep slope cut by submarine canyons (Fig. 3-7). Shelf areas have current and wave rippled sand with local coverings of silt and clay. Pebbles and cobbles occur nearshore at the plunge point, but are uncommon elsewhere. Slope sediments are distributed by slumps and slides. Mud, sand, and gravel occur as random, discontinuous, lenticular units, and matrix-supported conglomerates are significant (Westcott and Ethridge, 1980).

Alluvial fans prograding onto a shelf produce a seaward-fining sequence that overlies the normal shelf deposits. Cross-bedded and burrowed sands offshore from the transitional beaches, bars, or barriers grade into very fine-grained sands, silts, and clays. Progradation of the fan will produce a coarsening-upward sequence (Wescott and Ethridge, 1980).

In the Hornelen basin where debris flows were the principal depositional mechanism, subaqueous fan delta flows are interbedded with and laterally equivalent to subaerial flows and lacustrine deposits (Fig. 3-10). Subaqueous flows are lower in viscosity than their subaerial equivalents, are finer grained, and have a higher percentage of matrix and lower sediment concentration. Inverse and inverse-to-nor-

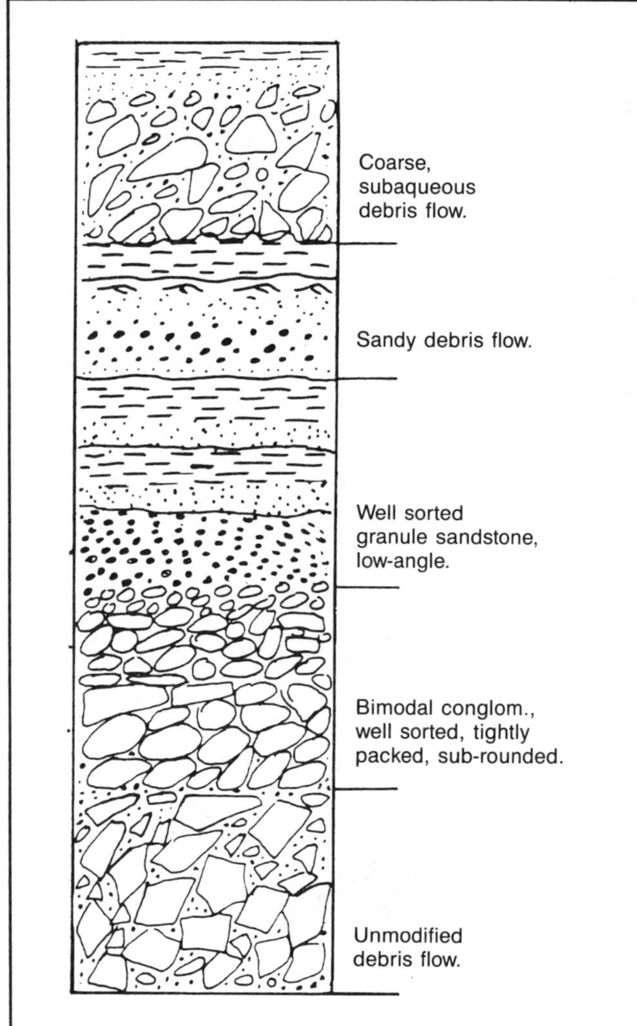

Fig. 3-9. Vertical sequence developed by transgressive, low-energy, lacustrine shoreline. (From Gloppen and Steel, 1981; permission to publish by SEPM).

mal graded bedding are common. Rippled or massive fine sandstones which grade up from the matrix of underlying conglomerate frequently cover the flows. The fine sandstones may be deposits from turbidity currents that were above and following the debris flows (Gloppen and Steel, 1981).

HYDROCARBONS IN ALLUVIAL FANS-FAN DELTAS

Quiriquire Field, Venezuela

Thousands of oil seeps from Quiriquire formation rocks prompted exploratory drilling in northeastern Venezuela. The field was discovered to the south and east of the seeps and became a major producer (Fig. 3-11). Production is from the Quiriquire formation, a Plio-Pleistocene sequence of coarse and fine clastics that unconformably overlies Miocene to Cretaceous rocks (Fig. 3-12). Outcrops and conventional cores show the reservoir to consist of alternating layers of clay and conglomerate with lignite and lignitic clay scattered throughout. Clays are sandy while the coarse interbeds are unsorted silty sands, conglomerates, and boulder beds. Fresh to brackish water macrofossils, crab remains, and plant fragments support a continental alluvial fan origin of the reservoir. Random occurrences of interbedded clays and fine sands along with muddy water molluscs indicate local areas of quiet water deposition (Borger, 1952).

Fan sediments were derived from the highlands to the northwest, and they are in close proximity and parallel to the mountain front. Structural contours on the pre-Pliocene surface show the highland with the field at the base of a steep slope (Fig. 3-13). Distribution of sediment sizes reflects a northwesterly source with coarse sediments dominant in the northern part of the field and clays and sandy clays predominant in the southern part (Borger, 1952).

Correlations in the field are extremely difficult. Beds are lenticular and show considerable lateral variability (Fig. 3-14). One laminated sandstone and clay bed with a characteristic SP curve has lateral continuity and allows correlations based on equivalent time of deposition rather than correlations of individual beds (Borger, 1952). Lack of bed continuity is based, in part, on variations in the gravity of the contained oil (Fig. 3-14).

Structurally, the Quiriquire reservoir is a southeast dipping homocline with no closure (Fig. 3-15). Hydrocarbons are trapped by permeability barriers and truncation against the underlying unconformity making the Quiriquire field a stratigraphic trap.

The limited descriptions of the rocks, the presence of fine-grained rocks, lignites and lignitic clay, and an assumed paleoclimate similar to today's climate suggest the Quiriquire alluvial fan was deposited in a humid climate. A modern analogue for the Quiriquire could be the Yallahs alluvial fan.

Brae Field, North Sea

The Brae field lies at the southern end of the Viking Graben and immediately east of the fault system forming the western boundary of the graben (Fig. 3-16). Its structure is a low amplitude anticline at the base of the fault system (Fig. 3-17). Sealing faults to the west provide most of the structural closure for the field (Fig. 3-18; Harms et al., 1981).

Principal reservoirs in the Brae field are Upper Jurassic conglomerates and sandstones occurring just below the Kimmeridge Clay (Fig. 3-19). Conventional cores show clasts in the conglomerates are angular pebbles, cobbles, and boulders up to three feet (1 m) across. Clast size, angularity, and close proximity to the fault zone suggest a nearby

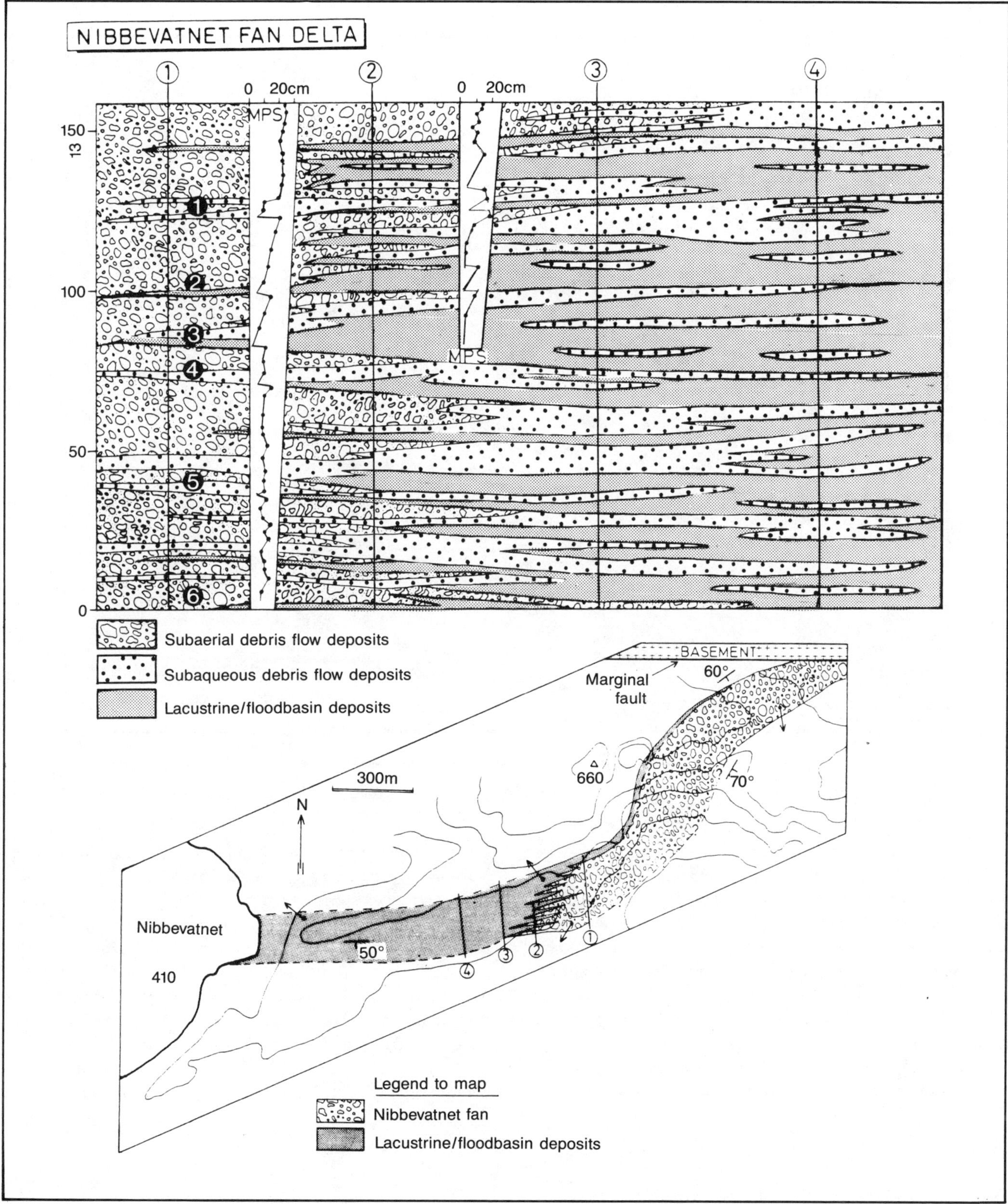

Fig. 3-10. Stratigraphic relationships of subaerial and subaqueous debris flow deposits, and lacustrine deposits, Nibbevatnet fan delta, Hornelen basin, Norway. Coarsening-upward sequence shown by MPS (maximum particle size) columns. (From Gloppen and Steel, 1981; permission to publish by SEPM).

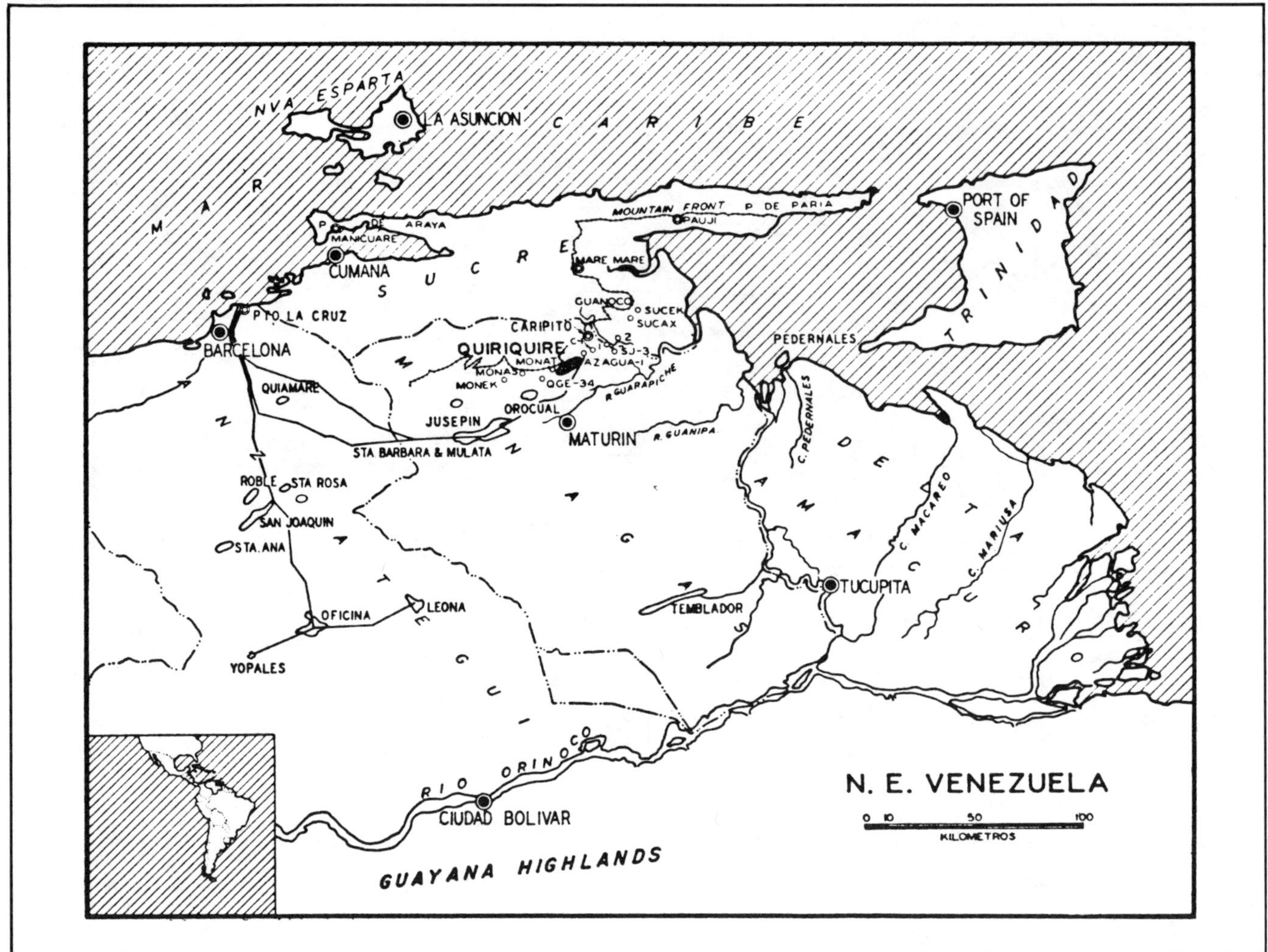

Fig. 3-11. Location of Quiriquire field in northeastern Venezuela. (From Borger, 1952; permission to publish by AAPG).

elevated source to the west. Conglomerate units are over 100 ft (30 m) thick, but consist of individual beds a few feet thick indicating deposition in shallow channels. Conglomerates and sandstones thin to the east and, in two to four miles (3 to 7 km), pass laterally into sandy or laminated siltstones. Variability of the Brae sediments along the trace of the fault system indicates contribution of detritus from numerous local drainages along the fault (Harms et al., 1981).

A series of coalescing fan deltas best describes the distribution of facies (Fig. 3-20). Conglomerates and sandstones were deposited on alluvial fans by streams coming from the highlands to the west. Laterally equivalent sandy or laminated siltstones were deposited in the adjacent sea. Primary dips were high (6° to 14°) and marine energy low. Different source areas for adjacent fans account for the changes in detrital character along the fault (Harms et al., 1981). A submarine fan environment of deposition is considered unlikely. Abrupt facies changes, steep primary dips of the siltstones, and sedimentary fabrics and structures (not described in the paper) seen in the different facies are not consistent with submarine fan deposition (Harms et al., 1981).

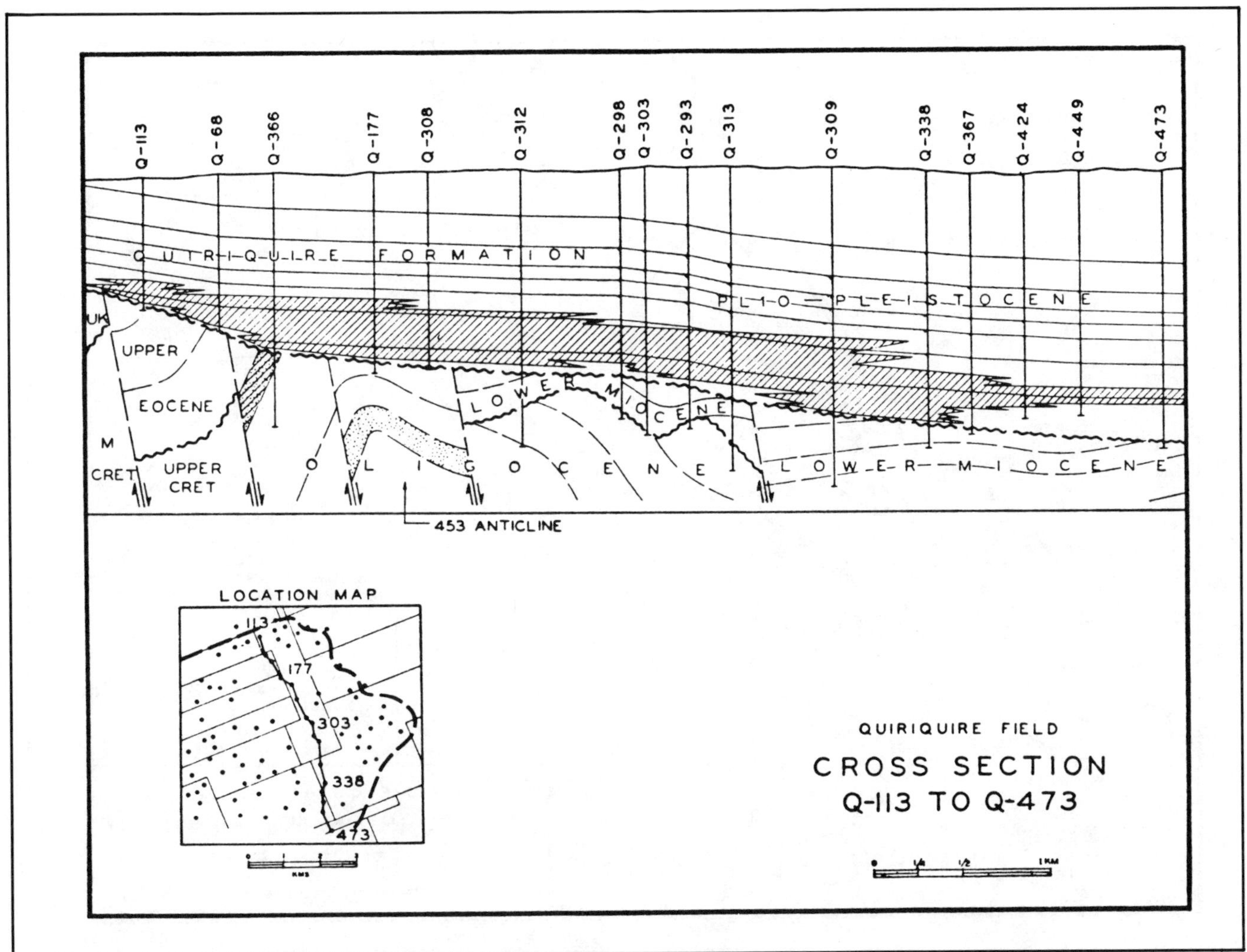

Fig. 3-12. North-south cross section through eastern portion of Quiriquire field. Quiriquire alluvial fan is shaded area above unconformity. (From Borger, 1952; permission to publish by AAPG).

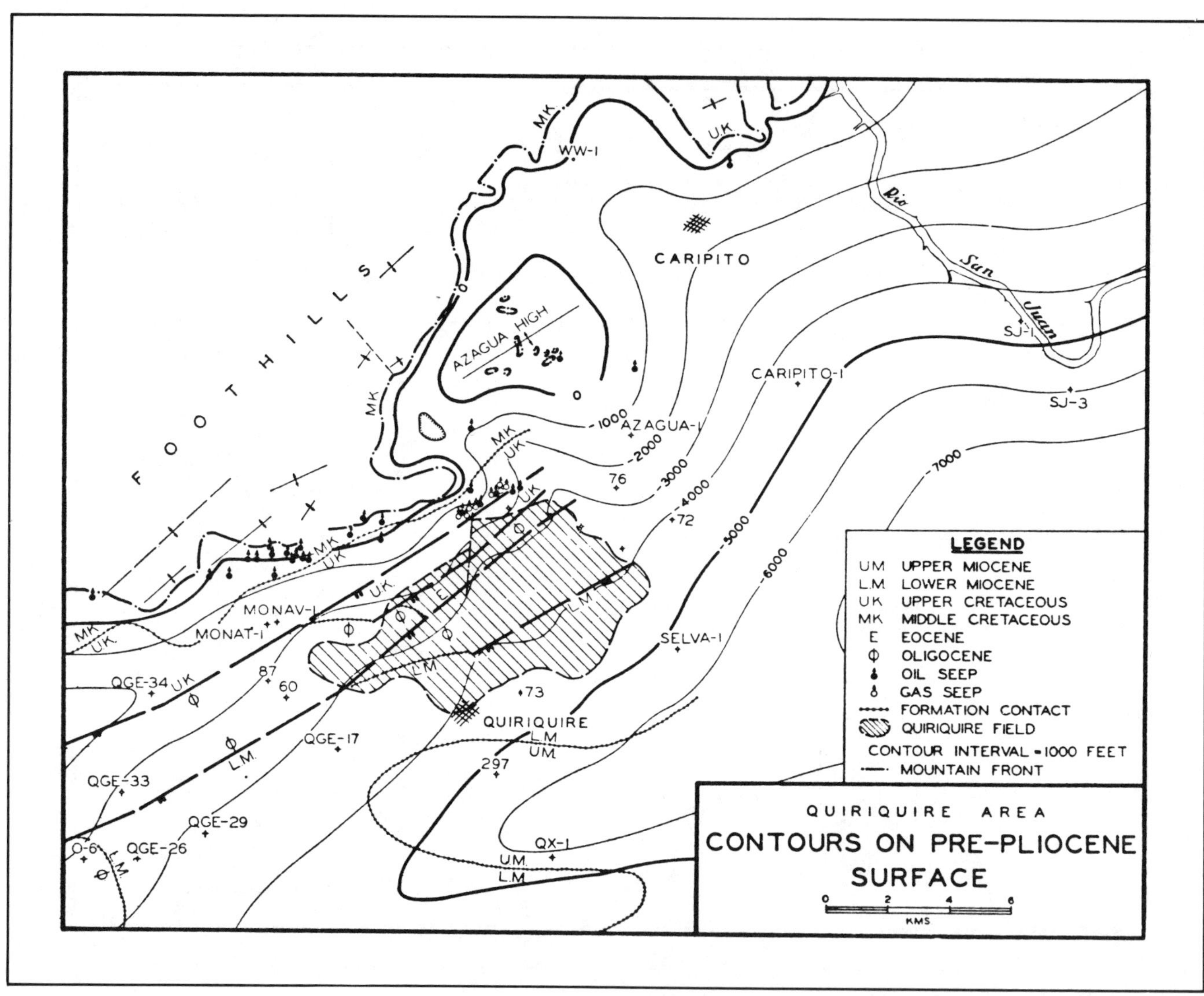

Fig. 3-13. Contours on pre-Pliocene surface and location of Quiriquire field. (From Borger, 1952; permission to publish by AAPG).

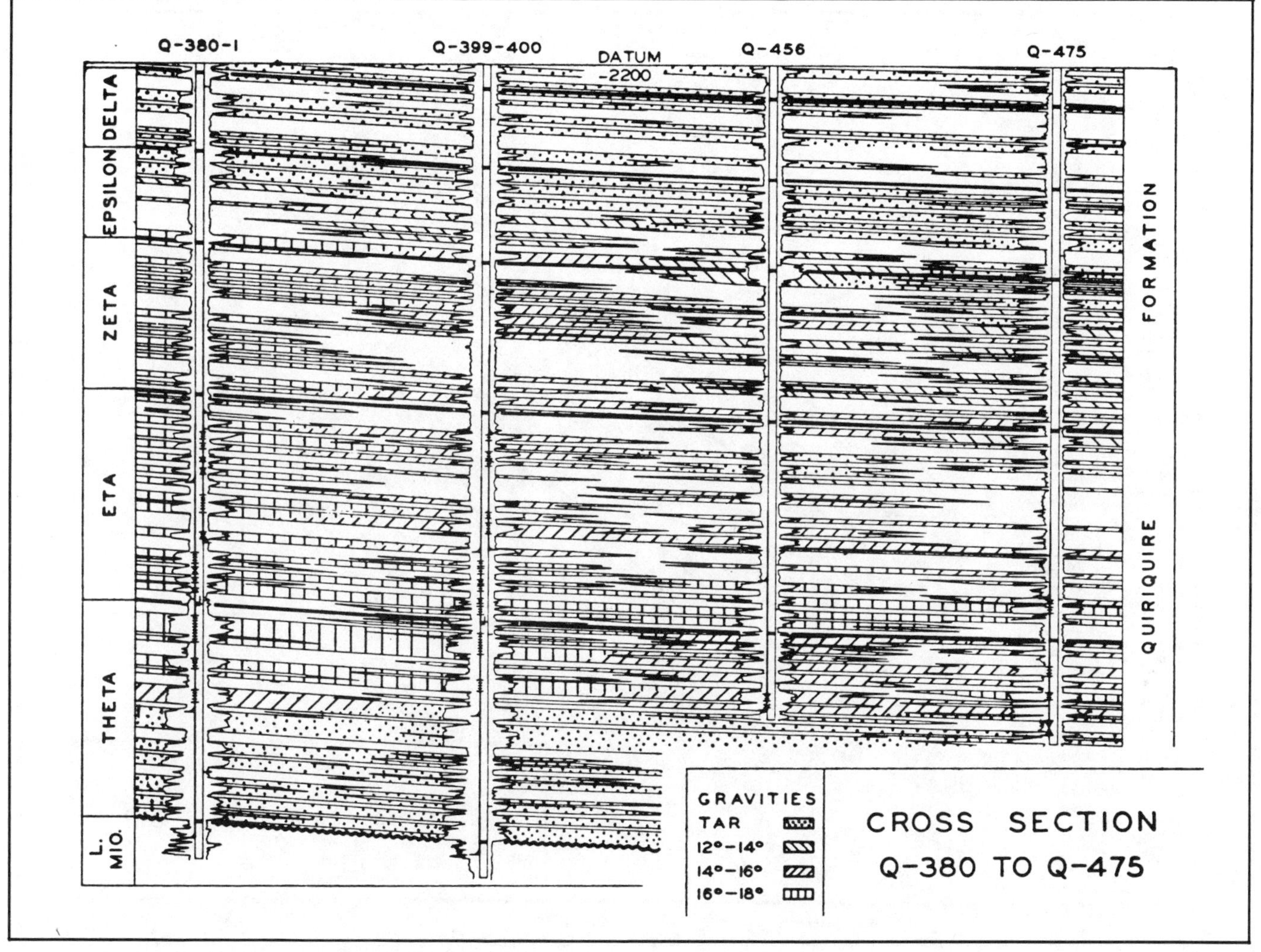

Fig. 3-14. Lateral discontinuity of beds within the Quiriguire reservoir as indicated by lateral changes in oil gravity. (From Borger, 1952; permission to publish by AAPG).

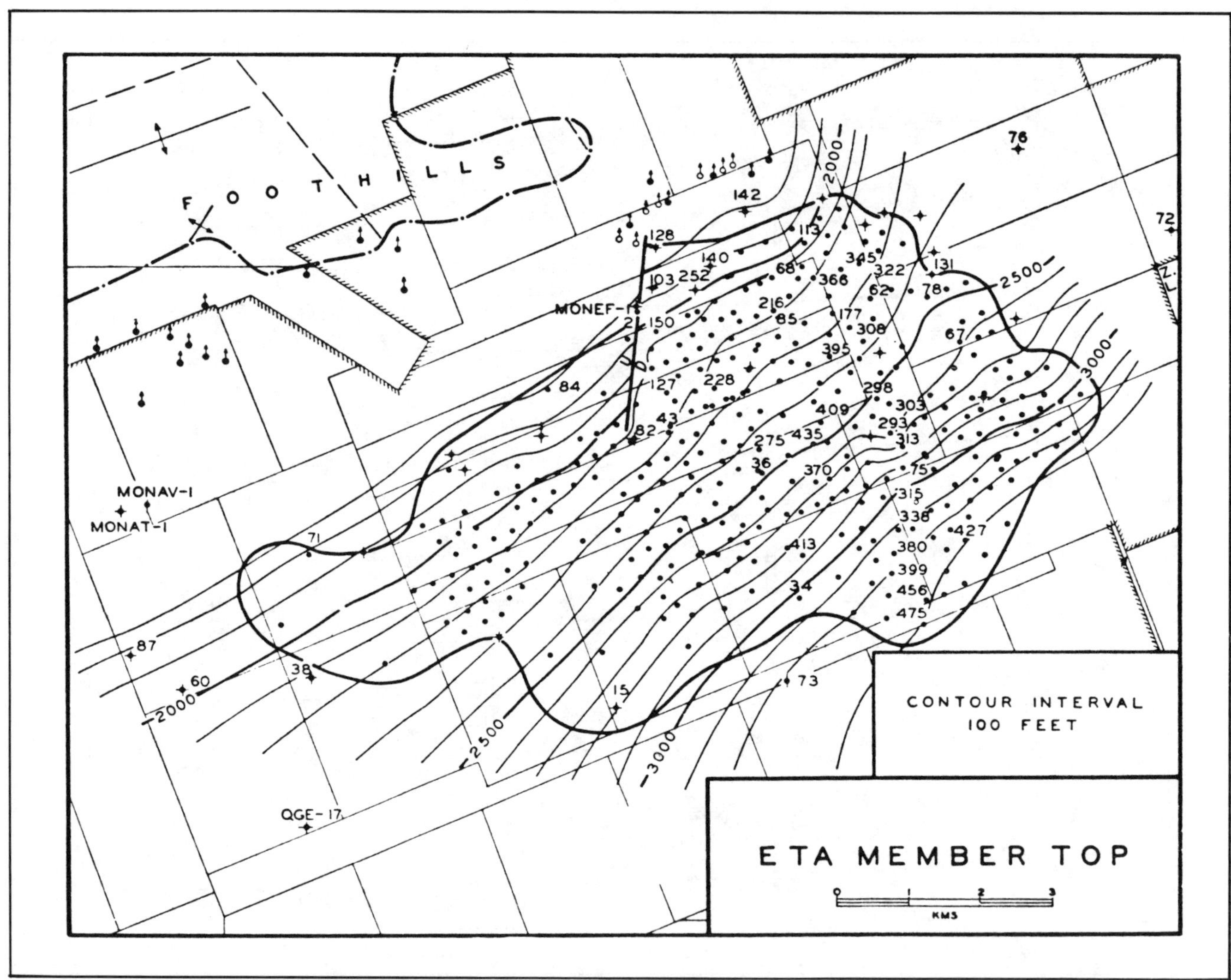

Fig. 3-15. Structure map on top of Eta member, Quiriquire formation. Note absence of any structural closure. (From Borger, 1952; permission to publish by AAPG).

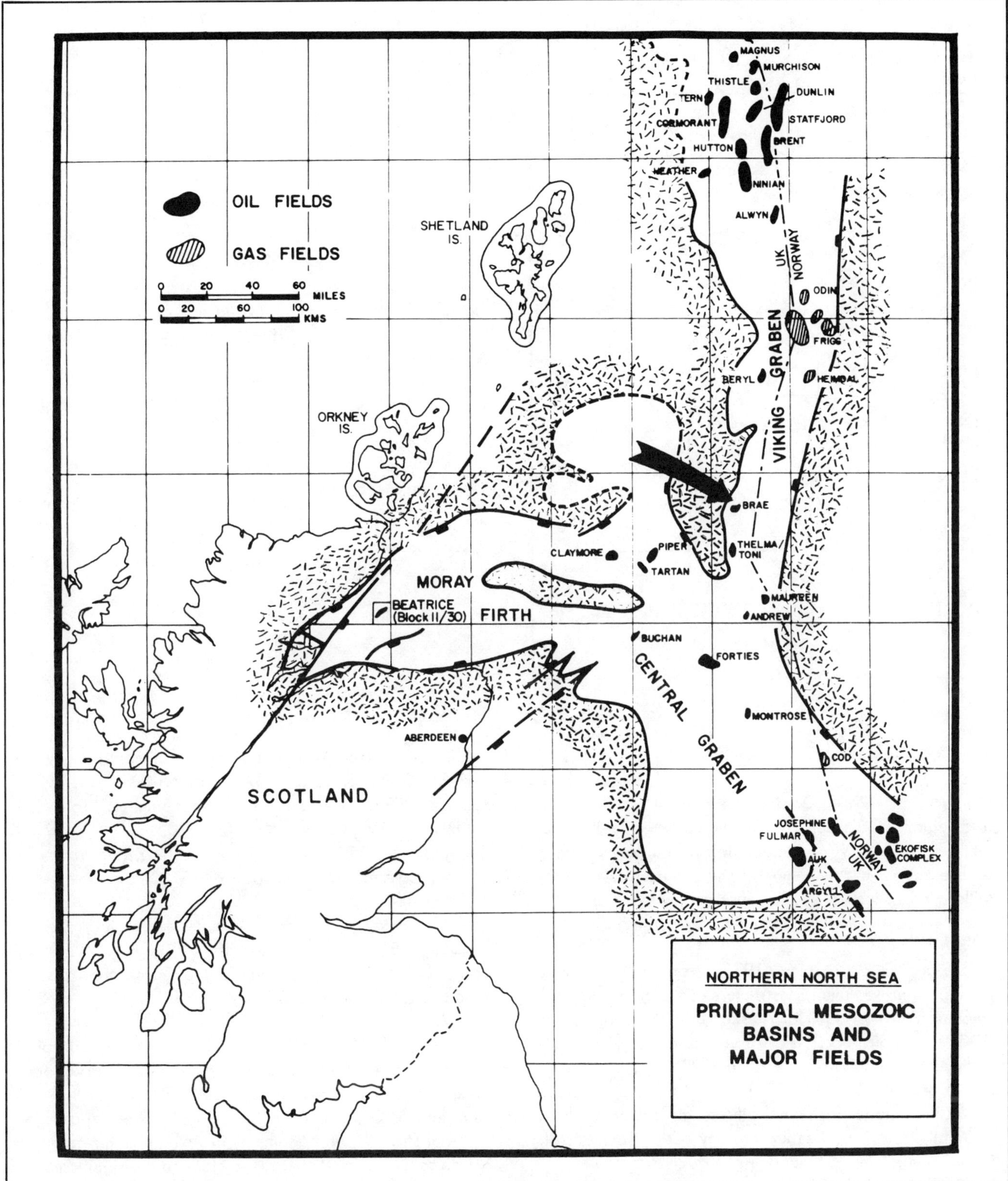

Fig. 3-16. Location map of Mesozoic basins and major oil fields, northern North Sea. (From Linsley et al., 1980; permission to publish by AAPG).

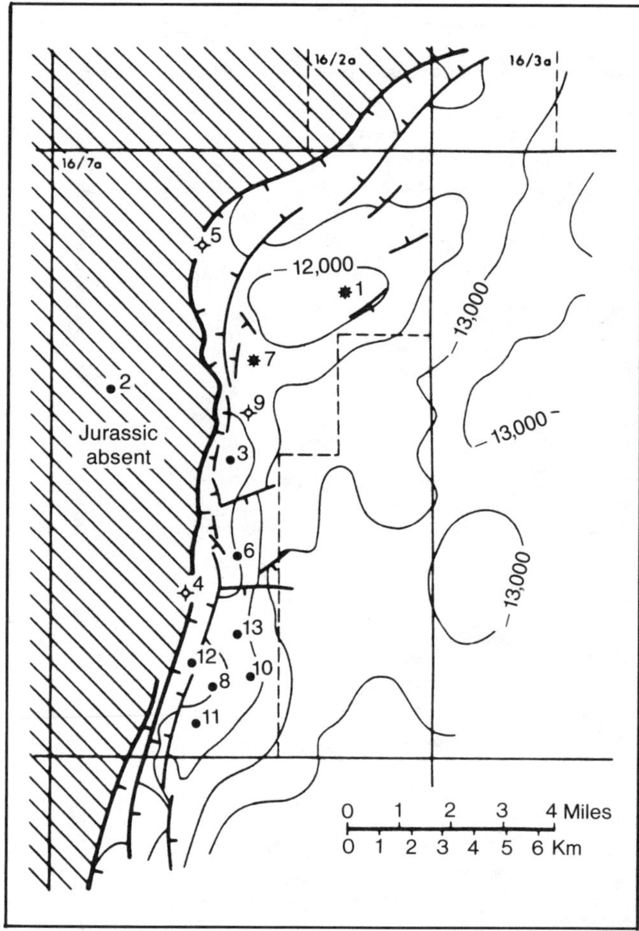

Fig. 3-17. Structure map on the top of the Jurassic Brae field. (From Harms et al., 1981; permission to publish from Institute of Petroleum, London).

SELECTED BIBLIOGRAPHY

Blissenbach, E., 1954, Geology of alluvial fans in semi-arid regions: GSA Bull., v. 65, no. 2, p. 175-190.

Boothroyd, J. C. , 1976, A model for alluvial fan-fan delta sedimentation in cold-temperate environments, *in* T. P. Miller, ed., Recent and ancient sedimentary environments in Alaska: Alaska Geol. Soc. Symp. Proc., p. N1-N13.

_______ and G. M. Ashley, 1975, Processes, bar morphology, and sedimentary structures on braided-outwash fans, northeastern Gulf of Alaska, *in* A. V. Jopling and B. C. McDonald, eds., Glaciofluvial and glaciolacustrine sedimentation: SEPM Sp. Pub. no. 23, p. 193-222.

_______ and D. Nummedal, 1978, Proglacial braided outwash: a model for humid alluvial fan deposits, *in* A. D. Miall, ed., Fluvial sedimentology: Can. Soc. Petrol. Geol. Mem. 5, p. 641-668.

Borger, H. D., 1952, Case history of the Quiriquire field, Venezuela: AAPG Bull., v. 36, no. 12, p. 2291-2330.

Bull, W. B., 1972, Recognition of alluvial-fan deposits in the stratigraphic record, *in* J. K. Rigby and W. K. Hamblin, eds., Recognition of ancient sedimentary environments: SEPM Sp. Pub. no. 16, p. 63-83.

_______, 1977, The alluvial fan environment: Progress in Phys. Geography, v. 1, p. 222-270.

Denny, C. S., 1965, Alluvial fans in the Death Valley region, California and Nevada: U. S. Geol. Survey Prof. Paper 466, 62 p.

_______, 1967, Fans and sediments: Amer. Jour. Sci., v. 265, p. 81-105.

Eckis, R., 1928, Alluvial fans of the Cucamonga district, southern California: Jour. Geol., v. 36, no. 3, p. 224-247.

Galloway, W. E., 1976, Sediments and stratigraphic framework of the Copper River fan delta, Alaska: Jour. Sed. Pet., v. 46, no. 3, p. 726-737.

Gloppen, T. G., and R. J. Steel, 1981, The deposits, internal structure and geometry in six alluvial fan-fan delta bodies (Devonian-Norway)—a study in the significance of bedding sequence in conglomerates, *in* F. G. Ethridge and R. M. Flores, eds., Recent and ancient nonmarine depositional environments: models for exploration: SEPM Sp. Pub. no. 31, p. 49-69.

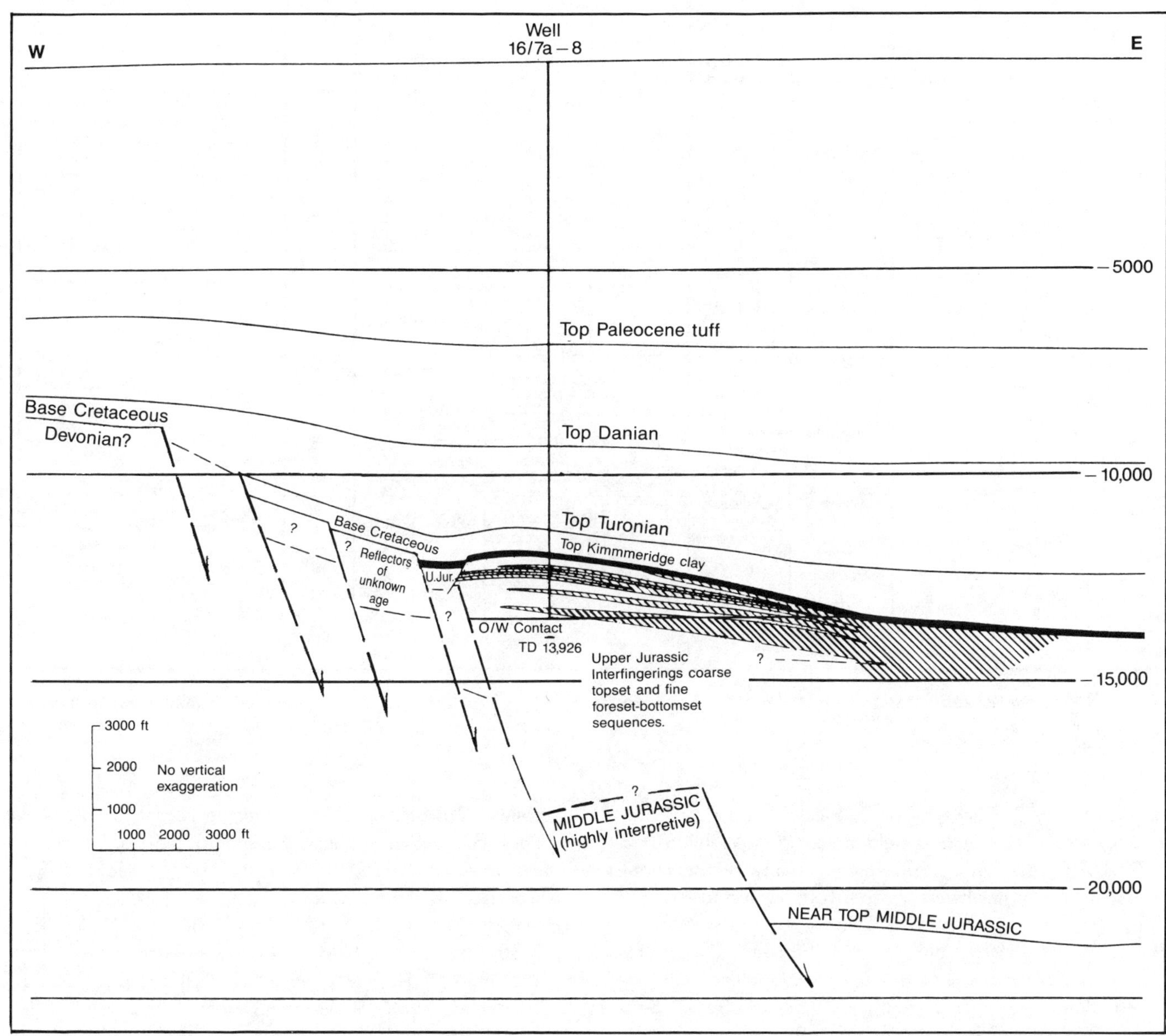

Fig. 3-18. East-west structural cross section, southern portion of Brae field. (From Harms et al., 1981; permission to publish by Institute of Petroleum, London).

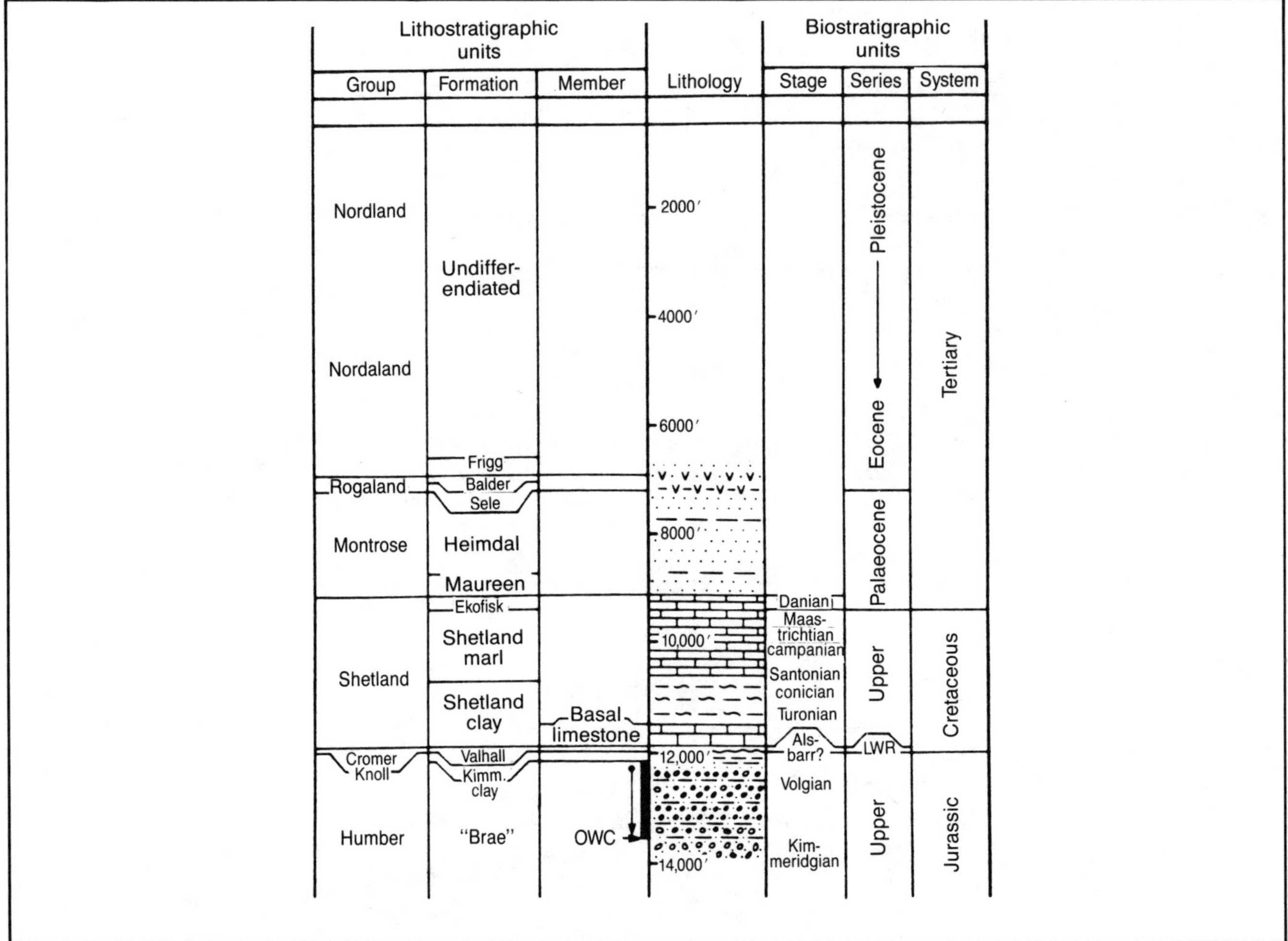

Fig. 3-19. Generalized stratigraphic column in the Brae area. (From Harms et al., 1981; permission to publish by Institute of Petroleum, London).

Harms, J. C., P. Tackenburg, E. Pickles, and R. E. Pollock, 1981, The Brae oilfield area, *in* L. V. Illing and G. D. Hobson, eds., Petroleum geology of the continental shelf of northwest Europe: London, Institute of Petroleum Geology, p. 352-357.

Heward, A. P., 1978, Alluvial fan and lacustrine sediments from the Stephanian A and B (La Magdalena, Cinera-Matallana and Sabero) coalfields, northern Spain: Sedimentology, v. 25, no. 4, p. 451-488.

Hooke, R. LeB., 1967, Processes on arid-region alluvial fans: Jour. Geol., v. 75, no. 4, p. 438-460.

_______, 1968, Steady-state relationships on arid-region alluvial fans in closed basins: Amer. Jour. Sci., v. 266, p. 609-629.

_______ and W. L. Rohrer, 1979, Geometry of alluvial fans: effect of discharge and sediment size: Earth Surface Processes, v. 4, p. 147-166.

Jahns, R. H., 1947, Geological features of the Connecticut Valley, Massachusetts, as related to recent floods: U.S. Geol. Survey Water Supply Paper 6, 158 p.

Johnson, A. M., 1970, Physical processes in geology: San Francisco, Freeman, Cooper and Co., 577 p.

Linsley, P. N., H. C. Potter, G. McNab, and D. Racher, 1980, The Beatrice field, inner Moray Firth, U.K. North Sea, *in* M. T. Halbouty, ed., Giant oil and gas fields of the decade 1968-1978: AAPG Mem. 30, p. 117-129.

Mack, G. H., and K. A. Rasmussen, 1984, Alluvial-fan sedimentation of the Cutler Formation (Permo-Pennsylvanian) near Gateway, Colorado: GSA Bull., v. 95, no. 1, p. 106-116.

Magleby, D. C., and I. E. Klein, 1965, Ground-water conditions and potential pumping resources above the Corcoran Clay—an addendum to the ground-water geology and resources definite plan appendix, 1963: U.S. Bur. Reclam. open-file report, 21 plates.

McGowen, J. H., and C. G. Groat, 1971, Van Horn sand-

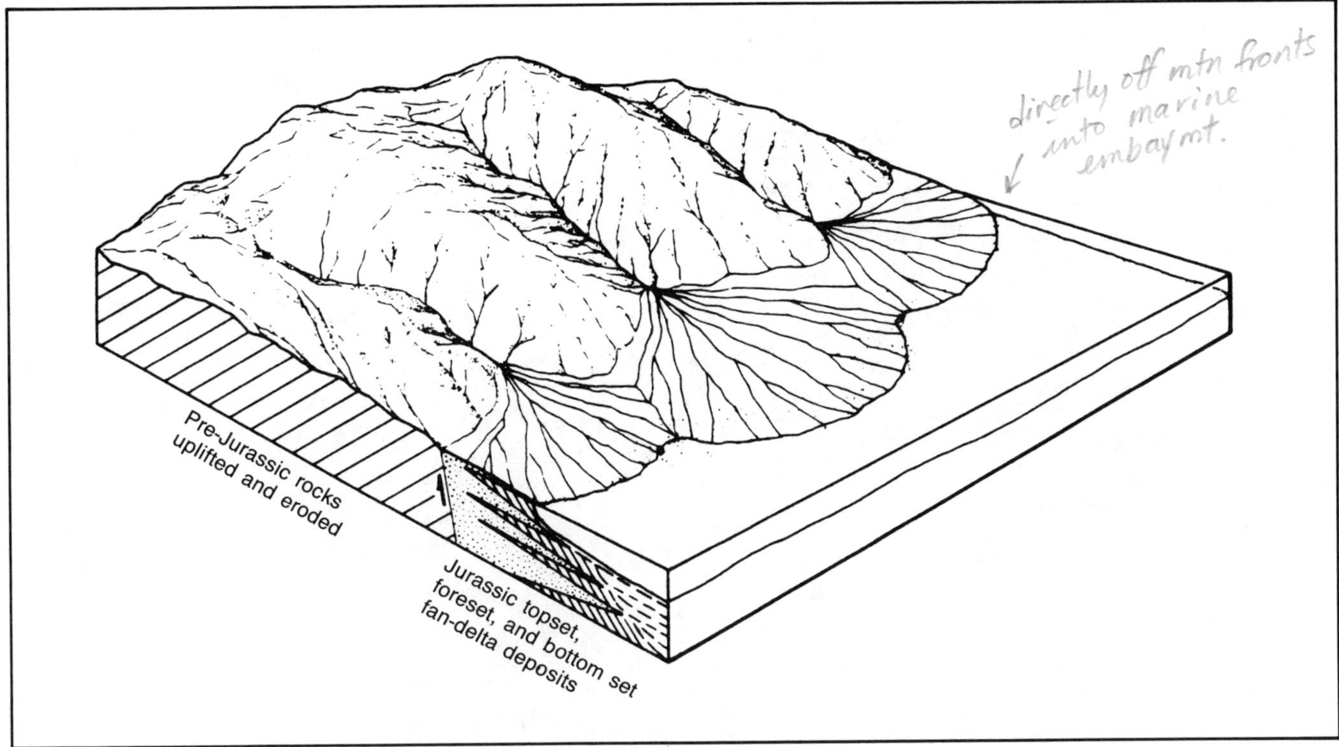

Fig. 3-20. Block diagram of topography and fan delta morphology as interpreted for the Brae area in Upper Jurassic time. (From Harms et al., 1981; permission to publish by Institute of Petroleum, London).

stone, west Texas, an alluvial fan model for mineral exploration: Texas Bur. Econ. Geology Report of Investigations no. 72, 57 p.

Nilsen, T. H., 1982, Alluvial fan deposits, *in* P. A. Scholle and D. Spearing, eds., Sandstone depositional environments: AAPG Mem. 31, p. 49-86.

Reid, J. C., 1974, Hazel Formation, Culberson and Hudspeth Counties, Texas: Master's thesis, Univ. Texas-Austin, 88 p.

Sharp, R. P. and L. H. Nobles, 1953, Mudflow of 1941 at Wrightwood, southern California: GSA Bull., v. 64, no. 5, p. 547-560.

Spearing, D. R., 1974, Alluvial fan deposits: Summary sheets of sedimentary deposits, Sheet 1: Boulder, Colorado, GSA.

Steel, R. J., S. Maehle, H. Nilsen, S. L. Roe, and A. Spinnangr, 1977, Coarsening-upward cycles in the alluvium of Hornelen Basin (Devonian) Norway: sedimentary response to tectonic events: GSA Bull., v. 88, no. 8, p. 1124-1134.

Wescott, W. A., and F. G. Ethridge, 1980, Fan-delta sedimentology and tectonic setting—Yallahs fan delta, southeast Jamaica: AAPG Bull., v. 64, no. 3, p. 374-399.

4 BRAIDED STREAMS

INTRODUCTION

A braided stream consists of numerous rapidly shifting channels of low sinuosity (Miall, 1977; Rust, 1978). At low and moderate flows, individual channels divide and rejoin around bars and islands (Cant, 1982). During high flows, most of the floodplain is underwater, although some elevated, well-established islands may remain dry (Williams and Rust, 1969; Smith, 1971; Rust, 1978). Channels are wide and shallow and generally floored by dunes and bars (Leopold and Wolman, 1957; Smith, 1970, 1974; Fahnestock and Bradley, 1973; Hein and Walker, 1977).

Several kinds of bars bound and separate the various channels of a braided stream. Three types (longitudinal, transverse, compound) are indicated by Miall (1977) as most likely to be preserved in the rock record. Longitudinal bars are elongate parallel to flow and bounded on both sides by active channels. Bar margins may be eroded (Rust, 1972; Miall, 1977). Transverse and linguoid bars are considered together by Miall (1977). They are lobate or rhombic in plan, the upper surface dips gently upstream, and the downstream end is an avalanche slope (Smith, 1974; Miall, 1977). Compound bars are point bars, side bars, and lateral bars formed in lower energy areas of braided streams (Collinson, 1970; Smith, 1974; Miall, 1977). They are larger than the other bars and form by the coalescing of small bars and dunes (Collinson, 1970; Miall, 1977).

CAUSES OF BRAIDING

About a dozen factors have been mentioned by numerous authors as requisites for the formation of braided streams. Leopold and Wolman (1957) and others have variously considered width, depth, slope, bed roughness, velocity, sediment caliber and quantity, and water discharge and variability. Other factors include lack of cohesion or the erodibility of river banks (Macklin, 1956; Leopold et al., 1964; Kessler and Cooper, 1970; Cant, 1982), climate (Doeglas, 1962; Schumm, 1968), and vegetation (Schumm, 1968; Rust, 1978).

There are, however, inconsistencies in some factors. Braided river deposits are frequently coarse-grained, e.g., the gravel deposits of the Kicking Horse River in British Columbia (Smith, 1974). Bijou Creek, Colorado, on the other hand, deposits only sand (McKee et al., 1967) and the Yellow River, China, deposits silt (Chien, 1961). Sediments deposited by these rivers apparently are a function of available source material and unrelated to braiding.

High gradient frequently is related to braiding (Doeglas, 1962; Walker, 1976). Gradients of modern braided rivers vary widely and may even be less than gradients of meandering streams. Kessler and Cooper (1970) question the importance of gradient in forming braided rivers, and a comparison of the gradients of the Kosi and Mississippi Rivers supports their view (Table 4-1).

BRAIDED STREAM DEPOSITS

Channels

Bed load moves down braided channels as ripples, sand waves, dunes (Cant, 1978a), gravel sheets (Hein and Walker, 1977), and bars (Smith, 1974; Hein and Walker, 1977). Sandy braided river channels are floored by ripples, sand waves, dunes, and bars while channels of gravelly braided rivers have gravel sheets and bars.

Ripples. At low water velocities, ripples may form on channel beds or on other larger bedforms such as sand waves. Ripples are small-scale bedforms a few centimeters high with wave lengths of tens of centimeters. Stoss sides

TABLE 4-1. Gradients of four braided rivers and the lower 125 mi of the Mississippi River.

River and location	Gradient	Reference
South Platte, Denver, Colorado, USA	.00200	Smith, 1970
Kicking Horse, Field, B. C., Canada	.00500	Smith, 1974
South Saskatchewan Outlook, Saskatchewan, Canada	.00030	Cant, 1978a
Kosi, India	.00006	Gole and Chitale, 1966
Mississippi, Louisiana, USA	.00007	Calculated from Thornbury, 1954

slope gently upstream and lee sides form steep, downstream avalanche faces (Harms et al., 1975; Cant, 1978b). Ripples are recognized in ancient sequences by ripple cross-lamination (Cant and Walker, 1976).

Sand waves. This bedform is a wedge-shaped or tabular sedimentary unit with a relatively straight crest and well-defined slip face (Fig. 4-1). It forms at moderate flow velocities and its size is related to depth of flow (Harms et al., 1975). In shallow water, sand waves have amplitudes of 2 to 40 cm and wavelengths of 1 to 20 m (Cant, 1978b). Much larger sand waves with heights from 8 to 15 m and wave lengths from 180 to 900 m occur in the Brahmaputra River (Coleman, 1969). Planar cross-bedding (Fig. 4-1) characterizes the internal structure of sand waves, and ripples are common on the stoss sides. Ripple cross-lamination may be lateral to or superimposed on the cross-bedding (Cant, 1978b).

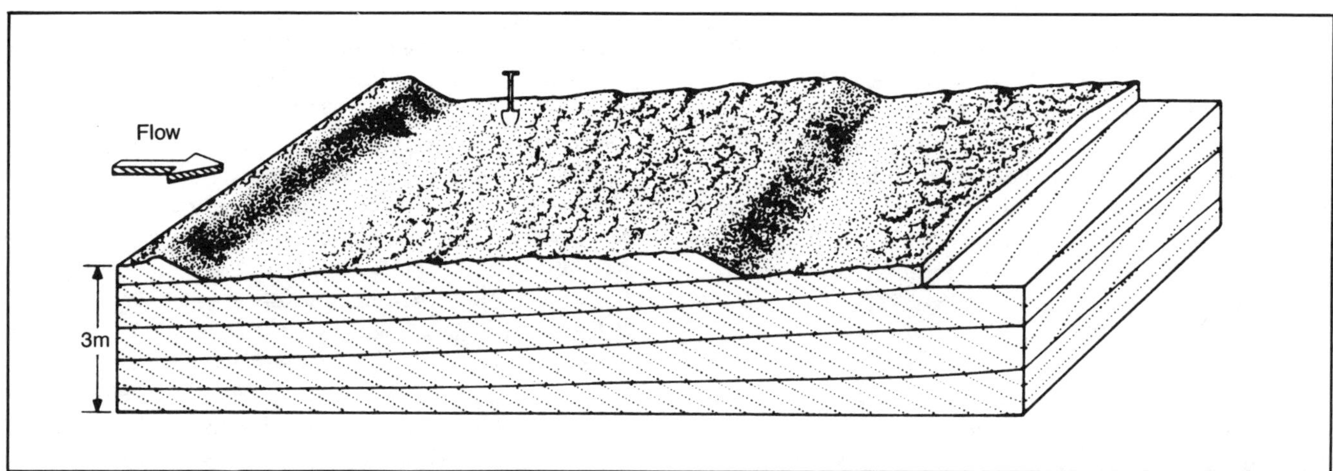

Fig. 4-1. Tabular cross-bedding formed by migrating sand waves. (From Harms et al., 1975; permission to publish by SEPM).

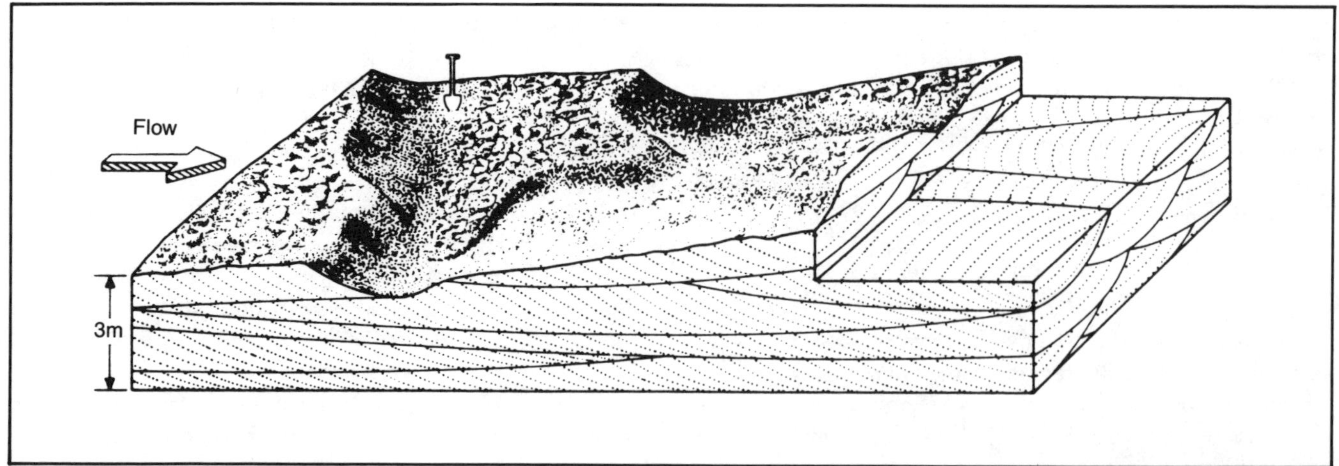

Fig. 4-2. Trough cross-bedding formed by migrating dunes. (From Harms et al., 1975; permission to publish by SEPM).

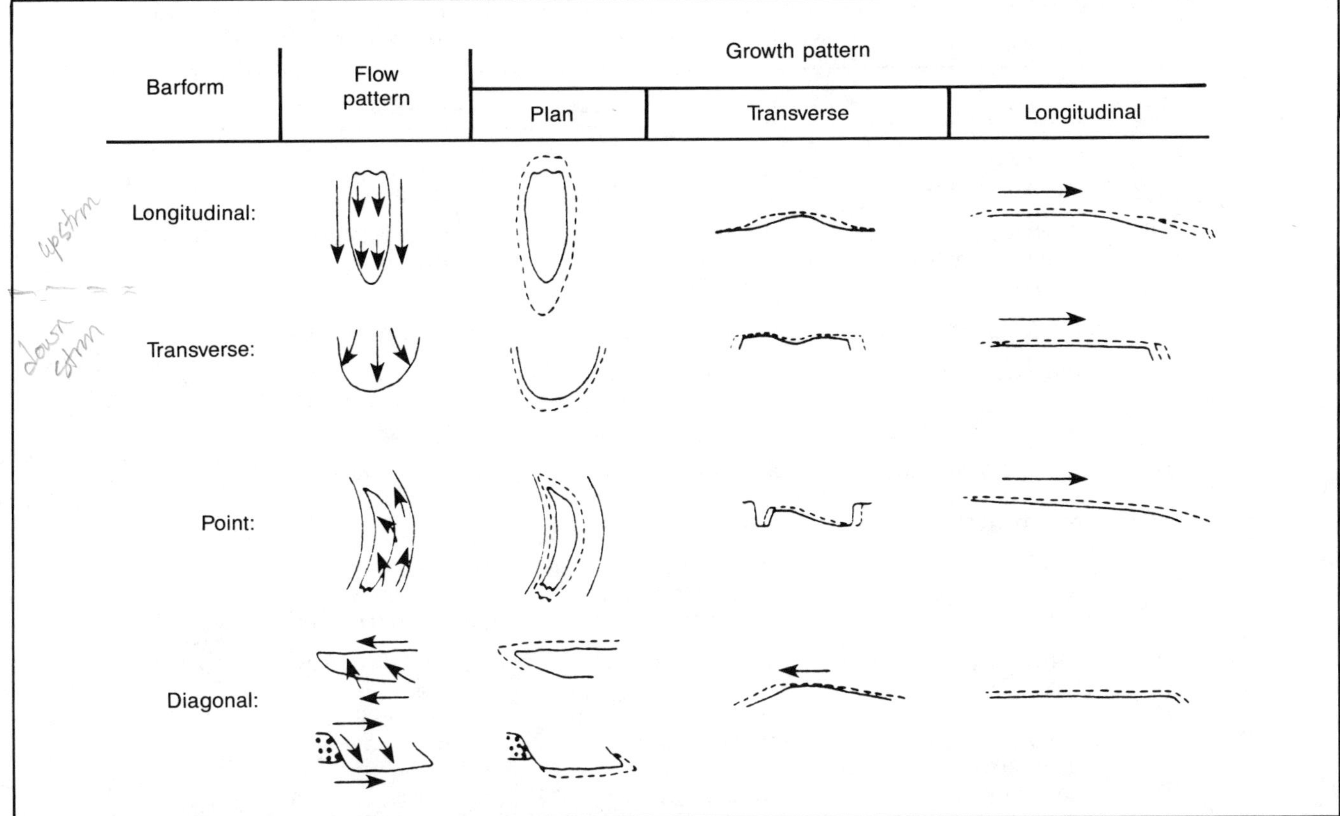

Fig. 4-3. Morphology, pattern of growth, and water flow over and near bars commonly found in braided channels. Dashed lines indicate accretion. (From Smith, 1974; permission to publish by Journal of Geology, University of Chicago).

Dunes. With increasing flow velocity, sand waves grade gradually into dunes. Crest patterns are irregular and scours occur in the trough areas downstream from the avalanche faces (Fig. 4-2). The irregularity of the bedform produces trough cross-bedding as the internal structure. Dune size varies directly with water velocity (Cant, 1978b) and water depth, and inversely with grain size. At high velocity and shallow depth, dunes are planed off and show a rounded profile. With further increases in velocity, the planed dunes become lower and grade into the upper flat bed phase (Harms et al., 1975).

Cant (1978b) studied the changes in dunes in a channel of the South Saskatchewan River during a flood using an echo sounder. During rising stage, the dunes averaged 0.5 m in amplitude and 5 to 10 m in wave length. Amplitudes reached 2 m with 40 m wave lengths at peak flow. Not long after peak flow during falling stage, both small and large du.ıes occurred together as the high flow forms had not yet been completely reworked by the lower flow. Later, at still lower flow, dune amplitude was 50 cm and wave length was 5 m.

Gravel sheets. Hein and Walker (1977) describe gravel sheet bedforms from the Kicking Horse River in British Columbia. They are low relief, coarse-grained deposits, one

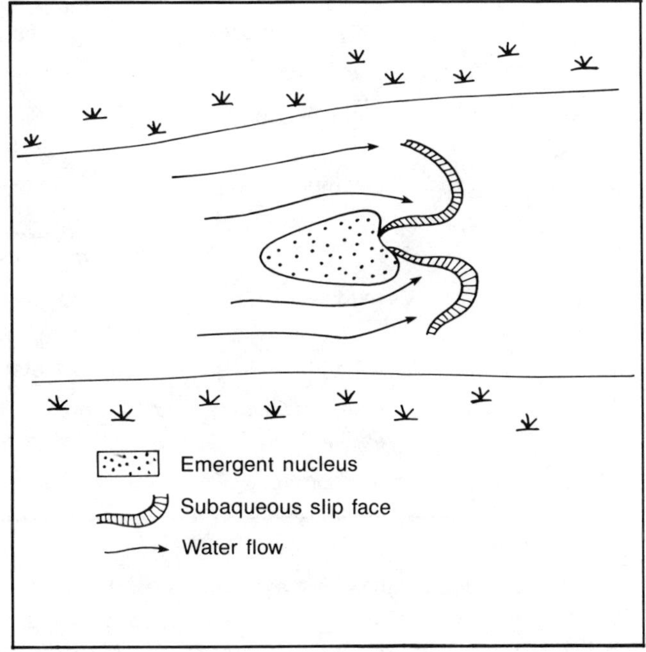

Fig. 4-4. Development of a sand flat in a sandy braided stream.

TABLE 4-2. Lithofacies and sedimentary structures of modern and ancient braided river deposits. (From Miall, 1978; permission to publish by Canadian Society of Petroleum Geologists).

Facies Code	Lithofacies	Sedimentary structures	Interpretation
Gms	massive, matrix supported gravel	none	debris flow deposits
Gm	massive or crudely bedded gravel	horizontal bedding, imbrication	longitudinal bars, lag deposits, sieve deposits
Gt	gravel, stratified	trough crossbeds	minor channel fills
Gp	gravel, stratified	planar crossbeds	linguoid bars or deltaic growths from older bar remnants
St	sand, medium to v. coarse, may be pebbly	solitary (theta) or grouped (pi) trough crossbeds	dunes (lower flow regime)
Sp	sand, medium to v. coarse, may be pebbly	solitary (alpha) or grouped (omikron) planar crossbeds	linguoid, transverse bars, sand waves (lower flow regime)
Sr	sand, very fine to coarse	ripple marks of all types	ripples (lower flow regime)
Sh	sand, very fine to very coarse, may be pebbly	horizontal lamination, parting or streaming lineation	planar bed flow (l. and u. flow regime)
Sl	sand, fine	low angle (<10°) crossbeds	scour fills, crevasse splays, antidunes
Se	erosional scours with intraclasts	crude crossbedding	scour fills
Ss	sand, fine to coarse, may be pebbly	broad, shallow scours including eta cross-stratification	scour fills
Sse, She, Spe	sand	analogous to *Ss, Sh, Sp*	eolian deposits
Fl	sand, silt, mud	fine lamination, very small ripples	overbank or waning flood deposits
Fsc	silt, mud	laminated to massive	backswamp deposits
Fcf	mud	massive, with freshwater molluscs	backswamp pond deposits
Fm	mud, silt	massive, desiccation cracks	overbank or drape deposits
Fr	silt, mud	rootlets	seatearth
C	coal, carbonaceous mud	plants, mud films	swamp deposits
P	carbonate	pedogenic features	soil

or two pebble diameters thick that extend most of the width of the channel. The sheets move at peak flows, and at low flows they may act as loci for bar growth. At high discharges of sediment and water, a gravel sheet grows downstream more than it aggrades, producing the massive or low-angle stratified deposits seen in flow parallel bars, e.g., longitudinal and diagonal bars (Fig. 4-3). At lower flows, a sheet aggrades vertically causing the bedform to have a slip face and producing the cross-bedded, fine gravel deposits of bars oriented normal to flow, e.g., transverse bars.

Bars. Longitudinal, transverse, point, and diagonal bars are recognized in modern braided channels on the basis of

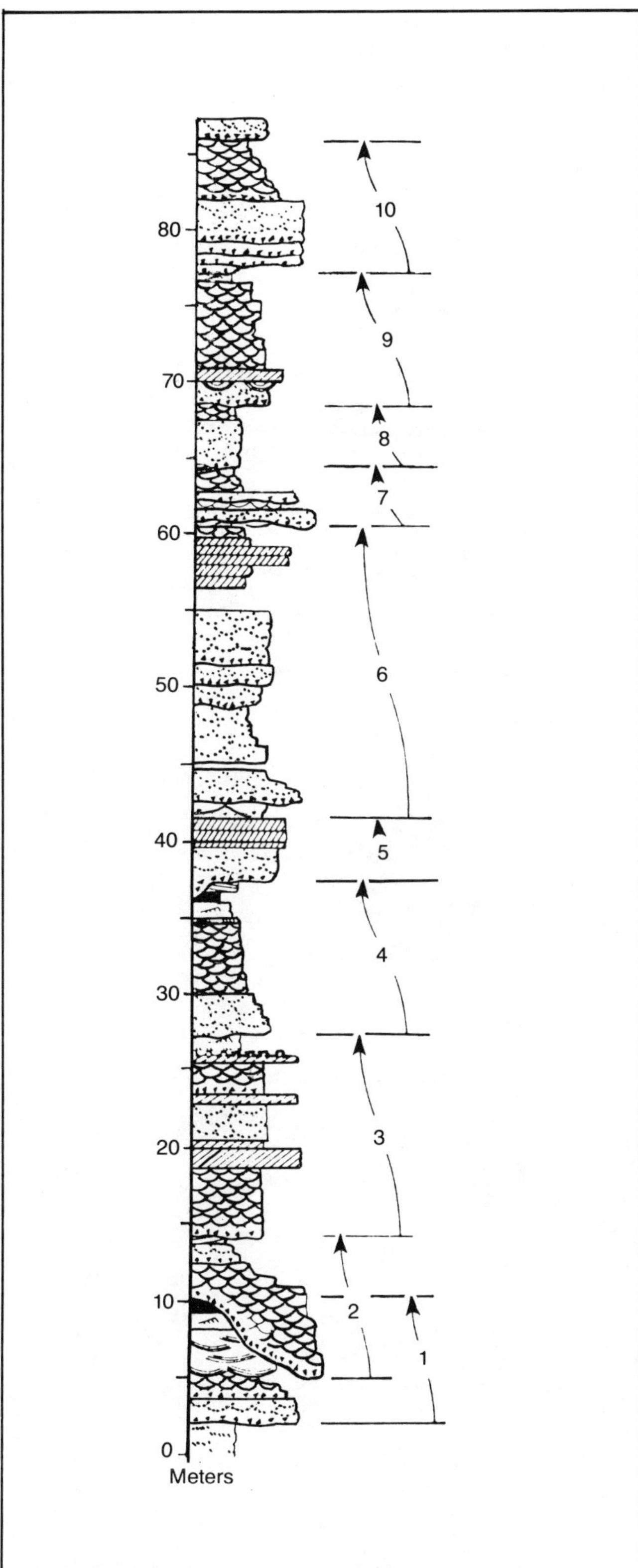

Fig. 4-5. Fining-upward sequences in the Battery Point Sandstone. See Figure 4-6 for facies symbols and descriptions. (Modified after Cant and Walker, 1976; permission to publish by the National Research Council of Canada).

their morphology (Fig. 4-3; Smith, 1974; Hein and Walker, 1977). Hein and Walker (1977) question the utility of a bar classification based on surface morphology and suggest the presence or absence of a slip face at or near the angle of repose is more important. The authors agree with Hein and Walker, especially since the presence or absence of cross-bedding is all with which the exploration geologist will have to work when studying cores. This is consistent with Miall's (1977) classification of bars and the production of internal stratification described in the section on gravel sheets.

Sand Flats

In sandy braided streams, large areas of sand called sand flats are exposed at medium to low river stage. They are combinations of bars and smaller features that are continually subject to reworking and destruction by flood events

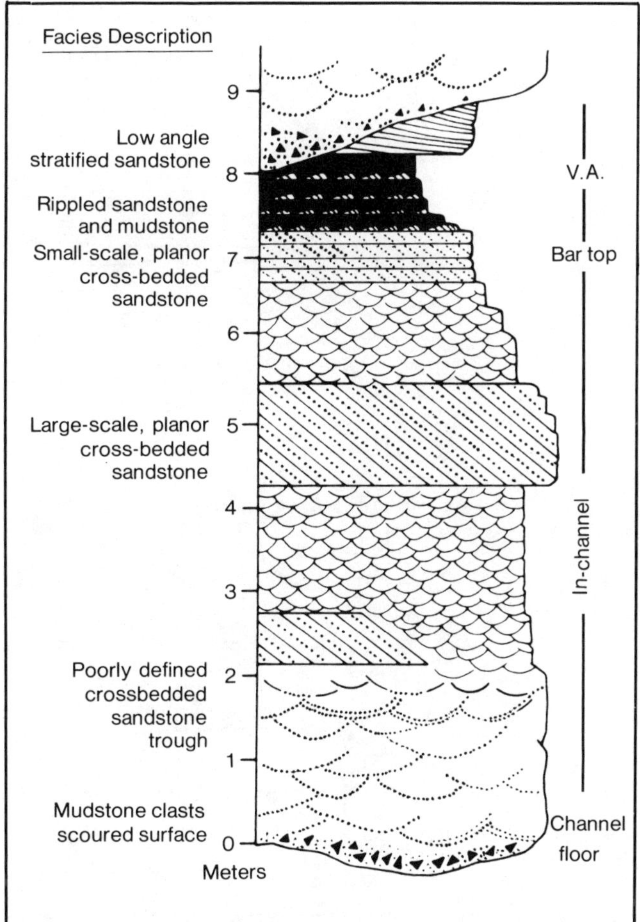

Fig. 4-6. Battery Point statistical summary sequence derived from the measured section shown in Figure 5. "V. A." signifies overbank vertical accretion deposits. (Modified after Cant, 1978a; permission to publish by Canadian Society of Petroleum Geologists).

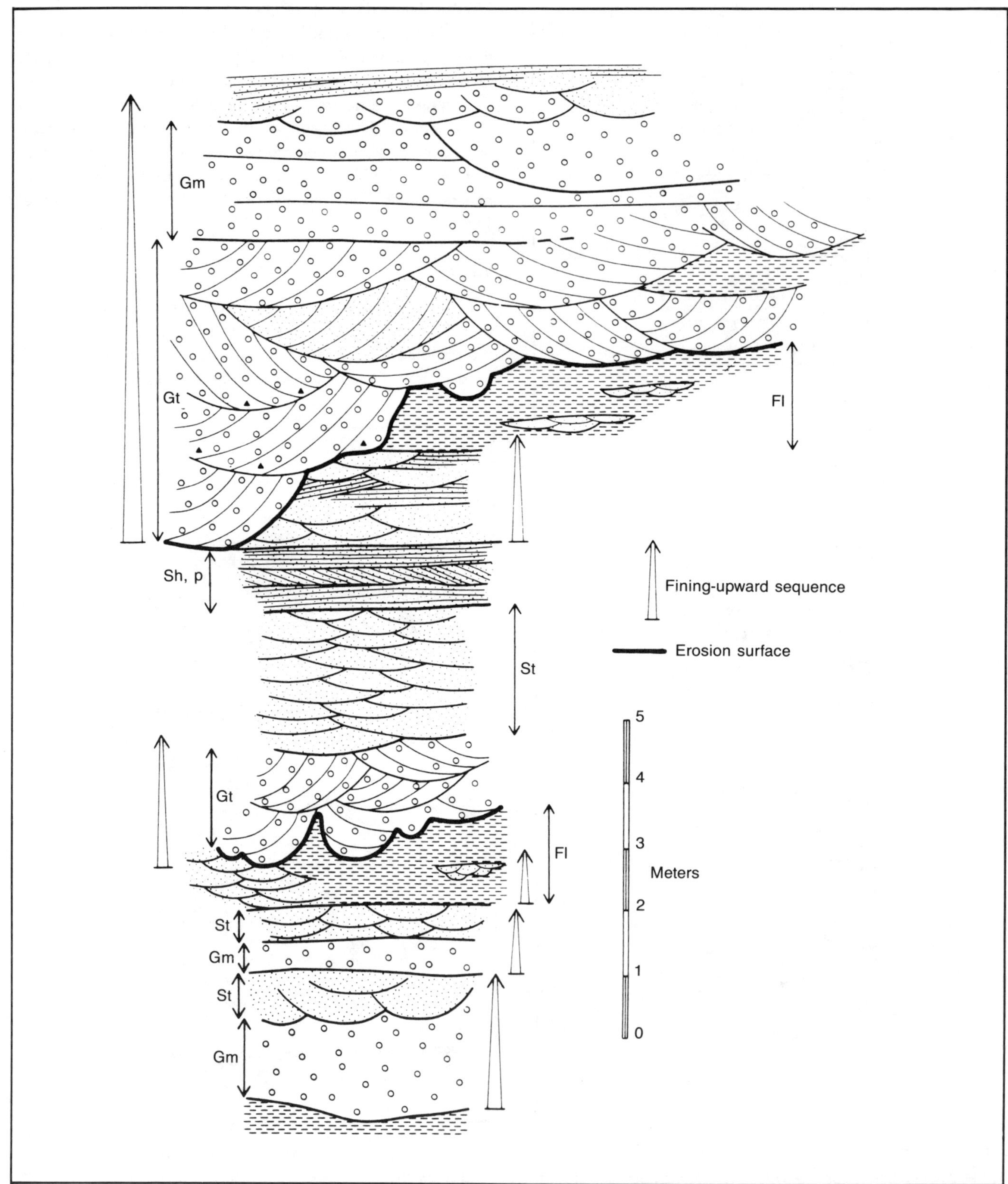

Fig. 4-7. Stratigraphic section of a part of the upper member of the Cannes de Roche Formation. See Table 4-2 for identification of abbreviations. (From Rust, 1978; permission to publish by Canadian Society of Petroleum Geologists).

(Cant and Walker, 1978). Sand flats in the South Saskatchewan River are essentially slip face bars that grow downstream from both sides of an emergent nucleus, such as the top of a diagonal bar, during falling river stage (Fig. 4-4). Downstream growth on either side of the nucleus creates a protected area in the lee of the nucleus where finer-grained sediments may accumulate. Sand flats build vertically during high river stages as bars and other bedforms override them. Floods may modify sand flats by dissecting them with small channels which may be refilled with sediment. Major channel shifts and erosion of upstream ends may also drastically alter or even eliminate sand flats (Cant and Walker, 1978).

The major bedform of a sand flat is a slipface bar with planar cross-stratification for its internal structure. Smaller-scale planar cross-beds from bars superimposed on the flats may overlie the original cross-beds. Parallel laminae deposited during floods may also occur. Small-scale trough and planar cross-bedding and ripple cross-lamination from small bedforms may occur in the upper parts of a sand flat but are less likely to be preserved than the larger-scale planar cross-bed sets (Cant and Walker, 1978).

Fine-grained Deposits

Silts and clays may be deposited in the braided river environment as overbank deposits and by settling from standing or slowly moving water left by receding floods. Deposits from standing water form lenticular mud drapes over low energy bedforms (Kelling, 1968; Miall, 1977). They

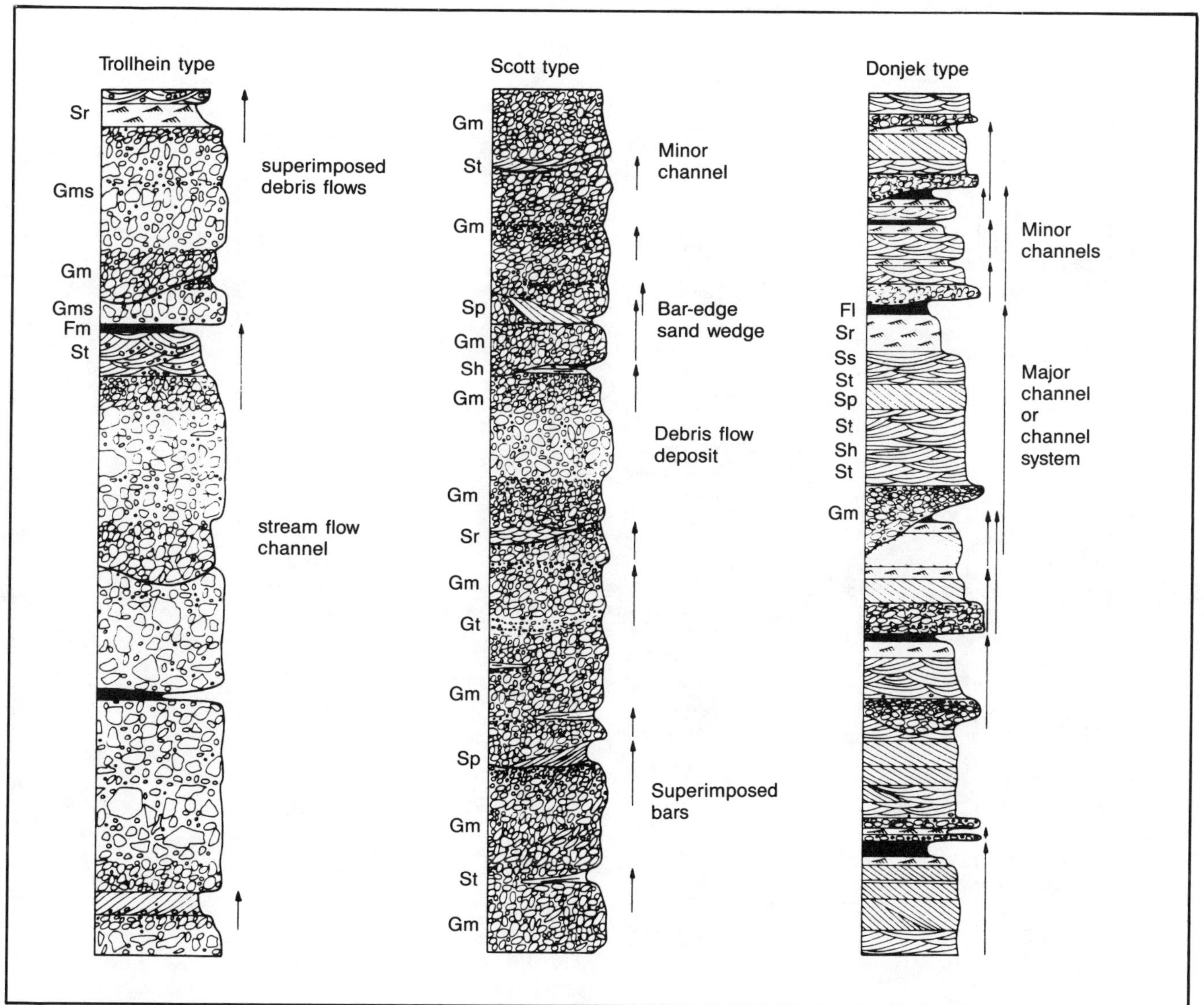

Fig. 4-8. Vertical profile models of gravelly braided stream deposits. Arrows show small-scale cyclic sequences. See Table 4-2 for facies codes. (From Miall, 1978; permission to publish by Canadian Society of Petroleum Geologists).

have low preservation potential due to erosion by subsequent floods (Kelling, 1968; Williams and Rust, 1969), but some do occur. Kelling (1968) describes silt banded mudstones with a few current ripples and rare oscillation ripples near the top of fining-upward sequences in the Rhondda Beds (Upper Pennsylvanian) of South Wales. Smith (1970) notes thin, lenticular shales and siltstones deposited in cut-off or plugged anabranches in the Shawangunk-Green Pond Conglomerates (Lower Silurian) in eastern Pennsylvania.

Areas with vegetation, such as high, semipermanent islands and floodplains, receive fine-grained sediments during the highest floods in braided streams as the fine material is trapped by the vegetation (Miall, 1977). Silty mudstones have ripple drift cross-lamination, draped lamination, and root burrows (Kelling, 1968; Boothroyd and Ashley, 1975).

Sedimentary Sequences

The potential for growth, modification, truncation, and complete erosion of bedforms in braided rivers is very high. This and the wide range of sediment sizes cause great variability in both vertical and lateral sedimentary sequences. There is, however, a fundamental order within the variability that permits identification of the braided stream environment.

Vertical Sequences. A single cycle of braided stream deposition results in a fining-upward sequence. In a flood or

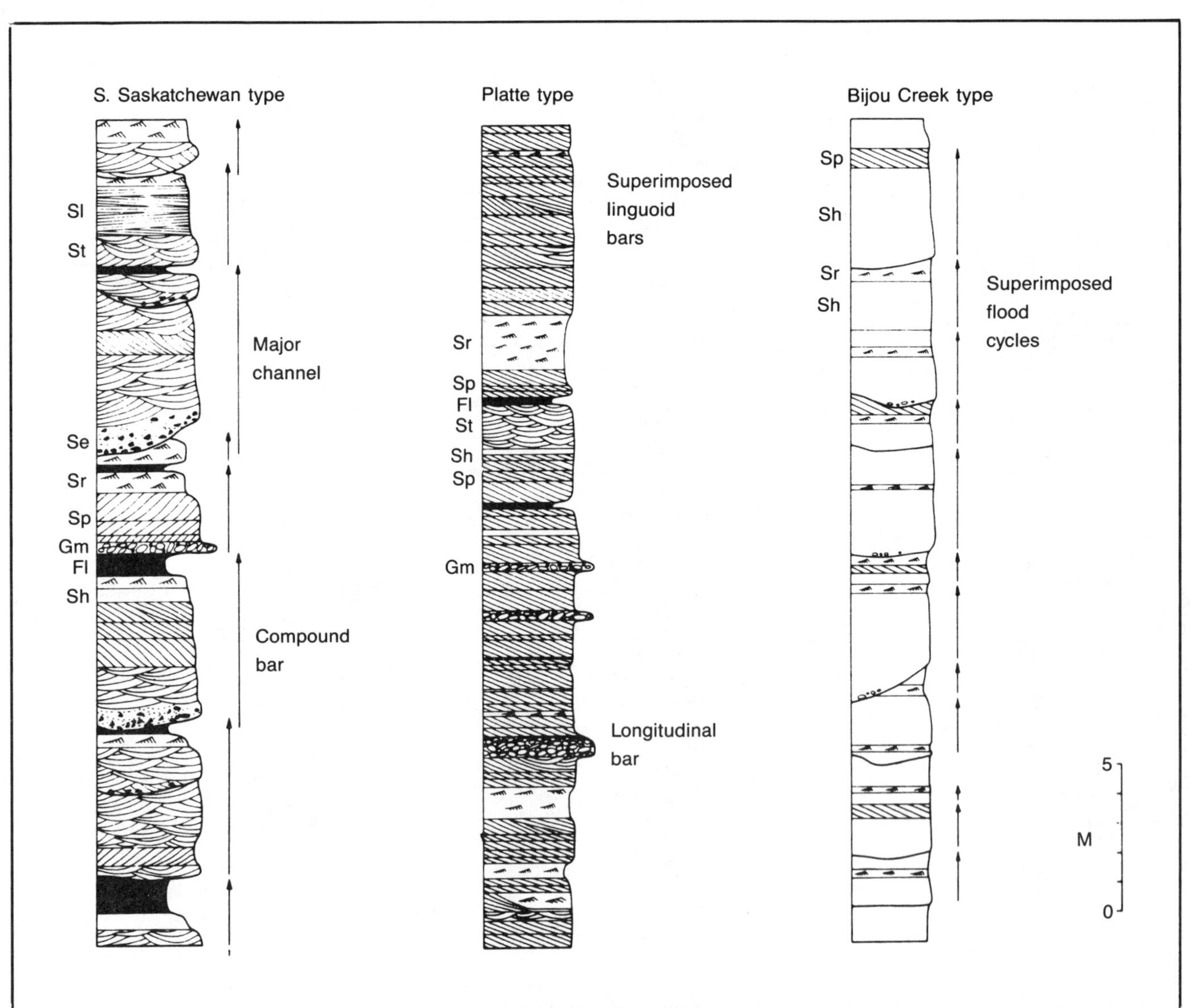

Fig. 4-9. Vertical profile models of sandy braided stream deposits. Arrows are small-scale cyclic sequences. See Table 4-2 for facies codes. (From Miall, 1978; permission to publish by Canadian Society of Petroleum Geologists).

TABLE 4-3. Facies assemblages and environments of the six principal gravel-dominated and sand-dominated braided river deposits (Figs. 4-8 and 4-9). See Table 4-2 for facies abbreviations. (From Miall, 1978; permission to publish by Canadian Society of Petroleum Geologists).

Name	Environmental setting	Main facies	Minor facies
Trollheim type (G_I)	proximal rivers (predominantly alluvial fans) subject to debris flows	*Gms, Gm*	*St, Sp, Fl, Fm*
Scott type (G_{II})	proximal rivers (including alluvial fans) with stream flows	*Gm*	*Gp, Gt, Sp, St, Sr, Fl, Fm*
Donjek type (G_{III})	distal gravelly rivers (cyclic deposits)	*Gm, Gt, St*	*Gp, Sh, Sr, Sp, Fl, Fm*
South Saskatchewan type (S_{II})	sandy braided rivers (cyclic deposits)	*St*	*Sp, Se, Sr, Sh, Ss, Sl, Gm, Fl, Fm*
Platte type (S_{II})	sandy braided rivers (virtually non cyclic)	*St, Sp*	*Sh, Sr, Ss, Gm, Fl, Fm*
Bijou Creek type (S_I)	Ephemeral or perennial rivers subject to flash floods	*Sh, Sl*	*Sp, Sr*

during channel aggradation, successively higher deposits form at progressively lower energy levels (Miall, 1977). In the Battery Point sandstone (Lower Devonian) in Quebec, Cant and Walker (1976) found ten, fining-upward, braided stream sequences (Fig. 4-5). Each starts at a scoured erosion surface that is overlain by coarse sandstone with mudstone intraclasts. The sequences then carry vertically into variously cross-bedded and rippled sandstones and mudstones. Statistical analysis of the different vertical facies changes in the Battery Point sandstone (Fig. 4-5) yielded the summary vertical section shown in Figure 4-6.

Williams and Rust (1969) identify about a dozen different vertical sequences of fining-upward channel fills in the Donjek River, Yukon. Each sequence starts with an initial gravel deposit on an erosion surface and ends with sand, silt, or silty clay deposited under differing circumstances and containing a variety of internal structures. An ancient, mixed grain size deposit, the upper member of the Cannes de Roche formation (Mississippian), Gaspe, Quebec (Fig. 4-7), has fining-upward sequences indicating the gradual filling of active channels (Rust, 1978). Trough cross-bedded gravels deposited by dunes are covered by massive gravels,

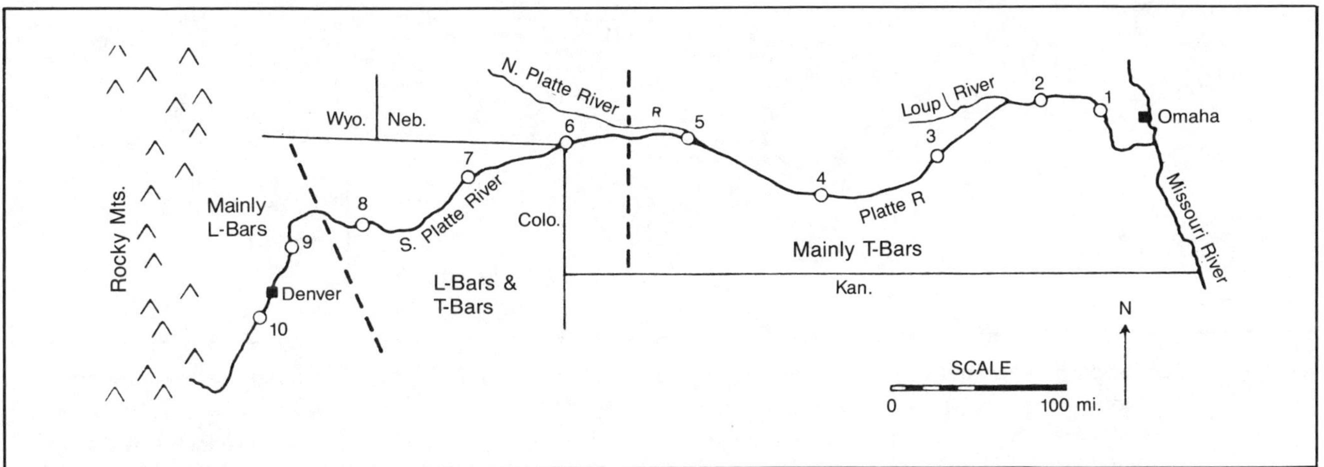

Fig. 4-10. Distribution of longitudinal bars (L-bars) and transverse bars (T-bars) along the South Platte-Platte Rivers. (From Smith, 1970; permission to publish by GSA).

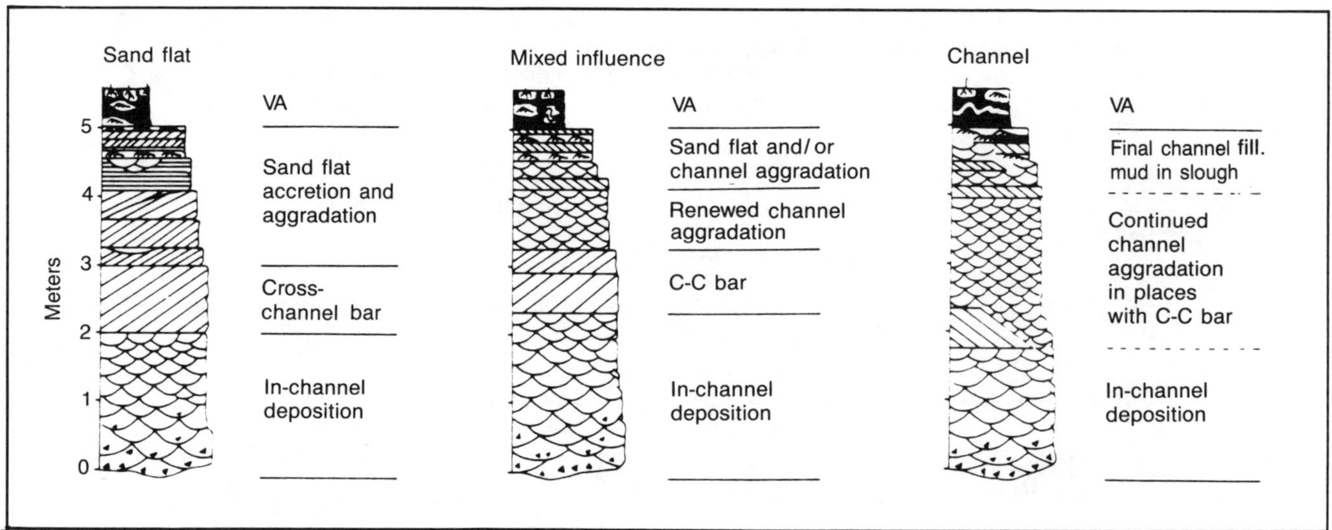

Fig. 4-11. Facies sequences in a sandy braided river. (From Cant, 1978a; permission to publish by Canadian Society of Petroleum Geologists).

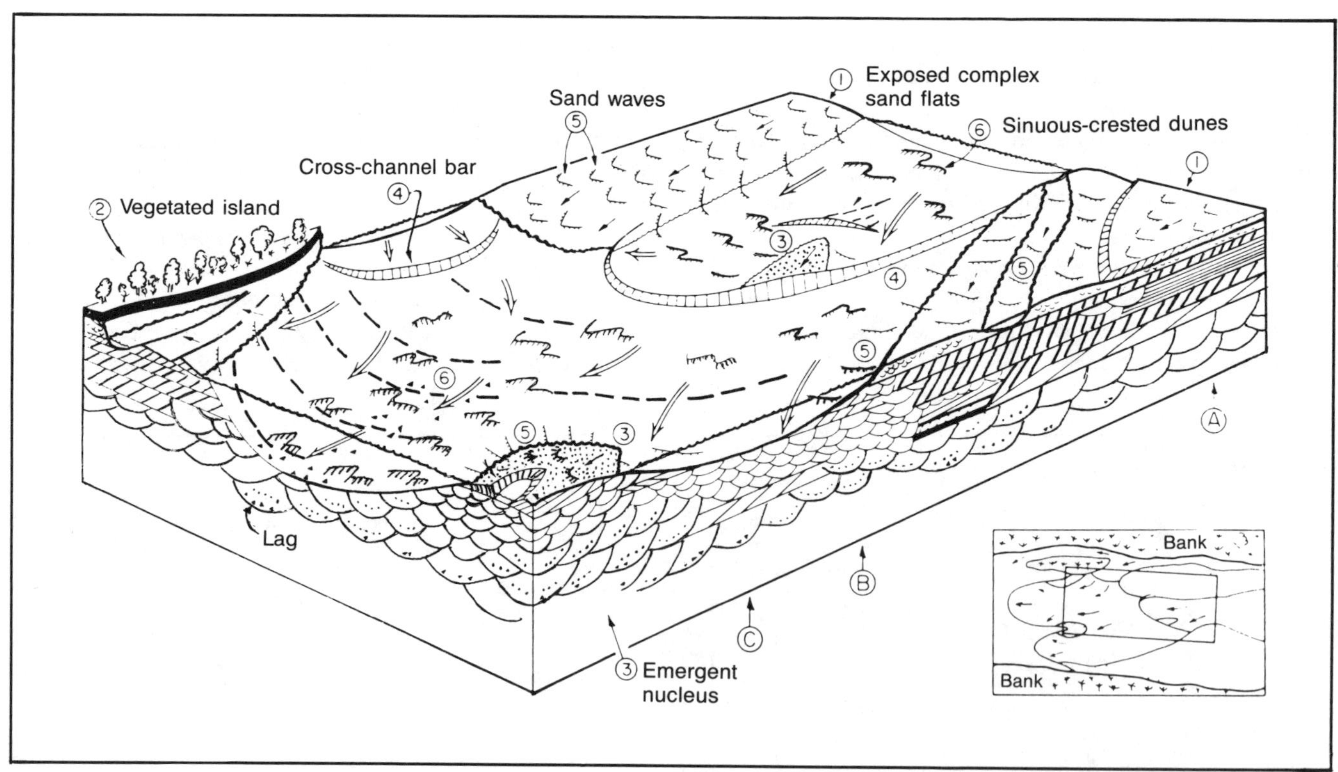

Fig. 4-12. Block diagram summarizing the major morphological elements and their associated bedforms and stratifications. The hypothetical reach is outlined by a rectangle in the inset. Stippled areas are emergent. Single shafted arrows indicate direction of bedform movement, and double-shafted arrows indicate flow directions. **A** *locates sedimentary sequence dominated by sand flat development (Fig. 4-11).* **B** *has mixed sand flat and channel influence, and* **C** *is dominated by channel aggradation. Numbers are explained on the diagram.* (From Cant and Walker, 1978; permission to publish from International Association of Sedimentologists).

trough cross-bedded sands, or horizontally bedded sands from migrating bars and/or channel aggradation. Migration of the active river channel ends the cycle with overbank deposits of laminated or cross-laminated very fine sand, silt, or mud (Rust, 1978). Limited representation of overbank deposits (Fl of Fig. 4-7) demonstrates their low preservation potential.

In the Kicking Horse River, British Columbia, a modern, gravel-dominant river, Smith (1974) finds that simple, single cycle gravel bars fine upward. The most likely bedding is massive, crudely horizontal, or low-angle inclination with imbricated pebbles. High angle, planar cross-stratification is not common, but where it does occur, it is confined to finer gravels. Simple bars are usually strongly modified by erosion or changing depositional patterns producing complex internal structures. Ancient deposits should be ". . . an assorted array of laterally adjacent and juxtaposed remnants of bar and channel-fill sediments fortuitously preserved by the shifting channels" (Smith, 1974).

Miall (1977, 1978) has constructed six vertical profile models for braided streams (Figs. 4-8 and 4-9). The Trollheim type (Fig. 4-8) is composed mostly of matrix-supported, debris-flow gravel and is more an alluvial fan deposit than a braided stream deposit. Facies and environmental summaries for each model appear in Table 4-3.

Horizontal Sequences. Braided stream deposits have considerable horizontal variation. The Slims River in Yukon, Canada, changes from a gravel-dominant river to a silt-dominant river in a distance of 22 km (Fahnestock, 1969). In the South Platte-Platte River system of Colorado and Nebraska, grain size decreases and bedforms change downstream (Smith, 1970). Upstream, longitudinal bars with coarse-grained, poorly sorted sediments are the dominant bedform (Fig. 4-10). Downstream, transverse bars composed of better-sorted, finer-grained sands occur. Between the two areas is a transition zone with both types of bars. The relationship between bar type and sedimentary texture confirms similar observations by Ore (1964). Downstream changes in bar type result in changing stratification with crude horizontal bedding related to longitudinal bars grading downstream to planar cross-bedding associated with transverse bars (Smith, 1970).

Miall's vertical profiles also indicate horizontal variations in braided stream deposits. Proximal and distal environments are mentioned for the Scott and Donjek models, re-

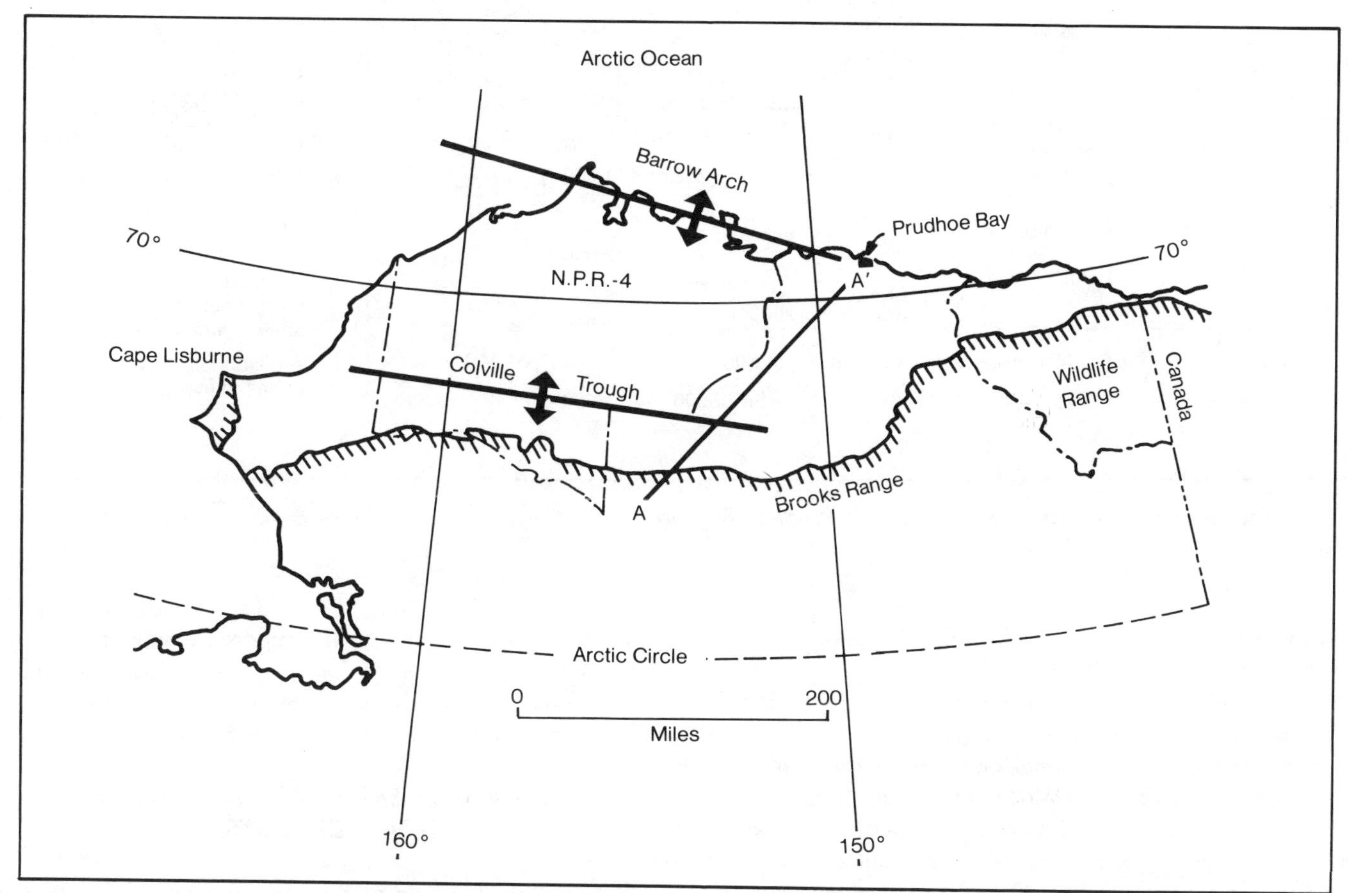

Fig. 4-13. Major structural features of northern Alaska. (From Jones and Speers, 1976; permission to publish by AAPG).

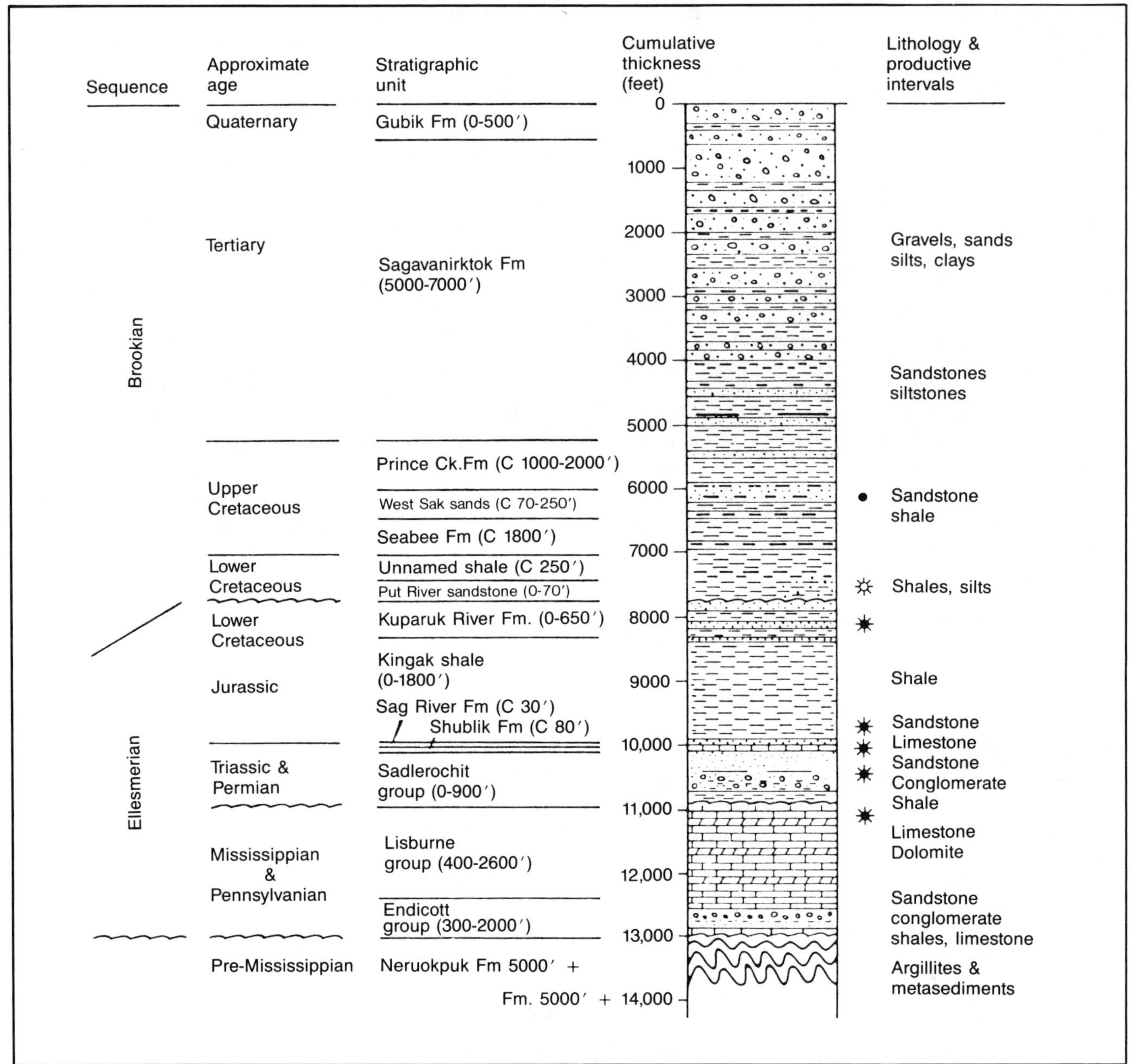

Fig. 4-14. Generalized stratigraphic column in Prudhoe Bay area. (From Jamison et al., 1980; permission to publish by AAPG).

spectively (Table 4-3), and Miall (1978) suggests that the Scott, Donjek, and South Saskatchewan profiles may constitute a proximal to distal sequence with a decreasing gravel to sand ratio (Figs. 4-8 and 4-9).

In developing a depositional model for the South Saskatchewan River, Cant and Walker (1978) illustrate the nature of local horizontal variations in sediments deposited in a sandy braided stream. Three vertical profiles through a sand flat, a channel, and the transitional area marginal to both show the primary difference between the profiles to be internal stratification (Figs. 4-11 and 4-12). All three profiles fine upward and have comparable sand development. Consequently, an SP or gamma-ray log in a well through one of the sands would not define the environment, whereas a conventional core would define it.

HYDROCARBON EXPLORATION IN BRAIDED STREAM DEPOSITS

The Prudhoe Bay field on the north coast of Alaska (Fig. 4-13) is the largest oil field in the United States (Jamison et al., 1980). Production is principally from braided stream

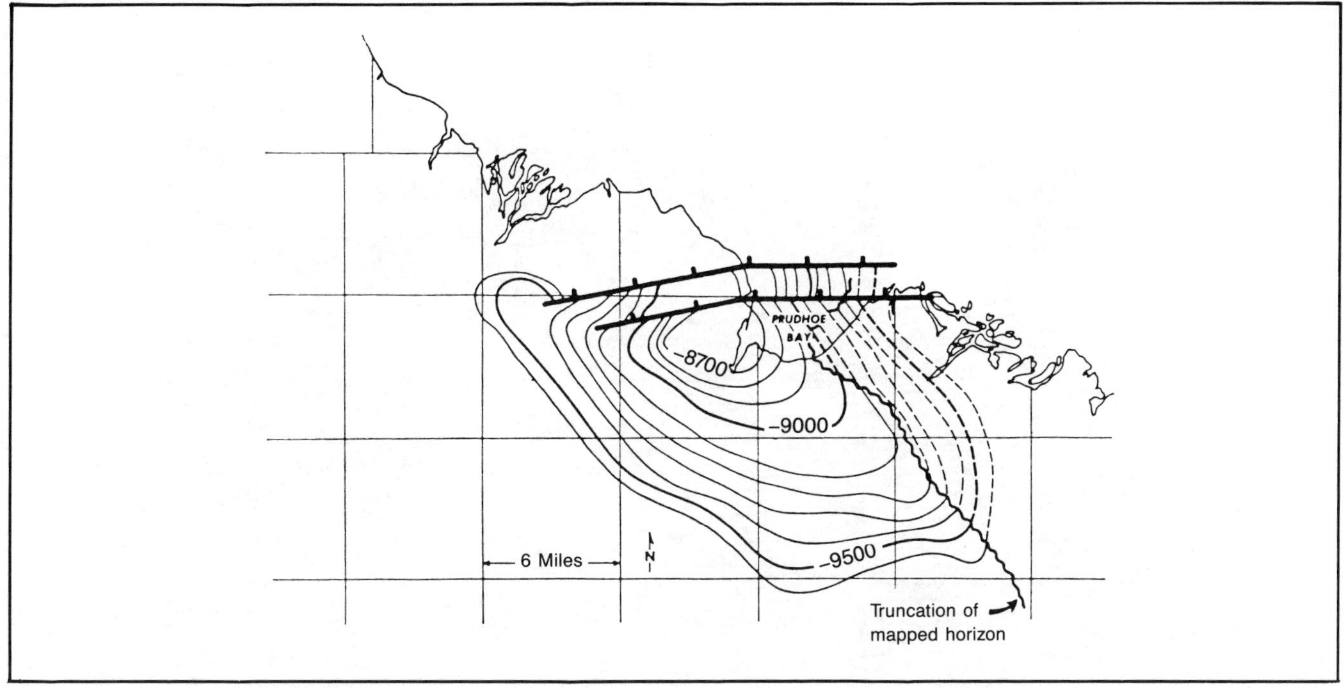

Fig. 4-15. Pre-sale (Spring 1965) seismic interpretation of Prudhoe Bay structure. Major elements of trapping are seen. Compare with Figure 4-16, more current map. (From Morgridge and Smith, 1972; permission to publish by AAPG).

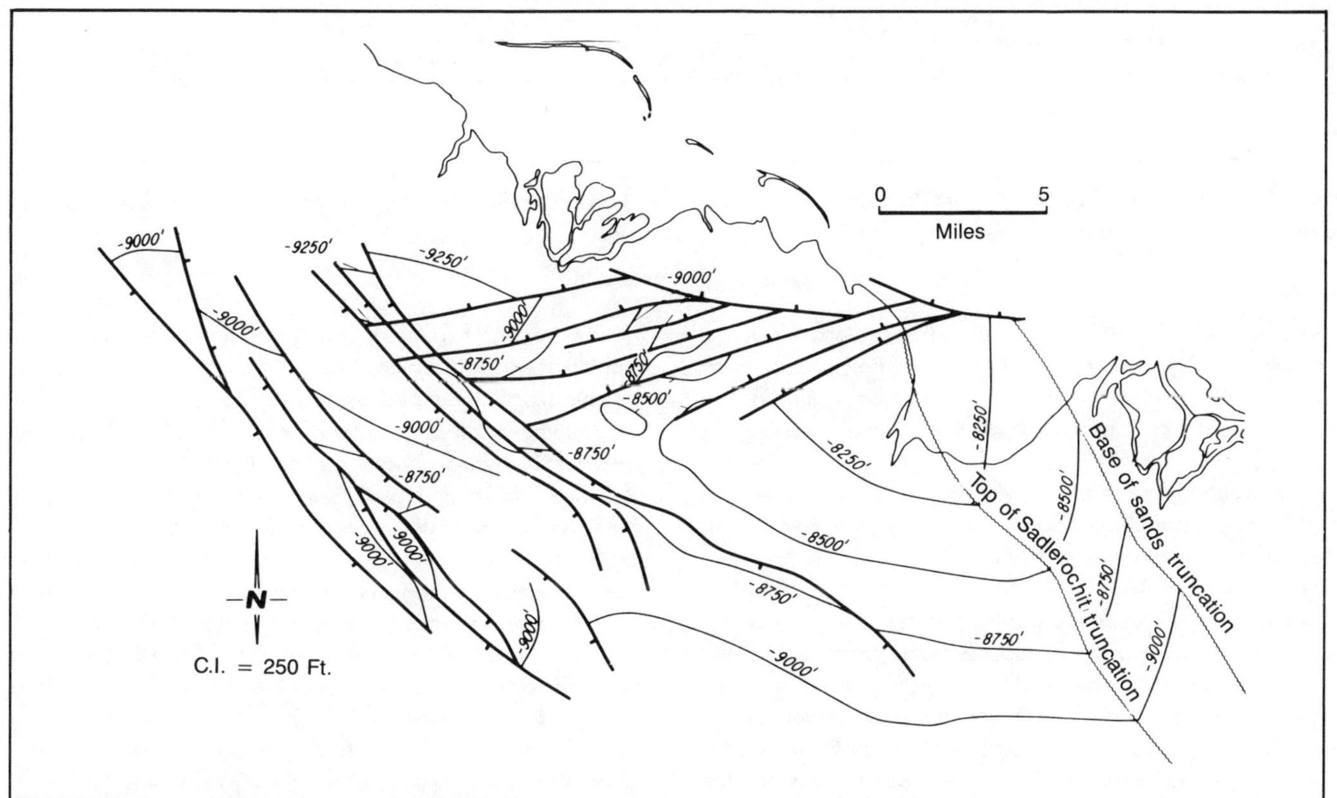

Fig. 4-16. Structure map of the top of the Sadlerochit. (From Jones and Speers, 1976; permission to publish by AAPG).

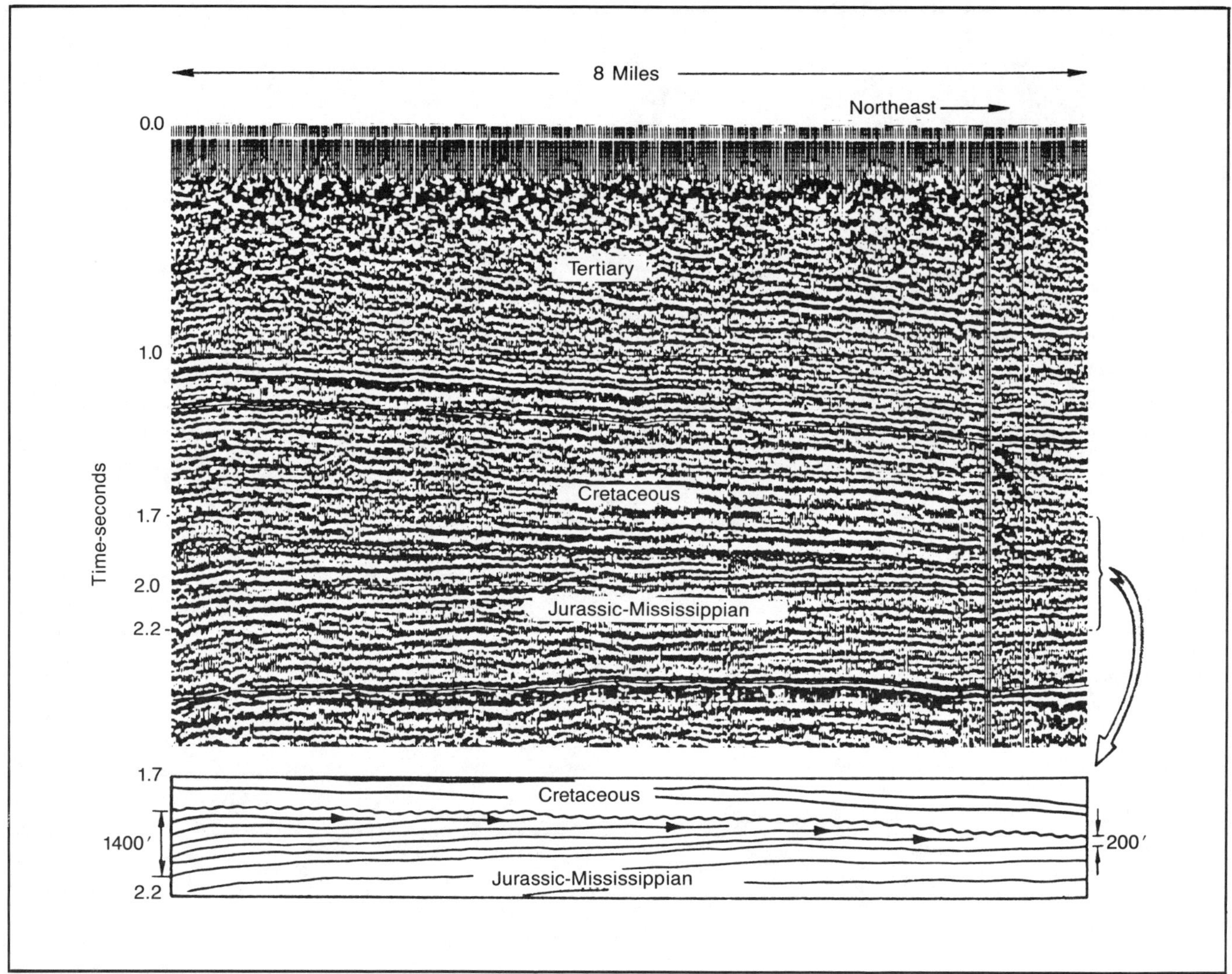

Fig. 4.17. Southwest-northeast-trending seismic profile showing unconformity truncating seismic events between 1.7 and 2.2 sec. (From Morgridge and Smith, 1972; permission to publish by AAPG).

sediments of the Ivishak sandstone in the Sadlerochit group (Permo-Triassic). Other formations from Pennsylvanian to Cretaceous age also produce but are not as prolific as the Ivishak (Fig. 4-14; Eckelmann et al., 1975; Jones and Speers, 1976).

The Prudhoe Bay structure was originally discovered and outlined by seismology (Fig. 4-15; Morgridge and Smith, 1972). It is a westerly plunging structural nose on trend with the Barrow Arch (Figs. 4-13, 4-15, and 4-16). Normal faults cut the structure on the northwest and the southwest. Pre-Cretaceous rocks dip generally to the south and are truncated on the northeast side of the structure by an unconformity (Figs. 4-15, 4-16, and 4-17). Lower Cretaceous shales lying on the unconformity are both the source rocks for the hydrocarbons and the sealing rocks for the Permo-Triassic reservoir (Figs. 4-17 and 4-18). Closure is a combination of faulting, sealing against an unconformity, and structural dip making the field a structural and stratigraphic trap (Jones and Speers, 1976).

Long-known oil seeps east of Point Barrow indicated the presence of hydrocarbons on the North Slope. Discovery of hydrocarbons in Naval Petroleum Reserve No. 4 (Fig. 4-13) further indicated potential for the area. In the late 1950's and early 1960's, surface geologic mapping of the North Slope was conducted by various companies on nonfederal lands. Seismic coverage was initiated in the early 1960's (Jamison et al., 1980). Data from 1964 shooting by Richfield and Humble defined the Prudhoe Bay structure (Fig. 4-15; Morgridge and Smith, 1972). After leasing the crestal portion of the structure in two separate state lease sales, ARCO and Humble drilled the Prudhoe Bay State No. 1 and found 400 ft (135 m) of gas column. A confirmation well, Sag River State No. 1, seven miles (10 km) southeast and well down the flank of the structure, encountered 400 ft (135 m) of oil column (Morgridge and Smith, 1972).

Conventional cores from the two wells were analyzed in detail for vertical sequences of sedimentary features. Comparisons between the sequences and modern analogues were made to identify environmental facies. As more wells were drilled, environmental facies calibrated to log patterns in cored wells were carried into uncored wells to further develop the reservoir model (Eckelmann et al., 1975). The first two cores showed the Sadlerochit group to have shallow marine deltaic deposits in the lower part and braided river deposits in the upper part (Fig. 4-19; Morgridge and Smith, 1972; Eckelmann et al., 1975). Fine-grained shallow marine and shoreline siltstones and sandstones of the Echooka formation rest uncomformably on the late Paleozoic carbonates of the Lisburne Group. The Kavik marine shale overlies the Echooka and grades upward into prodelta claystones and silty claystones. Delta-front silty claystones, siltstones, and very fine-grained silty sandstones of the lower Ivishak overlie the Kavik. All three units thicken to the south.

Stream mouth bars of very fine-grained sandstone overlie the delta-front sediments. Bar deposits thicken at the expense of delta-front fine clastics and grain size increases upward. South dipping cross-bedding occurs in oriented cores and on outcrops. Parallel lamination has been cited as indicating redistribution of stream mouth bars by waves (Eckelmann et al., 1975). Sheet-like delta plain deposits consisting of siltstones, silty claystones, and claystones deposited by flooding of braided streams overlie the bar deposits (Eckelmann et al., 1975).

Braided stream deposits constitute the bulk of the Ivishak sandstone reservoir (Fig. 4-19). In the lower part of the interval, complete and incomplete, stacked, fining-upward cycles start with large-scale, cross-bedded pebble conglomerates at the base and grade upward to medium-grained and fine-grained sandstones. Point bar deposits attributed to meandering streams, bars, and floodplain deposits also occur. The upper part of the braided stream interval is sandstone with scattered pebbles; conglomerates are rare (Morgridge and Smith, 1972; Eckelmann et al., 1975). Sedimentary structures and grain size distribution in the lower part of the braided stream section resemble those described earlier in this chapter from both modern and ancient rivers.

A regional interpretation of the facies distribution of the Sadlerochit group shows braided stream deposits prograd-

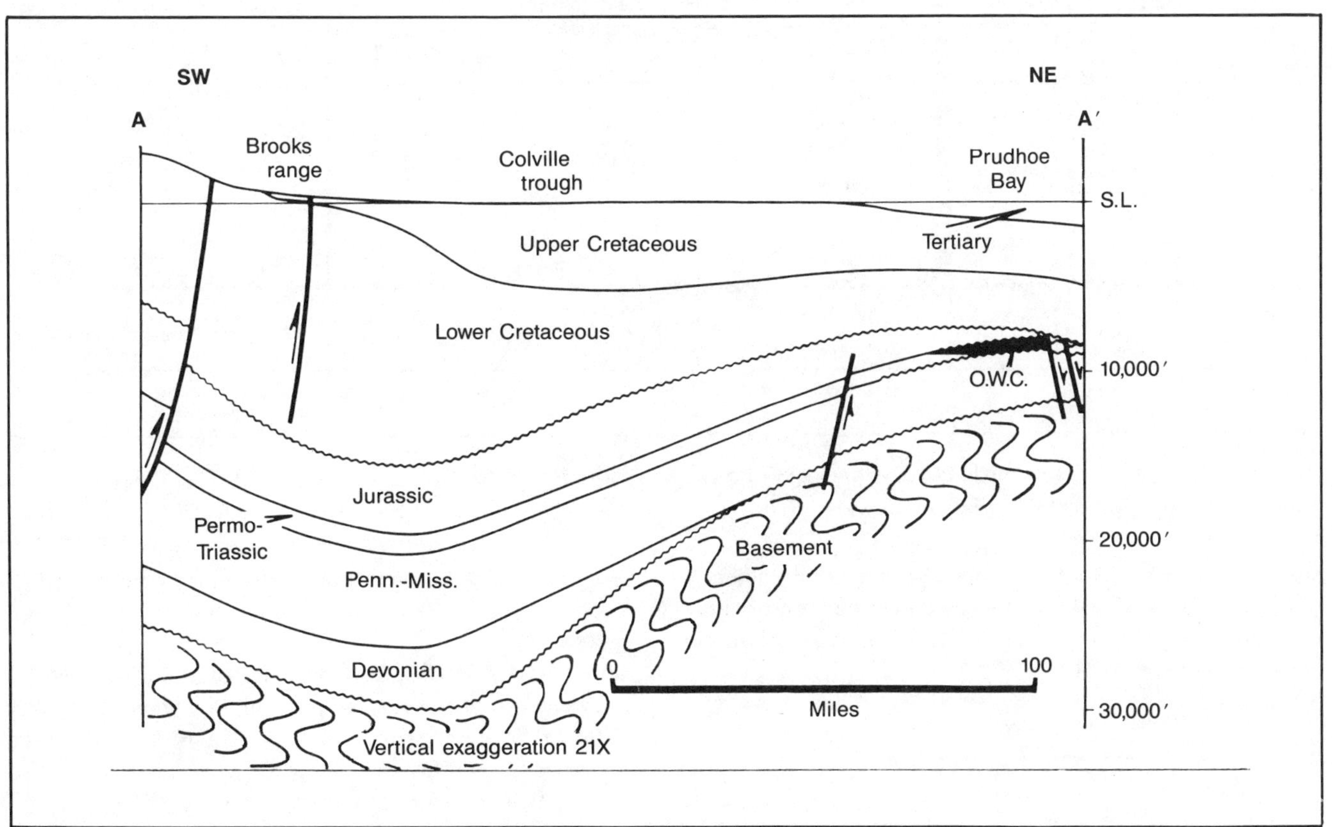

Fig. 4-18. Generalized cross section of the North Slope; see Figure 4-13 for location of cross section. (From Jones and Speers, 1976; permission to publish by AAPG).

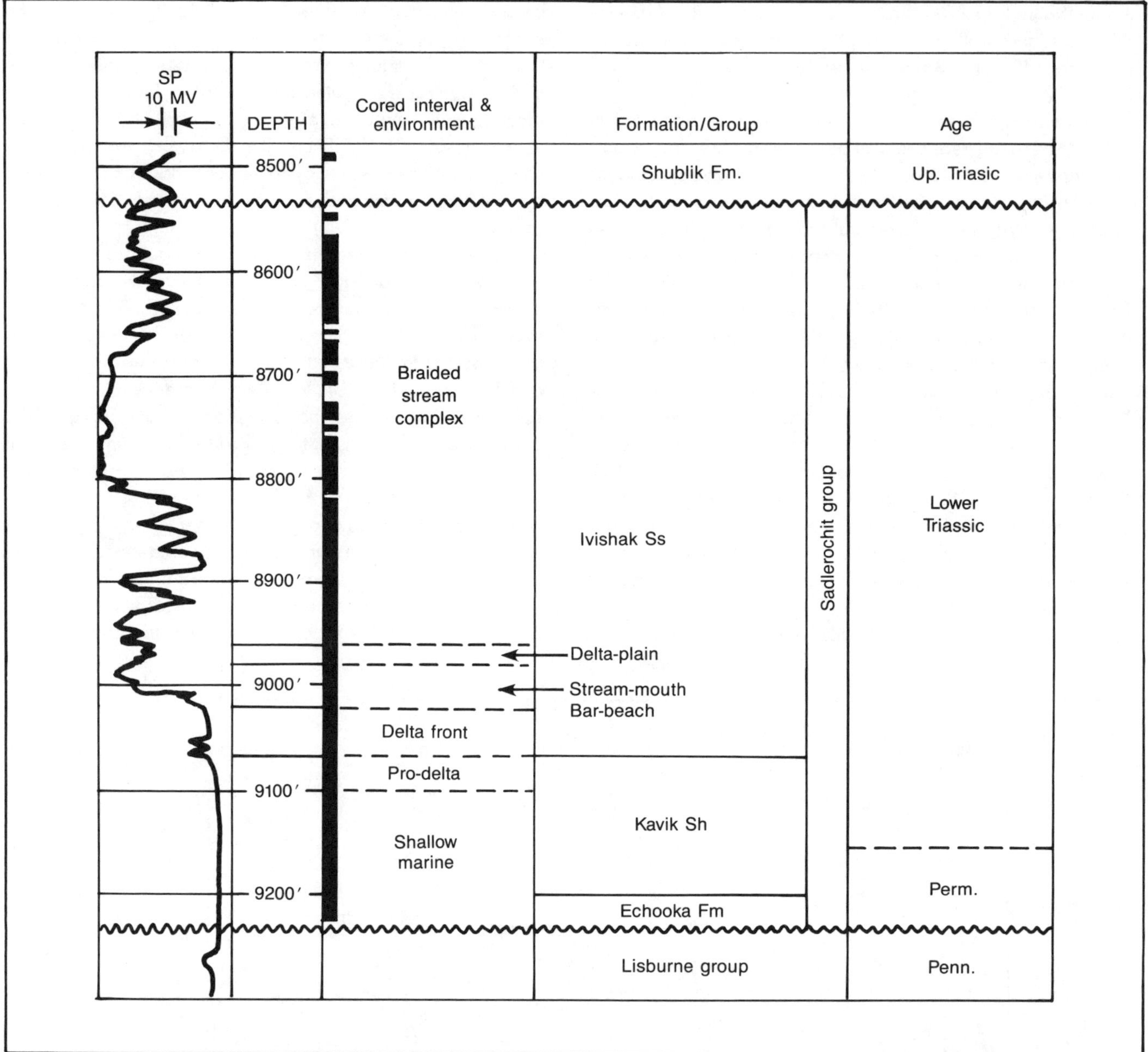

Fig. 4-19. Sadlerochit stratigraphy, environments and SP log character. (Modified after Eckelmann et al., 1975; permission to publish by Elsevier Applied Science Publishers).

ing over the delta-related sediments in a general southerly direction (Fig. 4-20). All of the braided stream and deltaic sediments were derived from a mountain source area to the north (Eckelmann et al., 1975).

SELECTED BIBLIOGRAPHY

Boothroyd, J. C., and G. M. Ashley, 1975, Process, bar morphology, and sedimentary structures on braided outwash fans, northeastern Gulf of Alaska, *in* A. V. Jopling, and B. C. McDonald, eds., Glaciofluvial and glaciolacustrine sedimentation: SEPM Sp. Pub. 23, p. 193-222.

Cant, D. J., 1978a, Development of a facies model for sandy braided river sedimentation: comparison of the South Saskatchewan River and the Battery Point Formation, *in* A. D. Miall, ed., Fluvial sedimentology: Can. Soc. Petrol. Geol., Mem. 5, p. 627-639.

———, 1978b, Bedforms and bar types in the South Saskatchewan River: Jour. Sed. Pet., v. 48, no. 4, p. 1321-1330.

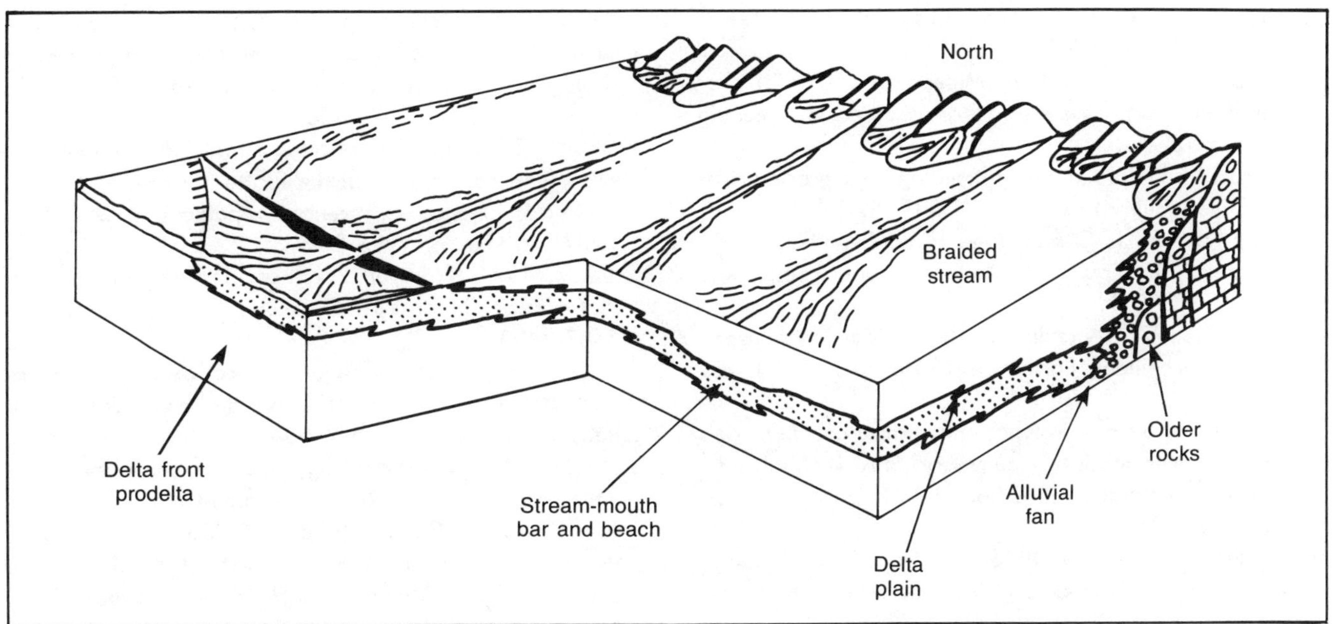

Fig. 4-20. Environmental facies distribution in and around Prudhoe Bay area. (From Eckelmann et al., 1975; permission to publish by Elsevier Applied Sciences Publishers).

_______, 1982, Fluvial facies models and their application, *in* P. A. Scholle, and D. Spearing, eds., Sandstone depositional environments: AAPG Mem. 31, p. 115-137.

_______, and R. G. Walker, 1976, Development of a braided fluvial facies model for the Devonian Battery Point Sandstone, Quebec: Can. Jour. Earth Sci., v. 13, no. 1, p. 102-119.

_______ and _______, 1978, Fluvial processes and facies sequences in the sandy braided South Saskatchewan River, Canada: Sedimentology, v. 25, no. 5, p. 625-648.

Chien, N., 1961, The braided stream of the lower Yellow River: Sci. Sinica, v. 10, p. 734-754.

Coleman, J. M., 1969, Brahmaputra River: channel processes and sedimentation: Sed. Geol., v. 3, no. 2-3, p. 129-239.

Collinson, J. D., 1970, Bedforms of the Tana River: Geografiska Annaler, v. 52A, p. 31-56.

Doeglas, D. J., 1962, The structure of sedimentary deposits of braided streams: Sedimentology, v. 1, no. 3, p. 167-190.

Eckelmann, W. R., R. J. De Witt, and W. L. Fisher, 1975, Prediction of fluvial-deltaic reservoir geometry, Prudhoe Bay field, Alaska: Proc. 9th World Petrol. Congr., Panel disc. 4 (2), p. 1-5.

Fahnestock, R. K., 1969, Morphology of the Slims River: Icefield Ranges Res. Proj., Sci. Res., 1, p. 161-172.

_______ and W. C. Bradley, 1973, Knik and Matanuska River, Alaska, a contrast in braiding, *in* M. Morisowa, ed., Fluvial geomorphology, SUNY Binghampton, Proc. 4th Geomorph. Symp., p. 220-250.

Gole, C. V., and S. V. Chitale, 1966, Inland delta building activity of the Kosi River: Proc. Amer. Soc. Civil Engineers, Jour. of Hydraulics Div., HY-2, p. 111-126.

Harms, J. C., J. B. Southard, D. R. Spearing, and R. G. Walker, 1975, Depositional environments as interpreted from primary sedimentary structures and stratification sequences: Dallas, SEPM Short Course 2, 161 p.

Haszeldine, R. S., 1983, Fluvial bars reconstructed from a deep, straight channel, Upper Carboniferous coalfield of northeast England: Jour. Sed. Pet., v. 53, no. 4, p. 1233-1247.

Hein, F. J, and R. G. Walker, 1977, Bar evolution and development of stratification in the gravelly, braided Kicking Horse River, British Columbia: Can. Jour. Earth Sci., v. 14, no. 4, p. 562-570.

Jamison, H. C., L. D. Brockett, and R. A. Mc Intosh, 1980, Prudhoe Bay: a 10-year perspective, *in* M. T. Halbouty, ed., Giant oil and gas fields of the decade 1968-1978, AAPG Mem. 30, p. 289-314.

Jones, H. P., and R. G. Speers, 1976, Permo-Triassic reservoirs of Prudhoe Bay field, North Slope, Alaska, *in* J. Braunstein, ed., North American oil and gas fields: AAPG Mem. 24, p. 23-50.

Kelling, G., 1968, Patterns of sedimentation in Rhondda Beds of South Wales: AAPG Bull., v. 52, no. 12, p, 2369-2386.

Kessler, L. G., II and F. G. Cooper, 1970, Channel sequences and braided stream development in the South Canadian River, Hutchinson, Roberts, and Hemphill Counties, Texas: Gulf Coast Ass'n. Geol. Soc. Trans., v. 20, p. 263-273.

Leopold, L. B., and M. G. Wolman, 1957, River channel

patterns, braided, meandering and straight: U. S. Geol. Surv. Prof. Paper 282-B, p. 39-85.

———, ———, and J. P. Miller, 1964, Fluvial processes in geomorphology: San Francisco, W. H. Freeman and Co., 522 p.

Macklin, J. R., 1956, Causes of braiding by a graded river (abs.): GSA Bull., v. 67, p. 1717-1718.

McKee, E. D., E. J. Crosby, and H. L. Berryhill, 1967, Flood deposits, Bijou Creek, Colorado: Jour. Sed. Pet., v. 37, p. 829-851.

Miall, A. D., 1977, A review of the braided river depositional environment: Earth Science Reviews, v. 13, no. 1, p. 1-62.

———, 1978, Lithofacies types and vertical profile models in braided river deposits: a summary, *in* A. D. Mial, ed., Fluvial sedimentology: Can. Soc. Petrol. Geol. Mem. 5, p. 597-604.

Morgridge, D. L. and W. B. Smith, Jr., 1972, Geology and discovery of Prudhoe Bay field, eastern Arctic Slope, Alaska, *in* R. E. King, ed., Stratigraphic oil and gas fields: AAPG Mem. 16, p. 489-501.

Ore, H. T., 1964, Some criteria for recognition of braided stream deposits: Wyoming Univ. Contrib. Geol., v. 3, p. 1-14.

Rust, B. R., 1972, Structure and process in a braided river: Sedimentology, v. 18, no. 3/4, p. 221-245.

———, 1978, Depositional models for braided alluvium, *in* A. D. Miall, ed., Fluvial sedimentology: Can. Soc. Petrol. Geol. Mem. 5, p. 605-625.

Schumm, S. A., 1968, Speculations concerning paleohydrologic controls of terrestrial sedimentation: GSA Bull., v. 79, no. 11, p. 1573-1588.

Smith, N. E., 1970, The braided stream depositional environment: comparison of the Platte River with some Silurian clastic rocks, north-central Appalachians: GSA Bull., v. 81, no. 10, p. 2993-3014.

———, 1971, Transverse bars and braiding in the lower Platte River, Nebraska: GSA Bull., v. 82, no. 12, p. 3407-3420.

———, 1972, Some sedimentological aspects of planar cross-stratification in a sandy braided river: Jour. Sed. Pet., v. 42, no. 3, p. 624-634.

———, 1974, Sedimentology and bar formation in the upper Kicking Horse River, a braided outwash stream: Jour. Geol., v. 82, no. 2, p. 205-233.

Steel, R. J., 1974, New Red Sandstone floodplain and piedmont sedimentation in the Hebridean Province, Scotland: Jour. Sed. Pet., v. 44, no. 2, p. 336-357.

Thornbury, W. D., 1954, Principles of geomorphology: New York, John Wiley and Sons, Inc., New York, 618 p.

Walker, R. G., 1976, Facies models 3: sandy fluvial systems: Geoscience Canada, v. 3, no. 2, p. 101-109.

Williams, P. F., and B. R. Rust, 1969, The sedimentology of a braided river: Jour. Sed. Pet., v. 39, no. 2, p. 649-679.

5 CHANNEL SANDSTONES

Channel sandstones are one of several types of "shoestring" reservoirs referred to by Rich (1923). They may trend for many miles in a given direction through a youthful valley or in multiple directions when they collectively fill the various tributaries that make up a given drainage system. The sandstone may be either continuous or discontinuous. In vertical sections the channel fill may consist of (1) coarse sandstone at the base, fining in texture upward, (2) well-sorted sandstone, either fine grained or coarse grained throughout, or (3) sandstone interstratified with shale "breaks." Thus, channel sandstones are quite variable in trends and composition.

Channel sandstones in many instances are deposited subaerially, but they also may be deposited within a restricted portion of a shallow marine environment in response to either a gradual or cyclic marine transgression over a previously eroded land surface. They are exposed abundantly in outcrops and, thus, have been studied in considerable detail by sedimentologists. Such studies, however, have limited application in identifying and tracing channel sandstones in the subsurface. When the explorationist attempts to apply surface studies involving cross-bedding, preferred alignment of *c*-axes of individual elongate sand grains, and upstream tilting of flattened pebbles to rotary cuttings or mechanical logs, he becomes frustrated with what the authors call "sedimentological overkill." These parameters, although signficant in outcrop studies, have little, if any, value in the subsurface. Even cores afford little help in determining these parameters because they are seldom oriented or sufficient in number to be meaningful. Thus, in working with channel sandstones, the explorationist is obliged to work with other criteria that must be used in combination. It is these other criteria that make up the main body of the following discussion.

DRAINAGE SYSTEMS

A background knowledge of the various types of drainage systems is paramount if the explorationist is to be effective in working with channel sandstones in the subsurface. Perhaps the most common type is the *dendritic* pattern of drainage, in which the majority of the tributaries bifurcate at acute angles. The overall pattern is analogous to the veins of a maple leaf. The dendritic drainage pattern generally is developed in rock strata that are essentially horizontal and offer a uniform degree of relative resistivity to weathering and erosion. Either a shale or sandstone terrane is conducive to the development of the dendritic drainage pattern.

The second most important drainage pattern is the *rectangular* in which the tributaries join their parent stream essentially at right angles. In this case the strata, being weathered and eroded, have a homoclinal tilt and consist of alternating hard and soft layers. The master stream *(consequent)* erodes across both types of strata and trends essentially normal to the strike of the strata. Its primary tributaries *(subsequent)* erode into and parallel with the strike of the softer strata. Tributaries to the subsequent are of two types. Those that flow down the dip slope are known as *resequent*. Short streams flow down the scarp slope, have comparatively steep gradients and are called *obsequent*. The subsurface significance and identification of these stream types were described by Martin (1966).

The third most important drainage pattern is the *trellis* in which subparallel streams follow the strike of the strata, which have been folded and truncated. Alternating hard and soft strata are subaerially exposed, and subparallel drainage courses are eroded into the softer strata.

A *barbed* drainage pattern is one in which tributaries

join the main stream at an angle greater than 90. This situation is not common and is, therefore, unimportant to the explorationist. Areas of ground, recessional, and terminal moraine have "dump" topography and the drainage may be *deranged*. In these areas there is no apparent systematic control other than the initial depressions on the original surface. This type of drainage has no significance in subsurface studies of channels. Other drainage patterns that are seldom encountered in subsurface studies are *centripetal, radial,* and *annular*. The three most common drainage patterns in the subsurface are the dendritic, rectangular, and trellis.

Individual channels may be straight, crooked, contorted, anastomosing, braided, meandering, or may consist of a deltaic system of distributaries. Braided streams and delta distributaries are of such importance as to merit separate treatments in this text. It generally is possible to determine from subsurface data not only the type of drainage pattern but also such related information as paleogradient and direction of paleoflow.

PALEODRAINAGE MAPPING

Little published material is available on methods of identifying and tracing channel deposits in the subsurface. One paper (Andresen, 1962), however, is particularly noteworthy. Andresen presented six methods of locating and tracing trends of valley axes—all of which are based on the depth a valley extends into the subjacent strata. Although Andresen set up his techniques on a theoretical basis, the author can vouch for their validity by virtue of having employed strikingly similar, if not the same, techniques for many years in subsurface studies.

Unconformity-Contour Method

Use of the unconformity-contour method requires that there be a recognizable lithologic distinction between the base of the channel fill and the subjacent strata. The elevation of the base of the channel fill is ascertained from either geophysical or sample logs, or from core data, and a structure map is drawn from these control points (Fig. 5-1). Axial lines drawn through the minimum radii of curvature of successive contour lines define the trends of lowest elevation on the unconformity surface.

Andresen (1962) pointed out three possible sources of error in interpretation of the paleodrainage by use of this technique (Fig. 5-1, A, B, C). In area A, a northwest-trending syncline is the result of structural deformation after valley carving. This axial trend, being genetically unrelated to erosion, will not contain valley fill. In area B, the topographically low area is elevated where a north-south-trending anticline was superimposed after valley carving. Topographic contour lines normally trend upstream, cross over abruptly, and then trend downstream on the opposite side.

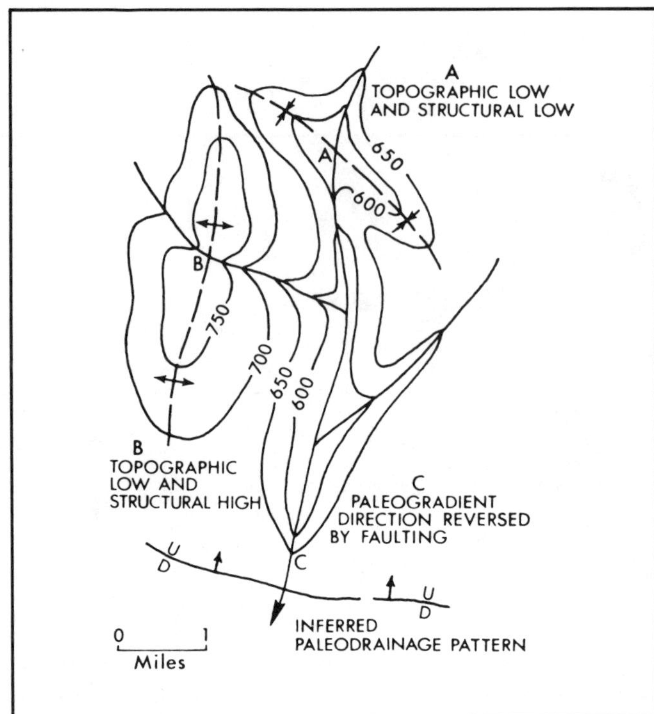

Fig. 5-1. Isopach method for interpreting paleodrainage; contours in feet. (From Andresen, 1962; permission to publish by AAPG).

This situation has been reversed in the headwaters portion of the stream (west of the anticlinal axis). In area C, faulting has produced an apparent reversal of the paleogradient. Because of these possible sources of error, Andresen recommended that this method be employed only where structural activity has been negligible since the time of valley formation. The authors find that these structural effects can be anticipated readily and, therefore, need not introduce errors in interpretation. By drawing a structural map on a marker bed a short distance either above or below the valley fill, one can anticipate the modifying effects of tectonic activity and make the proper allowance in reconstructing the paleodrainage (with its valley fill).

Superimposed structural effects can be eliminated almost entirely by constructing an isopach map of the strata between a marker bed above the channel fill and the base of the channel. Such a map provides a close approximation of the original paleotopography of the area with its drainage courses and the divides that separate them. A transparency of this map, when superimposed on a map such as that shown in Figure 5-1, furnishes the necessary data for removing the effects of post-erosional deformation.

Cross-Section Method

The cross-section technique, which is not new, involves the drawing of restricted stratigraphic cross sections by tracing electric or radioactivity logs of wells drilled on subsurface trends of channel fills. To plot a paleodrainage map, lines are drawn that connect the lowest point on each cross section with those on the other cross sections. Such lines will pass through progressively lower points in the direction of the paleogradient. Although Andresen (1962) did not illustrate this method, he pointed out its principal advantages: textures and lithologic types of valley fill can be plotted on the cross sections, and the manner in which the channel migrated across the valley can be observed readily. Andresen noted that it is a time-consuming method and that an advance knowledge of the valley trends is necessary for the proper selection of wells to be used. It is the authors' opinion that the principal value of such a map is to illustrate to management the ideas regarding paleodrainage that are gained by use of the other techniques reviewed here.

Paleogeologic-Map Method

The paleogeologic-map method involves the construction of a subcrop (or paleogeologic) map of an unconformity into which a drainage system has been eroded and later filled. In Figure 5-2, three formations (A, B, and C) have been partly eroded, and their respective subcrop distributions are shown. Valley axes are drawn in the middle parts of the subcrop of the oldest stratum. This technique has several limitations. For example, considerable subjectivity is involved in the establishment of formation boundaries; furthermore, downcutting must have been sufficient to expose more than one identifiable stratum. In the northern part of the area shown in Figure 5-2, the width of the valley exceeds 10 mi (16 km). The position of the valley axis is drawn in the middle of subcrop formation A. It is quite likely that a stream flowing in such a broad valley had pronounced meanders within a meander belt that was much less than 10 mi (16 km) wide. Thus, the position of the valley axis as generalized might be quite different from the last position of the stream that carved this channel.

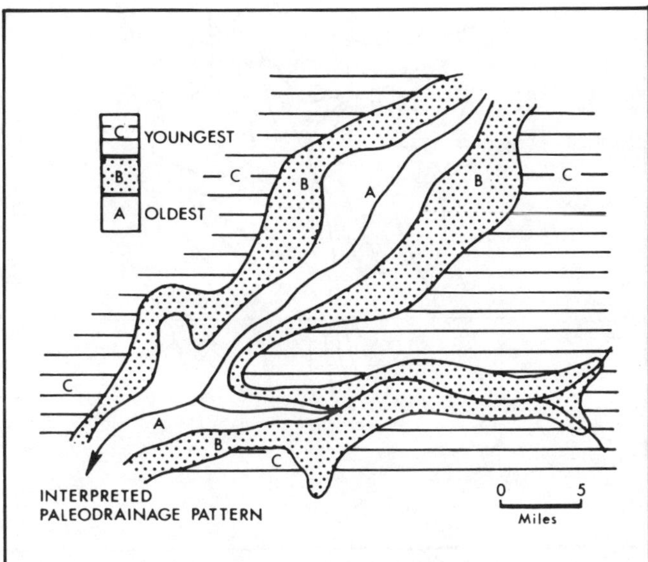

Fig. 5-2. Paleogeologic-map method for interpreting paleodrainage. (From Andresen, 1962; permission to publish by AAPG).

The principal application of this method is to sedimentary units deposited in a shelf environment. In this environment, eustatic changes in sea level cause extensive marine transgressions and regressions; as a result, vertical changes in lithology and thinner stratigraphic units are most numerous there. During a regressive phase of the shoreline, conditions are ideal for weathering and selective erosion of previously deposited thin beds. During a re-advance of the sea, the base level of deposition of the individual drainage systems moves landward, and coarser clastic material becomes concentrated in these paleodrainage courses. Because offshelf environments of deposition seldom are exposed to weathering and erosion, subaerial unconformities are unlikely to develop; also, the steady conditions of sedimentation cause thicker stratigraphic units to be deposited. It is for these reasons that this technique has almost no application to sedimentary units deposited in offshelf environments.

Marker-Bed Method

The marker-bed method, as illustrated in Figure 5-3, involves the construction of a modified paleogeologic map. The key to reconstructing paleodrainage is to map areas where a prominent lithologic marker bed is absent due to valley downcutting. Channel axes then are drawn through the middle of areas where the marker bed has been eroded away. Inasmuch as valleys commonly are asymmetrical in cross section, the paleodrainage pattern is generalized. Advantages of this technique are that it is easy and rapid, while accuracy is related directly to the amount of available subsurface control. This technique is primarily applicable to sediments deposited in the shelf environment for the same reasons as cited for the paleogeologic-map method.

Datum-Plane—Valley-Floor Isopach Method

The datum-plane—valley-floor isopach method involves the construction of an isopach map of a stratigraphic interval between the base of valley-fill material and a lithologic marker bed either above or below. Figure 5-4 illustrates an isopach map of a stratigraphic interval below the base of the channel fill. The minimum interval thickness is directly below the position of maximum downcutting of the stream that carved the channel. Such a map is very similar to a paleotopographic map. The paleogradient may be determined by dividing 5,280 ft by the number of feet of convergence along any one mile segment of an axial line.

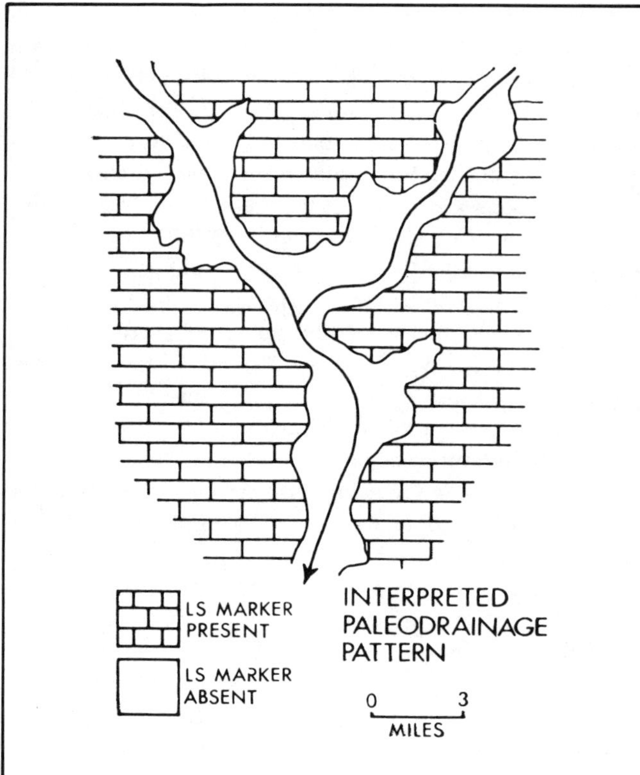

Fig. 5-3. Marker-bed method for interpreting paleodrainage. (From Andresen, 1962; permission to publish by AAPG).

The mapping of interval *L* is of limited value where the channel fill contains hydrocarbons. In this case, a 15–30 ft (4.6–9.1 m) "pocket" generally is drilled below the fill, and most of the producing wells fail to penetrate the lower datum. In tracing the reservoir facies that commonly make up the channel fill, a technique is needed that utilizes the maximum number of control points. The isopach contouring of interval *U* is such a technique. Interval *U* is actually a genetic increment of strata (GIS), whereas interval *L* is not. Trends of maximum thickness of this GIS are coincident with trends of maximum downcutting. The rate of thickening along axial lines, in feet per mile, is the approximate paleogradient.

The advantage of mapping interval *U* is that it uses the maximum number of control points. It also eliminates the problem of selecting a consistent top in cases where the explorationist might otherwise wish to isopach the channel fill. In many places the top of a channel fill is transitional with the overlying shales or consists of interbedded thin sandstones and shales. In an isopach map of interval *U*, the effects of tectonism that occurred after deposition of the upper datum are removed, and the strata are restored approximately to their original depositional attitude. The emphasis is solely one of reconstructing paleodrainage. This type of map, then, serves as a geologic framework for reconstructing the geometry of the channel fill. Once this has been done, a structure map of the upper datum should be drawn to determine the extent to which "post-upper datum" tectonism may have localized hydrocarbons in structurally high parts of the channel fill.

Valley-Fill Isopach Method

The construction of an isopach map of a channel fill generally is predicated on the assumption that most, if not all, of the fill is sandstone. In such cases, the maximum sandstone trends coincide with trends of maximum downcutting. Unfortunately, most channel fills are a mixture of conglomerate, sandstone, siltstone, and shale, and almost any combination of clastic sedimentary types is possible. Limestone, scattered thin streaks and fragments of coal, and fragments of igneous and metamorphic rocks may constitute parts of a channel fill. If the fill is shale lying on a shale base, it is practically impossible to identify the contact from geophysical logs alone. If it is shale lying on eroded limestone, the contrast is striking on electric and radioactivity logs.

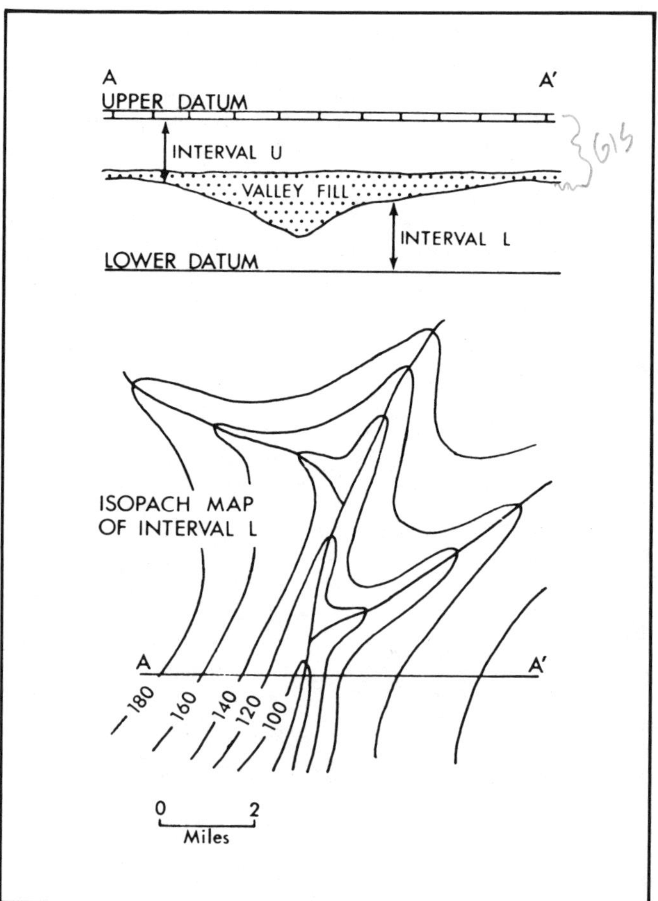

Fig. 5-4. Isopach method for interpreting paleodrainage; contours in feet. (From Andresen, 1962; permission to publish by AAPG).

The fill commonly consists of lenticular sandstones alternating with shale and siltstone. Several separate oil-reservoir sandstones may be present within a single channel fill, complicating the problem of tracing them in the subsurface.

Thus, an isopach map of a channel sandstone does not necessarily reveal the trend and position of maximum downcutting of the stream that carved the channel. Figure 5-5 shows the thickest part of the sand in a position that does not coincide with that of the true valley axis. Andresen (1962) pointed out that the silty-shale portion of the fill may not be recognizable on electric logs. This problem generally is not as serious as he suggests because a silty shale generally exhibits a 15–20 mv "shoulder effect" beyond the shale baseline on electric logs. In these instances, the base of the channel fill can be ascertained even though the lithology is not that of a reservoir facies.

The principal value of drawing an isopach map of a sandstone as illustrated in Figure 5-5 is one of making reserve estimates. Such a map seldom reveals channel-fill geometry and is of considerably less value in exploration and stepout drilling than the other types of maps discussed in the preceding paragraphs.

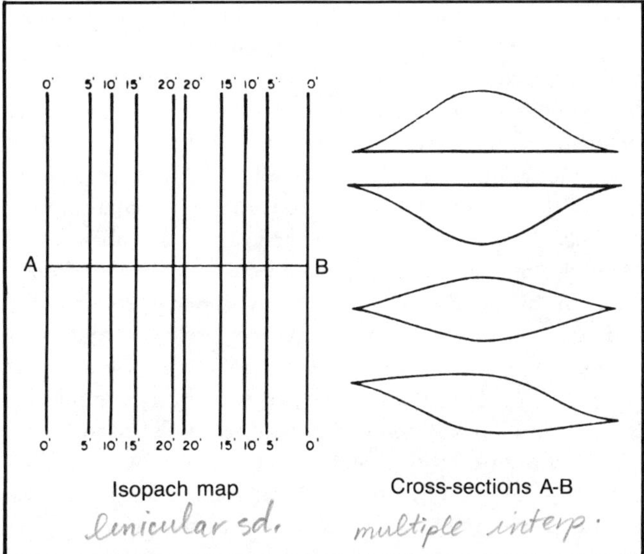

Fig. 5-6. Idealized isopach map of part of an elongate sandstone body and four of numerous possible cross sections that meet the requirements of this map. (From Rittenhouse, 1960; permission to publish by AAPG).

GEOMETRY OF CHANNEL SANDSTONES

To recognize channel sandstones in the subsurface, the exploration geologist must acquire a background knowledge of all the known shapes as a prerequisite for delineating and projecting them. The most significant single criterion for recognizing them from S.P.-G.R. log data is their downward thickening relative to one or more time marker beds or horizons. Geologists commonly draw an isopach map of an elongate lenticular sandstone and then assume that they have mapped a channel sandstone. This type of map falls far short of defining a channel sandstone because it shows only two dimensions of the sand body, namely, length and rate of thinning. There are at least nine different types of lenticular sandstone that can be illustrated by means of isopach contours, but such a map does not serve to positively identify it as a channel fill.

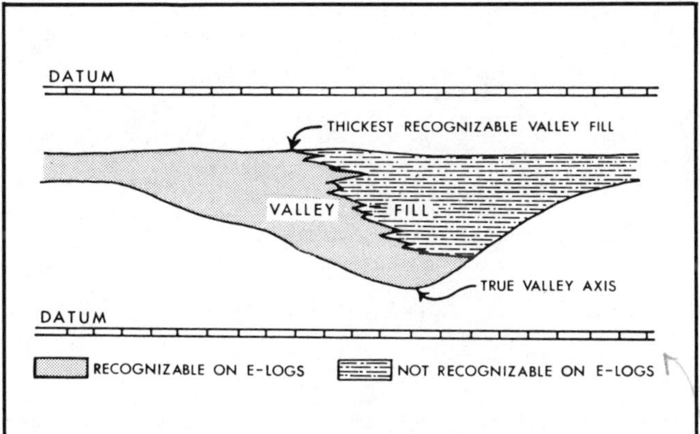

Fig. 5-5. Thickest sand development versus true valley-axis position. (From Andresen, 1962; permission to publish by AAPG).

Rittenhouse (1960) addressed this problem by stating, "Insofar as geometry is concerned, three major problems are (1) to reconstruct the geometry correctly, (2) to know what it implies concerning origin and (3) to know the distribution pattern of sediment of that origin in an analogous depositional situation." He further states,

> . . . various methods can be used to represent the geometry of individual sandstone bodies. Probably the most common are isopach maps and cross sections. These are commonly used alone, but they should be used in combination. For the isopach or thickness map, the first question is: thickness of what? of sand? sand and coarse silt? sand and gravel? sand, gravel, and coarse silt? all porous and permeable rock? or just porous and permeable rock that contains hydrocarbons, that is, net pay? The answer depends partly on how the isopach will be used and partly on the types of data available . . . For understanding the origin of the sandstone body, however, it seems desirable to use the thickness of genetically related rocks of sufficient coarse texture to have originally held hydrocarbons. If thicknesses are obtained from electric logs, the maps actually represent the thickness of rocks having relatively high resistivity and a self-potential equal to or exceeding a selected absolute or relative value. The map represents sand only to the extent that the relationship of resistivity and self-potential to sand is actually correct.

Figure 5-6 illustrates Rittenhouse's thoughts regarding the meaning of an isopach map of an individual sand body relative to its origin. This idealized isopach map of part of a lenticular sandstone body shows width, length, thickness, and trend. It does not, however, define the shape of the sand body. Any or all of the cross sections shown satisfy the requirements of the isopach map. Several other cross sectional shapes could be added to the four shown. This raises the question, which, if any, of these cross sections is correct? The answer lies in the construction of a cross section "hung" on a marker bed or horizon, preferably above the lenticular sand. The lower three cross sections of Figure 5-6 can be readily interpreted as channel sandstones, whereas, the uppermost cross section could be that of a barrier bar, chenier, offshore bar, tidal current ridge, sand wave, etc. All of the latter types of sand bodies should be "hung" on a marker bed or horizon *below* the sand body, rather than on a marker bed above, since they all thicken upward. Channel fill, on the other hand, thickens downward and, therefore, should be referred to a reference datum *above*. Downward thickening is the one criterion for a channel-fill that is nearly diagnostic.

Distortion of Cross Sections by Differential Compaction

Figure 5-7 is part of an illustration taken from Brown (1969) and illustrates a system of delta distributaries that spread in a general westerly direction across the east shelf of the Midland Basin of north-central Texas. The upper cross section of Figure 5-8 (profile A-B of Fig. 5-7) is "hung" on the Breckenridge limestone, which restores the various distributary channel sandstones to a close approximation of their original cross sectional shape. The lower cross section B is based on the same log data as cross section A, but it is "hung" on the Blach Ranch limestone below the channel sandstones. The result is that the cross sectional shapes of

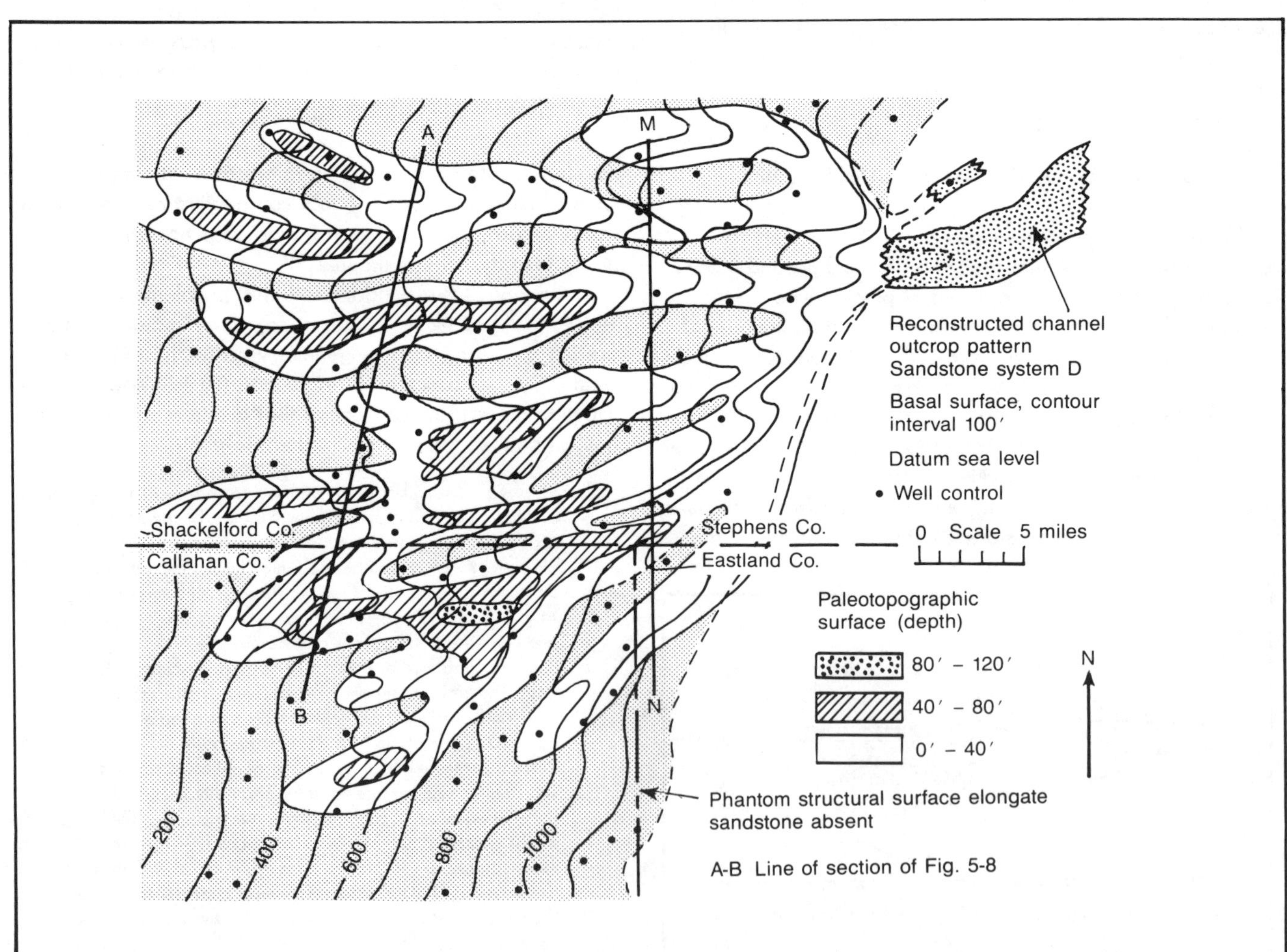

Fig. 5-7. Isopach map from base of a lenticular sandstone to overlying phantom structural surface near top of sandstone system, illustrating paleosurface configuration of sandstone system, eastern shelf of Midland Basin, Texas. (From Brown, 1969; permission to publish by Gulf Coast Association of Geological Societies).

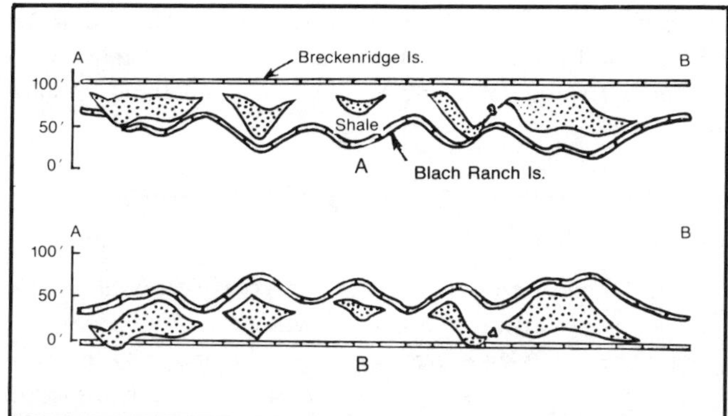

Fig. 5-8. Cross section of Blach Ranch-Breckinridge format (GIS). **A.** *Breckinridge limestone reference datum that restores sandstone profiles to close approximation of their original shapes.* **B.** *Same cross section as A, but drawn with Blach Ranch limestone as reference datum. Note extreme distortion of channel-sand profiles and "roller coaster" effect on Breckenridge limestone. For location of cross section see Figure 5-7.* (From Brown, 1969; permission to publish by Gulf Coast Association of Geological Societies).

the channel sands are badly distorted. In both cross sections A and B of Figure 5-8, it may be reasonably assumed that both limestones were deposited in an essentially horizontal position. Distortion of these limestone reference data is due almost entirely to differential compaction of the shales that occur laterally adjacent to the relatively noncompactible channel sandstones.

Three stages in the erosion, deposition, and differential compaction of sediments related to a channel sandstone are illustrated in Figure 5-9. Stage A shows a channel eroded into partially compacted clay and shale. In Stage B the channel is filled with sand to height (H), and the entire area is blanketed with a thin layer of overburden. The downwarped dashed lines are isochrons indicating that the upper part of the overburden may be deposited concurrently with the channel-fill sand. In Stage C the channel is completely filled with sand, and the laterally adjacent clay and shale are compacted to approximately two-thirds of their original thickness. The fact that differential compaction is concurrent with deposition of the channel fill and the overburden is indicated by the slight thinning of the overburden in the area directly above the position of maximum sand thickness. The most significant feature of this illustration is the thickening of the sand primarily at the expense of the underlying clay and shale and not at the expense of the overlying beds. This is a diagnostic criterion for identification of a channel sandstone in the subsurface and usually will serve to distinguish a channel-fill sandstone from an offshore bar in the subsurface. The latter type of sandstone body usually thickens primarily at the expense of the overburden. In making such a determination it is best to select several closely spaced geophysical logs that exhibit sharply contrasting thicknesses of sandstone. If a lithologic marker bed below the fill is selected as the datum of reference in constructing a stratigraphic profile, then any similar marker bed a short distance above the channel fill will be bowed upward (owing to differential compaction) if the fill consists principally of sandstone. However, the interval of shale that separates the top of the sandstone from the upper marker bed will remain fairly persistent in thickness except for a slight local thinning directly above the thickest portion of the lenticular sandstone.

It generally is more meaningful to restore the lens of sandstone to its original cross-sectional shape; this is accomplished by selecting a lithologic reference datum above the sandstone. By this means, the datum is restored to its original depositional attitude, and the top of the channel sandstone, likewise, is flattened. Such a restoration of the original sandstone profile, however, will cause any lithologic marker beds a short distance below the sandstone to be distorted out of their depositional attitude and to be bowed downward. Regardless of whether a reference datum is selected above the lens of channel sandstone or below it, the thickening of the sandstone is accomplished primarily at the expense of the subjacent strata.

In a delta environment, such as existed in Cisco time on the eastern shelf of the Midland Basin, there were numerous marine trangressions and regressions. With each regression, delta distributaries and delta plain fluvial channels were eroded. These channels were sites for sand accumulation during the following marine transgression. Brown (1969) states that during the "filling of earlier drowned channels,

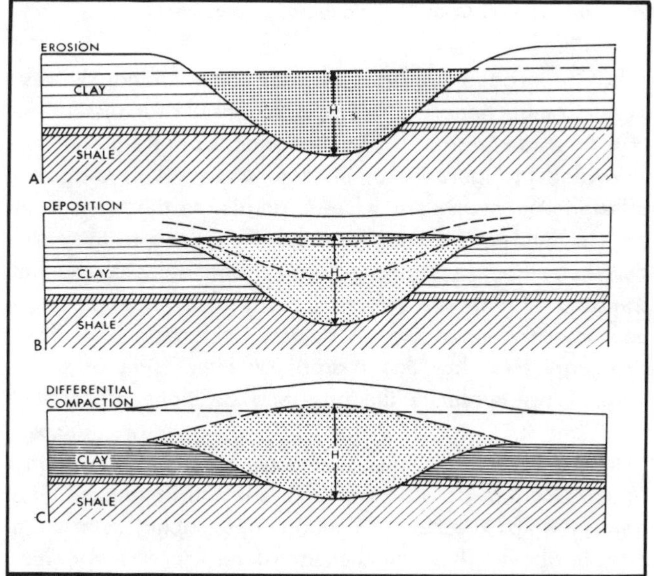

Fig. 5-9. Schematic stages of erosion, deposition, and differential compaction of sediments related to a channel sandstone. (From Busch, 1974).

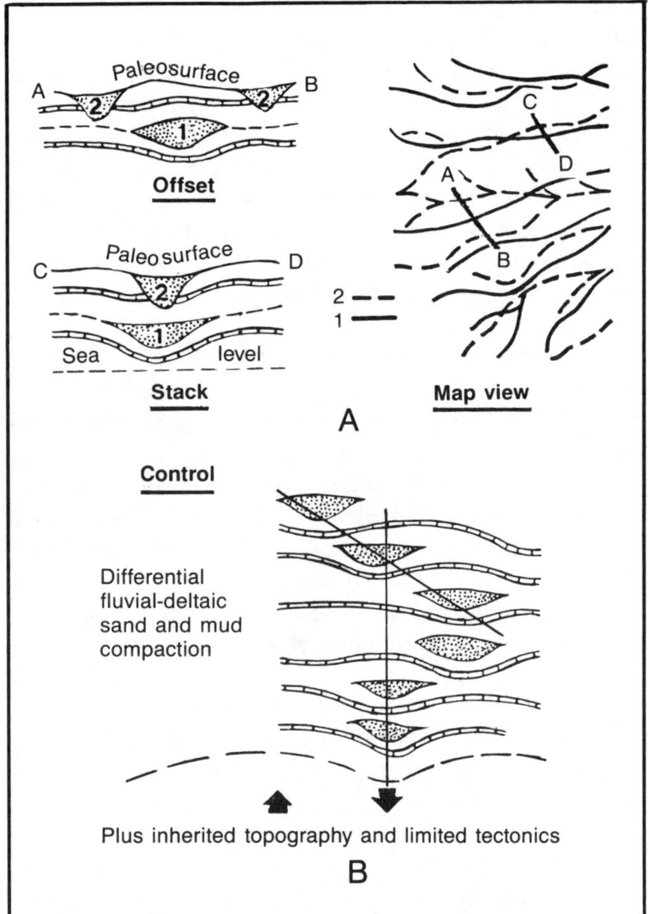

Fig. 5-10. Schematic relationship of sandstone geometry and inferred controlling factors. (From Brown, 1969; permission to publish by Gulf Coast Association of Geological Societies).

deltas prograded principally across compactional depressions. Compactional control was responsible for offset vertical sandstone patterns." This situation is diagrammatically illustrated in Figure 5-10. In cross section A-B, diagonal offsetting of channel sandstone 2 relative to the position of channel sandstone 1 may be noted. This is due to the early development of local depressions on both sides of channel sandstone 1 that are related directly to differential compaction of the shales laterally adjacent to sandstone 1. These topographically low linear areas on either side of sandstone 1 predetermine the offsetting positions of channel sandstone 2.

In profile C-D, channel sandstone 1 assumed a slightly concavo-convex downward profile early in its depositional history, which resulted in a slight topographic depression directly above. Thus, channel sandstone 2 occurs in a vertically stacked position relative to sandstone 1. The influence of inherited topography, related to differential compaction, and limited tectonics is illustrated in Figure 5-10B. Some of the channel sandstones are stacked, whereas others are offset. From the foregoing discussion it should be obvious that parallelism of marker beds is not likely to occur in a stratigraphic section containing multiple channel sandstones.

Cross Sections and Examples of Subsurface Channel Sandstones

It has been pointed out that the cross-sectional shapes of channel sandstones are quite variable. The S.P.-G.R. "signatures" likewise are variable. The majority of subaerial (continental) channel sandstones exhibit a bell-shaped S.P. and G.R. log "signature." This is due to a progressive decrease in grain size upward and an increase in clay content upward resulting in an abrupt basal contact and a transitional upper contact. The upper portion of a channel fill might even consist of thinly interstratified sandstone, siltstone, and shale. The bell-shaped "signature" should never be considered a diagnostic criterion because there are numerous exceptions. For example, delta distributary channel sandstones frequently exhibit a cylinder-shaped S.P.-G.R. log profile, which means a more uniform texture throughout and an abrupt base and top of the sand.[1] In the thinner (edge) portions of both types of channel sand there may be both a transitional top and base. Thus, a bell-shaped or cylindrical-shaped profile is only suggestive of a channel sandstone. A much more definitive criterion of both types is downward thickening relative to a marker bed occurring stratigraphically above the channel fill. Figure 5-11 illustrates seven types of channel fill commonly encountered in subsurface studies. Corresponding S.P. and G.R. log configurations also are shown for each.

Postcompaction Cut and Fill. Figure 5-11A illustrates the type of channel-fill profile that is formed where the substrata have been compacted thoroughly (by weight of overburden) prior to uplift and erosion. The channel is filled with noncompactible sand or with sand and gravel. Where the subjacent and lateral strata (usually shale) previously have been compacted, the fill exhibits a planoconvex-downward profile, even after considerable overburden is deposited. The electric log commonly will have a bell-shaped profile, indicating a sharp basal contact of the coarser detritus with the underlying rocks. There is a general decrease in grain size upward. The upper part of the fill may be a silty zone transitional with the overlying shale, or it may be interbedded fine-grained sandstone and shale. In either case, the upper boundary of the fill commonly is difficult to identify from the electric log. The electric log should be used as a supplemental tool and never as the sole criterion for identifying a channel sandstone in the subsurface.

The channel fill may consist of both horizontal and cross-

[1]Much of this discussion is taken directly from AAPG Memoir 21, but is considerably updated (Busch, 1974).

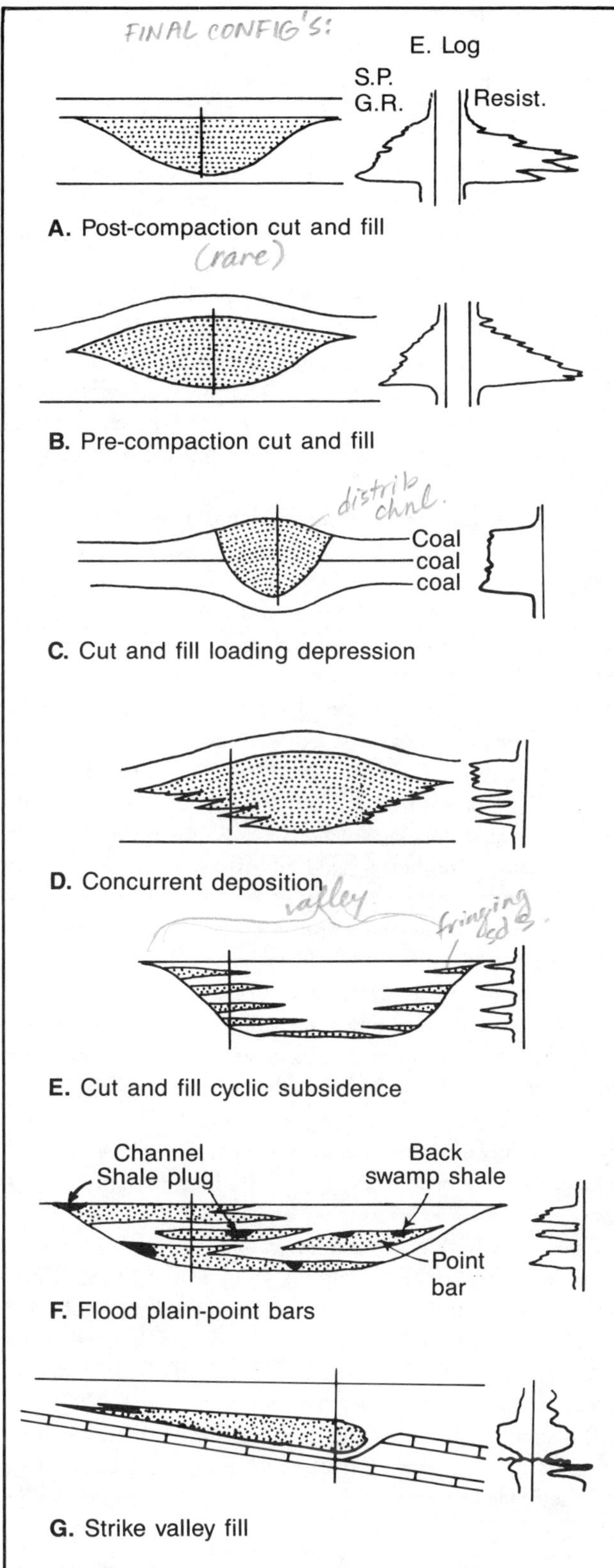

Fig. 5-11. Schematic cross sections of seven types of channel fill encountered in the subsurface. (Modified after Busch, 1974).

bedded sandstone. The texture may range from a pebbly conglomerate at the base to a very fine-grained sandstone elsewhere in the channel. Coarser-textured sandstone and pebbly conglomerate also may be present at various positions within the main body of the fill. Sorting generally is poor to fair. In many places the fill consists of discrete zones of well-sorted to poorly sorted sandstone interbedded and interlaminated with siltstone. Distinct shale "breaks" and "pockets" also may make up a part of the channel fill. It even is possible for all of the fill to consist of shale and silty shale. In this case, the electric log is of very limited value in subsurface identification, and the bell-shaped profile does not apply. A shale fill of a channel can be identified from the electric log, however, if the channel is in bedrock consisting principally of either sandstone or limestone.

Precompaction Cut and Fill. Figure 5-11B is the type of channel-fill profile that is created if the channel is eroded prior to or during compaction. The internal characteristics of the fill material generally are the same as those indicated for postcompaction cut-and-fill (Fig. 5-11A). The electric log, likewise, is similar to that of Figure 5-11A, provided the fill material is principally sandstone. The phenomenon of differential compaction is a complicating factor when analyzing this type of sandstone in the subsurface. Sand undergoes very little, if any, compaction due to the weight of the overburden, whereas the laterally adjacent shales ultimately can compact to 45–55 percent of their original depositional thickness.

A good example of precompaction cut-and-fill channel sandstone is illustrated in Figure 5-12. The stratigraphic profile A-A′ is constructed from gamma logs, supplemented by cores. The sandstone thickens downward at the expense of the underlying shale. The reference datum at the top of this GIS is a thin limestone marker bed. It can be identified readily on all gamma-ray logs and also was logged reliably by the drillers in those wells that were not mechanically logged. Thus, this limestone serves as a very useful reference datum not only over all of the 11 sq mi (28.5 sq km) area illustrated, but also over the surrounding area. Although the sandstone is nearly flat on top and convex downward at the base, it is, nevertheless, a biconvex lens in the subsurface. By "hanging" this profile on the thin limestone marker, the profile of the sandstone body has been distorted to a close approximation of its original cross sectional shape.

The stratigraphic interval between this reference datum and the base of the channel sandstone contains no disconformities and, therefore, constitutes a genetic increment of strata. A close inspection of the isopach map of this interval reveals that the interval increases in thickness from south to north; the increase is approximately 60 ft (18.3 m) in a distance of about 6 mi (9.7 km). The average paleogradient is about 10 ft/mi (1.9 m/km); the direction of stream

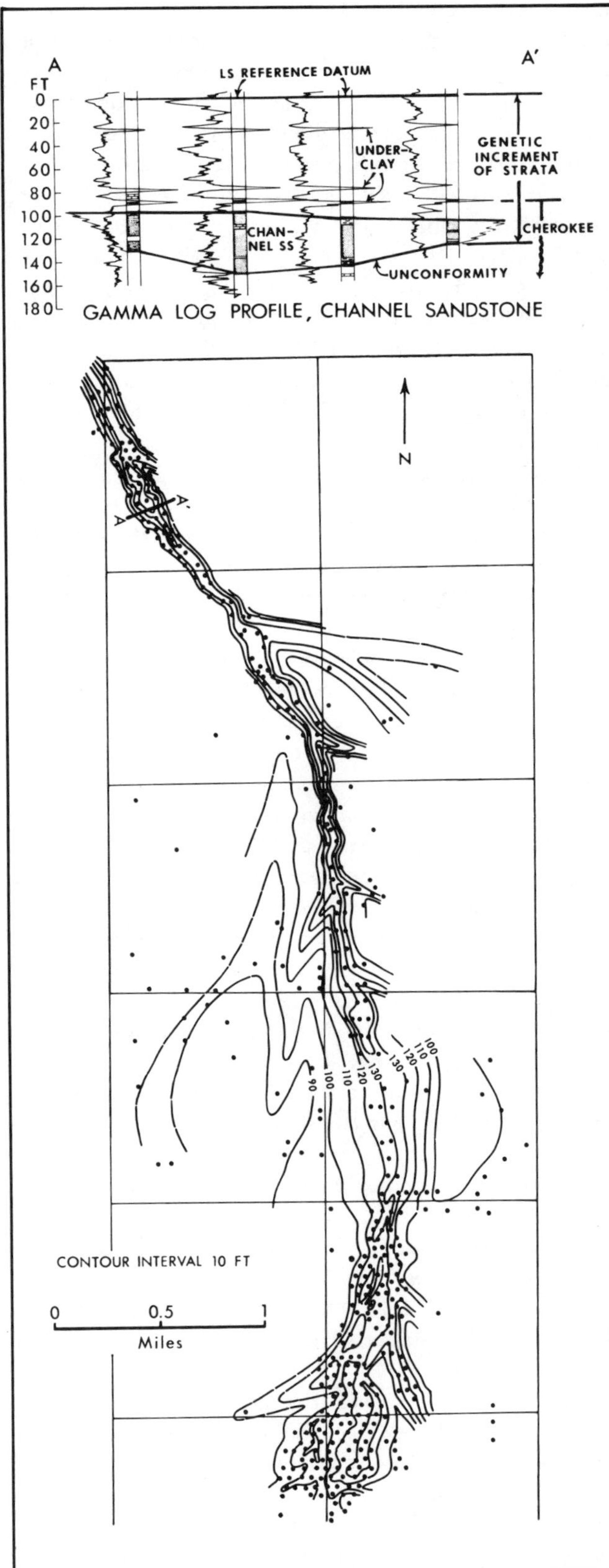

flow was to the north. The channel fill is predominantly sandstone, which is interrupted locally with shale and siltstone and scarce limestone "breaks." The sandstone is in sharp contact with shale both to the east and west; this is a true cut-and-fill channel.

This isopach map may be visualized in the same manner as a topographic map of a modern stream channel. The contour lines extend for considerable distances upstream, cross the channel abruptly, and then extend downstream on the opposite side. Thus, this map represents a topographic simulation of a stream channel in which a more-or-less continuous lenticular body of sandstone was deposited. The filling of the channel probably was effected by an upstream shift of the base level of deposition of the stream when the area was inundated by a north-to-south marine transgression.

Detailed stratigraphic studies of this area and the surrounding area indicate this channel was carved in a region of low topographic relief. The channel was cut into uniformly nonresistant, partially consolidated muds. Essential parallelism of thin limestones and coal marker beds present in the shales, both above and below the channel fill, indicates a horizontal attitude of the beds at the time of channel carving. Thus, any drainage pattern developed on this type of land surface should be dendritic. The ancient stream channel illustrated in Figure 5-12 represents only one of a series of such channels that must constitute a drainage system. To locate another channel fill of the dendritic network, of which this channel appears to be a component part, it is necessary to know the original direction of stream flow. Any possible tributaries will bear an acute, angular relationship to this ancient stream course.

To locate the most likely positions of tributary-stream "breakouts," it is important to examine closely the cores, logs, and productivity of all wells along the margin of this lenticular channel sandstone. An apparently localized or abrupt increase in sandstone thickness or an anomalously large cumulative production of a well along the margin might signal the site of a tributary "breakout."

Figure 5-13 is an isopach map of a genetic sequence of strata in which the oil-producing channel sandstone is present. It includes not only the area illustrated in Figure 5-12 but also additional territory to the east. The north-south-trending pool, shown in black, is the same one that occupies the sandstone-filled channel of Figure 5-12.

The stratigraphic relation of the oil-bearing sandstone to the genetic increment of strata of Figure 5-12 and the genetic sequence of strata of Figure 5-13 is illustrated in Fig-

Fig. 5-12. Isopach map of a genetic increment of strata (GIS) that includes an oil-saturated sandstone in lower one-third. Mapped interval is shown on inserted gamma-log profile A-A′. (From Busch, 1963).

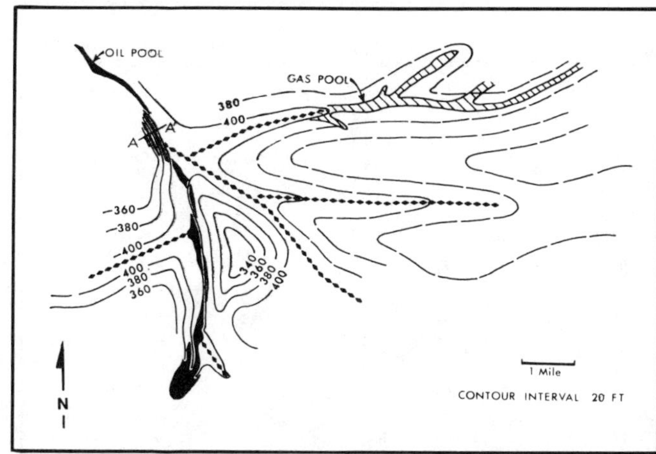

Fig. 5-13. Isopach map of Cherokee Group (Lower Pennsylvanian) showing oil and gas pools in an upper Cherokee sandstone. Trends of these elongate pools are coincident with the converging narrow trends of thicker Cherokee Group. Heavy stippled lines also indicate converging axial trends of thicker Cherokee. (From Busch, 1963).

ure 5-14. The presence of an unconformity at the base of the GSS was determined by constructing numerous stratigraphic profiles in the area illustrated in Figure 5-13. Abrupt changes in interval thickness below the lowest coal bed clearly indicate that the topographic lows of the unconformity surface were filled with sediment before marine deposition occurred on the topographic highs. After these hills were obliterated by filling of the intervening low areas, an alternating series of as many as five coal beds were deposited during the time represented by the GSS. This sequence indicates the following conditions: a monotonous mud-flat environment in which there was a cover of extensive vegetation, followed by minor subsidence and clastic deposition, and then repetitions of this series of conditions.

The oil pool of Figures 5-12 and 5-13 coincides for the most part with the trend and position of a linear area where the GSS has a maximum thickness. Furthermore, a lenticular gas-bearing sandstone is present along the north margin of the eastern half of this area. This lenticular sandstone is stratigraphically equivalent to the sandstone that produces oil in a cross trend to the west. The gas-bearing sandstone is also in a linear belt of maximum thickness of the GSS. This relation of oil-bearing and gas-bearing sandstone to maximum GSS thickness is more than a coincidence. Thin areas of the GSS coincide with paleotopographic highs on the underlying unconformity, and conversely, thick areas (or trends) represent paleotopographic lows on this surface. Areas where the GSS is thicker have undergone slightly more compaction than nearby areas where it is thinner. Thus, slight topographic depressions develop in the mud flats toward the end of deposition of the GSS most likely as a result of differential compaction of a predominantly shale section over buried hills. These depressions served as focal points for surface runoff of meteoric water; thus, erosion was initiated. Runoff over uniformly nonresistant, essentially horizontal strata resulted in a dendritic drainage pattern.

Several divergent thick trends of the GSS of Figure 5-13 are indicated by heavy stippled lines. It is along these lines that additional channel sandstones might be found. These trends, together with the known trends of two oil-bearing and gas-bearing channel sandstones, make up the complete dendrite of the drainage pattern, which is likely to have developed late in the period of GSS deposition.

Figure 5-15 is a structural map of a thin, persistent limestone marker bed, drawn with a contour interval of 20 ft

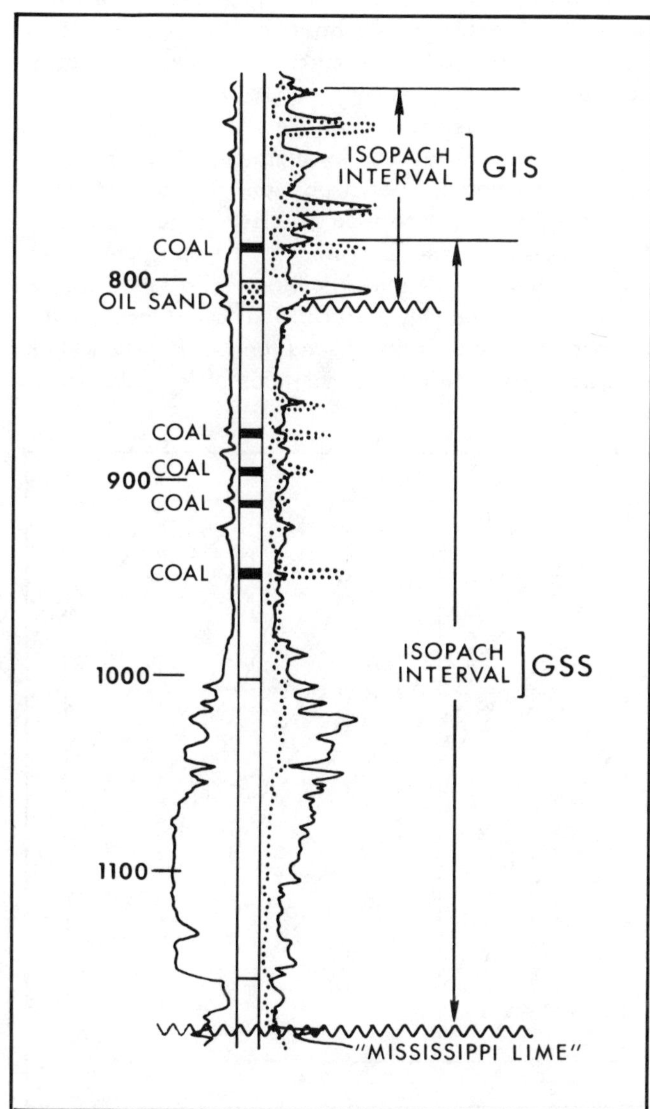

Fig. 5-14. Electric-log section of Lower Pennsylvanian, showing relation of stratigraphic intervals mapped in Figures 5-12 and 5-13. (From Busch, 1963).

(6.1 m). It shows a southwest homoclinal dip and affords no structural explanation for oil accumulation in pools X and Y. These pools produce oil from different sandstone lenses a short distance below the marker bed. This map was drawn shortly after the completion of a discovery well in pool Y, which had a natural production of 60 bbl of oil per hour from sandstone B of Figure 5-16. Sandstone A is the producing formation of pool X of Figure 5-15. Previous detailed studies of sandstone A, over an area of many townships, revealed that this reservoir is a channel-fill sandstone. In preparing a series of electric-log profiles through these two pools, it is immediately apparent that the predominant shale interval between limestones 1 and 2 thickens and thins abruptly. However, a persistent interval spacing is maintained between limestone 2 and coal and limestone marker beds both above and below limestone 1. Persistent interval thickness (or gradual basinward thickening) is considered a criterion in the selection of a marker-bed reference datum.

Limestone 1 is detrital and blankets an unconformity surface throughout the area of investigation. Although it represents contemporaneity of deposition, the same as limestone 2, it should not be used in selecting the upper limit of a GIS. If limestone 2 is used as the upper reference datum, the top of limestone 1 will present a subdued replica of the unconformity surface. The lowest part of the unconformity is at the base of sandstone A. The channel fill in Figure 5-16 is an alternating series of lenticular sandstones, shale, and limestone. Inasmuch as limestone 1 is present in all wells except one (in which sandstone B thickens downward and truncates it), the genetic increment of strata indicated in Figure 5-16 is ideal for reconstructing a subdued replica of the paleotopography of the unconformity.

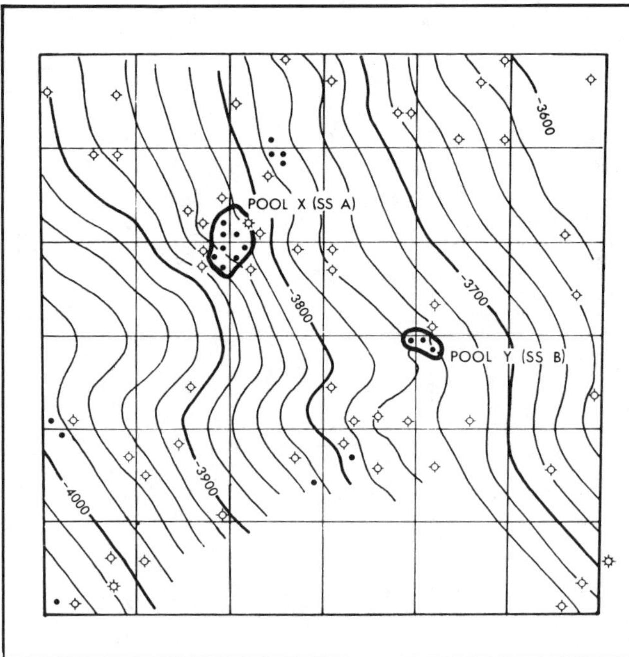

Fig. 5-15. Structure map of top of limestone 2 (shown on Figure 5-16). Contour interval 20 ft. (From Busch, 1974; permission to publish by AAPG).

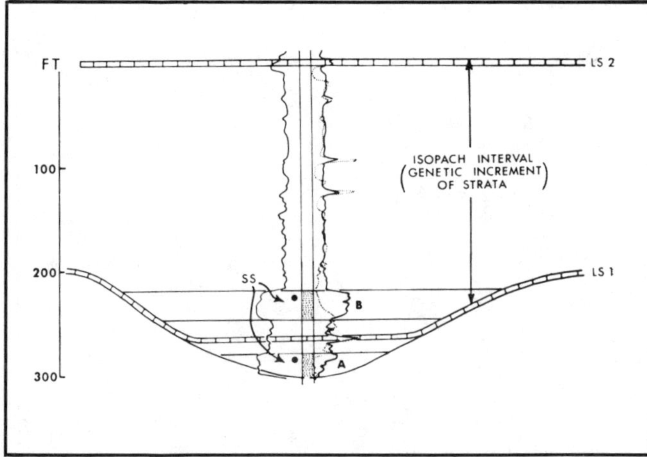

Fig. 5-16. Diagrammatic channel-fill profile through pool Y of Figure 5-15. (From Busch, 1974; permission to publish by AAPG).

Figure 5-17 is an isopach map of the GIS shown in Figure 5-16. The diagonally shaded area is an estimate of the trend and distribution of sandstone B. Although pool X (Fig. 5-15) produces from sandstone A, sandstone B also has good reservoir characteristics in these wells; however, it will not be produced until the reserves in sandstone A are depleted.

It may be seen that there is no similarity between Figures 5-15 and 5-17. Only when a sandstone-distribution map (shaded area, Fig. 5-17) is drawn should structural considerations be made. An isolated gas well 3 mi (4.8 km) east of pool Y had an initial production of 10 million cu ft of gas from sandstone B but is shut-in for lack of a pipeline connection to the market. This zone is approximately 150 ft (46 m) structurally higher in this well than the oil-producing zone of pool Y. Subsequent development drilling within the shaded area between pools X and Y has merged these two pools.

Cut-and-Fill Loading Depression. This type of channel fill is illustrated in Figures 5-11C, 5-18 and 5-19. The discussion of this example is borrowed largely from Burnham's (1956) unpublished paper. The stratigraphic interval of Figure 5-19 includes only two of the many Oficina sandstones, which were deposited on the shelf of the southern flank of the western Venezuela structural basin. The depositional environment during Oficina deposition was a combination of paludal, lagoonal, and brackish-marine conditions favorable for the development of extensive coal-forming swamps. It is an area in which there were numerous cyclic

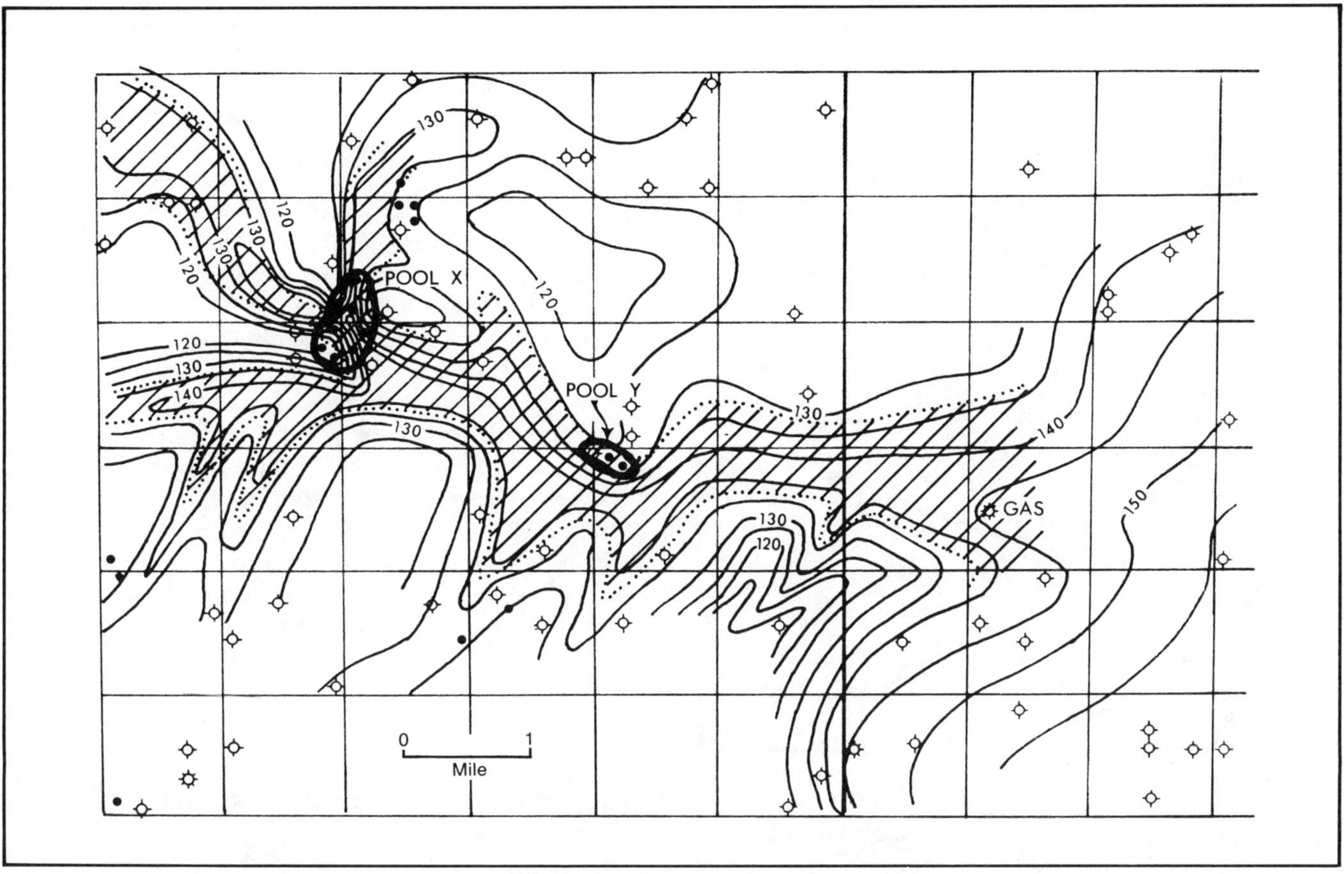

Fig. 5-17. Isopach map of the genetic increment between limestones 1 and 2 of Figure 5-16, showing a subdued replica of a paleodrainage course. Diagonally shaded area indicates trend and estimated width of sandstone B of Figure 5-16. Contour interval is 5 ft. (From Busch, 1974; permission to publish by AAPG).

changes in either sea level or land level, resulting in a very thick alternating series of shales and complex intervals of coarser clastic beds. This was the site of a large distributary-delta system that deposited its sediments partly in a coastal subaerial environment and partly in a very shallow subaqueous environment. Innumerable distributary channels were eroded into semiconsolidated muds. Burnham stated:

> With a slight rise in relative water level, the current velocity in the scour channels would be reduced and the channels would start back-filling with sand from behind sand bars at the mouths. This would result in "plugging with well-sorted sand channel fillings which make up narrow thick sand bodies characteristic of the shoal-water delta (Fisk)." The end result would be a channel almost choked with clean, well-sorted sand with a sluggish current depositing silts and silty shales in the upper part.

Burnham pointed out that these channel sands ". . . are younger than the missing correlation markers that they have cut out and replaced." He stressed the fact that key marker beds, such as lignites, are structurally depressed (i.e., they sag) beneath all trends of thick channel-sand development. This sag of the subjacent beds is accomplished without much thinning except where they have been eroded. In Figure 5-11C only the middle one of the three lignite beds is not deformed. The upper lignite bed is convex upward owing to differential compaction of the clays laterally adjacent to the sand-filled channel, whereas the lower lignite bed is depressed owing to the concentrated load of sand directly above. The stages in the development of this type of loading depression are illustrated in Figure 5-18. Burnham postulated ". . . that the underlying sediments were semi-consolidated when these were formed, and not underlying soft clays as described by Fisk. Semi-consolidated underlying sediments would spread the strain vertically and horizontally, giving sag rather than squeezing or flowage."

A 175 ft (53 m) stratigraphic profile involving six wells is illustrated in Figure 5-19. The reference datum is the I6 lignite zone at the top of the profile. The J3a siltstone and the top of the J3b zone are sharply depressed by loading compaction in areas directly beneath the two deep, sand-filled channels. It is the authors' opinion that this type of cut-and-fill loading depression of subjacent marker beds is

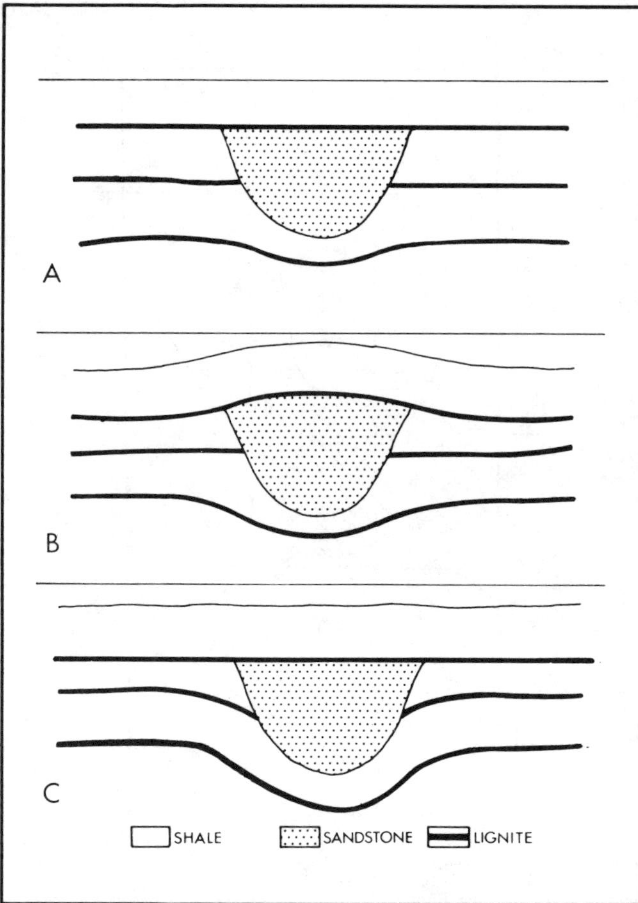

Fig. 5-18. Stages in development of cut-and-fill loading depression associated with a channel sandstone. (Modified after Burnham, 1956; personal communication).

much more likely to occur in a deltaic environment, where prodelta muds are deposited in abundance, than in a strictly subaerial depositional environment.

Concurrent Deposition. Figure 5-11D is an example of concurrent deposition in which the channel fills with sand at the same time that the laterally adjacent areas are receiving mud and silt. The diagnostic criteria for this type of channel fill are the same as those for precompaction cut-and-fill (Fig. 5-11B) plus the interbedding of silty sandstone with the laterally adjacent shale. The electric log exhibits interbedded sandstone and shale in the basal part of the fill and shows mostly sandstone in the upper part. It will exhibit only sandstone, however, if the well happens to be situated in the center of the channel.

This type of concurrent deposition occurs where terrestrial sediments are discharged into shallow-marine waters and deposited on an irregular submarine surface. As an example, growth structures (frequently deep-seated salt domes) on the shallow seafloor are known to have existed in abundance along the Gulf Coast area in Tertiary and Quaternary time. These structures produced submarine topographic irregularities that, in turn, caused a localization of discharge water between and around submarine "highs." Sediment transported through the submarine "lows" was sand, silt, and clay. Sand-size material was deposited in the lows, and it interfingered with clay-size and silt-size materials that were deposited concurrently in the quieter water along the margins of the lows and over the mounds separating the lows. Variations in the volumetric discharge of the sediments resulted in lateral interbedding of channel sandstone with clay and silt. The distributary pattern is not that of a true delta, but is somewhat haphazard owing to the irregular distribution of the submarine topographic highs.

An unpublished study by Slingerland (1967) contains an excellent example of the influence of growth structures in the development of the concurrent type of channel fill, as illustrated in Figure 5-20. The concurrent submarine channel fill occurs just below a *Bolivina perca* faunal zone, which was used as the reference datum for this illustration. This datum coincides perfectly with the base of a shale marker

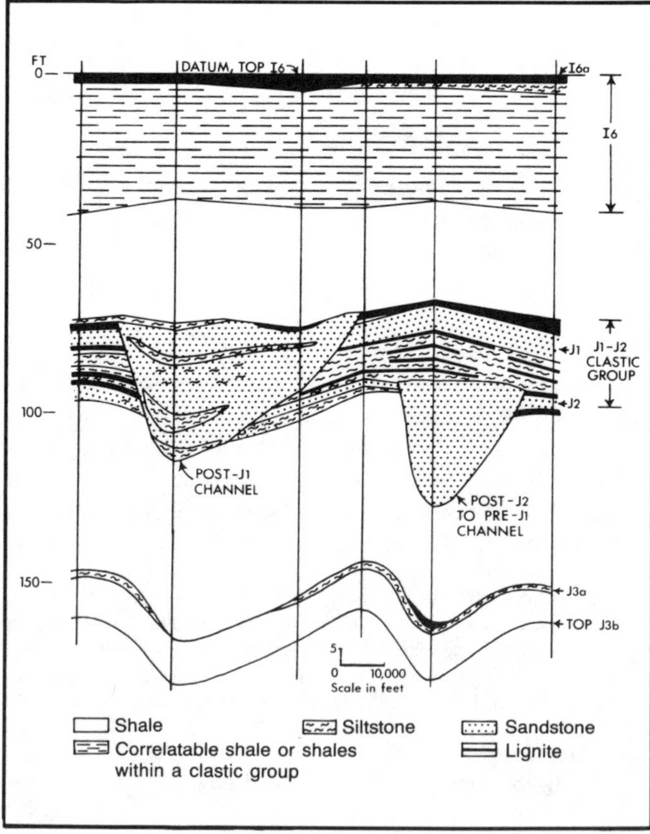

Fig. 5-19. Stratigraphic profile of part of Oficina Formation at East Mapiri, Maracaibo Basin, Venezuela, showing cut-and-fill loading depressions of J3a and top of J3b zones below position of maximum channel fill. (Modified after Burnham, 1956; personal communication).

bed. Interval thicknesses above the sandstone are remarkably uniform. The basal part of the sandstone interfingers with laterally adjacent shale, and the Anahuac reservoir sandstone increases in thickness at the expense of the subjacent shale. Because the *Bolivina perca* reference datum is restored arbitrarily to its original flattened condition, the two marker horizons below the lenticular sandstone are bowed downward. The lack of parallelism of the two marker horizons below the sandstone with the *Bolivina perca* reference datum is clear-cut proof of the differential compaction of the shales laterally adjacent to the channel sandstone. A plan view of the oil pool that produces from the Anahuac is shown in the upper part of Figure 5-20. It has the configuration of a channel, and the basinward direction is south.

This channel can be traced [illegible] a considerable distance farther south. An isopach map of a genetic increment in this area might be constructed using the *Bolivina perca* datum as the upper marker bed and either of the correlated marker horizons below the sandstone as the lower limit of the increment. On such a map, areas of maximum thickness will be coincident with maximum development of this Anahuac sandstone. The entire genetic increment consists of shale on both sides of the sandstone. This shale interval will have a minimum thickness above the growth structures that existed on the sea floor during this stage of Anahuac deposition. Mapping of the variations in thickness of this genetic increment makes it possible to reconstruct the submarine topography that existed during the time of this stage of Anahuac deposition.

The sandstone development can be inferred reasonably in the areas of the lows (thickest genetic increment). It is not to be expected that oil or gas will be present everywhere that the sandstone in the Anahuac is developed. Rather, it will be located in restricted updip parts of the sandstone that result from domal uplift and growth faulting. Thus, a structure map of the *Bolivina perca* datum should be constructed and superimposed on the map of predicted Anahuac sandstone distribution as a basis for defining updip stratigraphic traps in this formation.

Cut and Fill Cyclic Subsidence. Figure 5-11E illustrates paired fringing beach sands deposited on opposite sides of a linear trend of differentially compacted shale. This topographically low area is not a flood plain, but rather a trend of compacted shale that may be bordered on one side by a stratigraphically (noncompactible) sandstone and on the other side by a regional tilt. This situation localizes the position of runoff by meteoric water and predetermines the position and trend of downcutting by a stream. The Muddy sandstone of the Thermopolis shale of northeastern Wyoming is an example of such an area. The Muddy consists of a series of lenticular sandstone members and occurs in the upper portion of the Lower Cretaceous. It is overlain conformably by the Mowry shale, and it unconformably overlies irregularly eroded portions of the Skull Creek shale. This zone of sandstone lenses has been studied in a 576 sq mi (1,492 km^2) area, which includes all of Townships 47-54N and Ranges 67-70W, encompassing contiguous parts of Weston, Crook, and Campbell counties.

Easily identified siltstone and bentonite marker beds above and below the Muddy sandstones facilitate identification and correlation of the individual sandstone members composing

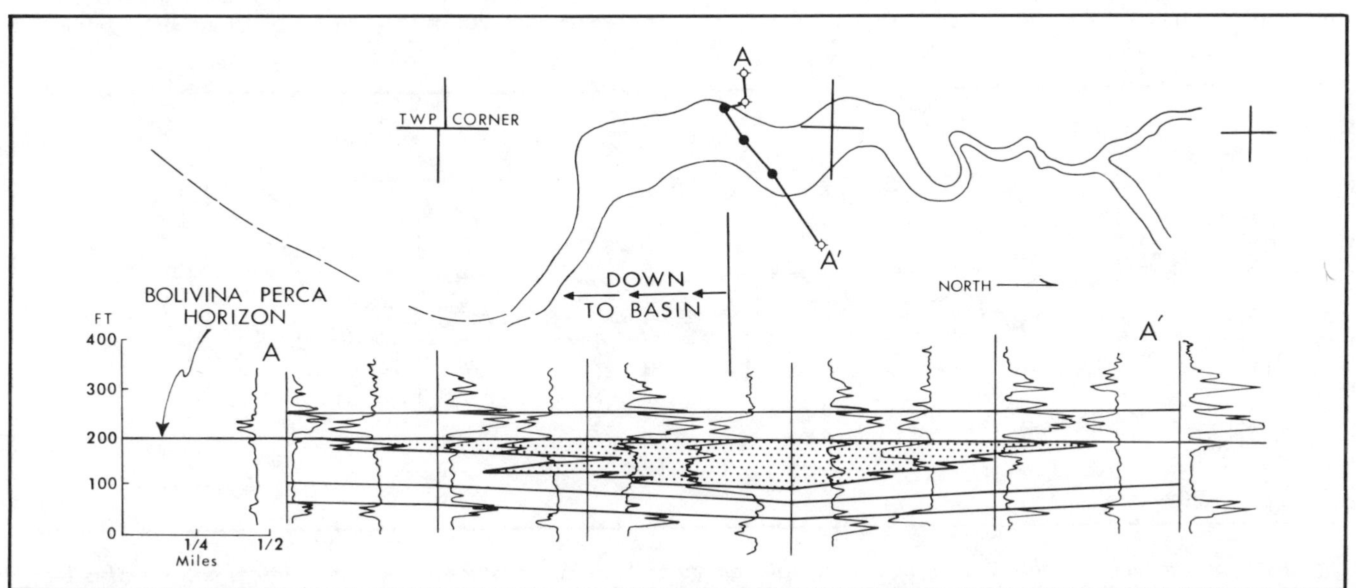

Fig. 5-20. Profile of an Anahuac (Oligocene-Miocene) channel sandstone, Acadia Parish, Louisiana, drawn to a **Bolivina perca** *reference datum.* (Modified after Slingerland, 1967; personal communication).

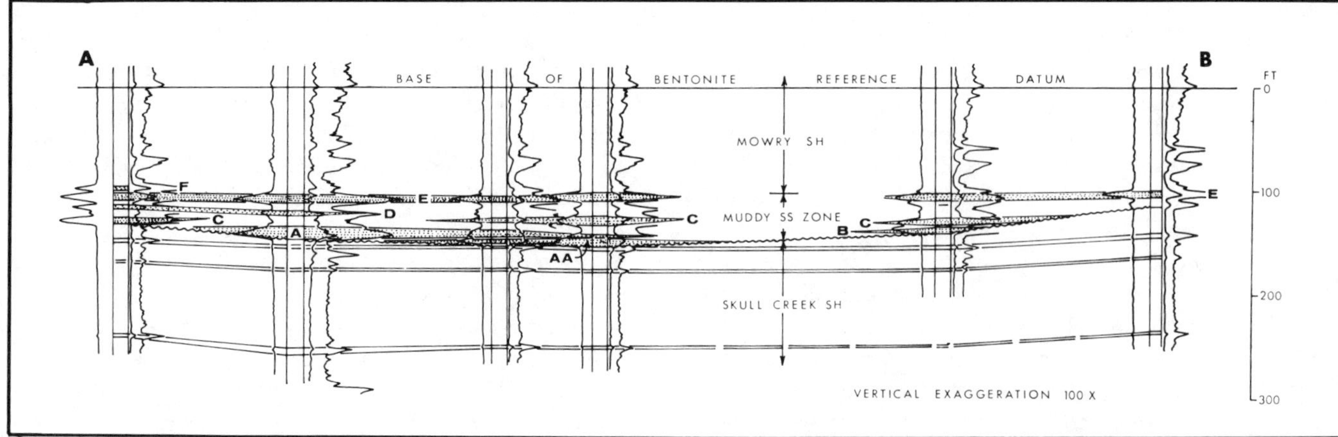

Fig. 5-21. Stratigraphic profile A-B, showing multiple lenticular Muddy sandstones. Reference datum is base of a bentonite marker bed. This illustration shows lenticular nature of sandstone members making up the Muddy zone, as well as downcutting of unconformity surface into Skull Creek shale. Location of profile is shown on Figure 5-23. (Modified after Busch, 1974; permission to publish by AAPG).

the Muddy zone. Figure 5-21 is a stratigraphic cross section that illustrates a bentonite marker bed (reference datum) in the Mowry shale, above the Muddy sandstones, and two siltstone zones in the Skull Creek shale, below the Muddy. The position of the unconformity at the base of the lowest development of Muddy sandstone is clearly indicated. The stratigraphically lowest units of the Muddy Sandstone Member occur where erosion into the underlying Skull Creek shale was deepest. The several sandstone units are designated alphabetically from the base up in the order of deposition.

Figure 5-22 is a stratigraphic profile of an area 7–10 mi (11.3–16.0 km) south of that illustrated in Figure 5-21. The sandstone units of the Muddy zone are lenticular. The "C" sandstone is the lowest unit present on Figure 5-22, whereas the "AA" member is the lowest unit present farther north (Fig. 5-21). Thus, the maximum downcutting at the unconformity increased from south to north. The maximum downcutting may be determined by preparing an isopach map of the genetic increment of strata between the bentonite reference datum and the unconformity as in Figure 5-23. It clearly shows a north-flowing drainage system that was eroded into the soft shales of the Skull Creek.

The Miller Creek, Donkey Creek, and Coyote Creek oil pools all produce principally from the stratigraphically lower Fall River sandstones. There is an abrupt sandstone-to-shale facies change along the eastern margins of all these pools. Approximately 200 ft (61 m) of Skull Creek shale separates

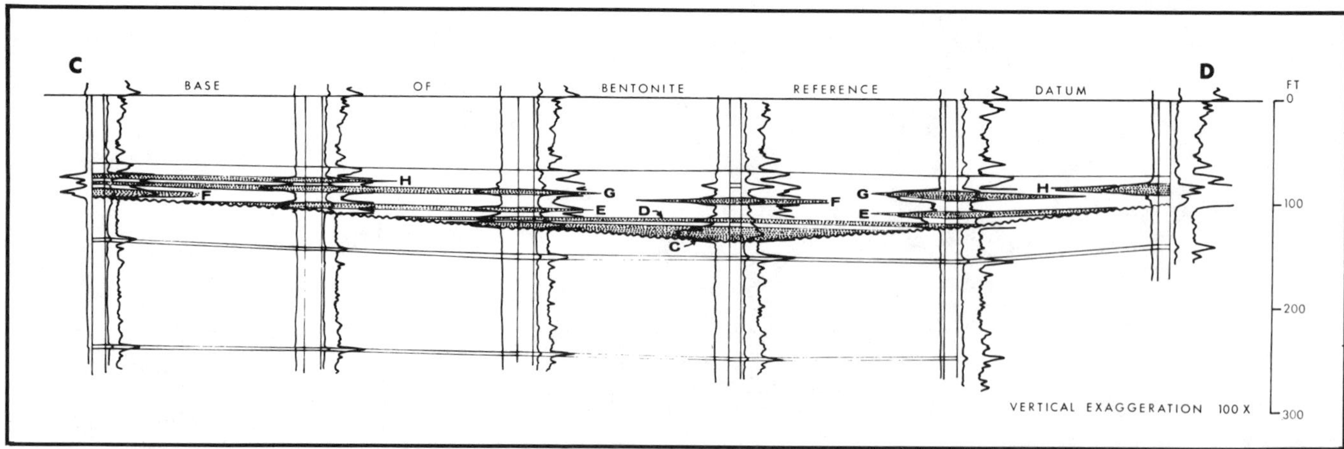

Fig. 5-22. Stratigraphic profile C-D, showing multiple lenticular Muddy sandstones. Reference datum is base of a bentonite marker bed. Sandstone members C and D of the Muddy zone occur along axis of the broad (low relief) channel rather than on opposite flanks, as in Figure 5-21. This is due to the fact that the profile is farther upstream than that of Figure 5-21. Location of the profile is shown on Figure 5-23. (Modified after Busch, 1974; permission to publish by AAPG).

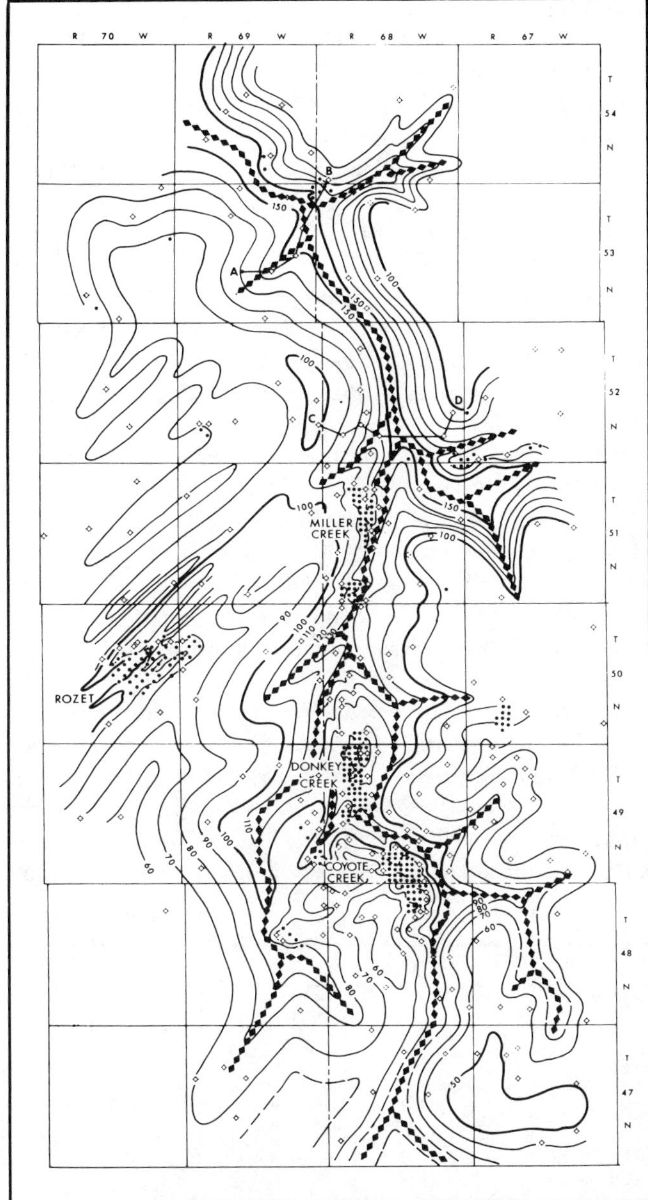

Fig. 5-23. Isopach map of genetic increment of strata (GIS) between bentonite marker bed and unconformity at base of Muddy sandstone zone. Map shows simulated paleotopographic surface of unconformity before deposition of Muddy sandstones. Contour interval 10 ft. (Modified after Busch, 1974; permission to publish by AAPG).

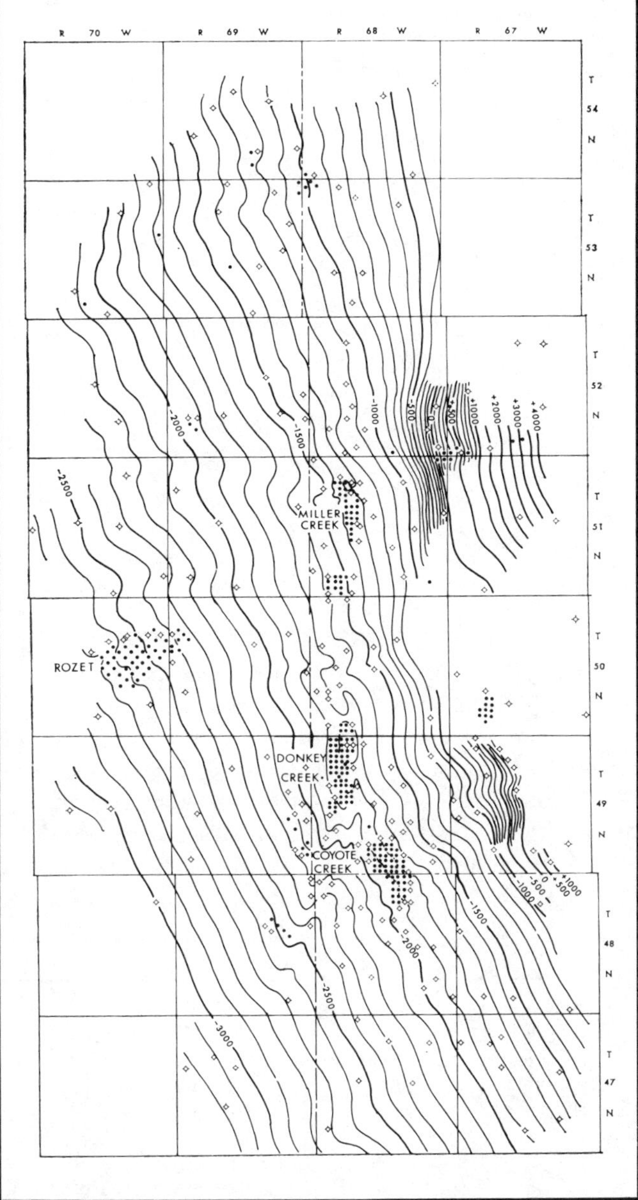

Fig. 5-24. Structure of bentonite marker bed, showing general homoclinal westward tilt into the Powder River basin. Contour intervals 100 ft and 500 ft. (Modified after Busch, 1974; permission to publish by AAPG).

the Fall River sandstones from the higher Muddy sandstones. Differential compaction of the Skull Creek shale directly east of the Miller Creek, Donkey Creek, and Coyote Creek pools was a determinative factor in the location of the main axial trend of this Muddy sandstone drainage course.

The structure of the bentonite marker bed is shown in Figure 5-24. There is slightly more than 6,500 ft (1,981 m) of basinward (Powder River basin) homoclinal dip to the west; a localized closure is present over the Donkey Creek oil pool. All of this westward tilting is related to the post-Cretaceous Laramide orogeny.

The combination of paleotopographic map (Fig. 5-23) and structure map (Fig. 5-24) serves as a geologic framework for more detailed studies of the oil possibilities of the sandstone units of the Muddy. The trends and geographic distribution of each of these sandstones can be determined readily by (1) identifying each of the alphabetical sand-

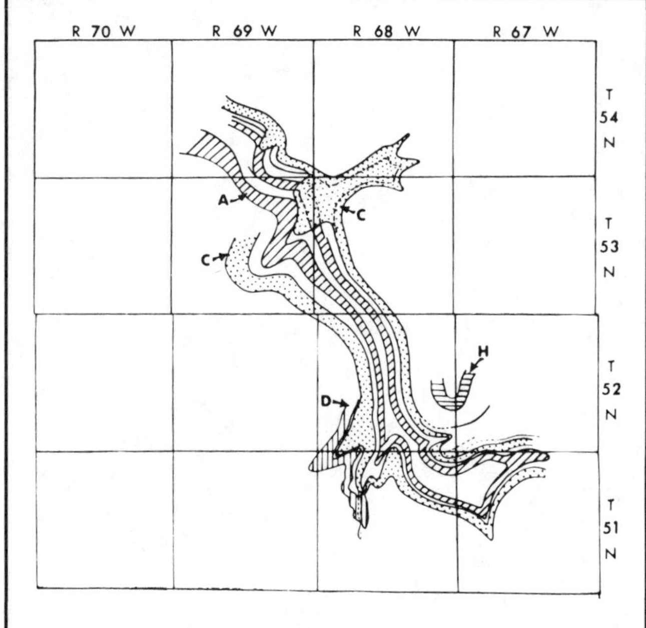

Fig. 5-25. Distribution of Muddy sandstone members A, C, D, and H. Members A and C are shown as paired bands on opposite sides of principal drainage courses. Stratigraphically higher sandstone members are progressively more widely separated and cross over main stream axes at successively more upstream positions. (Modified after Busch, 1974; permission to publish by AAPG).

stone members in each well, and (2) outlining each of the sandstones on an acetate overlay by following the contour lines of the paleotopography (Fig. 5-23). Because there are nine sandstone units of the Muddy in this area, there is considerable overlap. Thus, several acetate overlay maps must be drawn to avoid confusion. One such sandstone-distribution map is illustrated in Figure 5-25. It shows the respective distributions of sandstones A, C, D, and H. Sandstones A and C appear as narrow bands on either side of the axes of the main channel and its tributaries. If any one band of sandstone is followed in an upstream direction, it will cross the axis and extend downstream on the opposite side. Furthermore, the two bands of any one sandstone are farther from the channel axis the higher they are in the section. Thus, in profile, they have an en echelon arrangement on opposite sides of the channel. This pairing of sandstone bands on opposite sides of the channel is interpreted as the result of cyclic subsidence and marine transgression, which moved from north to south. By this postulation, each pair of sandstone bands is the result of a stillstand of the shoreline during which fringing beach sands were deposited.

To define drillable prospects in these sinuous bands of lenticular sandstones, it is necessary to use a combination of sandstone distribution (Fig. 5-25) and structural configuration (Fig. 5-24). Only the updip wedge edges of the individual sandstone members are likely to afford favorable conditions for the entrapment of oil and gas. No two pools

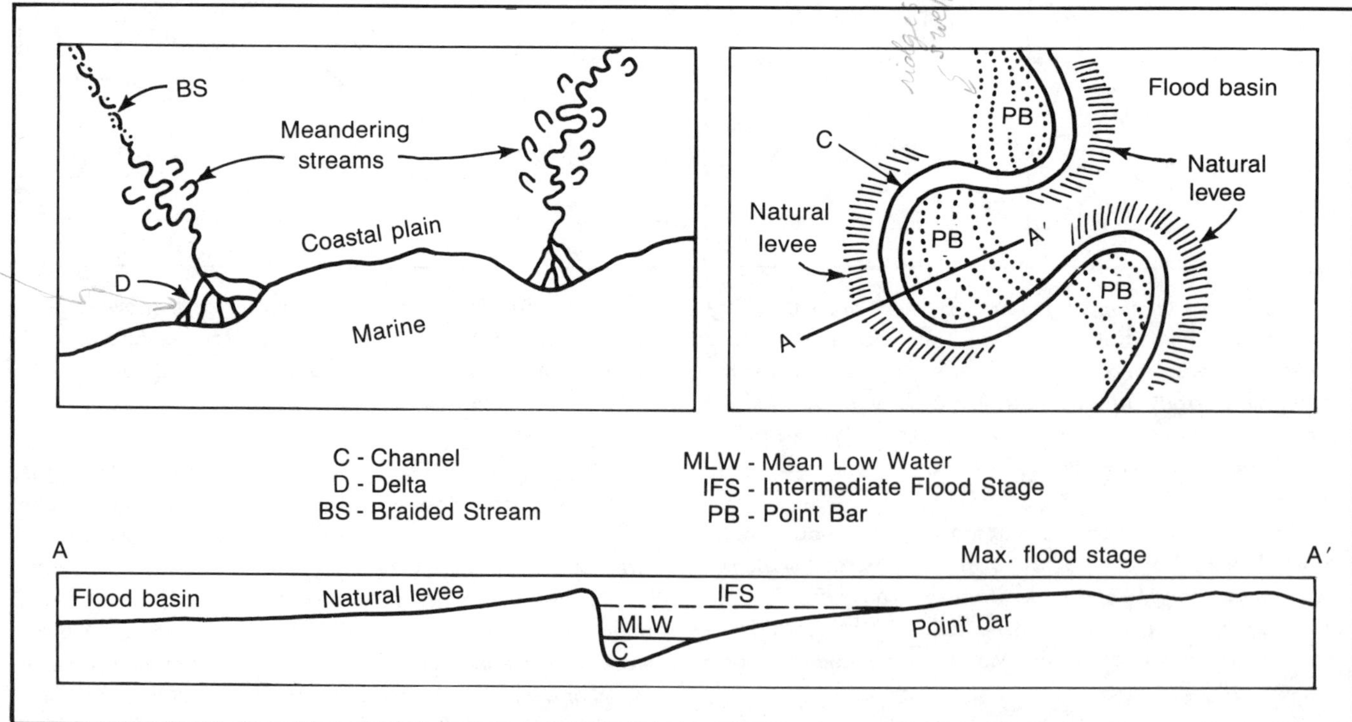

Fig. 5-26. Meandering stream model of clastic deposition. (Modified after LeBlanc, 1972; permission to publish by AAPG).

are likely to have the same outline, and each of the sandstone members must be considered as a separate potential source of supply.

Flood Plain Point Bars. Considerable study has been devoted to sand deposits associated with meandering streams. However, there is much still to be learned. Such deposits are called point bars and occur on the inside of meander loops. The most pronounced meander loops occur in the lower, downstream portions of drainage courses. It is in these areas that many streams have reached grade and simultaneously erode and deposit on opposite sides of the thalweg. Depositional models of both a flood basin and meander loops are illustrated in Figure 5-26. Meandering streams and associated point bars are by no means confined to the flood basin, as shown in this figure, but may also characterize the distributaries of wave-dominated deltas. Both the Niger delta of west-central Africa (Weber, 1971) and the Rhone delta of southeast France (Kruit, 1955; Oomkens, 1967) have meandering distributaries and numerous associated point bars.

River meanders tend to exhibit a constant ratio between the wave length and the radius of curvature, as illustrated in Figure 5-27. In the two illustrations of this figure Leopold and Langbein (1966) point out that "the value of this ratio for the meander that looks like a sine wave *(top)* is five for the wave length to one for the radius; the more tightly looped meander *(bottom)* has a corresponding value of three to one." By studying 50 typical meanders on many different rivers and streams Leopold and Langbein arrived at an average value of about 4.7 to 1, as plotted on Figure 5-28. It also may be noted from this figure that the meander length (L) averages 10.9 times the channel width (W). Sinuousity, or tightness of bend, may be expressed as the ratio of the length of the channel in a given curve to the wave length of the curve. Leopold and Langbein point out that in "the large majority of meandering rivers the value of this ratio ranges between 1.3 to one and four to one." These observations have considerable significance when estimating the size of a point-bar reservoir. They also influence the possible linear spacing of point bars within a single flood plain.

Cutoff meander loops (oxbow lakes) are a conspicuous feature of the flood plain. In most instances, these lakes silt and shale up in such a manner as to constitute clay plugs. In Figure 5-29A, both a neck cutoff and a chute cutoff are illustrated. Either may occur during a high water stage. A meander loop that bends in an upstream direction may be cut off abruptly during a high water stage, in which case only the proximal and distal ends of the lake are plugged with sand (Fig. 5-29B). Where a meander loops in a general downstream direction, abandonment is more gradual, and there is considerably more sand concentrated in the proximal and distal portions. Here the lake is much more restricted and only arcuate in shape.

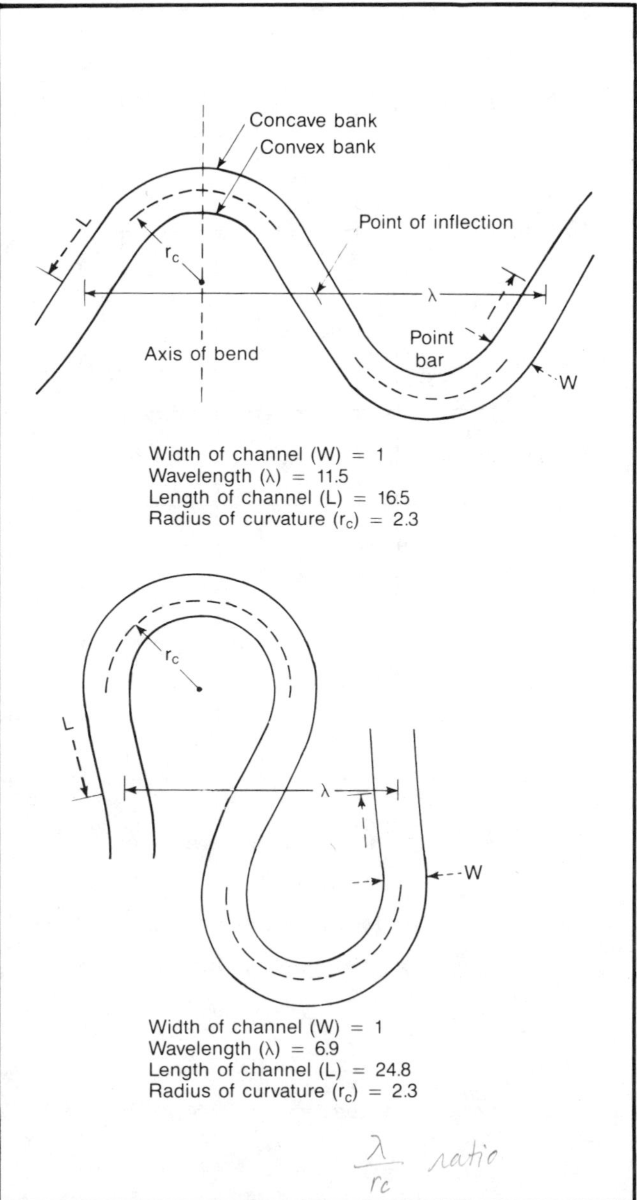

Fig. 5-27. Properties used to describe river meanders for two typical meander curves. A series of meanders has a regular appearance whenever there tends to be a constant ratio between the wave length (λ) of the curve and its radius of curvature (rc). The value of this ratio for the meander that looks rather like a sine wave **(top)** *is five to one; the tightly looped meander* **(bottom)** *has a corresponding value of three to one. An average value for this ratio is 4.7 to one. Sinuosity, or tightness of bend, is expressed as the ratio of the length of the channel (L) in a given curve to the wavelength of curve. The value of this ratio for the top curve is 1.4 to one and for the bottom curve 3.6 to one. On the average the value of this ratio ranges between 1.3 to one and four to one.* (Modified after Leopold and Langbein, 1966; from *Scientific American,* W. H. Freeman Co.).

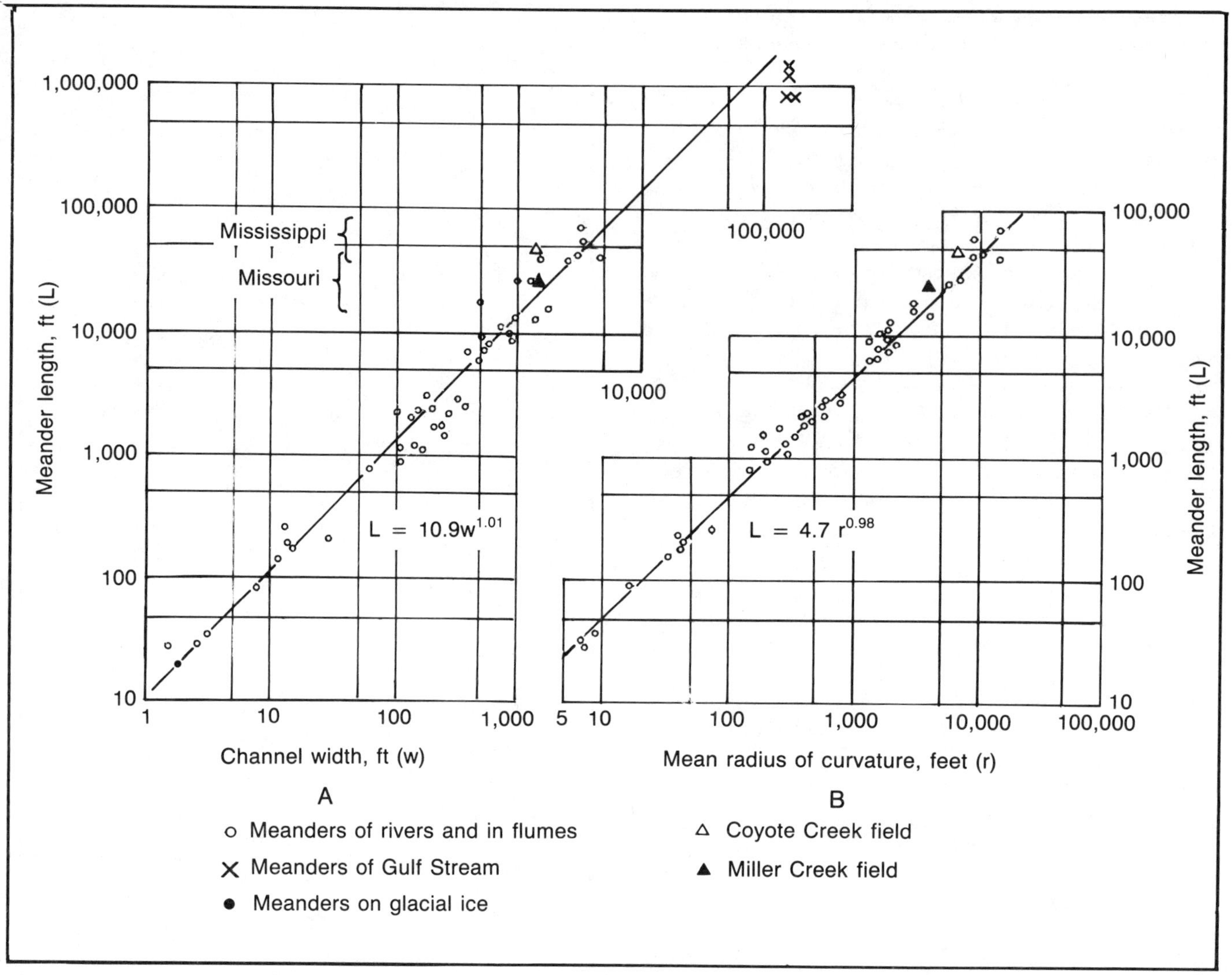

1,000,000
100,000
10,000
1,000
100
10
Meander length, ft (L)
Mississippi
Missouri
$L = 10.9w^{1.01}$
1 10 100 1,000
Channel width, ft (w)
A
100,000
10,000
1,000
100
10
Meander length, ft (L)
$L = 4.7\ r^{0.98}$
5 10 100 1,000 10,000 100,000
Mean radius of curvature, feet (r)
B

○ Meanders of rivers and in flumes
× Meanders of Gulf Stream
● Meanders on glacial ice
△ Coyote Creek field
▲ Miller Creek field

Fig. 5-28. Relations between channel parameters of modern streams. (Modified after Leopold and Wolman, 1960; permission to publish by GSA).

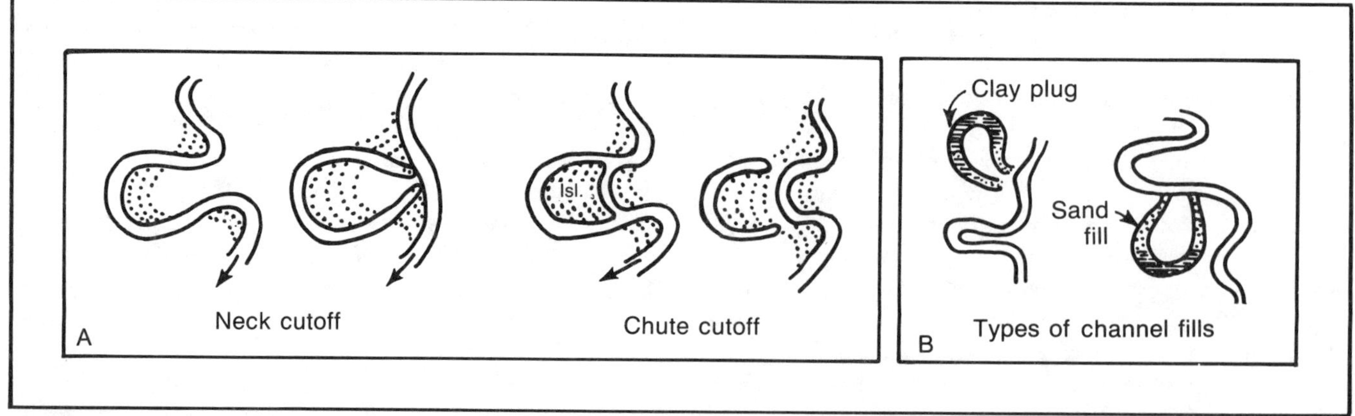

Fig. 5-29. Channel diversion, abandonment, and filling as a result of neck and chute cutoffs. (Modified after LeBlanc, 1972; permission to publish by AAPG).

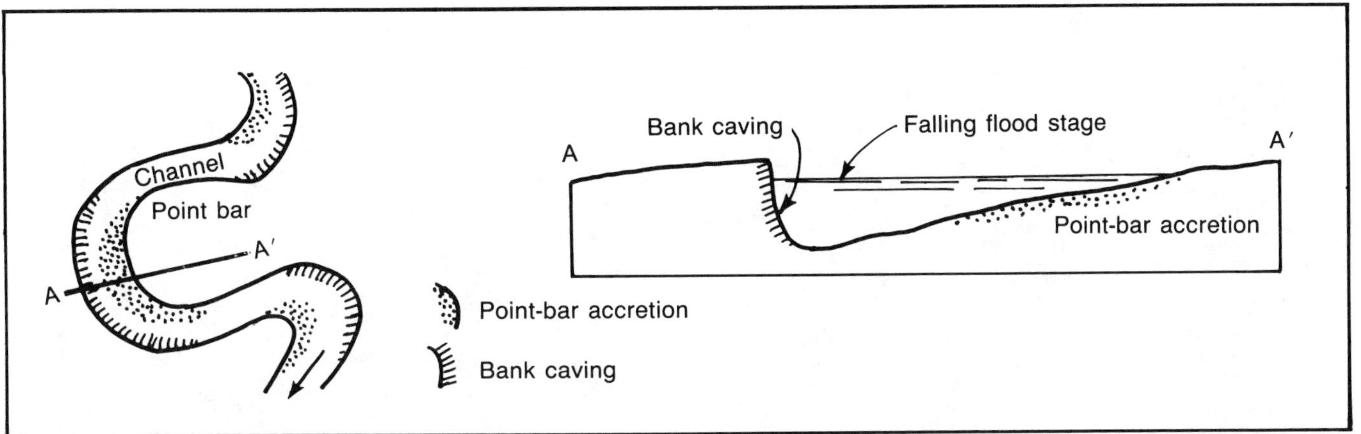

Fig. 5-30. Areas of bank caving and point-bar accretion along a meandering channel. (Modified after LeBlanc, 1972; permission to publish by AAPG).

LeBlanc (1972) points out that:

> The maximum rate of sediment deposition occurs during falling flood stages . . . The most important processes of sedimentation in the meandering-stream model are related to channel migration which occurs as a result of bank caving and point-bar accretion (Fig. 5-30) . . . Bank caving occurs at maximum rates in bends where the bed and bank materials are very sandy. Rates are much slower in areas where banks are characterized by clayey sediments. (Fisk, 1947).

The step-by-step accentuation of the meanders and the lateral accretion of point-bar sediments are illustrated in Figure 5-31. This figure, however, does not reflect that, in addition to lateral migration, the meanders also tend to migrate in a downstream direction. It should be noted that the main bulk of the deposited clastic material of the point bar occurs inside the meander loops and *not* in the channel. Point bars within a flood plain seldom are of the same size and orientation. Thicknesses of modern point bars range considerably. Those along the Mississippi River may exceed 150 ft (45.7 m), whereas, those along the Brazos of Texas are about 50 ft thick (Bernard et al., 1970). Generally speaking, the smaller the river, the thinner the point-bar sequence.

There are two principal types of sand bodies directly related to meandering streams, the point bar and the abandoned channel fill. LeBlanc (1972) states that, "The former, which are much more abundant than the latter, occur in the lower portion of the point bar sequence and constitute at least 75 percent of the sand deposited by a meandering stream." Sediments directly associated with point bars and abandoned channel fills are levees and flood plain deposits. Levees consist of unsorted fine-grained sands, silts, and clays. The flood basin deposits consist of blanket-like clays and silty clays deposited from suspension during short-term

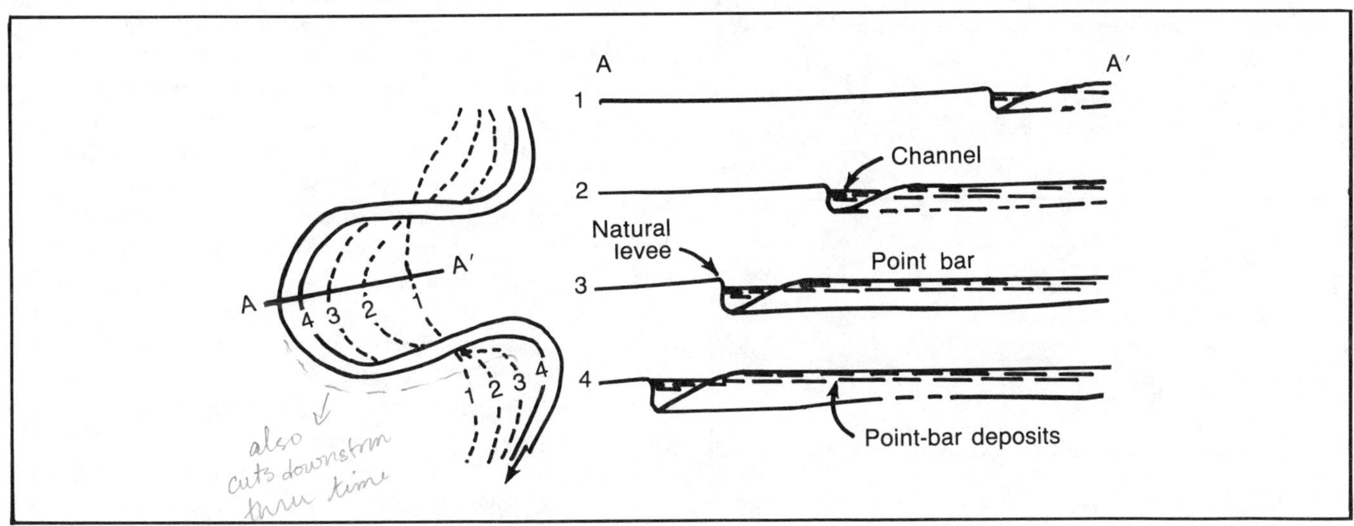

Fig. 5-31. Development of point-bar sequence of sediments. (Modified after LeBlanc, 1972; permission to publish by AAPG).

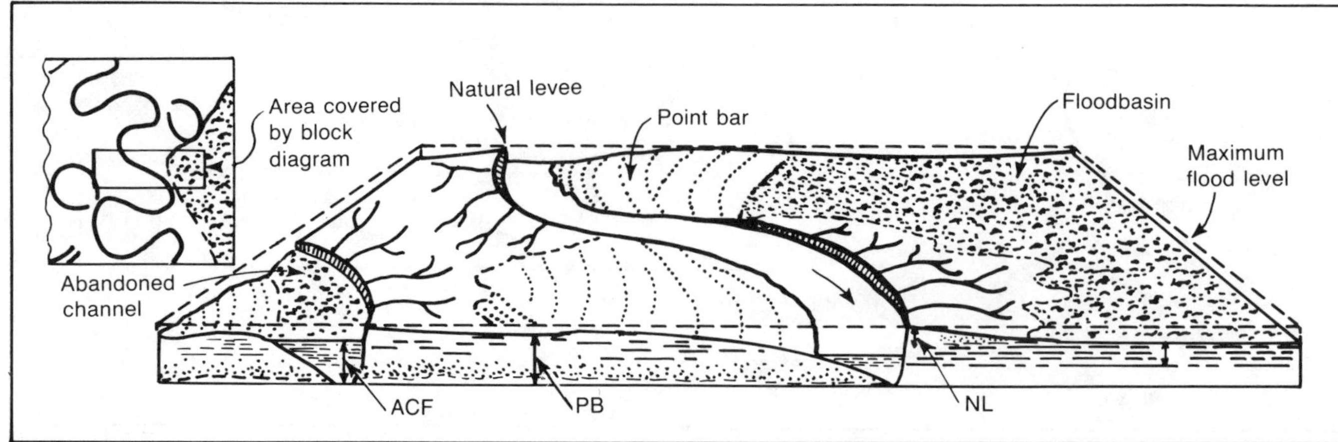

Fig. 5-32. Characteristics of meander belt and flood-plain deposits. (Modified after LeBlanc, 1972; permission to publish by AAPG).

floods that inundate the flood plain. The relationships of the point bar, channel, levee, and flood plain are illustrated in Figure 5-32. On the vertical face of this block diagram there is a generalized indication of a decrease in grain size upward in the point-bar portion. Actually, there is a quadripartite zonation within the vertical section of point bars that serves as a means of identifying this type of sedimentary body from a single unoriented core. This zonation is diagrammatically illustrated in Figure 5-33 in a coarse to fine upward sequence. LeBlanc describes these four zones, from the top down, as follows:

4. Highest part of bar (shallowest part of channel under flood conditions). Bed form consists mainly of small ripples.
3. Portion of bar (channel) covered by water during flood of maximum and intermediate magnitude. Bed form predominantly small ripples and horizontal. Deposits of this zone consist of cyclic units of horizontally-bedded and small scale cross-bedded silts and sandy silts, with some clays.
2. Portion of bar most frequently covered by water (from MLW to flood stages). Bed form consists of large ripples (scalloped). Sediments are mainly sands with some gravels that are festoon cross-bedded. Contains some thin units of horizontally-bedded and small scale cross-bedded sands.
1. Portion of channel generally always covered by water. Bed form not observed. Sediments are massive and very poorly sorted, usually containing clay balls.

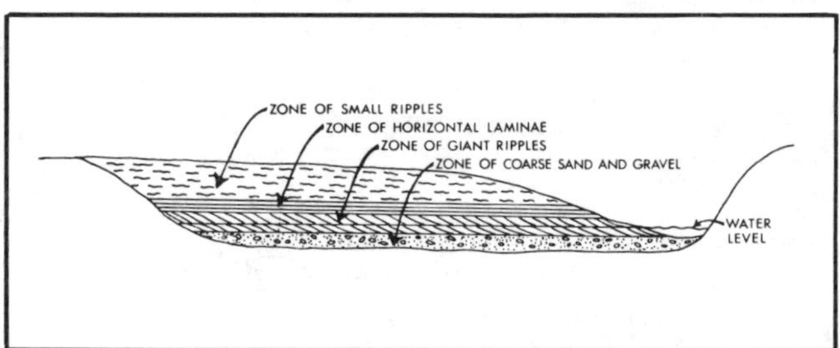

Fig. 5-33. Extremely generalized cross section through a point bar, showing quadripartite zonation. (Modified after Busch, 1974; permission to publish by AAPG).

In a few places, the normal sequence of zones in a point bar is interrupted by the presence of the basal, poorly-bedded sand and gravel zone lying unconformably on the zone of horizontally laminated beds or the uppermost zone of minute ripples. This interrupted sequence is interpreted as the result of partial erosion and deposition by the meandering stream. In other places, the complete sequence of zones may be developed several times within a single core. Such cores have been observed and studied from channel fills in the Maracaibo basin of Venezuela.

Point bars might actually coalesce to form a "blanket-like" body of sand of considerable geographic extent. Continuity of the sand, however, is interrupted by clay plugs that fill the abandoned meander loops in the final channel positions within the flood plain. Figure 5-34 is a hypothetical model of a meander belt showing diagrammatically the quadripartite zonation of point bars. This is a drawing of a plaster model constructed by Warne. The basal zone of coarse-textured, massive, poorly defined bedding probably is present throughout much of the area represented by the model; thus effecting fluid phase continuity within the interstitial pore spaces. A tilting of this model under natural conditions can result in updip migration and accumulation of hydrocarbons within each of the areas defined by clay-plugged paleodrainage courses. Thus, numerous oil pools, randomly oriented, irregularly spaced, and having quite dissimilar outlines are normally to be expected.

The S.P. and G.R. curves of geophysical logs run through

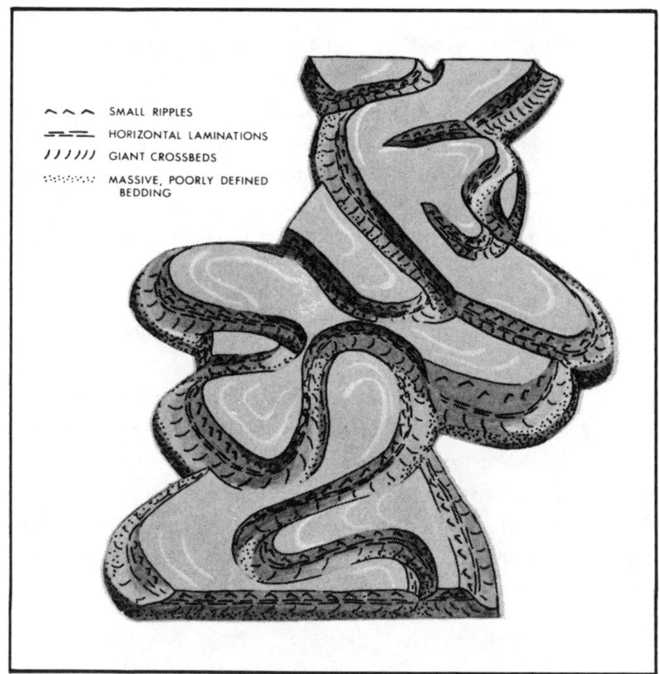

Fig. 5-34. Model of hypothetical meander belt, showing how a complex point-bar reservoir sandstone would appear with clay plugs and surrounding sedimentary material stripped away. Distribution of bedding structures is generalized. Zonation tends to follow this arrangement for a complete point-bar sequence. The uppermost zone(s) might be eroded away during later stages of meander development and a new sequence deposited on top. (Modified after Warne, personal loan).

a single point bar usually exhibit a bell-shaped profile. This profile is due to a general decrease in grain size upward and is accompanied by a general increase in the percentage of clay from the base up. These characteristics are consistent with the energy of the environment decreasing progressively upward through the four lithologic zones discussed.

Inasmuch as point bars make up only a part of the total sedimentary fill of a flood-plain deposit, they are difficult to interpret in the subsurface. Although there is still much to be learned about methods of prospecting for such reservoirs, several steps will assist in identifying them in the subsurface. A point bar can be identified from a single unoriented core by its four-part zonation. An isopach map this body of sand will exhibit an abrupt decrease in thickness at the edge where it is in contact with a clay-filled channel. The outline of a point bar will have one or more smooth-curving margins defined by dry holes directly offsetting good production. An isopach map of the shale interval between the top of the sandstone and a lithologic reference datum above the sandstone will reveal abrupt changes in thickness caused by differential compaction of the silt and clay in the meandering channels. A structure map of this reference datum will show a draping effect around the margins of the sandstone body. Other techniques currently are under study, but are still in the development stage.

The Little Creek pool of Lincoln and Pike counties, Mississippi, shows enough of the gross geometric features of a point bar to leave little doubt about the origin of the reservoir sandstone. This field, which produces from the lower part of the Tuscaloosa sandstone (Upper Cretaceous), is estimated to have an ultimate yield of 25,000,000 bbl of oil (Eisenstatt, 1960). Figure 5-35 is an isopach map of the producing sandstone of this pool and it illustrates the extremely irregular shape and thickness of the sandstone. The positions of dry holes that closely border the pool are especially significant. Sandstone 20–45 ft (6.1–13.7 m) thick grades completely into shale between wells drilled on 10-acre spacing. Eisenstatt stated, "The sand can be fairly thick in one test and absent in another only a few feet away. This phenomenon is important since it is affecting the development of the Field to the extent that many of the holes

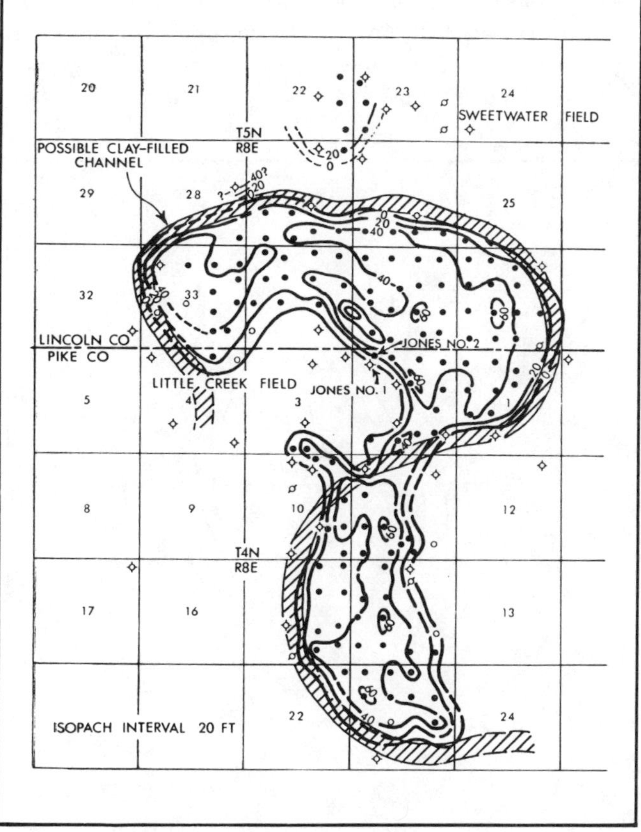

Fig. 5-35. Isopach of Denkman sandstone of Little Creek pool, Lincoln and Pike Counties, Mississippi, showing position and trend of a postulated clay-filled channel. (Modified after Eisenstatt, 1960; permission to publish by Gulf Coast Association of Geological Societies).

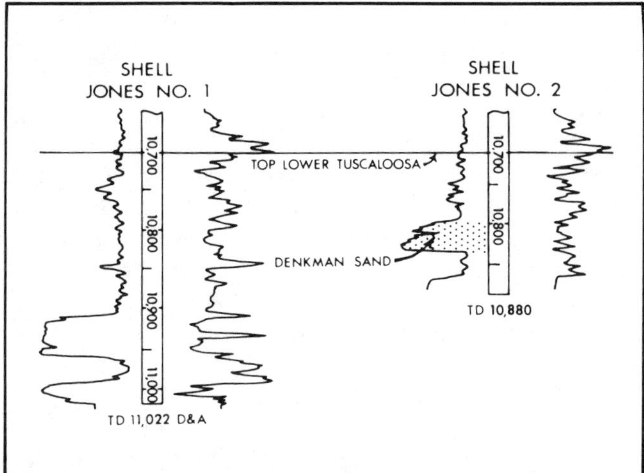

Fig. 5-36. Stratigraphic profile of upper portion of the Lower Tuscaloosa, showing abrupt pinchout of the Denkman point-bar sandstone in two wells approximately 250 ft apart; Little Creek pool, Lincoln and Pike Counties, Mississippi.

were drilled as exceptions to the field rules in order to find the edge of the sand body on a 40-acre unit." Logs of two wells spaced less than 660 ft (201 m) apart are shown in Figure 5-36. The "Denkman" sandstone (the producing zone of the Tuscaloosa) of the Jones No. 2 well is totally absent

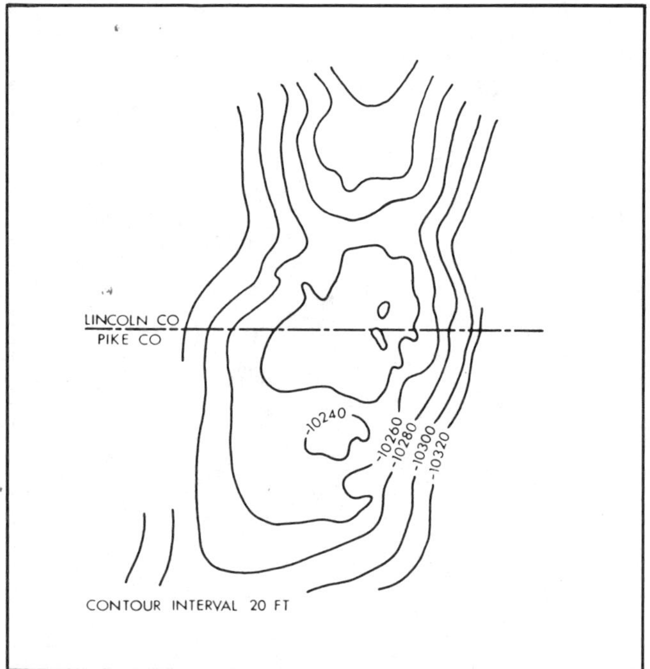

Fig. 5-37. Structure map of the top of the Lower Tuscaloosa, Little Creek pool, Lincoln and Pike Counties, Mississippi. Contour interval, 20 ft. (Modified after Eisenstatt, 1960; permission to publish by Gulf Coast Association of Geological Societies).

in the Jones No. 1 well. These two wells are labeled on Figure 5-35. The position and trend of a postulated clay-filled channel have been superimposed by the authors on Figure 5-35. Figure 5-37 is a structure map prepared by Eisenstatt depicting his interpretation of the configuration of the "Lower Tuscaloosa" sandstone. He stated that, "The data shows *(sic)* a gentle south plunging nose. Only about 30 feet (9 m) of counter-regional north dip is present from the highest well to the lowest north flank producer. However, an oil column of about 110 feet (33.5 m) is known to be present. Therefore, it is concluded that this is a combination structural-stratigraphic trap."

The Fall River sandstone on the northeast flank of the Powder River basin in northeastern Wyoming produces oil from eight point-bar reservoirs, as shown in Figure 5-38. Mettler (1966) first proposed a point bar reservoir for the Westmoorcroft pool, and Bolyard and McGregor (1966) suggested a similar origin for the Coyote Creek pool. Berg (1967) pointed out that:

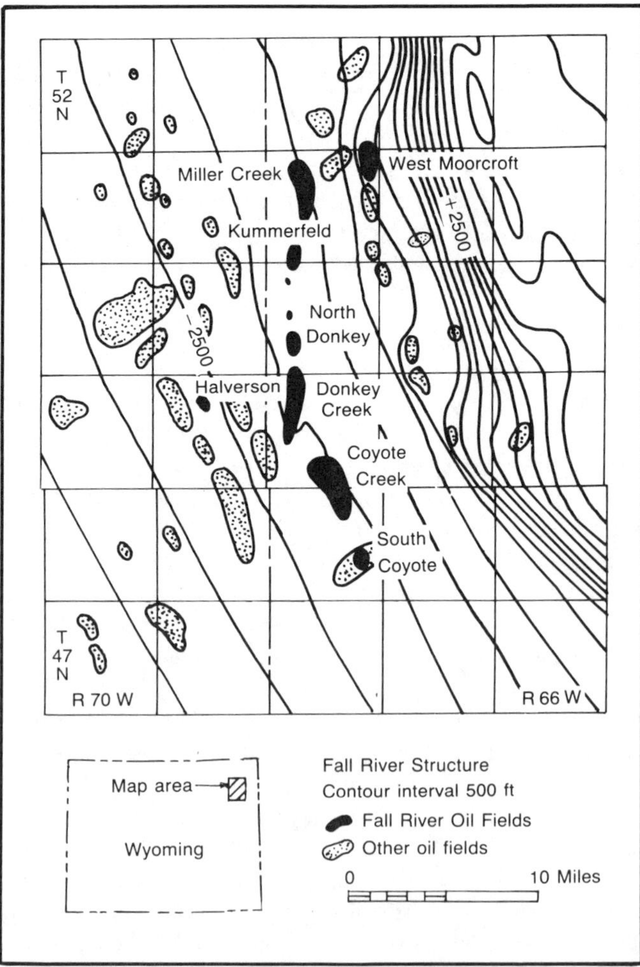

Fig. 5-38. Index map of Fall River oil fields in the northeast Powder River Basin, Wyoming. (Modified after Berg, 1967).

... all pt. bar deposits.

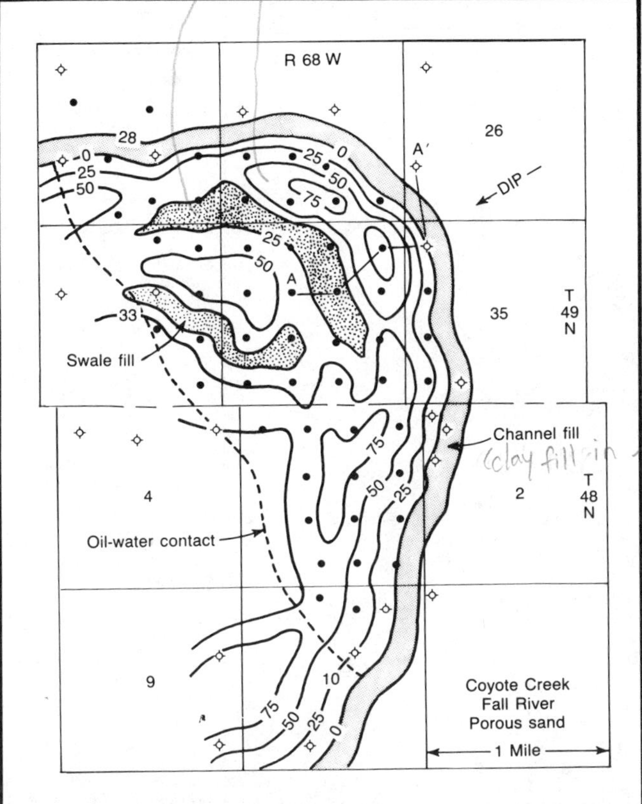

Fig. 5-39. *Isopach of porous Fall River sandstone, Coyote Creek pool, Crook and Weston counties, Wyoming. Contour interval 25 ft.* (Modified after Berg, 1967).

> Thick Fall River sandstone pinches out abruptly updip at Coyote Creek trapping an estimated 20 million barrels of oil . . . The pattern of porous and permeable sandstone shows linear and arcuate bands (Fig. 5-39) of thick sandstone that curve strongly updip. Porous sand thickness is variable and ranges up to a maximum of 90 feet . . . Downdip from the area of maximum thickness is an arcuate band of low permeability sandstone in which the upper Fall River is composed of tight sandstone, silt, and shale, but some permeable sandstone is found in the lower Fall River (Fig. 5-40). Greater but variable thicknesses of permeable sandstone occur farther downdip.

He further states that:

> Inside the point bar ridges are arcuate depressions or swales which may be filled with fine-grained sediment during subsequent high water stages. Clay filled swales may reach a thickness of 30 to 80 feet. As a result of intermittent flooding, a series of alternating ridges and clayey swales forms a pattern that generally follows the migrating of the river channel. When the river shortens its course during flood, it may cut across the floodplain between meander loops (neck cutoff) or seek a large swale as a new channel (chute cutoff). Consequently, an abandoned channel segment is left as an oxbow lake. In this pool there is approximately a 250 ft oil column.

The Miller Creek oil pool (Fall River sandstone) also has been interpreted by Berg (1967) as a point-bar reservoir similar to that at Coyote Creek. It has a conspicuous arcuate-trend updip and is bordered by a narrow shale-filled channel (Fig. 5-41). The maximum thickness of the reservoir sandstone is 52 ft (16 m) and thicknesses of 40 (12 m) to 50 ft (15 m) may be noted close to the channel fill. A clayfilled swell occurs just southwest of this thick, arcuate, sandstone trend. Two of the wells in this swale are dry holes.

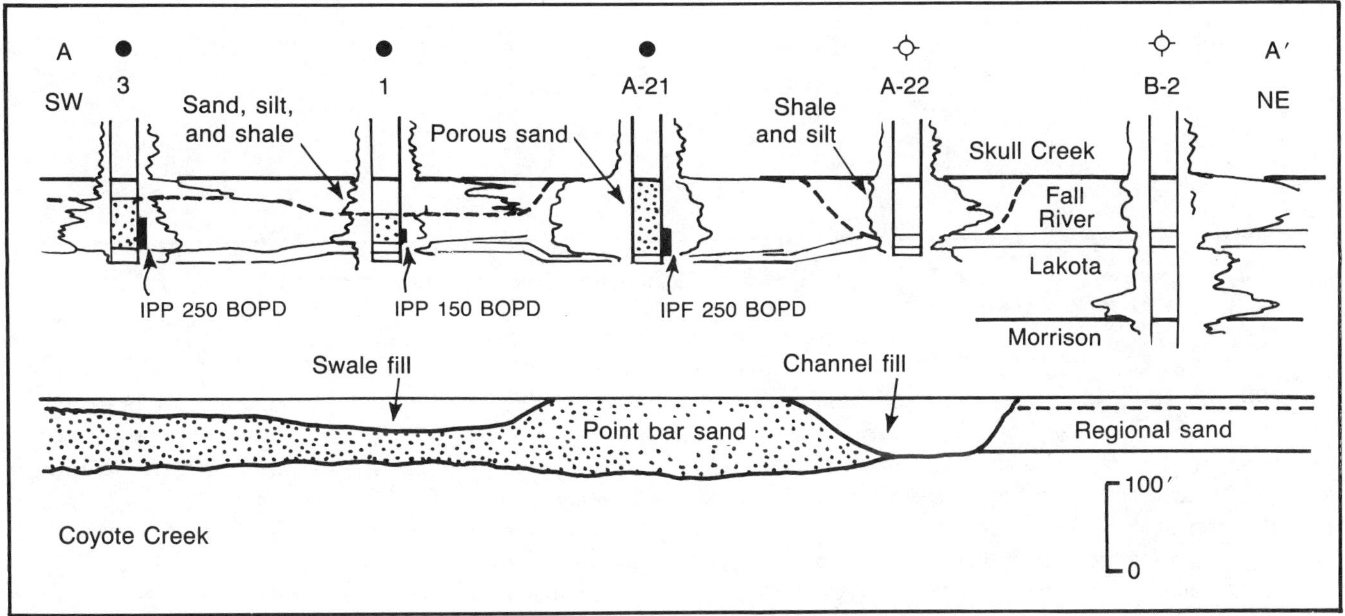

Fig. 5-40. *Electric log cross section A-A′ at updip edge of Coyote Creek pool and interpretation of meander belt facies. Location of section is shown on Fig. 5-38.* (Modified after Berg, 1967).

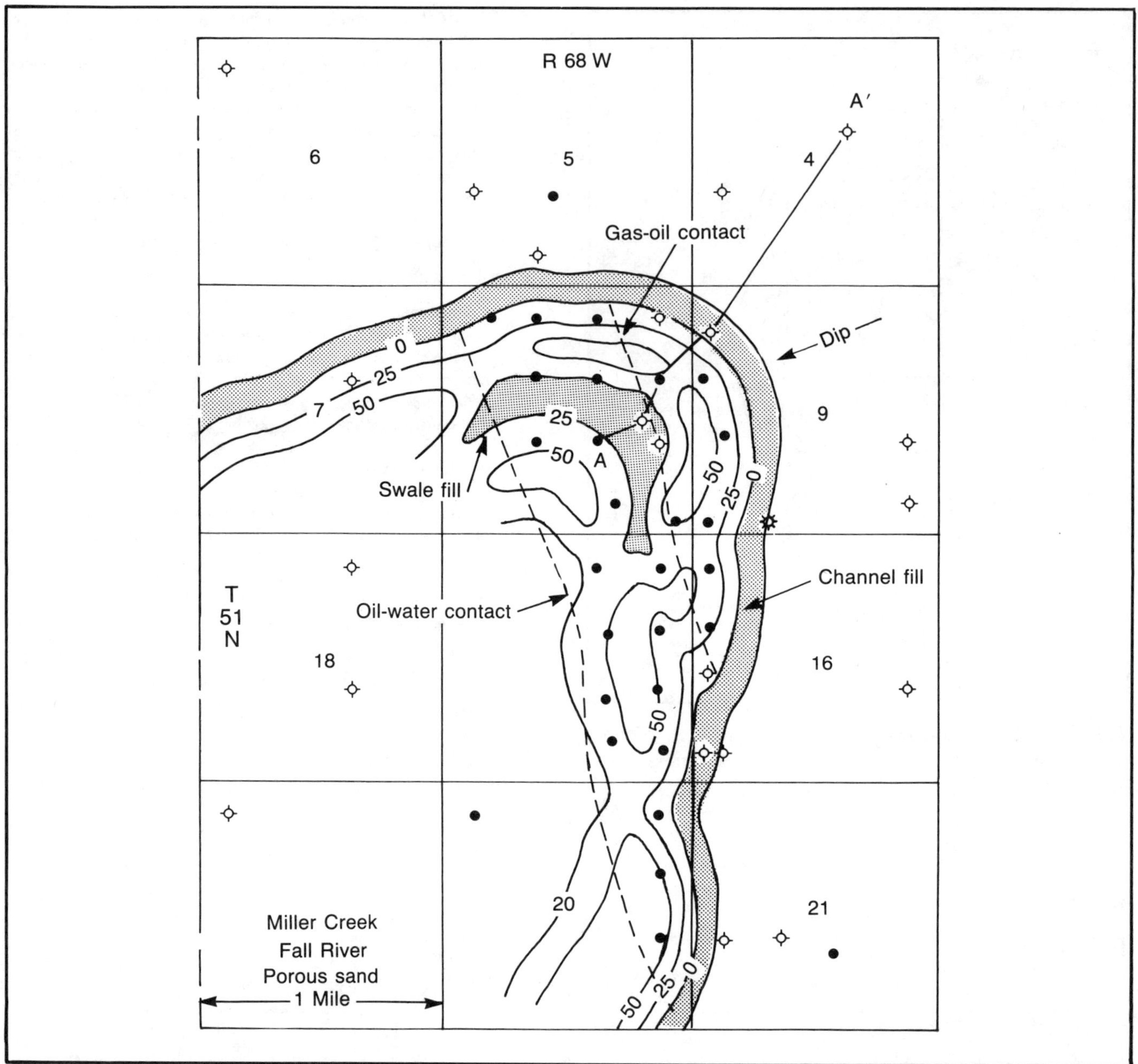

Fig. 5-41. Isopach map of porous Fall River sandstone, Miller Creek pool. Contour interval 25 ft. (From Truchot, 1963; modified by Berg, 1967).

Downdip from these wells oil is produced from wells that are stratigraphically higher than the two dry holes in the swales to the northeast. Electric logs along profile A-A′ (Fig. 5-42) show the distribution of meander belt facies at the north end of the pool.

In Figure 5-43 Berg (1967) illustrates the channel parameters for the Coyote Creek and Miller Creek pools. The ratio of wave length (W.L.) to radius of curvature (Rc) for Coyote Creek is 2.86 and that of Miller Creek is 3.33. These ratios compare favorably with the 3:1 ratio of the more tightly looped meander of Figure 5-27 (bottom). The sinuosities of the two meander curves of Figure 5-43 are quite similar. This property is determined as the ratio of the length of the channel in a given curve to the wave length. In the Coyote Creek pool it is 1.58 and in the Miller Creek pool it is 1.62.

Recognition of the point bar origin of a reservoir sandstone has a direct bearing on the selection of drill sites, drilling and coring procedures, estimation of recoverable reserves, and location of producing facilities. Ebanks and

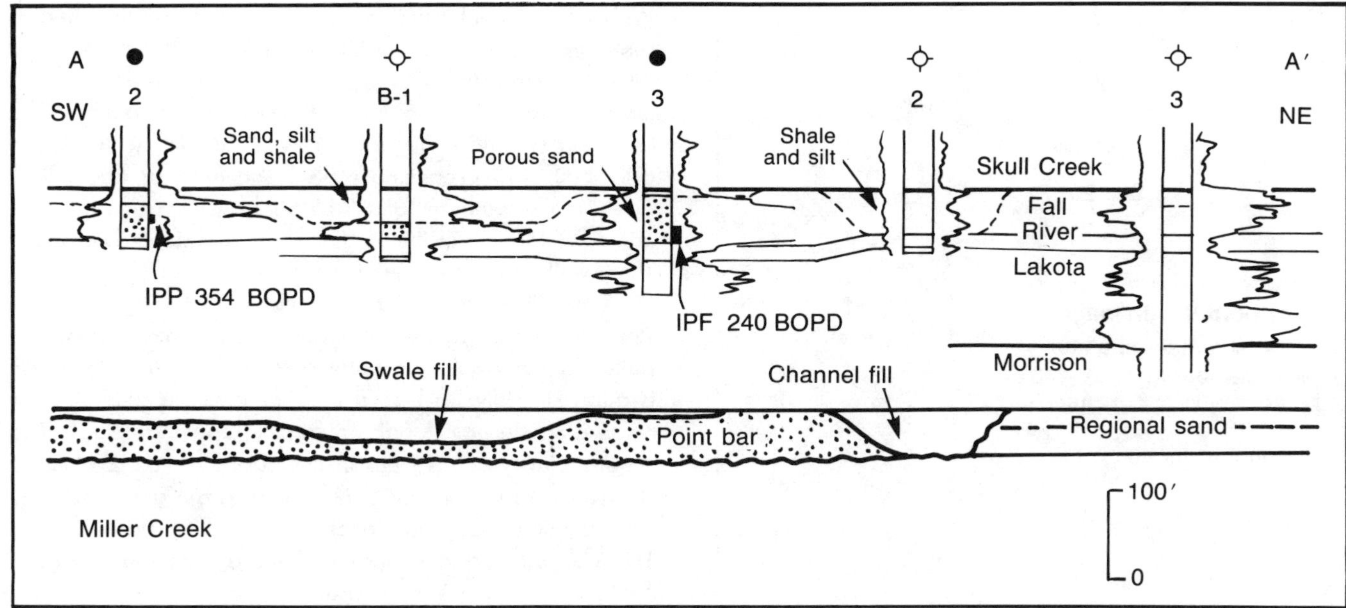

Fig. 5-42. Electric-log cross section A-A′ at Miller Creek pool and interpretation of meander belt facies. Location of section is shown on Fig. 5-40. (Modified after Berg, 1967).

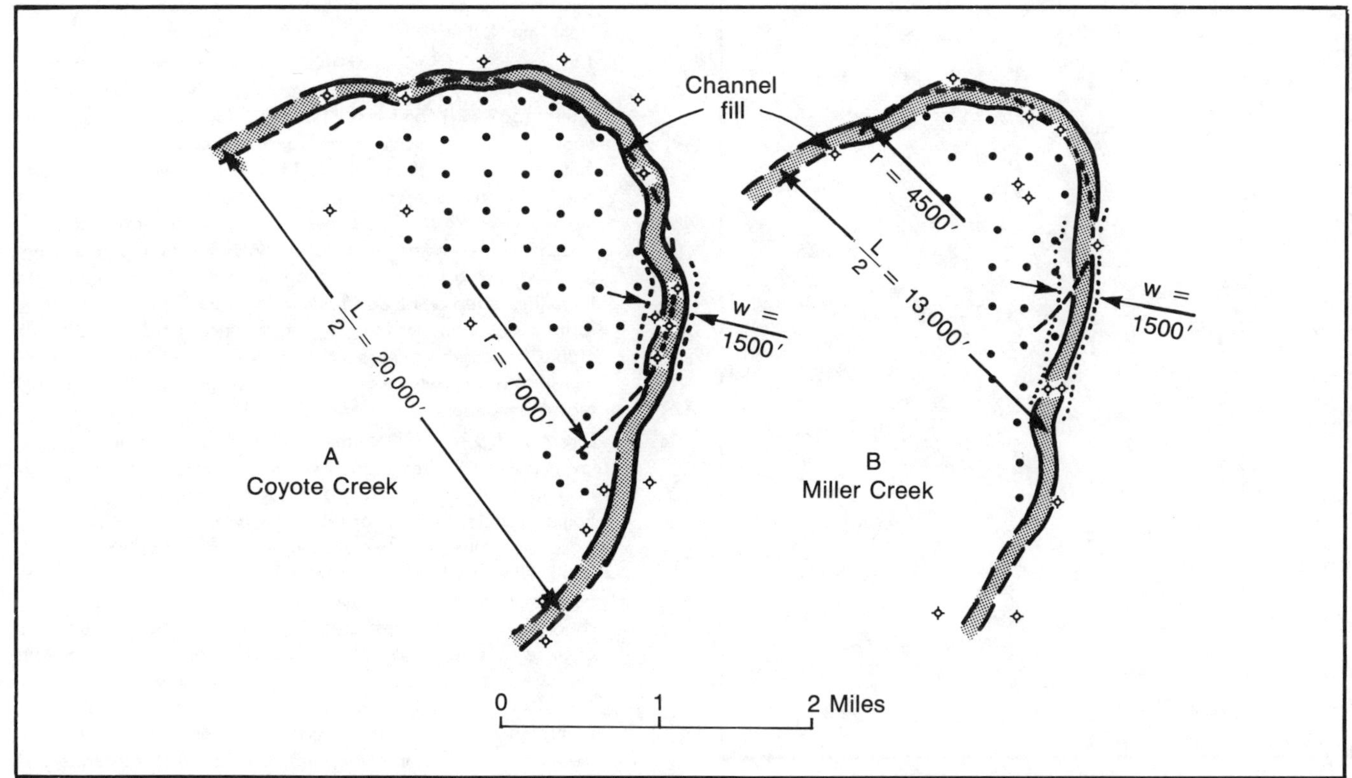

Fig. 5-43. Channel parameters for Coyote Creek and Miller Creek pools of northeastern Wyoming. (Modified after Berg, 1967).

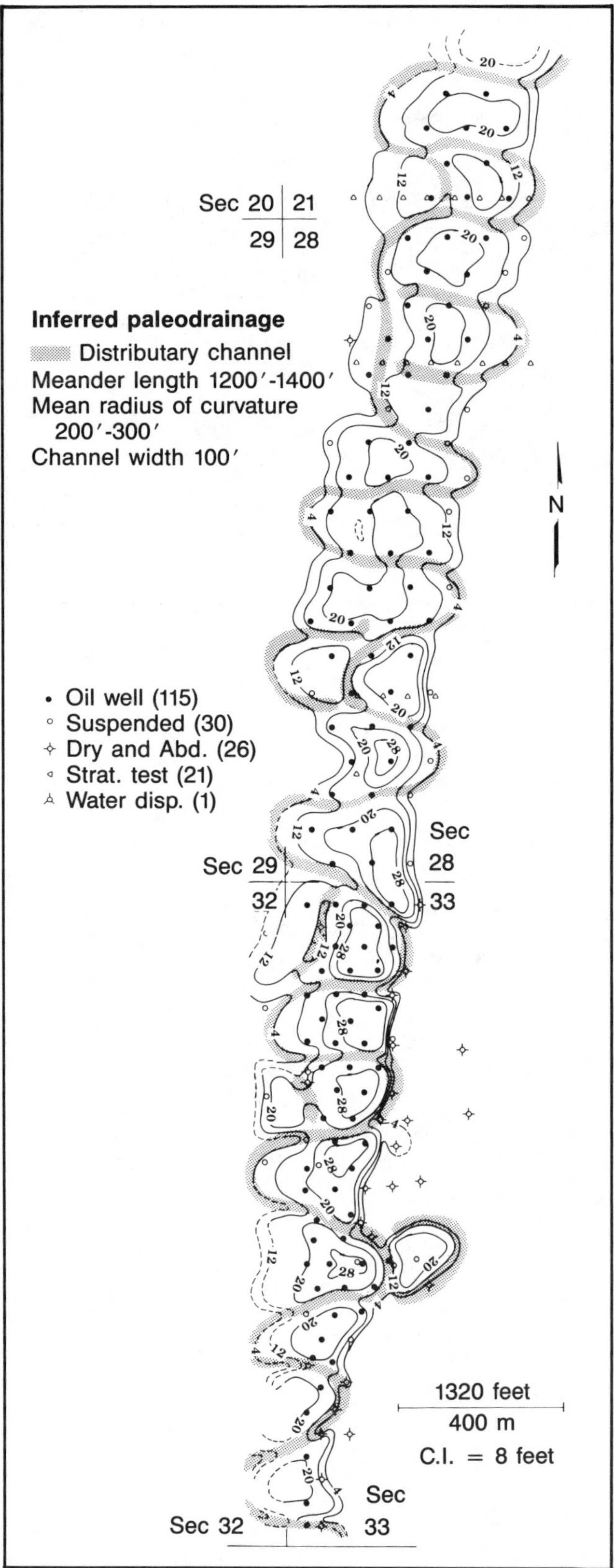

Fig. 5-44. Isopach map, gross thickness of Eastburn sandstone (Cherokee) in T35N-R33W, Eastburn pool, Vernon County, Missouri. (Modified after Ebanks and Weber, 1982; permission to publish by *Oil and Gas Journal*).

Weber (1982) published an excellent paper on the internal reservoir properties of such a sandstone body, the shallow (90–100 ft) Eastburn pool of Vernon County, in westernmost Missouri. This pool is slightly more than two miles (3.2 km) long and 400 to 1,200 ft (122–366 m) wide. There are about 115 producing wells in an area of about 200 acres. In the southern half of the pool, well spacing is about one acre; whereas in the northern half, it is about two acres. Thus, well spacing ranges from about 220 to 320 ft (67–97.5 m). Figure 5-44 is an isopach map of the gross sandstone thickness. The eastern and western margins of the pool shale out quite abruptly and the maximum thickness is approximately 28 ft (8.5 m). The contours exhibit local closures inside each of the meander loops (stippled). The great majority of the wells in this pool were cored; hence, detailed information was available on grain size, bedding, oil saturation, water saturation, permeability, porosity, etc. These parameters are illustrated for a type log in Figure 5-45. Ebanks and Weber state:

> Reservoir of the Eastburn sandstone is mainly influenced by grain size and sorting of the sand and by the amount and distribution of detrital clay and silt (Fig. 5-45). Other factors which are less important in this regard are the types and amounts of authigenic clays and cements. This relationship indicates that relict primary porosity is the most important kind of porosity and that permeability is strongly related to original patterns of sediment texture and distribution of facies. These relationships enable predictions about reservoir quality to be made from knowledge of facies in the reservoir sandstone.
>
> The lower point facies, or more coarse grained, well sorted section, with higher average porosity (26.3%) and permeability (562 md), and less abundant clay and silt than in the upper point bar facies, forms a continuous layer with fairly uniform thickness of about 12–16 ft in the lower one-third to one-half of the reservoir. The upper point bar facies, which is mostly ripple cross bedded, consists of more fine-grained sand, and has lower average porosity (24.9%) and permeability (372 md) and a larger proportion of silt and clay size particles than the lower point bar facies. The upper point bar facies forms a discontinuous layer whose thickness pinches and swells along the length of the sandstone body and whose areal dimensions are closely related to the somewhat predictable pattern of meanders of the paleochannel.
>
> Directional permeability within the sandstone is greatest in the lower part of the reservoir and tends to parallel the length of the sand body, or the orientation of the trough cross bedding. Some irregularity to this preferred pattern of fluid flow is introduced by the arcuate trend of these cross beds in plan view, and by the slump structures noted earlier. There should be few barriers to vertical communication within the sandstone. Occasional thin shale lenses, probably slough deposits, and streaks of carbonate cement are the only exceptions to this. Formation damage in a dirty sandstone such as this may present problems if the authigenic clays, particularly kaolinite, are caused to migrate as small particles into pore connections and to block them off. Introduction of non-native fluids, such as condensed steam can cause abrupt changes in pH that may aggravate this problem. Acidizing this formation to remove carbonate cement should not be necessary; furthermore, acidizing may cause partial dissolution of iron-bearing minerals, chlorite, and siderite.

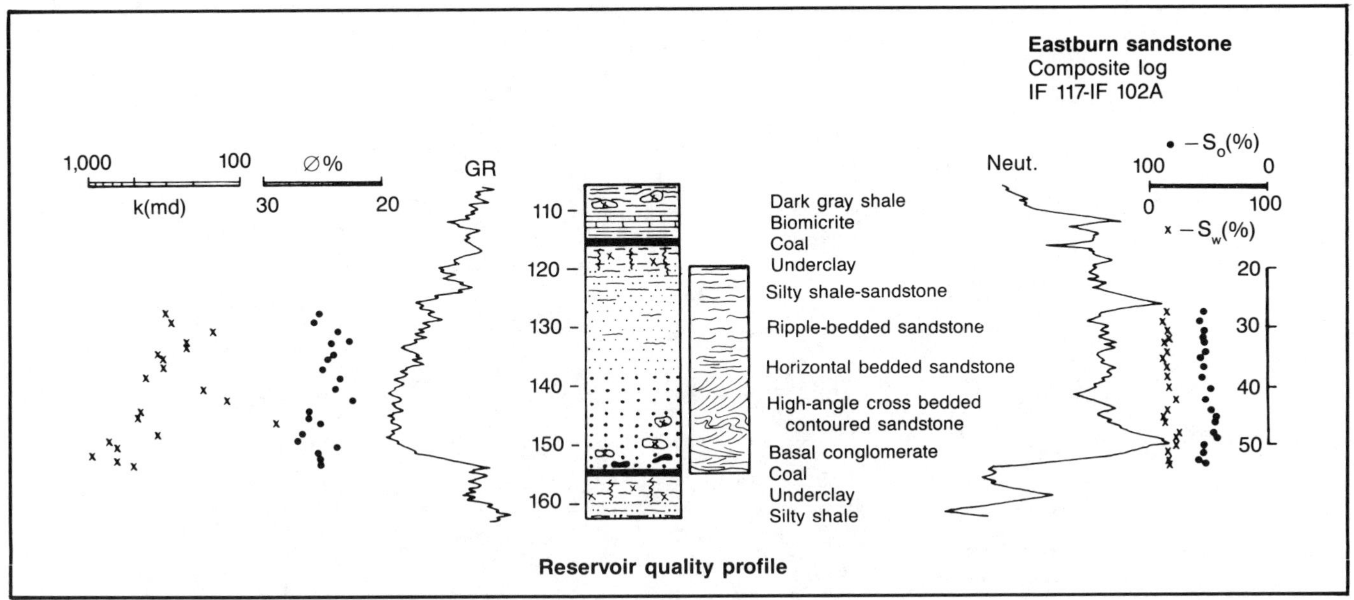

Fig. 5-45. Reservoir quality profile, Eastburn pool, Vernon County, Missouri. (Modified after Ebanks and Weber, 1982; permission to publish by *Oil and Gas Journal*).

The stream pattern of the Eastburn pool suggests that the stream course was regionally uplifted with the meanders entrenched in place. The meandering stippled band does not consist entirely of a clay plug but, rather, of alternating finely laminated, rippled sandstone and clay. (Ebanks, pers. comm.)

Another example of a point-bar reservoir is illustrated in Figure 5-46. It is the Selma pool (Burbank sandstone), situated in T22S-R21E, Anderson County, Kansas. This pool is approximately 1.5 mi (2.4 km) long, trends north-south, and is about 800 ft (240 m) wide. The interpretations shown in this figure are based on 131 drillers' logs (stippled) and three cores. In spite of the lack of quality data, the four interpretations were comparatively easy to make. Figure 5-46A is a structural interpretation of the top of the sandstone zone and clearly exhibits a sinuous trend of "highs" interpreted as coinciding with the thickest portions of the channel fill. It is structurally lowest along the east and west margins. This is attributed to differential compaction of the shales on either side and less compactability of the sand. East-west cross sections of the sandstone clearly exhibit biconvex profiles.

The structure of the base of the sandstone reflects the sinuous course of the erosional channel. The regional strike of the strata is north-south; hence, it has had very little effect on the structure of the base of the sand as shown in Figure 5-46B. The regional tilt of the strata is 30–50 ft (9–15 m) per mile to the west.

Figure 5-46C is an isopach map of the gross sand and exhibits the same meander pattern as shown in Figure 5-46A and B. The sinuous, thicker portions of the sand appear to occupy the deepest portions of the channel, shown in Figure 5-46B. This is contrary to the idea of the deepest portions of the channel being filled with a clay plug. The meanders appear to be entrenched in response to regional uplift, and the location of the thickest sand in the deepest portion of the channel may be related to this. Another possible explanation for failure to identify a clay plug might be the relatively inferior quality (drillers' logs) of the data on which this interpretation is based.

Figure 5-46D is an isopach map of net, floodable sandstone. The data on which this map is based were assembled by a petroleum engineer, and the contoured interpretation was made by a geologist without the benefit of Figures 5-46A-C. In fact, it was made one and one-half years later. In numerous instances the pore spaces of the basal several feet of the sandstone are filled with solid speciated hydrocarbon and hence are not included in floodable thickness. Likewise, the upper portion of the sandstone is variably shaly or silty. These intervals also were excluded in estimating thickness of floodable sandstone.

All four of the maps of Figure 5-46 were interpreted independently and, not withstanding this fact, they all display the same meander positions, similar wave lengths and radii of curvature. Such a map as Figure 5-46D is an absolute prerequisite for setting up a meaningful enhanced oil recovery program.

Strike Valley Sandstone. The term "strike-valley" sandstone was proposed by Busch (1959) to designate those sandstones deposited in low areas between cuestas at the time the land surface is inundated by a transgressive sea. Such cuestas may be either erosional escarpments or fault scarps. Erosional escarpments are the result of progressive truncation of a series of alternating resistant and nonresistant strata, all of which have been regionally tilted.

Strike-valley sandstones are unique in that they may be

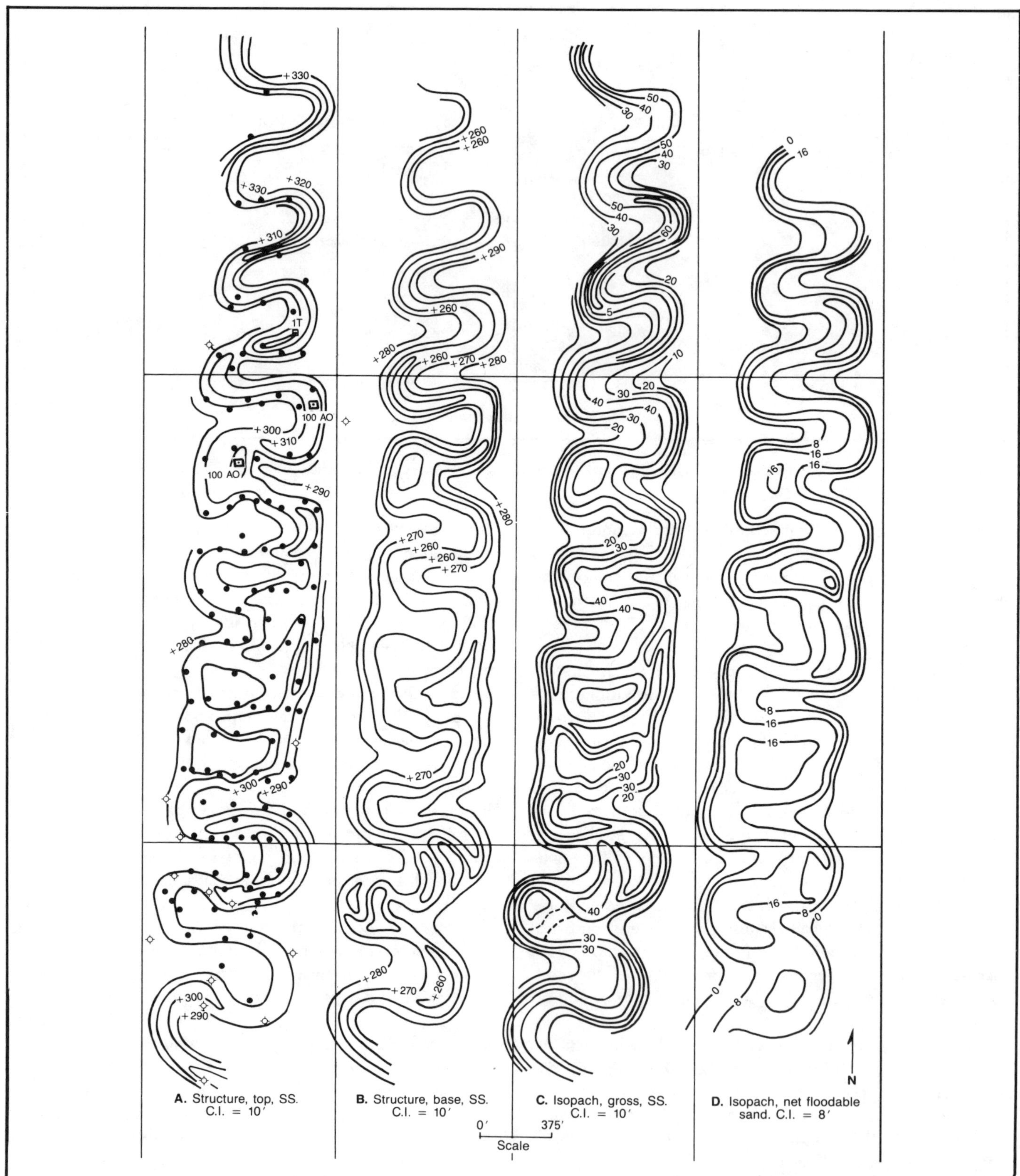

Fig. 5-46. Meandering-stream pattern of Selma pool, T22S-R21E, Anderson County, Kansas. ***A.*** *Structure, top of the Burbank sandstone,* ***B.*** *Structure of the base of the Burbank sandstone,* ***C.*** *Isopach of gross Burbank sandstone, and* ***D.*** *Isopach of net floodable Burbank sandstone.*

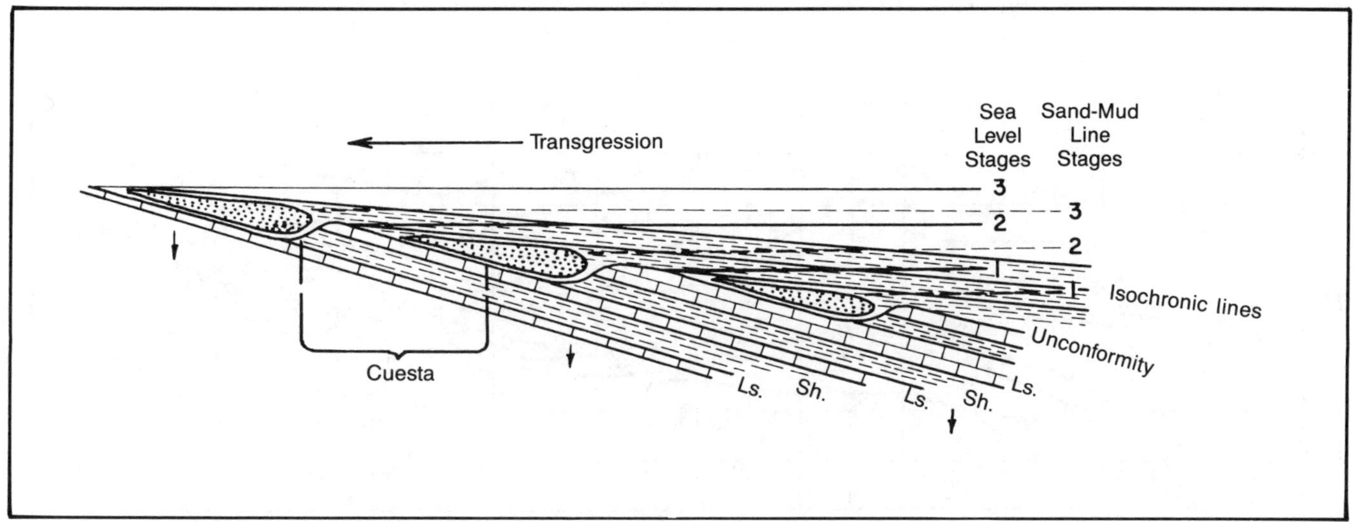

Fig. 5-47. Diagrammatic cross section showing several strike-valley sandstones occupying topographically low areas on the landward sides of escarpments (cuestas). Such sandstone bodies terminate abruptly toward the escarpments and taper gradually in a landward direction where they decrease in grain size and occasionally interfinger with siltstone and/or shale.

classified as either marginal-marine or channel sandstones. They were first described and illustrated from an example in the Midcontinent area by Busch (1959). Since then many examples have been identified by other geologists in other areas. Figure 5-47 is a diagrammatic cross section showing several strike-valley sandstones occupying topographically low areas on the landward sides of several escarpments (cuestas). As such, they terminate abruptly toward the escarpments and taper gradually in a landward direction where they decrease in grain size and sometimes interfinger with siltstone and shale. A distinct shale "break" separates the sandstone from the underlying unconformity. It is postulated that this shale constitutes a soil profile of the subjacent strata with admixtures of insoluble residues. It is typically thinnest under the thickest portion of the sandstone as shown in Figure 5-49. The controlling factor for the positioning of the sandstones is the *subsequent* valley next to the face of the escarpment. Tilting of the hard stratum of the escarpment may range from less than 1° to more than 45°.

Figure 5-48 shows the structural configuration of an unconformity surface in a pool area where there is abundant subsurface well control. Well spacing is one well per 40 acres, and all of the control consists of electrical log data. The contour interval is 10 ft (3 m). A narrow, west-plunging, asymmetric syncline underlies the producing strike-valley sandstone. This syncline is bordered on the south by a very narrow west-plunging structural nose, which is the position of the highest portion of the limestone escarpment. This is not a true structure map; rather it represents a combination of paleotopography and present-day structure, the latter imposed by postdepositional regional tilt. There is a direct relationship between the erosion surface and the lenticular sandstone that overlies it. The maximum thickness of this sandstone is 50–55 ft (15–18 m).

Two north-south cross sections show the body of sandstone that produces oil from the same pool illustrated in Figure 5-48. The cross section in the upper half of the figure is drawn with a limestone datum of reference; the limestone is a time marker that is persistent and easily recognized. The unconformity is indicated by a wavy line and is not parallel with the limestone datum. This unconformity shows progressive truncation northward, but a thin, resistant limestone escarpment is present just basinward from the southernmost producing well in the pool. The sandstone body (shaded black) is very asymmetric in cross section, having a teardrop profile. The relation of a strike-valley sandstone body to this combination of topographic and structural surface is more than a coincidence. The lower half of Figure 5-49 involves the same wells as those of the upper profile but it has been drawn with sea level as datum. The lenticular sandstone slopes at an angle similar to the slope of the unconformity surface, but it wedges out abruptly near the escarpment.

Figure 5-50 is a block diagram illustrating the positions of the strike-valley sandstone relative to the unconformity. During sea-level position 1, the shoreline was in the middle of the block diagram and the lower body of sandstone (shaded black) was accumulating on the escarpment side of the more seaward ridge. As the surface of deposition was transgressed, the shoreline shifted from right to left (south to north), and a second body of sandstone was deposited behind the next higher topographic escarpment. From a diagram of this type it is apparent that the top of the unconformity is not a true structural surface but is a combi-

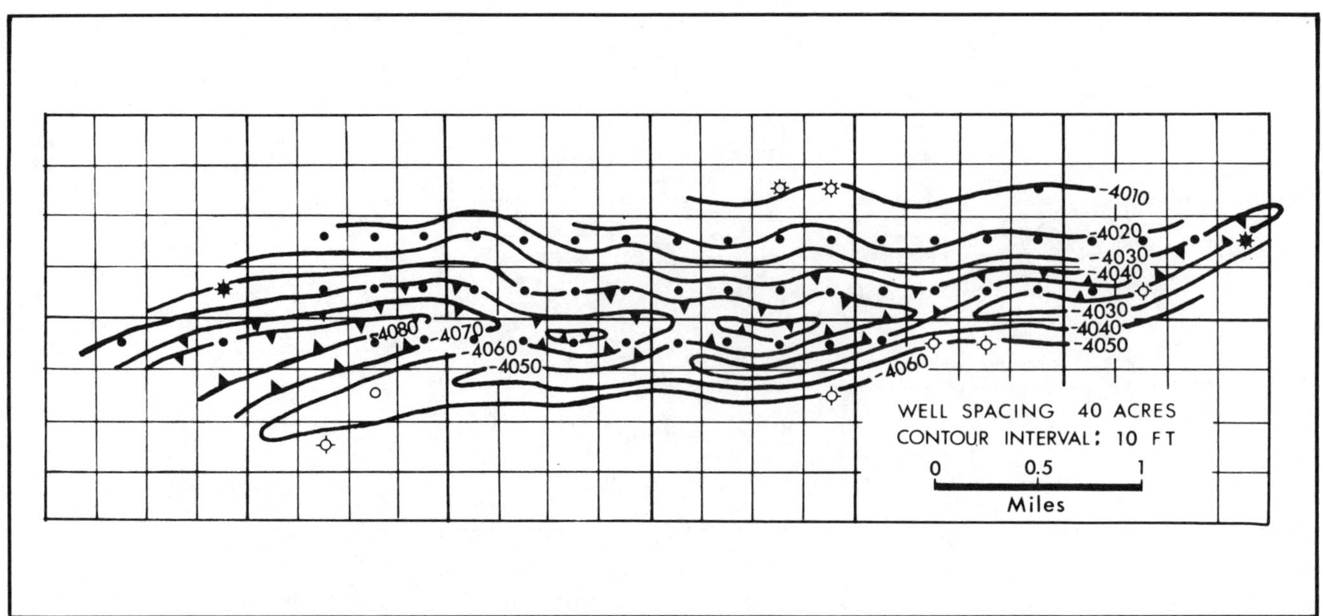

Fig. 5-48. Structural configuration of an unconformity surface, showing a west-plunging asymmetric syncline, bordered on the south by a very narrow west-plunging structural nose. Oil and gas wells (40 acre spacing) produce from a strike-valley sandstone that is positioned just above the unconformity surface of the syncline. (Modified after Busch, 1974; permission to publish by AAPG).

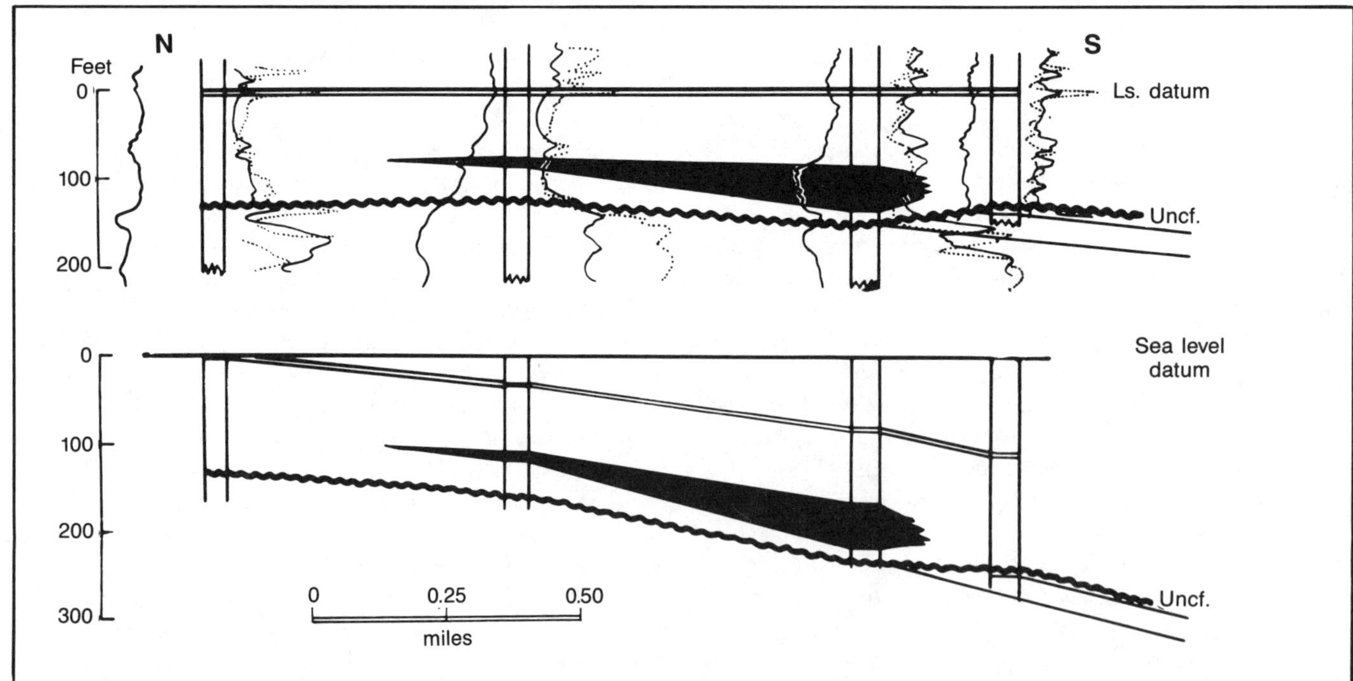

Fig. 5-49. North-south profiles, showing relation of a strike-valley sandstone to underlying unconformity surface. Note the occurrence of a shale "break" between the sandstone and the unconformity. (Modified after Busch, 1974; permission to publish by AAPG).

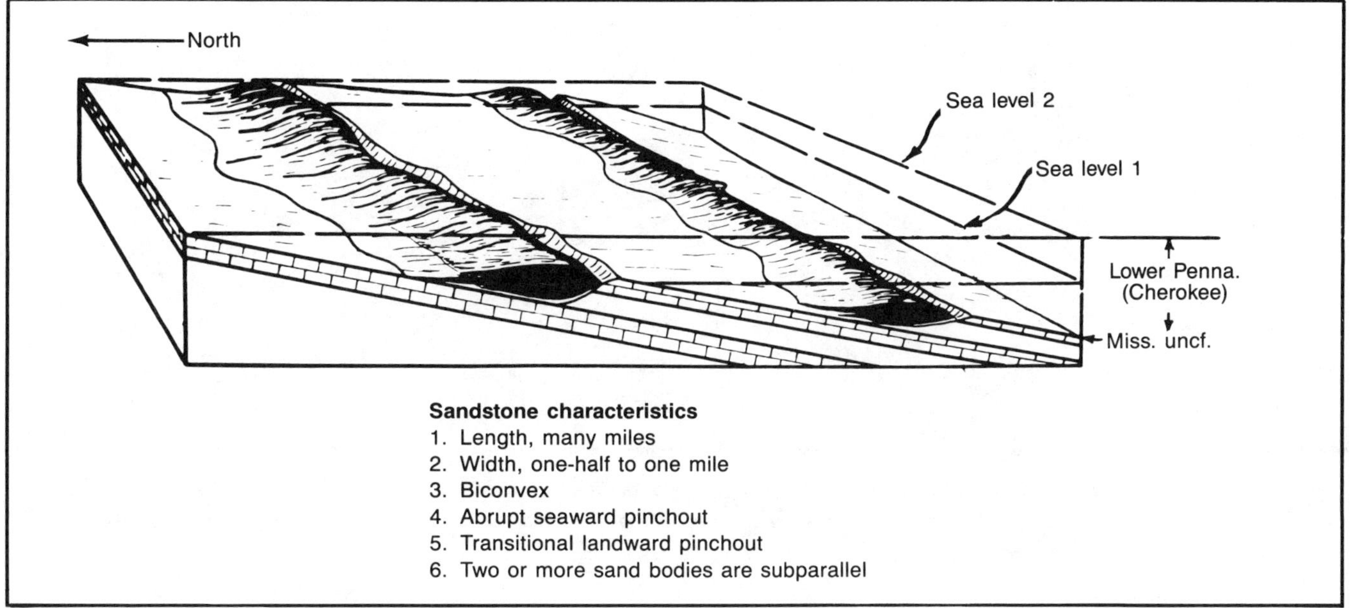

Fig. 5-50. Block diagram, illustrating relationship of strike-valley sandstones to erosional escarpments developed in a tilted and truncated series of alternating hard (limestone) and soft (shale) beds. (Modified after Busch, 1974; permission to publish by AAPG).

nation of structure and paleotopography. This type of surface, when inundated, controls the sites, trends, and linearity of the sands deposited.

The geometry of a sandstone of this type is summarized as follows:

1. The lengths of individual bodies of sandstone might be many miles.
2. The widths may vary from 0.5 to 1.0 mi (805–1610 m) or more, depending on the angularity of the depositional slope and the rate and volume of sand supply.
3. The cross section is asymmetrically biconvex with an abrupt seaward pinchout and a gradual landward pinchout.
4. Two or more such sand bodies are generally subparallel. Their trends are controlled by the paleotopography of the unconformity surface and not by the present structural grain of the strata beneath the unconformity.

Since a strike-valley sand is deposited in the drainage course of a subsequent stream, the sand body will have an initial slope in the direction of stream flow. The direction of flow and the paleogradient of a strike valley may be determined by constructing an isopach map of the genetic increment of strata between a marker bed (or horizon) above the sandstone and the unconformity below the sandstone. Such a GIS will thicken systematically in a downstream direction and the rate of thickening (in feet per mile) will approximate the paleogradient.

The first stream to develop on a newly uplifted coastal plain flows in a direction essentially perpendicular to the shoreline of the bordering embayment. Such a stream is a *consequent* stream. It will lengthen as a result of headwater erosion and, thereby, will cut a valley through both hard and soft rock. The first tributaries to develop will erode in the softer rock and flow into the master (consequent) stream essentially at right angles. These tributaries are subsequent streams that erode strike valleys. Thus, two subsequent streams may occupy the same strike valley behind an escarpment, and flow in opposite directions toward each other. Both of them will join the consequent stream at approximately the same point. When such a strike valley later becomes clogged with sand as a result of a marine transgression, the resulting strike-valley occurs within a genetic increment that exhibits thickening in opposite directions toward each other. At the point of maximum thickness of this type of genetic increment, it may be anticipated that a cross-trending channel sandstone will be present. This channel sandstone represents a clogging of the consequent stream channel. This clogging may be brought about by an upstream shift in the base level of deposition of the consequent stream as it gradually is inundated by the marine transgression. The origin of a sand-filled strike valley, however, is not explained as readily. Such a valley also may be filled as the base level shifts upstream during marine transgression. However, a more likely explanation for the origin of a strike-valley sand is transportation of marginal-

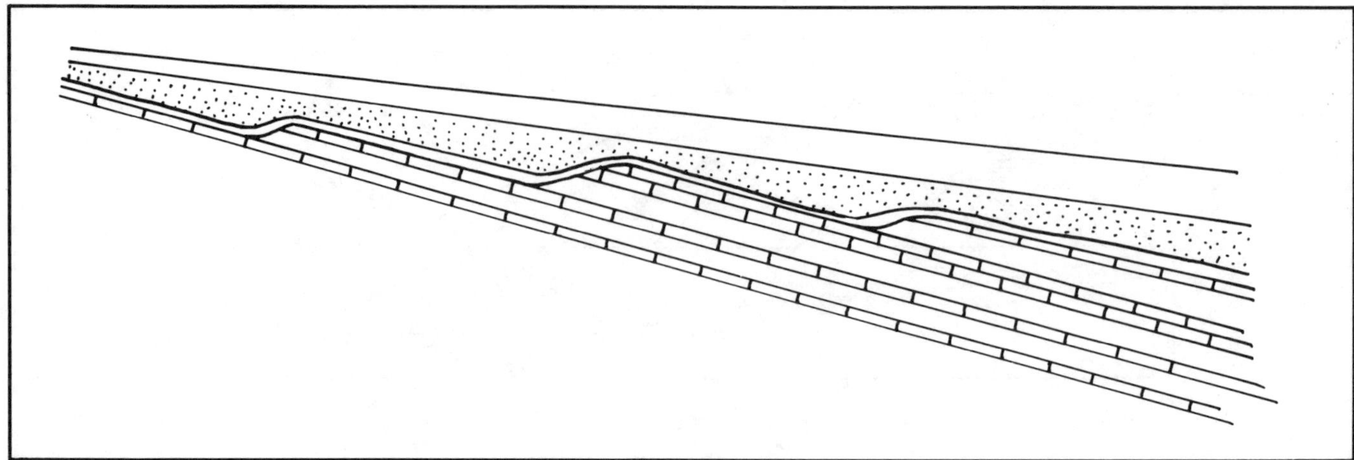

Fig. 5-51. Diagrammatic cross section of tilted and truncated alternately hard (limestone) and soft (shale) strata overlain by a sheet of sandstone that is flat on the top and escalloped on the base. Thicker portions of sandstone are due to compensatory deposition in topographically low areas.

marine sands by longshore currents after the bordering land mass is inundated.

The practical applications of reconstructing paleogeomorphology are discussed in detail by Martin (1966).

Figures 5-47 and 5-50 are quite diagrammatic and represent situations in which there is only a limited supply of sand. Thus, the sandstones are confined to topographically low areas between the escarpments. There are situations, however, in which an abundant supply of sand is available and the sand is not confined solely to the valleys. Such a case is illustrated diagrammatically in Figure 5-51 in which there was an abundant supply of sand under conditions of gradual marine transgression. The topographically low areas (strike valleys) were filled in first with sand that then spread across the escarpments so as to completely blanket the area. The upper surface of the sand is flat, whereas, the lower surface resembles a "wash board" upside down. Notice that a thin shale "break" occurs between the sand and the underlying unconformity.

Strike-valley sandstones are seldom isolated phenomena but generally occur in association with other types of fluvial sandstones. Figure 5-52 diagrammatically illustrates the rectangular drainage system that normally develops on a newly uplifted coastal plain where the underlying strata are alternately hard and soft and dip in a basinward direction. The first stream to develop on this newly uplifted coastal plain is the consequent stream. Its trend is essentially perpendicular to shoreline 1, and it increases in length by headward erosion. Its tributaries, subsequent streams, develop on the landward side of an escarpment (cuesta) and join the consequent stream approximately at right angles. *Obsequent* and *resequent* streams are tributaries that join the subsequent streams at nearly right angles.

The combination of these four streams constitutes a rectangular drainage system. As this systemmatically eroded surface is inundated in response to a gradual marine transgression, the base level of deposition of the consequent stream shifts in an upstream direction and becomes clogged with sand. The shoreline of sea level stage 2 is extremely irregular and occurs on both sides of the escarpment, resulting in extensive embayments on both sides of the consquent stream at the former sites of the two subsequent streams. It is in these embayments that strike-valley sands are deposited. At sea level stage 3 the escarpment of sea level stage 2 is completely inundated, and a second area of linear embayment occurs in subsequent valleys occurring in a more landward position. This is an ideal site for a second strike-valley sand to accumulate. These strike-valley sands are subparallel and merge at approximate right angles with the sand-filled channel of the consequent stream.

If two strike-valley pools that are miles apart are in the early development stages, it frequently is possible to ascertain the location of the consequent channel fill before the first well has been drilled into it. To do this it is necessary to construct a stratigraphic (longitudinal) profile of the strike valleys that join the consequent-stream channel at a common point, as shown in Figure 5-53. In this figure the subsequent-stream gradients are projected in opposite directions toward each other. The two streams have a common depth point where they flow into the consequent stream. It is through this point that the consequent-stream channel passes at right angles. If the strike-valley sandstones contain oil and/or gas, the probability is great that the consequent channel fill also will contain hydrocarbons. Acreage acquisition will involve a narrow trend normal to that of the strike-valley sandstones.

Diagrammatic illustrations, such as Figures 5-47, 5-50, 5-51, 5-52, and 5-53, fail to illustrate all of the stratigraphic complexities that are involved when one analyzes a rectangular drainage pattern. Mannhard and Busch (1974) made a detailed study of the Morrow (Penna) sandstones within a restricted area in southwestern Kansas and adjacent northwestern Oklahoma involving a rectangular drainage system. They analyzed a 16-township inlier area that embodies a complete rectangular drainage system. Figure 5-54 shows the location of the area as well as the position of six stratigraphic profiles (Figs. 5-55 and 5-56). Figure 5-54 illustrates two principal types of structures: (1) closures (such as the N. Buffalo field) caused by tectonism and (2) small scale noses and closures due to differential compaction of the shales within the Morrow section. In T27N-R21W and T28N-20W there is a well-defined southwest plunging structural nose on the Inola limestone. The structural expression of this feature, however, is several closures and southwest-plunging noses—all aligned in a southwesterly trend across these two townships. Most of the structures of the area are of the differential-compaction type. Many of these are subparallel southeast-plunging noses and northwest-southeast-trending closures. To analyze this structural relief, the Inola structure map (Fig. 5-54) was compared with the pre-Pennsylvanian topographic map (Fig. 5-57). The axes of structural lows on the Inola generally coincide with the axes of maximum Inola-to-Mississippian thickness (topographic lows) on the pre-Pennsylvanian paleotopographic map. Structural highs on the Inola, however, are above the axes of minimum thickness (buried hills). The Inola is downwarped along the axes of maximum (Morrow) sediment thickness as a result of the greater effect of compaction in these areas; consequently the Inola is draped over buried hills of the pre-Pennsylvanian erosional surface.

Mannhard and Busch state:

> In considering differential compaction it is assumed that the Mississippian sediments were lithified and compacted prior to Morrowan deposition. Cross sections show that the thicker sandstone section in the valley is accompanied by thicker shale sections, and by an increase in the number of shale beds deposited within the valley areas. In contrast, less shale was deposited above erosional remnants. Thus, there was more shale to compact in the valley areas than on erosional remnants. In addition, the individual sandstone beds in the valley-fill sec-

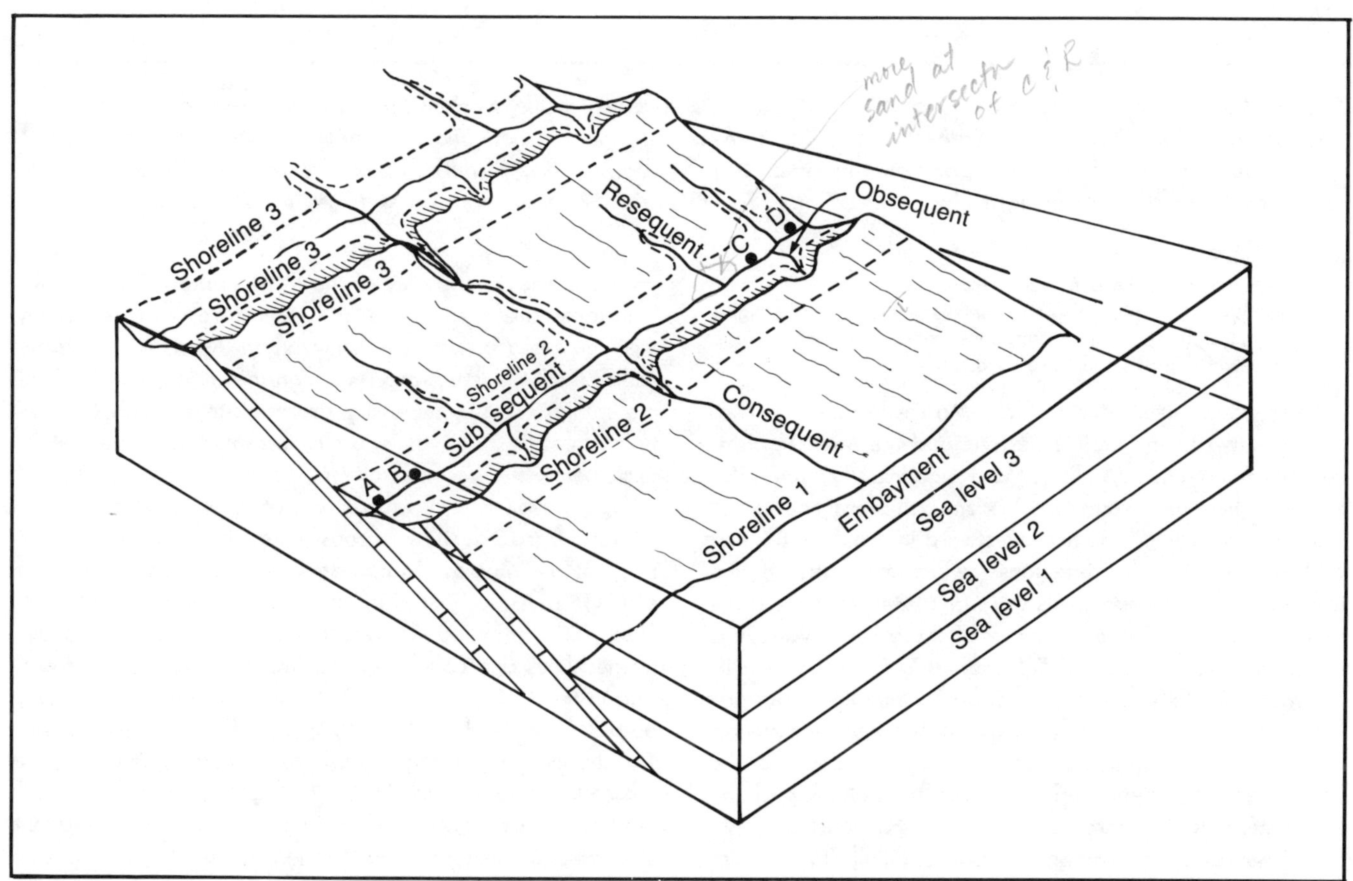

Fig. 5-52. Block diagram illustrating a rectangular drainage system developed on a newly uplifted coastal plain that is underlain by tilted, alternating hard and soft strata. With a gradual rise in sea level the **consequent** *and* **subsequent** *channels may become clogged with sand.*

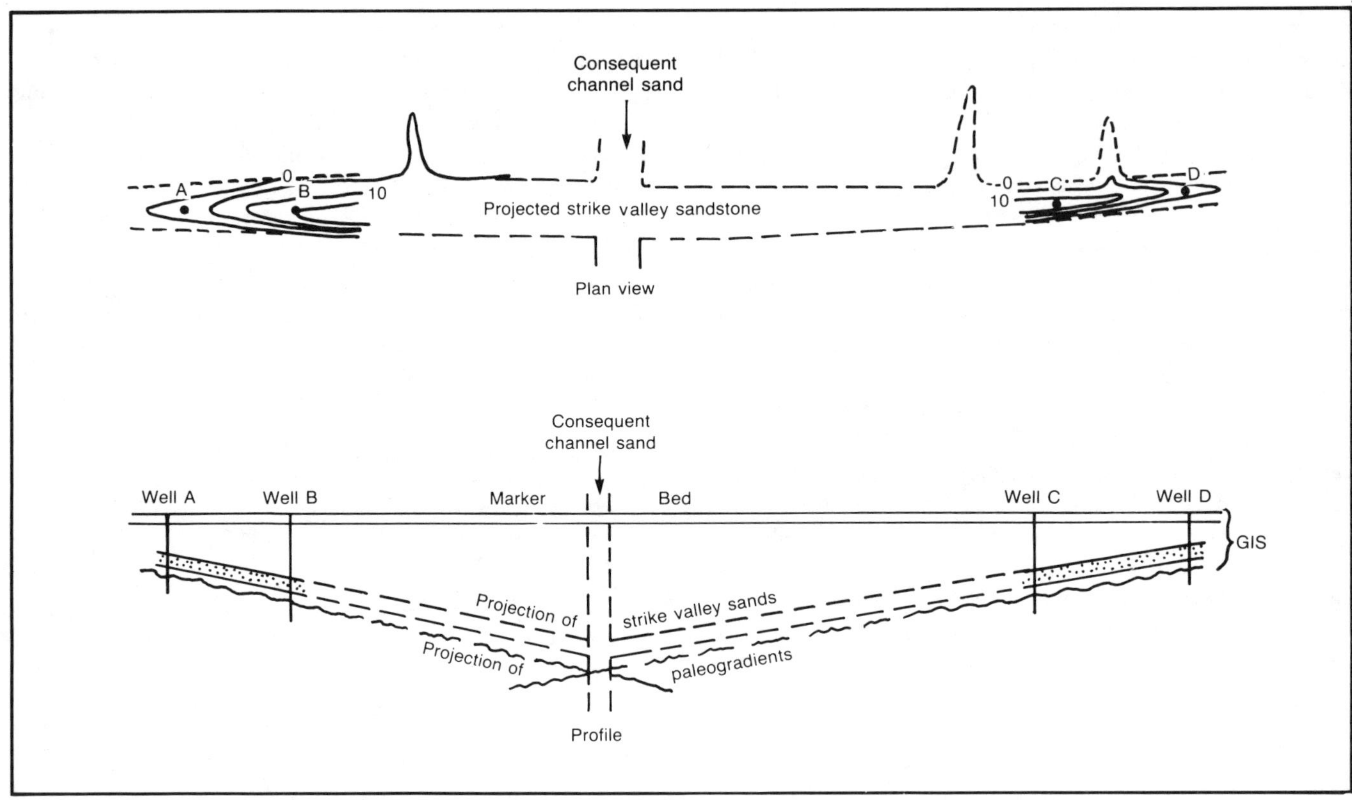

Fig. 5-53. Longitudinal stratigraphic profile along two **subsequent** *valleys, showing uniform rates of divergence in opposite directions (toward each other) of the GIS between the marker bed and the unconformity. The* **consequent** *channel fill occurs at the projected point of intersection of the opposing unconformity slopes. There is considerable vertical exaggeration of the* **subsequent** *channel gradients. See Figure 5-52 for location of wells A, B, C, and D.*

> tions generally have a higher shale content than "cleaner" sandstones which were spread across paleotopographic highs. As a result, the valley sandstones were slightly more 'compactible' than the sheet-like sandstones.

Figure 5-55 illustrates multiple Morrow sandstones occurring between the Inola limestone and the Mississippian-Pennsylvanian unconformity. More significantly, the GSS between these two reference horizons thickens southeastward in profile A-A' and northwestward in profile B-B'. This thickening in opposite directions toward each other is due to the fact that the sediments were deposited in strike (subsequent) valleys. The true nature of these strike valleys is well illustrated in Figure 5-56 (profiles D-D', E-E', and F-F'). Morrow sandstones were deposited initially in the paleotopographically low areas (valleys). Once these channels were filled with interbedded sand and shale, sheet-like sandstones were deposited over both the valleys and the escarpments that separated them. This is a good example of compensatory deposition in topographically low areas.

The key to correct interpretation of the Morrow formation in this area is the ability to identify the Mississippian-Pennsylvanian unconformity. In all of the cross sections of Figures 5-55 and 5-56 the Morrow rests on several different limestones and shales of the Mississippian. Some of the limestones are arenaceous and, therefore, can be readily misinterpreted as being Pennsylvanian (Morrowan) in age, rather than Mississippian. Thus, considerable core and sample studies were a necessary prerequisite to valid correlations. In such studies it is just as essential to correlate the beds below the unconformity as above.

Figure 5-57 is a reconstruction of the pre-Pennsylvanian paleotopography. It was constructed by drawing an isopach of the Morrow formation, which is a good example of a GSS. As such, it is a "cast" of this unconformity surface. Axes of paleodrainage courses are shown by heavy dashed lines and collectively illustrate a rectangular drainage pattern. The consequent stream, flowed southwesterly across T's 27 and 28N, R. 22W into the Anadarko Basin. The subparallel northwest-southeast trending tributaries are subsequent streams, all filled with sandstone and shale beds. Southwestward flowing streams flowing into the subsequent streams are resequents that eroded down the dip slopes of the Mississippian limestones and shales. Several very short obsequent streams (which flowed down the scarp slopes) flowed northeastward into the subsequent streams.

Mannhard and Busch state:

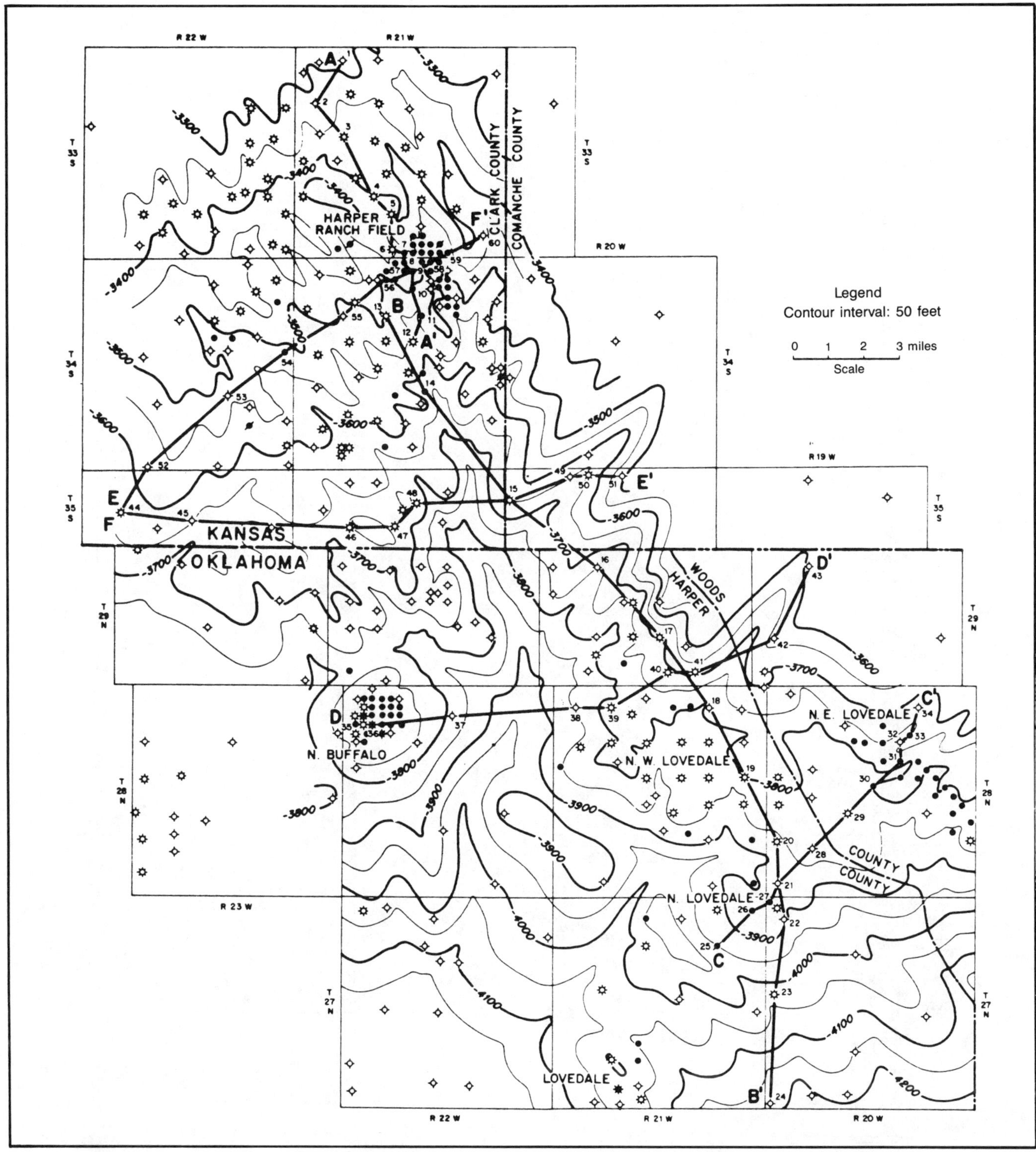

Fig. 5-54. Structure map of the Inola limestone in southcentral Kansas and adjacent northwestern Oklahoma, showing locations of six stratigraphic cross sections. (Modified after Mannhard and Busch, 1974; permission to publish by AAPG).

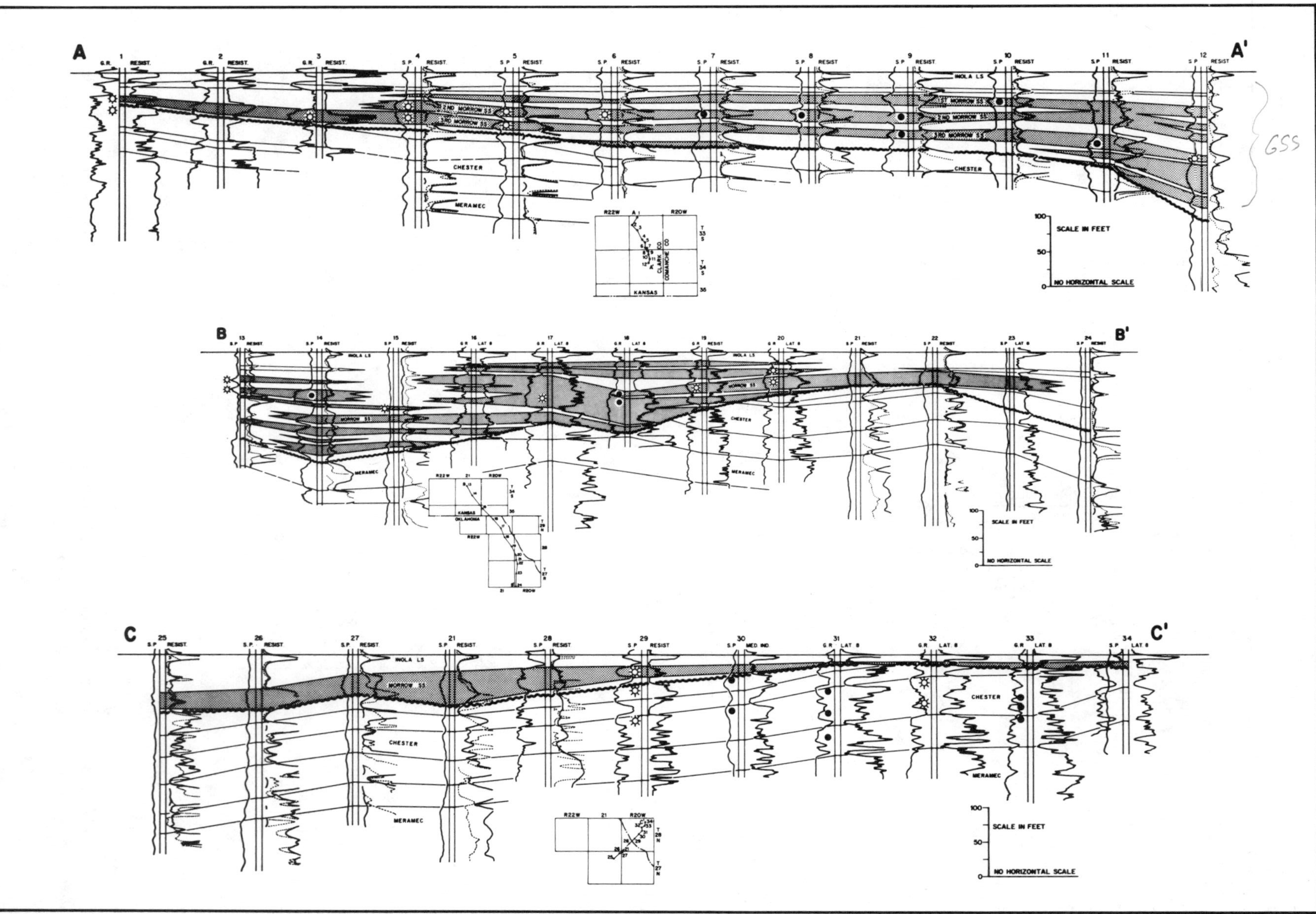

Fig. 5-55. Northwest-southeast-trending stratigraphic profiles A-A′ and B-B′, with the top of the Inola limestone as reference datum; no horizontal scale. For location of sections see Figure 5-54. (Modified after Mannhard and Busch, 1974; permission to publish by AAPG).

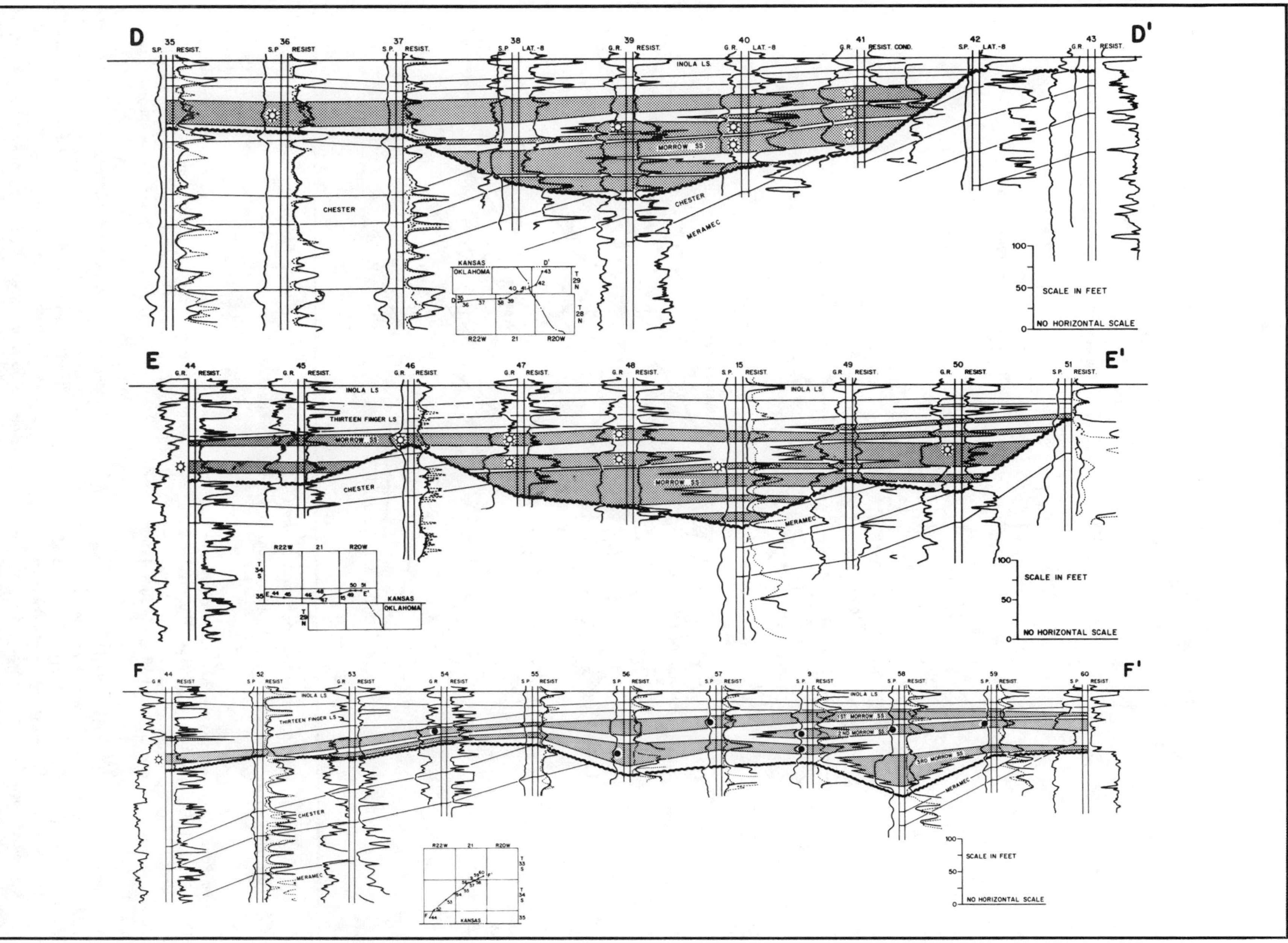

Fig. 5-56. Southwest-northeast-trending stratigraphic profiles C-C′, D-D′, E-E′, and F-F′, with the top of the Inola limestone as the reference datum; no horizontal scale. For location of sections see Figure 5-54. (Modified after Mannhard and Busch, 1974; permission to publish by AAPG).

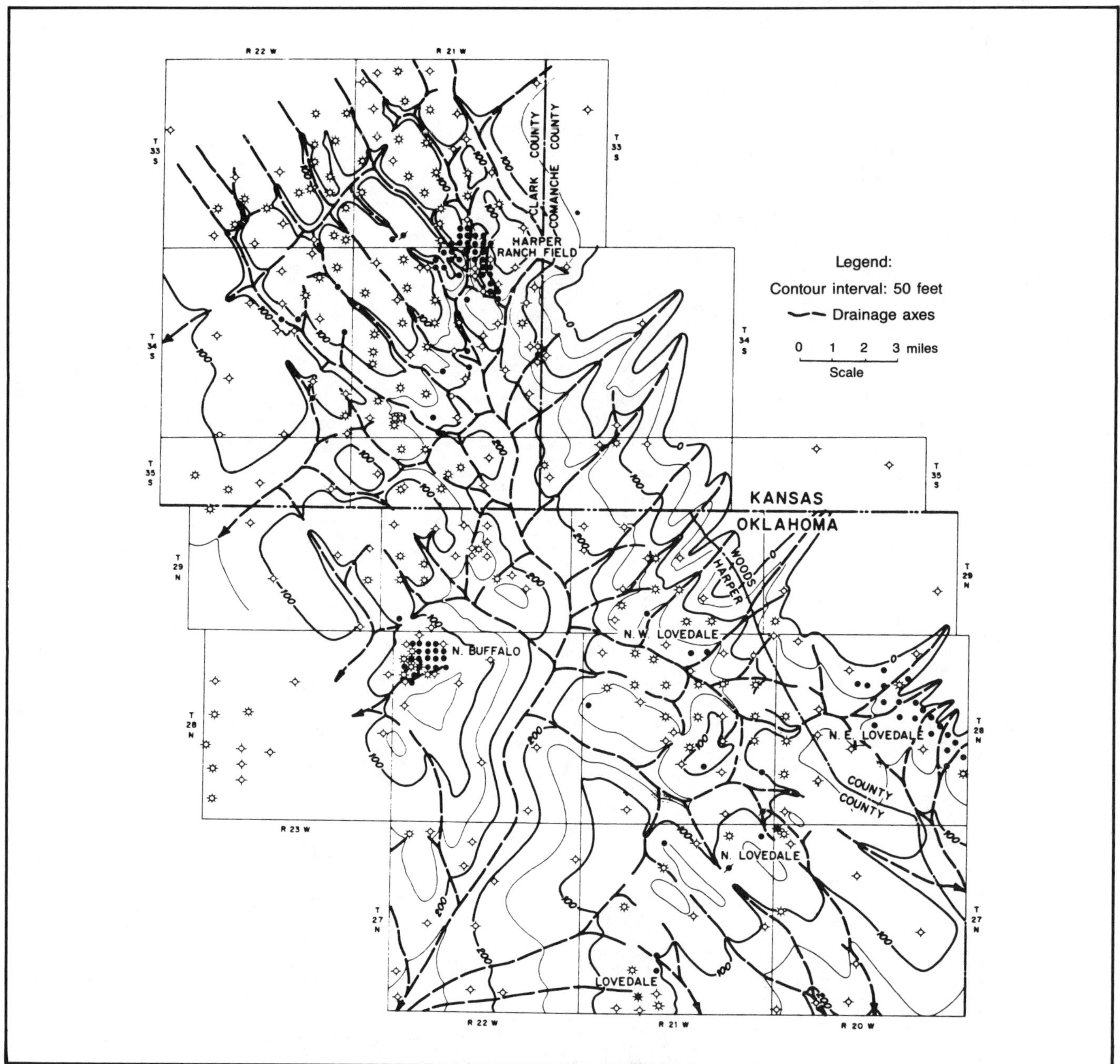

Fig. 5-57. Pre-Pennsylvanian topography (isopach map of the GSS from the top of the Inola limestone to the Mississippian unconformity). (Modified after Mannhard and Busch, 1974; permission to publish by AAPG).

The Harper Ranch field straddles the contiguous boundaries of T33, 34S, R21W, Clark County, Kansas. The discovery well, the United 1 Harper, in Sec. 9, T34S R21W, was completed in a Morrow sandstone for 6 MCF of gas in June 1953 (Waite, 1956).

Morrow sandstones of the Harper Ranch field have been zoned vertically into the First, Second, and Third Morrow sandstones, as shown on cross sections AA′ (Fig. 5-55) and FF′ (Fig. 5-56). The Second Morrow sandstone has the most favorable reservoir characteristics and the most widespread distribution. For this reason the Second sandstone accounts for more than 90 percent of the production within the area. Porosity calculations (from mechanical logs) range up to a maximum of 27 percent. On the basis of core analysis the porosity ranges from 10 to 26 percent and the permeability ranges from 25 to 350 md (Waite, 1956).

By detailed mapping an attempt was made to arrive at a logical explanation for each dry hole and producing well in the Second sand. Similar procedures should be applicable to the other Morrow sandstones within the study area.

Figure 5-58 is an isopach map of the interval from the top of the Inola limestone to the base of the Second sandstone. Such

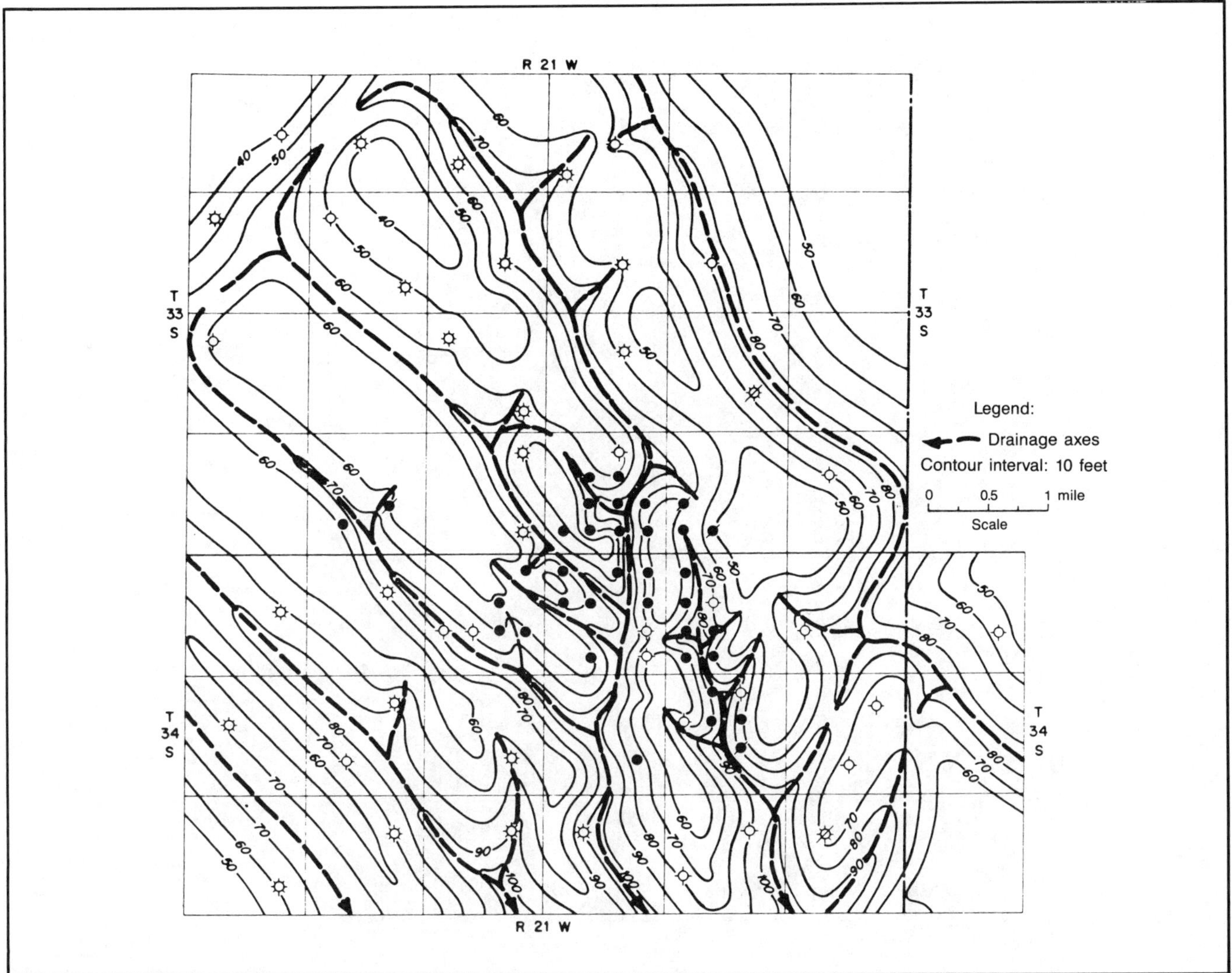

Fig. 5-58. Isopach map of the GIS from the top of the Inola limestone to the base of the Second Morrow sandstone. (Modified after Mannhard and Busch, 1974; permission to publish by AAPG).

a map provides a simulated reconstruction of the depositional surface on which the Second sandstone was deposited. The map shows an undulatory surface which is essentially a subdued replica of the pre-Pennsylvanian topography. There is a series of elongate, subparallel, southeast-trending ridges and intervening valleys. The surface probably was not exposed to subaerial erosion for any significant period of time, for the cross sections show no evidence of channeling into the shale section separating the Second and Third sands. The axes of maximum thickness on this map generally coincided with the axes of drainage courses developed on the pre-Pennsylvanian erosion surface. This is attributed to the fact that many pre-Pennsylvanian valleys were not filled entirely by compensatory deposition when the Second sand was deposited, and consequently the positions of the original valleys continued to be the sites of maximum sediment fill.

Figure 5-59 is an isopach map of the Second Morrow sandstone. The reliability of the Second sandstone isopach map was enhanced greatly by making a paleodrainage map first (Fig. 5-58). Maximum Second sandstone development is expected along the axes of paleodrainage.

An important aspect of the isopach map (Fig. 5-59), insofar as oil and gas accumulation is concerned, is the location of lines of zero thickness. Such lines represent the lateral limits of effective sandstone permeability.

Figure 5-60 is a structure map of the Second Morrow sandstone on which appropriate symbols have been superimposed to show both the known and probable distribution of oil, gas, and water. The zero sandstone thickness line was traced from the isopach map of the Second sandstone (Fig. 5-59).

Structural contours drawn on top of the Second Morrow sandstone bend in an upstream direction along the axes of maximum thickness (Inola to base of the Second sandstone interval) rather than in a downstream direction. This suggests that the Second sandstone was downwarped in response to compaction within the underlying section. Maximum compaction, and hence maximum downwarping, occurred over buried pre-Pennsyl-

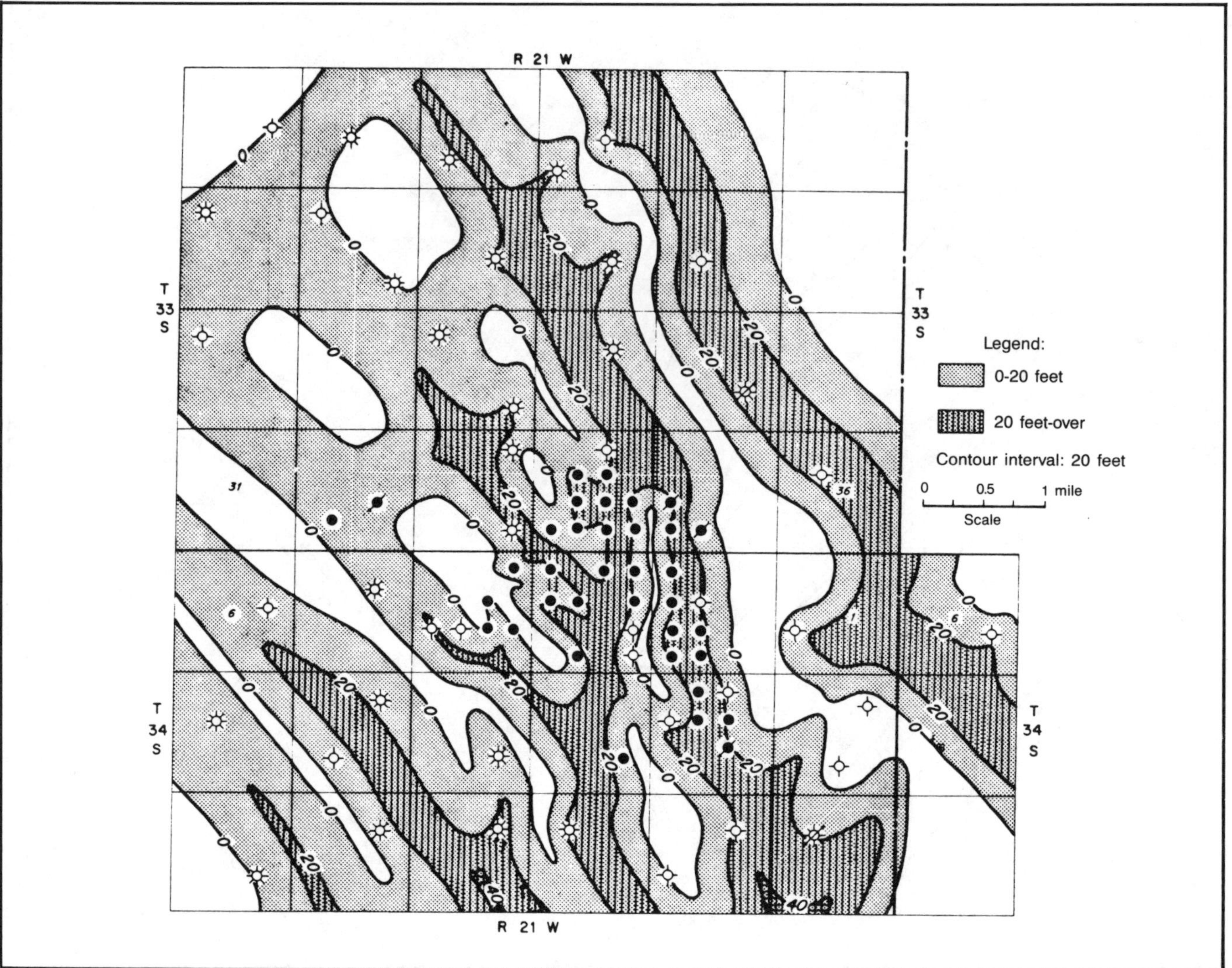

Fig. 5-59. Isopach map, Second Morrow sandstone. (Modified after Mannhard and Busch, 1974; permission to publish by AAPG).

vanian valleys which received relatively greater thicknesses of compactable shale than did surrounding paleotopographically high areas.

When production data for the Second sandstone are added to the structure map, it is apparent that the distribution of oil and gas is essentially independent of the structure. By superimposing the lines of zero-sandstone thickness from the Second Morrow sandstone isopach map (Fig. 5-59) onto the structure map, the necessary trapping mechanism for the oil and gas is provided. The updip wedge-outs and shale-outs of the Second sandstone served as effective barriers to the lateral migration of hydrocarbons. Vertical escape of hydrocarbons was prevented by the impermeable shale deposited on the Second sandstone.

In two areas (Fig. 5-60) gas is being produced from the Second Morrow sandstone at structurally lower elevations than the gas pool in the northwestern portion of T.33S, R.21W. One such area is in the southwestern part of the mapped area and includes parts of Secs. 7-9, 16-18, T34S, R21W. A much smaller area (occupied by a single abandoned gas well) is in the southeastern part of the mapped area and includes parts of Secs. 12, 13, T34S, R21W. The anomalous areas of gas production are separated by a larger updip trap with normal reservoir conditions in which gas is present at structurally higher elevations than oil, and oil at structurally higher elevations than water. The anomalous presence of gas in the downdip areas can be explained by Gussow's principle (Gussow, 1953), modified to apply to a stratigraphic trap. It is assumed that the downdip traps which now are producing only gas once were filled partly or entirely with oil. As gas migrated into these areas the oil was forced out at the bottom (spill points) of the traps and moved into the central updip trap. The positions where the oil escaped from the downdip traps into the updip trap are at the most downdip edges of the respective permeability barriers which separate the individual downdip traps from the central updip trap. The locations of the two spill points are shown on the structure map (Fig. 5-60). As gas entered the downdip traps all of the oil eventually was forced updip at the two spill points. The gas then began to spill out of the downdip traps and partly displaced the oil which had accumulated in the central updip trap.

Another seemingly anomalous situation exists along the eastern margin of the area. Wells in Secs. 15, 23, and 36, T33S, R21W, have tested salt water in the Second Morrow sandstone updip from oil and gas producing areas in the Second sandstone. The single abandoned gas well in Sec. 26 is anomalous and could result from a local lateral extension of the gas-producing area on the southwest. This condition suggests the presence of a permeability barrier restricting the migration of hydrocarbons into the area of salt-water accumulation. The existence of such a permeability barrier is confirmed by a gradual shaling out of the Second sandstone in wells at the northeast margin of the oil- and gas-producing area. The Second sandstone is completely absent in the two wells of Sec. 12, T34S, R21W. These are in the zone separating the oil- and gas-producing area from the area of water accumulation.

Shale-Filled Channels

In the previous discussion of channels the entire emphasis is on reservoir sandstones that make up a part, or all, of the channel fills. Channel fills, however, may consist of any type of sedimentary lithology. When, for example, a drainage system is filled with shale, there obviously is no reservoir potential within the individual channels. However, in instances of a dissected upland plateau, capped by a resistant sheet-like sandstone, the divides between the drainage courses may serve as reservoirs for hydrocarbons. Such a situation is described and illustrated by Nolte (1967) for a highly productive area in southwestern Saskatchewan, Canada. Figure 5-61 is a structural map of the Jurassic-Cretaceous unconformity in the Success Field. The three oil pools all produce from the Vanguard (Upper Jurassic) sandstone and occur in upland divide areas between a dendritic system of tributaries. The drainage courses (shaded) are filled with nonmarine lower Blairmore (Cantaur) carbonaceous shales, siltstones and lenticular clay-filled sandstones. These sediments are, for the most part, impermeable and serve as a seal for the underlying, middle-Vanguard reservoir sandstone. Nolte (1967) states that "the field is

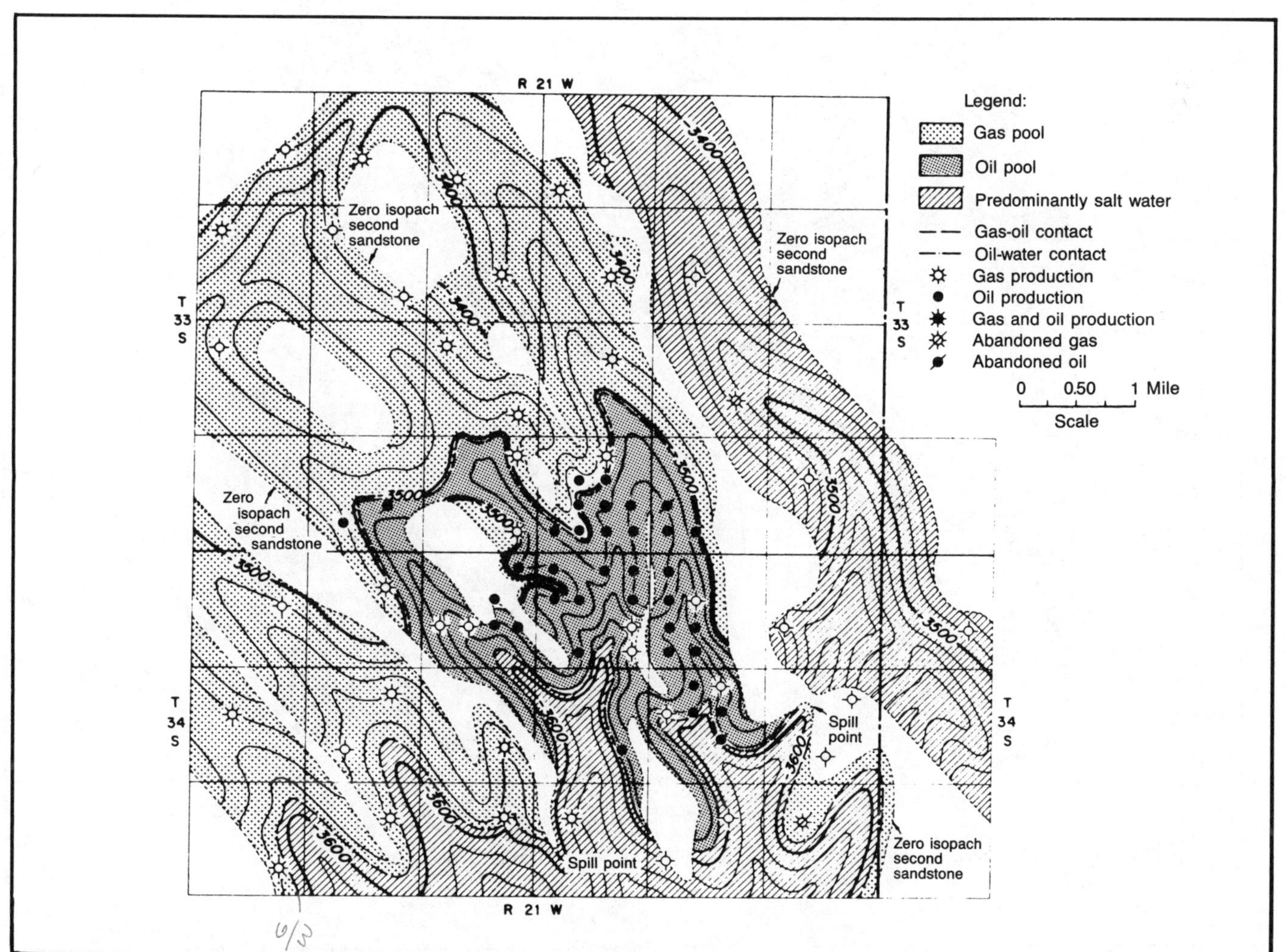

Fig. 5-60. Structure map, Second Morrow sandstone, showing anomalous oil and gas distribution; contour interval 20 ft. (Modified after Mannhard and Busch, 1974; permission to publish by AAPG).

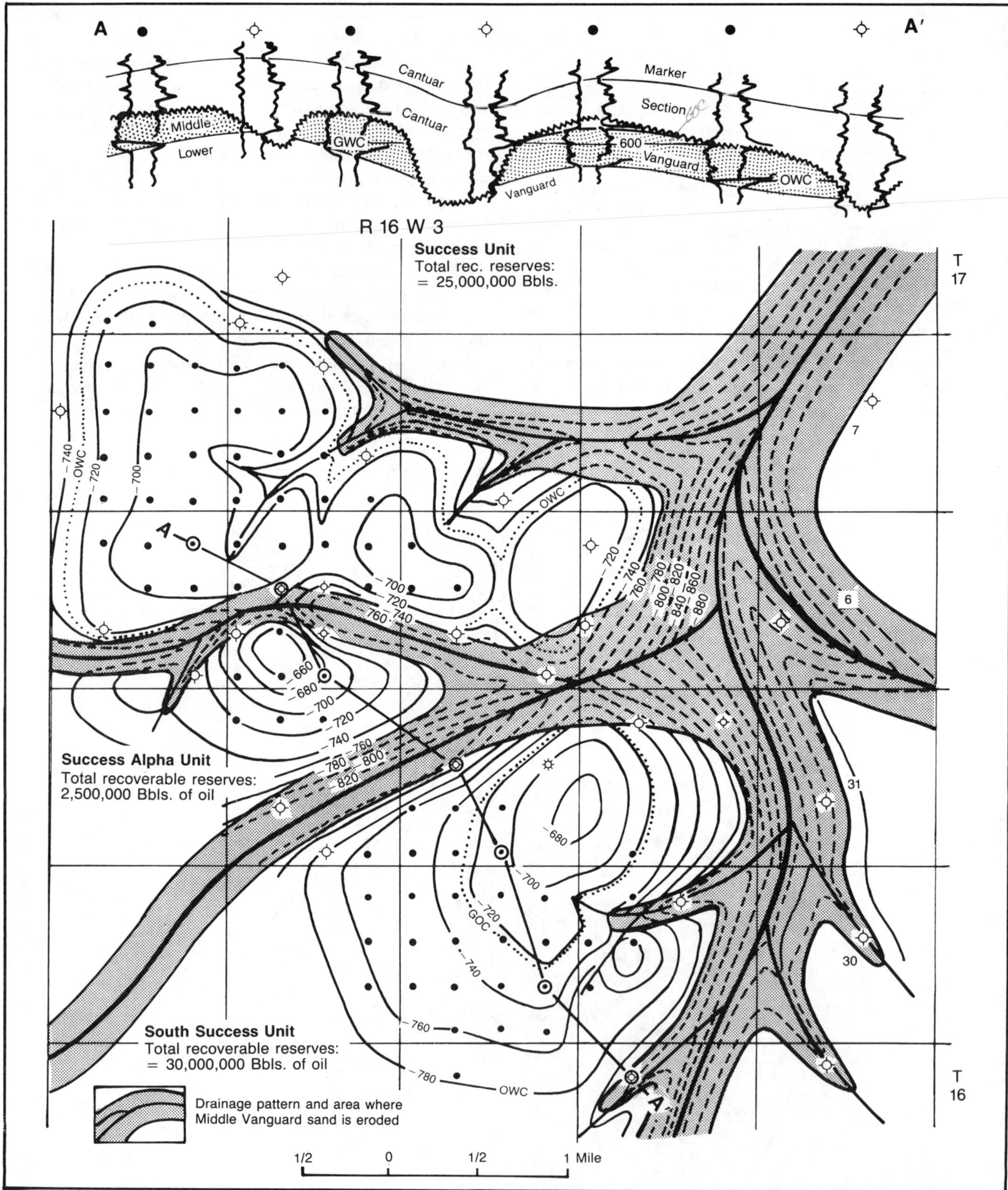

Fig. 5-61. Structure map of Jurassic unconformity at base of Cretaceous (Blairmore) strata, in the Success Field area, southwest Saskatchewan. Shaded area consists of impervious channel fill. (Modified after Nolte, 1967; permission to publish by Rocky Mountain Assoication of Geologists).

divided into three separate reservoirs with different oil-water and gas-oil interfaces. These reservoirs are separated by deep channels which cut through the Middle Vanguard sandstone and were subsequently filled with impervious Blairmore material." He notes that "the structure contours on the top of the surface of unconformity are primarily a function of topography rather than structure."

It is apparent to the authors that the Cantaur section of the basal Cretaceous Blairmore (see profile A-A' of Fig. 5-61) is truly a GIS. Thus, an isopach of this interval will present a "cast" of the Jurassic unconformity. The necessary data to construct such a map (Fig. 5-62) were loaned to the authors for this purpose. Figures 5-61 and 5-62 are quite similar, the principal difference being in the South Success unit. It is here that the structural relief is considerably greater than the paleotopographic relief. In fact, the topographic high is almost flat on top. The striking similarities of these two maps are attributed to minor post-Cretaceous tectonism in the immediate area. In areas where there has been considerable structural deformation either prior to or simultaneous with erosion, the GIS isopach map of the strata overlying the unconformity will present a much more accurate map of paleodrainage than a structure map of the unconformity.

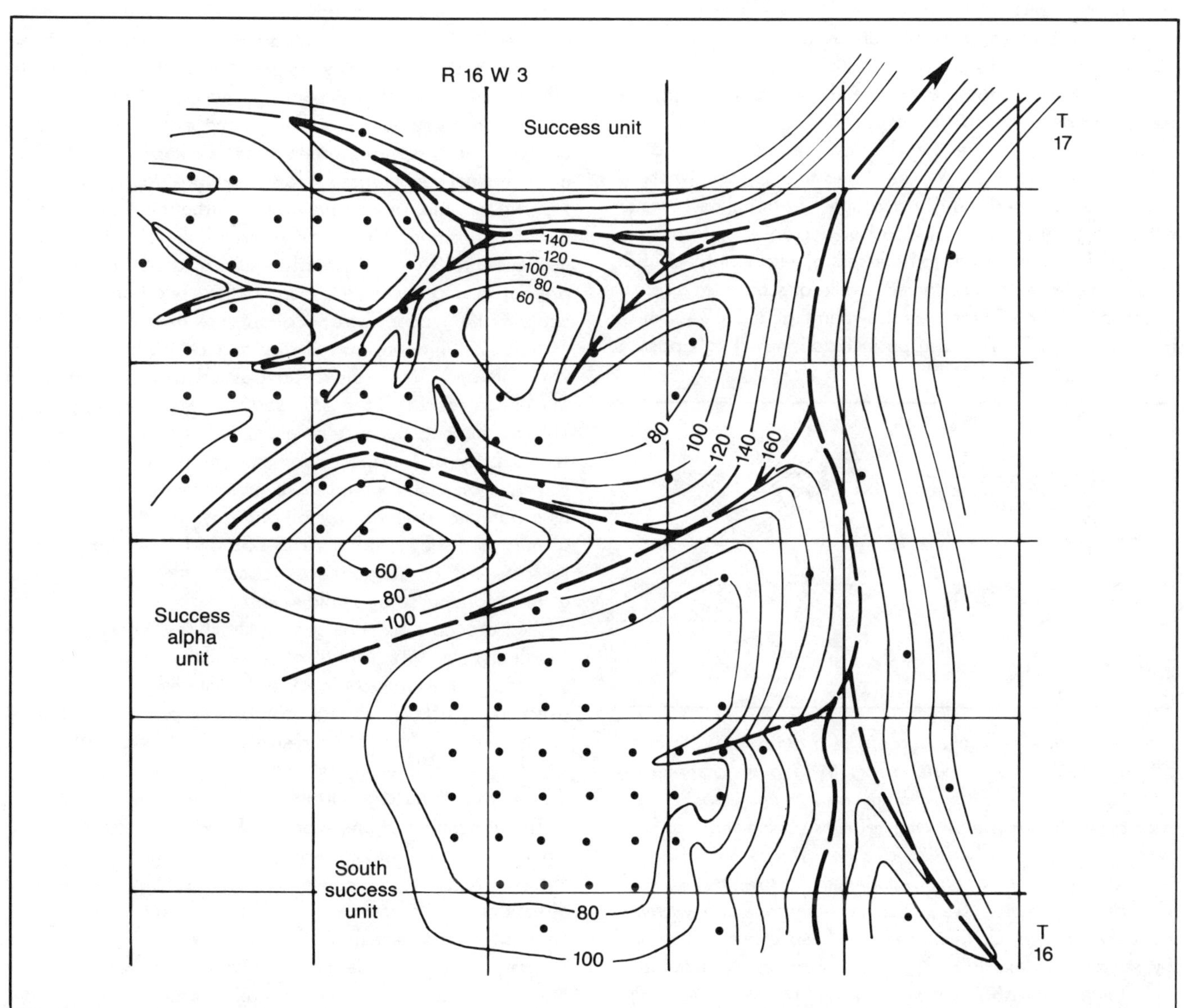

Fig. 5-62. Isopach map of the impervious shale and siltstone of the Cantaur section shown in profile A-A' of Figure 5-61. This genetic increment of strata, when isopached, presents a "cast" of the Jurassic unconformity that occurs at the base.

Internal Bedding

It has been pointed out that there is considerable variability in the types of sediment that might be deposited within a channel. In fact, practically every type of sedimentary rock, except evaporites, may make up the lithology of a channel fill. Grain size, degree of sorting, and cross-stratification are almost as variable as the lithology. In some flume experiments performed by McKee (1957), some information has been gained on the factors influencing stratification. His experiments involved the deposition of fill in channels cut by streams and submarine (submerged) currents. It was noted that ". . . channels cut by streams tend to be flat bottomed, steep sided to straight-walled, and shallow." Those cut by submarine currents ". . . are semi-circular in cross section as a result of constant slumping of the walls under water."

On the basis of these experiments, considerable significance can be attached to the position of water level as a determinative factor in the development of cross-stratification in a channel fill. McKee shows that, ". . . where a stream deposits sediment in a channel bottom, as a result either of increase in stream load or decrease in stream velocity, the layering tends to be essentially horizontal so that it conforms to the flat-bottomed profile of scour in a stream-cut channel." This situation is shown in Figure 5-63A, which is the tracing of a photograph included in McKee's paper.

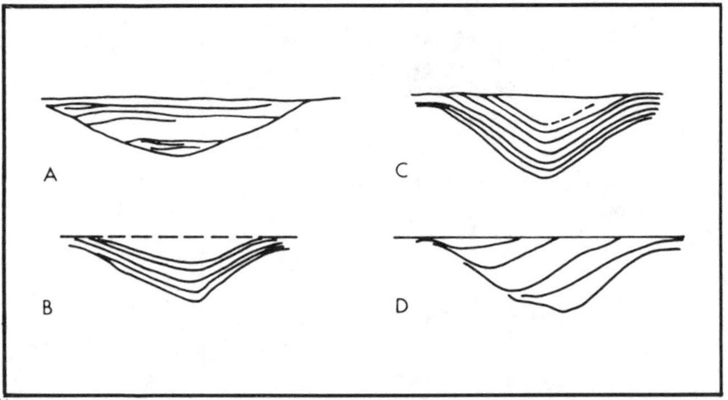

Fig. 5-63. Tracings of photographs of stratification of channel fill produced in flume experiments, **A**. *Trough cross-stratification formed by filling channel by stream deposits. Surface of water was below channel rims during most of process;* **B**. *Symmetrically-filled channel. Thickening of strata toward bottom of trough. Channel cut by stream and modified by rising water level; filled by submarine current moving down trough;* **C**. *Symmetrically-filled channel. Scoured by stream and modified by rising water level. Strata due to settling from above in quiet water;* **D**. *Asymmetrically-filled channel. Channel cut by stream and modified by rising water level; filled by submarine current moving in direction diagonal to channel.* (Modified after McKee, 1957; permission courtesy U.S. Geological Survey).

In a competely water-filled channel, the layering of channel deposits may be concave upward. McKee states that ". . . varying degrees of curvature result from differences in depth of water in the channel in which deposition occurs." This latter situation is illustrated in Figure 5-63B, in which the individual strata thicken toward the axis of the channel fill.

In another experiment, McKee allowed sediment to settle vertically through a column of quiet water occupying an eroded channel. In this case (Fig. 5-63C) there is very little thickening of the individual strata toward the axis. Being more or less of equal thickness, the strata tend to restrict the width and depth of the channel progressively as sedimentation proceeds. In a last experiment involving channel fill, McKee caused a submarine current to pass diagonally over the channel, and the result was a very asymmetric channel fill (Fig. 5-63D). Such asymmetry of the cross-stratification might be expected to develop on the inside of submarine meanders of a stream around growth structures.

In order to relate laboratory flume experiments to natural conditions, several points should be kept in mind. First, most channel sandstones (except for those deposited in deltas and around submarine growth structures) are of subaerial deposition. Thus, variations of the situation shown in Figure 5-63A should apply where cross-stratification is present in a channel fill. Second, the conditions set up to achieve the type of stratification shown in Figure 5-63C would result in deposition of clay and silt, rather than sand. Only silt and clay could be deposited from suspension in the upper reaches of a *still* body of water occupying an eroded channel, because a velocity of one-half to one knot is necessary to transport sand-size material. Thus, the resulting channel fill would have no commercial significance as a potential reservoir for oil and gas. Third, stratification of the types shown in Figures 5-63B and D will be confined to some deltaic distributaries (but not the bar-finger type of sandstone) and the shallow-water channel sandstones deposited around submarine growth structures. In both of these situations, the subjacent strata are likely to be clays and silty clays that will compact differentially in response to overburden. Thus, any depositional dip of the channel sandstones is likely to be distorted (flattened out) to the extent that the laterally adjacent shales are compacted by the weight of the overburden.

The direction and angle of tilt of bedding, as determined from a dipmeter, should never be considered as diagnostic criteria when working with channel sandstones. The tilts and directions of tilt are much too variable and, also, other types of sandstone exhibit the same variability.

Many channel sandstones exhibit a biconvex profile in the subsurface due to differential compaction of the laterally adjacent shales. Thus, dipmeter readings taken in the shale directly above a channel sandstone in many cases exhibit low angles of tilt in opposite directions away from the axis of channel fill.

SUMMARY OF CHANNEL SANDSTONE CRITERIA

Many geologists tend to employ only one or two criteria in tracing a channel sandstone in the subsurface. For example, the sand body might appear to be elongate and lenticular. There are other types of sandstone that also are elongate and lenticular, i.e., barrier bars, offshore bars, cheniers, tidal current ridges, sand waves, etc. Thus, it is paramount that multiple criteria be employed in working with subsurface channel sandstones.

Geometric Criteria

Although channel sandstones are linear and lenticular these criteria are, at best, somewhat superficial. The linearity should be related to the trend of the contour lines which represent its GIS. In most instances the linearity is normal to the GIS contours that represent the shoreline trends resulting from either marine transgression or regression. This is predicated on the fact that many streams take the shortest (steepest gradient) route to the sea. The GIS contours will exhibit a pronounced reentrant angle in an upstream (landward) direction.

The one criterion that is almost diagnostic of a channel fill is downward thickening. This phenomenon can be noted most readily by constructing several stratigraphic cross sections at right angles to the linear trend, all "hung" on the same time marker *above*. In so doing the usual shale section between the channel sandstone and the marker bed maintains a fairly uniform thickness. Slight thinning of this shale above the thickest portion of the channel sandstone is not unusual. When this does occur it is due to very early differential compaction of the clay-shales laterally adjacent to the channel sandstone. This causes a slight ridge effect on the sea floor directly above the sandstone and slight lateral creep of the muds (above the sand) away from the thickest part of the sand.

Internal Criteria

It is well established that flattened pebbles in the basal portion of a channel fill have a preferential tilt upstream. Also, elongate sand grains are preferentially oriented parallel with the long axis of the channel sandstone. Dip measurements on outcropping channel sandstones "point" in the prevailing direction(s) of sediment transport. Although all three of these criteria are of real significance in studying outcrops, they are of little value when working in the subsurface, because of the paucity of oriented cores. Rotary cuttings are of no value in determining these criteria. Thus, any attempt to use these well-established internal features when working in the subsurface is considered to be sedimentological overkill.

Unoriented cores, rotary cuttings, and mechanical logs are the only tools available in studying channel sandstones. From this combination of data it is possible to recognize a decrease in grain size upward. Many subaerial channel sandstones exhibit this aspect. It is not, however, diagnostic. Some subaerial channel sandstones are fine-grained from top to bottom. Deltaic distributary sandstones also frequently fail to exhibit a "fining" upward. The decrease in grain size upward, however, is a characteristic feature of point bars.

Close inspection of rotary cuttings frequently reveals a sandstone body consisting of 85%± of quartz sand and 15%± of rock fragments. The presence of rock fragments is indicative of a comparatively low energy environment (such as a fluvial channel) and, thus, rules out barrier bar, offshore bar, and chenier possibilities. Scattered carbonized plant remains frequently may be observed in cores, but they also may be noted in barrier bars.

E-logs and gamma logs frequently are used in identifying channel sandstones. When they have a bell-shaped profile they certainly are suggestive of a channel sandstone—but are not diagnostic. Such curves are due to both a decrease in grain size upward and an increase in siltstone and shale content in the same direction. Many, if not most, deltaic distributaries fail to exhibit a bell-shaped profile and have an equally abrupt basal and upper limit, indicating good potential reservoir sandstone throughout.

As indicated, the nearest thing available to a diagnostic criterion for a subsurface fluvial sandstone is downward thickening. In any study of a channel sandstone, multiple criteria should be employed. There is a synergetic relationship of criteria; that is, as the number of criteria increases, the degree of certainty increases exponentially.

SELECTED BIBLIOGRAPHY

Allen, J. R. L., 1965, A review of the origin and characteristics of recent alluvial sediments: Sedimentology (spec. issue), v. 5, no. 2, 191 p.

Andresen, M. J., 1961, Geology and petrology of the Trivoli sandstone in the Illinois Basin: Illinois Geol. Survey Circ. 316, 31 p.

_______ 1962, Paleodrainage patterns: their mapping from subsurface data, and their paleogeographic value: AAPG Bull., v. 46, no. 3, p. 398-405.

Asquith, D. O., 1966, Geology of late Cretaceous Mesaverde and Paleocene Fort Union oil production, Birch Creek Unit, Sublette County, Wyoming: AAPG Bull., v. 50, no. 10, p. 2176-2184.

Berg, R. R., 1967, Point bar origin of Fall River sandstone reservoirs, northeastern Wyoming: SPE of AIME, Paper No. 1953, 6 p.

Bernard, H. A. and C. F. Major, Jr., 1963, Recent meander belt deposits of the Brazos River: an alluvial "sand" model (abs.): AAPG Bull. v. 47, no. 2, p. 350.

———, ———, B. S. Parrot, and R. J. LeBlanc, Jr., 1970, Recent sediments of southeast Texas, a field guide to the Brazos alluvial and deltaic plains and the Galveston barrier island complex: Texas Univ. Bur. Econ. Geology Guidebook 11, 16 p.

Beutner, E. C., L. A. Gleckinger, and T. M. Gard, 1967, Bedding geometry in a Pennsylvanian channel sandstone: GSA Bull., v. 78, no. 7, p. 911-916.

Bloomer, R. R., 1977, Depositional environments of a reservoir sandstone in west-central Texas: AAPG Bull., v. 61, no. 3, p. 344-359.

Bolyard, D. W., and A. A. McGregor, 1966, Stratigraphy and petroleum potential of Lower Cretaceous Inyan Kara Group in northeastern Wyoming, southeastern Montana, and western South Dakota, AAPG Bull., v. 50, p. 2221-2244.

Brown, L. F., Jr., 1969, Geometry and distribution of fluvial and deltaic sandstones (Pennsylvanian and Permian) north-central Texas: Gulf Coast Assoc. Geol. Socs. Trans, v. 19, p. 23-47.

Buckley, A. B., 1922-23, The influence of silt on the velocity of flowing water in open channels: Inst. Civil Engineers Proc., v. 216, p. 183-211.

Burnham, W., 1956, Personal commun. and unpub. report of W. Burnham.

Busch, D. A., 1959, Prospecting for stratigraphic traps: AAPG Bull., v. 43, no. 12, p. 2829-2843.

———, 1963, Methods of prospecting for stratigraphic oil and gas traps: Assoc. Francaise Tech. Pétrole Bull. 160, pt. 1, p. 459–464; Bull. 161, pt. 2, p. 633–643.

———, 1974, Stratigraphic traps in sandstones— exploration techniques: AAPG Mem. 21, 174 pp.

Cant, D. J., and R. G. Walker, 1976, Development of a braided-fluvial facies model for the Devonian Battery Point Sandstone, Quebec: Can. Jour. Earth Sci., v. 13, no. 1, p. 102-119.

Coleman, J. M., 1969, Brahmaputra River: channel processes and sedimentation: Sed. Geol., v. 3, no. 2-3, p. 131-239.

Charles, H. H., 1941, Bush City oil field, Anderson County, Kansas, *in* A. I., Levorsen, ed., Stratigraphic type oil fields, a symposium: Tulsa, AAPG, p. 43-56.

Collinson, J. D., 1978, Vertical sequence and sand body shape in alluvial sequences, *in* A.D. Miall, ed., Fluvial sedimentology: Canadian Soc. Pet. Geologists Mem. 5, p. 577-586.

Conybeare, C. E. B., 1976, Geomorphology of oil and gas fields in sandstone bodies: New York, Elsevier Pub. Co., 341 p.

Das, I., 1950, Theory of the flow of water and universal hydraulic diagrams: Jour. Central Board Irrigation (India), v. 7, no. 2, p. 151-162.

Ebanks, W. J., Jr., and J. F. Weber, 1982, Development of a shallow heavy-oil deposit in Missouri: Oil and Gas Jour. (Sept. 27), p. 222-234.

Eckelman, W. R., R. J. DeWitt and W. L. Fisher, 1975, Prediction of fluvial-deltaic reservoir geometry, Prudhoe Bay Field, Alaska: Proc., 9th World Petroleum Congress, Tokyo, Panel Discussion 4 (2), p. 1-5.

Einstein, H. A., 1950, The bed-load function for sediment transportation in open channel flow: USDA Tech. Bull. 1026.

Eisenstatt, P., 1960, Little Creek field, Lincoln and Pike Counties, Mississippi: Gulf Coast Assoc. Geol. Soc. Trans., v. 10, p. 206-213.

Fisk, H. N., 1944, Geological investigation of the alluvial valley of the lower Mississippi River: Vicksburg, Mississippi, U.S. Army Corps of Engineers, Mississippi River Comm.

———, 1947, Fine-grained alluvial deposits and their effects on Mississippi River activity: Vicksburg, Mississippi, U.S. Army Corps of Engineers, Mississippi River Comm.

Frazier, D. E., and A. Osanik, 1961, Point-bar deposits, Oil River locksite, Louisiana: Gulf Coast Assoc. Geol. Socs. Trans. v. 11, p. 121-137.

Friedman, S. A., 1960, Channel-fill sandstones in the middle Pennsylvanian rocks of Indiana: Indiana Geol. Survey Rept. Prog., no. 23, 59 p.

Gilbert, G. K., 1914, Transportation of debris by running water: U.S. Geol. Survey Prof. Paper 86, 263 p.

Hartman, J. A., 1972, "G" Channel Sandstone, Main Pass Block 35 Field, Offshore Louisiana: AAPG Bull., v. 56, no. 3, p. 554-571.

Herbaly, E. L., 1974, Petroleum geology of Sweetgrass Arch, Alberta: AAPG Bull., v. 58, no. 11, p. 2227-2244.

Horton, R. E., 1945, Erosional development of streams and their drainage basin—hydrophysical approach to quantitative morphology: GSA Bull., v. 56, no. 3, p. 275-370.

Harms, J. C., D. B. MacKenzie and D. G. McCubbin, 1963, Stratification in modern sands of the Red River, Louisiana: Jour. Geol. v. 71, no. 5, p. 566-580.

———, J. B. Southard, D. R. Spearing and R. G. Walker, 1975, Depositional environment as interpreted from primary sedimentary structures and stratification sequences: SEPM Short Course 2, 161 p.

Hicklin, E. J., and G. C. Nanson, 1975, The character of channel migration on the Beatton River, northeast British Columbia, Canada: GSA Bull., v. 86, no. 4, p. 487-494.

Jackson, R. G., 1976, Depostional models of point bars in the lower Wabash River: Jour. Sed. Petrology, v. 46, no. 3, p. 579-595.

Kruit, C., 1955, Sediments of the Rhone delta, I. grain size and microfauna: Verh. Kon. Ned. Geol. Mijnb. Gen. Geol. Ser. 15, p. 357-514.

Le Blanc, R. J., 1972, Underground waste management and environmental implications *in* J. A. Peterson and J. C. Osmond, eds., Geometry of sandstone bodies, AAPG Mem. 18, p. 133-189.

Leopold, L. B., and W. B. Langbein, 1966, River meanders: Scientific American, June.

———, and T. Maddock, Jr., 1953, The hydraulic geometry of stream channels and some physiographic implications: U.S. Geol. Survey Prof. Paper 252, 57 p.

———, and M. G. Wolman, 1960, River meanders: GSA Bull., v. 71, no. 6, p. 769-794.

———, ———, and J. P. Miller, 1964, Fluvial processes in geomorphology: San Francisco, W. H. Freeman and Co., 522 p.

Mannhard, G. W., and D. A. Busch, 1974, Stratigraphic trap accumulation in southwestern Kansas and northwestern Oklahoma: AAPG Bull., v. 58, no. 3, p. 447-463.

Martin, R., 1966, Paleogeomorphology and its application to exploration for oil and gas (with examples from western Canada): AAPG Bull., v. 50, no. 10, p. 2277-2311.

Mettler, D. E., 1966, West Moorcraft Dakota field, Crook County Wyoming: The Mountain Geologist, v. 3, no. 2, p. 89-92.

McKee, E. D., 1957, Flume experiments on the production of stratification and cross-stratification: Jour. Sed. Pet., v. 27, no. 2, p. 129-134.

Nolte, C. J., 1967, The middle Vanguard sandstone: a profilic Jurassic objective in southwestern Saskatchewan, Canada: The Mountain Geologist, v. 4, no. 2, p. 41-51.

Oomkens, E., 1967, Depositional sequences and sand distribution in a deltaic complex, a sedimentological investigation of the post-glacial Rhone delta complex: Geologie en Mijnbouw, v. 46, p. 265-278.

Potter, P. E. and F. J. Pettijohn, 1963, Paleocurrents and basin analysis: New York, Academic Press, 296 p.

Putnam, P. E., 1982, Fluvial channel sandstones within upper Mannville (Albian) of Lloydminster area, Canada—geometry, petrography, and paleogeographic implications: AAPG Bull., v. 66, no. 4, p. 436-459.

Rich, J. L., 1923, Shoestring sands of eastern Kansas: AAPG Bull., v. 7, no. 2, p. 103-113.

Rittenhouse, G., 1961, Problems and principles of sandstone body classification: *in* J. A. Peterson and J. C. Osmond, eds., Geometry of sandstone bodies, Tulsa, AAPG, p. 3-12.

Schlee, J. S. and R. H. Moench, 1961, Properties and genesis of "Jackpile" sandstone, Laguna, New Mexico, *in* J. A. Petersen and J. C. Osmond, eds., Geometry of sandstone bodies: AAPG p. 134-150.

Schumm, S. A., and H. R. Khan, 1972, Experimental study of channel patterns: GSA Bull., v. 83, no. 6, p. 1755-1770.

Slingerland, R. E., 1967, Personal communication.

Steinmetz, R., 1972, Sedimentation of an Arkansas River sand bar in Oklahoma: a cautionary note on dipmeter interpretation: Shale Shaker, v. 23, no. 2, p. 32-38.

Stevens, J. C., 1937, Discussion of E.W. Lane, stable channels in erodible materials: Am. Soc. Civil Engineers Trans., no. 102, p. 145-149.

Thomas, A. R., 1946, Slope formulas for rivers and canals: Jour. Central Board Irrigation (India), v. 3, no. 1, pp. 40-49.

Sangree, J. B., and J. M. Widmer, 1977, Interpretation of clastic depositional facies: *in* C.E. Payton, ed., Seismic stratigraphy—applications to hydrocarbon exploration: AAPG Mem. 26, p. 168-184.

Shelton, J. W., 1967, Stratigraphic models and general criteria for recognition of alluvial, barrier bar, and turbidity-current sand deposits: AAPG Bull., v. 51, no. 12, p. 2441-2461.

Tinler, K. J., 1971, Active valley meanders in south-central Texas and their wider implications: GSA Bull., v. 82, no. 7, p. 1783-1800.

Truchot, Jr., J. F., 1963, The Miller Creek field, Crook County, Wyoming: Wyoming Geol. Assoc.—Billings Geol. Soc., Joint Field Conf. Guidebook, p. 129-132.

Vanoni, V. A., 1946, Transportation of suspended sediment by water: Am. Soc. Civil Engineers Trans., no. 111, p. 67-133.

Vigrass, L. W., 1977, Trapping of oil at intra-Mannville (Lower Cretaceous) disconformity in Lloydminster area, Alberta and Saskatchewan: AAPG Bull., v. 61, no. 7, p. 1010-1028.

Walker, R. G., 1976, Facies models 3, Sandy fluvial systems: Geosci. Canada, v. 3, p. 101-109.

Warne, J. R., personal loan.

Weber, K. J., 1971, Sedimentologic aspects of oil fields in the Niger delta: Geologie en Mijnbouw, v. 50, p. 559-576.

Wilson, C. W., Jr., 1948, Channels and channel-filling sediments of Richmond age in south-central Tennessee: GSA Bull., v. 59, no. 8, p. 733-765.

6 EOLIAN SANDS

INTRODUCTION

Eolian environments are often characterized as being relatively simple and as having good reservoirs with good continuity (LeBlanc, 1977). However, recent research shows the environment to be fairly complex with laterally and vertically discontinuous reservoirs (Ahlbrandt and Fryberger, 1982). Even dunes with their relatively simple, high-angle, cross-bedded internal structure may have porosity and permeability barriers related to the cross-bedding (Emmett et al., 1972; Lupe and Ahlbrandt, 1979).

Much of the eolian environment's complexity is due to three subenvironments that may be found in it (Fig. 6-1). A vertical buildup of wind-blown sand forms a dune; flat-lying sediments in low areas between dunes constitute an interdune; and flat to low-angle, nondune, eolian sediments marginal to a dune complex form a sand sheet (Ahlbrandt and Fryberger, 1982). Sedimentary characteristics of each subenvironment are different. Extradunes are noneolian environments marginal to an eolian environment but related to it in time and possibly by source (Lupe and Ahlbrandt, 1979).

DUNES

Three kinds of dunes are recognized by McKee (1979): simple, compound, and complex. Simple dunes are single sand bodies identified by body morphology and the number of slip or avalanche faces. Compound and complex dunes are simple dunes with the same or different dunes superimposed on them.

Exploration and development geologists need to be familiar with the variability of sedimentary structures that may be seen in cores and on dipmeter logs transecting ancient dunes. Such familiarity may help to identify dune type and orientation and aid in more effective exploration for and development of an eolian dune reservoir. Descriptions of the dunes that follow are taken from McKee (1979).

Dune Types

Dune type is a function of sand supply, wind strength and direction, vegetation, physical barriers to dune migration, and distance from source. Effects of three of the factors are demonstrated in the development of transverse dunes (Fig. 6-2), barchanoid ridges (Fig. 6-3), and barchan dunes (Fig. 6-4). All three types are free to migrate, develop where the wind blows primarily in one direction, and each is oriented normal to the wind. Unidirectional winds are indicated by each dune having a single slipface (Fig. 6-5). Transverse dunes have straight, continuous crests (Fig. 6-2), require the most sand to form, and are located closest to the sand source. Barchan dunes require the least amount of sand to form and are far from the sand source relative to transverse dunes. Barchanoid ridges have sinuous crests and resemble a series of end-to-end barchan dunes (Fig. 6-3). They are intermediate between barchan and transverse dunes with respect to sand supply and distance from source.

Other dunes are partially stabilized by vegetation or moisture. Unstabilized parts of parabolic (Fig. 6-6) and blowout dunes (Fig. 6-7) are free to migrate, and one or more slipfaces may be present.

Winds blowing alternately from two directions that form an acute angle produce linear dunes that are elongate sand ridges oriented parallel to the probable resultant of the two wind directions (Fig. 6-8). Linear dunes, also called seif or longitudinal dunes, have two slipfaces dipping in opposite directions (Fig. 6-9) and grow in the direction of the resultant wind. Individual dunes may be many kilometers long

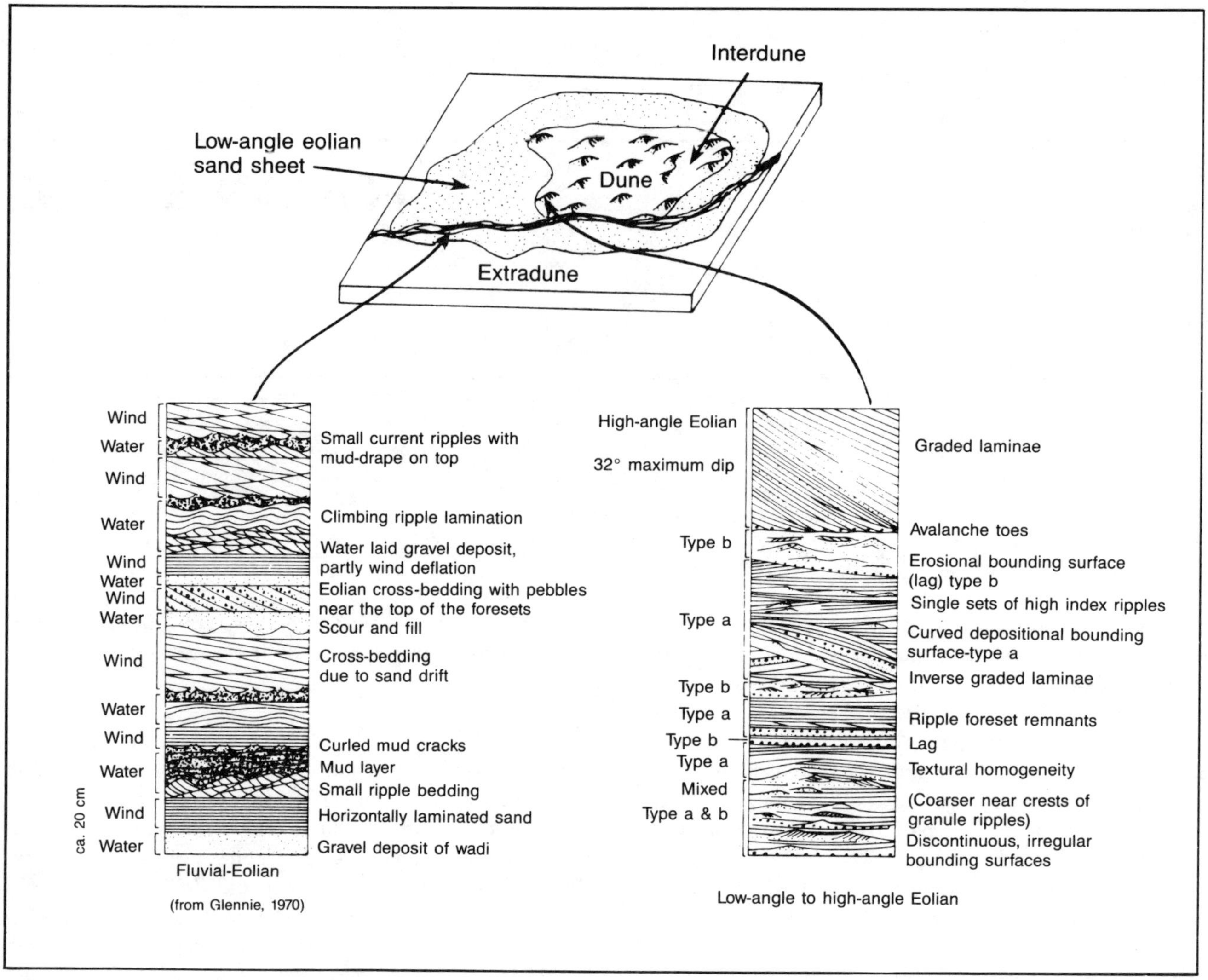

Fig. 6-1. Schematic diagram of subenvironments in the eolian environment with cross-cutting stream and hypothetical cross sections. (Redrawn from Fryberger et al., 1979; permission to publish by the Rocky Mountain Association of Geologists).

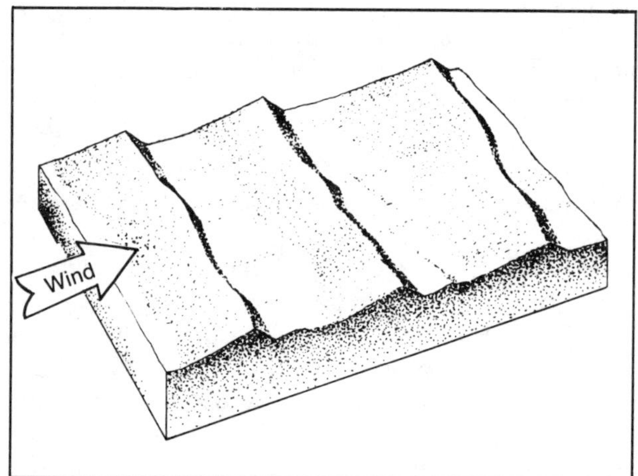

Fig. 6-2. Transverse dune with relatively straight crest and single slipface. (From McKee, 1979; permission courtesy U.S. Geological Survey).

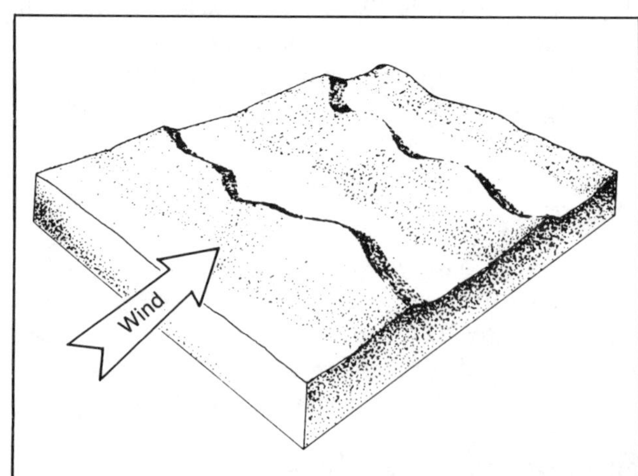

Fig. 6-3. Barchanoid ridge with sinuous crest and single slipface. (From McKee, 1979; permission courtesy U.S. Geological Survey).

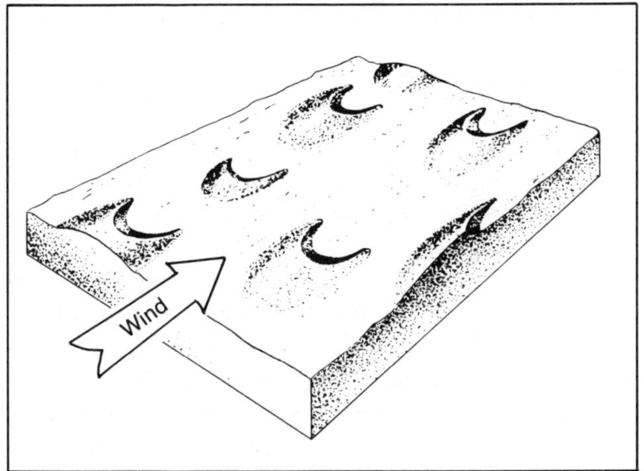

Fig. 6-4. Barchan dune with single slipface. (From McKee, 1979; permission courtesy U.S. Geological Survey).

and a dune field may cover hundreds of square kilometers.

Reversing dunes (Fig. 6-10) also result from bidirectional winds, but in this case the winds blow in opposite or nearly opposite directions with equal intensity. A second slipface oriented opposite to the primary slipface forms periodically producing sedimentary structures that dip in opposite directions (Fig. 6-11).

Star dunes (Fig. 6-12) have three or more arms radiating out from a high central area and as many slipfaces. Multidirectional winds form the dunes that tend to accrete vertically rather than migrate.

Dome dunes (Fig. 6-13) are circular to elliptical in plan and have no slipfaces. However, internal structures consist of unidirectional slipfaces suggesting a possible origin as a barchan dune. Dune form suggests truncation of the top and flattening of the lee slope by strong winds.

Basic dune forms also combine to form compound and complex dunes. In compound dunes, basic dunes of the same type overlap or are superimposed on one another (Fig. 6-14). In complex dunes (Fig. 6-15), basic dunes of different types coalesce or grow together.

Dune Structures

Few sedimentary structures are available for an exploration geologist to identify the eolian dune environment. McKee and Bigarella (1979) indicate the principal structures are sets of high-angle foreset beds (Figs. 6-1, 6-5, and 6-9), planar-tabular cross-bed sets, which may thin upward (Fig. 6-5), and horizontal or gently leeward dipping surfaces separating sets of cross-beds (Figs. 6-1, 6-5, and 6-9). In high dunes, some downwind bounding surfaces may dip more steeply. Of these structures, high-angle foresets are most apparent in conventional cores and on dipmeter logs. The orientation of various high-angle sets in a vertical section (Fig. 6-11) could give an indication of the types of

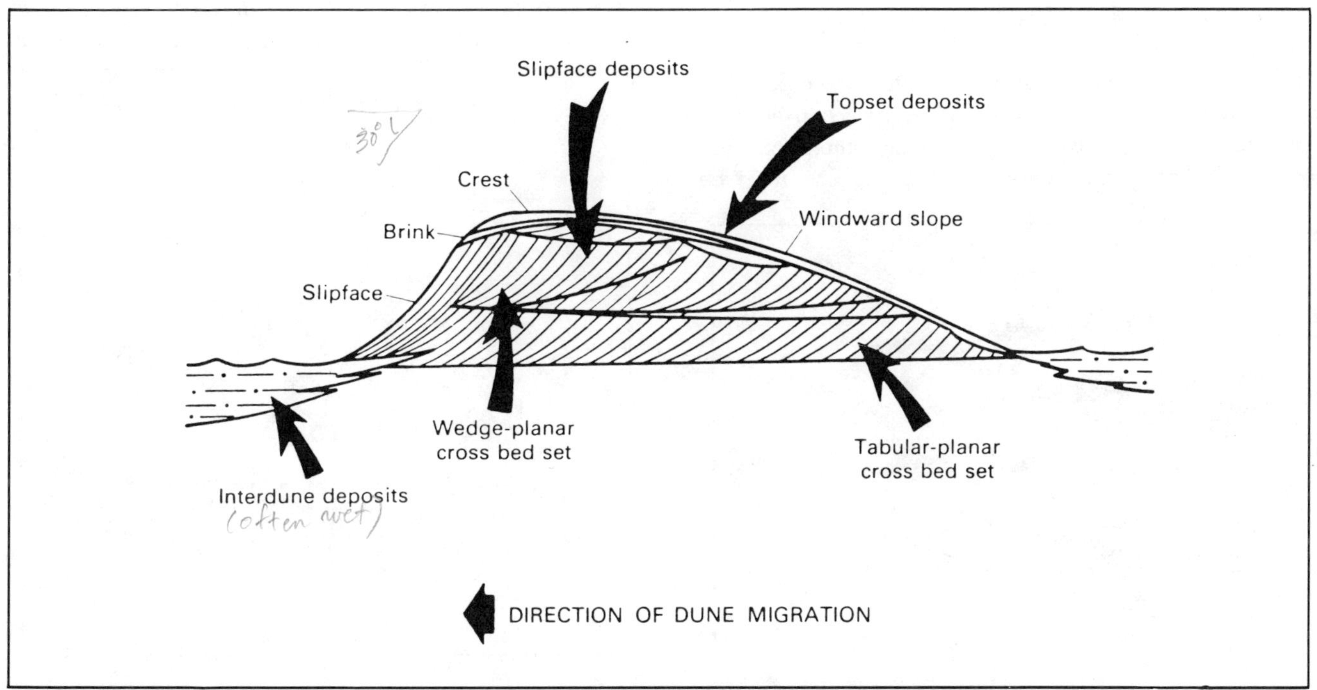

Fig. 6-5. Schematic cross section of single slipface dunes, such as barchan and transverse dunes. (From Ahlbrandt and Fryberger, 1980; permission courtesy U. S. Geological Survey).

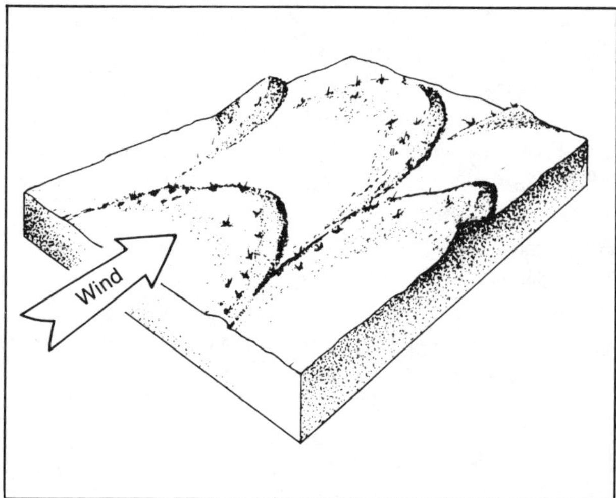

Fig. 6-6. Parabolic dunes. (From McKee, 1979; permission courtesy U.S. Geological Survey).

dunes encountered in a well by revealing the number of slipfaces of the dunes.

Deformational processes in dunes disrupt internal stratification and produce deformational structures in slipface deposits (Fig. 6-16). Avalanching down a slipface creates tensional features at the tops of slipfaces and compressional features at the bases (McKee and Bigarella, 1979). Additional disruption of dune bedding is caused by vegetation and burrowing organisms such as insects and reptiles (Ahlbrandt and Fryberger, 1982).

INTERDUNES

Separating individual dunes are low interdune areas. Deposition may occur in these lows before being covered by migrating dunes. A succession of dune and interdune beds increases the complexity of eolian deposits and must be considered when developing an eolian reservoir.

Migrating dunes control the thickness of interdune deposits by limiting the time an interdune is able to receive sediments. Rapid dune migration results in thin interdune deposition, whereas with more stationary dunes, interdunes have time to accumulate deposits of greater thickness and lateral extent (McKee and Bigarella, 1979; Ahlbrandt and Fryberger, 1981).

In the absence of an interdune classification, Ahlbrandt and Fryberger (1981) use descriptive terminology to identify several interdune environments. They recognize deflationary and depositional interdunes, and within the depositional group they consider wet, dry, and evaporite interdunes.

Deflationary Interdunes

Formation of deflationary interdunes is self-explanatory. Sediments consist of coarse lags, deposits related to the presence of vegetation, and features connected with deflation to or near the water table. Coarse lags tend to be thin horizons between dune cross-bed sets and may be rippled (Ahlbrandt and Fryberger, 1981). Vegetation effects include disruption of internal stratification by root growth, remnant calcified root molds and casts (Glennie and Evamy, 1968), sand deposition in the lee of a plant, and scour and fill around the base of a plant (Ahlbrant and Fryberger, 1981). Deflation to moisture may produce adhesion ripples that form by transported sand grains adhering to moist areas and growing upwind. Such deposits are thin unless the water table is rising, in which case they may be up to a meter thick. Adhesion structures produce an irregular interdune surface at the centimeter scale that causes any lag deposits to be wavy and uneven (Ahlbrandt and Fryberger, 1981).

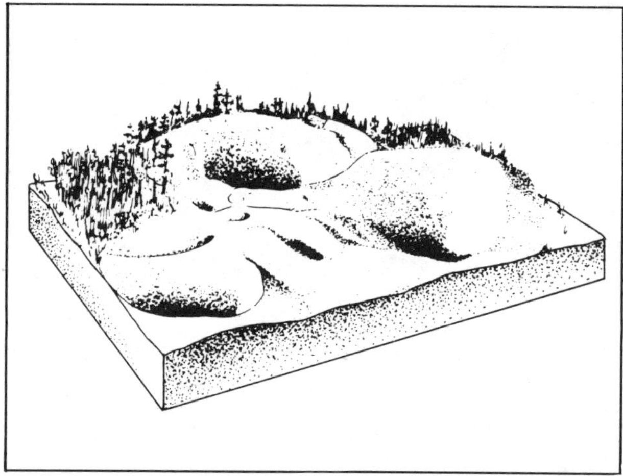

Fig. 6-7. Blowout dunes. (From McKee, 1979; permission courtesy U.S. Geological Survey).

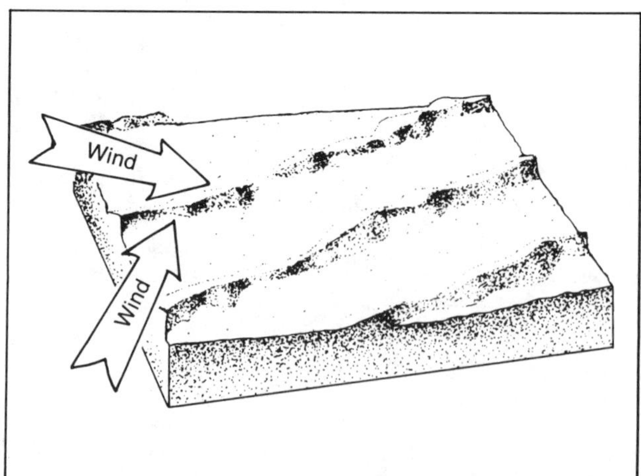

Fig. 6-8. Linear dunes grow downwind and have two slipfaces due to two prevailing wind directions. (From McKee, 1979; permission courtesy U.S. Geological Survey).

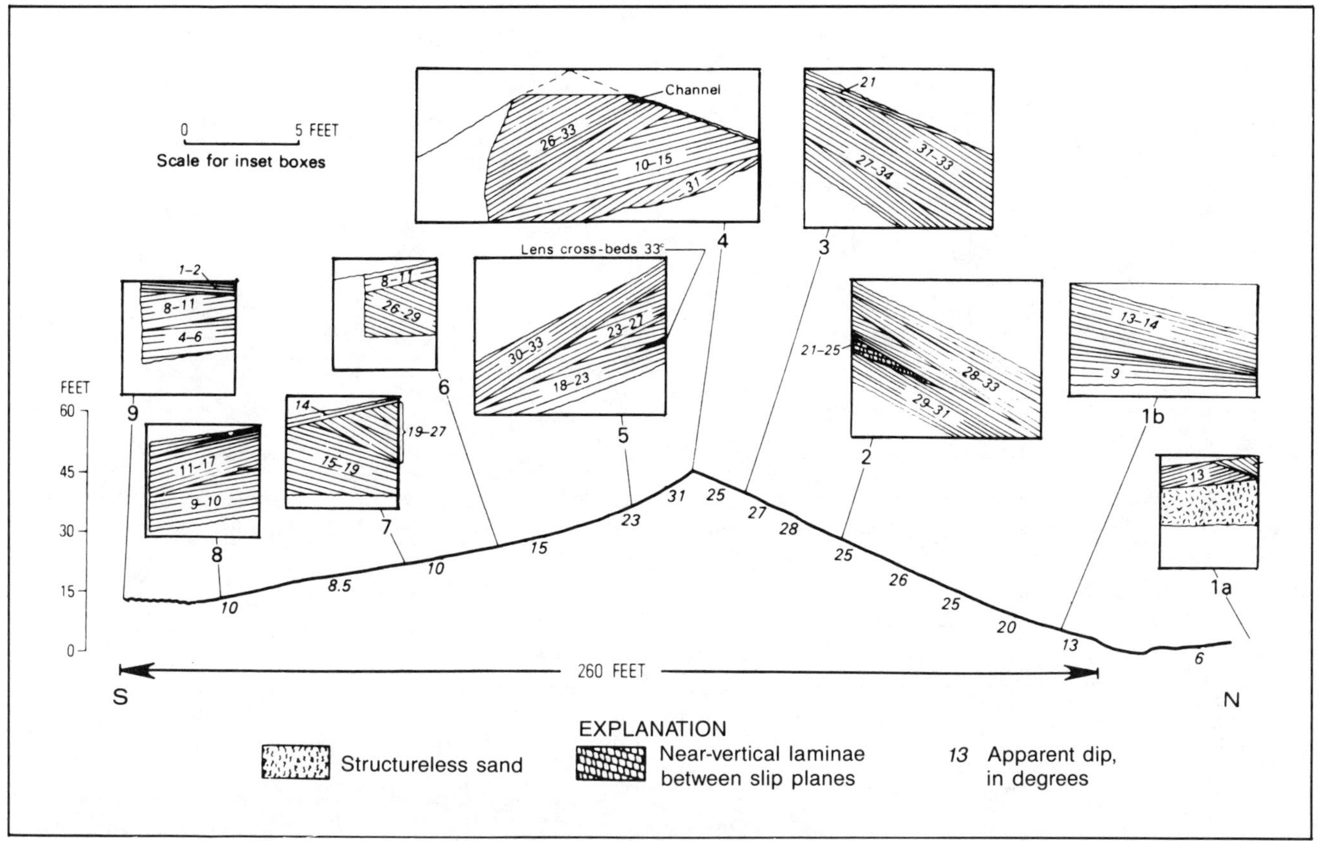

Fig. 6-9. Cross section of linear dune near Sebha, Libya, based on a series of test pits. (From McKee and Tibbits, 1964; as published in McKee and Bigarella, 1979; permission courtesy U.S. Geological Survey).

Depositional Interdunes

Dry Interdunes. Within this group, Ahlbrandt and Fryberger (1981) include interdunes that are periodically wetted by rains or water from ephemeral streams. They also indicate this may be the most common interdune.

Sediments of dry interdunes are silts, clays, and poorly

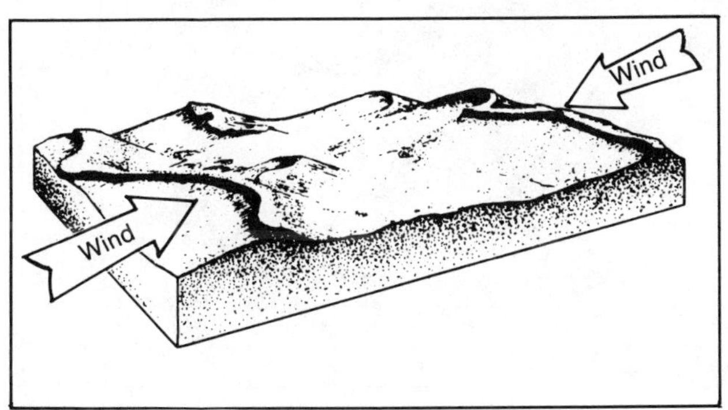

Fig. 6-10. Reversing dunes with two slipfaces due to two diametrically opposed wind directions. (From McKee, 1979; permission courtesy U.S. Geological Survey).

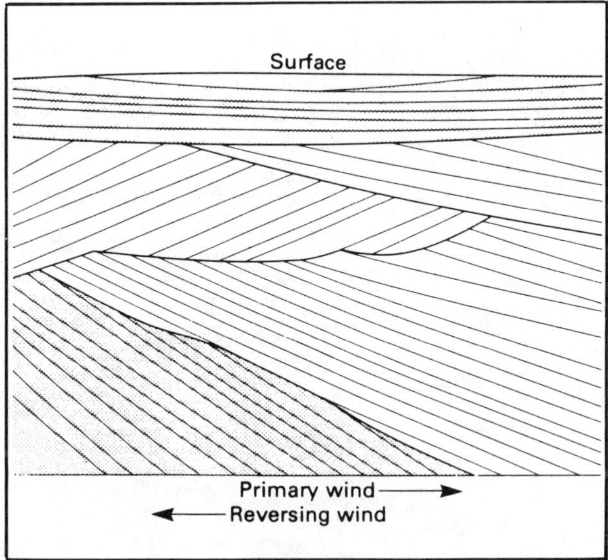

Fig. 6-11. Internal structure of reversing dune parallel to the opposing dominant wind directions. (Based on photo by Mark, from McKee, 1979; permission courtesy U.S. Geological Survey).

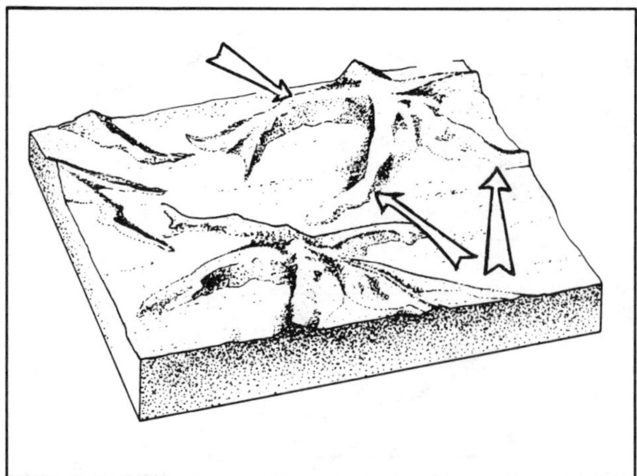

Fig. 6-12. Star dunes with numerous slipfaces. Some of the several wind directions indicated by arrows. Dune accretes vertically rather than migrates. (From McKee, 1979; permission courtesy U.S. Geological Survey).

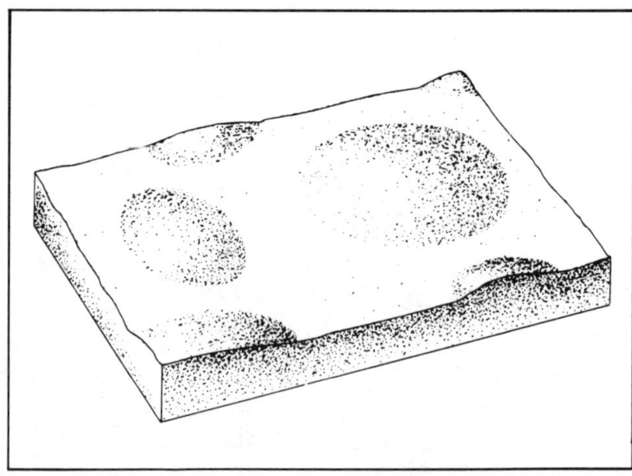

Fig. 6-13. Dome dunes with no slipfaces. (From McKee, 1979; permission courtesy U.S. Geological Survey).

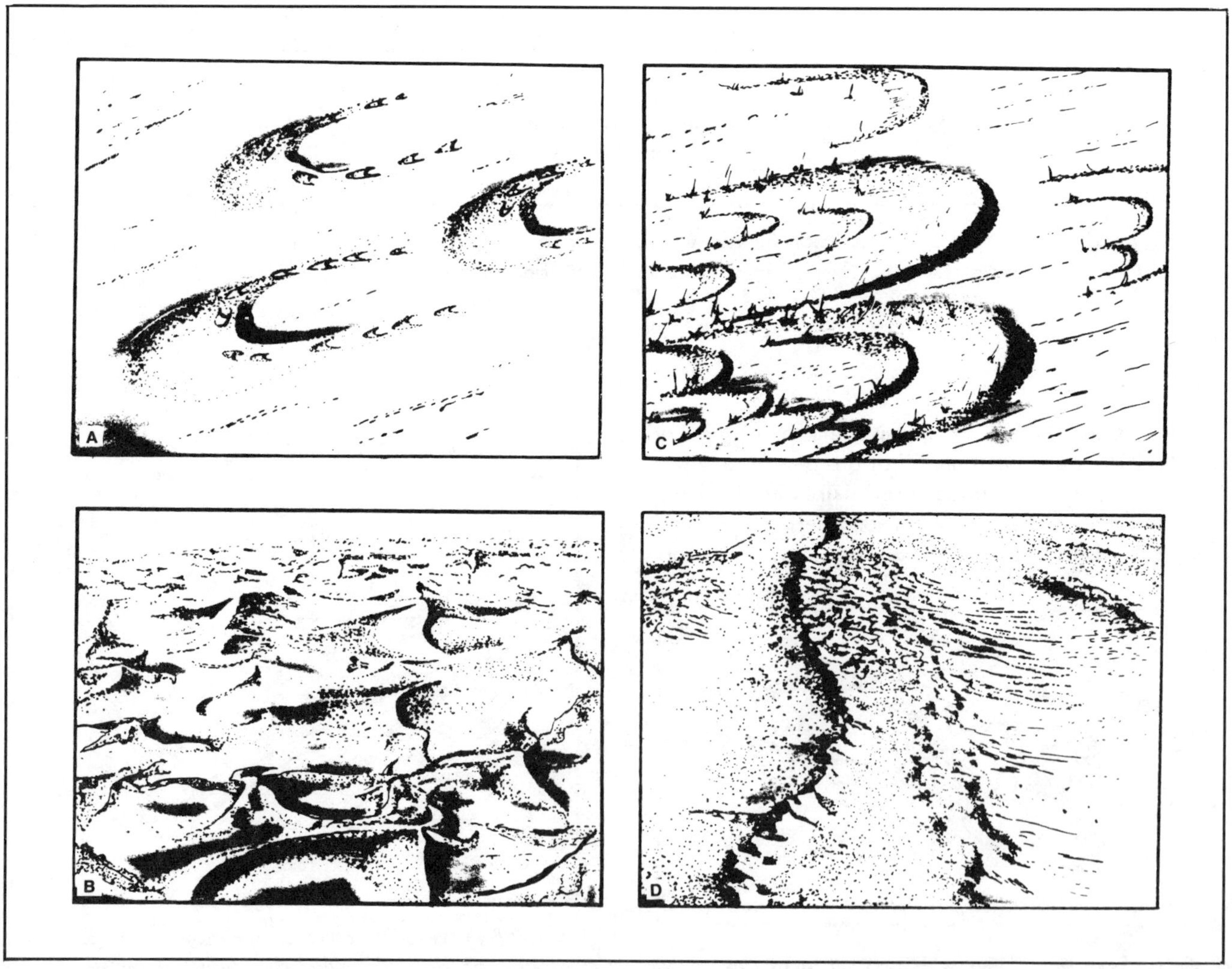

Fig. 6-14. Compound dunes. **A**. *Small barchan dunes on a large barchan.* **B**. *Coalescing star dunes.* **C**. *Parabolic dunes within parabolic dunes.* **D**. *Superimposed linear dunes.* (From McKee, 1979; permission courtesy U.S. Geological Survey).

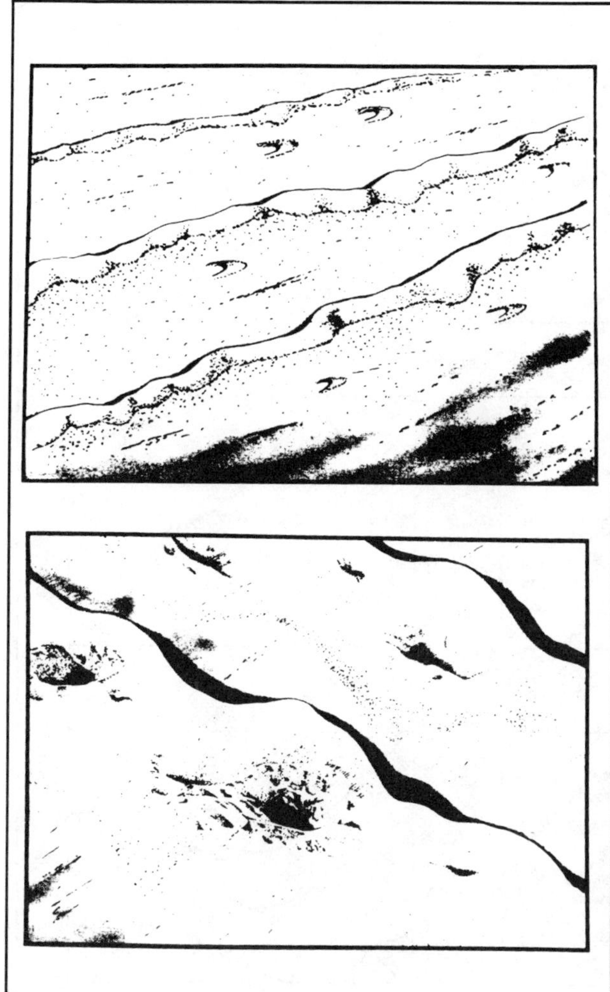

Fig. 6-15. Complex dunes. **Top.** *Linear dunes with barchans in interdunes.* **Bottom.** *Blowouts on transverse dunes.* (From McKee, 1979; permission courtesy U.S. Geological Survey).

sorted, bimodal sands (McKee and Bigarella, 1979; Ahlbrandt, 1979). Sedimentary features include ripples in sand and granule sizes, grainfalls on the lee side of vegetation and dunes, avalanching at the bases of dunes adjacent to the interdune, curved and gently dipping erosion surfaces, centimeter scale normal and inversely graded laminae, and interbedded high-angle eolian deposits from small dunes migrating across the interdune (Fryberger et al., 1979). Disruption of the sediment by bioturbation and cementation may make it structureless (Ahlbrandt et al., 1978).

Wet Interdunes. Perennial lakes or ponds in interdunes receive sediments that differ from those of dry interdunes. Fine-grained sediments, which might otherwise be lost to deflation, are trapped by the water, and large amounts of silt and clay are retained (Ahlbrandt and Fryberger, 1981). Floral and faunal activity is so enhanced by the water that fossils (Gradzinski and Jerzykiewicz, 1974; Hanley, 1980) and potential hydrocarbon source beds (Ahlbrandt and Fryberger, 1980) occur in wet interdunes. Other characteristics include easy contortion of incompetent, saturated, interdune sediments by loading from encroaching dune sediment, associated cements and sand crystals of salts of sodium, calcium, and potassium, and dissipation structures (Ahlbrandt and Fryberger, 1981). Dissipation structures are dark, wavy, discontinuous bands of mostly clay derived from soil-forming processes (Ruhe, 1975) and infiltration of water containing suspended clays (Bigarella, 1975; Ahlbrandt and Andrews, 1978).

Wet interdune deposits, then, are fossiliferous sands and organic silts and clays that are several meters thick. They are bioturbated and have dissipation structures and salts of calcium, sodium and potassium. The deposits are barriers to migrating fluids and may serve as limited sources of hydrocarbons for associated, reservoir-quality, dune deposits (Picard, 1975).

Evaporite Interdunes. These interdunes are primarily beds of chemically precipitated salts derived from interdune ponds and sabkhas. Trenches dug in a coastal sabkha in Saudia Arabia had a wide range of internal characteristics. One trench showed ". . . a complex of silty sandstone, sandstone, gypsum crystal mounds that affected overlying sediments, mudstone, gypsum, anhydrite and brine in stratigraphic succession from top to bottom . . . A second trench revealed low angle eolian sands with bioturbation features comparable to low angle eolian features described (for sand sheets). A third trench revealed a nearsurface silty sandstone and organic mud with bioturbating organisms actively forming oxidized trails in the reducing mud, anhydrite and brine . . . These trenches were dug within several hundred meters of one another with little apparent surficial differences" (Ahlbrandt and Fryberger, 1981). Evaporite interdune deposits consist of evaporite minerals and clastics that are structureless or have discontinuous flat bedding, wavy nonparallel lamination, bioturbation, and disrupted bedding from evaporite mineral growth (Ahlbrandt and Fryberger, 1981).

Interdune deposits may be considered a double-edged sword with regard to eolian reservoirs. On one hand, the fine-grained deposits act as permeability barriers to migrating fluids and increase the complexity of eolian reservoirs. On the other hand, some interdunes have deposits sufficiently rich in organic matter to act as sources of hydrocarbons that might not otherwise be available.

SAND SHEETS

Marginal to the relatively high-relief, dune-interdune environment is the relatively low-relief sand sheet (Fig. 6-1), a low-angle eolian deposit with lamination dips ranging up to about 20°. At such low dips, there are no slipfaces. Deposits consist of remnants of dunes, grassy sand mounds,

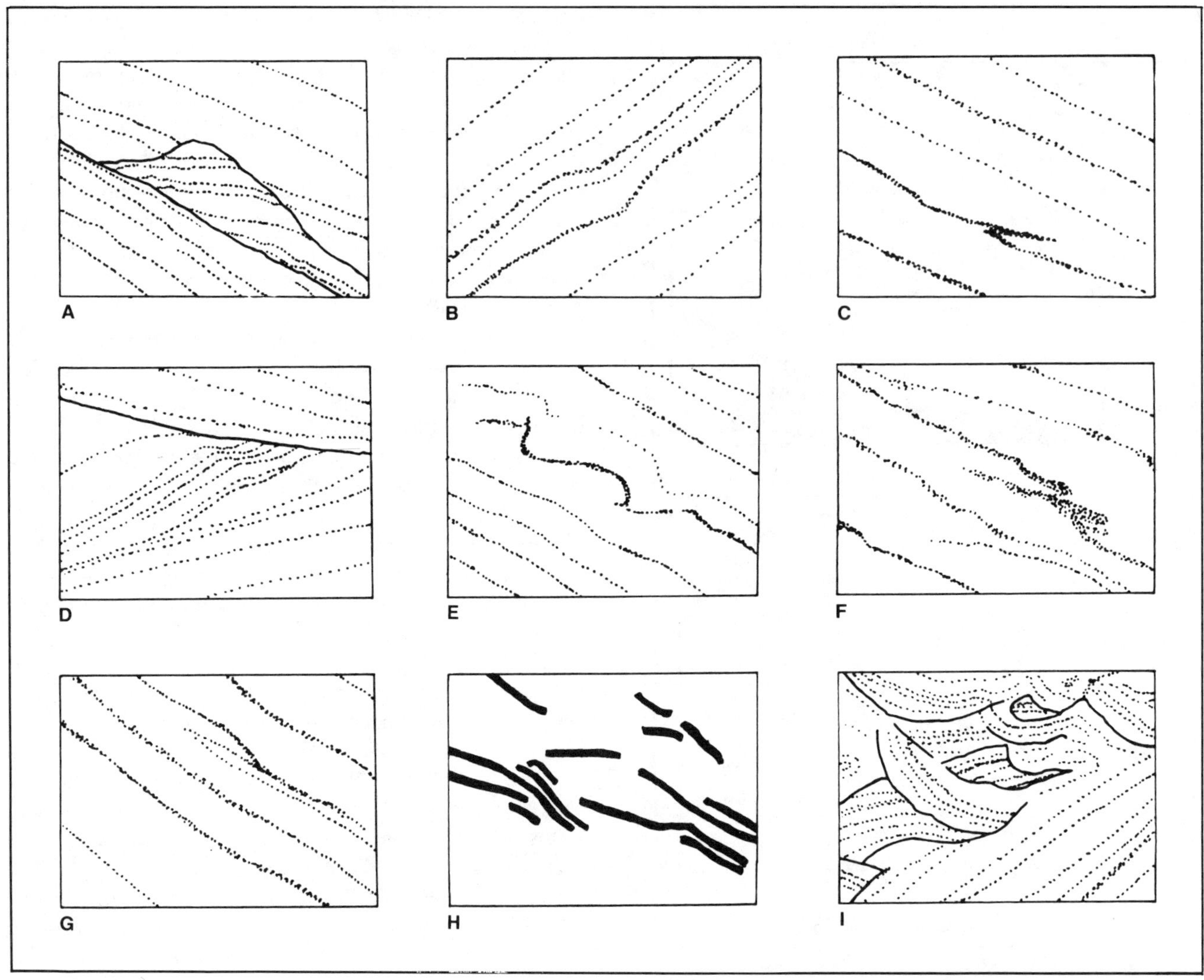

Fig. 6-16. Principal types of deformational structures in avalanche deposits of dunes: **A**. *Rotated structures;* **B**. *warps or gentle folds;* **C**. *flame structure;* **D**. *drag fold;* **E**. *high-angle asymmetrical folds;* **F**. *folds;* **G**. *overthrust;* **H**. *break-aparts;* **I**. *breccias. Each block represents an area approximately 6 x 4 inches (15 x 10 cm).* (From McKee and Bigarella, 1979; permission courtesy U.S. Geological Survey).

and low-angle eolian deposits possibly produced by low-energy winds (Fryberger et al., 1979).

Two principal types of low-angle deposits occur, separated by a gently dipping, curved, erosion surface. Type "a" deposits are ". . . thinly laminated (≤2 mm) to laminated (2 mm to 1 cm), with graded laminations common. They consist of better-sorted sand than in type "b" sets, and have few, if any, preserved ripples or ripple foresets. Laminations are more continuous . . . (compared to "b" sets), and are commonly convex upward. Type "b" sets result from alternation of deposition and erosion which combine to produce stacked fine and coarse beds or laminae of sand. Thus, type "b" sets are poorly sorted (commonly bimodal) . . ., although individual coarse and fine beds and laminae may be well-sorted. A common feature in type "b" sets is isolated, or single horizons of coarse-grained, high index ripples (as defined by Sharp, 1963; Tanner, 1967). The coarse layers and coarse ripples observed in type "b" sets form by deflation of type "a" material, or in some instances by direct deposition of coarser material over finer material" (Fryberger et al., 1979).

The low-angle to high-angle cross section in Figure 6-1 illustrates a hypothetical sand sheet overlain by a dune. Type "a" and "b" sands are identified on the left of the section. Ripples in the "b" horizons and the curved erosion surfaces between "a" and "b" layers are shown. A section elsewhere through the sand sheet, but outside the dune field, would be similar except for the high-angle cross-beds at the top.

As sand sheets are eolian deposits marginal to dune fields,

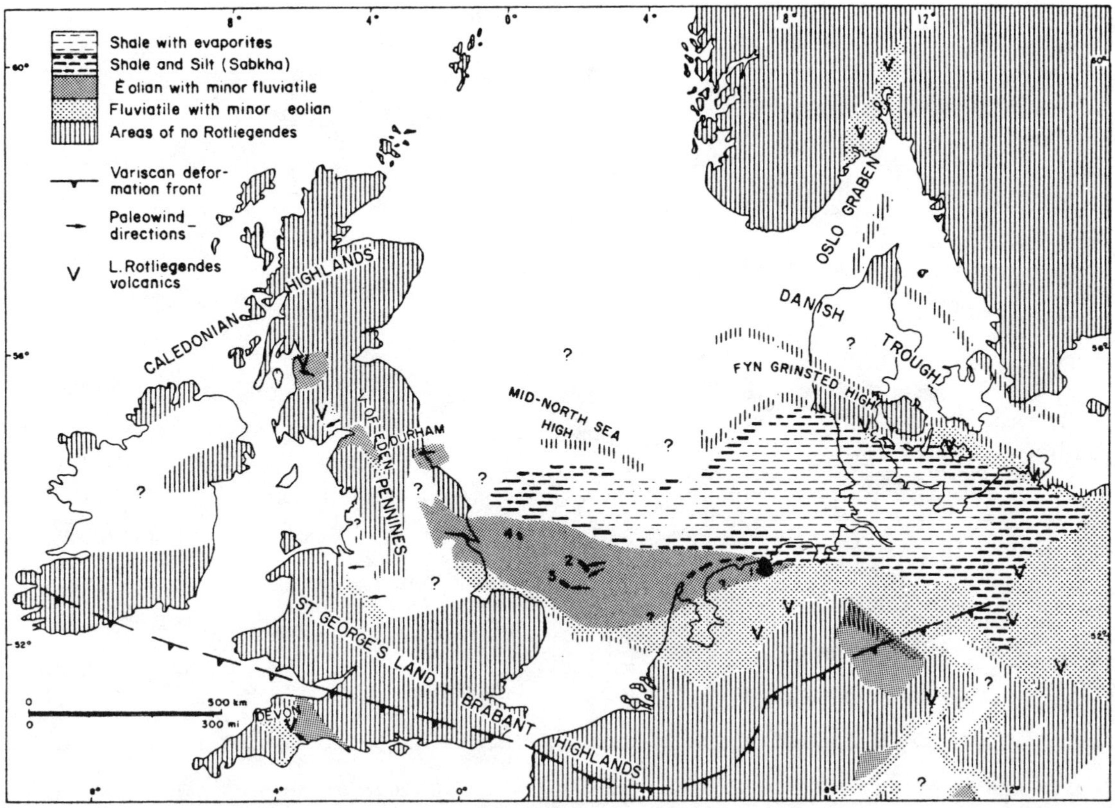

Fig. 6-17. General distribution of selected Lower Permian tectonic elements and Rotliegendes facies in northwestern Europe. (From Glennie, 1972; permission to publish by AAPG).

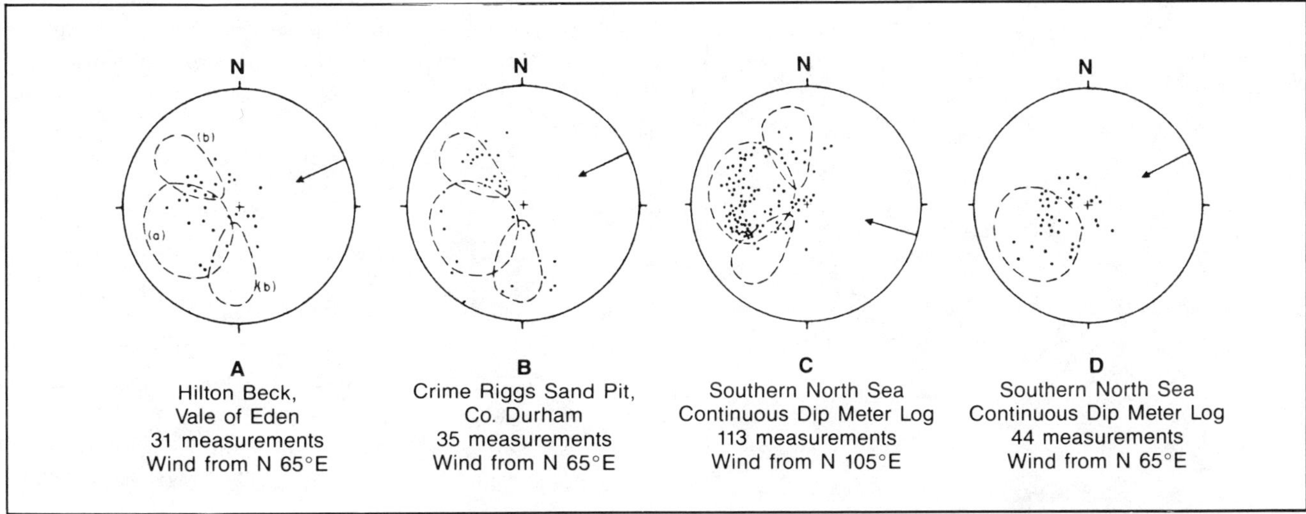

Fig. 6-18. Upper hemisphere stereographic polar nets of poles to Rotliegendes dune bedding. **A** *and* **B** *are from outcrops in northern England;* **C** *and* **D** *are from wells in the southern North Sea.* **Dashed Lines** *outline areas of transverse* **(a)** *and longitudinal* **(b)** *dunes;* **arrow** *is deduced paleowind direction.* (From Glennie, 1972; permission to publish by AAPG).

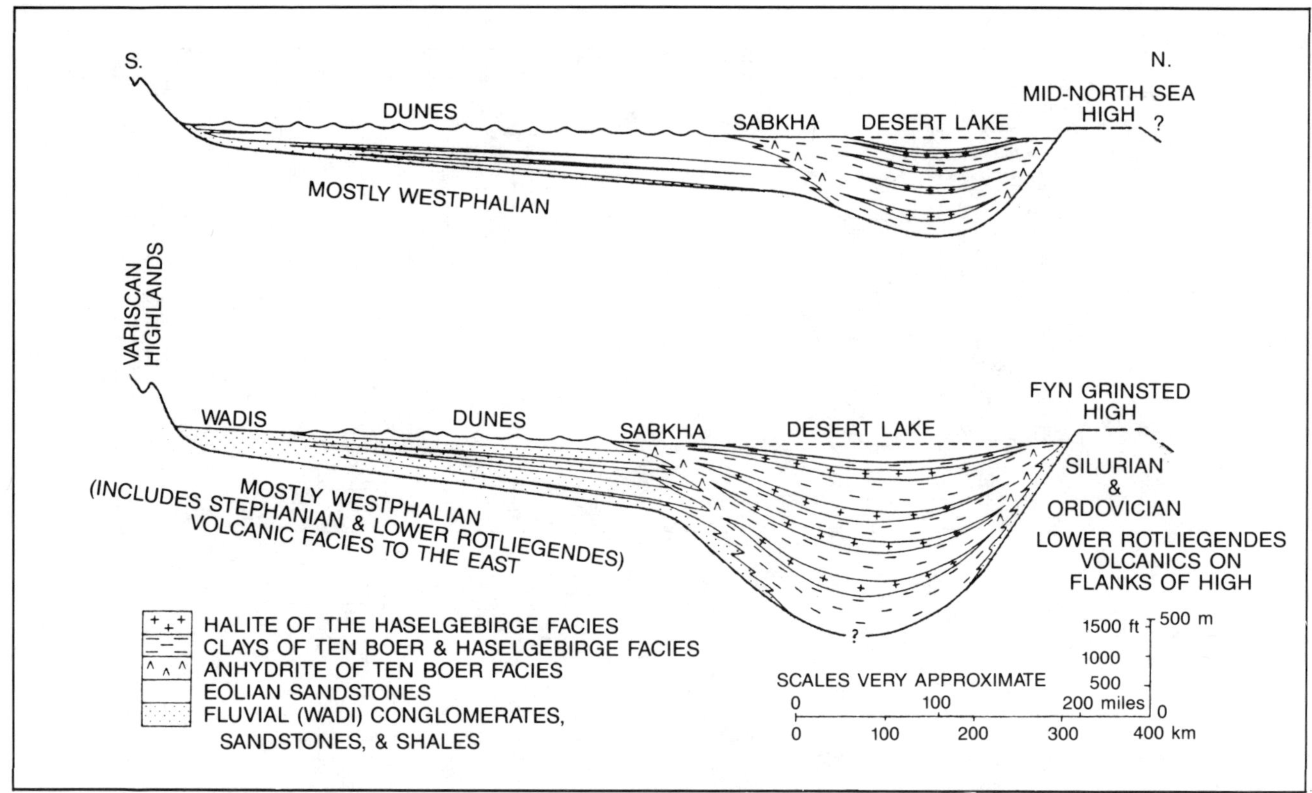

Fig. 6-19. North-south schematic cross-sections showing distribution of facies in the Rotliegendes depositional basin. Sections are in the southern North Sea and eastern Netherlands south of the Mid-North Sea high and Fyn Grinsted high, respectively. (From Glennie, 1972; permission to publish by AAPG).

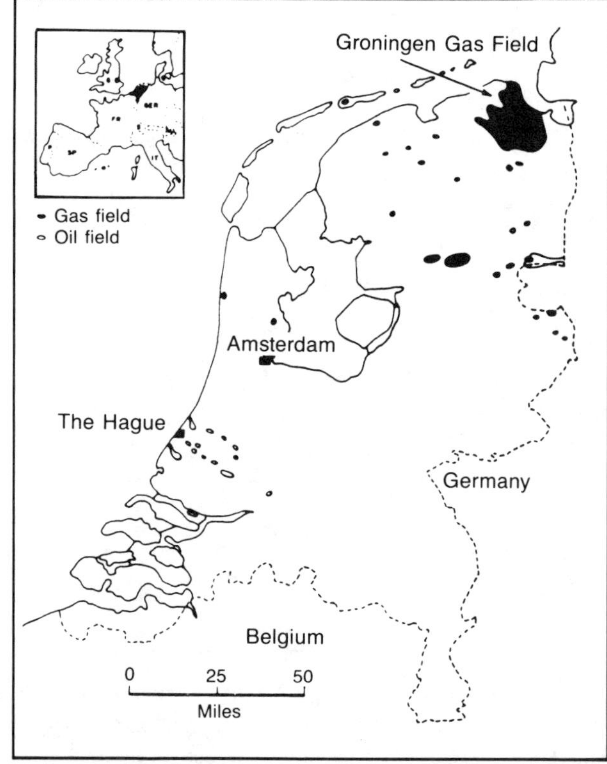

Fig. 6-20. Location map for Groningen gas field, northeastern Netherlands. (From Stauble and Milius, 1970; permission to publish by AAPG).

they should be intercalated with dune-interdune and extradune deposits. They should also show features of other eolian environments such as bioturbation, grainfall deposits, and scour and fill around obstructions such as plants. Recognition of a sand sheet in an exploratory well core would indicate probable proximity to a dune field and should encourage further exploration for a possible eolian dune reservoir.

HYDROCARBON EXPLORATION IN EOLIAN DEPOSITS

Oil and gas are found in eolian deposits of Paleozoic and Mesozoic age in the United States and northwestern Europe. Gas fields at Groningen (Netherlands) and in the southern North Sea produce from the Rotliegendes (Lower Permian) sandstone. Oil and gas are produced from the Tensleep-Weber (Pennsylvanian-Permian) sandstones and the Nugget (Triassic-Jurassic) sandstone of Colorado and Wyoming in the western U.S. Rotliegendes fields seem to be relatively simple sedimentologically, while some reservoirs in the Tensleep are so complex that they are developed on ten or even five acre spacing (Emmett et al., 1972; Ahlbrandt and Fryberger, 1982).

Rotliegendes Fields

In Late Carboniferous-Early Permian time, the Variscan orogeny created a belt of highlands running from southern England, across northern France, into central Germany. North of the mountains was a basin, which was bounded on the north by the Mid-North Sea high and the Fyn Grinsted high and on the west by the Pennines in central England (Fig. 6-17). A large lake extending approximately 1000 km east-west and 300–400 km north-south covered the northern part of the basin. In Early Permian time, sediments derived from the denudation of the Variscan highlands were carried north by ephemeral streams across a desert toward the lake. Most of the coarse clastics were deposited prior to reaching the lake, and deflation by easterly winds produced extensive dune fields south of the lake (Glennie, 1972). Arid conditions prevalent during deposition of the Lower Permian Rotliegendes formation persisted through the rest of the Permian when the Zechstein evaporites were deposited. (Glennie, 1972).

Outcrop studies in England, Scotland, Norway, Poland, Germany, and France, and conventional cores from wells in Germany, the Netherlands, and the southern North Sea indicate five desert facies occur in the upper Rotliegendes. Lower Rotliegendes rocks are nonproductive and not considered herein. Characteristics of each of the facies have all been seen in the modern desert environment (Glennie, 1970, 1972). The facies are described by Glennie (1972) as wadi, dune, interdune sabkha, sabkha associated with temporary desert lakes, and sabkha marginal to a desert lake. His descriptions are summarized in the following discussion.

Rotliegendes Facies. Around the southern rim of the basin north of the Variscan Mountains are conglomerates, sandstones and clays of the wadi facies. Fluvial sandy conglomerates contain subround to round quartzite pebbles and angular to round Carboniferous shale pebbles. Red clays are cracked and curled indicating subaerial dessication. Sandstones are fine- to medium-grained with some coarse-grained zones. Sorting is poor but individual laminae may be well sorted. Dips vary from subhorizontal to 27°, with

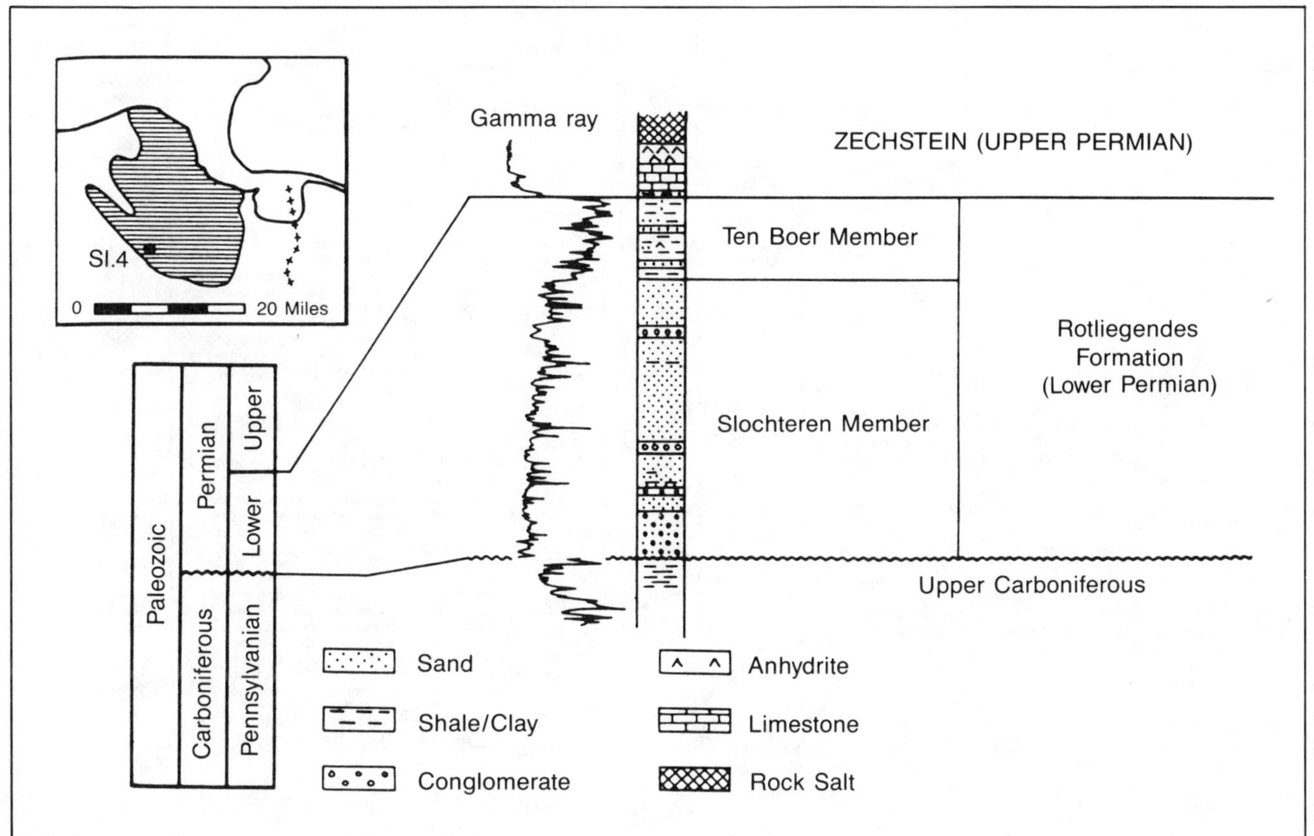

Fig. 6-21. Type section of the Rotliegendes formation, Groningen gas field, in Slochteren No. 4 well. (From Stäuble and Milius, 1970; permission to publish by AAPG).

steeply dipping sandstones usually being truncated by subhorizontal, coarser-grained sandstones. Dips of the coarser sandstones increase upsection until they are truncated. Dip direction is quite uniform and suggests a principal wind direction from the east.

Some sandstones are structureless, poorly sorted, very fine-grained to medium-grained and contain clay pebbles and flakes. They were probably reworked by wind or water, and the clay pebbles are rip-ups from clay drapes on a wadi floor.

Combinations of fluvial and eolian processes are seen in modern wadis. Fluvial wadi deposits are frequently eroded and partially overlain by eolian sands, with some of the sand being derived from the wadi. The fluvial-eolian cross section in Figure 6-1 depicts the sediments and structures of the wadi environment.

Rotliegendes sand dunes have thick sets of laminated sandstones like those of the wadi eolian sandstones. Because coarse sands accumulate preferentially at the base of a migrating dune's avalanche slope, a set starts with subhorizontal, relatively coarse sand lying on an intraformational unconformity. Upsection, grain size decreases and dip of laminae increases to a maximum of 27°. Compaction causes the maximum dip to be less than the angle of repose of modern sands. Stereonet plots of dips of dune laminae from outcrops in England and wells in the North Sea (Fig. 6-18) show uniform wind direction and differentiate between transverse (or barchan) and longitudinal dunes. Longitudinal dunes seem to be more common in England and Germany close to the Variscan Mountains, while transverse dunes seem to be more common away from the mountains.

Interdune sabkhas occur in the upper parts of the dune sands. Adhesion ripples of fine- to coarse-grained, poorly sorted sand with some silt and little clay directly overlie high- and low-angle dune sandstones. Some of these interdunes have crystals and cement of anhydrite and halite.

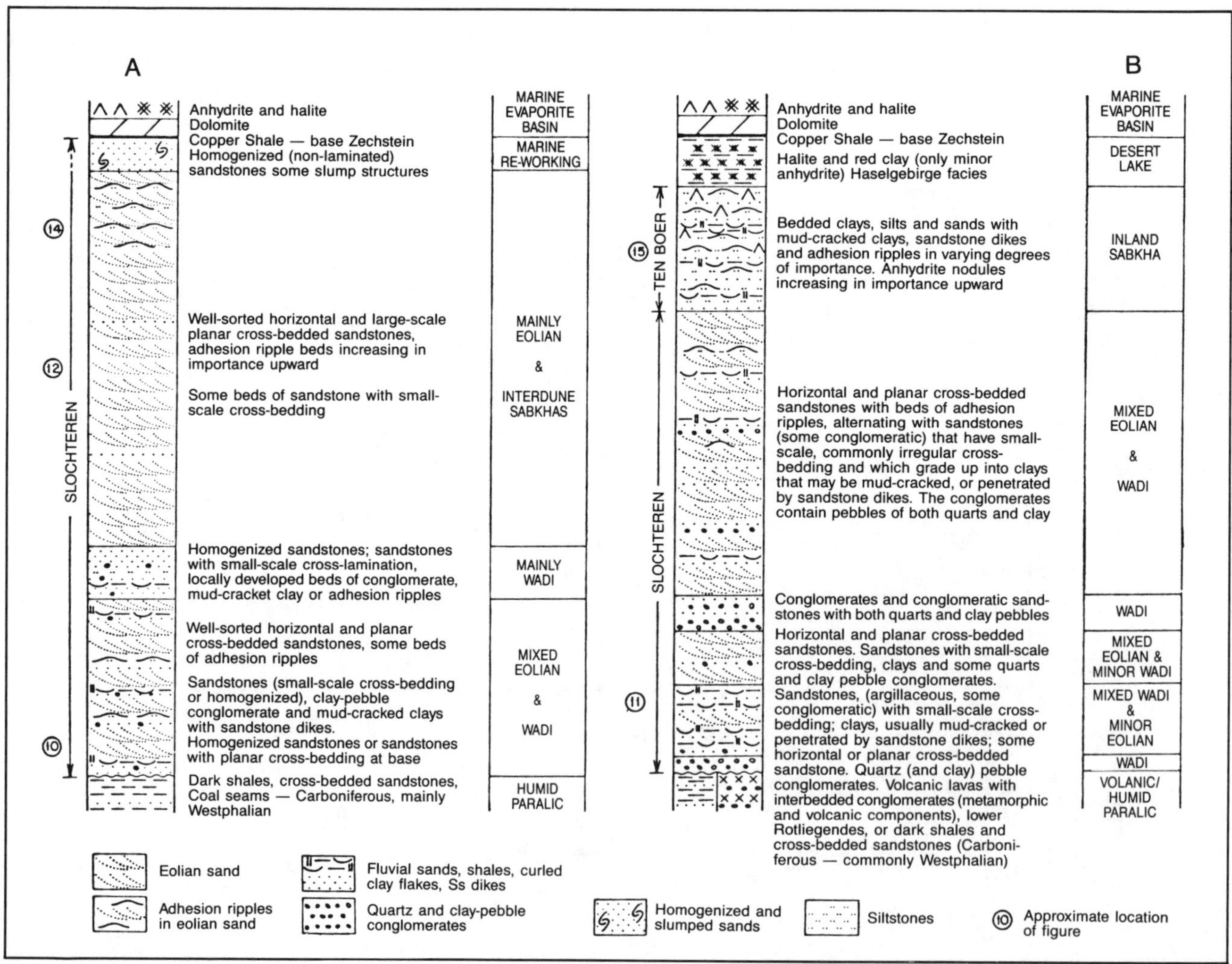

Fig. 6-22. Schematic cross section of Rotliegendes facies in southwestern (**A**) *and south-central* (**B**) *Rotliegendes basin.* (From Glennie, 1972; permission to publish by AAPG).

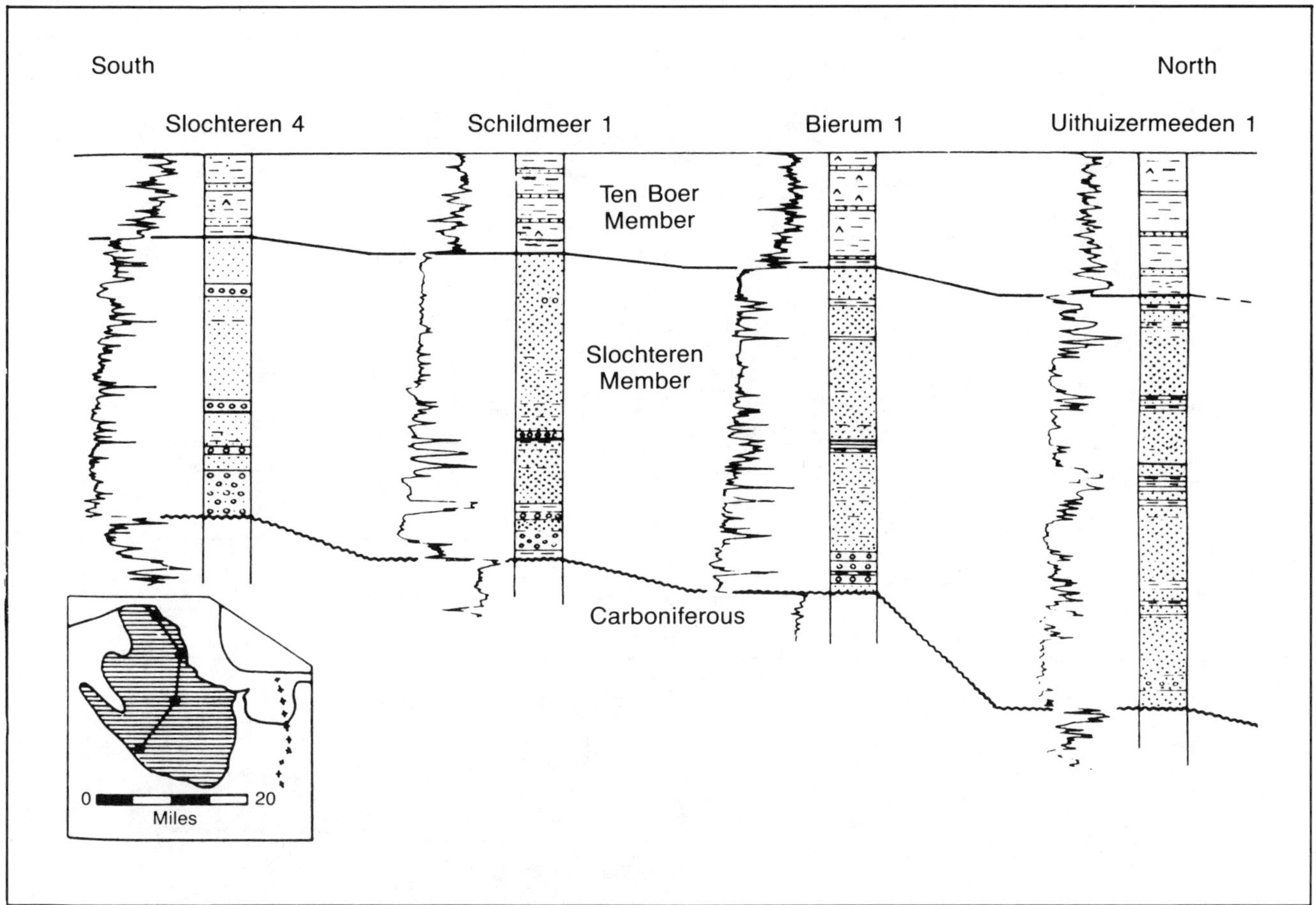

Fig. 6-23. North-south stratigraphic cross section through the Rotliegendes formation in Groningen gas field. (From Stäuble and Milius, 1970; permission to publish by AAPG).

Interdune sabkhas probably correspond to deflationary interdunes (Ahlbrandt and Fryberger, 1981).

Other sabkhas are associated with ephemeral desert lakes. Deposits consist of horizontally bedded clays with nodular anhydrite and little sand. Some of the clays are cracked and curled indicating subaerial dessication. Sedimentary structures in the sandstones indicate both eolian and fluvial processes were active. Environments suggested by the limited amount of anhydrite are temporary lakes supplied by ephemeral streams. The deposits seem to correspond to dry interdunes (Ahlbrandt and Fryberger, 1981).

In the southeastern portion of the Rotliegendes basin is the best development of a sabkha marginal to the large desert lake (Fig. 6-17). Deposits of eolian sandstone are overlain by red claystone with minor amounts of sandstone and siltstone. Nodular anhydrite occurs locally in sandy layers. Structures in the sandstones are both water and wind produced. Subaerial exposure and dessication are indicated by cracking and curling in some horizontally bedded clays. Adhesion ripples are common and associated with nodular anhydrite. The facies grades laterally, and sometimes vertically, into the Haselgebirge facies, a clayey sequence with up to 30% halite deposited in the desert lake. A nonmarine body of water in the Rotliegendes basin is indicated by deposits with no marine fossils or other salts associated with marine evaporites.

In addition to the distribution of facies shown in Figure 6-17, a general model for Rotliegendes deposition may be illustrated by schematic north-south cross sections (Fig. 6-19). A section to the west in the southern North Sea runs from the Variscan Mountains to the mid-North Sea High, and an eastern section runs from the mountains north to the Fyn Grinsted high. Both show a south to north progression of fluvial to dune to lakeshore sabkha to desert lake sedimentation.

Some of the northwest European gas fields are considered in light of Glennie's (1972) Rotliegendes model and facies.

Groningen. This giant gas field is located in the Netherlands (Fig. 6-20). The first exploratory well, a dry hole, drilled in the Groningen area tested Zechstein carbonates. When the well was deepened, 600 ft (183 m) of water-

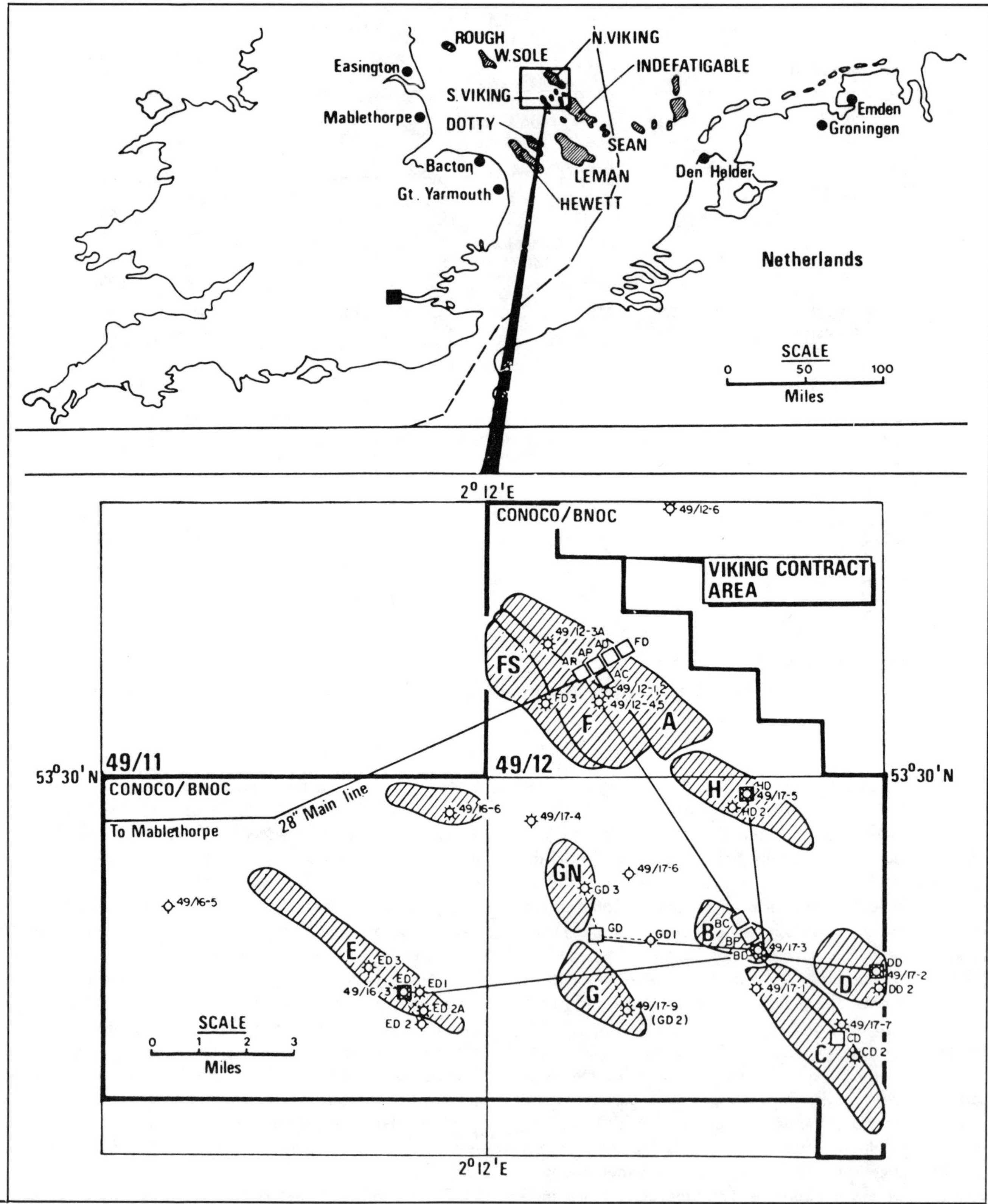

Fig. 6-24. Location maps for Viking gas field, southern North Sea. (From Gage, 1980; permission to publish by AAPG).

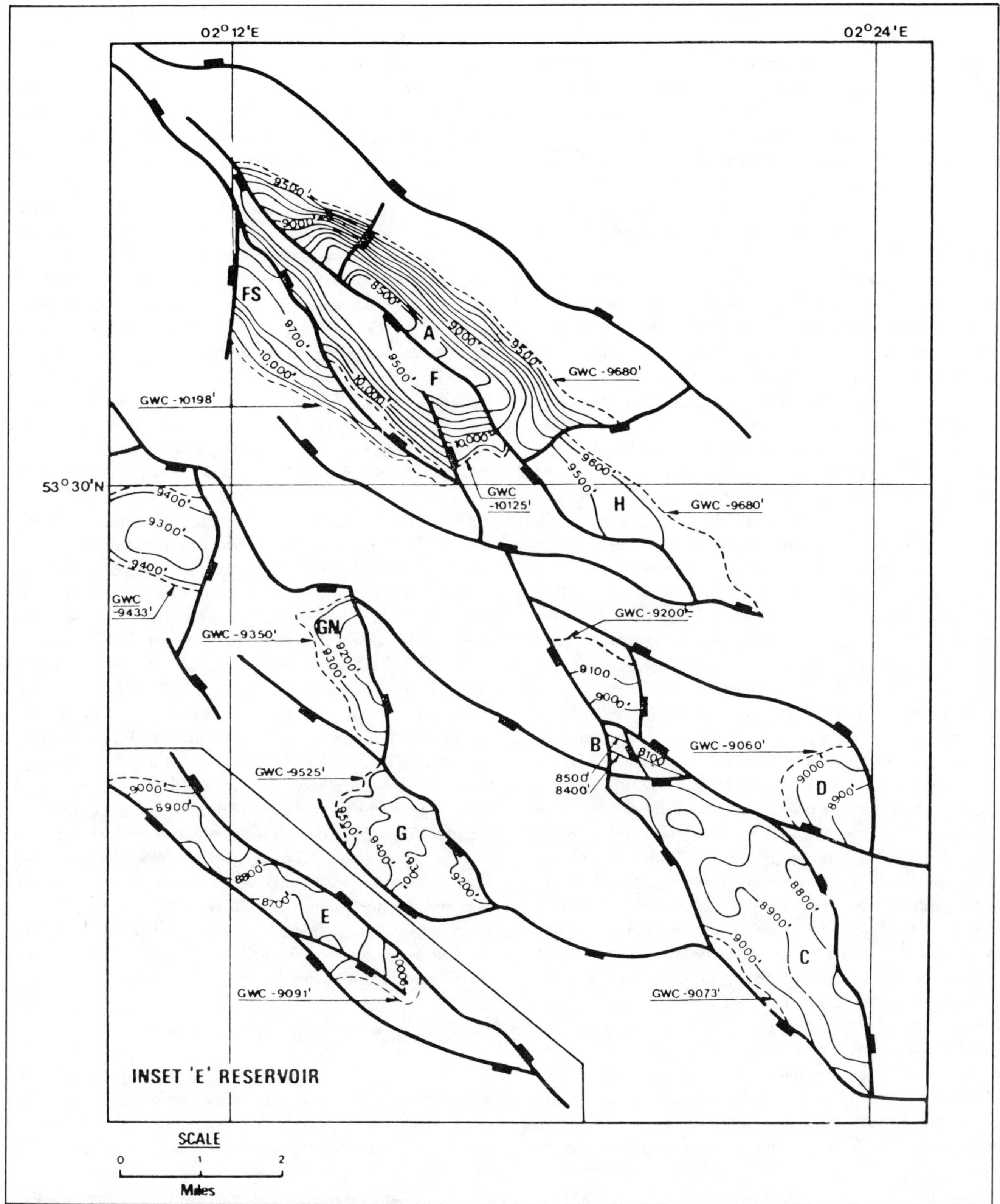

Fig. 6-25. Structure map of the Rotliegendes, Viking gas field. Note ten gas pools with different gas-water contacts. (From Gage, 1980; permission to publish by AAPG).

bearing, reservoir quality Rotliegendes sandstones were encountered. A later well, Slochteren 1, discovered gas in the Rotliegendes (Stäuble and Milius, 1970). Although not specifically stated, Stäuble and Milius (1970) imply that the original seismic maps of Groningen were substantially incorrect. They write, "The next well, Delfzijl 1, revealed only the upper part of the sequence to be gas-bearing on what then was thought to be a separate structure. Seismic detailing, additional appraisal wells, and the use of improved velocity data led to drastically altered maps showing the separate structures to be part of one giant gas field" (Stäuble and Milius, 1970).

Analysis of cores from Groningen wells led to a two member division of the Rotliegendes. Conventional cores of the entire formation taken in the Slochteren No. 4 well serve as the type section for the two members, the Slochteren and Ten Boer (Figs. 6-21 and 6-22). Eolian dune sandstones and wadi conglomerates, sandstones, and clays constitute the Slochteren member (Fig. 6-22). Generally, wadi beds are more common low in section with eolian deposits dominating the upper part. Slochteren sandstones are poorly sorted to well sorted, fine to pebbly, and brown-gray. Conglomerates are reddish with quartz and clay pebbles. More frequent occurrences of conglomerate in the southern part of Groningen field indicate the source for coarse clastics was to the south (Stäuble and Milius, 1970).

Overlying the Slochteren is the Ten Boer member of the Rotliegendes. It consists of reddish-brown, silty to fine sandy claystone, anhydrite nodules, and interbedded siltstones and sandstones (Fig. 6-22). The member thickens to the north (Fig. 6-23) and is said by Stäuble and Milius (1970) to be a sabkha deposit marginal to a subsiding coastal or continental basin. An equivalent environment is the sabkha facies marginal to the desert lake (Glennie, 1972).

Source rocks for the gas contained in the Rotliegendes eolian dune reservoirs are thought to be the Carboniferous coals below the Variscan unconformity (Stäuble and Milius, 1970; van Veen, 1975). Sealing is provided by the thick, overlying, Zechstein evaporites (Stäuble and Milius, 1970).

Viking Gas Field. Ten commercial gas pools constitute this field located in the British sector of the southern North Sea (Fig. 6-24). Each pool occurs on one of a series of northwest-southeast trending horsts and has its own gas-water contact (Fig. 6-25; Gray, 1975; Gage, 1980). A typical Rotliegendes succession of rocks occurs with eolian dune sandstones forming the major reservoirs (Fig. 6-26).

Lateral complexity in the Rotliegendes is shown in two northwest-southeast cross sections (Figs. 6-27 and 6-28). Section A-A′ (Fig. 6-27) is in the North Viking area passing through pools "A" and "H" (Fig. 6-25). Two sabkha zones separate the principally eolian section into three, correlatable units (Fig. 6-27). Either temporary desert lake interdunes or temporary expansions of the main desert lake from the north produced the sabkha horizons (Gage, 1980). Slochteren and Ten Boer rocks seen in the Groningen area are not identified here. Section B-B′ (Fig. 6-28), south of section A-A′, passes through pools "B" and "C" (Fig. 6-25). A wadi environment in the northwest changes to a dominantly eolian environment in the southeast. Eolian and wadi facies in the Viking field were identified by sedimentological analysis of conventional cores. Both facies were extended to uncored wells using log and dipmeter characteristics.

Initial seismic mapping of Rotliegendes fields was difficult. Major difficulties related to high horizontal seismic velocity gradients arose early in the Viking field's history when few velocity surveys were available (Gray, 1975). Variations in the thickness of Zechstein salts added to the problems (Butler, 1975; Gage, 1980). Extensive appraisal drilling and detailed seismic work alleviated many of the problems in Rotliegendes fields (Butler, 1975; van Veen, 1975). In the Viking field, "Continuous velocity analysis of most of the seismic lines over the structure enabled a satisfactory velocity gradient to be calculated, after much trial and error, and the seismically calculated depth picture is now in good accord with the well results." (Gray, 1975).

To summarize exploration in the Rotliegendes, descriptions of the facies and recognition of deposition under arid conditions came from examination of outcrops scattered around northwestern Europe. The significance of facies changes within the formation were not understood, however, until compared with modern desert facies (Glennie, 1970, 1972). Outcrop studies along with drilling in Germany, Denmark, the Netherlands, and the North Sea demonstrated the existence of a large Permo-Triassic basin in northwestern Europe with the Rotliegendes being the basal formation (Glennie, 1972). Discovery of the Groningen gas field in 1959 established the prospectiveness of the area. Although structures in the Rotliegendes could be generally located seismically, details were obscure and seismic maps were often incorrect. Geophysical problems were related, in part, to variations in thickness of the Zechstein evaporites and strong, horizontal, seismic velocity gradients in units overlying the Rotliegendes. Not until sufficient appraisal wells were drilled and accurate velocity gradients calculated, were the seismic problems overcome.

Uses Of Logs In Eolian Environments

Wireline logs may be useful in distinguishing various eolian environments where there are sufficient petrological differences among the deposits. However, the petrological and depositional characteristics of the various environments must be known so their differences may be exploited.

With preservation of original porosity and permeability, it may be possible to distinguish between different facies in dune-extradune deposits (Lupe and Ahlbrandt, 1979). Dune sediments are relatively well-sorted, well-rounded, and

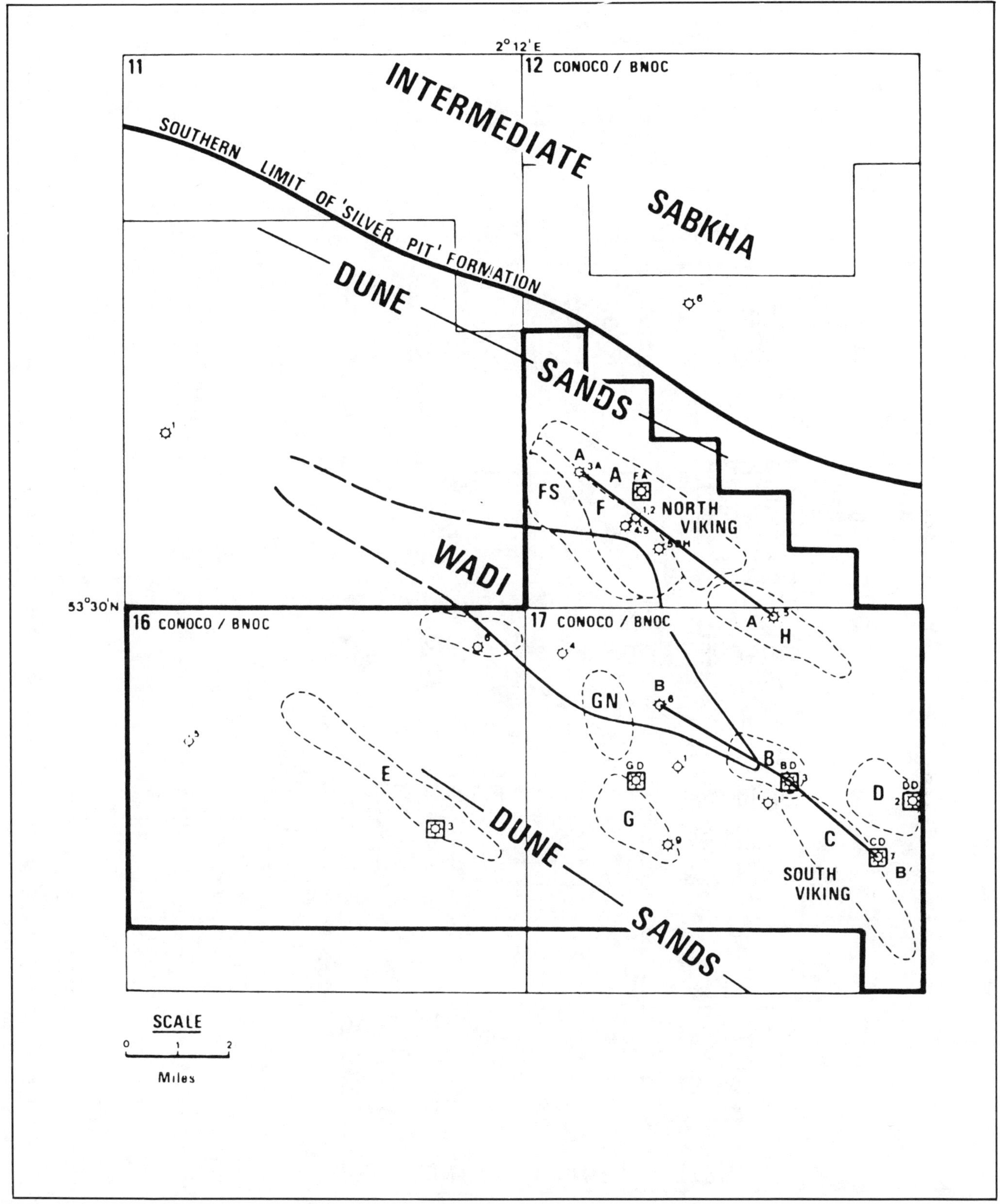

Fig. 6-26. Distribution of desert environments in the Rotliegendes formation, Viking gas field. A-A' *and* B-B' *are approximate lines of cross sections in Figures 6-27 and 6-28.* (Modified after Gage, 1980; permission to publish by AAPG).

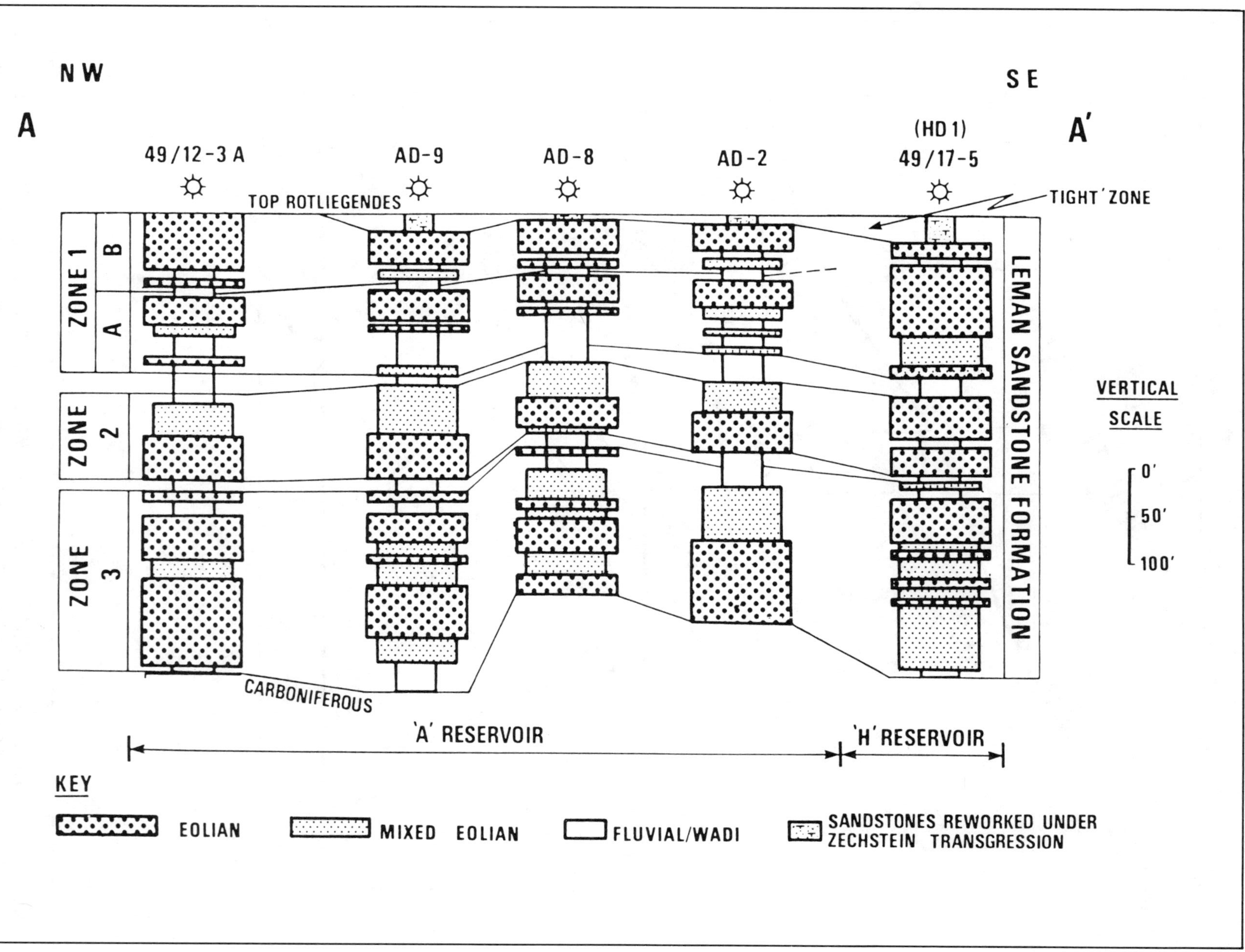

Fig. 6-27. Lateral facies variation of Rotliegendes formation, North Viking gas field. Eolian dunes are dominant facies. See Figure 6-26 for approximate location of cross section. (From Gage, 1980; permission to publish by AAPG).

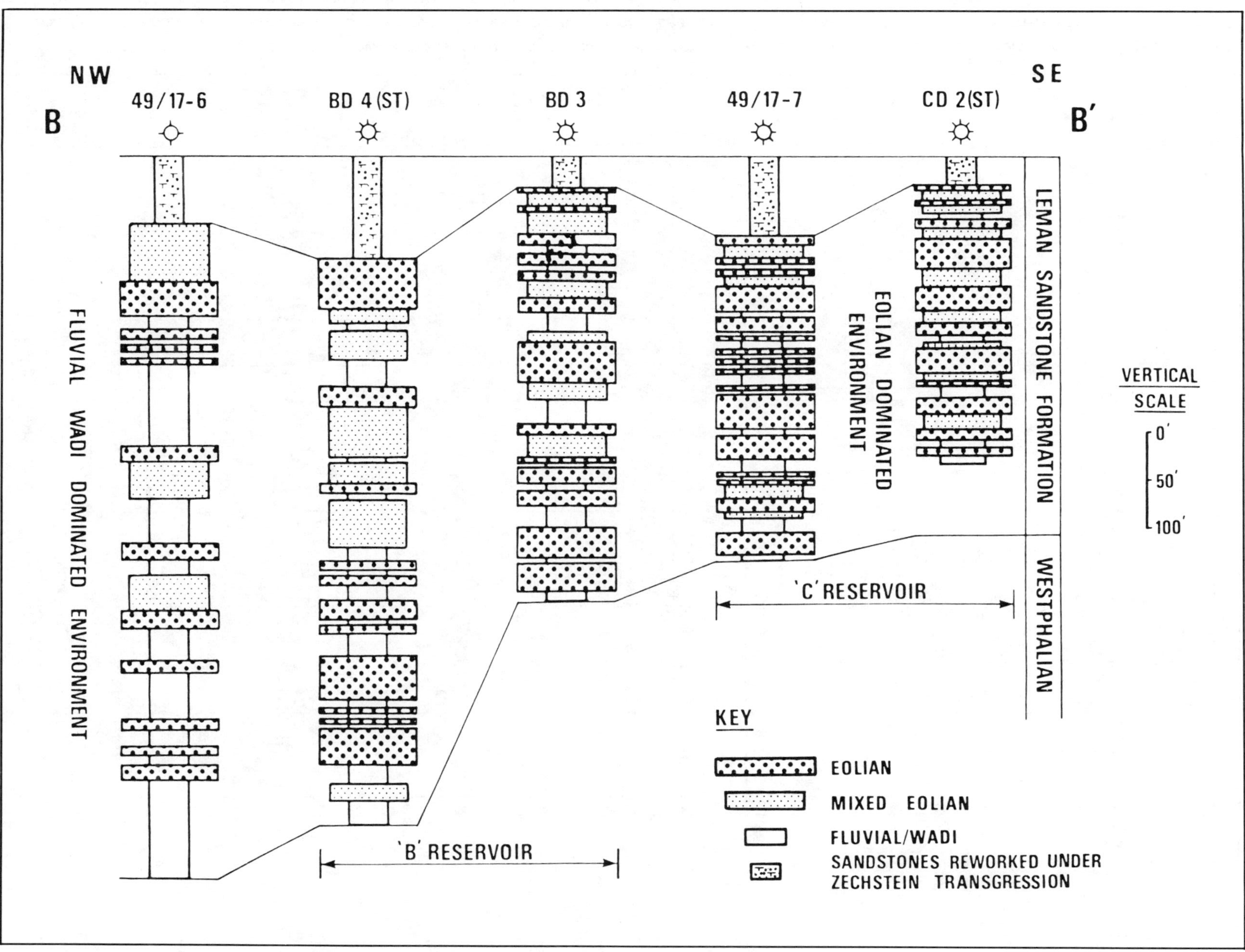

Fig. 6-28. Lateral facies variation of Rotliegendes formation around South Viking gas field. Wadi-dominated environment to northwest becomes eolian dominated environment to southeast. Approximate location of cross section shown in Figure 6-26. (From Gage, 1980; permission to publish from AAPG).

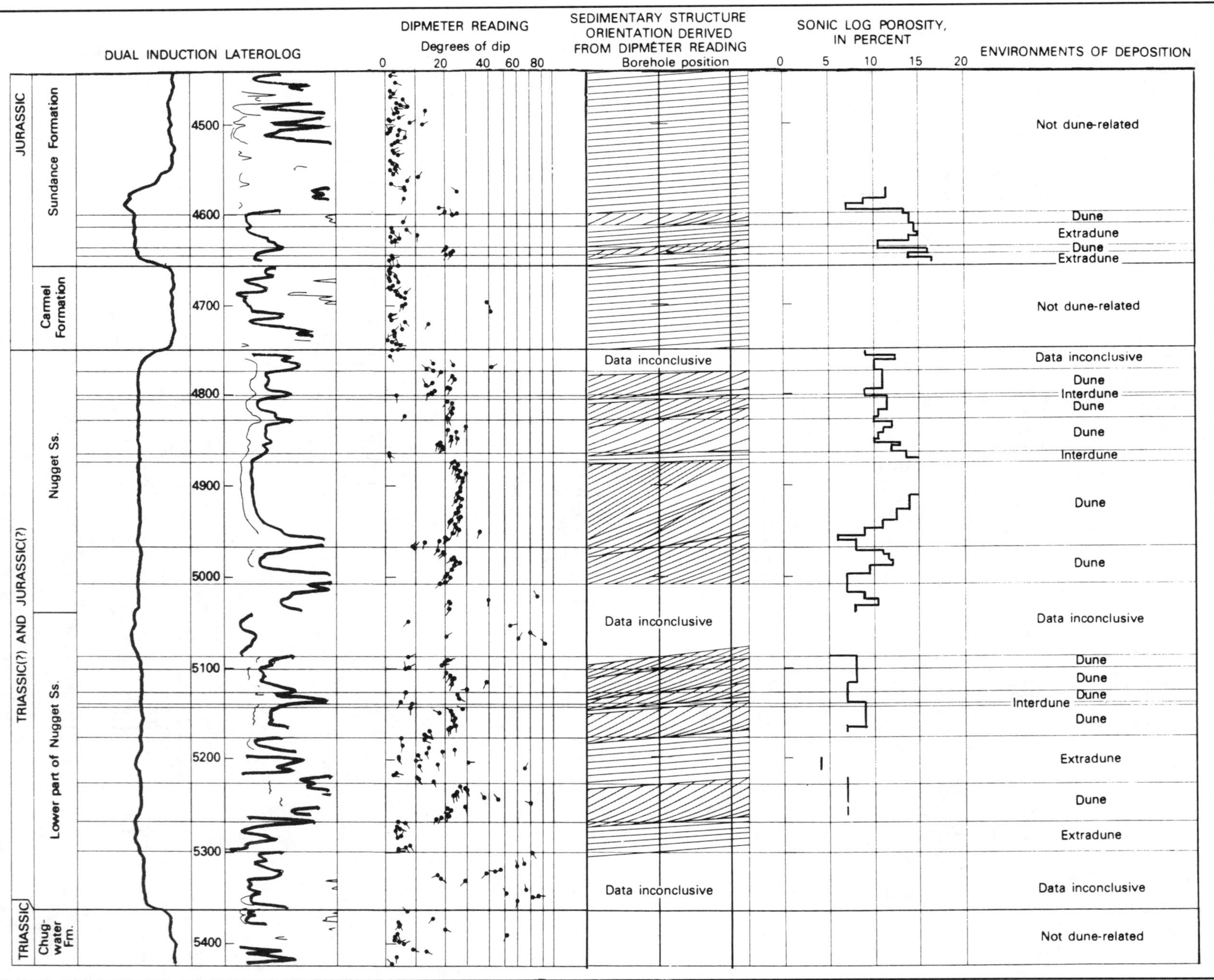

Fig. 6-29. Identification of depositional environments from sedimentary structures and petrological properties as seen on electric logs. Note correlation of porosity with sedimentary structures and distinction between dune and nondune facies with dipmeter and resistivity curves. Logs from Hilliard Oil and Gas Co., Joyce Creek 1, Sweetwater County, Wyoming. (From Lupe and Ahlbrandt, 1979; permission courtesy U. S. Geological Survey).

mineralogically mature. Extradune and interdune sediments are most likely deposited by water and less mature texturally and compositionally than dune sands. Therefore, more round and better-sorted dune sediments should be more porous and permeable than the interdune and extradune sediments (Lupe and Ahlbrandt, 1979). Dune sediments, being slipface deposits, should also have higher initial dips than the other sediments.

Conventional cores and electric logs often detect the indicated differences. Porosity and permeability may be measured directly from cores and calculated from logs. Depositional environments may be interpreted by observing sedimentary structures and textures in conventional cores, from electric logs calibrated with cores, and from dipmeter data (Lupe and Ahlbrandt, 1979).

In the Hilliard Oil and Gas Co. Joyce Creek 1 well in Wyoming, logs effectively differentiate between dune and nondune facies (Fig. 6-29). High-angle slipface dune deposits and the low-angle interdune or extradune beds are readily distinguished by the dipmeter. Generally responding inversely to porosity, resistivity is low in high porosity zones and high in low porosity zones. Depositional environment is implied by resistivity as porous zones should be dunes and less porous zones should be nondune beds (Lupe and Ahlbrandt, 1979). This example illustrates how knowledge of depositional environments and their petrological characteristics may be used to gain additional information from wireline logs.

SELECTED BIBLIOGRAPHY

Ahlbrandt, T. S., 1975, Comparison of textures and structures to distinguish eolian environments, Killpecker dune field, Wyoming: The Mountain Geologist, v. 12, no. 2, p. 61-73.

_______, 1979, Textural parameters in eolian deposits, *in* E. D. McKee, ed., A study of global sand seas: U.S. Geol. Survey Prof. Paper 1052, p. 21-52.

_______ and S. Andrews, 1978, Distinctive sedimentary features of cold climate eolian deposits, North Park, Colorado: Palaeogeography, Palaeoclimatology, Palaeoecology, v. 25, p. 327-351.

_______, _______, and D. T. Gwynne, 1978, Bioturbation in eolian deposits: Jour. Sed. Pet., v. 48, no. 3, p. 839-848.

_______, and S. G. Fryberger, 1980, Eolian deposits in the Nebraska Sand Hills: U. S. Geol. Survey Prof. Paper 1120 A, p. 1-24.

_______ and _______, 1981, Sedimentary features and significance of interdune deposits, *in* F. G. Ethridge and R. M. Flores, eds., Recent and ancient nonmarine depositional environments: models for exploration: SEPM Spec. Pub. No. 31, p. 293-314.

_______ and _______, 1982, Introduction to eolian deposits, *in* P. A. Scholle and D. Spearing, eds., Sandstone depositional environments: AAPG Mem. 31, p. 11-47.

Bagnold, R. A., 1941, The physics of blown sand and desert dunes: London, Methuen, 265 p.

Bigarella, J. J., 1972, Eolian environments: their characteristics, recognition and importance, *in* J. K. Rigby and W. K. Hamblin, eds., Recognition of ancient sedimentary environments: SEPM Sp. Pub. No. 16, p. 12-62.

_______, 1975, Lagoa dune field (State of Santa Catarina, Brazil), a model of eolian and pluvial activity: Int. Symp. Quarternary, Bol. Paran. Geoci., v. 33, p. 133-167.

Butler, J. B., 1975, The West Sole gas field, *in* A. W. Woodland, ed., Petroleum and the continental shelf of northwest Europe, 1. Geology: London, Applied Sci. Pubs., p. 213-219.

Emmett, W. R., K. W. Beaver, and J. A. McCaleb, 1972, Pennsylvanian Tensleep reservoir, Little Buffalo Basin oil field, Big Horn Basin, Wyoming: The Mountain Geologist, v. 9, no. 1, p. 21-31.

France, D. S., 1975, The geology of the Indefatigable gas field, *in* A. W. Woodland, ed., Petroleum and the continental shelf of the northwest Europe, 1. Geology: London, Applied Sci. Pubs., p. 233-239.

Fryberger, S. G., 1979, Eolian-fluviatile (continental) origin of ancient stratigraphic trap for petroleum in Weber Sandstone, Rangely oil field, Colorado: The Mountain Geologist, v. 16, no. 1, p. 1-36.

_______, T. S. Ahlbrandt, and S. Andrews, 1979, Origin, sedimentary features, and significance of low-angle eolian "sand sheet" deposits, Great Sand Dunes National Monument and vicinity, Colorado: Jour. Sed. Pet., v. 49, no. 3, p. 733-746.

Gage, M., 1980, A review of the Viking gas field, *in* M. T. Halbouty, ed., Giant oil and gas fields of the decade: 1968-1978: AAPG Mem. 30, p. 39-57.

Glennie, K. W., 1970, Desert sedimentary environments: Developments in sedimentology 14: Amsterdam, Elsevier, 222 p.

_______, 1972, Permian Rotliegendes of northwest Europe interpreted in light of modern desert sedimentation studies: AAPG Bull., v. 56, no. 6, p. 1048-1071.

_______ and B. D. Evamy, 1968, Dikaka: plants and root structures associated with eolian sand: Palaeogeography, Palaeoclimatology, Palaeoecology, v. 23, p. 77-87.

_______, G. C. Mudd, and P. J. C. Nagtegaal, 1978, Depositional environment and diagenesis of Permian Rotliegendes sandstones in Leman Bank and Sole Pit areas of the UK southern North Sea: Jour. Geol. Soc. London, v. 135, p. 25-34.

Gradzinski, R., and T. Jerzykiewicz, 1974, Dinosaur- and

mammal-bearing aeolian and associated deposits of the Upper Cretaceous in the Gobi Desert (Mongolia): Sed. Geol. v. 12, no. 4, p. 249-278.

Gray, I., 1975, Viking gas-field, *in* A. W. Woodland, ed., Petroleum and the continental shelf of northwest Europe, 1. Geology: London, Applied Sci. Pubs., p. 241-247.

Hanley, J. H., 1980, Paleoecology of nonmarine Mollusca from some paleointerdune deposits in the Nebraska Sand Hills: U. S. Geol. Survey Prof. Paper 1120B, p. 25-28.

Hunter, R. E., 1981, Stratification styles in eolian sandstones: some Pennsylvanian to Jurassic examples from the western interior U. S. A., *in* F. G. Ethridge and R. M. Flores, eds., Recent and ancient nonmarine depositional environments: models for exploration: SEPM Sp. Pub. no. 31, p. 315-329.

Lamb, C. F., 1980, Painter Reservoir field: giant in the Wyoming thrust belt, *in* M. T. Halbouty, ed., Giant oil and gas fields of the decade: 1968-1978: AAPG Mem. 30, p. 281-288.

LeBlanc, R. J., Sr., 1977, Distribution and continuity of sandstone reservoirs—Part 1: Jour. Pet. Tech., v. 29, July, p. 776-792.

Lupe, R., and T. S. Ahlbrandt, 1979, Sediments of the ancient eolian environment—reservoir inhomogeneity, *in* E. D. McKee, ed., A study of global sand seas: U. S. Geol. Survey Prof. Paper 1052, p. 241-252.

Marie, J. P. P., 1975, Rotliegendes stratigraphy and diagenesis, *in* A. W. Woodland, ed., Petroleum and the continental shelf of northwest Europe, 1. Geology: London, Applied Sci. Pubs., p. 205-210.

McKee, E. D., 1979, Introduction to a study of global sand seas, *in* E. D. McKee, ed., A study of global sand seas: U. S. Geol. Survey Prop. Paper 1052, p. 1-19.

_______ and J. J. Bigarella, 1979, Sedimentary structures in dunes, *in* E. D. McKee, ed., A study of global sand seas: U. S. Geol. Survey Prop. Paper 1052, p. 83-134.

_______ and G. C. Tibbits, Jr., 1964, Primary structures of a seif dune and associated deposits in Libya: Jour. Sed. Pet., v. 34, no. 1, p. 5-17.

Merk, G. P., 1960, Great sand dunes of Colorado, *in* Guide to the geology of Colorado: Geol. Soc. Amer., Rocky Mtn. Assoc. Geol., and Colorado Sci. Soc., p. 127-129.

Moiola, R. I., and A. B. Spencer, 1979, Differentiation of eolian deposits by discriminant analysis, *in* E. D. McKee, ed., A study of global sand seas: U. S. Geol. Survey Prop. Paper 1052, p. 53-60.

Morgan, J. T., F. S. Cordiner, and A. R. Livingston, 1978, Tensleep reservoir, Oregon Basin field, Wyoming: AAPG Bull., v. 62, no. 4, p. 609-632.

Oele, J. A., A. C. P. J. Hol, and J. Tiemens, 1981, Some Rotliegendes gas fields of the K and L blocks, Netherlands offshore (1968–1978)—a case history, *in* L. V. Illing and G. D. Hobson, eds., Petroleum geology of the continental shelf of northwest Europe: Institute of Petroleum Geology, London, p. 289-300.

Picard, M. D., 1975, Facies, petrography and petroleum potential of Nugget Sandstone (Jurassic), southwestern Wyoming and northeastern Utah, *in* D. W. Bolyard, ed., Deep drilling frontiers of the central Rocky Mountains: Rocky Mountain Assoc. Geologists, p. 109-127.

Ruhe, R. V., 1975, Geomorphology: Boston, Houghton Mifflin Co., 246 p.

Sharp, R. P., 1963, Wind ripples: Jour. Geol., v. 71, no. 5, p. 617-636.

Sneh, A., 1983, Desert stream sequences in the Sinai Peninsula: Jour. Sed. Pet., v. 53, no. 4, p. 1271-1279.

Stäuble, A. J., and G. Milius, 1970, Geology of Groningen gas field, Netherlands, *in* M. T. Halbouty, ed., Geology of giant petroleum fields: AAPG Mem. 14, p. 359-369.

Tanner, W. R., 1967, Ripple mark indices and their uses: Sedimentology, v. 9, no. 2, p. 89-104.

van Veen, F. R., 1975, Geology of the Leman gas-field, *in* A. W. Woodland, ed., Petroleum and the continental shelf of northwest Europe, 1. Geology: London, Applied Sci. Pubs., p. 223-231.

Walker, T. R., and G. V. Middleton, 1977, Facies models 9, eolian sands: Geoscience Canada, v. 4, no. 4, p. 182-189.

7 LACUSTRINE SEDIMENTS

INTRODUCTION

Lakes exist in a variety of settings on the earth's surface. They occur in polar to equatorial regions, in humid to arid climates, and in sizes from small ponds to major water bodies covering hundreds of thousands of square kilometers. Lake origins are varied and include formation in tectonic depressions, damming of drainages by landslides and volcanics, and interruption of drainage by glaciation. Several modern lakes occur in lows along the East African Rift, suggesting ancient rifts and passive margins as likely sites for lakes (Veevers, 1977; Veevers and Cotterill, 1978). Examples of ancient rift basin deposits are the Lockatong formation, a Triassic lake deposit in the eastern U.S., and the Lagoa Feia formation, a Lower Cretaceous lacustrine deposit in the Campos Basin, Brazil (Bacoccoli et al., 1980). Ancient lakes (pre-Pleistocene) seem to be primarily tectonic in origin. Lacustrine deposits in structural basins include the Miocene of the block faulted Zhanhua Basin in eastern China (Sizhong and Ping, 1980) and the Green River formation (Eocene) in the structurally low Uinta Basin, northeastern Utah, U.S.A. (Ryder et al., 1976).

Lake deposits are of interest to the exploration geologist as they may be prolific producers of hydrocarbons. Production from lacustrine sediments occurs in China (Meyerhoff and Willums, 1976; Sizhong and Ping, 1980), the Green River formation, Utah, U.S.A. (Chatfield, 1972), and the Lagoa Feia formation, Campos Basin, Brazil (Bacoccoli et al., 1980). The Brazilian example suggests it is not unreasonable to expect future production from other passive margin lake beds on either side of the Atlantic and elsewhere in the world.

IDENTIFICATION OF LACUSTRINE DEPOSITS

Lacustrine sediments are deposited under a wide range of conditions. Lake waters vary from fresh to hypersaline; bottom waters may be agitated and oxidizing or stagnant and reducing; biological activity may be little or great; and deposits include clastics, organics, and evaporites. Variations in the conditions under which lake sediments are deposited may make identification of the deposits difficult.

Additional complications arise in trying to distinguish between inland seas and lakes. Both have wave action and nearshore currents that produce similar facies patterns, stratigraphic sequences, lithologies, sedimentary structures, bedding, and paleocurrent patterns (Picard and High, 1972). Offshore, below wave base, a lake floor may be less disturbed by physical processes while a sea floor may have more current or organic activity. However, oxygenated lake bottoms may be reworked by organisms (Picard and High, 1972), and turbidity currents may flow along lake floors, making the lake deposits resemble marine deposits.

Separation of fluvial and fresh water lake deposits also has its pitfalls. Stream deposits show unidirectional currents, which assist in their identification. However, extensive overbank deposits laid down by slow-moving, receding flood waters may closely resemble lacustrine sediments deposited from standing water (Picard and High, 1972).

Summaries of characteristics of lacustrine deposits are given by Fouch and Dean (1982) and Picard and High (1972, 1981). Characteristics mentioned by the former authors are more applicable to recent north temperate lakes while those of the latter seem to be more oriented toward ancient lakes and more appropriately considered here. The following summary of lake deposit characteristics is taken from Picard and High (1972, 1981); the interested reader is referred to their papers for more detailed discussions.

Stratigraphic Characteristics

Lacustrine deposits may cover small or large areas (Figs. 7-1 and 7-2), but even the most extensive ancient lake, the 130,000 km^2 Popo Agie (Fig. 7-2), is substantially smaller than an epicontinental sea. Thus, lacustrine deposits will not be as laterally extensive or as linear as marine rocks with which they may be confused.

Compared with marine and inland sea deposits, shore and nearshore lacustrine deposits form narrow bands because lakes have shallow wave bases and virtually no tides to spread the sediments normal to the shoreline. Transition from the fluvial-deltaic environment to the offshore is the same as in marine rocks (refer to Chap. 8) but it occurs more rapidly. However, changing lake level, especially over a gently sloping bottom, can spread the transition deposits substantially. Narrow bands of shore and nearshore deposits suggest lake depositior. but wide zones of these deposits do not preclude a lake environment.

Vertical sedimentary sequences may serve to identify lacustrine deposits. As a lake fills with sediment, the ideal sedimentary sequence should be regressive and coarsen upward (Fig. 7-3; Twenhofel, 1932; Visher, 1965). Sediments

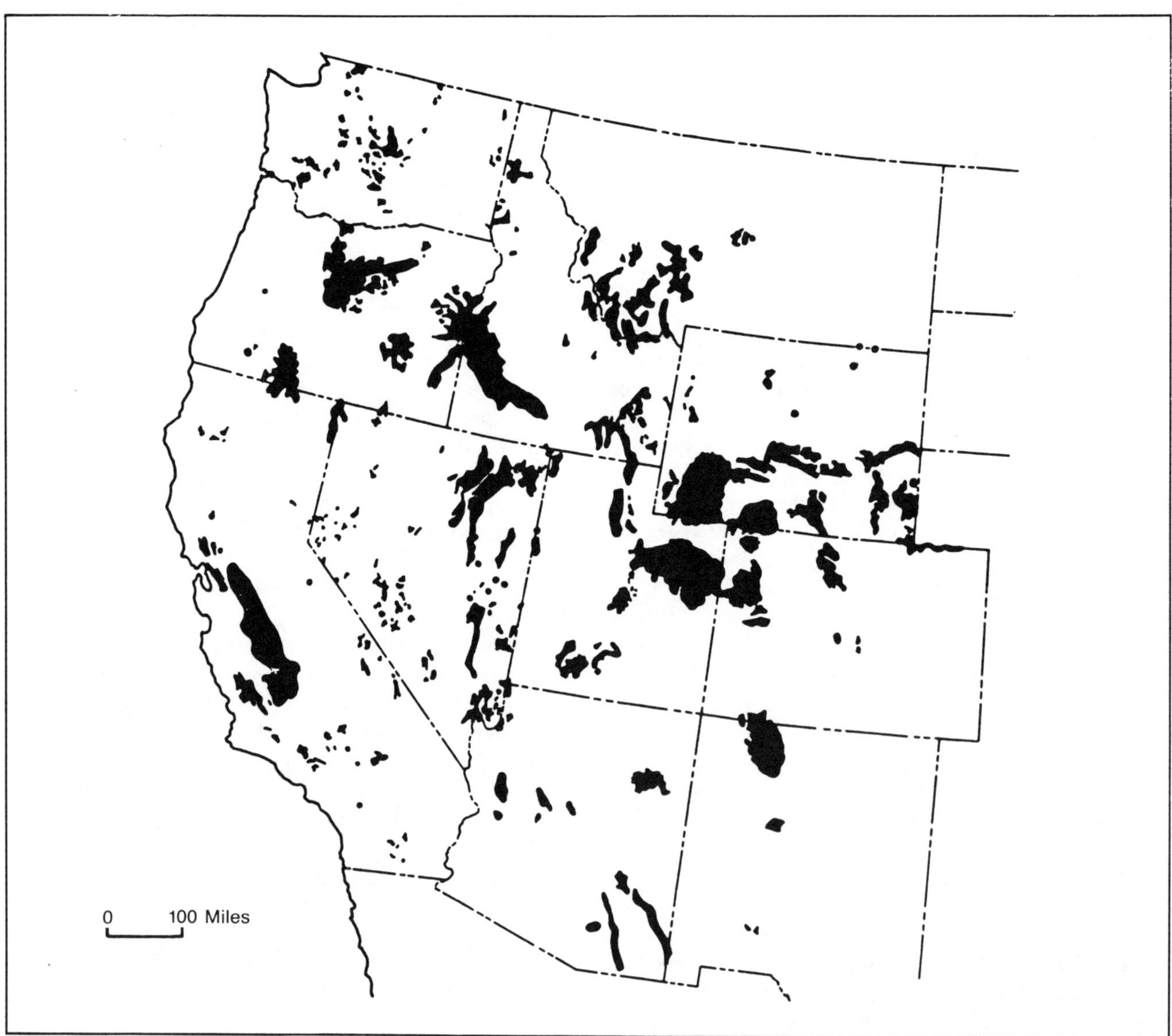

Fig. 7-1. Distribution of Tertiary lake deposits in western U.S. (From Feth, 1963, as published in Picard and High, 1972; permission to publish by SEPM).

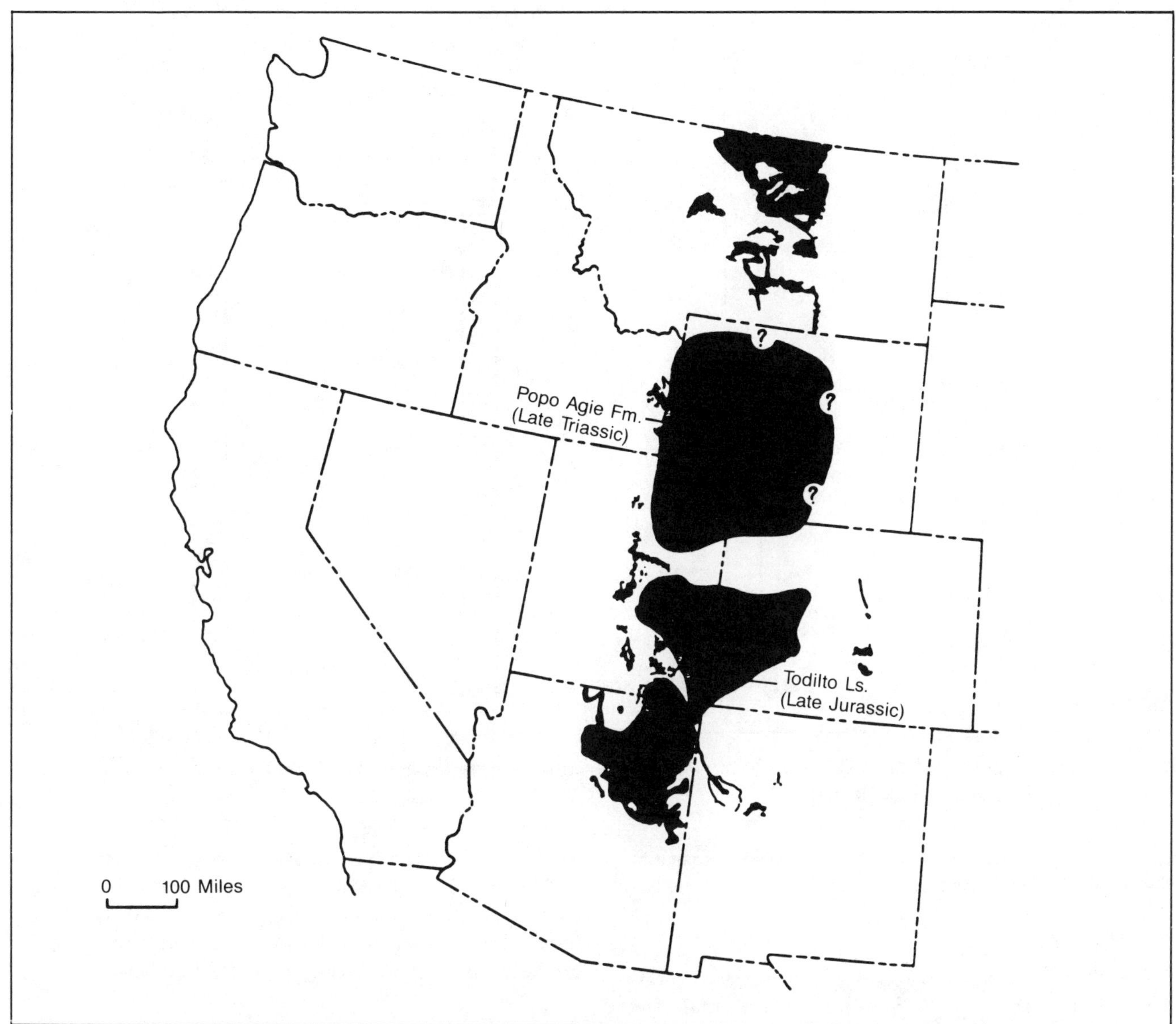

Fig. 7-2. Distribution of pre-Tertiary lake deposits in western U.S. (From Feth, 1964, as published in Picard and High, 1972, permission to publish by SEPM).

in Lake Uinta approximate the ideal (Fig. 7-3). However, cyclic sedimentation in the Lockatong formation and the absence of any obvious depositional trend in the Popo Agie (Fig. 7-3) indicate lacustrine sedimentary sequences are generally complex and do not conform to the ideal.

Lateral stratigraphic associations, on the other hand, may distinguish lake deposits. Lake sediments, being continental features, should be associated with other continental sediments, soil zones, and unconformities. Any or all of these continental characteristics should surround the suspected lacustrine deposits. However, syndepositional marine rocks on one or more sides would suggest coastal lagoon rather than lake deposition.

Paleontology

Organisms and organic debris preserved in sediments aid in identifying lacustrine sediments but do not do so strictly on their own. Fresh water lake deposits may be distinguished from marine units on the basis of certain fossils, but the fresh water organisms may also occur in fluvial deposits. Spores and pollen are readily transported into lacustrine and marine environments and are of limited utility in differentiating between the two environments. A purely terrestrial assemblage of spores suggests nonmarine conditions but does not distinguish between stream, swamp, and lake deposits. Other likely nonmarine fossils are insects

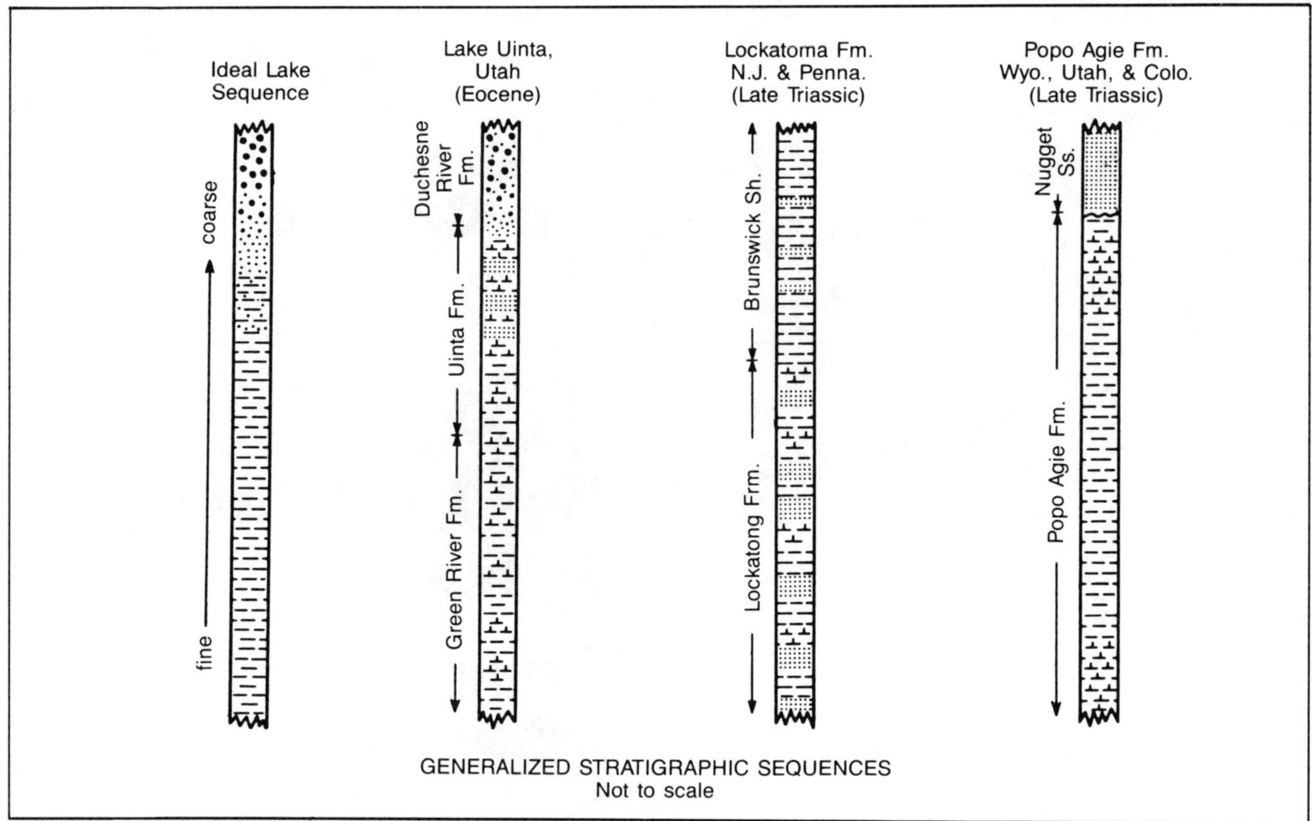

Fig. 7-3. Comparison of "ideal" lacustrine sedimentary sequence with three ancient lacustrine sequences. (From Picard and High, 1972, permission to publish by SEPM).

and vertebrates except for fish. Insect larvae in particular indicate lacustrine environments.

Geochemistry

Evaporite and authigenic mineral suites are useful in differentiating between deposits from marine and nonmarine bodies of water. Marine evaporites are chemically simple and consist of common minerals such as calcite, dolomite, gypsum, anhydrite, halite, and chlorides of potassium and magnesium. Lacustrine evaporites are more complex and consist of bicarbonates, carbonates, hydrous carbonates, sulfates, and chlorides of sodium and calcium.

Lacustrine authigenic mineral assemblages tend to be more sodium-rich and complex than marine assemblages because lake chemistry is responsive to source area chemistry. Suites, such as those in Table 7-1, could only occur in lacustrine deposits, and some authigenic minerals in association with iron minerals may indicate Eh-pH conditions that are found only in nonmarine waters. Unusual mineral suites should also be useful in differentiating between lacustrine and fluvial deposits.

Cyclic Deposition

As lake levels rise and fall, transgressive and regressive sedimentary sequences are deposited in nearshore environments. In Lake Uinta, such sequences correspond to channel cutting and infilling as the lake margin moved back and forth across floodplains peripheral to the lake. Cyclic deposition also occurs in marine deposits but lacustrine cycles are likely to be more widespread and significant.

Ancient Lake Examples

Three examples of ancient lake deposits illustrate the characteristics used to identify the deposits. In each case, it is the sum of the characteristics that best serves to specify the lacustrine environment.

Popo Agie formation. The Popo Agie formation was deposited in a large, late Triassic lake in the western United States (Fig. 7-2). It consists of four informal units, a purple and an ocher unit between a lower and an upper carbonate. The lower carbonate, a carbonate pebble conglomerate, is overlain by the purple unit, a silty, montmorillonitic mudstone. Both are fluvial in origin and contain other sediments deposited in local ponds or isolated lakes. Overlying the purple unit are lake deposits of the ocher unit and upper carbonate that consist of silty, analcimic, montmorillonitic mudstone and argillaceous dolomite to dolomitic mudstone, respectively. Erosion truncates the Popo Agie to the north and east but correlative fluvial deposits mark the southern and western limits. Fossils are not common but include amphibians, reptiles, and freshwater pelecypods.

Paleontology, mineralogy, geochemistry and stratigraphic associations indicate the lacustrine nature of the Popo Agie. Fossils are nonmarine, and analcime and carbonate minerals indicate a basic pH. That, plus high alkali content, suggest a lacustrine rather than a fluvial environment as streams tend to be more acidic and have lower concentrations of alkalis. Additional support for a lacustrine rather than a marine environment comes from the lateral transition to fluvial beds and the lack of any horizontally equivalent marine units.

Green River and Uinta formations. Mineralogy, paleontology, cyclic deposits, and enclosing fluvial rocks indicate lacustrine environments for the Green River and Lower Uinta formations. Eocene lacustrine strata were deposited in Lake Uinta in northeastern Utah and northwestern Colorado as equivalent rocks were deposited in Lake Gosiute in southwestern Wyoming (Fig. 7-1). Sediments deposited in central areas of Lake Uinta include oil shale, dark calcareous and dolomitic shales, and sandy to argillaceous limestone and dolomite. Shoreline deposits are red and green claystones, calcareous and argillaceous siltstone and sandstone, intraformational and chert pebble conglomerate, and assorted oolitic, algal, and clastic carbonates. These rocks form a lenticular deposit that is encased in fluvial deposits. Fossils include well-preserved fish, bone fragments of birds and turtles, gastropods, algae, and ostracods. Laterally equivalent fluvial beds contain a mammalian fauna.

Depositional cycles related to fluctuating lake level occur along with corresponding episodes of lake margin fluvial channel cut and fill. Formation mineralogy (Table 7-1) consists of many unusual carbonates and silicates.

Lockatong formation. This Upper Triassic formation is the lacustrine facies of the nonmarine Newark Group deposited in the Newark rift basin in New Jersey and Pennsylvania (U.S.A). Publications by Van Houten (1964, 1965) serve as the basis for the following discussion.

The Lockatong formation is a lens of sediment 90 miles

TABLE 7-1. Authigenic minerals found in deposits from Searles Lake, California, and the Green River formation. (From Picard and High, 1972; permission to publish from SEPM).

Searles Lake (late Quaternary)	Green River Formation (Eocene)	
Carbonates	*Carbonates*	*Zeolites*
Aragonite $CaCO_3$	"Alstonite-bromlite" (?) $CaBa(CO_3)_2$	Analcime $NaAlSi_2O_6 \cdot H_2O$
Burkeite $2Na_2SO_4 \cdot Na_2CO_3$	Barytocalcite $CaBa(CO_3)_2$	Clinoptilolite $NaAlSi_{4.2-5}O_{10.4-12} \cdot 3.5–4H_2O$
Calcite $CaCO_3$	Brayleyite $Na_3PO_4 \cdot MgCO_3$	Mordenite $NaAlSi_{4.5-5}O_{11-12} \cdot 3.2–3.5H_2O$
Dolomite $CaMg(CO_3)_2$	(carbonate-phosphate)	
Gaylussite $Na_2Ca(CO_3)_2 \cdot 5H_2O$	Burbankite $Na_2(Ca,Sr,Ba,Ce)_4(CO_3)_5$	*Sulfides*
Nahcolite $NaHCO_3$	Calcite $CaCO_3$	Marcasite FeS_2
Pirssonite $Na_2Ca(CO_3)_2 \cdot 2H_2O$	Dawsonite $Na_3Al(CO_3)_3 \cdot 2Al(OH)_3$	Pyrite FeS_2
Trona $Na_2CO_3 \cdot NaHCO_3 \cdot 2H_2O$	Dolomite $CaMg(CO_3)_2$	Pyrrhotite $Fe_{1-x}S$
Tychite $2Na_2CO_3 \cdot 2MgCO_3 \cdot Na_2SO_4$	Eitelite $Na_2Mg(CO_3)_2$	Wurtzite ZnS
	Gaylussite $Na_2Ca(CO_3)_2 \cdot 5H_2O$	
Silicates	Magnesite $MgCO_3$	*Sulfates*
Adularia $KalSi_3O_8$	Nahcolite $NaHCO_3$	Anhydrite $CaSO_4$
Searlesite $NaBSi_2O_6 \cdot H_2O$	Pirssonite $Na_2Ca(CO_3)_2 \cdot 2H_2O$	Barite $BaSO_4$
	Shortite $Na_2Ca_2(CO_3)_2$	(?) Bassanite $CaSO_4 \cdot 1/2\ H_2O$
Zeolites	Siderite $FeCO_3$	Gypsum $CaSO_4 \cdot 2H_2O$
Analcime $NaAlSi_2O_6 \cdot H_2O$	Thermonatrite $Na_2CO_3 \cdot H_2O$	
Phillipsite $KCa(Al_3Si_5O_{16}) \cdot 6H_2O$	Trona $Na_2CO_3 \cdot 2H_2O$	*Chlorides*
	Witherite $BaCO_3$	Halite NaCl
Sulfates		Northupite $Na_2CO_3 \cdot MgCO_3 \cdot NaCl$
Aphthitalite $K_3Na(SO_4)_2$	*Silicates*	
Mirabilite $Na_2SO_4 \cdot 10H_2O$	Acmite $NaFe^{3+}Si_2O_6$	*Phosphates*
Thenardite Na_2SO_4	Albite $NaAlSi_3O_8$	Collophane $Ca_{10}(PO_4)_6CO_3 \cdot H_2O$
	Clay Minerals	Fluorapatite $Ca_{10}(PO_4)_6F_2$
Chlorides, Fluorides	Elpidite $Na_2ZrSi_6O_{15} \cdot 3H_2O$	
Galeite $Na_2SO_4 \cdot Na(F,Cl)$	Garrelsite $(Ba,Ca,Mg)B_3SiO_6(OH)_3$	*Hydrocarbons*
Halite NaCl	Labuntsovite (K,Ba,Na,Ca,Mn)	albertite
Hanksite $9Na_5SO_4 \cdot 2Na_2CO_3 \cdot KCl$	$(Ti,Nb)(Si,Al)_2-(O,OH)_7H_2O$	coal
Northupite $Na_2CO_3 \cdot MgCO_3 \cdot NaCl$	Leucosphenite $CaBaNa_3BTi_3Si_9O_{29}$	gilsonite
Schairerite $Na_2SO_4 \cdot Na(F,Cl)$	Loughlinite $(Na_2,Mg)_2Si_3O_6(OH)_4$	ingramite
Sulfohalite $2Na_2SO_4 \cdot NaCl \cdot NaF$	Quartz SiO_2	ozokerite
Teepleite $Na_2B_2O_4 \cdot 2NaCl \cdot 4H_2O$	Reedmergnerite $NaBSi_3O_8$	tabbyite
	Riebeckite-magnesioriebeckite	uintahite
Borates	$(Na_2(Mg,Fe^{2+})_3(Fe^{3+},Al)_2Si_8O_{22}(OH)_2$	utahite
Borax $Na_2B_4O_7 \cdot 10H_2O$	Searlesite $NaBSi_2O_6 \cdot H_2O$	wurtzilite
Tincalconite $Na_2B_4 \cdot 5H_2O$	Sepiolite $Mg_2Si_3O_6(OH)_4$	

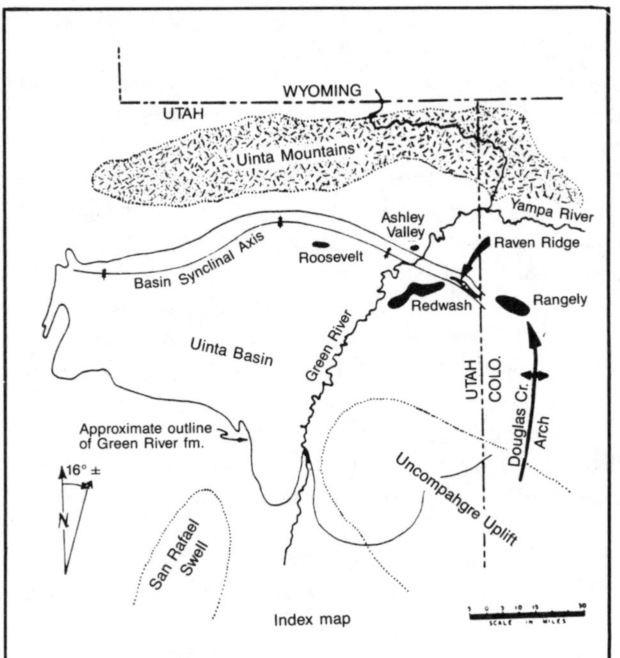

Fig. 7-4. Area map of Uinta basin, northeastern Utah. (Redrawn from Chatfield, 1972; permission to publish by AAPG).

(145 km) long, 30 miles (48 km) wide, and up to 3750 feet (1143 m) thick. Fluvial deposits limit the formation to the east and southeast; fanglomerate occurs to the northwest. Clastic and chemical cycles characterize the lacustrine strata. A lower black shale, mudstone and marlstone, and an upper massive mudstone with lenticular feldspathic siltstone make up the clastic cycles. Chemical cycles have increasing analcime and decreasing carbonate upward. Cycles result from the basin being alternately open and closed to the sea, producing clastic and chemical cycles, respectively.

A number of characteristics of Lockatong strata point to various lacustrine features. Quiet water deposition is indicated by thin bedding, varves, small-scale ripple and cross-stratification, and lateral persistence of beds. Evidence of shallow water is the presence of subaerial dessication cracks and salt casts. A stratified water body is suggested by abundant pyrite and a lack of sessile bottom dwelling or-

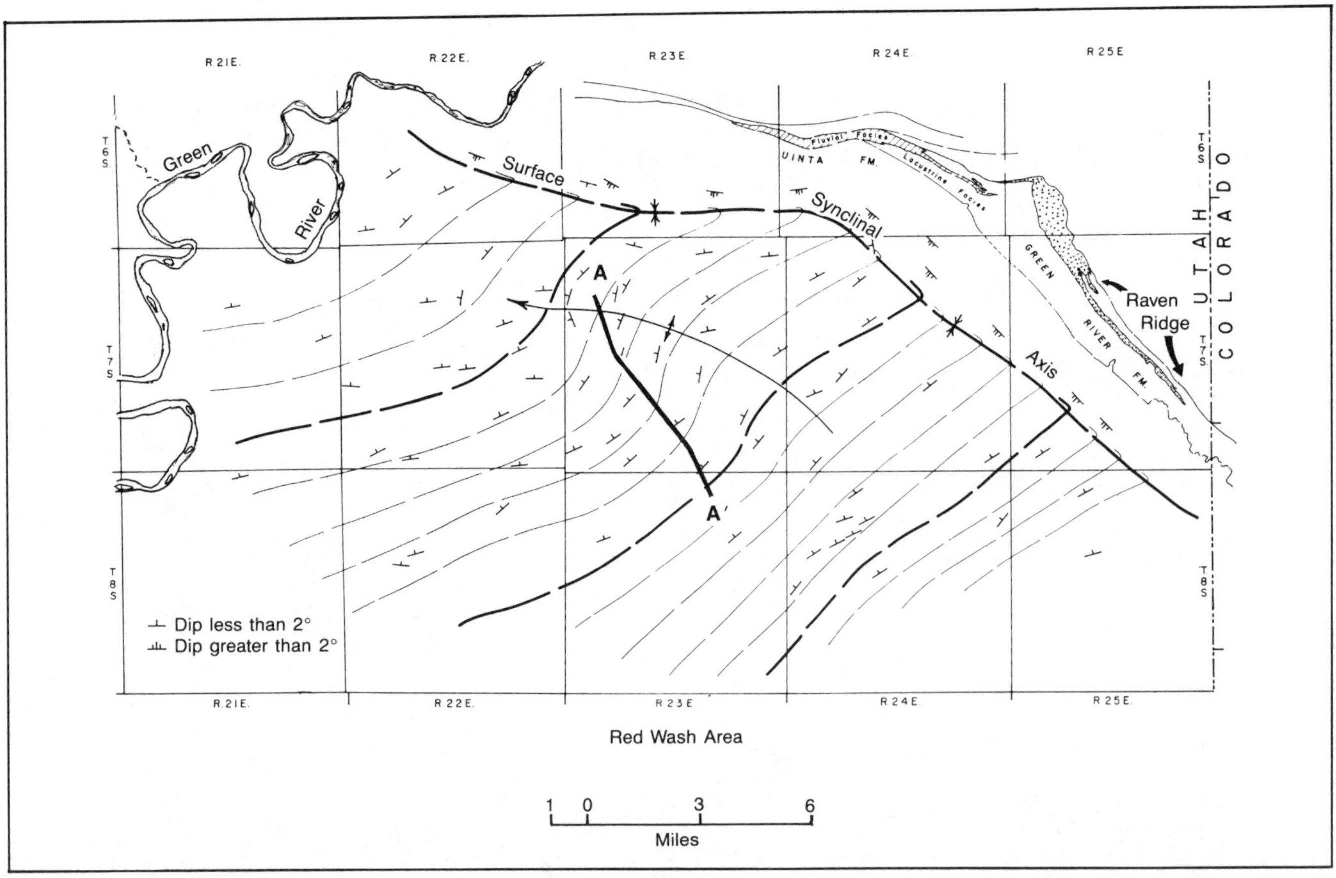

Fig. 7-5. Surface structure map of Red Wash area (T7S, R22-23E). Green River outcrop and sand pinchout area at Raven Ridge shown to east. (From Chatfield, 1972; permission to publish by AAPG).

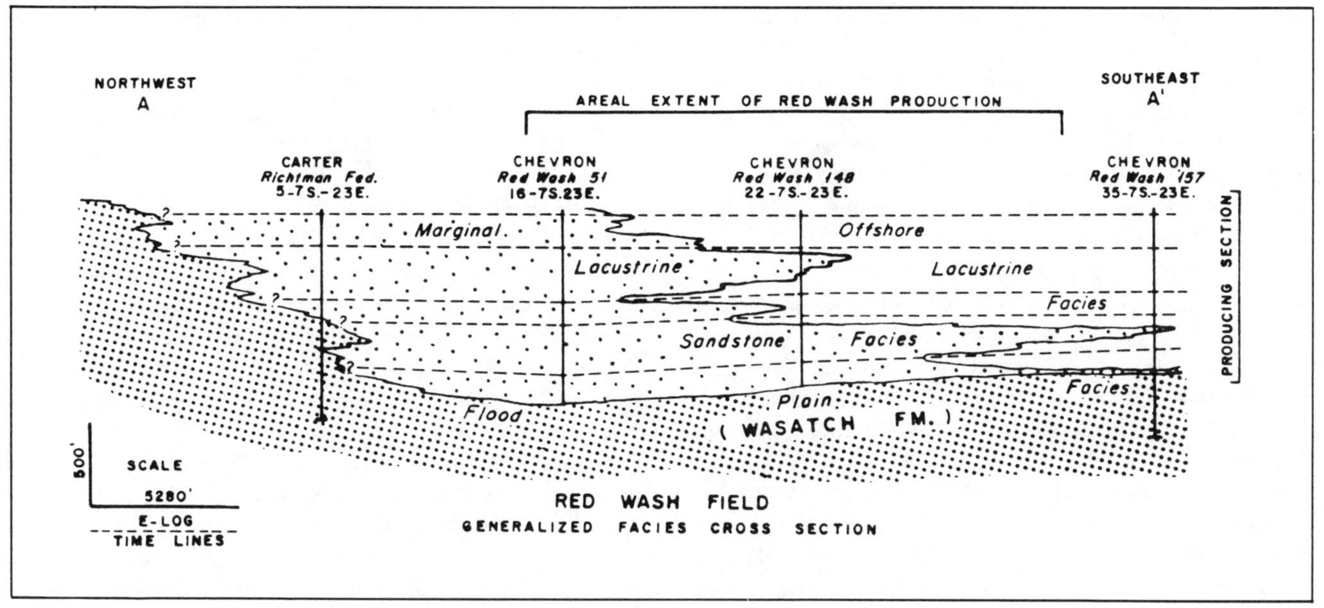

Fig. 7-6. Generalized stratigraphic cross section across Red Wash area showing facies changes. See Figure 7-5 for location. (From Chatfield, 1972; permission to publish by AAPG).

RED WASH AREA
BOUGUER GRAVITY MAP
1945 & 1949 DATA
C.I.=1 MILLIGALL

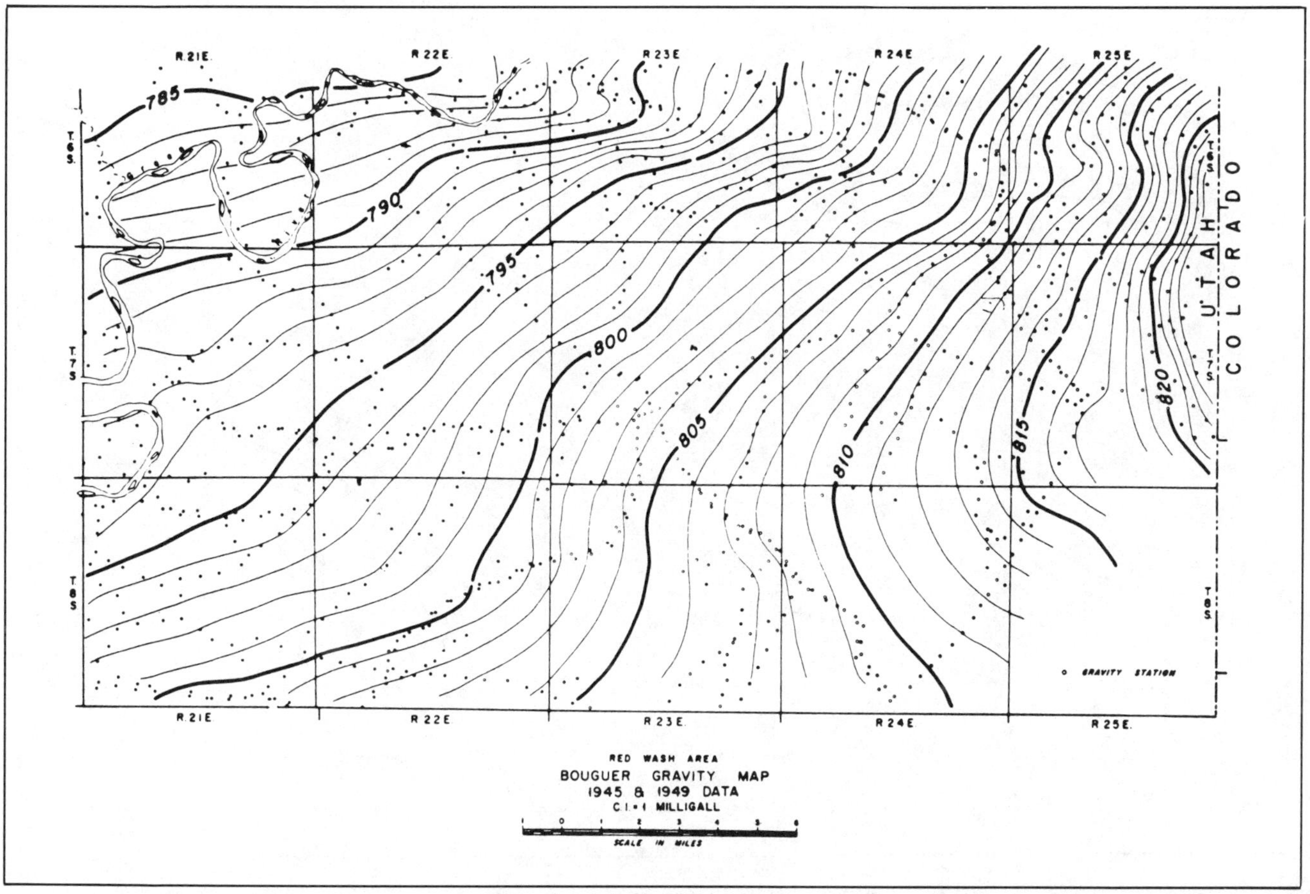

Fig. 7-7. Bouguer gravity map, Red Wash area. (From Chatfield, 1972; permission to publish by AAPG).

ganisms. Nonmarine conditions are demonstrated by amphibian and fish fossils.

Based on the preceding characteristics, the lacustrine nature of the Lockatong is interpreted from paleontology, mineralogy, sedimentary structures, cyclic deposits, and regional relationships. The data were obtained from outcrops, a luxury generally not available to the exploration geologist. However, much of the same information is available in drill cuttings and conventional cores.

In working with lacustrine or suspected lacustrine rocks, Picard and High (1972) recommend a dual approach: determine if the deposit was formed in marine or nonmarine water through the use of paleontology (marine or nonmarine fossils), mineralogy (conventional or unusual suites of authigenic or evaporite minerals), and regional stratigraphy (lake beds have limited lateral extent and are enclosed by fluvial or other nonmarine beds). If the rocks are nonmarine, then determine if deposition was from flowing or standing water by examining bedding, sedimentary structures, grain size, sorting, paleocurrent patterns (unimodal for streams), and carbonates (deposited from standing water). One must, however, keep in mind that slack water deposition on floodplains may confuse the interpretation.

HYDROCARBON EXPLORATION IN LACUSTRINE SEDIMENTS

Prior to the discovery of major hydrocarbon accumulations in lacustrine deposits in China, the Green River formation (Eocene) of Utah produced the greatest volume of hydrocarbons from lake deposits. Much Green River production comes from the Red Wash field in northeastern Utah (Fig. 7-4). Discovery of the Red Wash field is a good example of a multifaceted approach to oil exploration using both geology and geophysics. It is also, however, a good example of letting a prospective area lie dormant for an extended period of time. Chatfield's (1972) paper is the basis for the following exploration history of Red Wash field.

In the early 1920's and 1930's, field work in the Uinta basin resulted in the description of the Green River oil shales and established the regional stratigraphy, hydrocarbon potential, and structural history of the basin. In spite of this work, no exploration programs were undertaken until 1945.

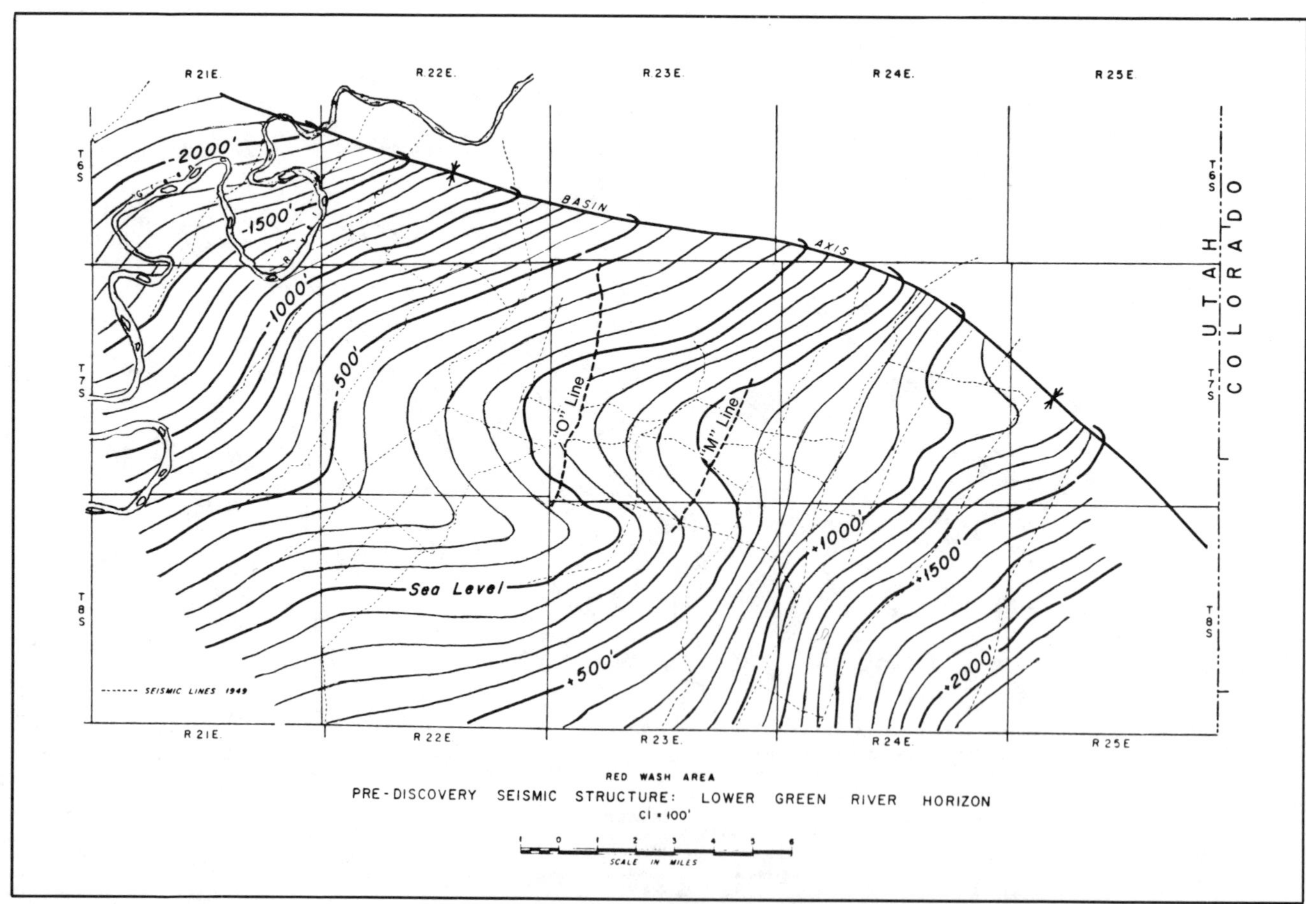

Fig. 7-8. Seismic structure map on a horizon in the lower Green River formation. (From Chatfield, 1972; permission to publish by AAPG).

In that year, a program was started to compile geologic information of the basin, study outcrops around the basin edge, and conduct reconnaissance gravity and air photo surveys. Any indicated surface anomalies were examined in detail by surface field parties. One of the anomalies detected from aerial photographs was a subtle anticlinal feature in T7S, R22-23E, Utah. Surface mapping (Fig. 7-5) confirmed a west-plunging anticlinal nose in fluvial rocks of the Uinta formation (Fig. 7-3).

With the structural feature confirmed, Green River formation outcrops at Raven Ridge 12 mi (19 km) to the east were measured and mapped in detail. Facies distributions showed a northwest to southeast progression from fluvial to nearshore lacustrine to open water lacustrine paleoenvironments. Lithologies varied with the facies and ranged from more than 50% sandstone in the north to open lacustrine shales and marlstones in the south. A south-prograding delta with sediments derived from the Uinta Mountains to the north is represented (Fig. 7-6).

Also observed at Raven Ridge were oil-stained and oil-saturated Green River sandstones with the oil occurring primarily at the southern (basinward) ends of the sandstones. Projection of beds with reservoir potential westward into the subsurface indicated the beds could wedge out near the axis of the surface anticlinal nose which was confirmed at Red Wash. No well control existed to substantiate the projection.

Geologists had taken the prospect as far as they could and geophysical confirmation was needed. Additional gravity data in the Red Wash area were obtained to supplement the original reconnaissance gravity survey (Fig. 7-7). A broad, west-plunging nose slightly to the south of the similar surface feature mapped from aerial photos and by field parties was evident. Disparity in the location of the structural axis of the nose between the gravity and the geological maps was attributed to inaccuracies caused by topog-

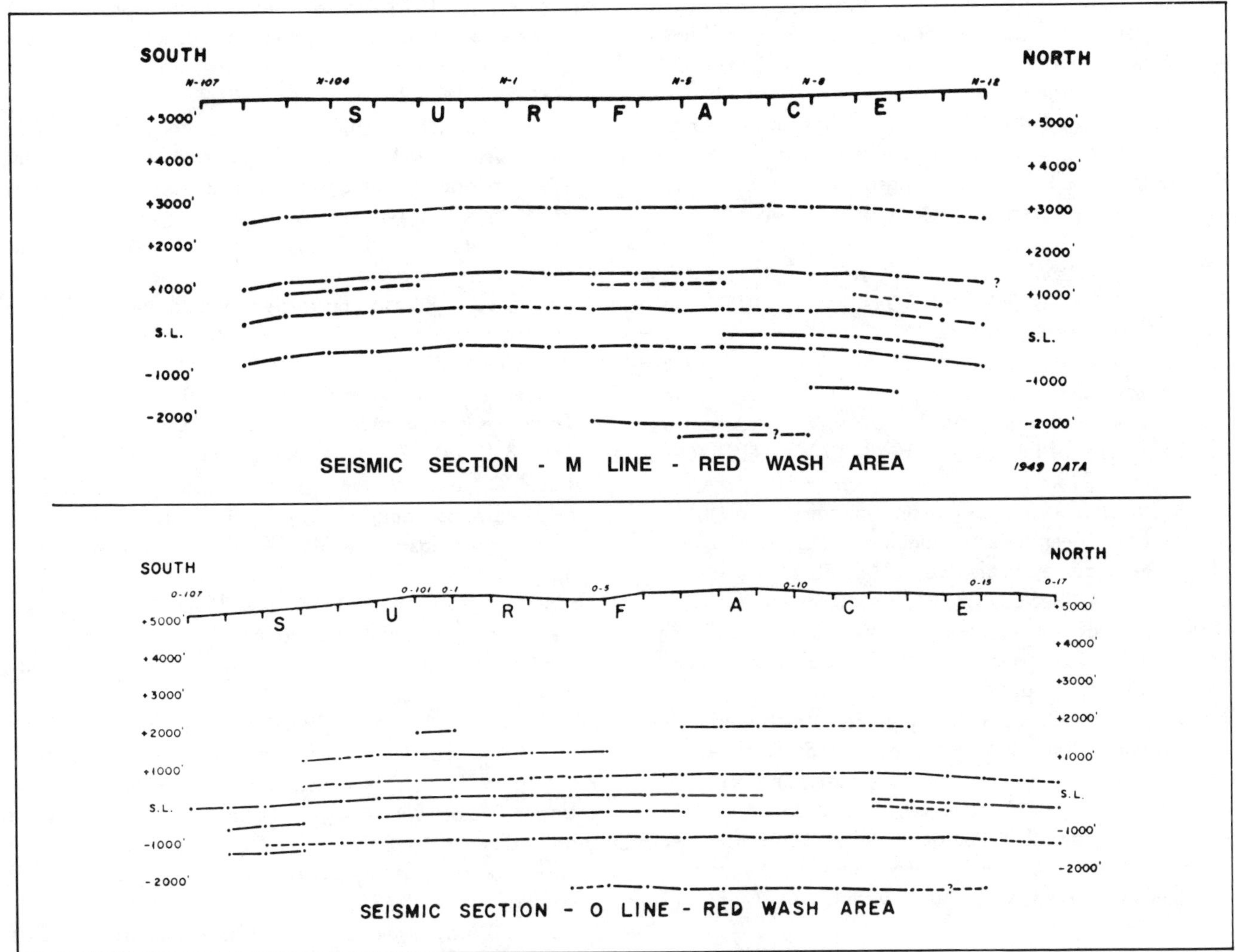

Fig. 7-9. Seismic depth sections across Red Wash anticlinal nose. See Figure 7-8 for locations of sections. (From Chatfield, 1972; permission to publish by AAPG).

raphy, and seismic confirmation was scheduled.

Seismic structure mapping on a horizon within the Green River formation confirmed the presence of the Red Wash structure (Fig. 7-8). Dip reversal in seismic sections normal to the axis of the nose is subtle but present (Fig. 7-9).

The concept of the prospect, therefore, was Green River deltaic sands derived from the north wedging out to the south and being replaced by lacustrine shales and marlstones. Pinch out of the coarse clastics coincided approximately with the axis of a west-plunging anticlinal nose in the Red Wash area according to projections from outcrops 12 mi (19 km) to the east. Absence of thinning in Green River rocks over the structural feature indicated the structure postdated deposition and would aid in trapping any hydrocarbons. For two reasons the exploratory well was located on the crest of the structure. First, this was the area of projected pinching out of the sands. Second, the Roosevelt field to the northwest (Fig. 7-4) produced from fractured Green River shales and siltstones. Should production from similar reservoirs occur at Red Wash, the crest of the structure, being the area of maximum extension, should have the greatest fracturing. The discovery well came in at 339 barrels of oil per day.

Development of the Red Wash field has confirmed the accuracy of the concept of the prospect. Discovery of the field was the result of a well-integrated exploration program using many facets of geology and geophysics.

SELECTED BIBLIOGRAPHY

Bacoccoli, G., R. G. Morales, and O. A. J. Campos, 1980, The Namorado oil field: a major oil discovery in the Campos basin, Brazil, *in* M. T. Halbouty, ed., Giant oil and gas fields of the decade: 1968–1978: AAPG Mem. 30, p. 329-338.

Bradley, W. H., 1931, Origin and microfossils of the oil shale of the Green River formation of Colorado and Utah: U.S. Geol. Survey Prof. Paper 168, 58 p.

______, 1964, Geology of the Green River formation and associated rocks in southwestern Wyoming and adjacent parts of Colorado and Utah: U. S. Geol. Survey Prop. Paper 496-A, 86 p.

Chatfield, J., 1972, Case history of Red Wash field, Uintah County, Utah, *in* R. E. King, ed., Stratigraphic oil and gas fields—classification, exploration methods, and case histories: AAPG Mem. 16, p. 342-353.

Eugster, H. P., and R. C. Surdam, 1973, Depositional environment of the Green River formation of Wyoming: a preliminary report: GSA Bull., v. 84, no. 1, p. 115-120.

Feth, J. H., 1963, Tertiary lake deposits in western coterminous United States: Science, v. 139, no. 3550, p. 107-110.

______, 1964, Review and annotated bibliography of ancient lake deposits (Precambrian to Pleistocene) in the Western States: U. S. Geol. Survey Bull. 1080, 119 p.

Fouch, T. D., 1975, Lithofacies and related hydrocarbon accumulations in Tertiary strata of the western and central Uinta basin, Utah, *in* D. W. Bolyard, ed., Symposium on deep drilling frontiers in the central Rocky Mountains: Rocky Mtn. Assoc. Geologists, p. 163-173.

______ and W. E. Dean, 1982, Lacustrine and associated clastic depositional environments, *in* P. A. Scholle and D. Spearing, eds., Sandstone depositional environments: AAPG Mem. 31, p. 87-114.

Matter, A., and M. E. Tucker, eds., 1978, Modern and ancient lake sediments: Internat'l Assoc. Sedimentologists Sp. Pub. no. 2, 290 p.

Meyerhoff, A. A., and J. O. Willums, 1976, Petroleum geology and industry of the People's Republic of China: United Nations ESCAP, CCOP Tech. Bull., v. 10, p. 103-212.

Picard, M. D., 1972, Paleoenvironmental reconstruction in an area of rapid facies change, Parachute Creek member of Green River formation (Eocene), Uinta basin, Utah: GSA Bull., v. 83, no. 9, p. 2689-2708.

______ and L. R. High, Jr., 1972, Criteria for recognizing lacustrine rocks, *in* J. K. Rigby and W. K. Hamblin, eds., Recognition of ancient sedimentary environments: SEPM Sp. Pub. No. 16, p. 108-145.

______ and ______, 1981, Physical stratigraphy of ancient lacustrine deposits, *in* F. G. Ethridge and R. M. Flores, eds., Recent and ancient nonmarine depositional environments: models for exploration: SEPM Sp. Pub. No. 31, p. 233-259.

Ryder, R. T., T. D. Fouch, and J. H. Elison, 1976, Early Tertiary sedimentation in the western Uinta basin, Utah: GSA Bull., v. 87, no. 4, p. 496-512.

Sizhong, C., and W. Ping, 1980, Geology of Gudao oil field and surrounding areas, *in* M. T. Halbouty, ed., Giant oil and gas fields of the decade: 1968-1978: AAPG Mem. 30, p. 471-486.

Surdam, R. C., and K. O. Stanley, 1979, Lacustrine sedimentation during the culminating phase of Eocene Lake Gosiute, Wyoming (Green River formation): GSA Bull., v. 90, no. 1, Part I, p. 93-110.

Twenhofel, W. H., 1932, Treatise on sedimentation, 2nd ed.: Baltimore, The Williams and Wilkins Co., 926 p.

Van Houten, F. B., 1964, Cyclic lacustrine sedimentation, Upper Triassic Lockatong formation, central New Jersey and adjacent Pennsylvania, *in* D. F. Merriam, ed., Symposium on cyclic sedimentation: Kansas Geol. Survey Bull., no. 169, p. 497-531.

______, 1965, Composition of Triassic Lockatong and associated formations of Newark Group, central New Jersey and adjacent Pennsylvania: Amer. Jour. Sci., v. 263, no. 10, p. 825-863.

Veevers, J. J., 1977, Rifted arch basins and post-breakup rim basins on passive continental margins: Tectonophysics, v. 41, p. T1-T5.

_______ and D. Cotterill, 1978, Western margin of Australia: evolution of a rifted arch system: GSA Bull., v. 89, no. 3, p. 337-355.

Visher, G. S., 1965, Use of vertical profile in environmental reconstruction: AAPG Bull., v. 49, no. 1, p. 41-61.

Wolfbauer, C. A., 1973, Criteria for recognizing paleoenvironments in a playa-lake complex—the Green River formation of Wyoming, *in* Wyoming Geol. Assoc. guidebook to the geology and mineral resources of the greater Green River basin: Wyo. Geol. Assoc., p. 87-91.

8 DELTAS

INTRODUCTION

Deltaic deposition occurs where a river enters a receiving body of water. Sediments and water from a drainage basin are carried through an alluvial valley to the receiving basin. At the point of entry, the river ceases to be an agent of transportation, becomes an agent of dispersal, and deposits its sedimentary load. Deltaic depositional environments include distributary channels, interdistributary bays, distributary mouth bars, tidal flats and ridges, marshes, swamps, beaches, and dunes (Coleman and Prior, 1980, 1982).

The diversity of depositional environments makes deltas attractive exploration targets for hydrocarbons. Fine-grained clastics rich in organic matter and coarse-grained deposits vary abruptly both horizontally and vertically frequently producing porous reservoirs encased in impermeable sealing rocks that also may be sources of hydrocarbons.

FACTORS CONTROLLING DELTA DEVELOPMENT

For a delta to form and grow, sediment must accumulate at the basin margin more rapidly than it is dispersed by processes active in the receiving basin. Processes, some of which are interrelated, operating in the drainage basin, at the river mouth, and in the receiving basin, control the accumulation and dispersal of deltaic sediments. Coleman and Prior (1980, 1982) describe these factors for modern rivers and their studies serve as the basis for the following discussion of Recent deltas.

Drainage Basin Factors

Climate. Climate is the prime factor operating in the drainage basin and on the delta. It controls physical and chemical weathering in the basin, thereby regulating the composition and quantity of sediment available to the rivers for transport. Seasonal distribution and quantity of precipitation control runoff, which affects sediment volume and caliber, channel morphology in the alluvial valley, and delivery of water and sediment to the river mouth. Climatic factors on the delta regulate organic deposits and may lead to deposition of evaporites under arid conditions.

Discharge. Sediment and water discharge are related to climate. With a uniform distribution of annual runoff, the volume of sediment and water carried by a river is relatively constant, producing stable meandering streams. Alluvial valley deposits are laterally restricted, relatively well sorted, and consist of fining-upward sequences. Deltaic sands tend to be long, linear, and oriented at high angles to the coast. Under conditions of erratic runoff, sediment discharge is highly variable. Alluvial valley channels tend to be braided and migrate rapidly. Alluvial valley deposits are widespread, coarse- to fine-grained, and poorly sorted. Deltaic sands are reworked by receiving basin processes (waves, tides, and longshore drift) during periods of low flow, and deltaic sand body orientation will be parallel or subparallel to the coast.

River Mouth Factors

Interaction between river water effluent and the water of the receiving basin causes diffusion and loss of competence of the effluent. Due to the loss of competence, sedimentation patterns are related to inertia of the effluent and associated turbulent diffusion, friction between the effluent and the seabed just beyond the river mouth, and buoyancy from any density contrast between the effluent and basin water. With high effluent velocity, relatively deep water basinward from the river mouth, and a small difference in

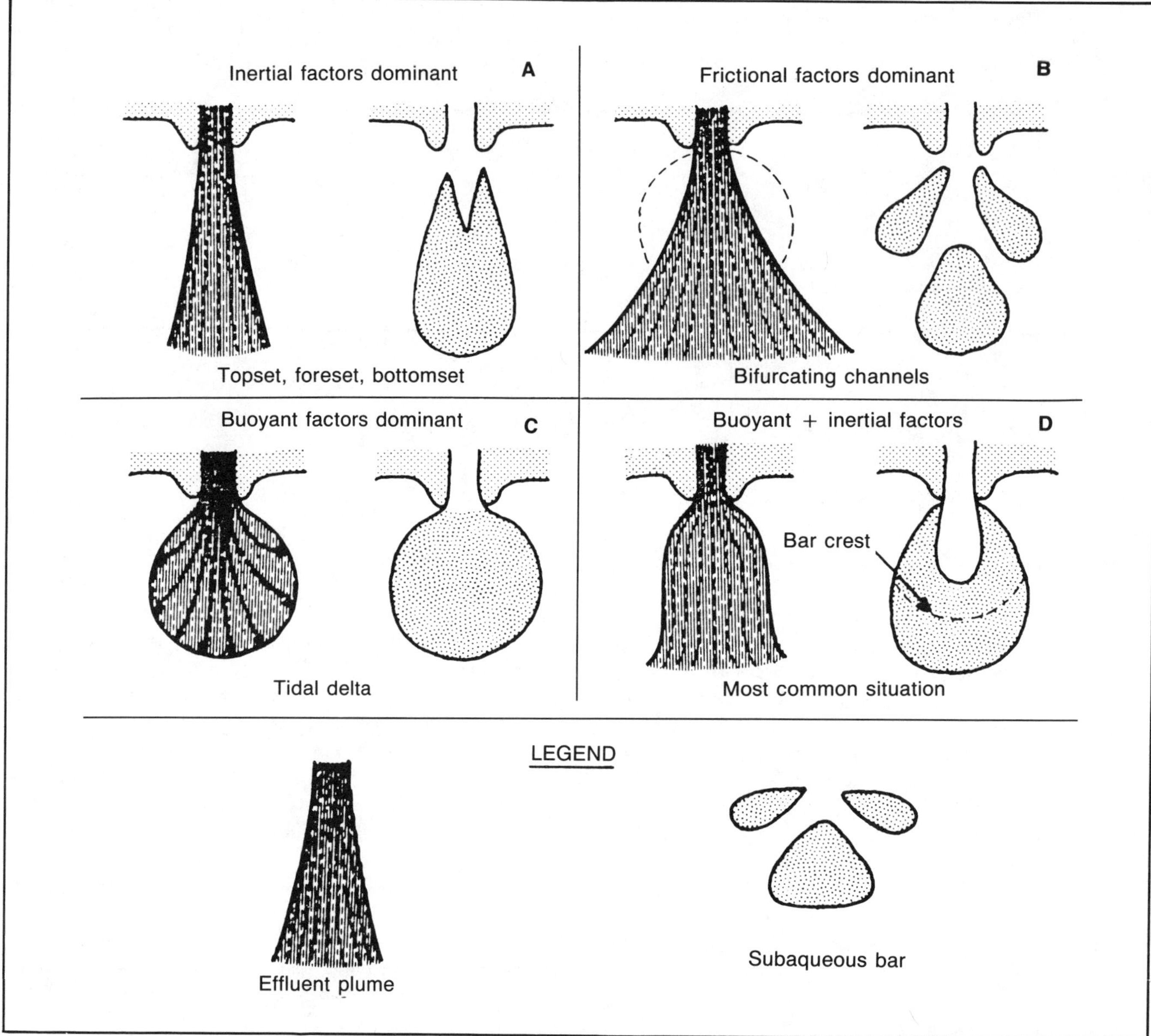

Fig. 8-1. River mouth factors affecting delta development. (From Coleman and Prior, 1980; permission to publish by AAPG).

density between effluent and basin waters, inertial forces dominate, and the effluent behaves and diffuses as a turbulent jet. The resulting river mouth bar is a thick, linear sand body of limited horizontal extent (Fig. 8-1A). Vertical transition from fine-grained shelf clays to overlying bar sands is rapid.

If there is shallow water seaward of the river mouth, turbulent diffusion of the effluent is essentially horizontal due to friction between the effluent and the bottom. Initially, the river mouth bar is a broad, arcuate, radial bar that grows laterally as the effluent expands laterally (Fig. 8-1B). Eventually, subaqueous natural levees develop along the edges of the expanding effluent and restrict lateral expansion of the river mouth bar. With continued accretion of the center of the bar, channels develop along zones of maximum turbulence near the natural levees leaving a triangular shoal between two diverging channels. Coalescing of distributary mouth bars from adjacent bifurcating channels may form a continuous sand strip around the periphery of the delta.

Density differences between effluent and basin waters affect river mouth bar morphology. Low density, fresh, sediment-laden river water overrides heavier, basin salt water. Deceleration of the effluent is radial from the river mouth and sands are deposited around the mouth. Bars are thin and have lateral continuity (Fig. 8-1C). Most river mouth

TABLE 8-1. Comparison of mean annual wave energy and discharge for seven deltas. (From Wright and Coleman, 1973; permission to publish from AAPG).

Delta	Mean Annual Deepwater Wave Power ft-lb/sec	Mean Annual Nearshore Wave Power ft-lb/sec	Mean Discharge $ft^3/sec \times 10^3$	Mean Annual Discharge Effectiveness Index	Mean Annual Attenuation Ratio
Mississippi	237.4	0.03	624.6	5477.0	7913.3
Danube	51.7	0.03	222.0	1171.0	2585.0
Ebro	168.8	0.11	19.5	267.8	1299.5
Niger	152.1	1.48	384.8	4.4	102.8
Nile	306.2	7.49	52.2	3.2	42.5
Sao Francisco	834.6	22.40	110.2	1.3	37.2
Senegal	351.9	84.60	27.2	0.3	4.2

processes are varying combinations of the above factors and the most common bar configurations are those in Figures 8-1B and 8-1D.

Receiving Basin Factors

Waves. Wave energy at the shoreline is the most important factor in coastline development. Deep-water wave energy does not vary drastically in major bodies of water world-wide, but the power of the waves near shore does (Table 8-1). Waves sort and distribute sediments along the coast forming beaches and barriers. Deltaic deposits alter the coastline, while waves act on the deposits to restore the coast. Deltaic geometry is a balance between the ability of a river to deliver sediment to the delta and the ability of waves to redistribute the sediment. Table 8-1 quantifies the two opposing processes for seven modern deltas. Nearshore wave power is a measure of the waves' capability to alter the coastline. Discharge effectiveness index is the ratio of average discharge per unit channel width to average nearshore wave power per unit crest width. No physical meaning is attributed to the index, but it does serve to rank deltas by relative degree of river versus wave dominance. Attenuation ratio is the frictional loss of energy sustained by waves in moving over the shelf from deep to shallow water. Comparison of Table 8-1 with Figure 8-2 shows that attenuation varies inversely with shelf slope. Greater attenuation results from large energy losses on shallow shelf slopes.

A spectrum of delta morphology exists, therefore, depending on the relative importance of river and wave forces. When river forces are dominant, the delta shoreline is irregular and characterized by extended distributaries sep-

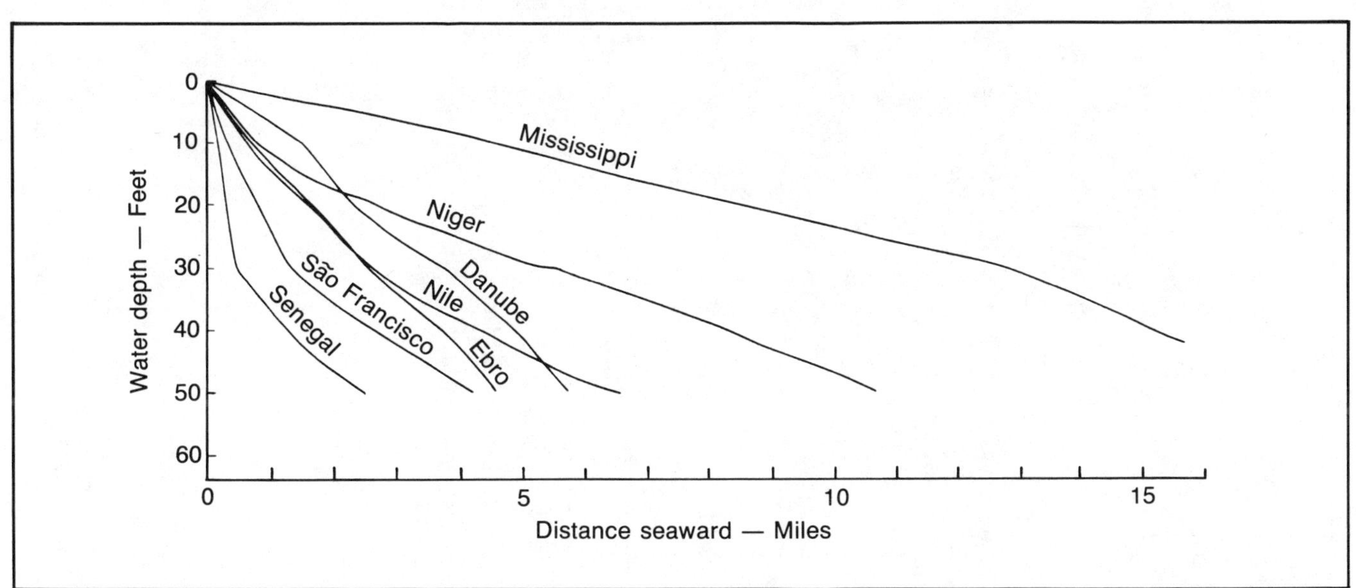

Fig. 8-2. Average submarine profiles off seven deltas. (From Wright and Coleman, 1973; permission to publish from AAPG).

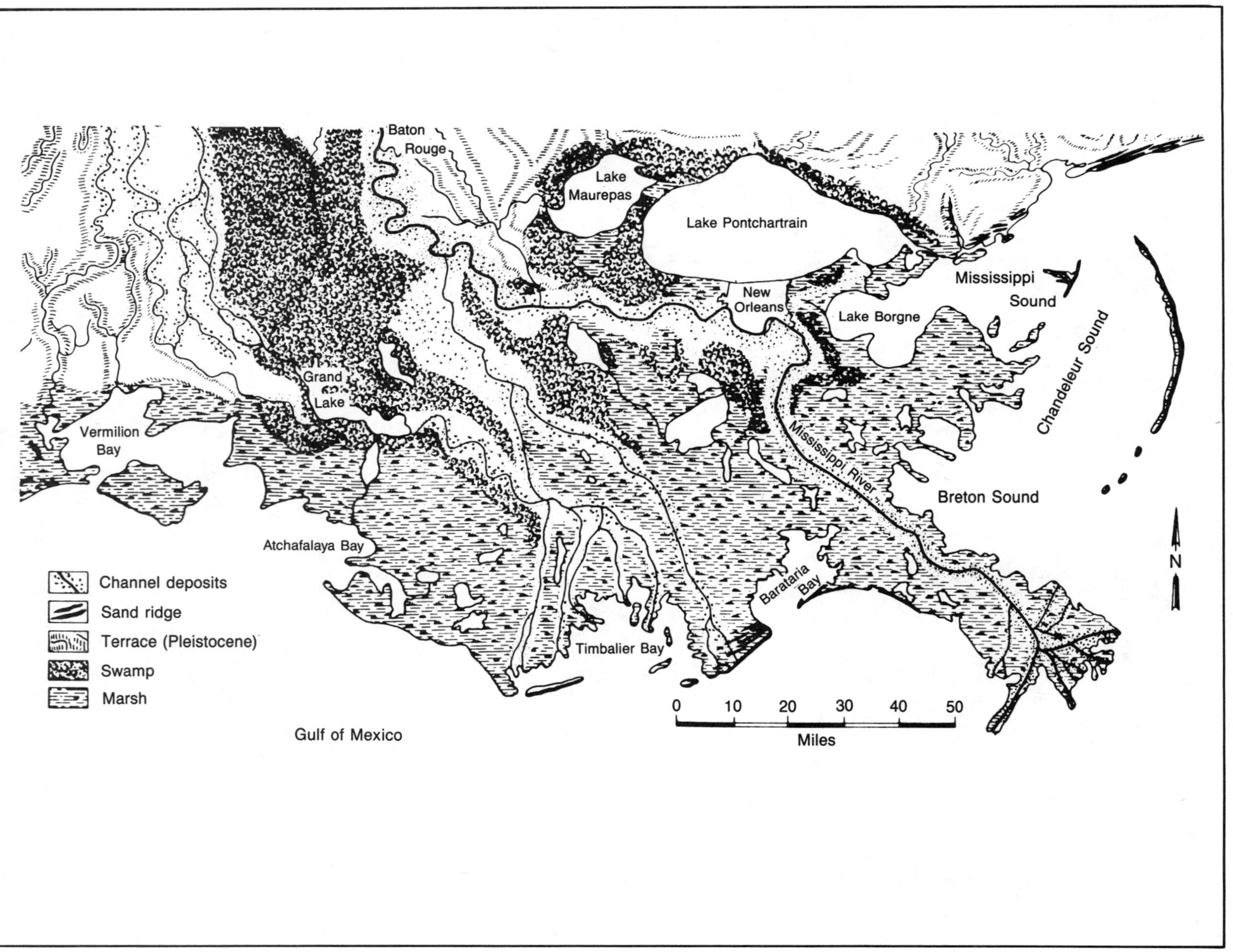

Fig. 8-3. Geomorphology of the Mississippi (USA) delta. (From Wright and Coleman, 1973; permission to publish by AAPG).

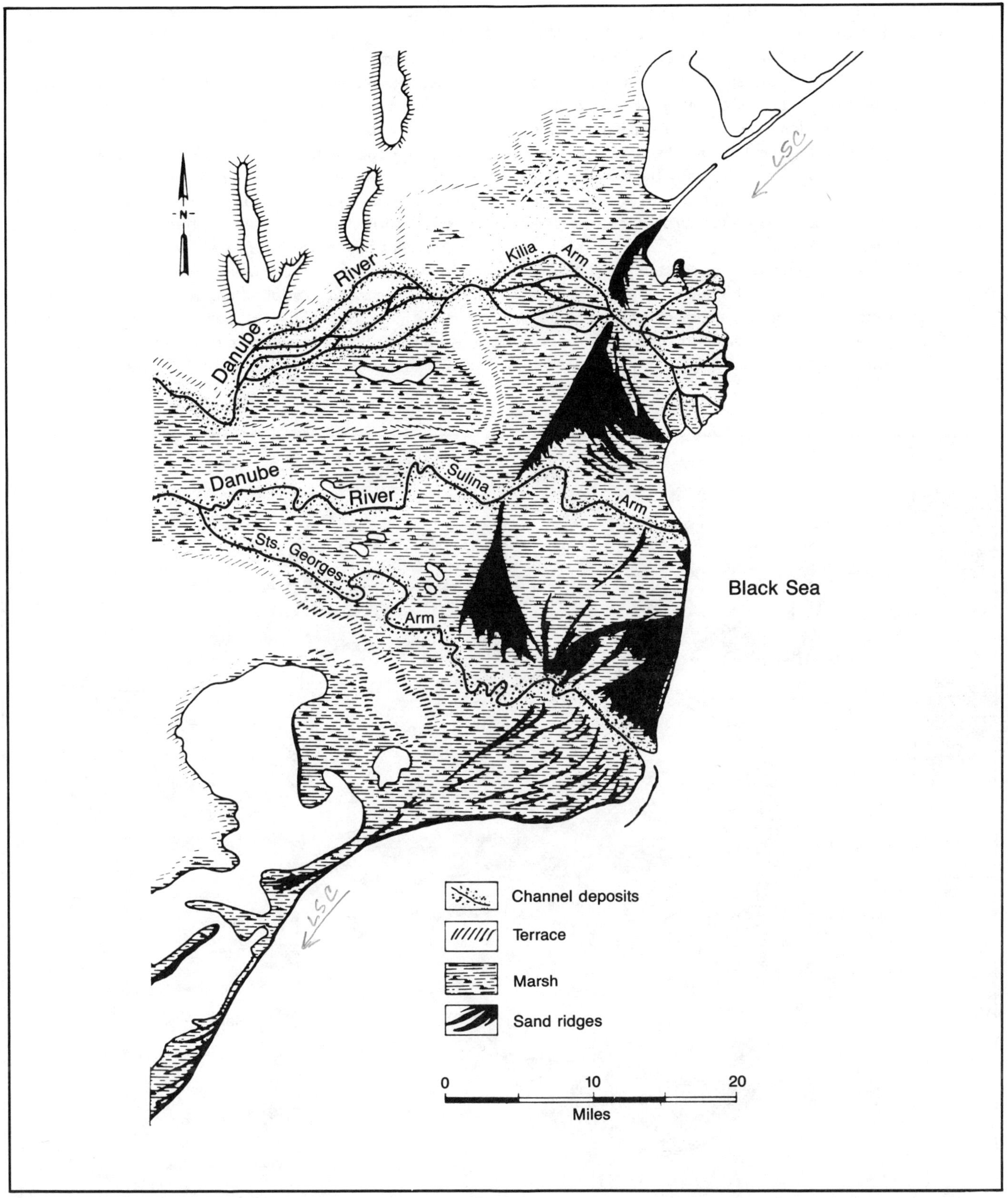

Fig. 8-4. Geomorphology of the Danube (Rumania) delta. (From Wright and Coleman, 1973; permission to publish by AAPG).

arated by bays and marshes (Fig. 8-3). As wave forces increase relative to river forces, areas between protruding distributaries become smoother (Figs. 8-4 and 8-5). With an even balance between river and wave forces, the delta has a smooth, sandy, arcuate shoreline, although distributary mouths may protrude slightly, (Figs. 8-6 and 8-7). With increasing wave dominance, the shoreline becomes smoother, the delta protrudes less into the basin, and deltaic sediments become increasingly sandy (Figs. 8-8 and 8-9). Table 8-2 relates variations in river versus wave dominance and offshore slope to delta morphology and sedimentological characteristics.

Tides. High tide range on a delta causes strong bidirectional sediment transport, loss of effluent buoyancy through destruction of the vertical density gradient by tidal mixing, and horizontal and vertical extension of marine-fluvial interactions. Through these processes, tides may modify delta and sand body morphology to a considerable degree, such as on the Mekong delta of Vietnam (Fig. 8-10). Its distributary mouths tend to be funnel shaped, and distributary mouth bar sands are formed into elongate, subaqueous ridges oriented parallel to the tidal currents. Tidal ridges may be up to 30 m thick, hundreds of meters wide, and a few to over 25 km long.

Strong flood tides with large tidal ranges produce a number of deltaic features. Bedload is transported upstream causing abandoned distributaries to fill with sand. In active distributaries, backwater effects enhance point bar development causing greater lateral migration of channels and increased meandering of distributaries upstream from the area of maximum tidal influence. Well-developed sand units are deposited within the delta plain by the meandering channels. Backwater effects at high tide also cause extensive overbanking and numerous crevasse splays.

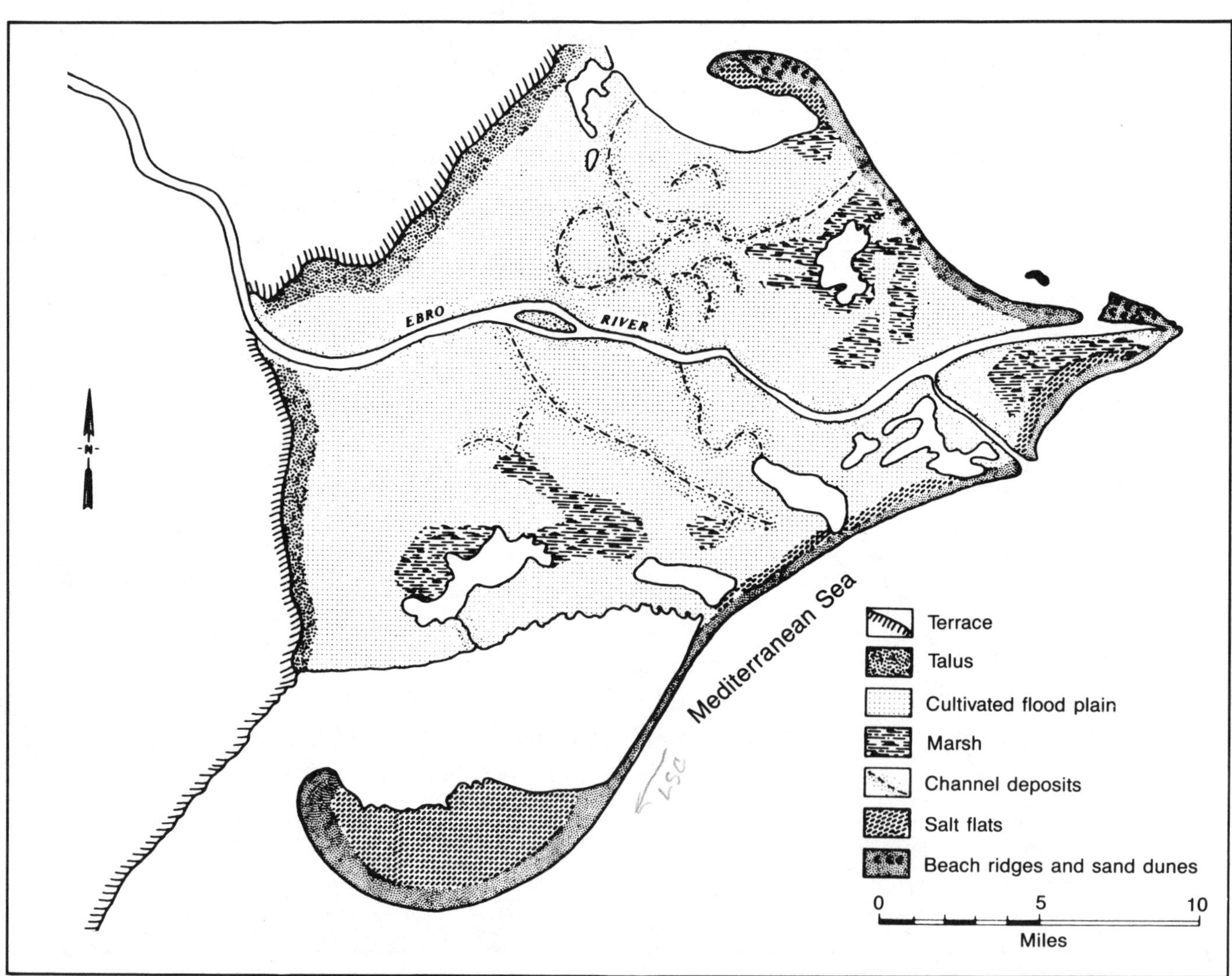

Fig. 8-5. Geomorphology of the Ebro (Spain) delta. (From Wright and Coleman, 1973; permission to publish by AAPG).

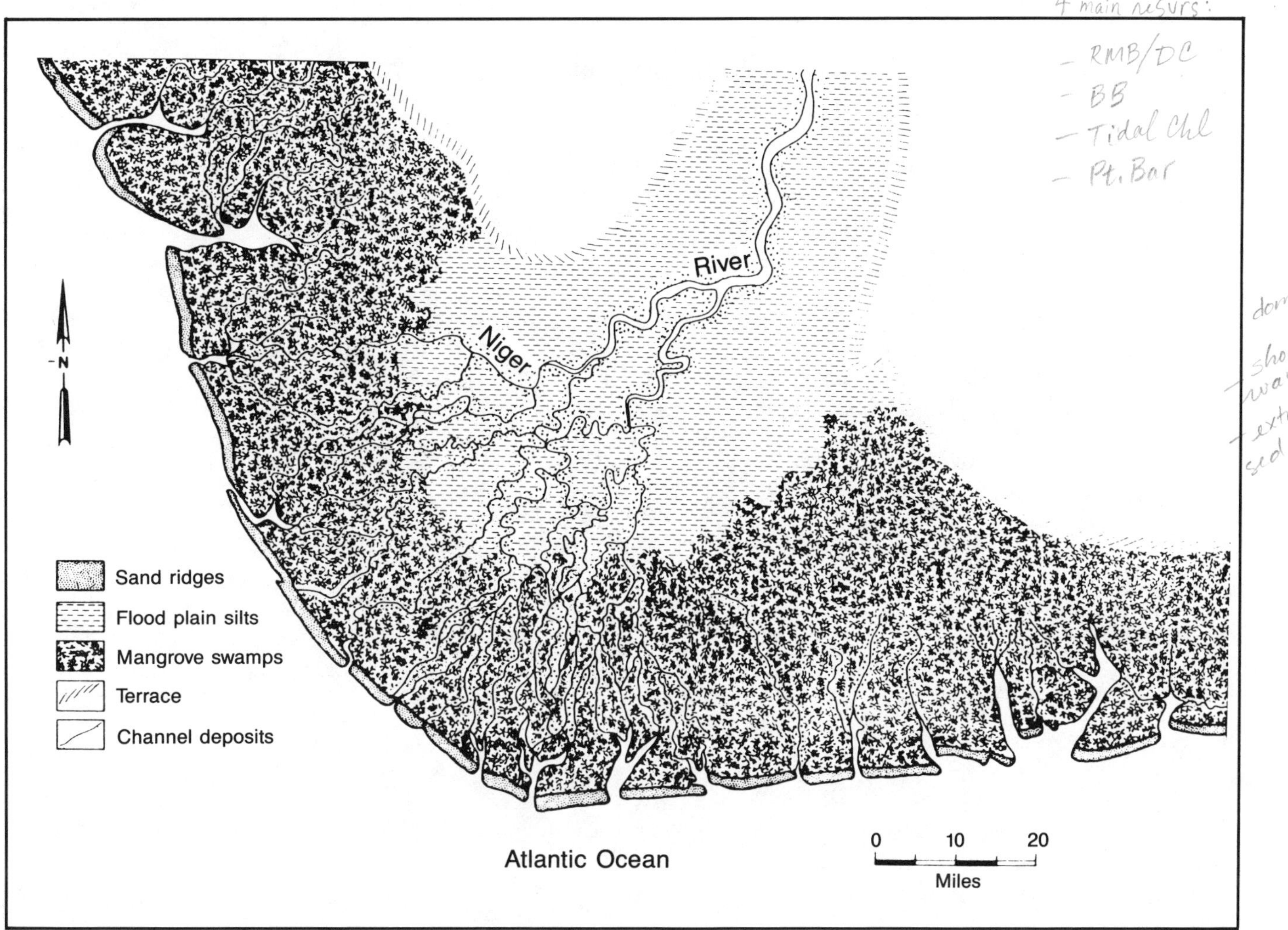

Fig. 8-6. Geomorphology of the Niger (Nigeria) delta. (From Wright and Coleman, 1973; permission to publish from AAPG).

Currents. Wind, waves, density gradients, and tides produce currents that modify deltas. Tidal currents are strong in narrow embayments and partially enclosed basins. Wind-driven currents are more effective in large open basins with low tidal influence. Sand bodies produced by currents are oriented parallel or subparallel to depositional strike and may be relatively far from shore.

Shelf slope. Shelf slope controls the power of incoming waves at the shoreline. Low slopes attenuate wave energy by friction between the wave and the sea bottom. Less wave drag occurs on steep slopes and more wave power reaches the shore. High rates of deltaic sedimentation may reduce the shelf slope, thereby reducing incoming wave energy and allowing further build-up of the delta.

Shelf slope is also a factor in the switching pattern of a delta. Lobe switching occurs on a delta, such as the Mississippi, where tidal range, wave energy, and shelf slope are all low. A delta lobe progrades by a system of distributary channels until it is abandoned and a new lobe forms elsewhere (Fig. 8-11A). Overlapping lobes and coalescing distributary mouth bars form extensive sands.

Delta growth by channel switching occurs on a shelf with an intermediate slope, high wave energy, and high tidal range. Upstream channel shifts on the delta plain produce a new river course and a new area of delta growth (Fig. 8-11B). The Ganges-Brahmaputra delta switches in this manner.

Growth by alternate channel extension occurs on the Danube delta as it crosses a shelf with a steep slope and variable wave and tidal energy. Two or more major distributaries originate at a nearly common point and flow, without dividing, to the basin (Fig. 8-11C). One distributary will carry most of the water and sediment and actively prograde while the sediments of the relatively inactive distributary are reworked by waves into beach ridges. When the active distributary progrades sufficiently to lose its favorable gradient, the sediment and water will go down the other distributary, causing it to prograde. Beach ridges are left behind by the newly active distributary as it progrades, and ridges form on the relatively inactive distributary. Mul-

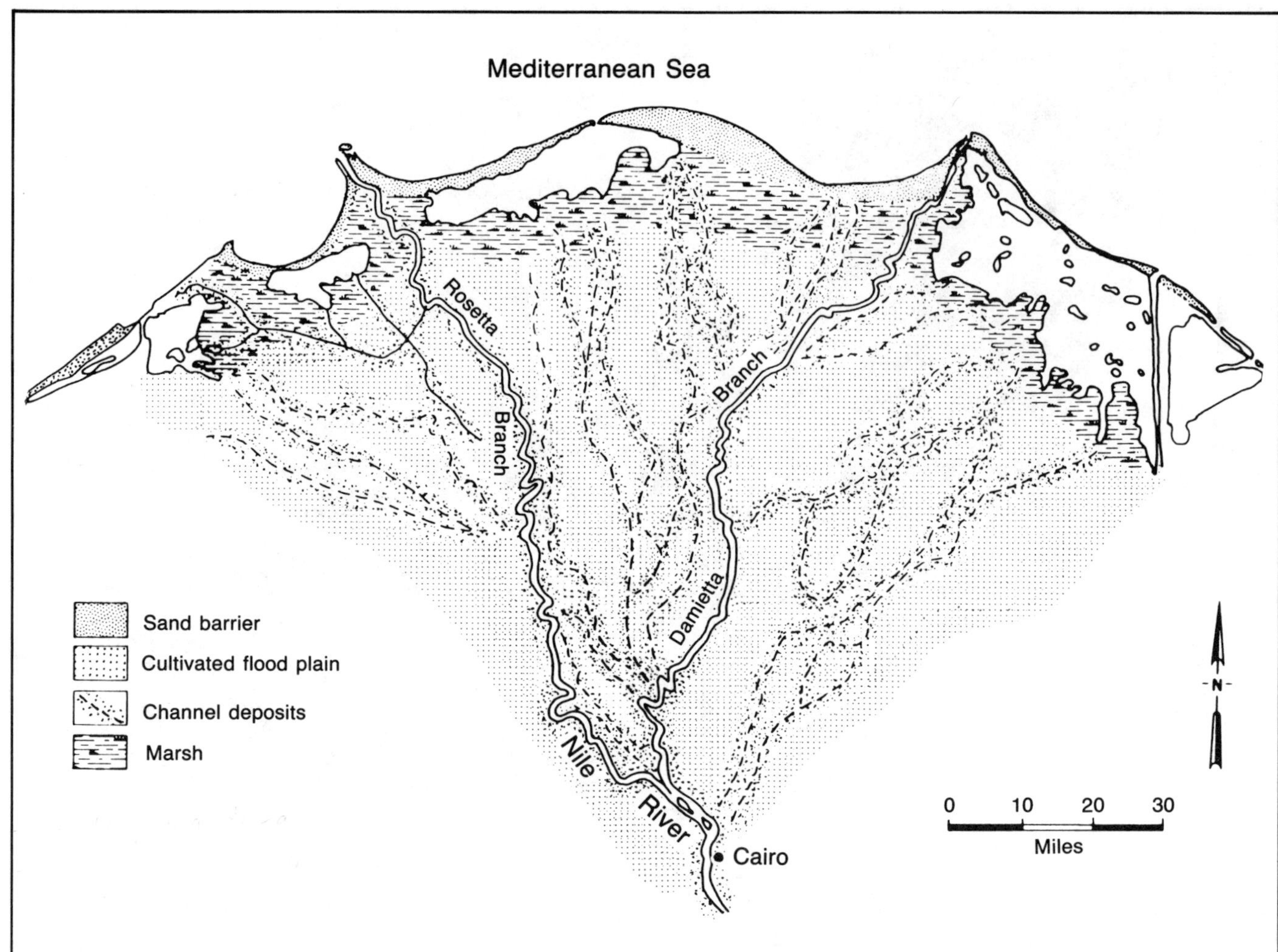

Fig. 8-7. Geomorphology of the Nile (Egypt) delta. (From Wright and Coleman, 1973; permission to publish by AAPG).

tiple sequences of beach ridges characterize a delta plain formed by channel extension (Coleman and Wright, 1975).

Depositional basin geometry and tectonics. Receiving basin geometry affects delta development by controlling processes that modify deltas. Elongate, narrow basins may have strong tidal currents but weak waves and permanent currents. Large, open basins, however, usually have strong waves and currents and low tidal ranges. Basin shelf characteristics (shelf slope and width) are also important.

Depositional basin tectonics control the thickness and lateral continuity of deltaic deposits. In a subsiding basin, deposits are thick and limited laterally due to the sinking basin floor. Additional room for vertical accretion is made when fluids are expelled from buried, undercompacted, fine-grained sediments. Failure of unstable sedimentary buildups transfers large quantities of deltaic sediment offshore into deeper water and permits still more nearshore deposition. Vertical accretion is limited in tectonically stable basins resulting in thin, laterally extensive deltaic deposits.

Coleman and Prior's (1982) observations suggest that waves, tides, and fluvial sediment influx are the immediate controls on delta development. Fisher's (1969) delta classification is based on the same three factors. He recognizes two types of high-constructive deltas, lobate and elongate, and two types of destructive deltas, tide-dominated and wave-dominated (Fig. 8-12). Galloway (1975) restructures Fisher's classification into a triangular diagram (Fig. 8-13) with fluvial-, wave-, and tide-dominated deltas as end members in a three-fold scheme.

DELTAIC ENVIRONMENTS AND FACIES

Studies of modern deltaic sediments provide models for interpreting ancient deltas. The most studied modern delta is that of the Mississippi River. Although the birdfoot shape is the least typical of all deltas, many of its the subenvironments and facies occur on other modern deltas. Reviews by Coleman and Prior (1980, 1982) of the environments and facies of the Mississippi and five other deltas serve as the basis for the following discussion.

A delta consists of three principal parts, upper and lower subaerial delta plains, and a subaqueous delta (Fig. 8-14). The upper plain is essentially a continuation of the alluvial valley and is acted upon by fluvial processes. It is the oldest part of the delta and above significant marine or tidal influence. Extending from the highest area of marine or tidal influence to the low tide line is the lower delta plain. It is subject to both fluvial and marine processes and is most extensive on deltas where delta topography and surface dips are low and tidal range is high. The subaqueous delta extends from the low tide line seaward and includes those shelf areas receiving fluvial sediments. It may be tens of kilometers wide and extend to depths of 300 m. In a regressive sequence, the subaqueous delta is the base across which the subaerial delta progrades.

Upper Delta Plain

Depositional environments of the upper delta plain are essentially the same as those of the alluvial valley. Deposits of braided and meandering streams, lakes, swamps, and floodplains all occur here.

Braided streams. Braided streams are characterized by large variations in water discharge, relatively steep gradients, high bed load to suspended load ratios, high width-to-depth ratios, and rapid lateral migration of channels. Deposits are widespread, thin, frequently truncated, fining-upward sequences containing cross-bedding, ripple marks, lenticular and horizontal laminations, and convolute bedding. Individual units contain laterally discontinuous and truncated sands resting on a scoured surface of great lateral extent (Fig. 8-15A). Fine-grained sediments between sands reduce reservoir continuity. Substantial vertical variation occurs in cross-bedding, grain size distribution, direction and angle of dip, and relative porosity (Fig. 8-15B). Refer to Chapter 4 for a more complete discussion of braided streams outside the deltaic environment.

Meandering streams. Meandering rivers passing through alluvial valleys generally have high suspended loads, low gradients, fine-grained bed loads, relatively uniform water

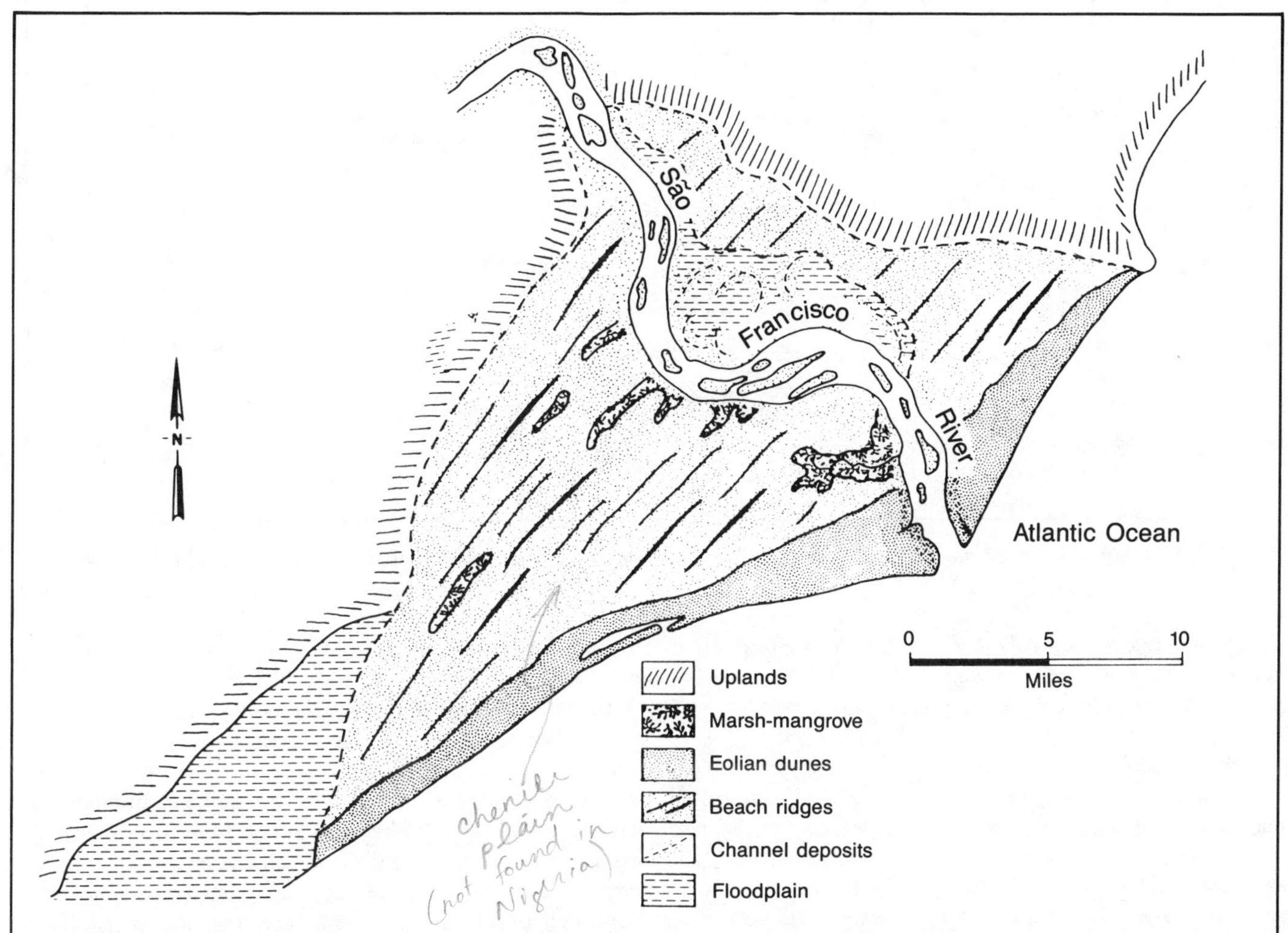

Fig. 8-8. Geomorphology of the São Francisco (Brazil) delta. (From Wright and Coleman, 1973; permission to publish by AAPG).

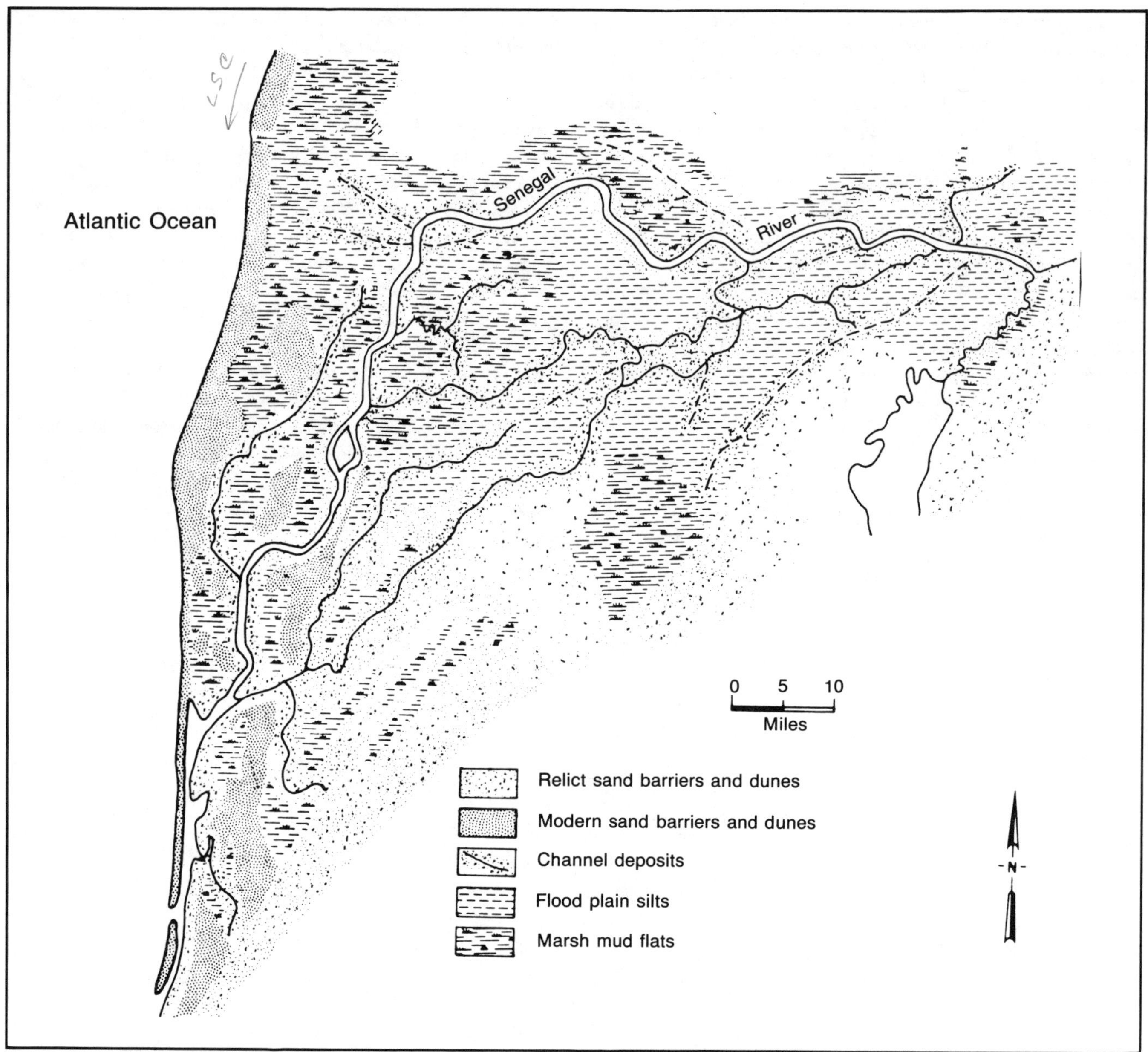

Fig. 8-9. Geomorphology of the Senegal (Senegal) delta. (From Wright and Coleman, 1973; permission to publish by AAPG).

discharge and flood regularly (Chapter 5). Meandering distributaries, however, occur on deltas with high tidal ranges and when the rivers have relatively coarse-grained bed loads. Backwater effects at high tide and bed load deposition require the distributaries to migrate laterally to maintain their gradients. Meander belt width varies directly as river discharge and channel depth and may exceed 15 km. Sediments rest on an eroded surface (Fig. 8-16A) and include point bars and abandoned channels filled with overbank and slumped-in muds. Sedimentary structures reflect deposition in various flow regimes and soft sediment deformation. Vertical variations in the sediments include fining-upward sequences and variable directions and amounts of dip. Sandstone porosity is generally good (Fig. 8-16B).

Lake fills. Formation of upper delta plain lakes is not well understood but does seem related to local subsidence and compaction. Lake areas vary from a few to many tens of square kilometers and depths are shallow, perhaps up to 20 m. Sediments in the lakes range from fresh water to evaporite deposits depending on local climate. Unless cut and infilled by a distributary, lake sediments are usually deposited under reducing conditions and consist of fine-grained silts and clays derived from flooding. Organic matter is abundant and some of it may occur as shelly layers

TABLE 8-2. Relationships between river, wave, and shelf slope characteristics and delta morphology and sedimentology. (From Wright and Coleman, 1973; permission to publish from AAPG).

Offshore slope	Flat convex →		Steep concave
Relative nearshore wave power	Low →		High
Relative riverine influence	High ←		Low
Configuration charactertistics	Highly indented, extended distributaries →	Smooth, arcuate, slightly protruding river mouths →	Straight, coastline, no protrusions
Delta plain landforms	Marsh, bays —	Subtle beach ridges and dunes alternating with marsh and mangrove →	Barriers, beach ridge and dunes dominate
Sand body	Shoestring sands, low lateral continuity →	Moderate lateral continuity along localized bands →	High lateral continuity, thick sand sheets
Sediment composition and sorting characteristics	Fine-grained, poorly sorted silts and sands laterally, well-sorted channel sands →	Alternating bands of moderately sorted dirty sands and poorly sorted silts and clays →	Well-sorted clean sands

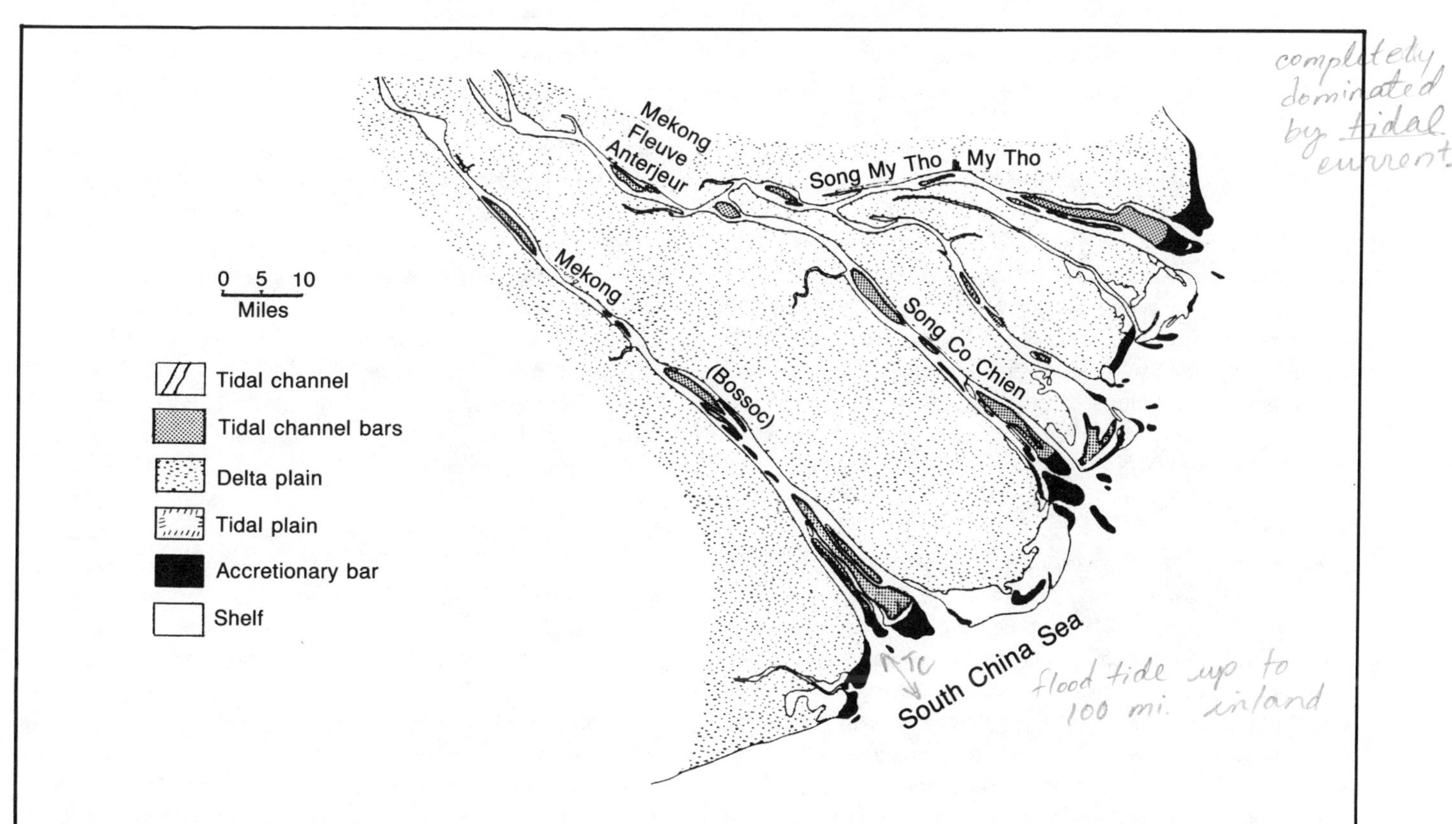

Fig. 8-10. Funnel-shaped distributary mouths of the Mekong delta. (From Fisher, 1969; permission to publish by Texas Bureau of Economic Geology).

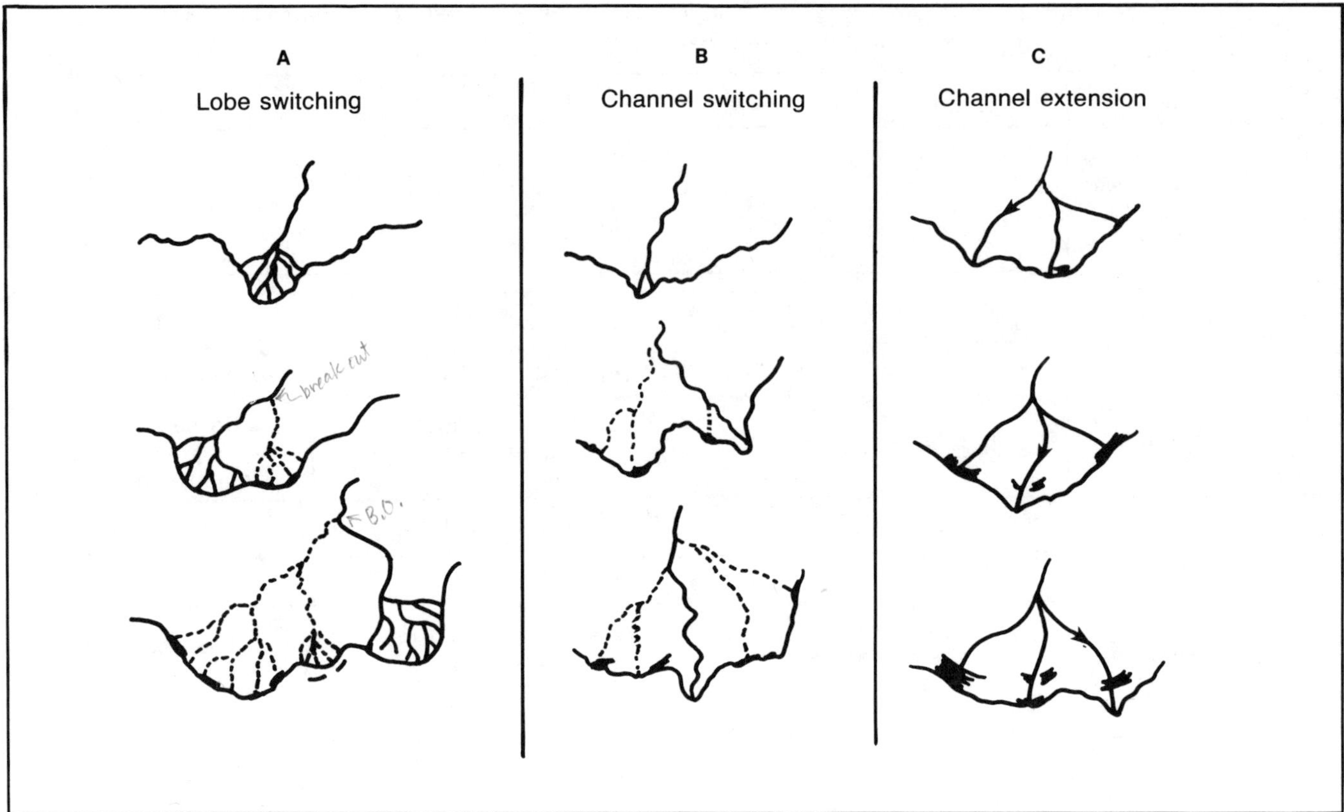

Fig. 8-11. Delta switching patterns. (From Coleman and Wright, 1975; permission to publish by Houston Geological Society).

if there is seasonal turnover in the lake.

Infilling of a delta plain lake may be rapid if the filling mechanism is a crevasse splay or small deltaic deposit (Fig. 8-17A). Lamination and bed thickness increase and depositional slope rises as the fill progrades. When infilling is complete, root burrows penetrate the top of the fill, and the whole sequence is buried by fine-grained, reducing, backswamp deposits. Lake fill sediments coarsen upward, dip directions are more consistent than those of meander point bars, and dip angles and porosity generally increases upward (Fig. 8-17B).

Lower Delta Plain

Lower delta plain environments consist of distributary channels and interdistributary areas. Channels are more numerous than on the upper delta plain due to anastomosing and bifurcation of the distributaries. Interdistributary areas are extensive and consist of migrating tidal channels, interdistributary bays, swamps, and marshes. Principal sand bodies are stacked, thin, crevasse splay sands deposited in bays separated by fine-grained bay fill and marsh deposits. In arid regions, deltas with high tidal ranges have interdistributary evaporites.

Bay fill deposits. Interdistributary bays cover the greatest area of many lower delta plains. They are hundreds of meters to kilometers in length and width and up to 10 m deep. As bays may be connected to the sea or surrounded by marshes and distributary channels, bay waters range from marine to brackish. Sedimentary infilling of interdistributary bays is by crevasse splays that extend seaward across the bays through a network of bifurcating channels (Fig. 8-18; D, E, and F). The filling process is cyclic and, on the Mississippi delta, one cycle lasts for 100 to 150 years. A crevasse forms during a flood and sediments enter the bay. During successive floods, sediments build to the bay-water surface and prograde into the bay. Sediment and water discharged into the bay reach a maximum in several years and then gradually wane. When sediment build-up either ceases or cannot keep pace with general subsidence, the bay fill gradually sinks below sea level and an open bay is re-established.

A cycle of filling of West Bay on the Mississippi delta is recorded in a series of maps (Fig. 8-19A). West Bay crevasse originated in 1839, but hydrographic charts from 1845 still showed open water in the bay. By 1875, the crevasse splay had built to sea level and had prograded into West Bay. In 1922, the crevasse splay lost its gradient advantage and was abandoned. By 1958, there had been substantial loss of land due to subsidence, and in 1978 the bay fill was completely submerged. A future splay could repeat the process if the birdfoot delta remains active.

The sedimentation pattern for a bay fill is one of shallow, brackish water to marine clays encroached upon by a coarsening-upward deltaic sequence (Fig. 8-19B). Prograding splay deposits are covered, in turn, by overbank sands, silts, and clays. With reversion to an open bay environment, a cap of burrowed, organic, marsh clays and/or open bay silts and clays is laid down. Dip is principally seaward at low angles, and porosity varies directly as grain size in the splay deposit.

Abandoned distributary deposits. Distributaries on the lower delta plain are quite stable and do not tend to migrate laterally. The exception is on deltas where the tide range is high and meander belt deposits develop. Distributary channels are a few meters to over a kilometer wide and one to tens of meters deep (Fig. 8-20A). Channel depth decreases toward the distributary mouth bar where there may be three meters of water or less.

Abandoned distributaries fill with locally derived materials. Where tidal range is high, sands may be brought into the channel by the flood tide. On deltas with low tidal range, poorly sorted sands, silts and organic debris are initially deposited in an abandoned channel. As filling progresses and any existing currents wane, the fill becomes increasingly fine-grained and rich in organic matter. Slumping of infilling sediments produces variable dip angles and directions (Fig. 8-20B).

Subaqueous Delta

Extending seaward from the low tide line and including the area actively receiving fluvial sediments is the subaqueous delta. If sediment accumulation is faster than reworking by marine processes and exceeds the rate of subsidence,

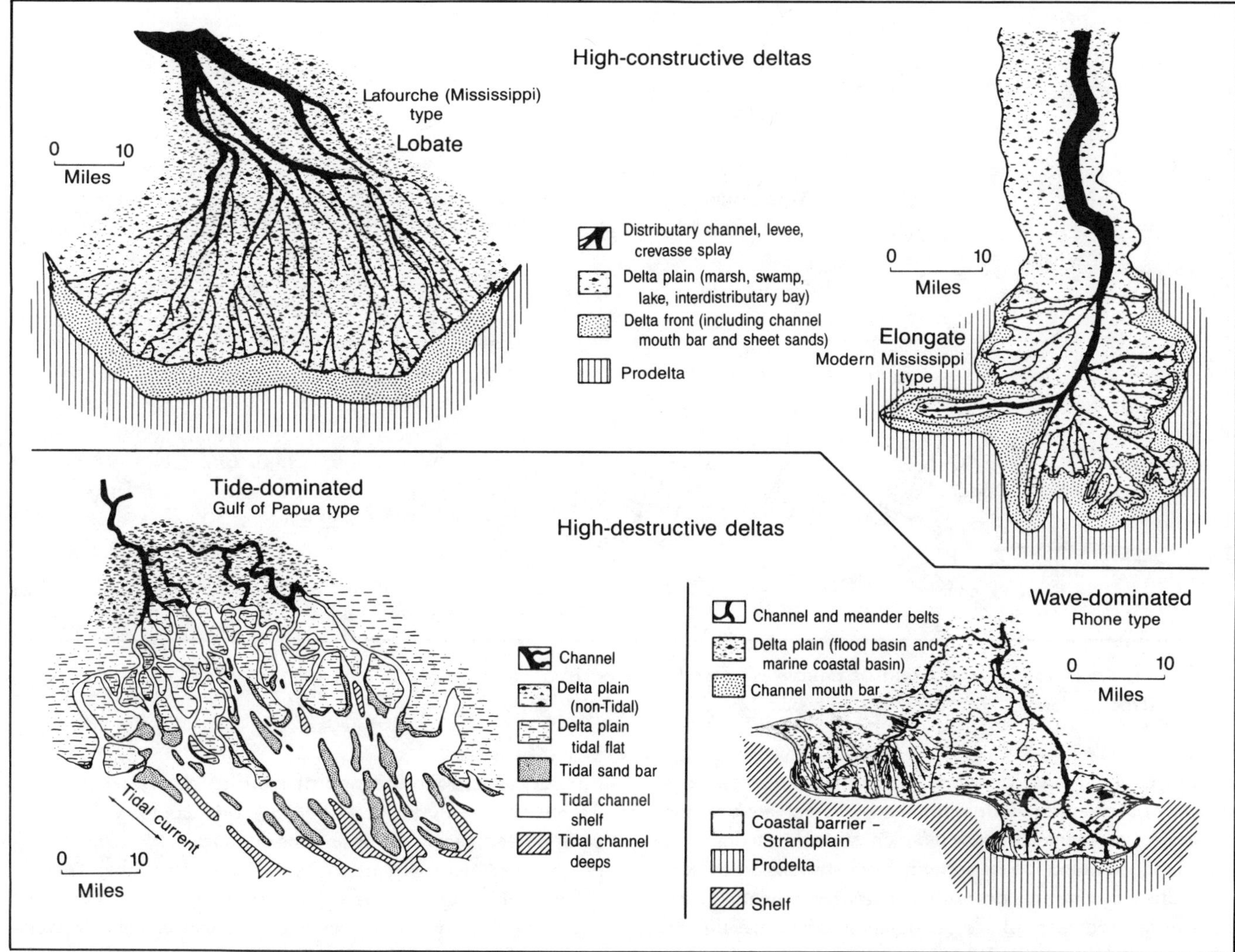

Fig. 8-12. Morphologies of high-constructive and high-destructive deltas. (From Fisher, 1969; permission to publish by Texas Bureau of Economic Geology).

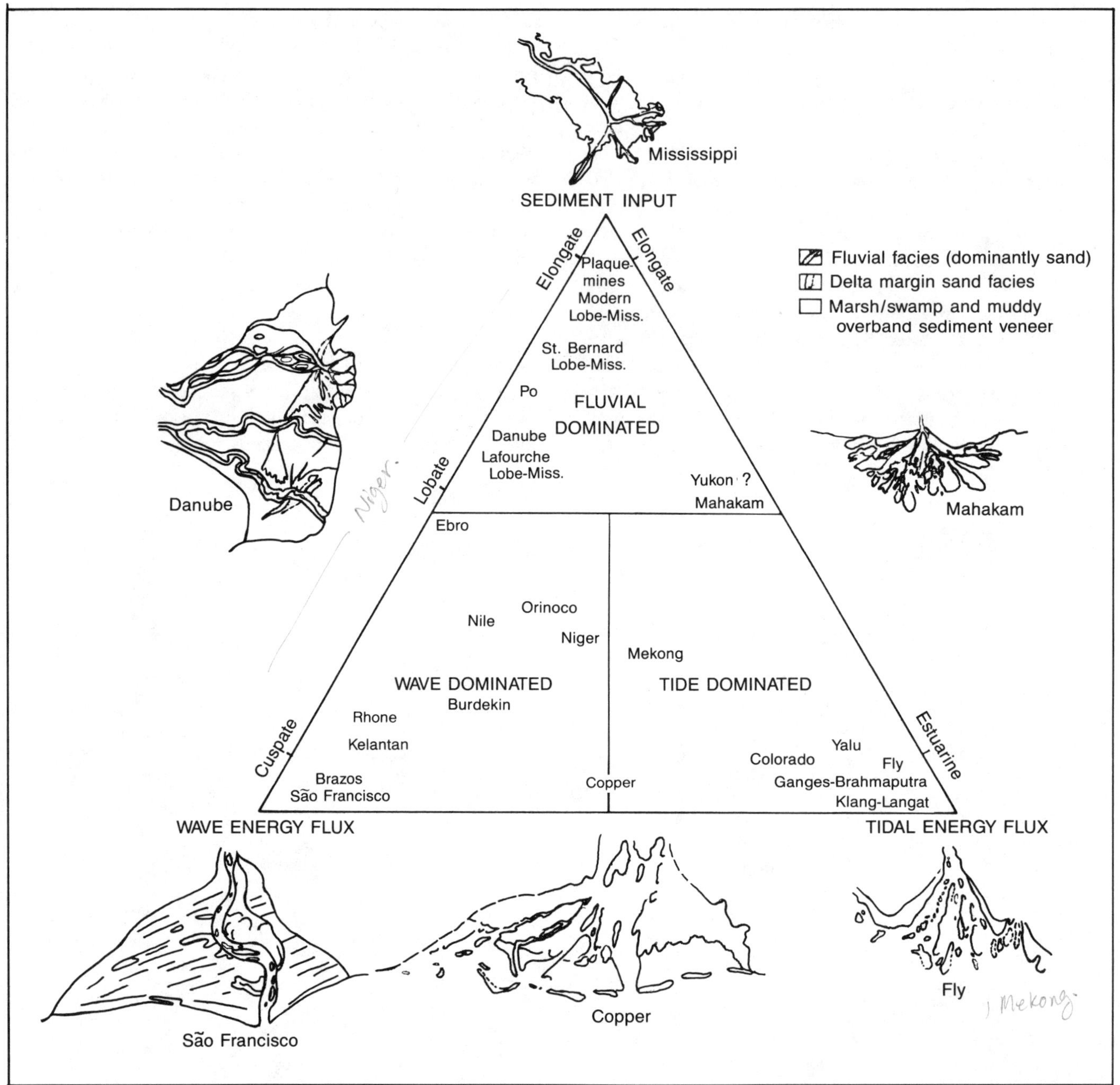

Fig. 8-13. Three-fold classification of deltaic systems based on relative importance of sediment influx, wave and tidal forces. (From Galloway, 1975; permission to publish by Houston Geological Society).

the subaerial delta will prograde across the subaqueous delta and form a complete deltaic sequence. Sediments decrease in grain size seaward, and sedimentation far from shore may be a combination of settling from suspension through the water column and downslope mass movements. Depositional environments of the subaqueous delta are the distributary mouth bar, the distal bar or delta front, and the prodelta (Figs. 8-21 and 8-22).

Distributary mouth bar. Shoaling at the mouth of a distributary marks the location of the distributary mouth bar. As the river effluent loses velocity and competence upon entering the receiving basin, the bedload and coarse suspended sediment are deposited to form the bar. Deltaic deposition occurs at the highest rate here. The mouth of Southwest Pass on the Mississippi delta receives one to three meters of sediment per year, causing the distributary to prograde approximately 100 m per year.

Distributary mouth bar geometry is a function of effluent

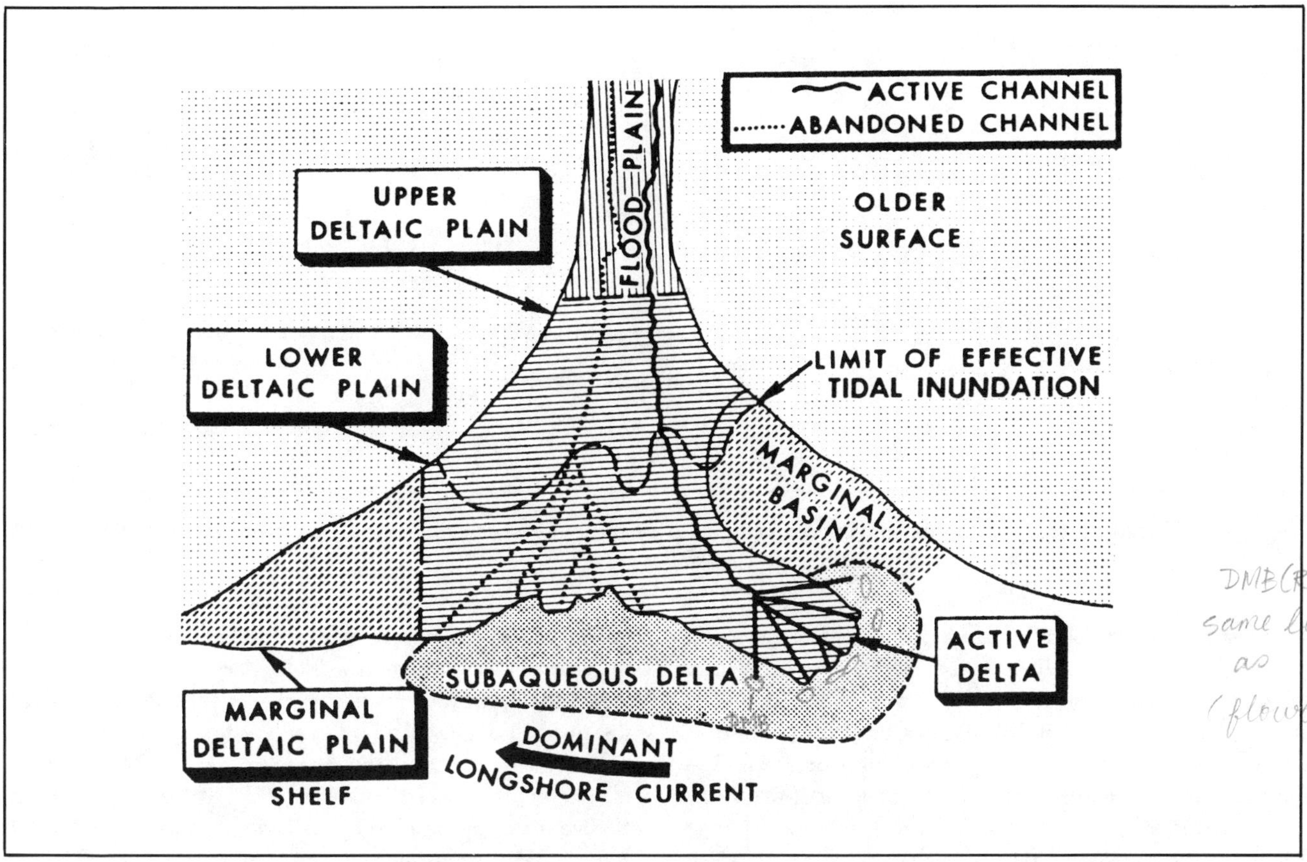

Fig. 8-14. Components of a delta. (From Coleman and Prior, 1982; permission to publish by AAPG).

dynamics and marine processes. River flow conditions, density differences between effluent and basin waters, shelf slope seaward of the river mouth, tidal range and currents within the lower river channel, and redistribution by fluvial sediments of marine processes all affect the bar. In a river-dominated delta such as the Mississippi delta, prograding distributaries produce elongate distributary mouth bars (the bar-fingers of Fisk, 1961; Fig. 8-23). Sand bodies are tens of kilometers long, over six kilometers wide, and 20 to 80 m thick (Figs. 8-23 and 8-24). Tide-dominated deltas differ

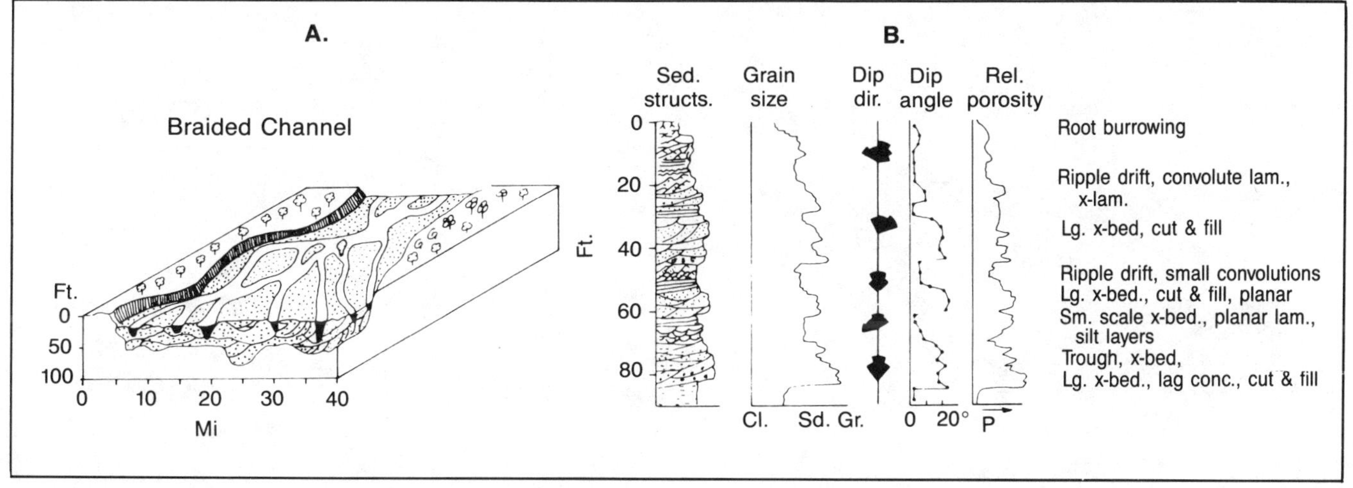

Fig. 8-15. Summary diagram of the major characteristics of braided channel deposits on the upper delta plain. (Modified after Coleman and Prior, 1982; permission to publish by AAPG).

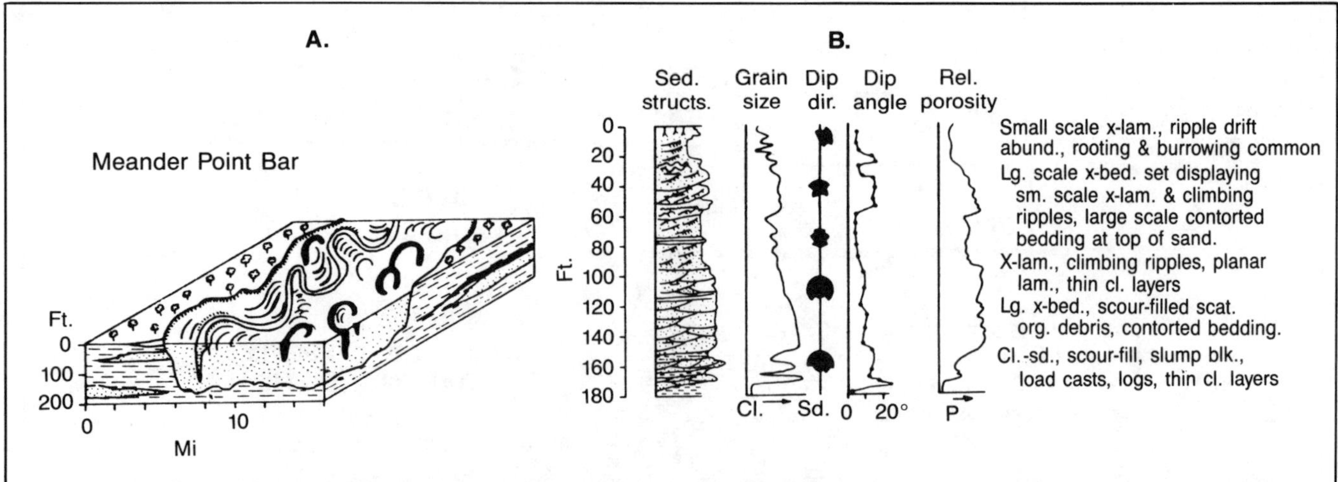

Fig. 8-16. Summary diagram of the major characteristics of meandering stream deposits on the upper delta plain. (Modified after Coleman and Prior, 1982; permission to publish by AAPG).

from river-dominated deltas by tending to have funnel-shaped mouths (Fig. 8-10) and river mouth sands formed into a series of subparallel ridges oriented parallel to the tidal current (Figs. 8-25 and 8-28). Ridges may attain heights of over 20 m, lengths of 15 km, and widths of 2 km. The deltas of the Ord River (Australia), Klang River (Malaysia), and Colorado River (Gulf of California) have such ridge systems. Wave-dominated deltas, such as the Sao Francisco (Fig. 8-8) and Senegal (Fig. 8-9) seem to have poorly defined distributary mouth bars (Fig. 8-26). High wave energy, in possible combination with strong longshore drift, tends to rework the river sands into beach ridges that, in turn, may supply sands for eolian dunes.

Distributary mouth bar sediments are the coarsest materials moved by the river. Fine-grained sediments may be deposited on the bar during periods of low river flow, but these are likely to be reworked and removed by marine processes and subsequent high river flows. Progradation of the distributary mouth bar over the fine-grained sediments of the distal bar and the prodelta (Fig. 8-27A) produces a coarsening-upward sequence (Fig. 8-27B). Laterally extensive sand deposits may be formed if the bars from bifurcating distributaries coalesce. Sedimentary structures are related to currents and waves and include cross-lamination, larger-scale trough cross-bedding, current ripple drift, and climbing ripples (Fig. 8-27B). Slumping is common, especially in the lower parts of the bar where over-pressured sediments are more likely to occur.

Overlying the distributary mouth bar are overbank splays (natural levees) and shallow bay sediments. Splays are pro-

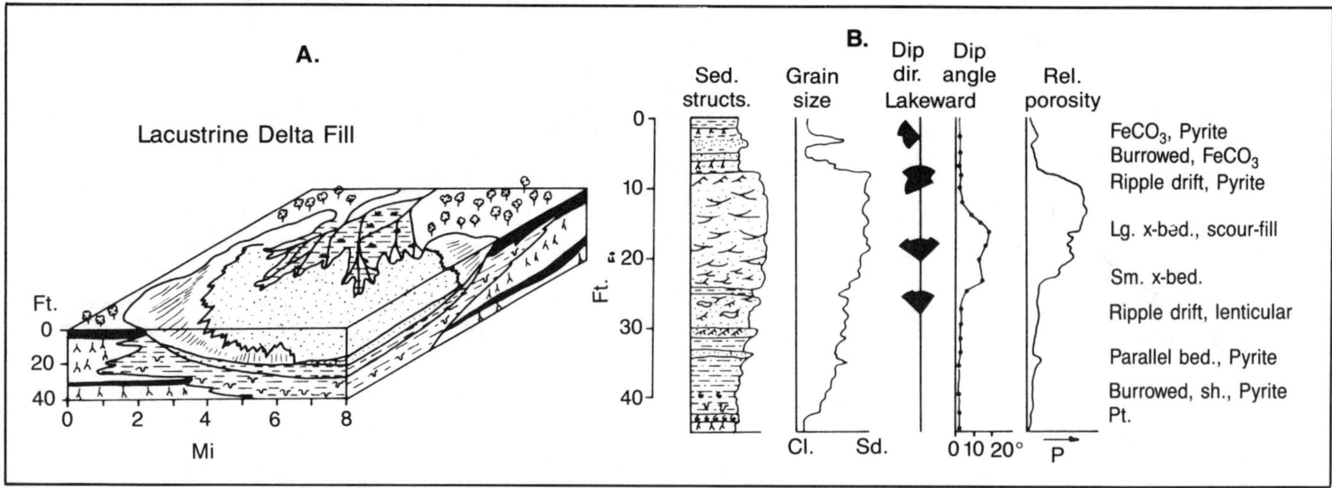

Fig. 8-17. Summary diagram of the major characteristics of a crevasse splay filling a lake on the upper delta plain. (Modified after Coleman and Prior, 1982; permission to publish by AAPG).

duced when floods top the distributary channels. Deposits build up until they reach flood level and consist of coarsening-upward sequences with small-scale cross-laminations and climbing ripples.

Sedimentological data on the ridges of tide-dominated deltas indicate that ridge sands are coarse-grained, well-sorted, and thick. Sands in the swales are more poorly developed and are associated with shell-lag deposits, organic debris, and clay clasts. In or near the distributary mouth, the sand may rest on a scoured surface and have the charateristics of channel sand. Cross-stratification is large-scale, small-scale, and herringbone, indicating both uni-

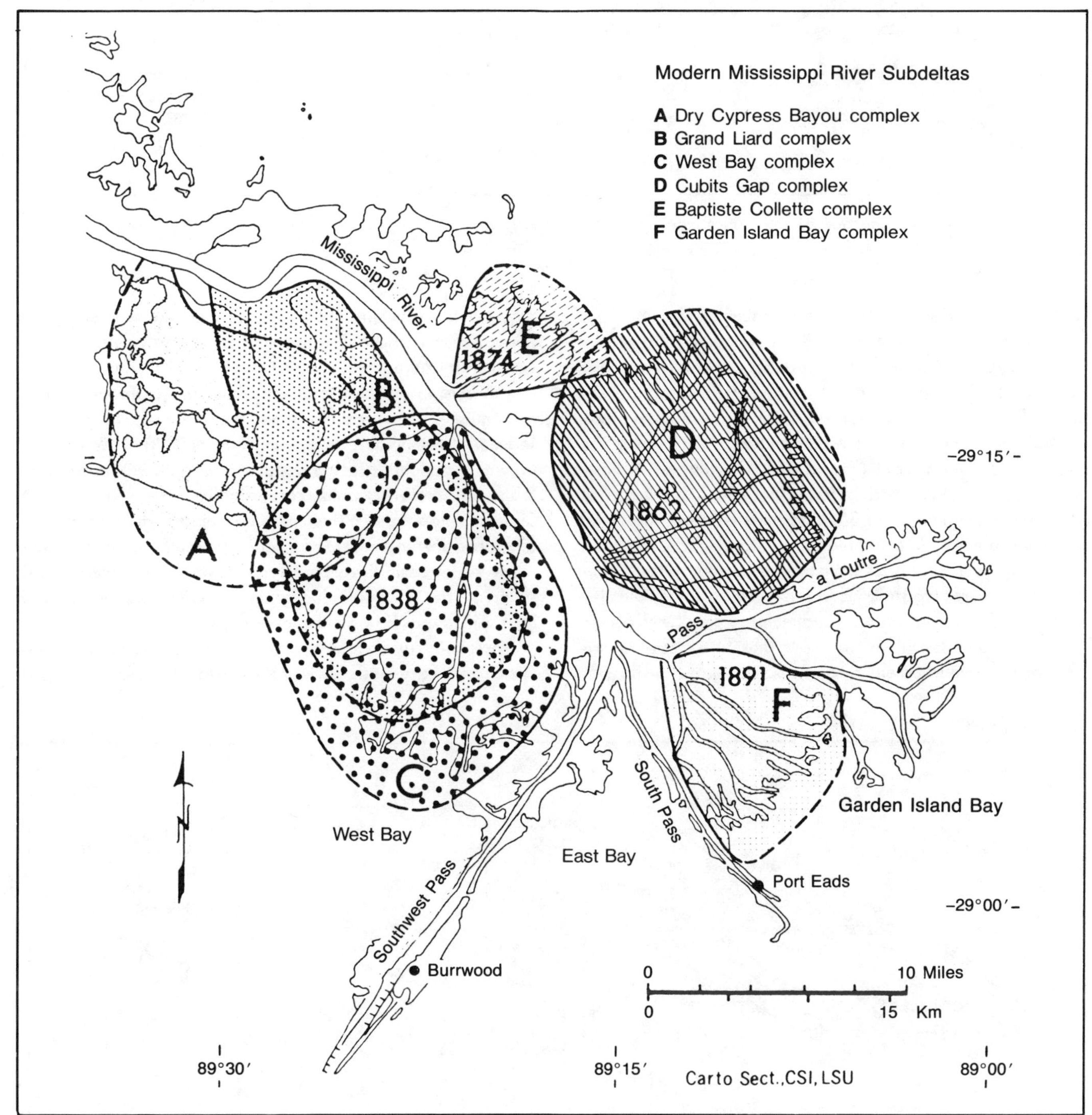

Fig. 8-18. Crevasse splays filling bays of the modern bird-foot delta of the Mississippi River. (From Coleman and Prior, 1980; permission to publish by AAPG).

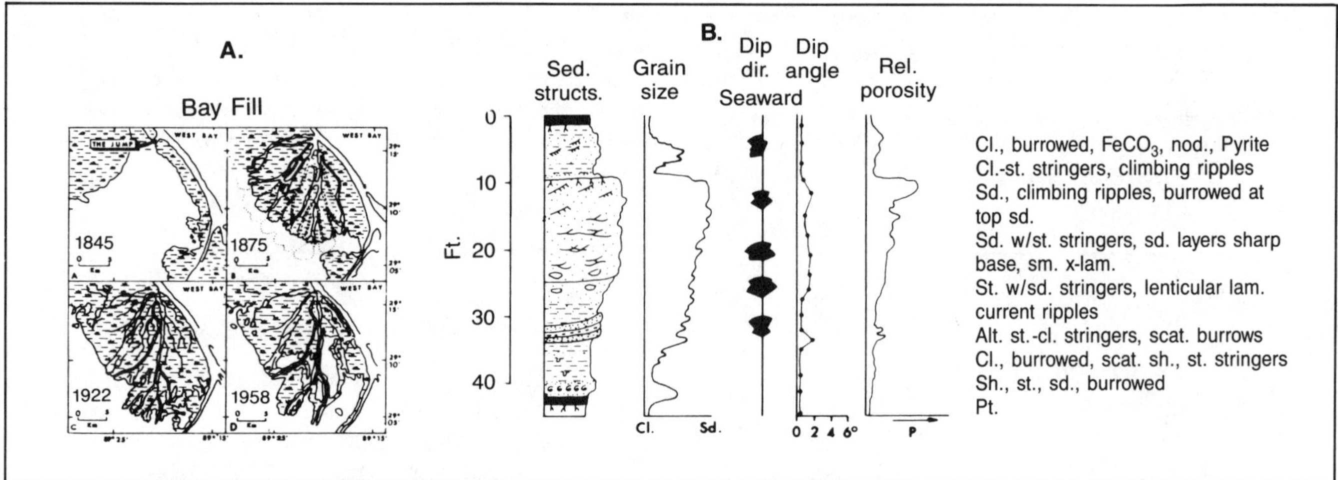

Fig. 8-19. Summary diagram of the major characteristics of bay-fill deposits on the lower delta plain. (Modified after Coleman and Prior, 1982; permission to publish by AAPG).

modal and bimodal transport of the sediments (Fig. 8-28B). Dip angles range from 1-10° and the range of relative porosities is great. Parallel lamination in the sands is probably caused by deposition in the upper flow regime resulting from high tidal current velocities in shallow water.

Distal bar. Seaward of the distributary mouth bar is the distal bar or delta front (Fig. 8-21). The highest depositional slope on the delta is found here with an angle of generally less than 1/2° to 3/4°. Sediments low in the distal bar are lenticular and laminated silts and clays with thin layers of graded and cross-laminated sands and silts. Higher in the bar where deposition is in shallower water, grain size increases and current-induced sedimentary structures occur, such as graded sands, minor scour and fill, cross-lamination, and starved current ripples. Sands of the distal bar may be of sufficient quality to be reservoirs. Pleistocene distal bars of offshore Texas and Louisiana contain gas in commercial quantities.

Prodelta. A higher sedimentation rate from suspension distinguishes the prodelta from the open shelf. Prodelta sediments are widespread laminated silts and clays containing occasional graded bedding and an open marine microfauna. Bioturbation is rare due to the relatively high sedimentation rate. When present, bioturbation is generally found low in the section near open shelf sediments where rates of deposition are lower. Unstable sediments and slumping are common due to high fluid pressures caused by rapid deposition of fine-grained sediments.

Variability in deltaic deposits. Deltaic sedimentary environments vary both horizontally and vertically. Deposits

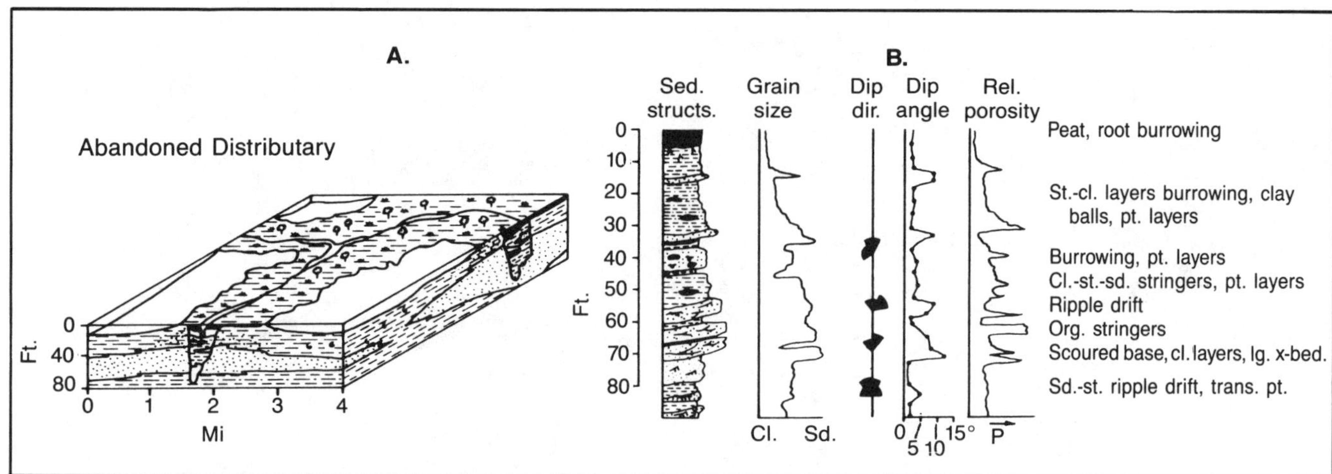

Fig. 8-20. Summary diagram of the major characteristics of abandoned distributary deposits on the lower delta plain. (Modified from Coleman and Prior, 1982; permission to publish by AAPG).

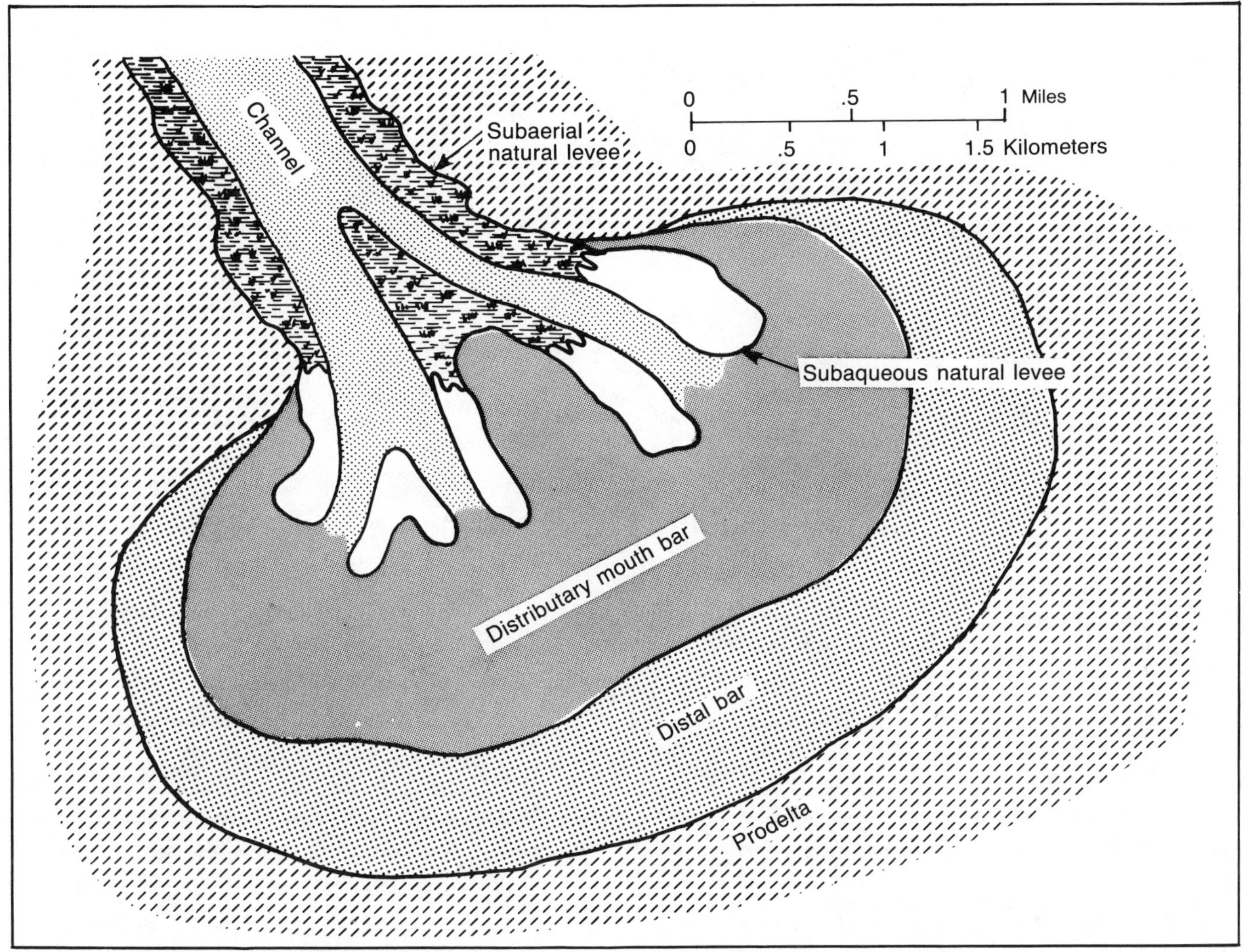

Fig. 8-21. Depositional environments at the mouth of a distributary. (From Coleman and Prior, 1980; permission to publish by AAPG).

of individual environments range from hundreds of meters to kilometers in width and from a few to many tens of meters in thickness depending on the size of the distributary. In a prograding delta (Fig. 8-29), marine sediments are overlain by increasingly proximal and subaerial deltaic deposits. Marine deposits are laterally continuous whereas those of the subaerial delta are laterally restricted. Sediments encountered in a borehole are a consequence of lateral position relative to the distributary. Proximity to the distributary is shown by high sand content (Fig. 8-29, Logs D and E) while increasing distance from the distributary is shown by decreasing sand (Fig. 8-29, Logs C, B, and A).

Synsedimentary deformation of deltaic deposits. Substantial redistribution and modification of deltaic sediments may occur by contemporaneous mass movement. Such movement is the result of high rates of deposition of fine- and coarse-grained sediment, which produces deposits with high water content and excess pore-water pressure. Additional fluid pressure and loss of internal strength comes from biogenic gases produced by decomposing organic matter in the sediments. Failure occurs when the internal strength of the sediment is reduced and/or external stresses rise and exceed the internal strength. Some modern deltas with recurrent, contemporaneous slumping are the Magdalena (Columbia), Amazon (Brazil), Nile (Egypt), Niger (Nigeria), Hwang-Ho (China), and Mississippi (USA).

Studies off the Mississippi delta indicate numerous types of failures occur around the periphery of the modern delta (Fig. 8-30). Peripheral slumps, retrogressive slides, and growth faults are failures of interest in exploration. Peripheral slumps are failures of distributary mouth bar deposits along curved shear planes. Resembling rotational slumps, they produce a series of stairstep scarps up to eight meters high on the sea floor and tensional cracks upslope from the main scarp. Dip of the shear planes quickly decreases downslope, becoming parallel to the bedding and

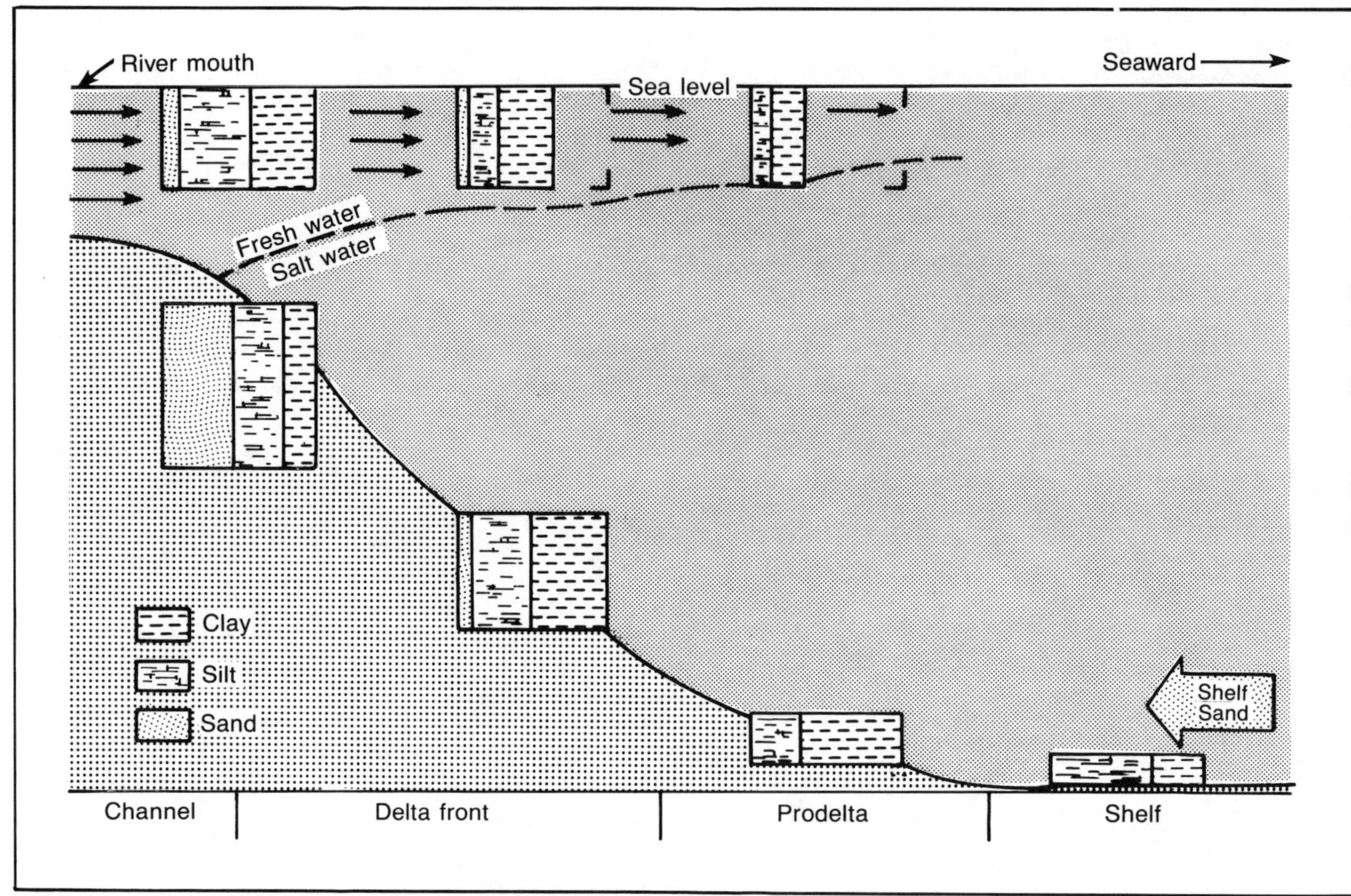

Fig. 8-22. Distribution of subaqueous delta environments and sediments. (From Scruton, 1960; permission to publish by AAPG).

making the primary movement of peripheral slumps translational. Individual blocks are from hundreds of meters to a few kilometers wide and up to 35 m thick. Downslope slide rates vary from a few hundred meters to a kilometer per year. With the involvement of distributary mouth bar sediments in the slumps, coarse, shallow water sediments are transferred into deeper water offshore.

Elongate retrogressive slides originate in shallow water (10–100 m) and consist of three parts: a shallow water, bowl-shaped depression bounded by scarps; an elongate gully oriented parallel to dip emanating from the downslope side of the depression; and a deep-water area of accumulation below the gully consisting of overlapping depositional lobes. Retrogressive slides originate in the depression, the center of which is hummocky due to disorganized masses of blocks in remobilized and disrupted sediments. The gully is a winding feature from a few to 20 m deep and up to 10 km long. Width varies, alternating between a few tens of meters in constricted areas to several hundred meters in more bulbous areas. Overbanking of the gully occurs periodically as shown by splays resembling natural levees. At the seaward end of a gully there is a series of overlapping lobes, each deposited by a separate slide event. Lobe surfaces are irregular and have mud volcanoes. A scarp up to 25 m high marks the downslope extent of each lobe. Forward of the lobes are small pressure ridges and volcanoes emitting mud and fluids indicating relief of over-pressure in the substrate caused by loading from the lobes. Coalescing lobes from adjacent slides may form complexes many kilometers wide along strike.

Growth faults are failures along the outer shelf in front of prograding deltas. An individual fault may be up to several kilometers long, but many faults arranged end to end may extend along strike for tens or hundreds of kilometers. Growth faults step down to the basin and may involve sediments up to 750 m deep. Downthrown sides are sites of additional sediment accumulation causing beds to thicken across the faults (Fig. 8-31). Continuous movement of the faults, combined with continuous sedimentation, results in increasing displacement with depth, hence the name growth fault. Like the shear planes in peripheral slumps, growth fault planes flatten with depth and become bedding plane faults.

Subaqueous slumps consist of multiple slump blocks that rotate landward, reversing their original depositional dip (Fig. 8-32A). Such dip reversals may reach 20° (Fig. 8-32B). Bedding and sedimentary structures are disrupted near shear planes but preserved in the central portions of the slump blocks. Grain size may be variable, especially if distributary mouth bar sands have slumped basinward where they rest on and are covered by fine-grained prodelta muds (Fig. 8-32B).

EXPLORATION FOR HYDROCARBONS IN DELTAS

Deltas are good hydrocarbon prospects because of the great variety of reservoir rocks they contain. Identification of a delta in the subsurface is frequently difficult, but application of a number of exploration techniques should result in the correct identification of the environment and enable the geologist to better find the fairways of sand deposition. Some of the successful methods used to explore in deltas are discussed in this section.

Oklahoma

Booch Delta of McAlester basin. The Booch delta, covering approximately 2,000 sq mi (5180 sq km), was de-

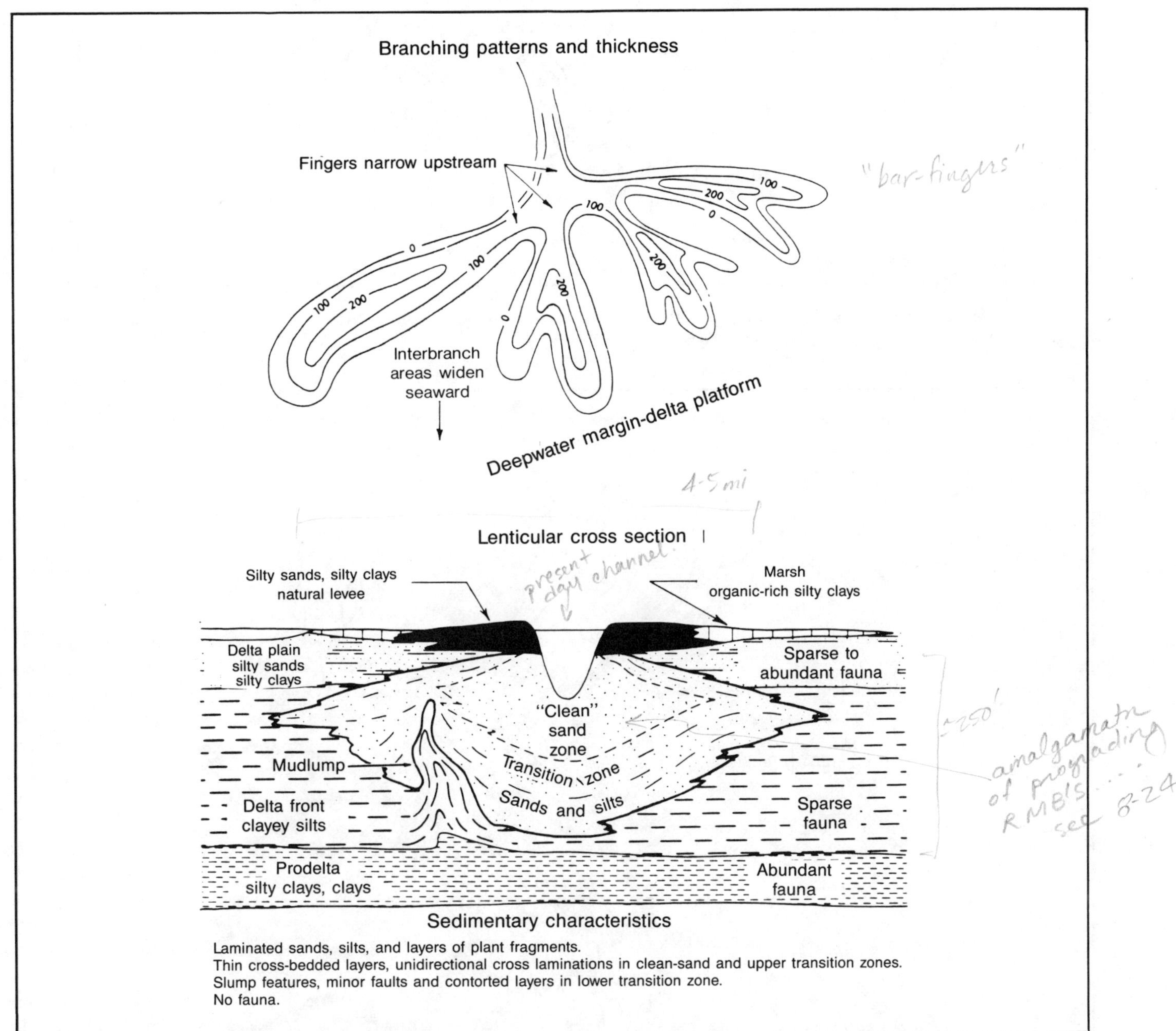

Fig. 8-23. Geometry and sedimentary characteristics of elongate distributary mouth bar sands. (From Fisk, 1961; permission to publish by AAPG).

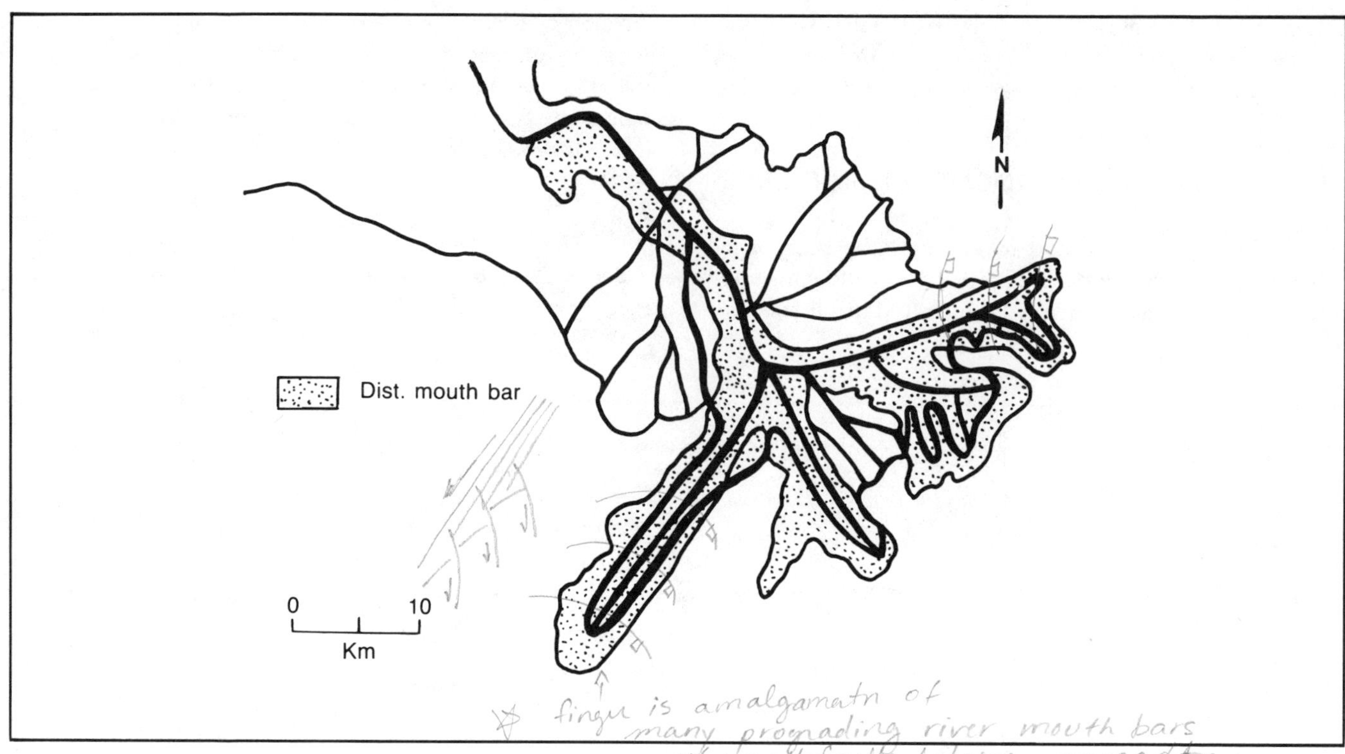

Fig. 8-24. Morphology of a distributary mouth bar of a river-dominated delta (Mississippi River, USA. (Modified after Coleman and Prior, 1980; permission to publish by AAPG).

finger is amalgamatn of many progading river mouth bars w/ growth-faults helping progradtn. ← (helped out by seismic)

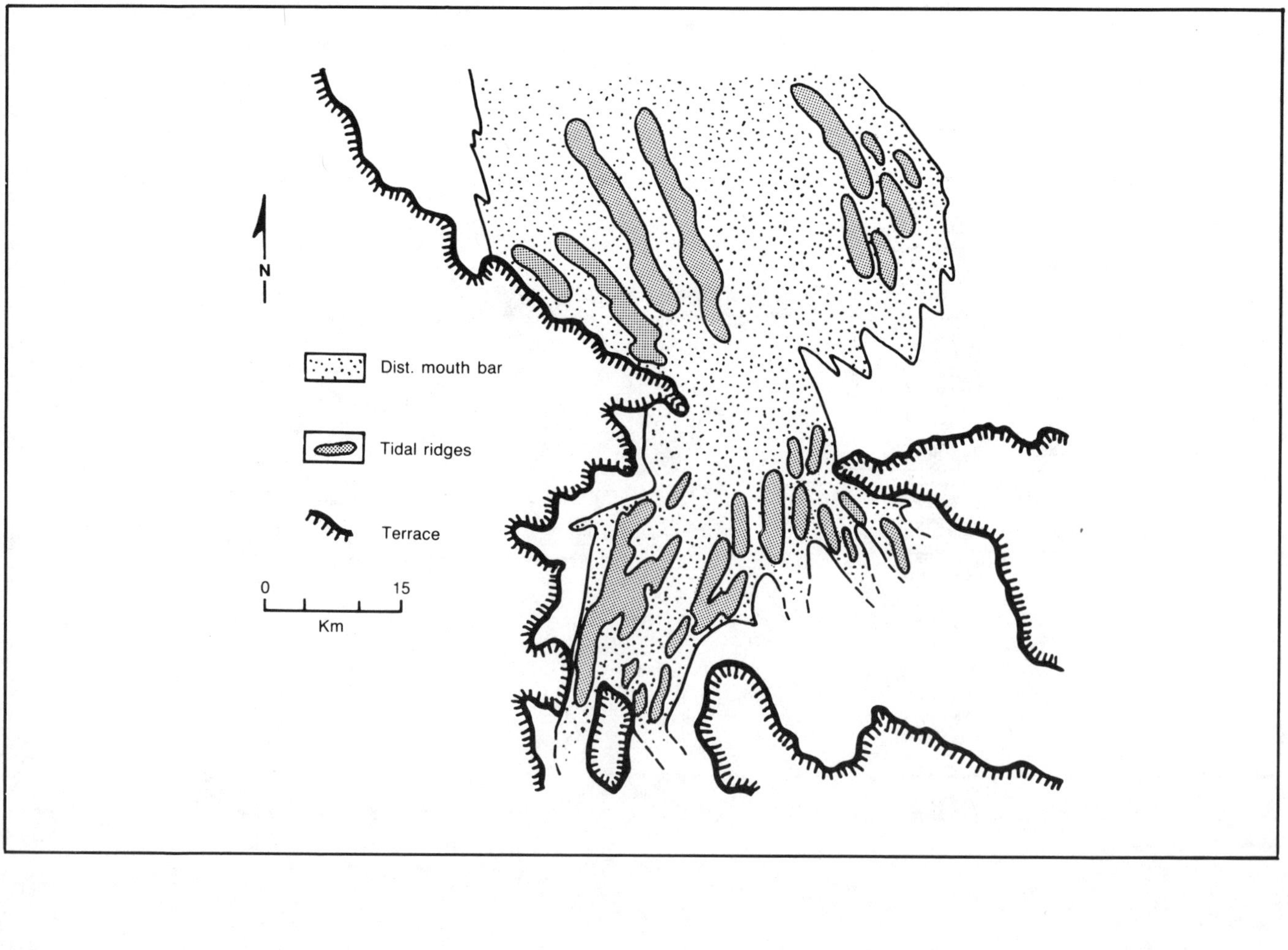

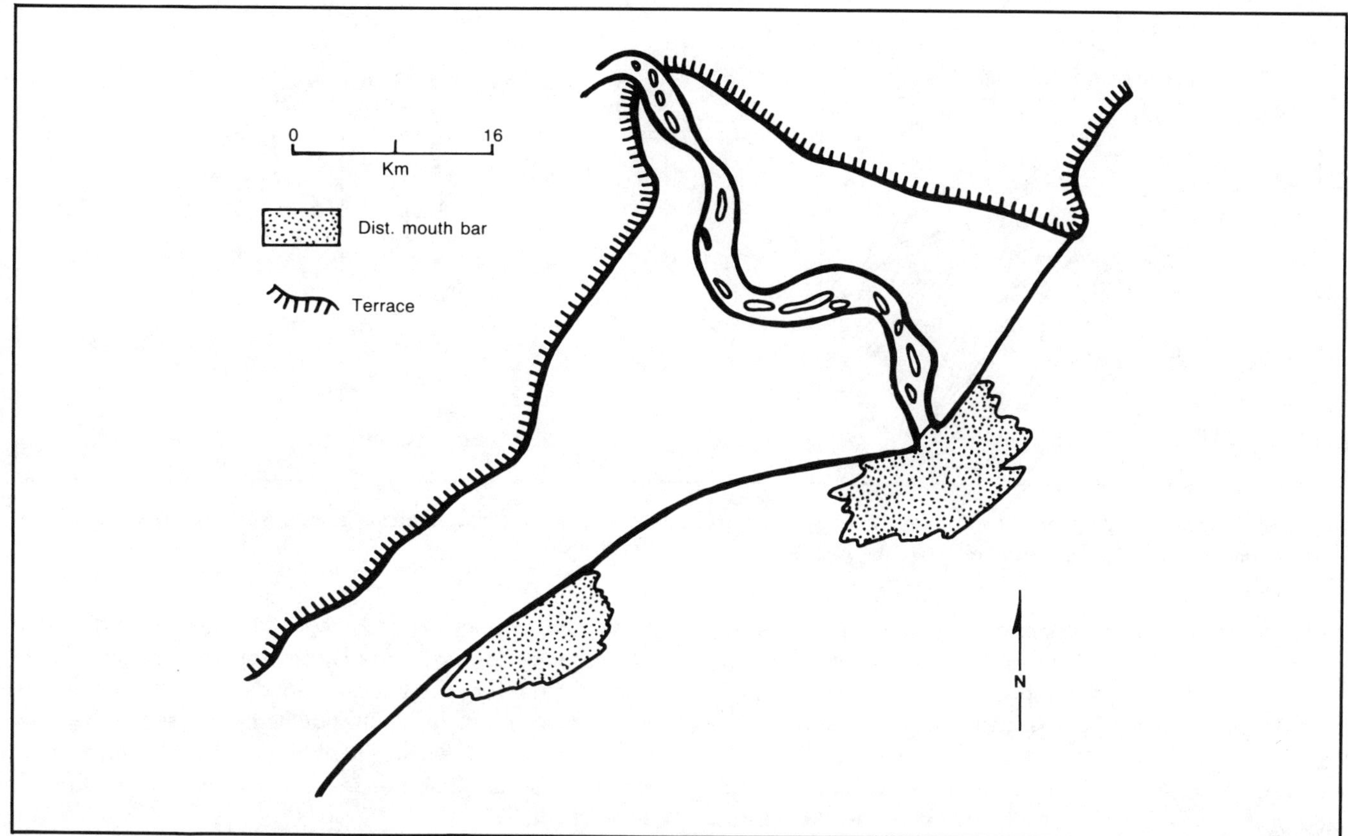

Fig. 8-26. Morphology of a distributary mouth bar of a wave-dominated delta (São Francisco River, Brazil). (Modified from Coleman and Prior, 1980; permission to publish by AAPG).

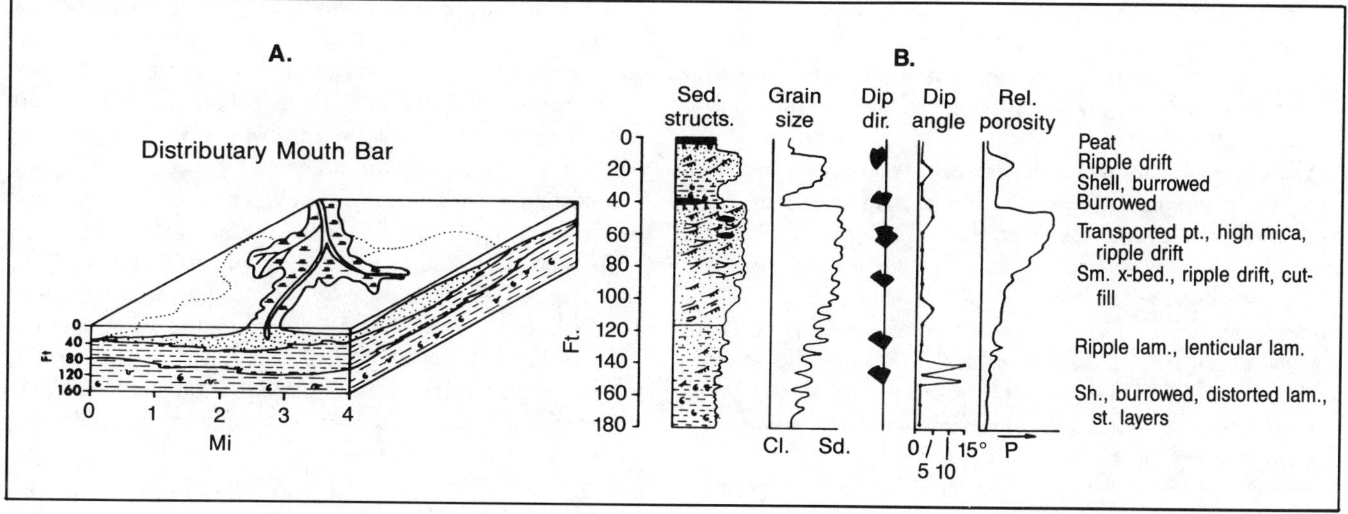

Fig. 8-27. Summary diagram of the major characteristics of distributary-mouth bar deposits in the subaqueous delta plain. (From Coleman and Prior, 1982; permission to publish by AAPG).

←

Fig. 8-25. Morphology of a distributary mouth bar of a tide-dominated delta (Ord River, Australia). (Modified from Coleman and Prior, 1980; permission to publish by AAPG).

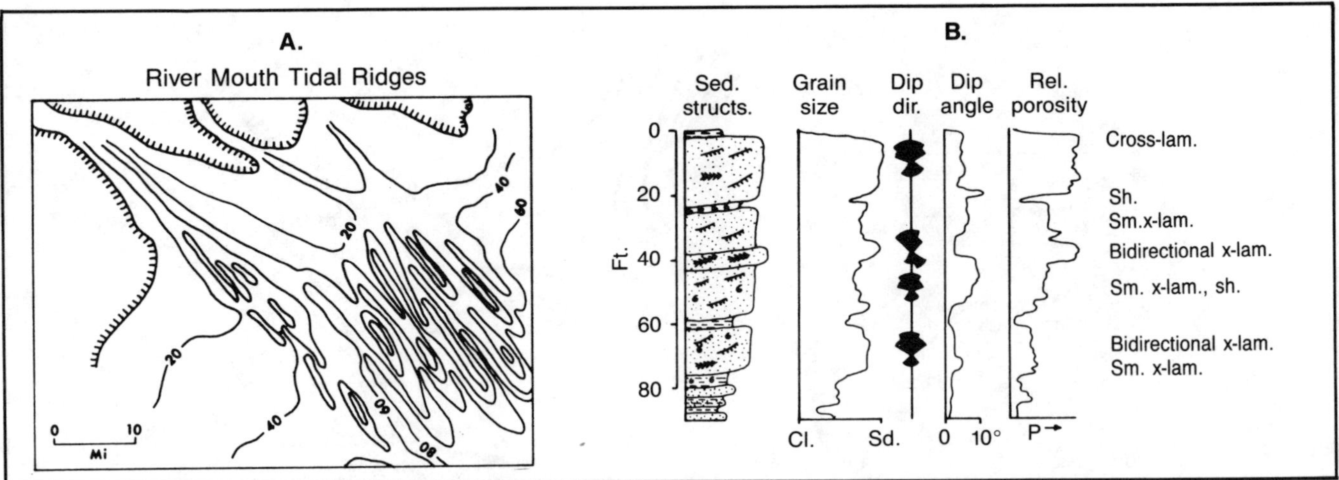

Fig. 8-28. Summary diagram of the major characteristics of river mouth tidal ridges in the subaqueous delta plain. (From Coleman and Prior, 1982; permission to publish by AAPG).

posited on a shelf in the eastern Oklahoma part of the Arkoma basin in early Pennsylvanian time. To analyze the sandstones making up the delta, it is best to first consider the GSS of which they are a part. North-south and east-west stratigraphic cross sections were constructed for the McAlester formation, which contains the delta. Wells not used in the correlation grid were correlated to the nearest wells in the grid.

Sediments of the McAlester formation are predominantly shale except for the four members of the Booch sandstone (Fig. 8-33). In some wells, the middle member of the Booch sandstone increases abruptly in thickness to as much as 250 ft (76 m). In so doing, it merges vertically with the lower middle member of the Booch and occasionally with the lower member of the Booch. Thickening is principally downward at the expense of the shale encasing the lower middle and lower Booch.

Shales of the Savannah formation conformably overlie and are lithologically indistinguishable from shales of the McAlester. The contact between them is arbitrarily placed at a re-entrant in the short normal resistivity curve, which is easily recognized on over 580 of 650 electric logs available for the study (Fig. 8-33). In the deeper part of the Arkoma basin, the McAlester rests conformably on the Atoka shale. On much of the shelf, however, the McAlester unconformably overlies successively older units to the west in the direction of the Hunton arch.

Electric logs and samples of the McAlester show that basinward (southeast) the formation consists almost entirely of shale. On the shelf, however, it is characterized by alternating beds of shale, sandstone, limestone, and coal. Deep water never covered the shelf as indicated by the coal beds.

An isopach map of the McAlester formation (Fig. 8-34) was drawn to: (1) reconstruct the shape of the northwestern part of the Arkoma depositional basin; (2) determine the trend and position of the hinge line that existed during McAlester deposition; and (3) ascertain the principal directions of transport of McAlester sands and mud. Formation thickness increases gradually southeast across the shelf and then rapidly into the basin after crossing the hinge line. Maximum accumulation of sediments in the mapped area is 1600 ft (488 m).

In the northeastern part of the study area (T13-15N, R14-16E), the upper Booch sandstone is the principal unit present and is a series of delta distributary channel fills (Fig. 8-35). In the central part of the study area, thick sandstones represent vertical merging of all four Booch sandstones. West of R12E where the Booch is less than 60 ft (18 m) thick, the middle Booch is the principal sandstone. Other Booch units are correlated only as mappable zones rather than as lithologic units. Lower middle and lower Booch sandstones are not present except where merged vertically with the middle Booch in the channel areas.

Sandstone is prominently developed in a deep channel trending southeast from T15N, R11E, to T11N, R14E (Fig. 8-35). Channel width ranges from three to five miles (5–8 km), mapped length is over 35 mi (56 km), and maximum depth exceeds 240 ft (73 m). Undoubtedly the channel extends farther southeast, but the data for mapping it in this direction were not available. Where the channel fill is thickest, sandstone comprises over 50% of the total thickness of the McAlester with the rest of the section being predominantly shale. Compaction of McAlester shales away from the channel is sufficient to show the location and trend of this sandstone on the isopach map of the McAlester formation (Fig. 8-34). As sandstone is relatively noncompactible, the northwest convexity of the 300, 400, and 500 ft contours is due primarily to the presence of the channel sandstone. Numerous smaller channel sandstones combined with the major channel sandstone reveal a distributary net-

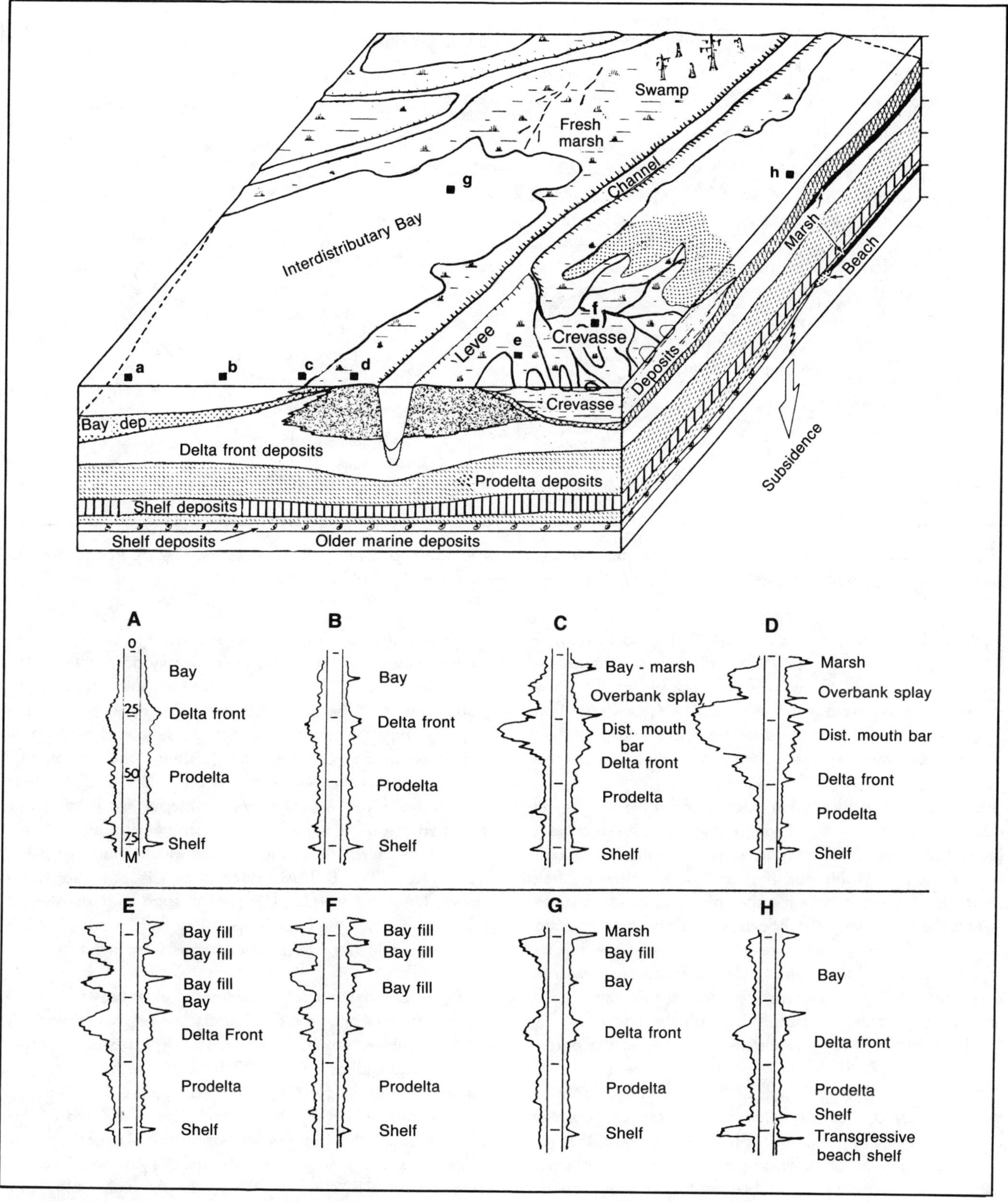

Fig. 8-29. Block diagram and schematic electric logs showing lateral and vertical facies relationships in deltaic deposits. No horizontal or vertical scale. (From Coleman and Prior, 1980; permission to publish by AAPG).

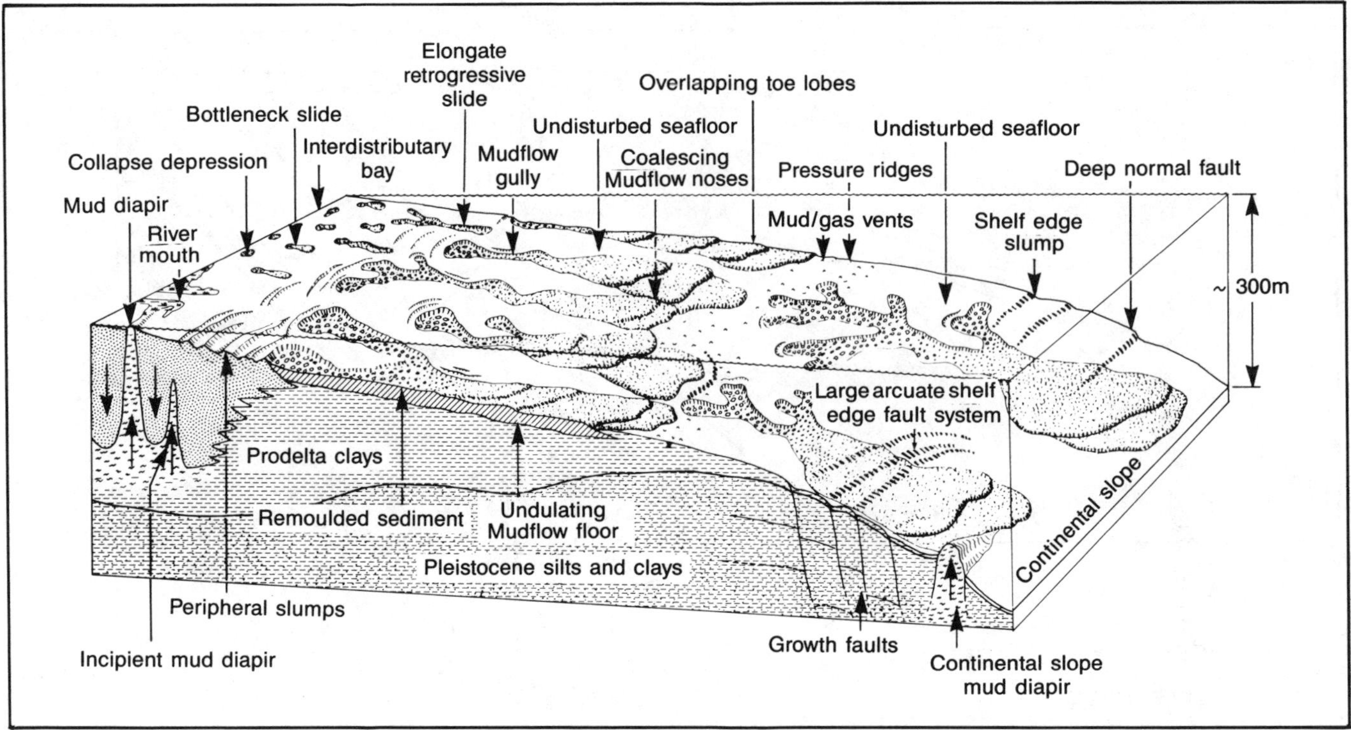

Fig. 8-30. Block diagram illustrating diapirism, contemporaneous faulting, and failures in submarine deltaic sediments in the Mississippi delta. (From Coleman and Prior, 1982; permission to publish by AAPG).

work of a large delta with its apex to the north (Fig. 8-35). It is not likely that all the distributaries were formed simultaneously, and the map is a composite of all stages of distributary development during McAlester deposition. Some of the smaller distributaries close to the largest channel sandstone may represent sand accumulation as crevasse splays.

Based on several thousand electric and drillers' logs, the isopach map of the Booch delta (Fig. 8-35) serves as a geological framework from which other studies may be made. For example, in T6N, R8E (Fig. 8-35), the Hawkins field produces from the middle member of the Booch sandstone. A coal bed overlying the Booch with a characteristic signature on the long normal resistivity curve of the electric logs serves as a local datum. Structure contours describe a sinuous, arcuate high over the pool to the south and a narrow, bifurcating structural ridge on the north (Fig. 8-36A). Maximum relief over the southern half of the pool is approximately 75 ft (23 m). To the north, relief exceeds 120 ft (37 m). When the trends of the structural highs (heavy dashed lines) are compared with the axes of maximum sandstone thickness (Fig. 8-36B), it is apparent that the structure is due in part to variations in sandstone thickness. Logs of wells outside of and marginal to the pool show the Booch is lenslike and grades into shale in all directions. The coal datum serves as the means of identifying the Booch interval outside the productive area.

An isopotential map of the Booch (Fig. 8-36C) shows trends of maximum initial potential (heavy dashed lines) coincide with trends of maximum sandstone thickness (Fig. 8-36B). This is to be expected since the initial potential is partly a function of relative permeability and amount of sandstone at any one borehole position. The isopotential map also shows the deltaic aspect of the Booch better than any of the other maps. Local postdepositional tectonism modified the area, with the most pronounced effect being a local structural depression in contiguous parts of Secs. 16 and 21 (Fig. 8-36A). Isopach and isopotential maps define the sand trends in the depression and assisted in carrying out a systematic step-out drilling program. Drill sites were selected where axial trends were most likely to extend.

Significant petrographic differences exist between distributary channel and interdistributary sandstones, especially in the southwestern half of the delta (Fig. 8-35). Where channel sandstones exceed 20 ft (6 m) in thickness, they generally have larger mean grain diameters than those of the interchannel areas, the averages being 99 and 86 microns, respectively. Channel sandstones are also better sorted than those of the interchannel and have a lower average clay content, 15% versus 20%. Both the higher clay content and smaller sandstone grain size of the interchannel areas may be responsible for the relative absence of oil in these areas.

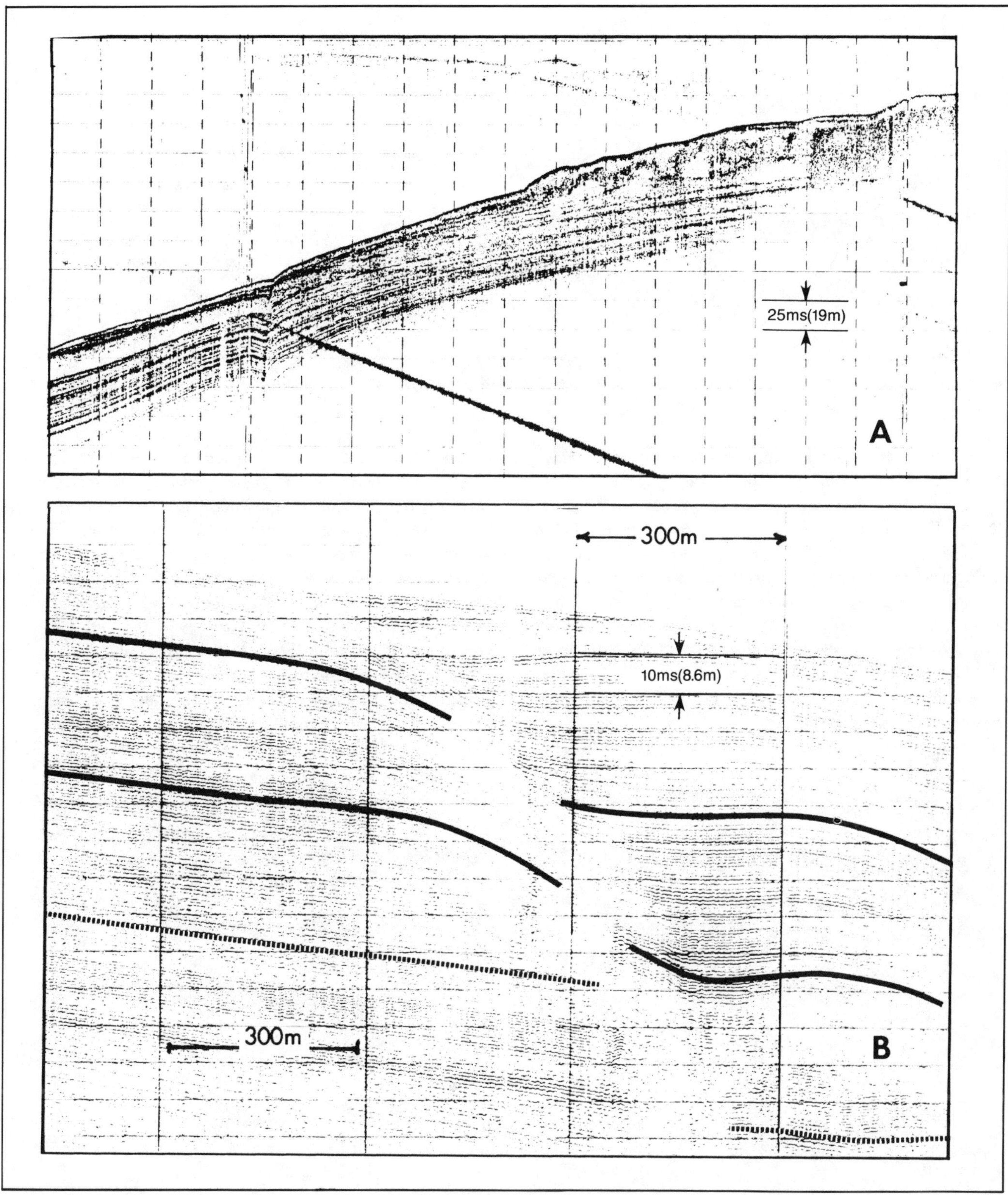

Fig. 8-31. High resolution seismic record of active growth fault seaward of mouth of South Pass, Mississippi delta. Mudflow is upslope from fault in A. (From Coleman and Prior, 1982; permission to publish by AAPG).

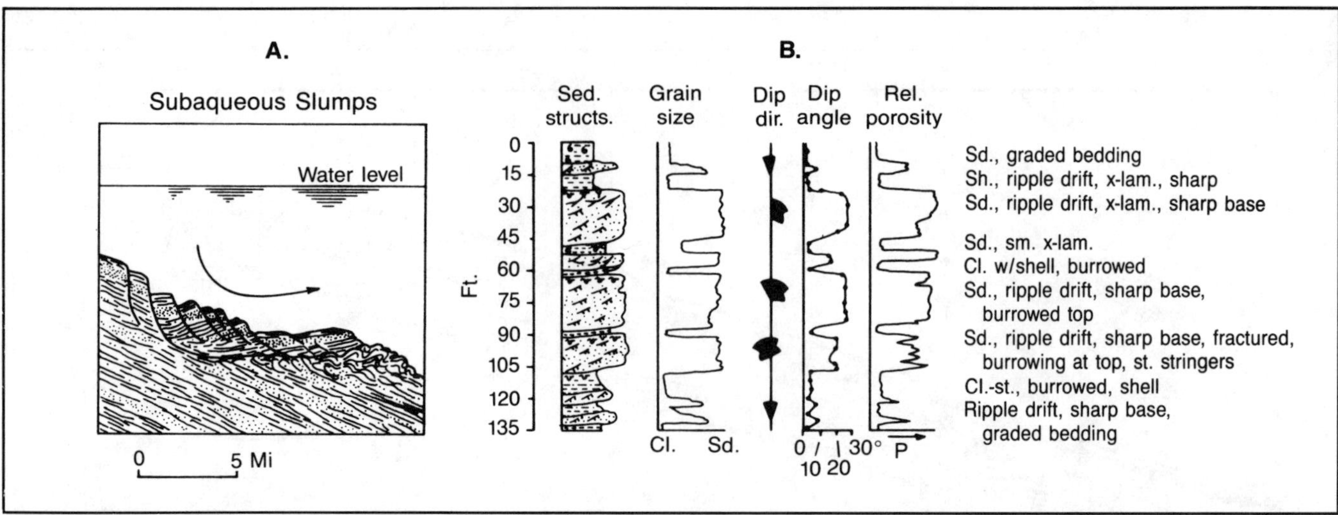

Fig. 8-32. Summary diagram of the major characteristics of slump deposits in the subaqueous delta plain. (Modified from Coleman and Prior, 1982; permission to publish by AAPG).

Most oil fields of the Booch sandstone are not in thick channel sandstones, but in thinner channel sandstones. A cluster of oil pools of irregular outline is present in T13 and 14N, R14E (Fig. 8-35) about midway between two main distributary systems. If sufficient log data were available, the pools probably could be shown to occur in thinner distributary channel sandstones close to a backswamp environment. Other elongate pools are in distributary channel sandstones to the southwest (Fig. 8-35). All of these bear little relation to structure and probably are the result of permeability seals surrounding more porous and permeable parts of channel sandstones.

Several significant oil pools to the northeast outside of the mapped area occupy selected parts of thick channel sandstones. There the channel sandstones are medium- to coarse-grained with little interstitial clay. Pools occur where axes of west-plunging structural noses intersect axes of major distributary channel sandstones. Oil-water contacts are readily determined on the north and south sides of the fields, and abrupt loss of permeability occurs on the east and west margins where the channel sandstones merge with the silty, very fine-grained interdistributary sandstones.

Upper Tonkawa delta, Anadarko basin. An excellent example of a deltaic sandstone reservoir is the upper member of the Tonkawa sandstone (Pennsylvanian) in the Oklahoma panhandle part of the Anadarko basin. Khaiwka's (1968)

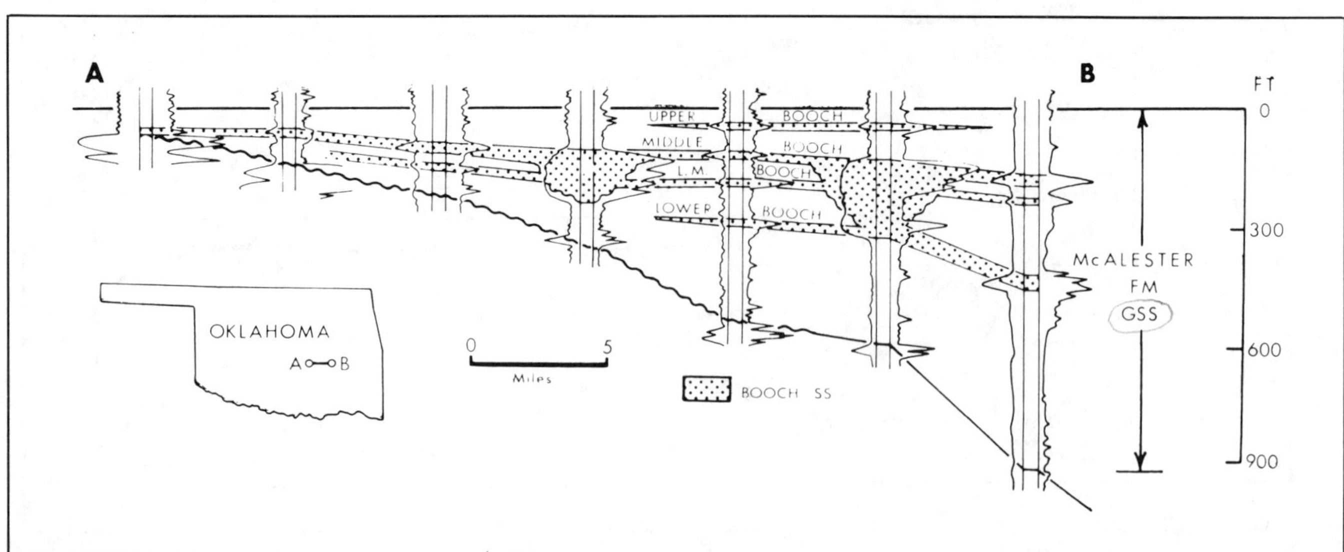

Fig. 8-33. Correlation of Booch sandstone members of McAlester formation. (From Busch, 1974; permission to publish by AAPG).

fluvial channel sands, from downwd cutting, discontin.
thick sands ... in some areas, downcut thru unc.

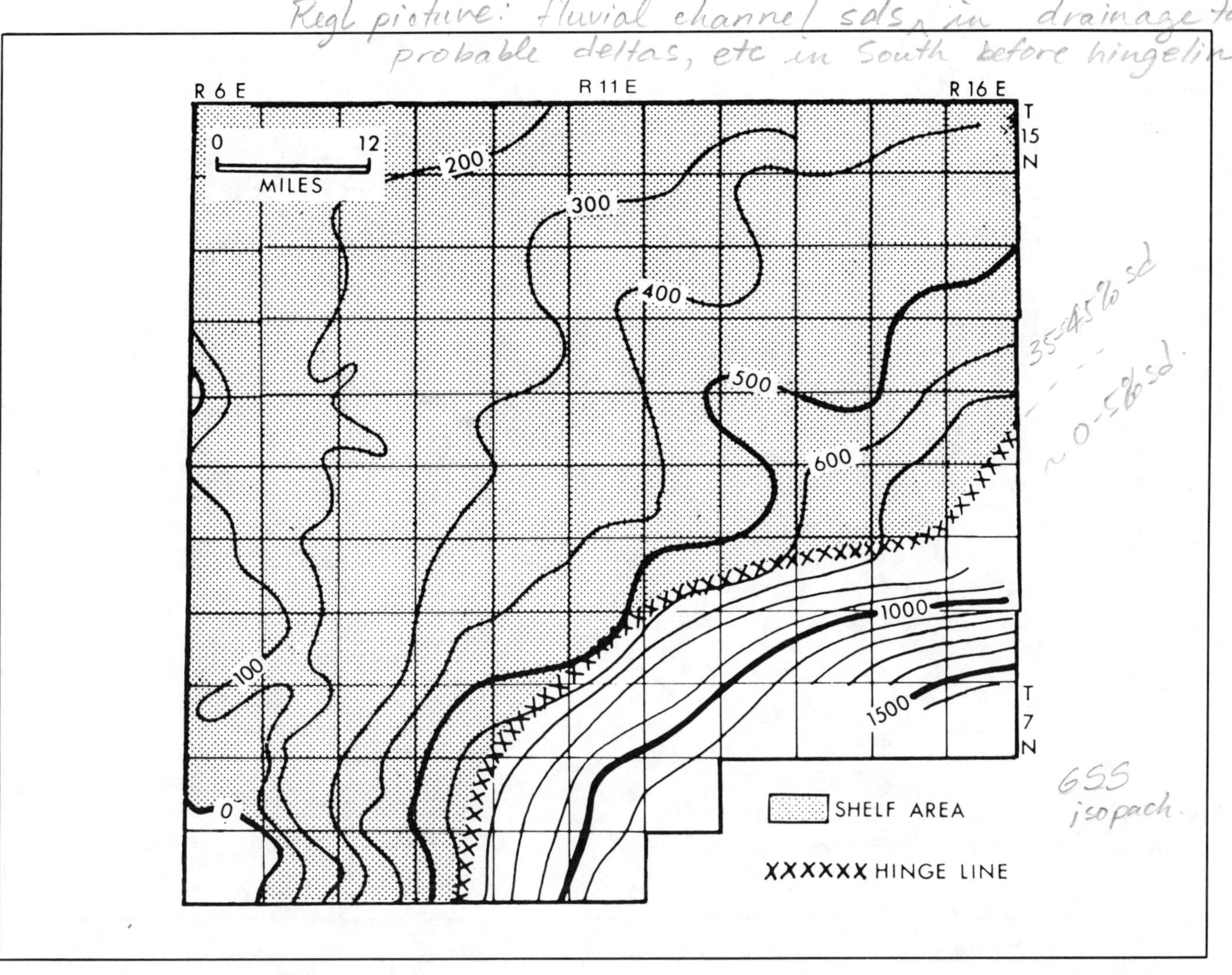

Fig. 8-34. Isopach map of McAlester formation. Hinge line separates shelf environment on northwest from more rapidly subsiding basin area to southeast. (From Busch, 1974; permission to publish by AAPG).

dissertation is the reference for this example.

The Tonkawa sandstone consists of three members (Fig. 8-37A). A thin persistent limestone, the "Haskell", overlies the upper member and is a convenient datum for stratigraphic analysis. Downward thickening of the upper member relative to the limestone datum (Fig. 8-37B) is typical for the area and demonstrates the channel aspect of the sandstone. Channeling is supported by GIS and upper Tonkawa sandstone isopach maps (Figs. 8-38 and 8-39), which are readily interpreted as distributary channel systems. Areas of maximum GIS thickness (Fig. 8-38) correspond to avenues of maximum downcutting that are the axes of distributary channels. With a well density of one well per section, only a generalized interpretation is possible.

Using the GIS isopach map as a guide, Khaiwka made an interpretation of the thickness variations of the upper Tonkawa sandstone (Fig. 8-39). Trends of maximum sandstone thickness do not always coincide with trends of maximum downcutting which Khaiwka attributes to lateral migration of distributaries. Interdistributary areas commonly have thinner sandstone of reservoir quality.

Gas distribution in the upper member of the Tonkawa sandstone is controlled by a combination of sandstone distribution and structure (Fig. 8-40). Gas migrated north-northwest until it reached the updip extent of the sandstone. A gas-water contact occurs at approximately -3,270 ft (-997 m) to the southeast.

Regional stratigraphic analysis of this kind is beneficial to the exploration geologist during development of a field. Preferred locations with the thickest sandstone above the hydrocarbon-water contact are predictable once the depositional nature of the sandstone is ascertained.

Endicott deltas, Anadarko basin

Another excellent example of a subsurface deltaic sandstone is the Endicott formation (Pennsylvanian) in north-

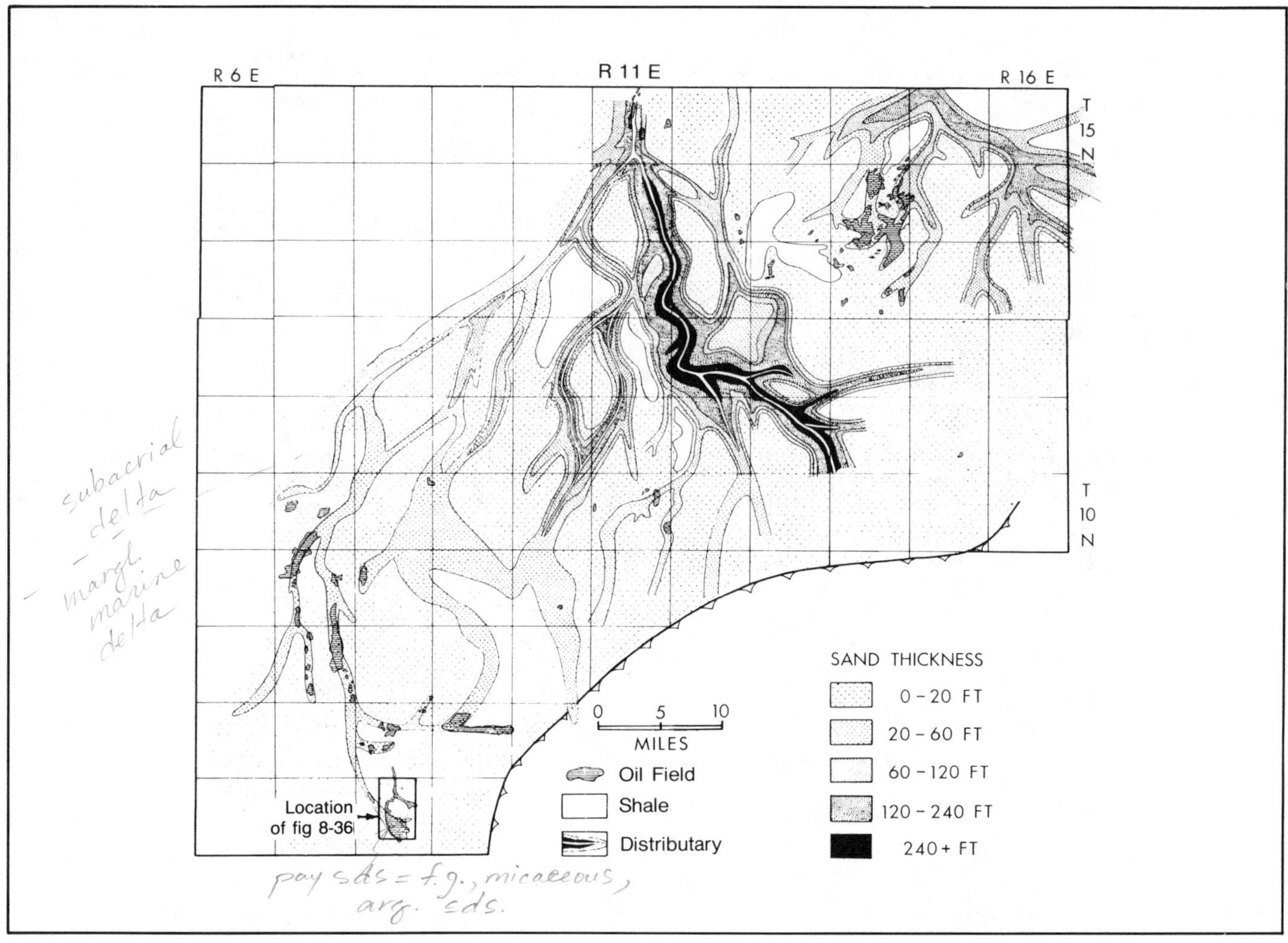

Fig. 8-35. Isopach map of Booch sandstone. (From Busch, 1974; permission to publish by AAPG).

western Oklahoma. It was deposited in a series of bifurcating distributary channels eroded into the north flank of the Anadarko basin and unconformably overlies either the Toronto limestone or the unnamed shale directly below the Toronto (Fig. 8-41). Thickening of the Endicott is clearly downward. Under areas of maximum Endicott channel fill, differential compaction has caused the underlying Lovell sandstone, "Haskell" limestone, and Tonkawa sandstone to bow downward (Fig. 8-41).

An excellent marker bed in this area is the Oread limestone, and it serves as the upper limit of a GIS that extends down to the unconformity at the base of the Endicott channel fill. An isopach map of the GIS (Fig. 8-42) presents a cast of the erosion surface that immediately preceded deposition of the Endicott sandstone. Axes of maximum downcutting, indicated by bifurcating dashed lines, largely controlled the positions of thick Endicott sandstone trends. Overall, the Endicott sandstone forms a deltaic complex consisting of two laterally merging distributary systems that fan out southward.

If only the generally smooth upper surface is considered, the Endicott resembles a blanket sandstone. However, the thickness of this formation ranges from zero to more than 180 ft (55 m), and it may change of as much as 100 ft (30 m) in a distance of one mile (1.6 km). Khaiwka (1968) prepared an unpublished isopach map of the Endicott that is almost identical to Figure 8-42 insofar as trends of maximum thickness are concerned. Thalwegs of the bifurcating distributaries on the two isopach maps are coincident, indicating there was no lateral migration of the distributaries.

Endicott structure is a south-dipping homocline devoid of any significant structural noses or closures. Although the formation has good reservoir characteristics, it lacks a trapping mechanism, such as an updip permeability seal or structural closure to retain hydrocarbons. The only known Endicott production occurs from several wells in the Oklahoma and Texas panhandle parts of the Anadarko basin.

Red Fork delta, McAlester basin

For many years geologists have debated the nature of the Red Fork sandstone (Pennsylvanian) of Noble County,

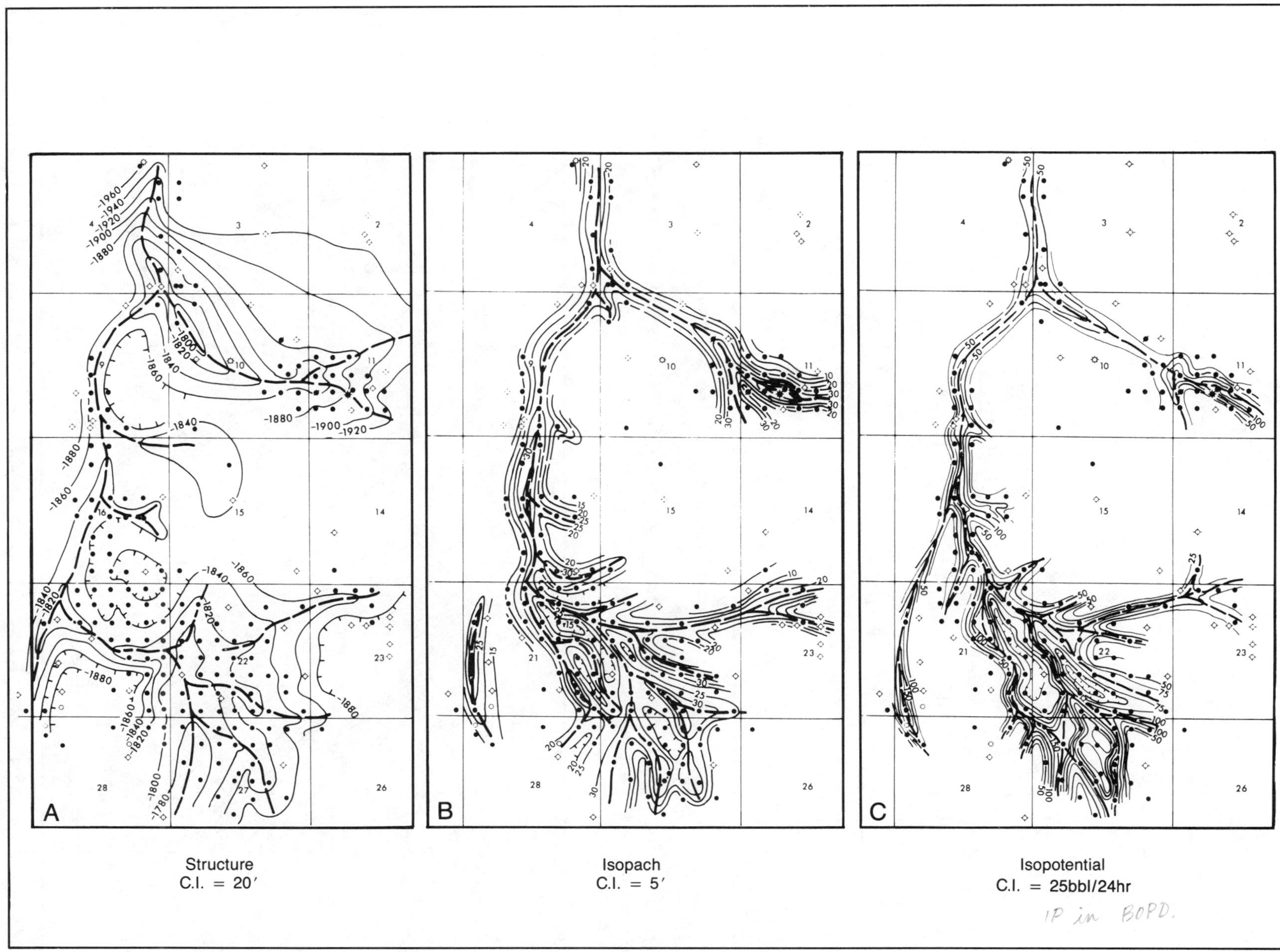

Fig. 8-36. Hawkins field, T6N, R18E, Hughes County, Oklahoma. **A**. *Structure, top of Booch sandstone,* **B**. *Isopach map of Booch sandstone,* **C**. *Isopotential map of Booch sandstone. See Figure 8-35 for location.* (From Busch, 1974; permission to publish by AAPG).

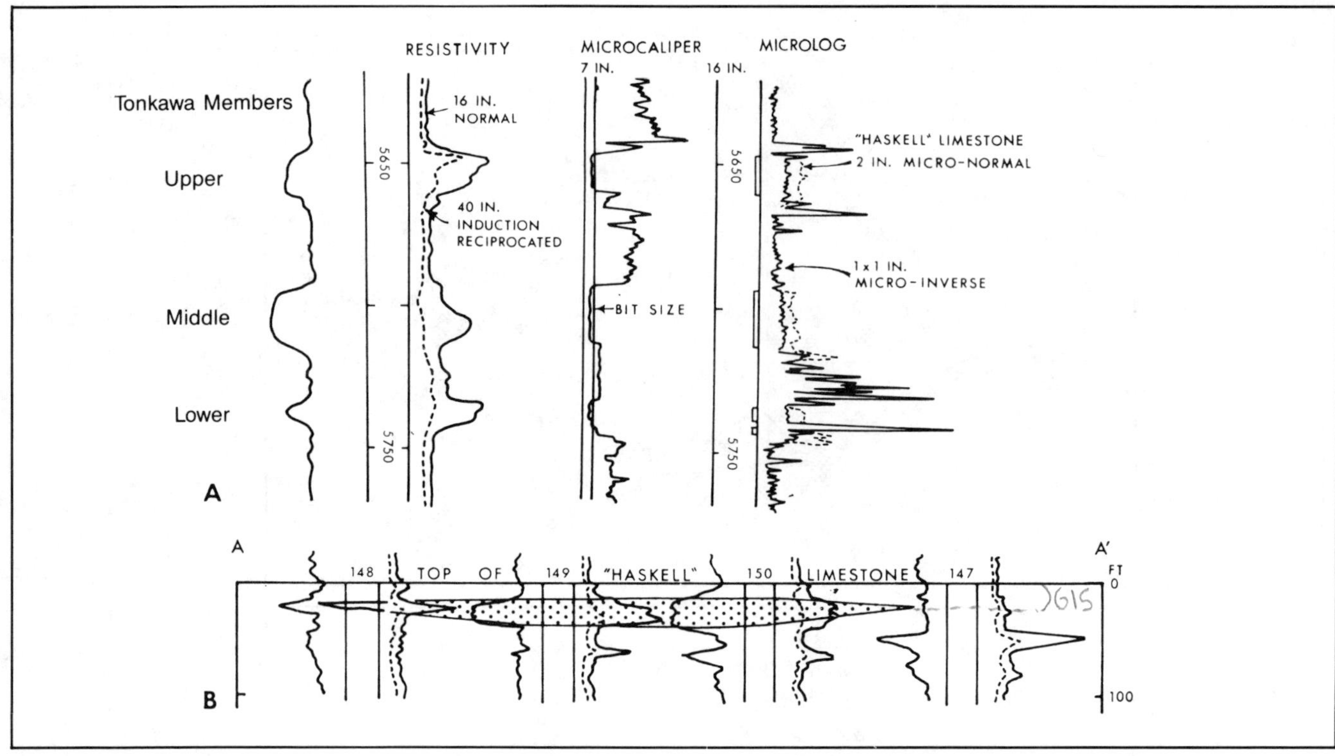

Fig. 8-37. **A.** *Electric logs showing three members of Tonkawa sandstone and their stratigraphic positions relative to the "Haskell" limestone.* **B.** *Stratigraphic cross section A-A' of upper Tonkawa sandstone showing downward thickening at expense of underlying shale. Datum is top of "Haskell" limestone. See Figure 8-38 for location of cross section.* (From Busch, 1974; permission to publish by AAPG).

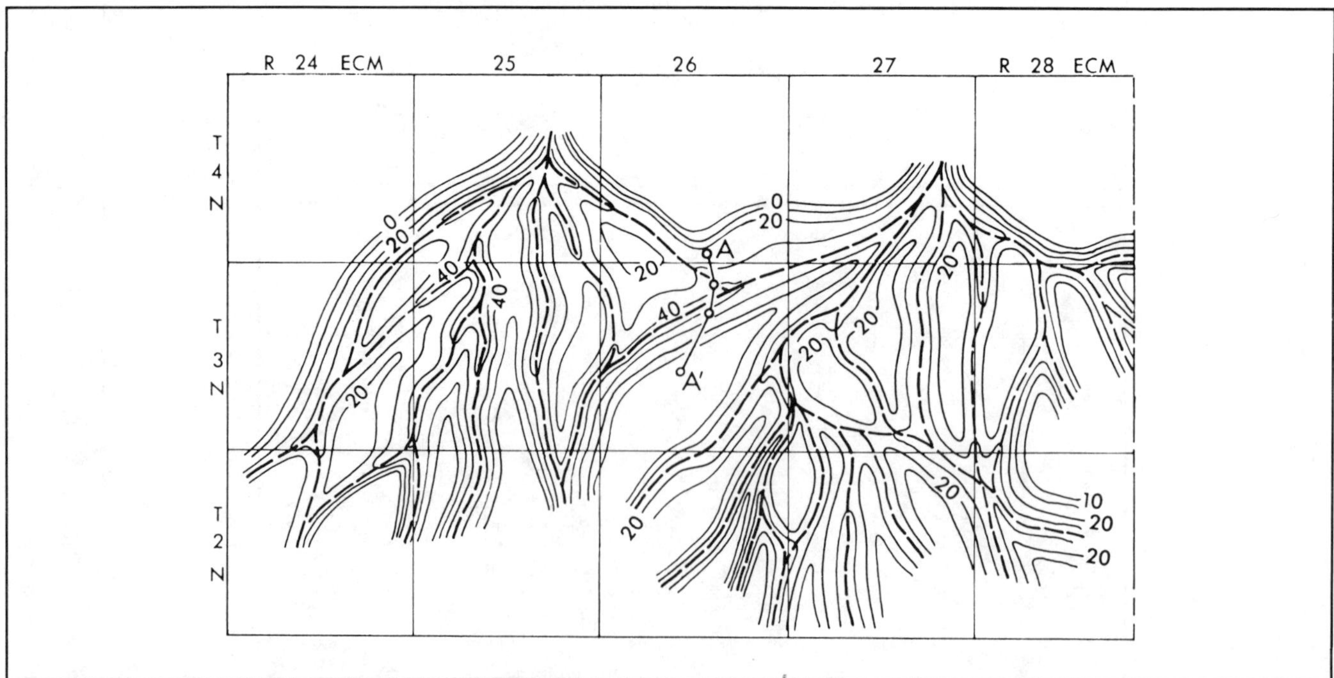

Fig. 8-38. Isopach map of GIS between top of "Haskell" limestone and base of upper Tonkawa sandstone. Map shows two laterally coalescing distributary systems. Contour interval 10 ft. (From Khaiwka, 1968; permission to publish by The University of Oklahoma).

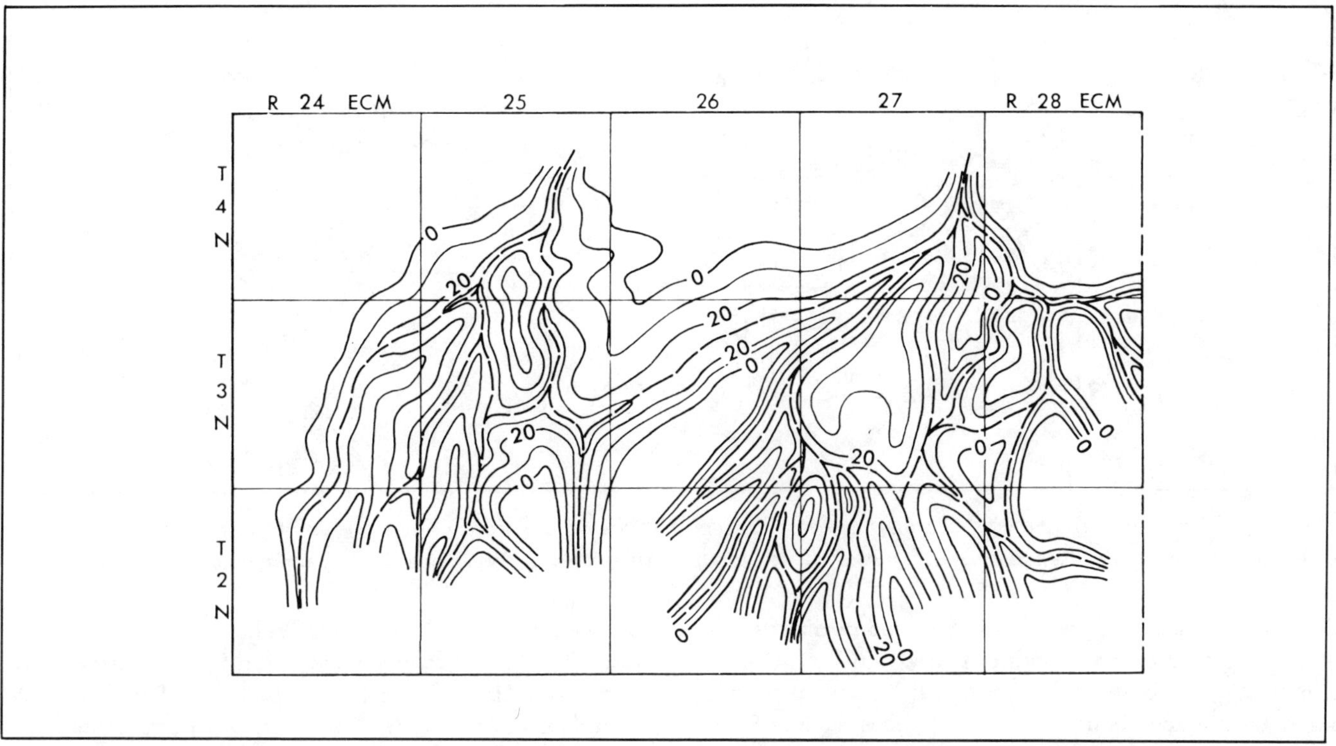

Fig. 8-39. Isopach map of upper Tonkawa sandstone, Beaver County, Oklahoma. Contour interval 10 ft. (From Khaiwka, 1968; permission to publish by The University of Oklahoma).

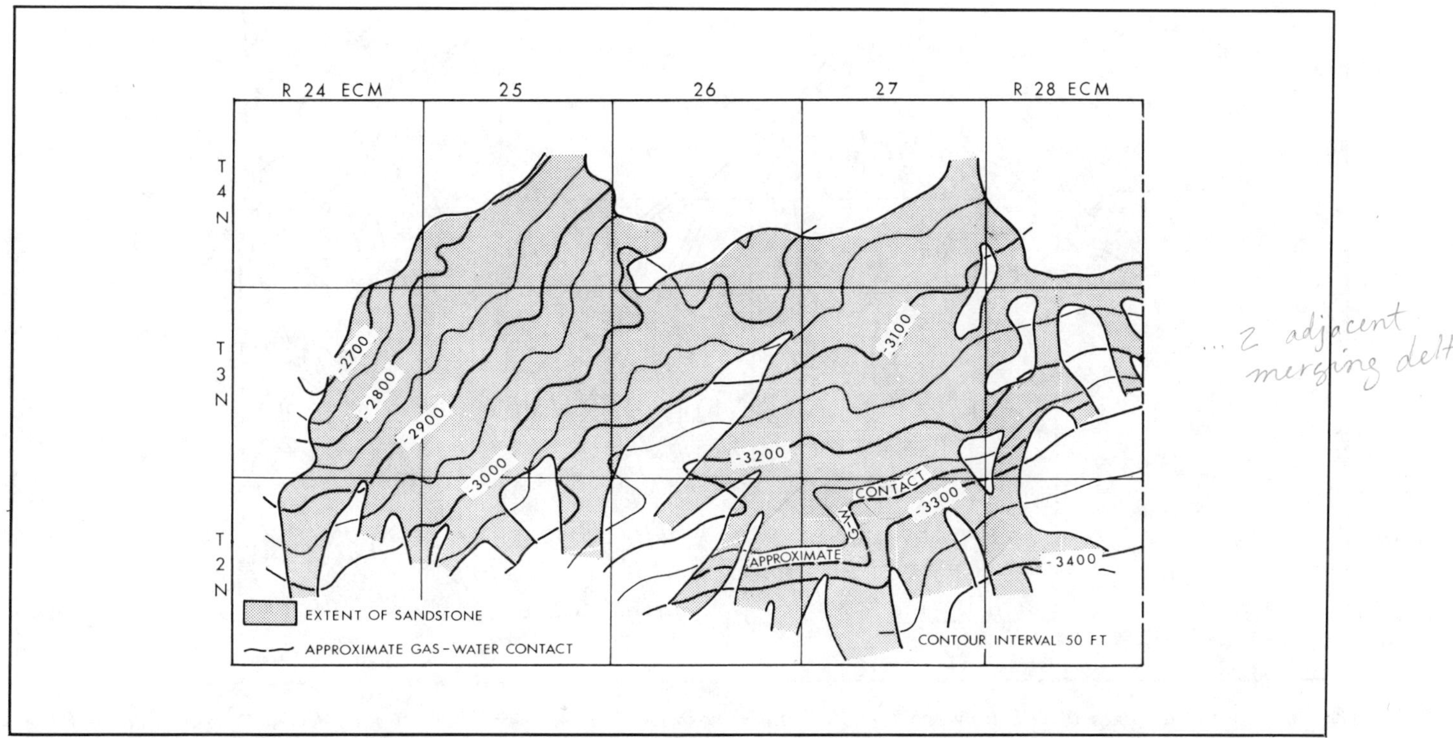

... 2 adjacent merging deltas.

Fig. 8-40. Structure map, top of Tonkawa sandstone, Beaver County, Oklahoma. (From Khaiwka, 1968; permission to publish by The University of Oklahoma).

(strat. trap)

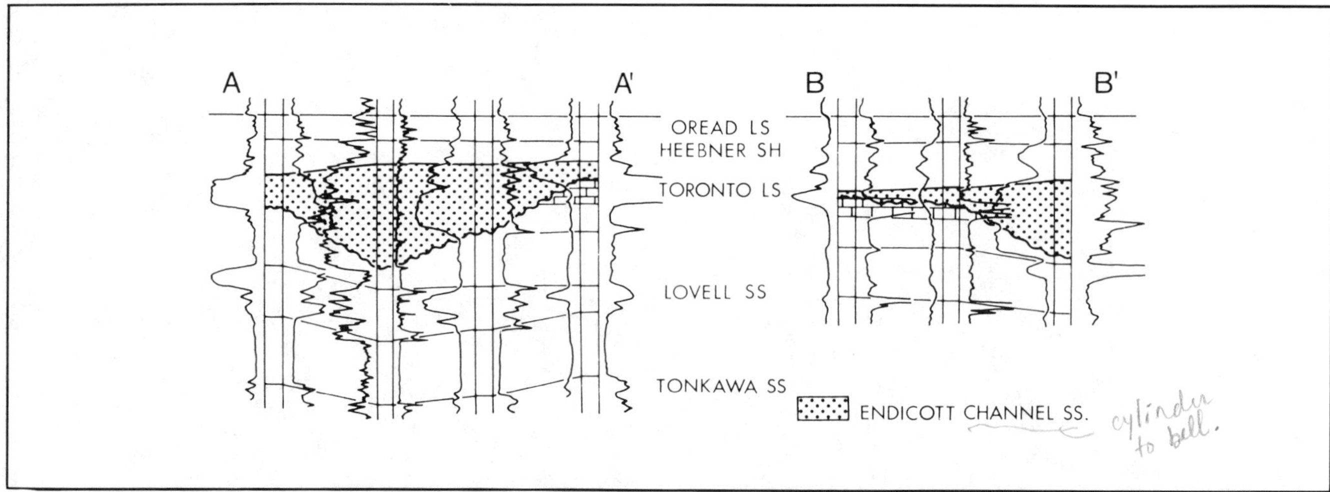

Fig. 8-41. Cross sections showing channel aspect of Endicott sandstone. Toronto limestone is eroded away at positions of maximum sandstone thickness. See Figure 8-42 for location of profiles. (From Busch, 1974; permission to publish by AAPG).

Oklahoma. It has been variously interpreted as an offshore bar, an isolated channel sandstone, and a delta. A structure map of the base of the Oswego limestone, a readily mappable horizon about 175 ft (53 m) above the Red Fork, shows an irregular homocline sloping toward the west (Fig. 8-43). A southwest-plunging syncline interrupts this slope to the west and is related to the Tonkawa fault in the northwest corner of the mapped area. Red Fork sandstone production describes a horseshoe-shaped area (Fig. 8-43). Oil is produced in the western and northern parts of this horseshoe-shaped area, whereas the eastern limb produces primarily gas. All production is from the lenticular Red Fork

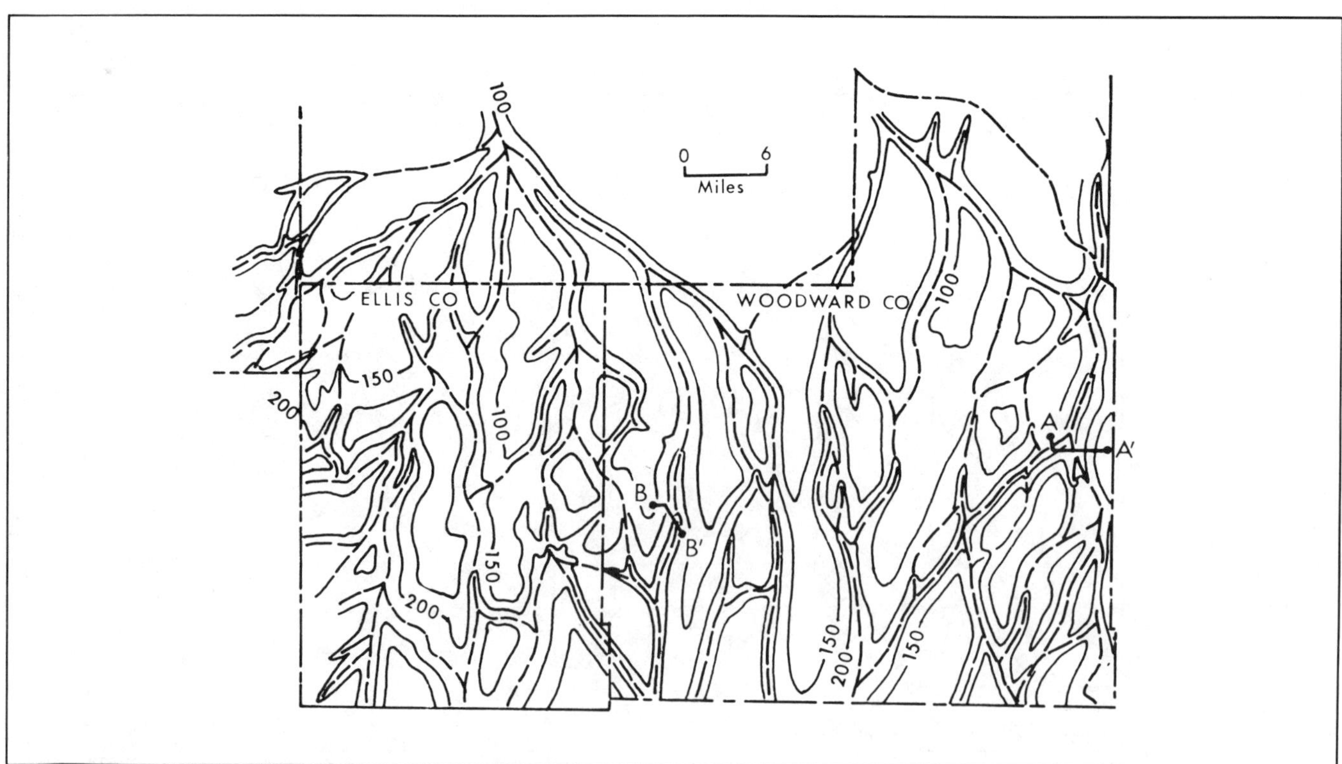

Fig. 8-42. Isopach map of GIS between top of Oread limestone and base of Endicott sandstone. Map is a cast of the eroded surface upon which the Endicott was deposited in an area of two laterally coalescing deltaic distributary systems. Contour interval 50 ft. (From Busch, 1974; permission to publish by AAPG).

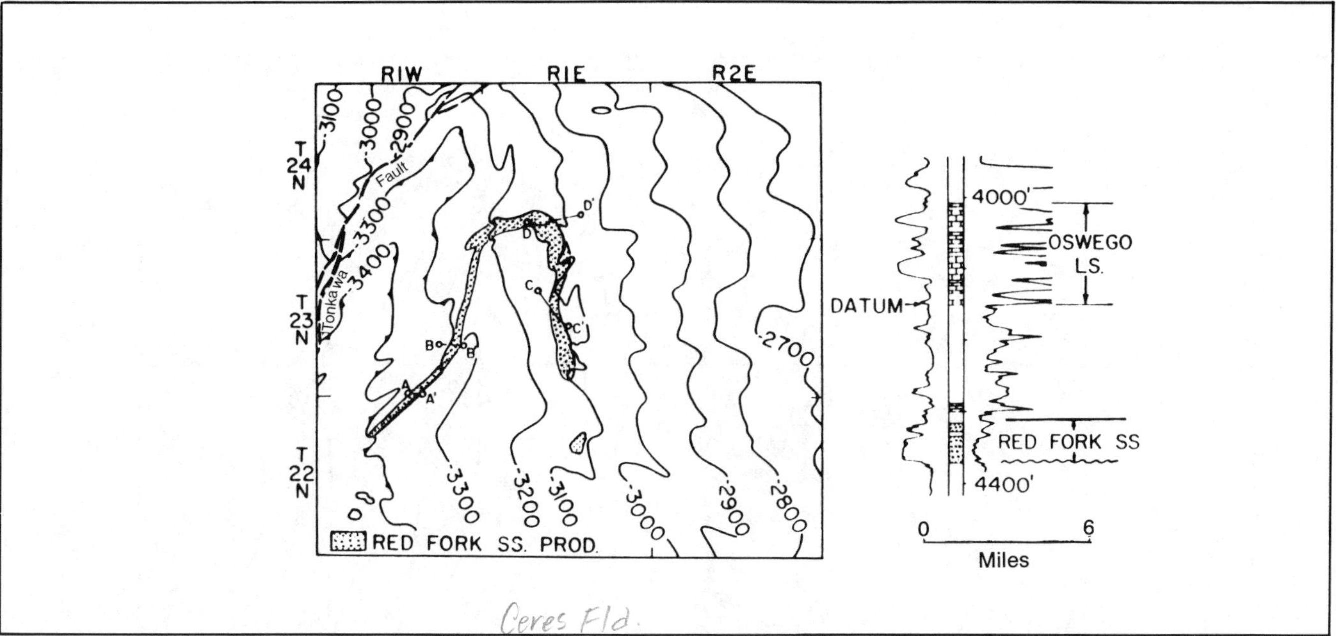

Fig. 8-43. Structure of base of Oswego limestone showing location of four cross sections. (Modified from Scott, 1970; permission to publish by The University of Oklahoma).

sandstone. Differential compaction of shales laterally adjacent to the Red Fork is apparent along the west limb of the horseshoe where structure contour lines nose southwestward as they cross the Red Fork sandstone (Figs. 8-43 and 8-44).

A published cross section, using the base of the Oswego limestone as a datum (Neal, 1951), shows the Red Fork sandstone increasing in thickness at the expense of the shale directly overlying it (Fig. 8-45). On this basis, an offshore bar interpretation was made for the Red Fork. Scott (1970) made a different interpretation using the same four electric logs as Neal. A marker bed near the base of the Red Fork sandstone is the key to the correct interpretation (Fig. 8-46). It is present in wells 1 and 4 and eroded out in wells 2 and 3 where the Red Fork has its maximum thickness. Viewed this way, the Red Fork is clearly a channel sandstone. Other stratigraphic cross sections also show downcutting (Figs. 8-47, 8-48, and 8-49). In each case, the marker bed is truncated by the Red Fork, confirming the channel aspect of the sandstone.

In his study, Scott (1970) considered the interval from the Pink limestone to the base of the Red Fork sandstone as a GIS and reconstructed a paleodistributary system of a delta (Fig. 8-50). To date, oil and gas production is restricted to the two principal distributaries of this delta. Other distributaries may be crevasse splays in which the sandstones generally are thinner, more argillaceous, and nonprospective.

Lyons and Dobrin (1972) demonstrate that the Red Fork channel sandstone of the Ceres field can be identified and traced by seismic profiling. From a seismic profile (Fig. 8-51) that crosses the Ceres field at right angles, a plot of the travel time difference between the top of the Oswego limestone and the top of the Mississippian limestone was made. The heavy isochron at the top of Fig. 8-52 represents the running average of five points and the greatest time differential occurs at the known position of the thickest Red Fork channel fill. By using only the travel time interval between the top of the Oswego and the Mississippian surface, elevation effects and any other seismic errors are eliminated.

A regional, seismic and subsurface, isopach map of the interval between the Pink limestone and the Mississippian limestone (Fig. 8-53) offers considerable support to the idea that the two major distributaries of the Ceres delta occur directly above Mississippian channels. The top of the Pink limestone is an excellent marker horizon and is parallel to the top of the Oswego limestone. Similarities in the southward divergent trends of this mapped interval (Fig. 8-53) and the two major Red Fork distributaries (Fig. 8-50) suggest that channeling in the Mississippian unconformity influenced the positions of later channels in the Red Fork.

Texas

Cook deltas, Midland basin. The Midland basin in west Texas is bordered on the east by the Eastern shelf (Fig. 8-54). In Wolfcamp time (Early Permian), streams flowed from east to west across the shelf into the Midland basin. By examining several thousand logs, Bloomer (1977) was able

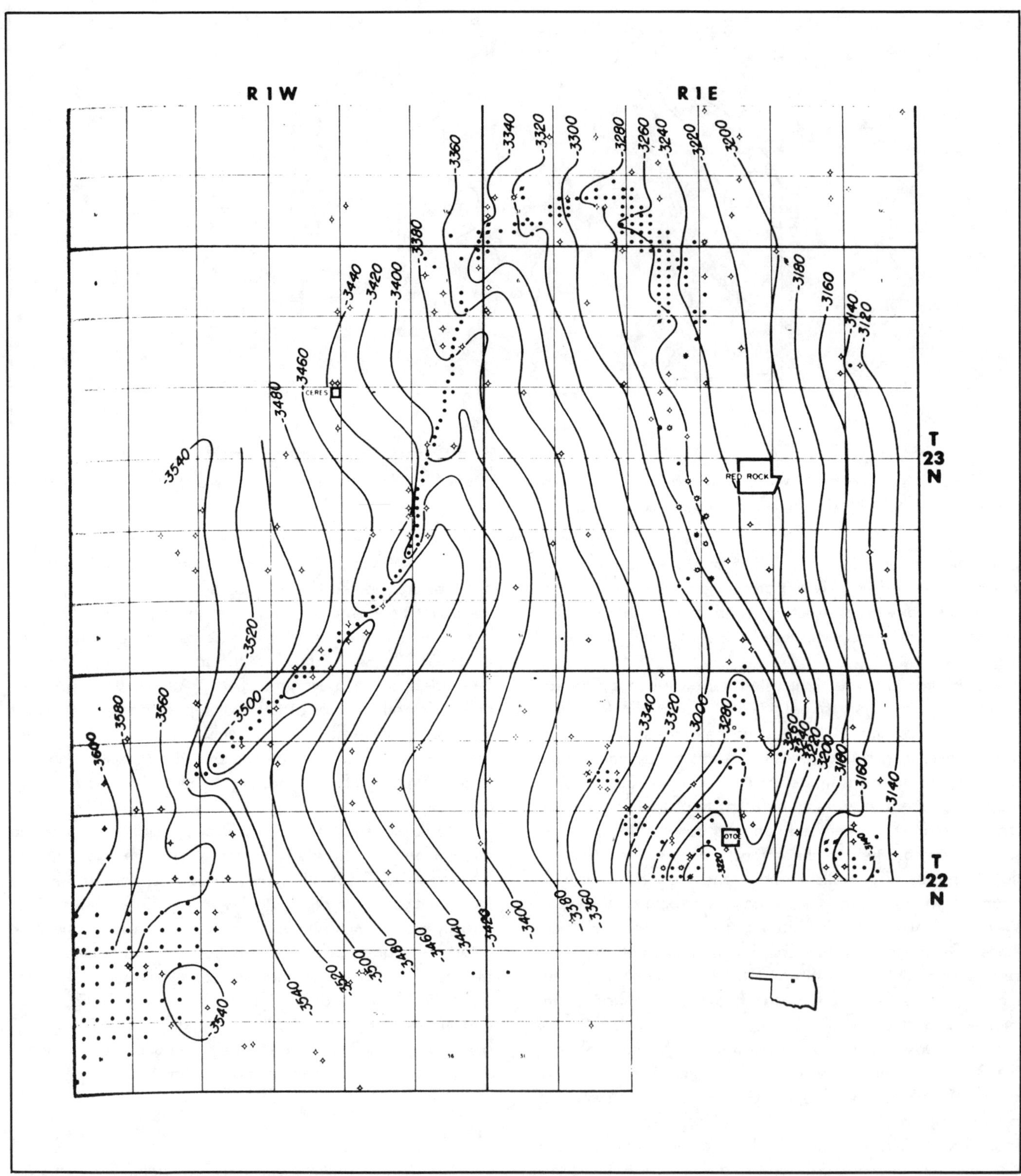

Fig. 8-44. Structure of the Pink limestone, showing conspicuous drape effect with southwest structural nosing where contours cross the Ceres oil pool, Noble County, Oklahoma. (From Lyons & Dobrin, 1972; permission to publish by AAPG).

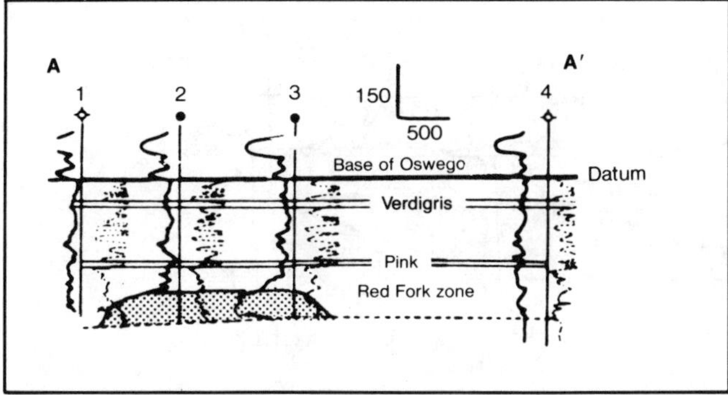

Fig. 8-45. Sand bar interpretation of Red Fork Sandstone, South Ceres field, Noble County, Oklahoma. See Figure 8-43 for location of profile. (From Neal, 1951; permission to publish by World Oil).

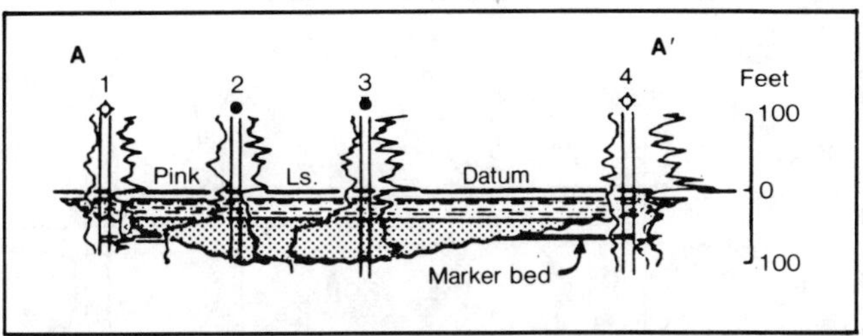

Fig. 8-46. Channel interpretation of Red Fork Sandstone, South Ceres field. See Figure 8-43 for location of profile. (Modified after Scott, 1970; permission to publish by The University of Oklahoma).

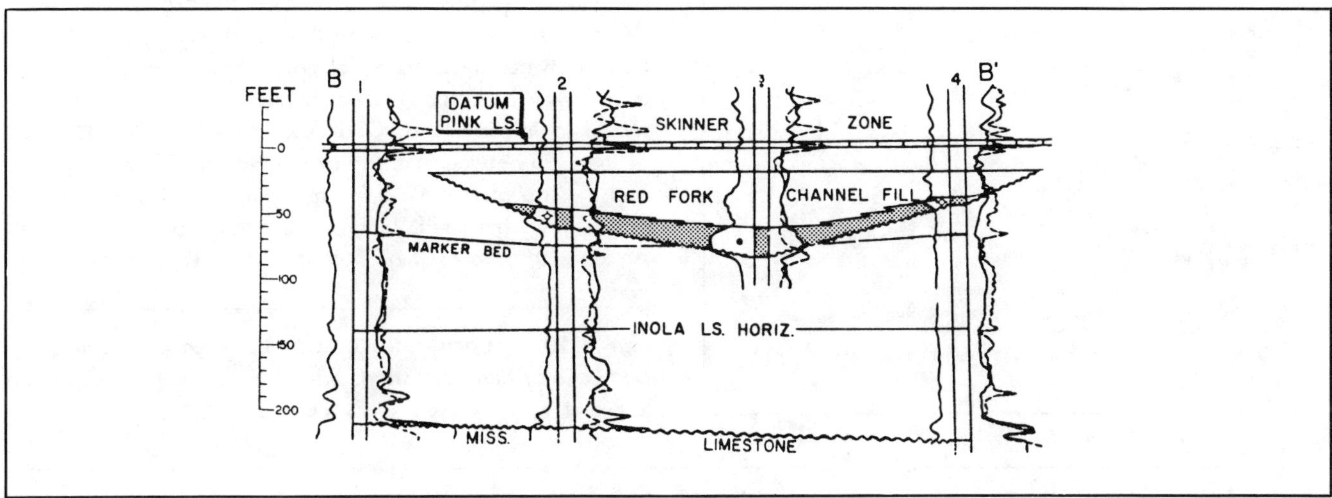

Fig. 8-47. Cross section B-B′, showing Red Fork channel fill in which over half of fill is silty shale. See Figure 8-43 for location of profile. (Modified after Scott, 1970; permission to publish by The University of Oklahoma).

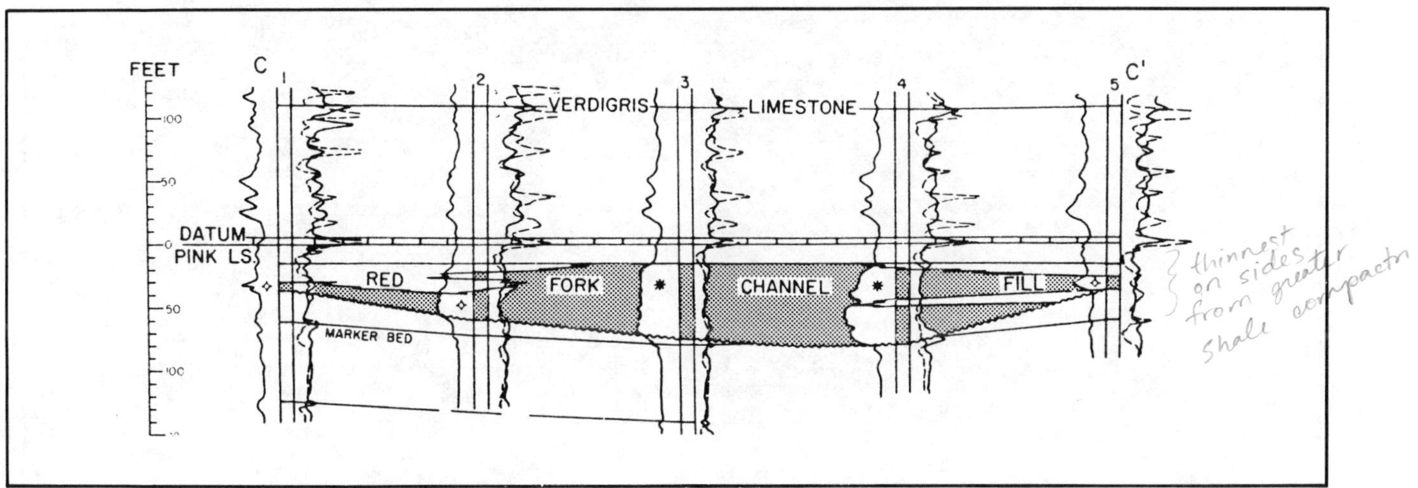

Fig. 8-48. Cross section C-C″, showing Red Fork channel fill, predominantly sandstone. See Figure 8-43 for location of profile. (From Scott, 1970; permission to publish by The University of Oklahoma).

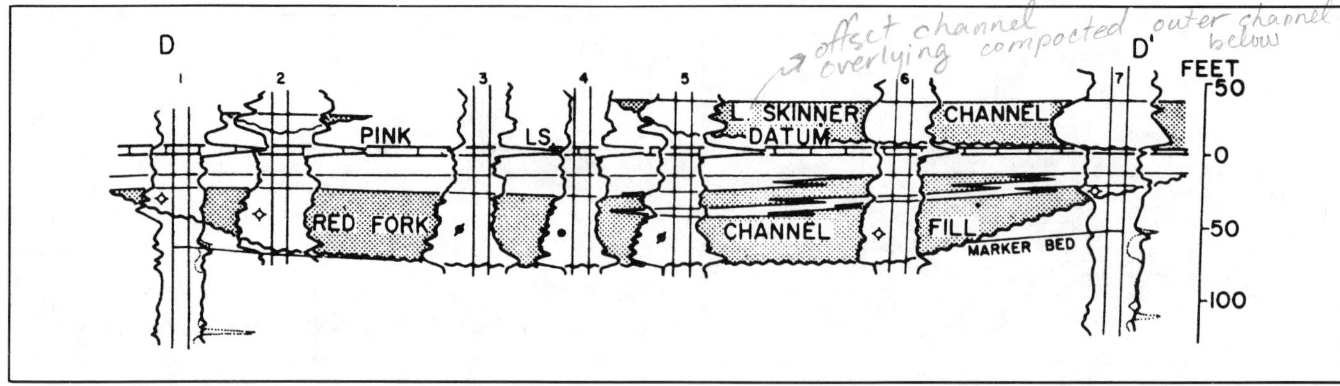

Fig. 8-49. Cross section D-D', showing Red Fork channel fill where channel has maximum width. See Figure 8-43 for location of profile. (From Scott, 1970; permission to publish by The University of Oklahoma).

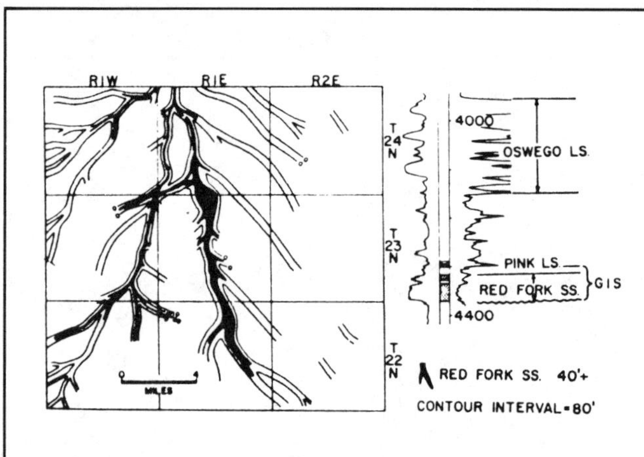

Fig. 8-50. Isopach map, of GIS between Pink Limestone and base of Red Fork sandstone. (From Scott, 1970; permission to publish by The University of Oklahoma).

to trace two fluvial systems from outcrops in the east, across the shelf in the subsurface to deltas in the basin to the west. Of interest to Bloomer was that portion of the Waldrip Shale between the Crystal Falls and Flippen limestones (Fig. 8-55). Any sandstone between these two limestones is referred to informally as Cook sandstone. An isopach map of the GSS between the top of the Noodle Creek limestone and the top of the Waldrip limestone II at the edge of the basin shows the shelf, hinge, and basin slope (Fig. 8-56). An isopach map of the Cook sandstone (Fig. 8-57) shows

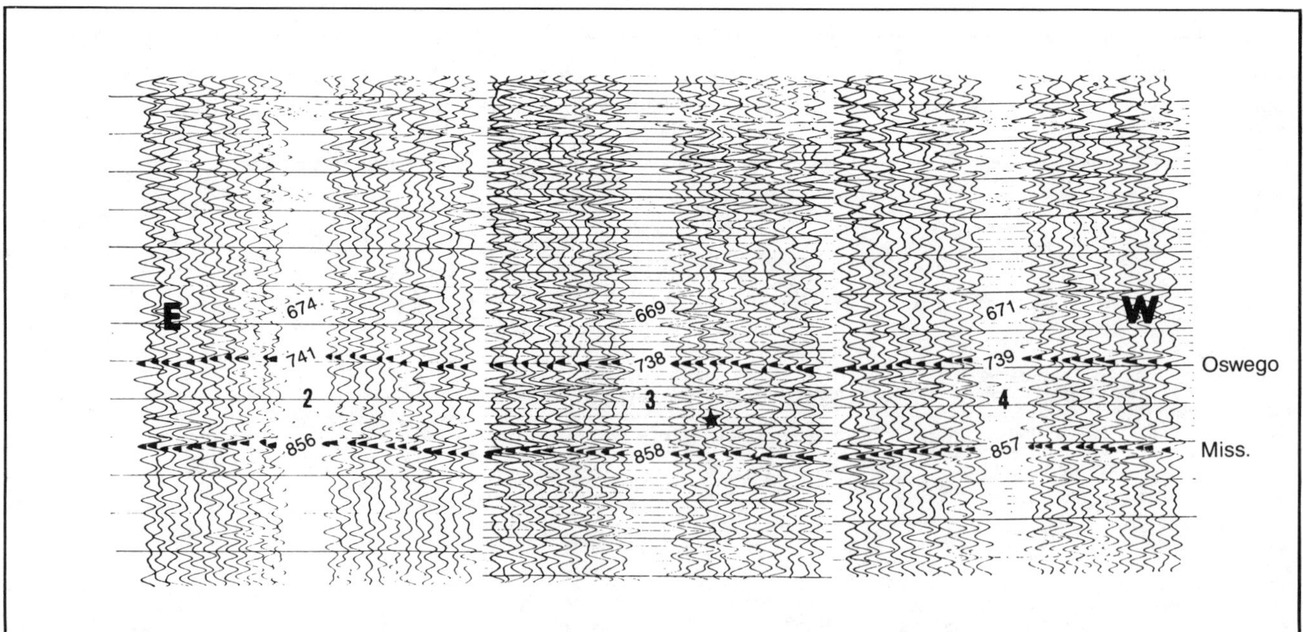

Fig. 8-51. Oswego and Mississippian limestone reflections that determine channel defined in Figure 8-52. (From Lyons and Dobrin, 1972; permission to publish by AAPG).

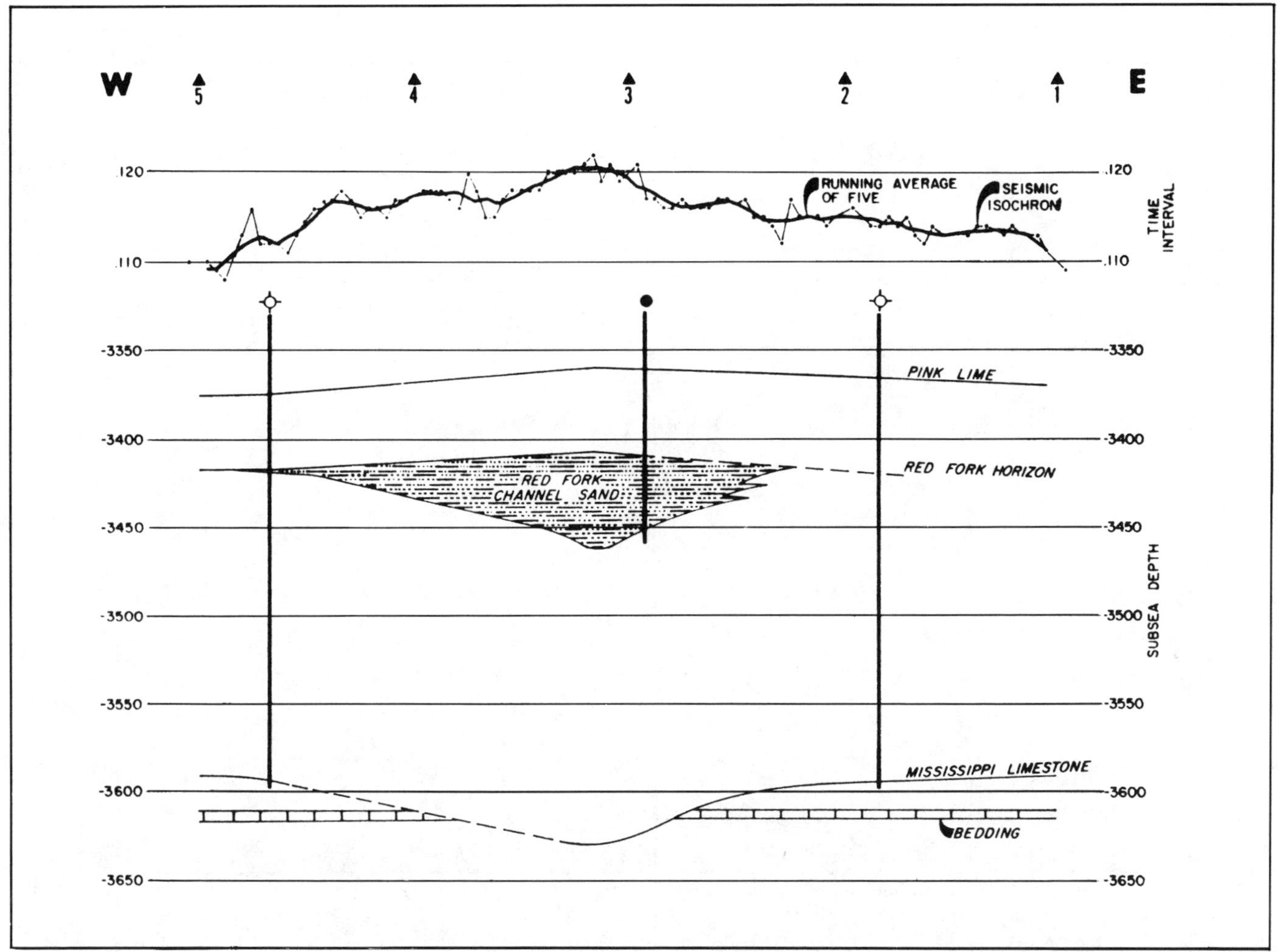

Fig. 8-52. Cross section made from seismic records defines a narrow subsurface channel filled with productive Red Fork sandstone. (From Lyons and Dobrin, 1972; permission to publish by AAPG).

the two fluvial systems coming from the east and their respective deltas beyond the edge of the shelf.

East-west cross sections through the two deltas (Figs. 8-58 and 8-59) show the shelf, hinge, basin slope, and the prograding clastic wedges of Cook sandstone and Waldrip shale between the base of the Flippen limestone and the top of the Waldrip limestone II. In the southern delta, Cook strata in wells 6-8 (cross section C-C; Fig. 8-58) consist of 50 ft (15 m) of thin, lenticular, fine-grained sandstones with interbedded carbonaceous shale and lenticular coal. Bloomer interprets these rocks as upper delta floodplain and swamp deposits. Upper Cook sandstones in wells 2-5 (Fig. 8-58) are about 40 ft (12 m) thick and have high amplitude, cylindrical to funnel-shaped SP curves on electric logs. At this same horizon, most other wells in the southern delta west of the upper delta plain facies have the same log characteristics. Bloomer interprets this facies as delta front and distributary-channel sandstone. Between the sandstone facies and the Waldrip limestone II are prodelta shales and sands that formed the platform across which the submarine delta prograded. Delta plain strata overlie the delta front and distributary channel sandstones to complete the prograding deltaic sequence.

Two coalescing crevasse splays are seen on the north side of the southern delta (Fig. 8-57). A double lobate feature just south of wells 7-9 on line D-D′ consists of about 10 ft (3 m) of shaly Cook sandstone. Both morphology and petrography suggest crevasse splays.

Northern delta strata are similar to those of the southern delta. Wells 7-9 in cross section D-D′ (Fig. 8-59) have 50 ft (15 m) of upper delta plain Cook rocks. Upper Cook sandstones in wells 2–6 average 20 ft (6 m) in thickness and have the same SP character as the delta front and distributary channel facies of the southern delta. The thick sands in well 4 may be a distributary mouth bar.

Oil and gas are produced from Cook sandstone in the Group 4000 field in the southern delta. An isopach map of the GIS between the bases of the Flippen limestone and

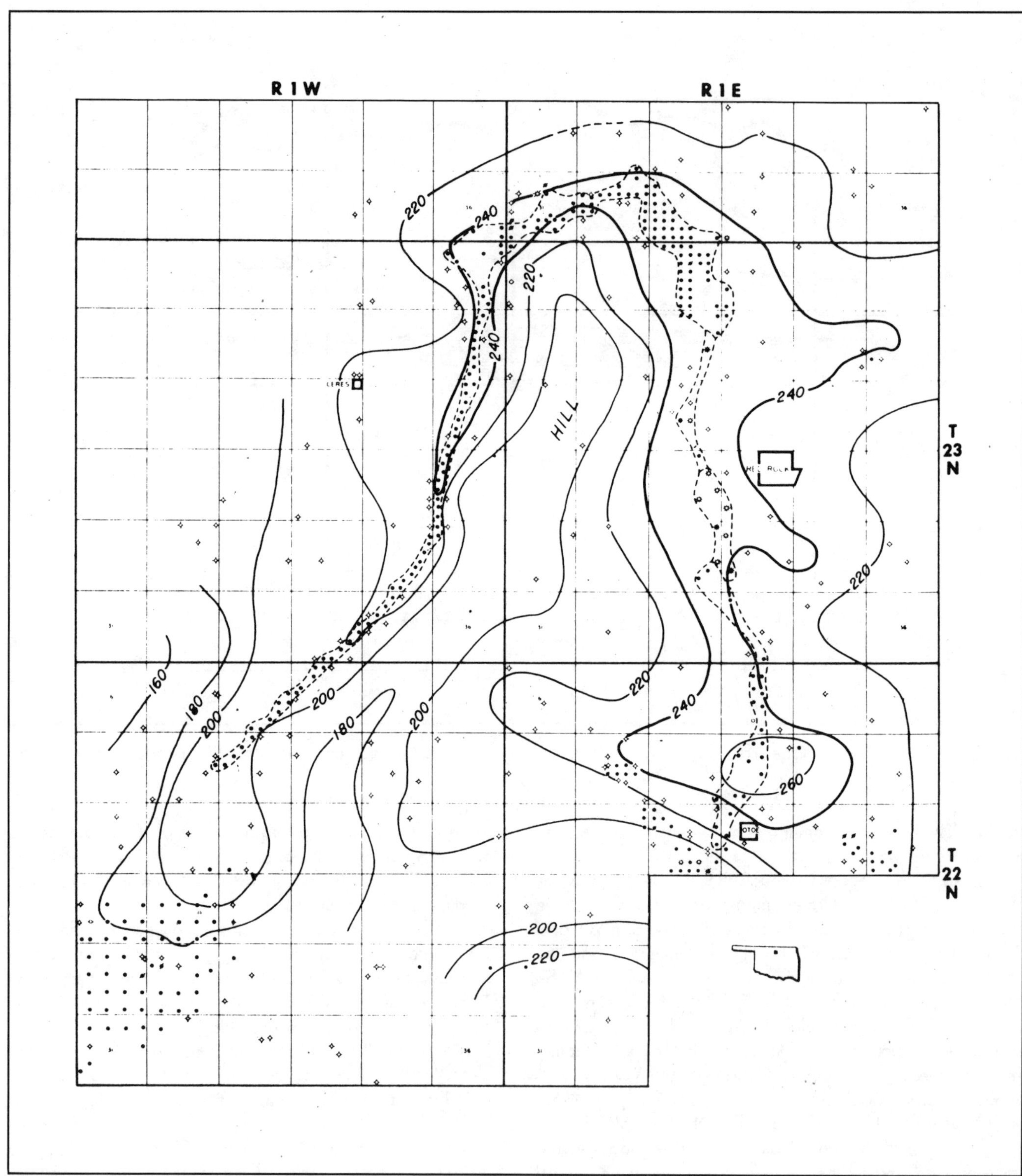

Fig. 8-53. Seismic and subsurface isopach map of the stratigraphic interval between the Pink limestone and the Mississippian limestone, Ceres field area, Noble County, Oklahoma. (From Lyons and Dobrin, 1972; permission to publish by AAPG).

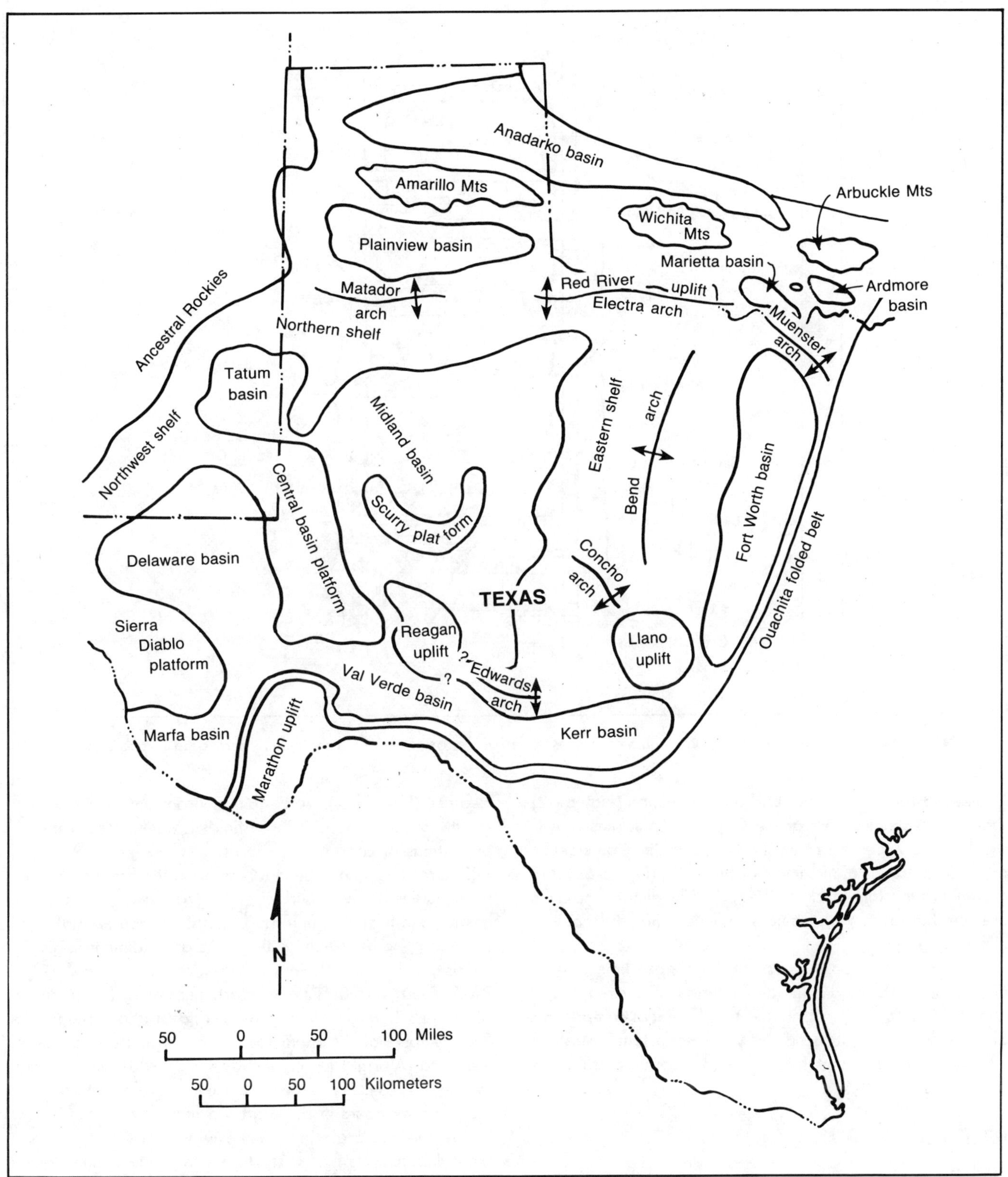

Fig. 8-54. Principal tectonic elements in west Texas.

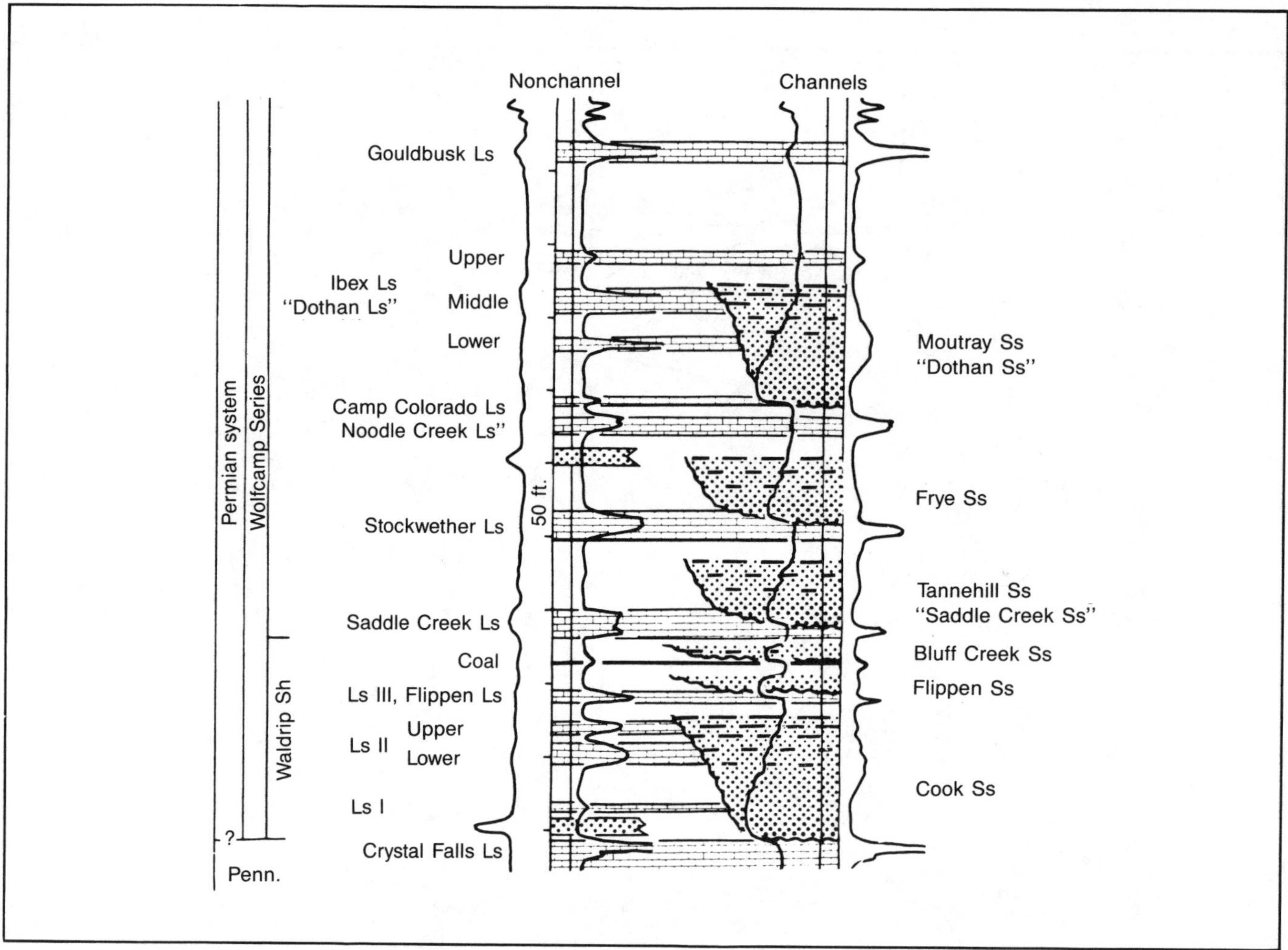

Fig. 8-55. Lower Permian stratigraphic section for west-central Texas. (From Bloomer, 1977; permission to publish by AAPG).

Cook sandstone (Fig. 8-60) and a cross section (Fig. 8-62) show the channel aspect of the Cook as it thickens downward at the expense of the underlying shale. Coincidence of the axial trends of maximum downcutting (Fig. 8-60) and maximum sandstone thickness (Fig. 8-61) indicates the axes are the thalwegs of distributary channels that did not migrate laterally.

A structure map on the base of the Flippen limestone in the Group 4000 field (Fig. 8-63) shows a low amplitude domal structure that is attributed to differential compaction of shale within and around the Cook sandstone. Most of the oil has come from thin, lenticular, upper delta plain sandstones (Fig. 8-62).

North Sea

Viking Graben fields. Sandstones of the Brent formation (Middle Jurassic) are major reservoir rocks in the Viking Graben of the northern North Sea. Among the fields producing from the Brent are Statfjord, Brent, Ninian, and Heather (Fig. 8-64). A deltaic environment of deposition for the Brent in the first three fields is interpreted from examinations of conventional cores.

Five members have been identified in the Brent sandstone from studies of cores and wireline logs. Descriptions of the members in Statfjord field (Fig. 8-65) are representative of those in the other fields as the rocks are sedimentologically consistent. Log correlations between and within fields (Figs. 8-65, 8-66, and 8-67) are good, permitting identification of the members. In Ninian, Brent and Statfjord fields, the five members of the Brent formation constitute a prograding delta (Albright et al., 1980; Kirk, 1980; Simpson and Whitley, 1981). The environmental interpretation made for the Brent members in the Statfjord field (Fig. 8-65) is similar to the interpretations in the two other fields. A complete deltaic cycle exists, starting with prodelta fine clastics in the Broom and ending with delta-plain channel sandstones, interdistributary bay, lake, and swamp muds, and crevasse splay deposits in the Ness and Tarbert members. Gross geometry of the members (Fig. 8-68) and an increasing marine aspect of the Ness to the north indicate the Brent delta prograded from south to north (Kirk, 1980).

Not all of the Brent formation is considered to be deltaic. A barrier island-lagoon environment is interpreted for the Etive and Ness members in South Cormorant and Heather Fields (Fig. 8-64; Budding and Inglin, 1981; Gray and Barnes, 1981). At South Cormorant, the Ness member is thought to have some delta plain sediments in the southern portion of the field (Budding and Inglin, 1981). The western limit of the Brent delta or the edge of a delta lobe may be established by the South Cormorant and Heather fields.

SELECTED BIBLIOGRAPHY

Albright, W. A., W. L. Turner, and K. R. Williamson, 1980, Ninian field, U.K. sector, North Sea, *in* M. T. Halbouty, ed., Giant oil and gas fields of the decade 1968-1978: AAPG Mem. 30, p. 173-193.

Allen, J. R. L., 1970, Sediments of the modern Niger delta, *in* J. P. Morgan, ed., Deltaic sedimentation: modern and ancient: SEPM Sp. Pub. I5, p. 138-151.

Avbovbo, A. A., 1978, Tertiary lithostratigraphy of Niger delta: AAPG Bull., v. 62, no. 2, p. 295-306.

Axelson, V., 1967, The Laiture delta, a study of deltaic morphology and processes: Geografiska Annaler, v. 49, p. 1-127.

Bagans, B. P., J. C. Horne, and J. C. Ferm, 1975, Carboniferous and Recent Mississippi lower delta plains: Gulf Coast Assoc. Geol. Soc. Trans., v. 25, p. 183-191.

Bates, Charles C., 1953, Rational theory of delta formation: AAPG Bull., v. 37, no. 9, p. 2119-2162.

Belt, E. S., 1975, Scottish Carboniferous cyclothem patterns and their paleoenvironmental significance, *in* M. L.

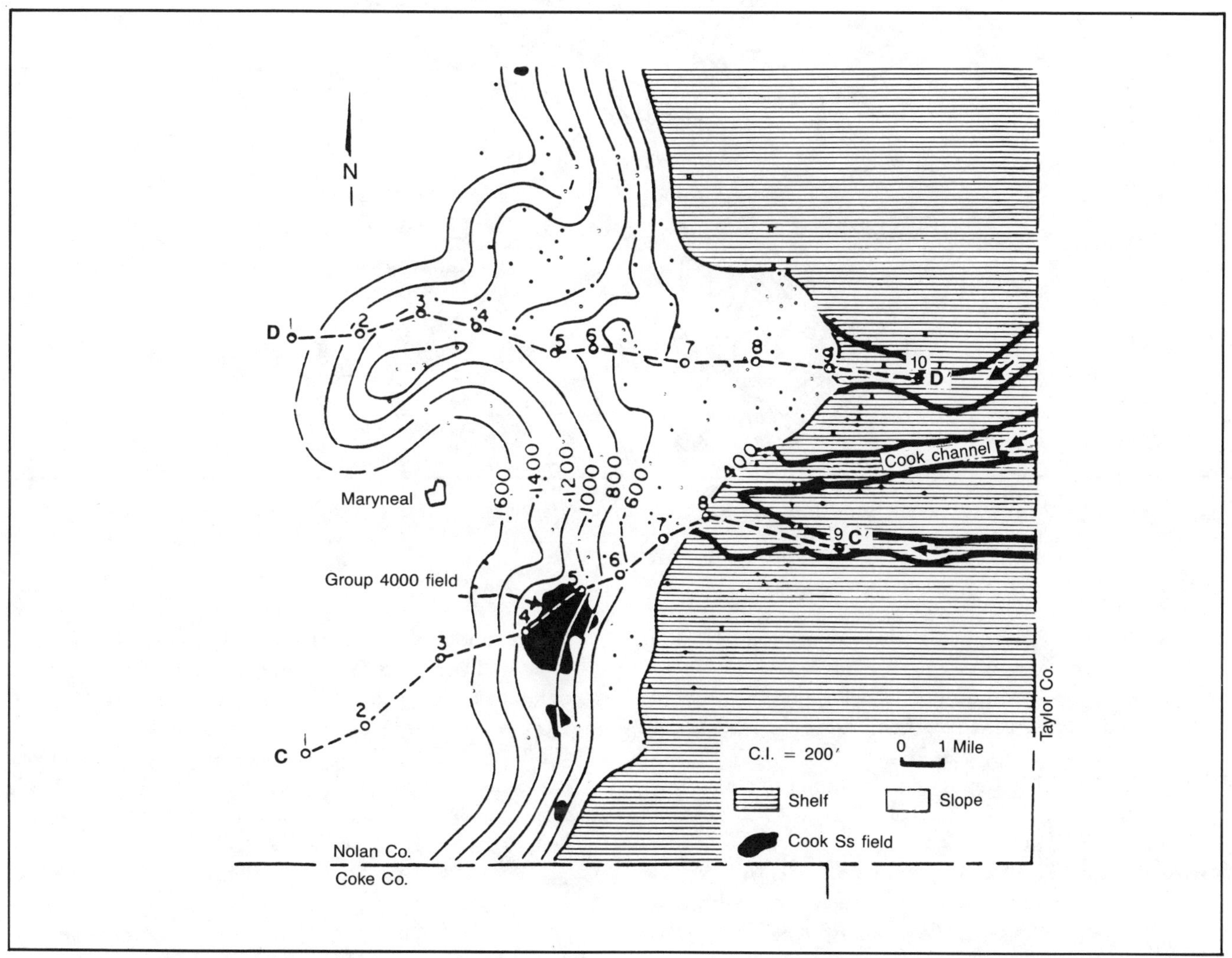

Fig. 8-56. Isopach map of GSS between top of Noodle Creek limestone and top of Waldrip limestone II, Nolan County, Texas. Map represents paleoslope of Waldrip limestone II. (From Bloomer, 1977; permission to publish by AAPG).

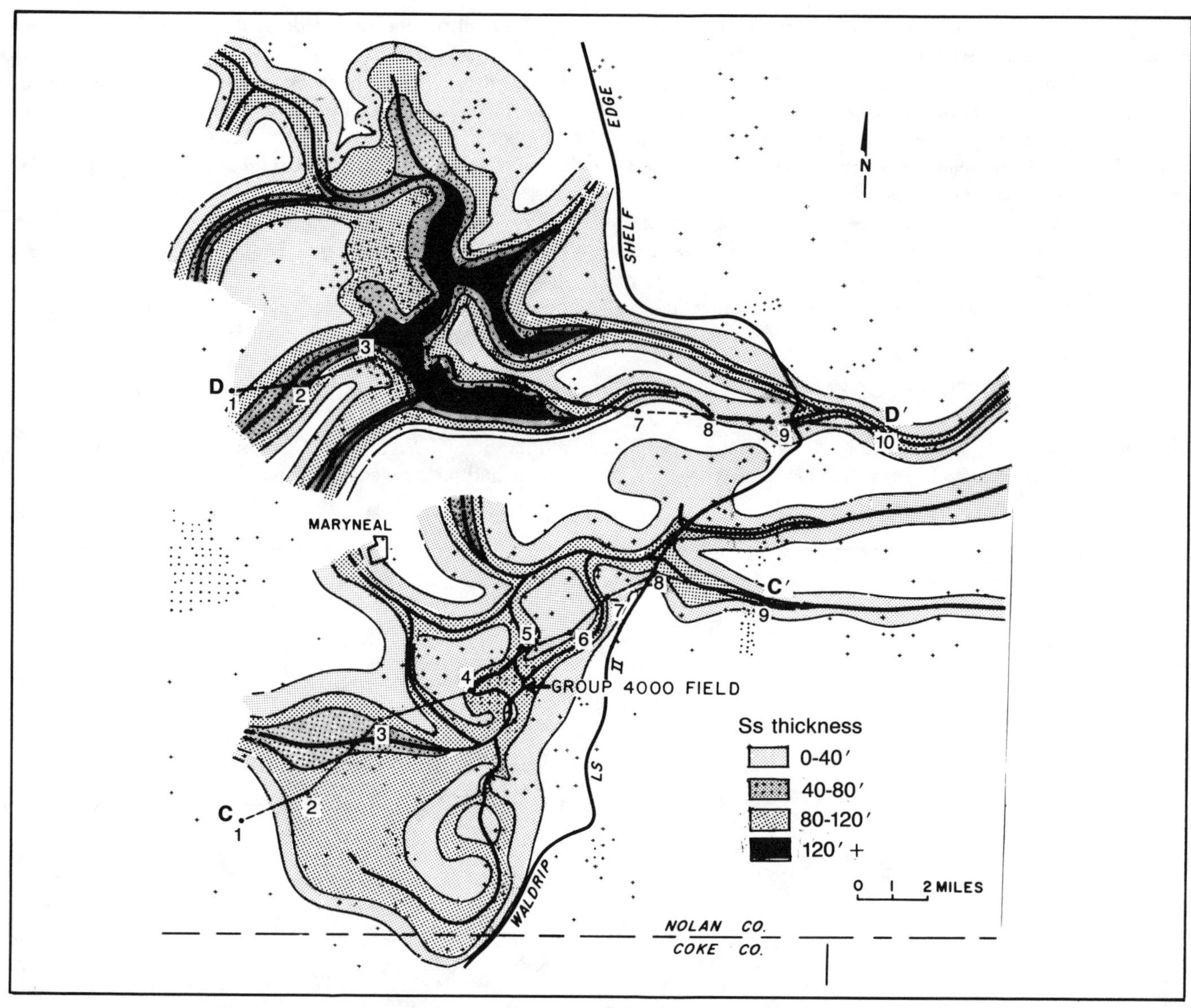

Fig. 8-57. Isopach map of Cook sandstone fluvial and deltaic deposits, Nolan County, Texas. (From Bloomer, 1977; permission to publish by AAPG).

Broussard, ed., Deltas, models for exploration: Houston Geol. Soc., p. 427-449.

Berg, R. R., 1979, Oil and gas in delta-margin facies of Dakota sandstone, Lone Pine field, New Mexico: AAPG Bull., v. 63, no. 6, p. 886-904.

Bloomer, R. R., 1977, Depositional environments of a reservoir sandstone in west-central Texas: AAPG Bull., v. 61, no. 3, p. 344-359.

Bowen, J. M., 1975, The Brent oil field, *in* A. W. Woodland, ed., Petroleum and the continental shelf of northwest Europe, 1, Geology: London, Applied Sci. Pubs., p. 353-361.

Broussard, M. L., ed., 1975, Deltas, models for exploration: Houston Geol. Soc., 555 p.

Budding, M. C., and H. F. Inglin, 1981, A reservoir geological model of the Brent sands in southern Cormorant, *in* L. V. Illing, and G. D. Hobson, ed., Petroleum geology of the continental shelf of north-west Europe: Inst. of Petroleum Geology, London, p. 326-334.

Busch, D. A., 1953, The significance of deltas in subsurface exploration: Tulsa Geol. Soc. Digest, v. 21, p. 71-80.

———, 1959, Prospecting for stratigraphic traps: AAPG Bull., v. 43, no. 12, p. 2829-2843.

———, 1971, Genetic units in delta prospecting: AAPG Bull., v. 55, no. 8, p. 1137-1154.

———, 1974, Stratigraphic traps in sandstones—exploration techniques: AAPG Mem. 21, 174 p.

Coleman, J. M., and D. B. Prior, 1980, Deltaic sand bodies: AAPG Education Course Note Series No. 15, 171 p.

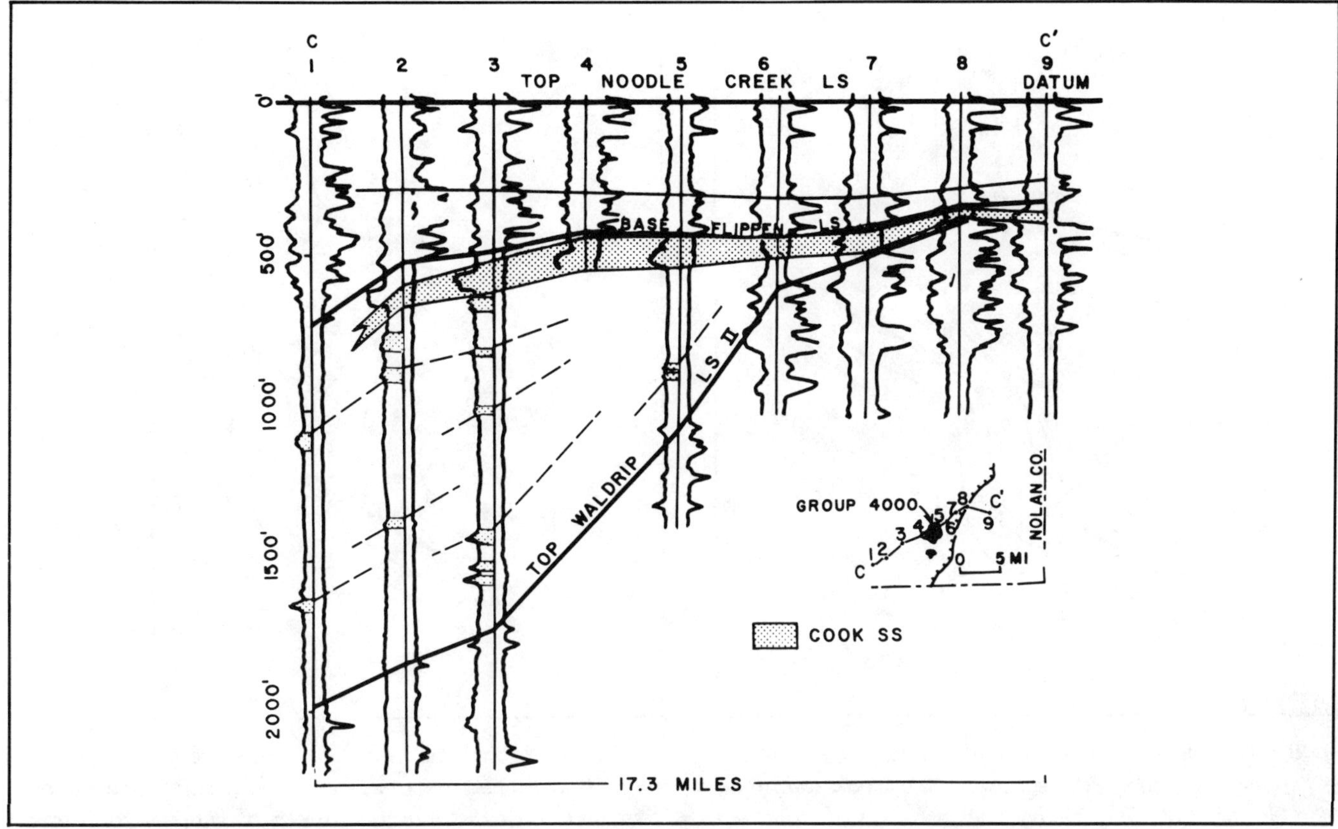

Fig. 8-58. Stratigraphic cross section C-C′ through southern Cook delta. No horizontal scale between wells. See Fig. 8-55 or 8-56 for location. (From Bloomer, 1977; permission to publish by AAPG).

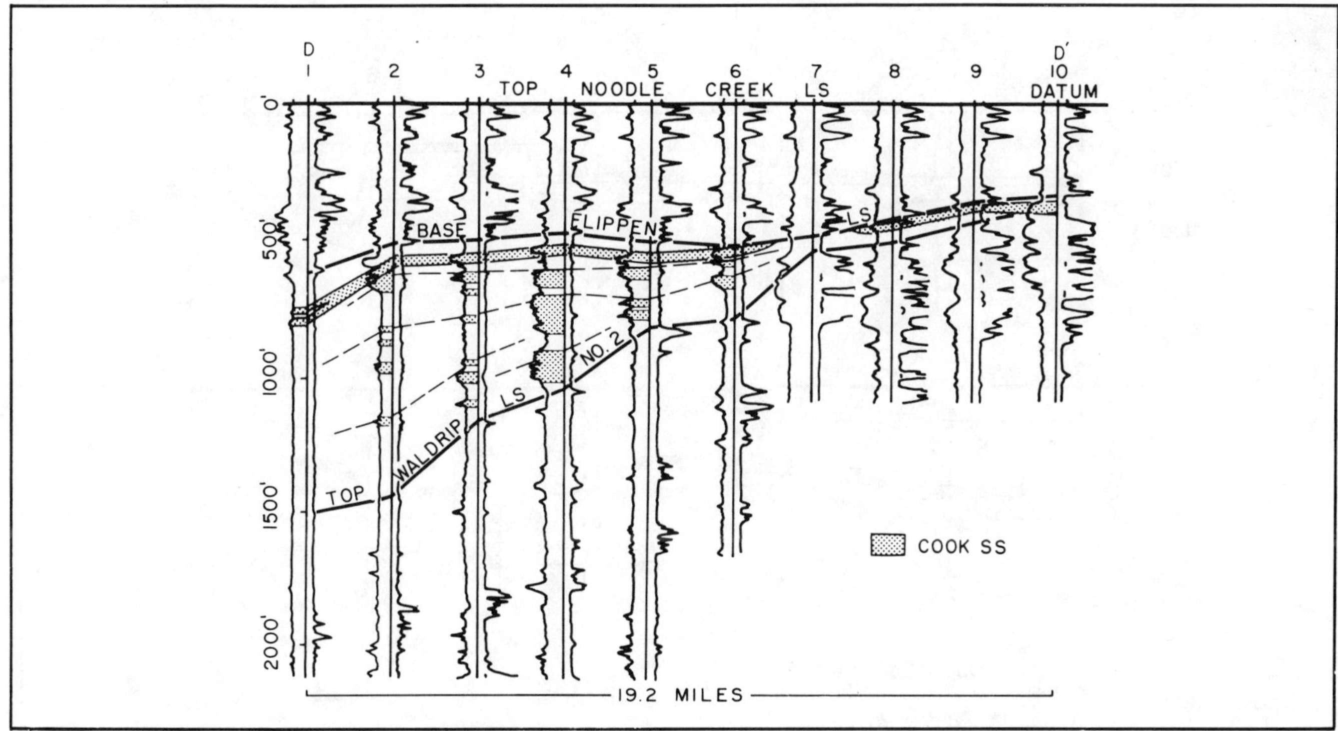

Fig. 8-59. Stratigraphic cross section D-D′ through northern Cook delta. No horizontal scale between wells. See Fig. 8-55 or 8-56 for location. (From Bloomer, 1977; permission to publish by AAPG).

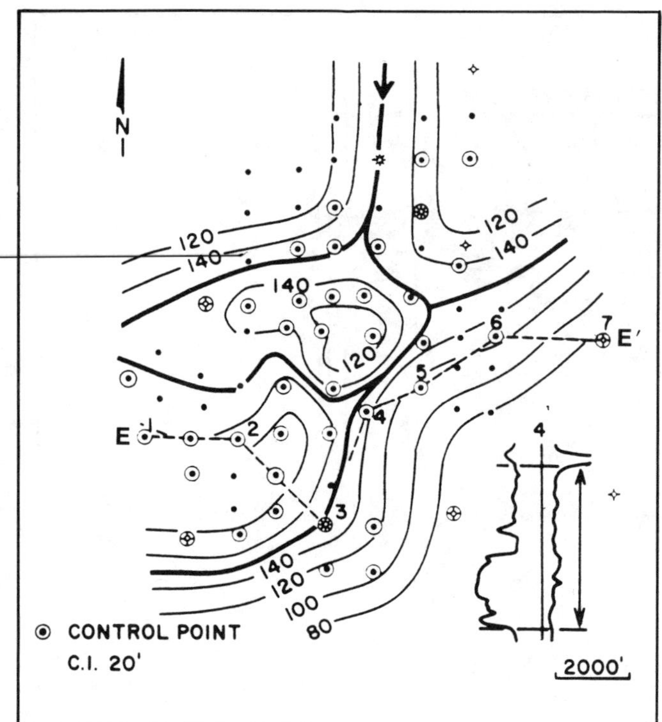

Fig. 8-60. Isopach map of the GIS between the base of the Flippen limestone and the base of the Cook sandstone in the Group 4000 field. Heavy lines are axes of maximum thickness. (From Bloomer, 1977; permission to publish by AAPG).

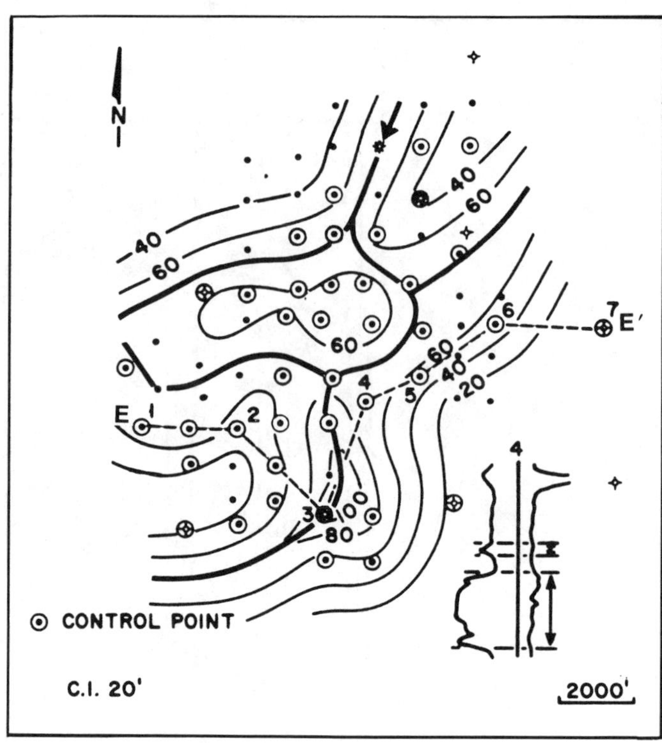

Fig. 8-61. Isopach map of Cook sandstone in Group 4000 field. Heavy lines are trends of thickest sandstone deposition and show distributary pattern of channel sandstone. (From Bloomer, 1977; permission to publish by AAPG).

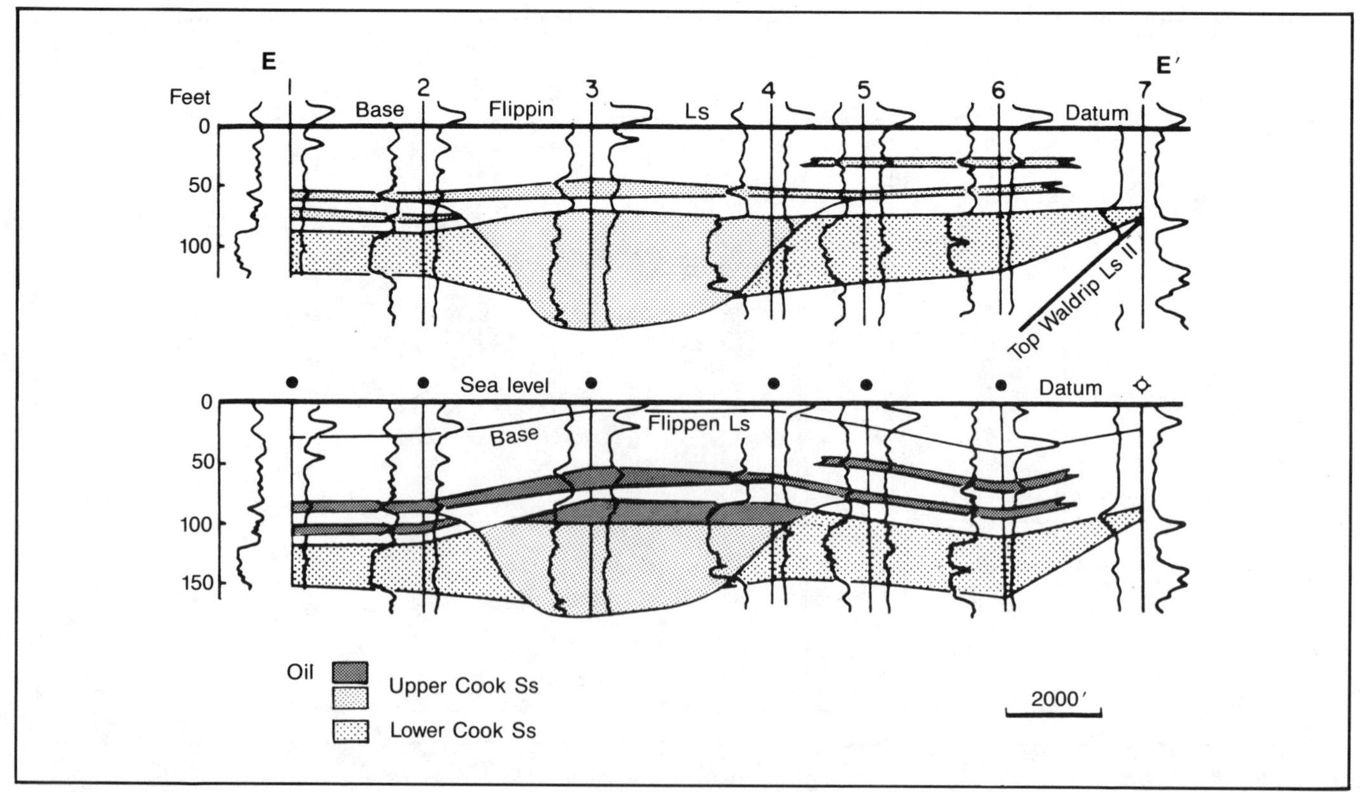

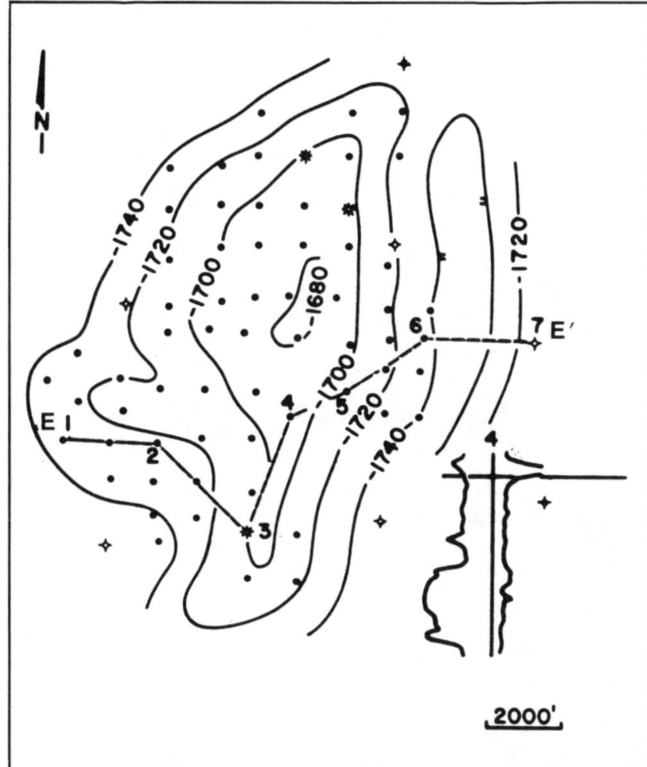

Fig. 8-63. Structure map of base of Flippen limestone in Group 4000 field. (From Bloomer, 1977; permission to publish by AAPG).

_______ and _______, 1982, Deltaic environments of deposition: *in* P. A. Scholle and D. Spearing, eds., Sandstone depositional environments: AAPG Mem. 31, p. 139-178.

_______ and L. D. Wright, 1975, Modern river deltas: variability of processes and sand bodies, *in* M. L. Broussard, ed., Deltas, models for exploration: Houston Geol. Soc., p. 99-150.

Combaz, A., and M. de Matharel, 1978, Organic sedimentation and genesis of petroleum in Mahakam delta, Borneo: AAPG Bull., v. 62, no. 9, p. 1684-1695.

Curtis, D. M., 1970, Miocene deltaic sedimentation, Louisiana Gulf Coast, *in* J. P. Morgan, ed., Deltaic sedimentation, modern and ancient: SEPM Sp. Pub. 15, p. 293-308.

Dailly, G. C., 1976, Pendulum effect and Niger delta prolific belt: AAPG Bull., v. 60, no. 9, p. 1543-1549.

De Matharel, M., P. Lehmann, and T. Oki, 1980, Geology of the Bekapai field, *in* M. T. Halbouty, ed., Giant oil and gas fields of the decade 1968-1978: AAPG Mem. 30, p. 459-469.

Edwards, M. B., 1981, Upper Wilcox Rosita delta system of south Texas: growth-faulted shelf-edge deltas: AAPG Bull., v. 65, no. 1, p. 54-73.

Ejedawe, J. E., 1981, Patterns of incidence of oil reserves in Niger delta basin: AAPG Bull., v. 65, no. 9, p. 1574-1585.

Evamy, D. D., J. Haremboure, P. Knaap, W. A. Molloy, and P. H. Rowlands, 1978, Hydrocarbon habitat of Tertiary Niger delta: AAPG Bull., v. 62, no. 1, p. 1-39.

Ferm, J. C., 1970, Allegheny deltaic deposits, *in* J. P. Morgan, ed., Deltaic sedimentation, modern and ancient: SEPM Sp. Pub. 15, p. 246-255.

Fisher, W. L., 1969, Facies characterization of Gulf Coast basin delta systems, with some Holocene analogues: Gulf Coast Assoc. Geol. Soc. Trans., v. 19, p. 239-261.

_______ and J. H. McGowen, 1969, Depositional systems in the Wilcox Group (Eocene) of Texas and their relationship to occurrence of oil and gas: AAPG Bull., v. 53, no. 1, p. 30-54.

Fisk, H. N., 1961, Bar-finger sands of the Mississippi delta, *in* J. A. Peterson and J. C. Osmond, eds., Geometry of sandstone bodies: AAPG Sp. Pub., p. 29-52.

Galloway, W. E., 1975, Process framework for describing the morphologic and stratigraphic evolution of deltaic depositional systems, *in* M. L. Broussard, ed., Deltas, models for exploration: Houston Geol. Soc. p. 86-98.

Gray, W. D. T. and G. Barnes, 1981, The Heather oil field, *in* L. V. Illing and G. D. Hobson, eds., Petroleum geology of the continental shelf of north-west Europe: Inst. of Petroleum Geology, London, p. 335-341.

Hancock, N. J., and M. J. Fisher, 1981, Middle Jurassic North Sea deltas with particular reference to Yorkshire, *in* L. V. Illing and G. D. Hobson, eds., Petroleum geology of the continental shelf of north-west Europe: Inst. of Petroleum Geology, London, p. 186-195.

Khaiwka, M. H., 1968, Geometry and depositional environments of Pennsylvanian reservoir sandstones, northwestern Oklahoma: Ph.D. Diss., Univ. of Oklahoma, 126 p.

Kirk, R. H., 1980, Statfjord field—a North Sea giant, *in* M. T. Halbouty, ed., Giant oil and gas fields of the decade 1968-1978: AAPG Mem. 30, p. 95-116.

LeBlanc, R. J., 1972, Geometry of sandstone reservoir bodies, *in* T.D. Cook, ed., Underground waste management and environmental implications: AAPG Mem. 18, p. 133-190.

Lyons, P. L., and M. B. Dobrin, 1972, Seismic exploration

Fig. 8-62. Stratigraphic and structure cross sections E-E' of Cook sandstone in Group 4000 field. Delta-front sandstones are cut by distributary channel and filled with channel sandstone. See Fig. 8-59 or 8-60 for location. (From Bloomer, 1977; permission to publish by AAPG).

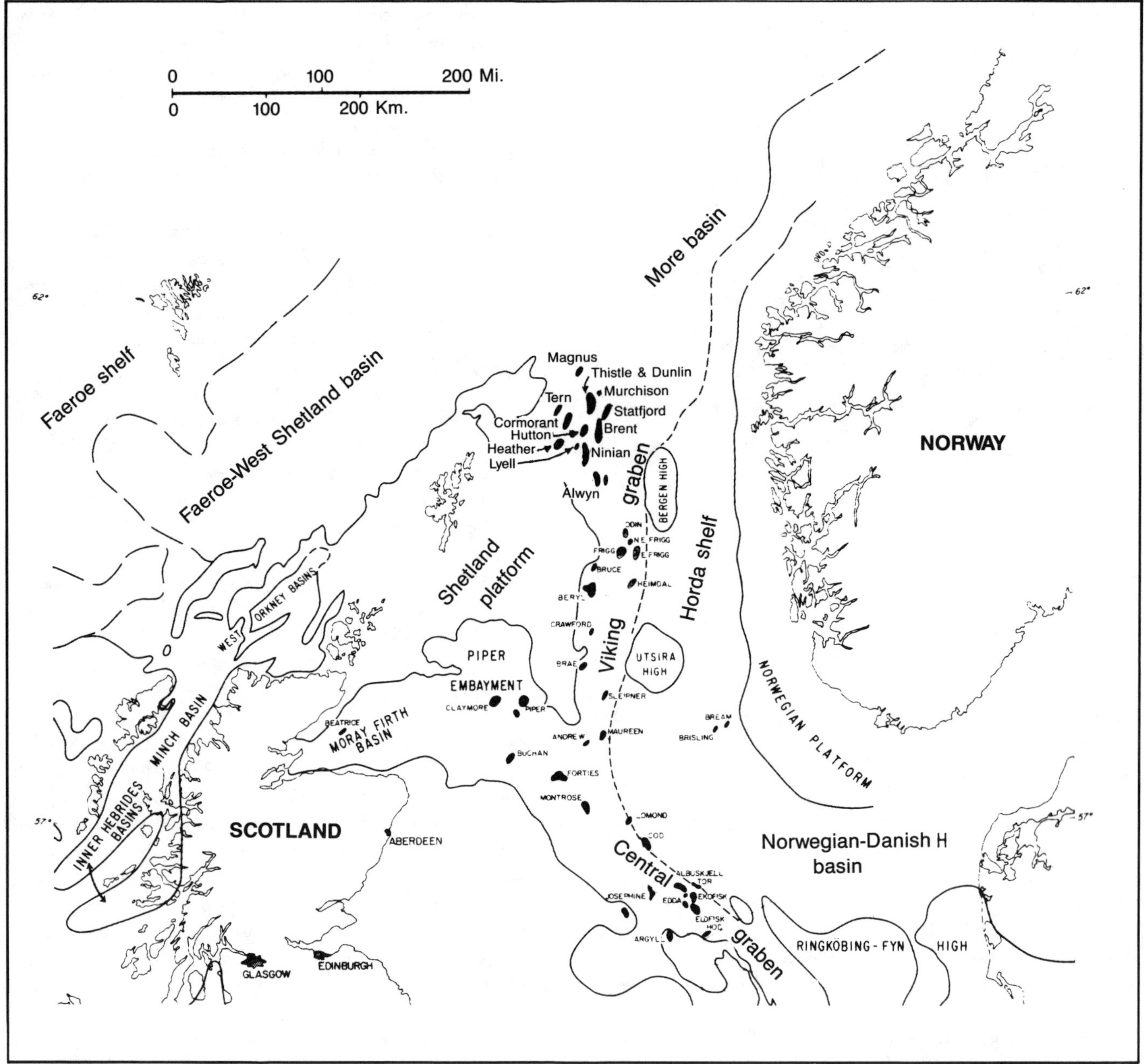

Fig. 8-64. Area map showing northern North Sea fields and major features. (From Albright et al., 1980; permission to publish by AAPG).

for stratigraphic traps, *in* R. E. King, ed., Stratigraphic oil and gas fields—classification, exploration methods, and case histories, AAPG Mem. 16, SEG Sp. Pub. No. 10, p. 225-243.

Magnier, P., T. Oki, and L. Kartaadiputra, 1975, The Mahakam delta, Kalimantan, Indonesia: 9th World Petroleum Cong., Tokyo, Proc., p. 239-250.

Michelson, J. E., 1976, Miocene deltaic oil habitat, Trinidad: AAPG Bull., v. 60, no. 9, p. 1502-1519.

Morgan, J. P., 1970, Depositional processes and products in the deltaic environment, *in* J. P. Morgan, ed., Deltaic sedimentation, modern and ancient: SEPM Sp. Pub. 15, p. 31-47.

Neal, E. P., 1951, South Ceres, Oklahoma's oddest shoestring field: World Oil, v. 133, no. 7, p. 92, 94, 98.

Oomkens, E., 1967, Depositional sequences and sand distribution in a deltaic complex; a sedimentological investigation of the post-glacial Rhone complex: Geologie en Mijnbouw, v. 46, p. 265-278.

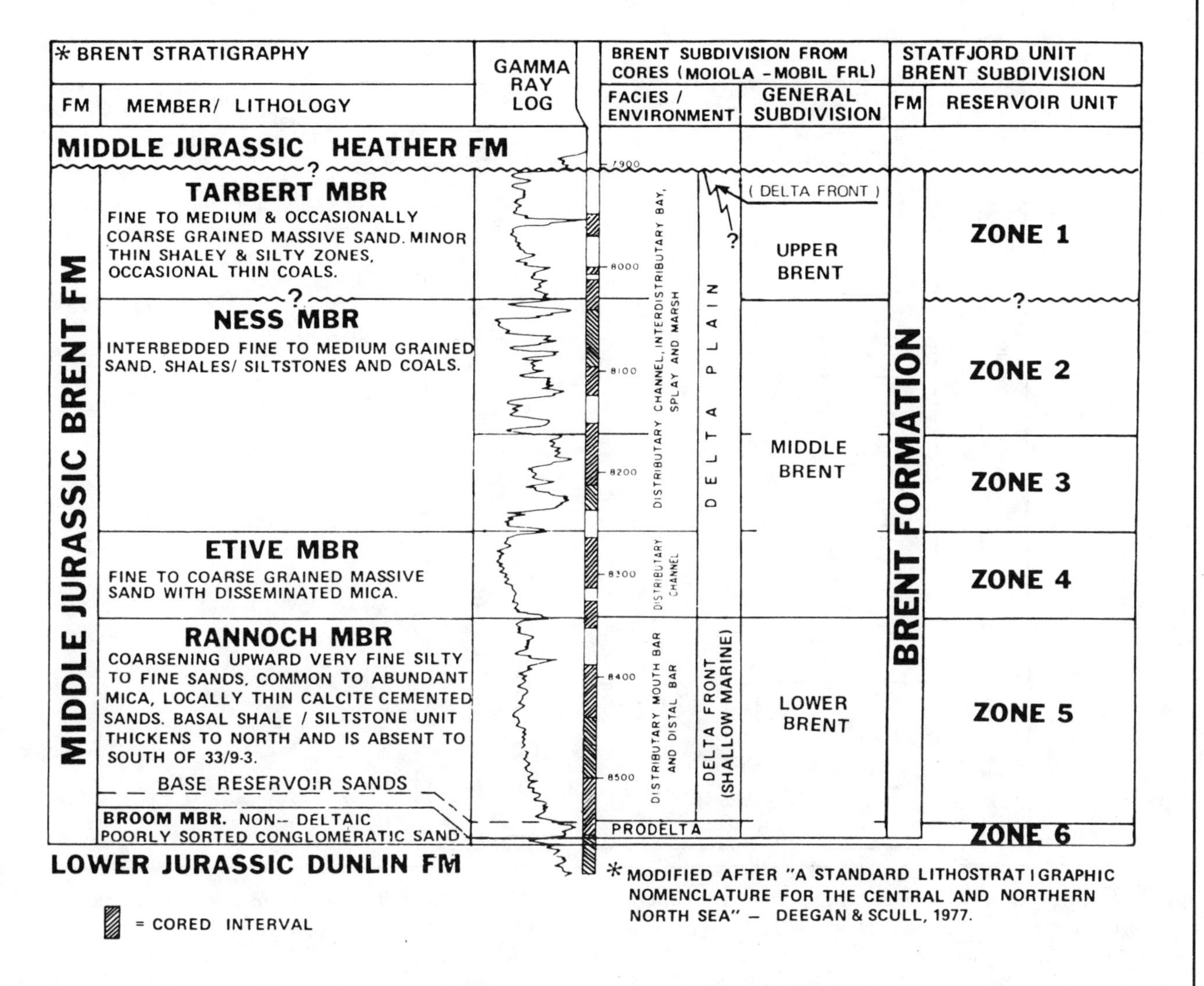

Fig. 8-65. Nomenclature, zonation, and description of sediments, Brent formation, Statfjord field. (From Kirk, 1980; permission to publish by AAPG).

Rice, D. D., 1980, Coastal and deltaic sedimentation of Upper Cretaceous Eagle sandstone: relation to shallow gas accumulations, north-central Montana, AAPG Bull., v. 64, no. 3, p. 316-338.

Scott, A. J., and W. L. Fisher, 1969, Delta systems and deltaic deposition, *in* Delta systems in the exploration for oil and gas, a research symposium: Texas Univ. Bur. Econ. Geology, p. 10-39.

Scott, J. D., 1970, Cherokee sandstones of a portion of Noble County, Oklahoma: Master's Thesis, Univ. of Oklahoma.

Scruton, P. C., 1960, Delta building and the deltaic sequence, *in* F. P. Shepard et. al., eds., Recent sediments, northwest Gulf of Mexico, 1951-1958: AAPG Sp. Pub. p. 82-102.

Shepard, F. P., 1973, Sea floor off Magdalena delta and Santa Marta area, Colombia: GSA Bull., v. 84, no. 6, p. 1955-1972.

Short, K. C., and A. J. Stäuble, 1967, Outline of geology of Niger delta: AAPG Bull., v. 51, no. 5, p. 761-779.

Simpson, R. D. J., and P. K. J. Whitley, 1981, Geological input to reservoir simulation of the Brent Formation, *in* L. V. Illing and G. D. Hobson, eds., Petroleum geology of the continental shelf of north-west Europe: Inst. of Petroleum Geology, London, p. 310-3l4.

Verdier, A. C., T. Oki, and A. Saurdy, 1980, Geology of the Handil Field (East Kalimantan–Indonesia), *in* M. T. Halbouty, ed., Giant oil and gas fields of the decade 1968-1978: AAPG Mem. 30, p. 399-42l.

Wanless, H. R., J. R. Baroffio, J. C. Gamble, J. C. Horne,

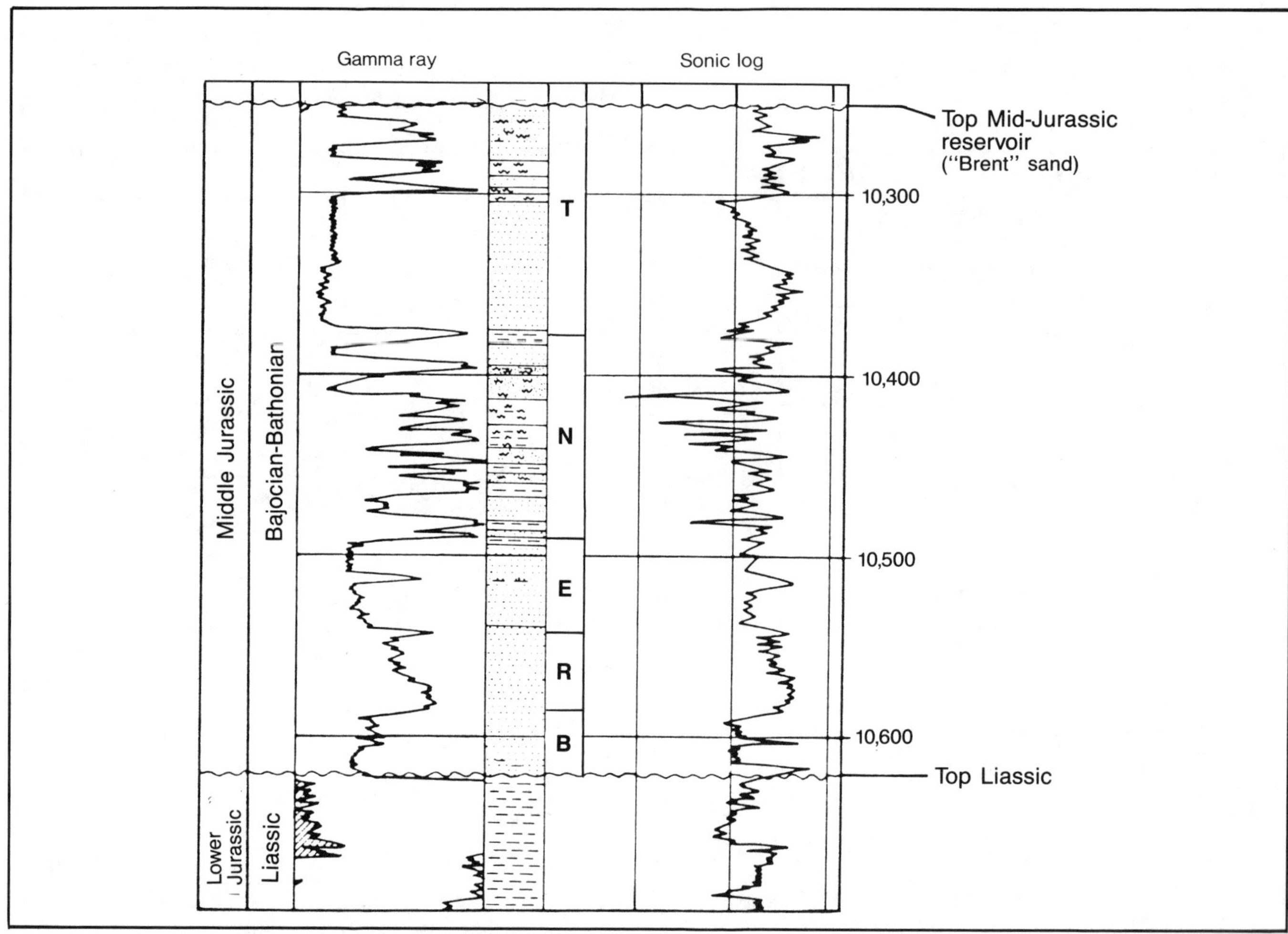

Fig. 8-66. Log character and zonation, Brent formation, well 3/3-3, Ninian field. Letters to right of lithology track indicate members: T = *Tarbert;* N = *Ness;* E = *Etive;* R = *Rannoch;* B = *Broom.* (Modified after Albright et al., 1980; permission to publish by AAPG).

D. R. Orlopp, A. Rocha-Campos, J. E. Souter, P. C. Trescott, R. S. Vail, and C. R. Wright, 1970, Late Paleozoic deltas in the central and eastern United States, *in* J. P. Morgan, ed., Deltaic sedimentation, modern and ancient: SEPM Spec. Pub. 15, p. 215-245.

Weber, K. J., 1971, Sedimentological aspects of oil fields in the Niger Delta: Geol. en Mijnbouw, v. 50., p. 559-576.

——— and E. Daukoru, 1975, Petroleum geology of the Niger Delta: Proc. 9th World Petrol. Congr., Panel Disc. 4, p. 1-14.

Wermund, E. G., and W. A. Jenkins, Jr., 1970, Recognition of deltas by fitting trend surfaces to Upper Pennsylvania sandstones in north central Texas, *in* J. P. Morgan, ed., Deltaic sedimentation, modern and ancient: SEPM Sp. Pub. 15, p. 256-269.

Whitbread, T., and G. Kelling, 1982, Mrar formation of western Libya—evolution of an early Carboniferous delta system: AAPG Bull., v. 66, no. 8, p. 1091-1107.

Wright, L. D., and J. M. Coleman, 1973, Variations in morphology of major river deltas as functions of ocean wave and river discharge regimes: AAPG Bull., v. 57, no. 2, p. 370-398.

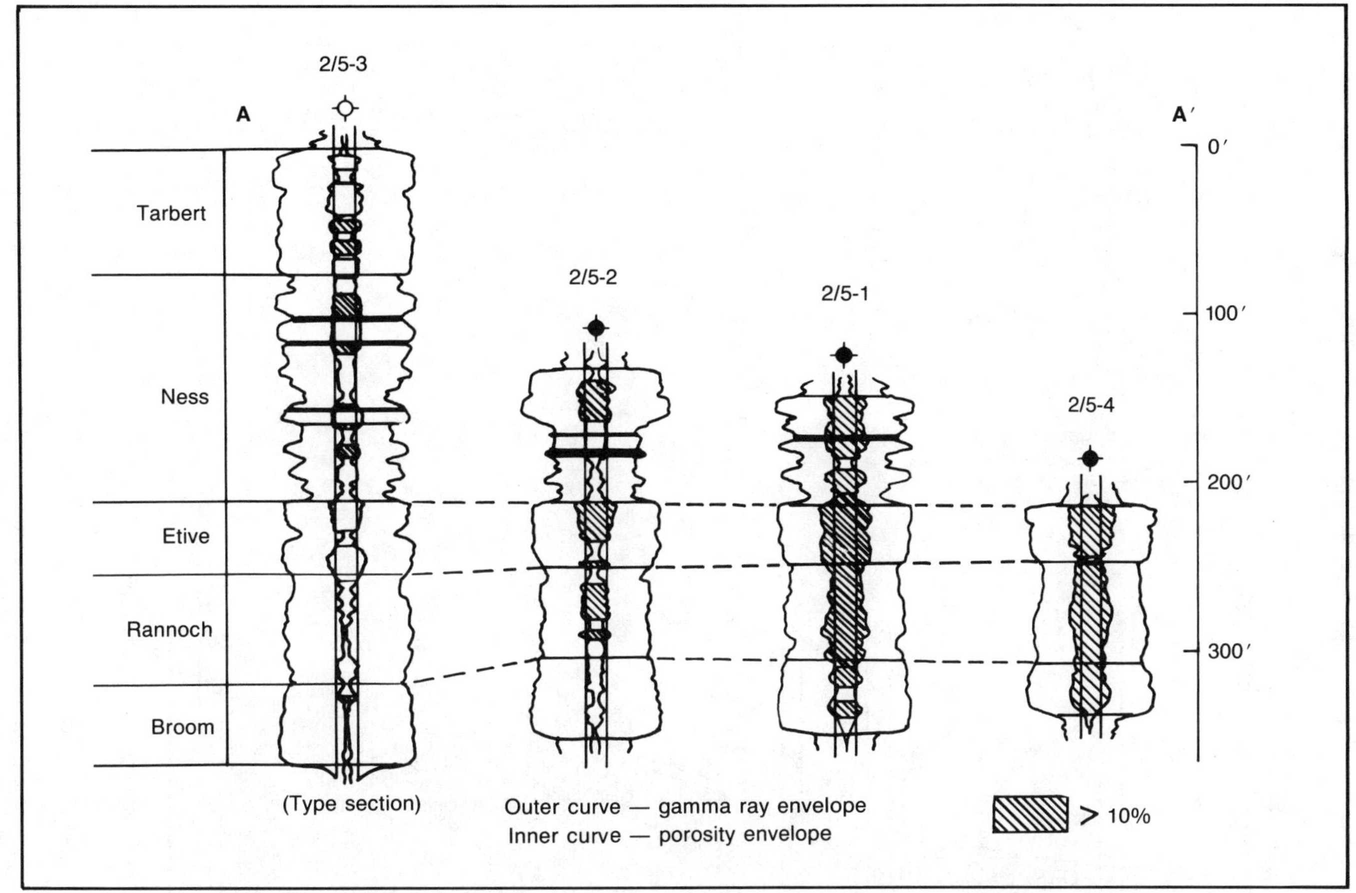

Fig. 8-67. Log character and zonation, Brent formation, Heather field. (From Gray and Barnes, 1981; permission to publish from Institute of Petroleum, London).

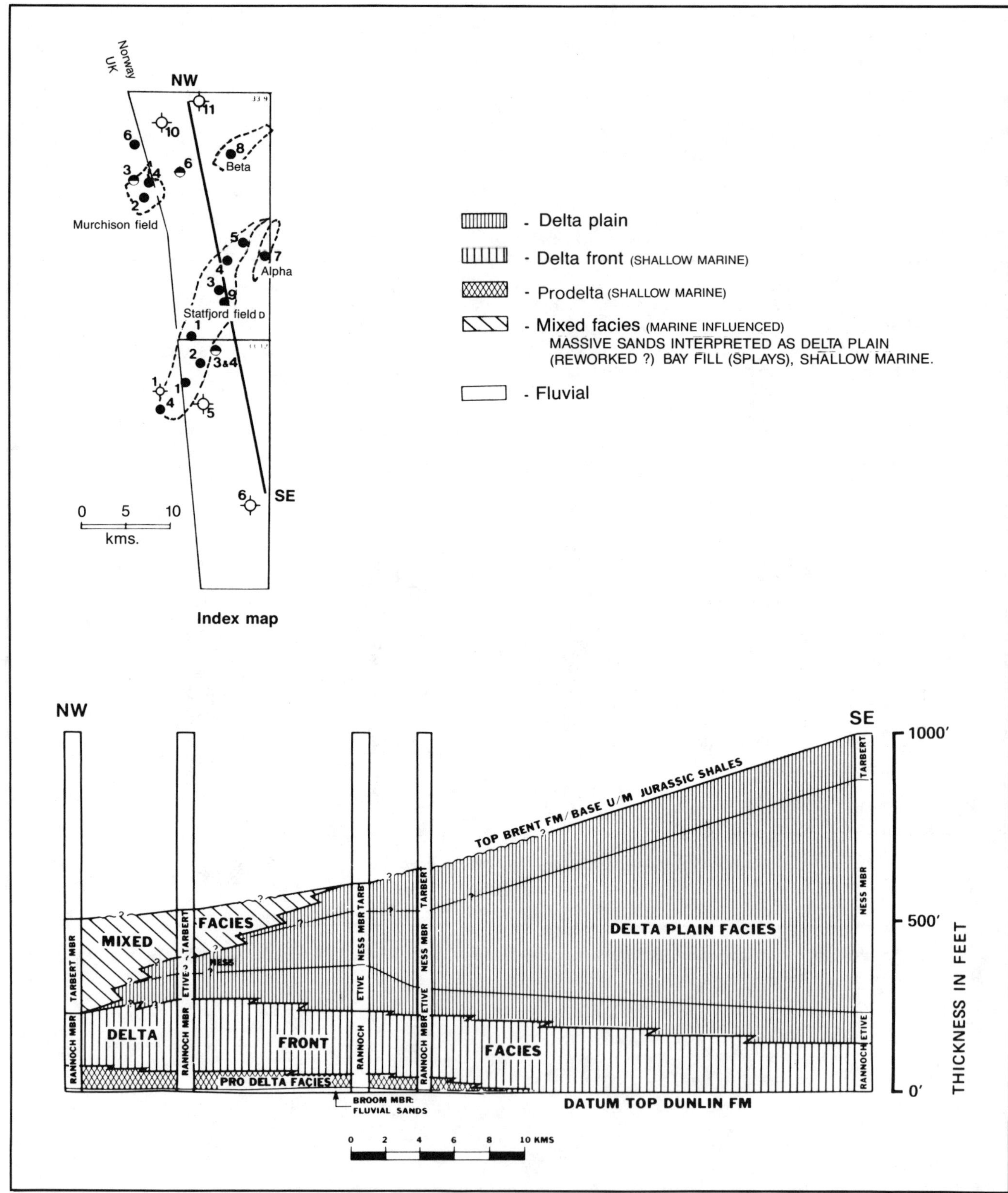

Fig. 8-68. NW-SE cross section in Statfjord field area illustrating southeast to northwest thinning and facies change of Brent formation. (From Kirk, 1980; permission to publish by AAPG).

9 COASTAL-INTERDELTAIC RESERVOIRS

The coastal-interdeltaic environment includes most of the genetically related subenvironments that occur along the margin of a major marine embayment. These subenvironments were first referred to as *paralic* environments by Tercier (1940). They occur within the shallower portions of what Krumbein et al. (1949) later referred to as the *epineritic* environment.

There are seven distinct but related types of sediment that are deposited in the paralic environments: mud flat, chenier, barrier island (bar), lagoon, tidal channel, tidal delta, and offshore bar. All of these sediments are initially transported away from river mouths, or deltas, by long shore drift (currents) for considerable distances along the coast. Clays and fine silts are carried in suspension, whereas coarse silt and sand are transported either as bed load or by wave energy along the beach and shallow water zone. LeBlanc (1972) points out that, "The suspended silt and clay load is dispersed at a rapid rate and is most significant in the development of the mud flats of the chenier plain. Lateral movement of the sand bed load occurs at a relatively slow rate and is most significant in the development of the cheniers and the barrier-island complex." The buildup of offshore bars, however, is related primarily to varying intensity of wave energy conditions.

CHENIERS

General Characteristics

Cheniers were described first along the coast of southwest Louisiana by Russell and Howe in 1935. They later were described by Fisk (1955), Fisk and McFarlan (1955), Byrne et al. (1959), and Gould and McFarlan (1959). Cheniers consist of long, narrow, sandy ridges that are subparallel with the coastline. They rise slightly above the surrounding marshes, lakes, and water courses. Individual cheniers rise from a few inches to 10 ft (3 m) above sea level along selected portions of the Louisiana coast (see Figs. 9-1 and 9-2). In the same area they are 2 to 15 ft (0.6 to 4.5 m) thick, 100 to 1,500 ft (30 to 457 m) wide, and up to 30 mi (48 km) in uninterrupted length. They have a gentle seaward slope. In plan view, they generally are slightly concave seaward, except near the mouths of rivers where they tend to converge and curve sharply landward. Cheniers occur in multiples and collectively make up a small percentage of the sediments of the chenier plain. The principal lithology of the chenier plain is silty clay with much lesser amounts of peat. The individual sandy cheniers are pronounced interruptions in this overall chenier plain material. A significant feature of cheniers is that all such relict sands are deposited in a regressive shoreline situation and, thus, will occur at the same stratigraphic level relative to a marker bed above or below. This generally is not the case with ancient multiple barrier bars.

The base of an individual chenier may be either flat or slightly convex downward. The bottom commonly is downwarped a few feet below sea level as a result of compaction of the underlying silts and clays. The cross sectional shape of an individual chenier (Pecan Island) is illustrated in the upper half of Figure 9-3, which has a vertical exaggeration of 200 times.

Lithologically, cheniers consist principally of sand and shells with minor quantities of silt and clay. Byrne et al. (1959) state, "The shells, which make up 22 percent of the facies, occur as distinct layers up to several inches thick and as fragments disseminated throughout the sand."

Figure 9-4 is a generalized illustration of the internal structure of a chenier and its associated mud flat. A single

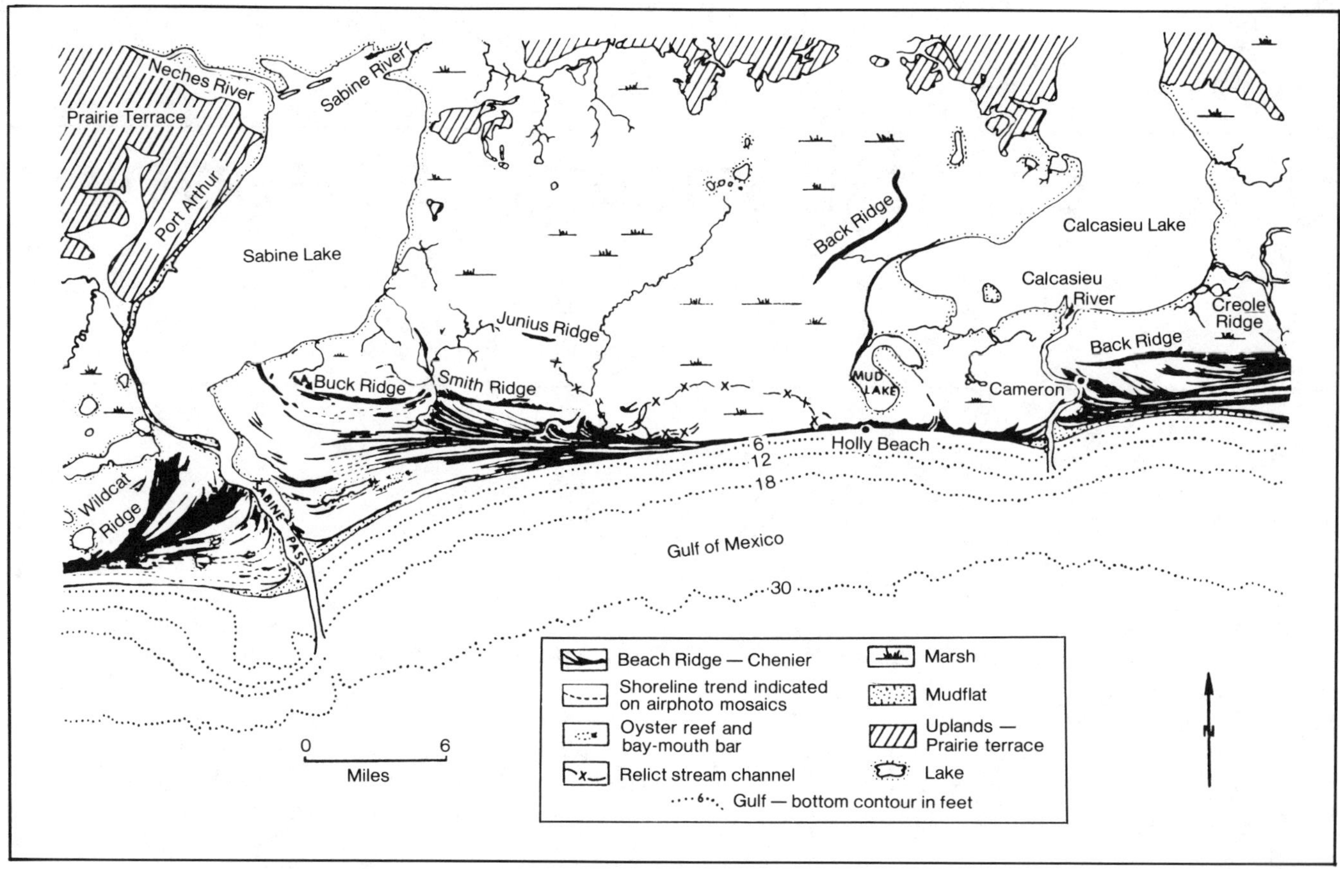

Fig. 9-1. Chenier plain of southwestern Louisiana (in two parts; see Fig. 9-2). (Modified after Byrne et al., 1959; permission to publish by Gulf Coast Association of Geological Societies).

unoriented core reveals eolian cross stratification in the upper one-third, horizontal bedding in the middle one-third, and a zone of bioturbation in the lower one-third. This tripartite zonation is not unique to cheniers but also can be noted in the subaerial portions of barrier bars. An unoriented core taken from the mud flat consists predominantly of brackish-water shale with an occasional thin layer of sand, which was washed into the swamp during an intensive storm. The same situation may occur in the lagoonal sediments behind a barrier bar.

Figure 9-5 illustrates the stages in development of a chenier plain and its lateral position relative to the direction of longshore drift away from a delta. All of the sediments of the chenier plain are regressive in nature.

Origin of Cheniers

The origin of cheniers is somewhat conjectural. Available information indicates that the five following environmental conditions apparently are necessary to their development:

1. Low-lying marshy or swampy coastal plain adjacent to the delta of a river that transports a large load of sediment
2. Longshore drift that transports delta (land-derived) sediments to the bordering coastal plain
3. Periods of abundant supply of fine-grained sediment (to form mud flats) alternating with periods of slight sediment supply (during which beach ridges or cheniers are constructed)
4. Predominance of clay and silt with a minor amount of sand
5. Moderate wave action along the shore

Gould and Morgan (1962) point out:

> . . . alternations in coastal outbuilding and relative stability reflect pulsations in the supply of Mississippi River sediments carried into the area of longshore currents. Such pulsations have resulted not from changes in the load of the Mississippi but from wide lateral shifts in the position of the mouth during construction of the deltaic plain. Periods of rapid progradation resulted when the river discharged into the western part of the deltaic plain, whereas intervals of coastal stability and local retreat followed shifts in discharge to more easterly positions.

LeBlanc (1972) attributes mud flat silts and shales to major floods that result in the sudden large influx of sediments

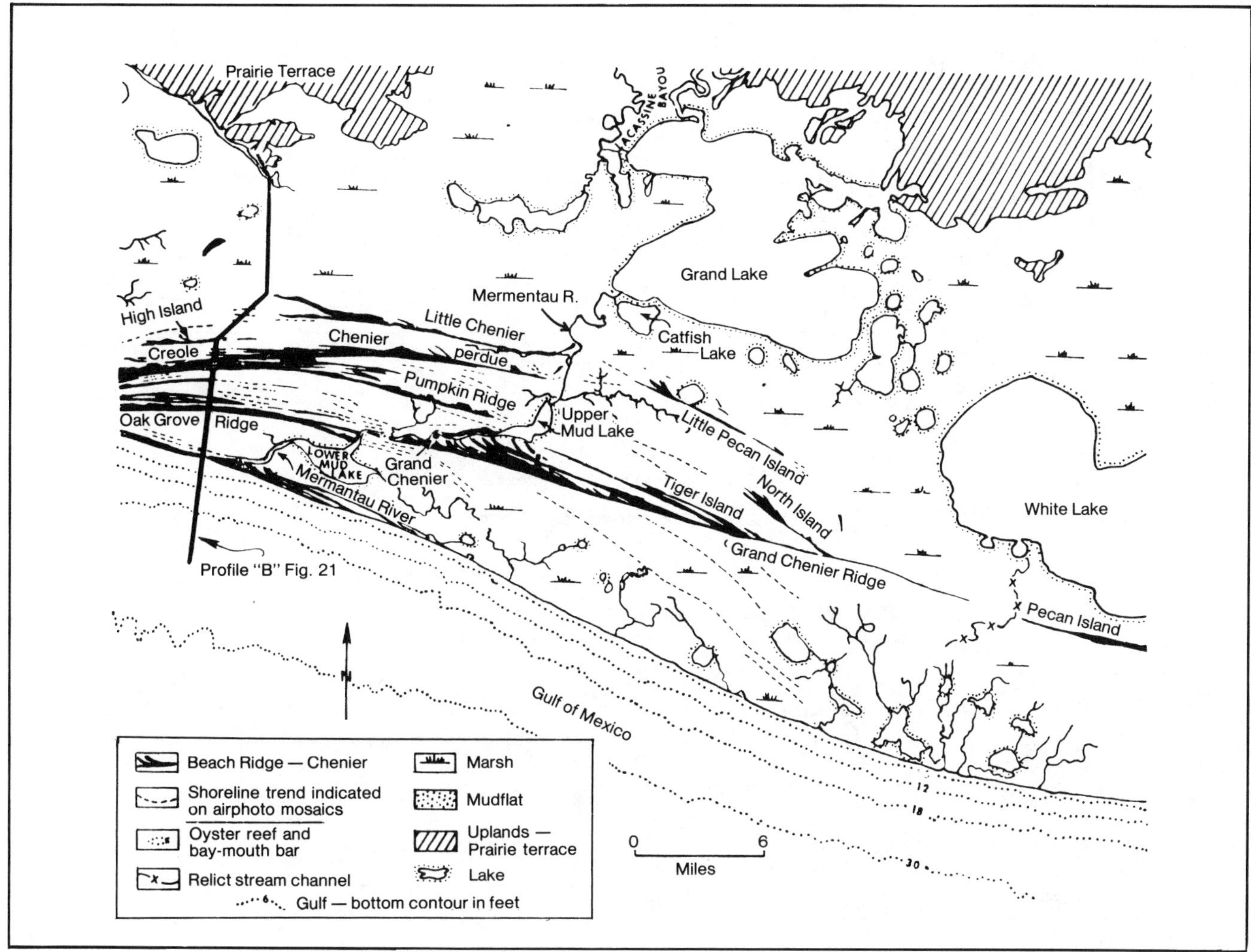

Fig. 9-2. Chenier plain of southwestern Louisiana (in two parts; see Fig. 9-1). (Modified after Byrne et al., 1959; permission to publish by Gulf Coast Association of Geological Societies).

at river mouths. Much of the suspended load introduced to the coastal marine environment is rapidly dispersed laterally along the coast by the predominant longshore drift. A considerable portion of this suspended load is deposited along the shoreline (on the delta flank) as extensive mud flats. This period of regressive sedimentation (progradation or offlap) occurs in a relatively short period when rivers are at flood stages.

During long periods when rivers are not flooding, the supply of sediment to the coast is reduced considerably or is nil. Coastal marine currents and wave action rework the seaward edge of the newly formed mud flat, and a transgressive situation develops. A slight increase in sand supply can result in a regressive situation, and the initial transgressive beach accumulation will grow seaward by regressive beach accretion to form a long, narrow, well-defined chenier on the seaward edge of the extensive mud flat.

It is apparent that Gould and Morgan relate chenier plain deposition to wide lateral shifts in the mouth of the Mississippi River, whereas LeBlanc emphasizes flood stages versus non-flood stages of the river supplying the sediment.

Kaczorowski (1980) has compared the carbon 14 age-dating of the Mississippi delta lobe positions with a growth rate curve from the midwestern chenier plain and finds they are not consistent. He therefore believes local river influx is more significant in the building of the chenier plain than the Mississippi River discharge direction. Obviously, more research is necessary on the origin of the sediments of the chenier plain.

From the foregoing description of cheniers along the southwest Louisiana coast, it is apparent that this type of relict beach sand has considerable potential significance as a reservoir sandstone. Cheniers are genetically related to the deltaic environment. Additional sites where cheniers have been reported are in the vicinity of the Orinoco delta (Surinam and French Guiana) and adjacent to the deltas of the

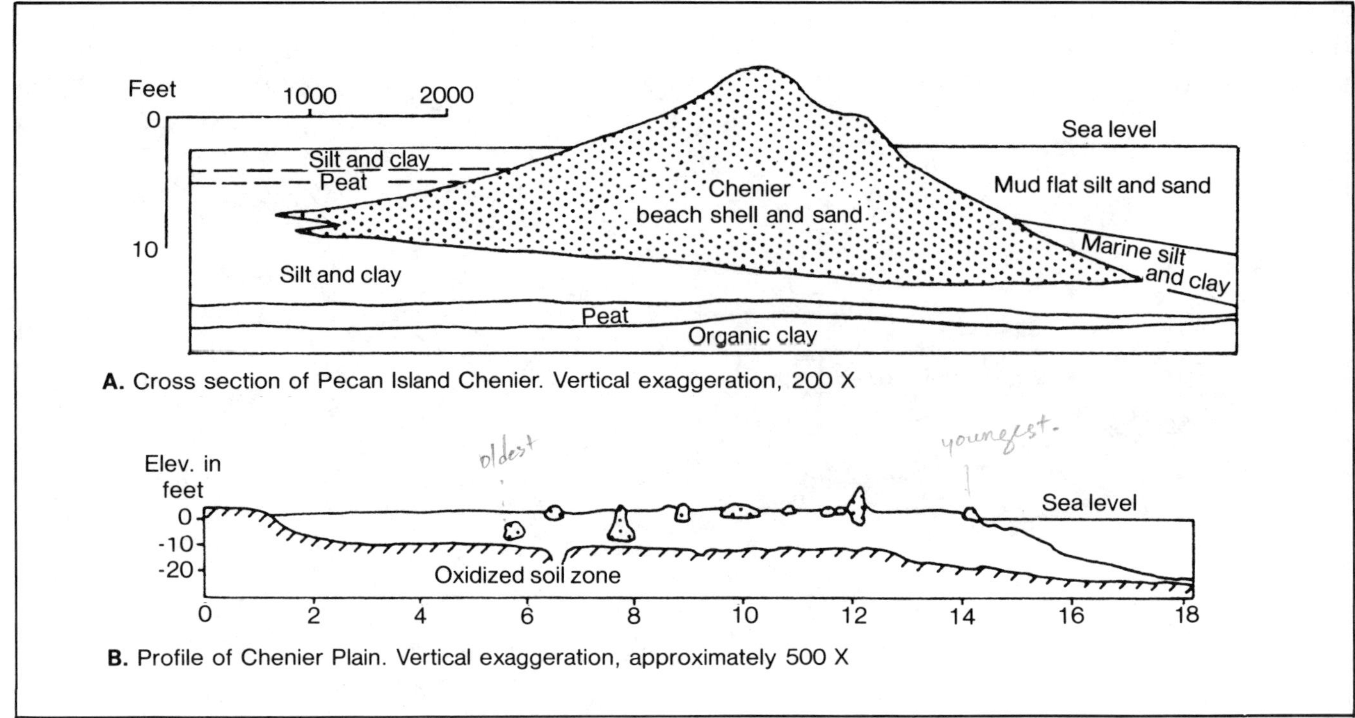

Fig. 9-3. Cross sections of **(A)** *Pecan Island chenier (right side of Fig. 9-2).* (Modified after Gould and Morgan, 1962; permission to publish by Houston Geological Society). **(B)** *chenier plain (Fig. 9-2).* (Modified after Byrne et al., 1959; permission to publish by Gulf Coast Association of Geological Societies).

Amazon and São Francisco Rivers of Brazil. They are also thought to be present adjacent to the deltas of the Rhone (France) and Po (Italy) Rivers, as well as along the Gulf of Venezuela.

Although they rarely exceed 15 ft (4.6 m) in thickness along the southwest Louisiana coast, cheniers may be much thicker, depending on such variables as the available supply of sand and shell material and the intensity of waves and longshore drift. Modern cheniers adjacent to the delta

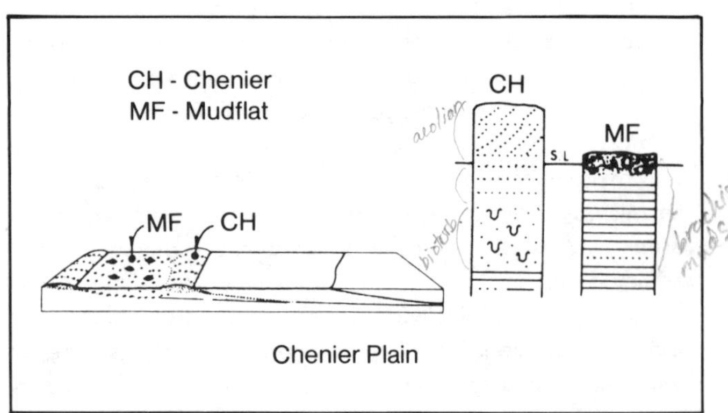

Fig. 9-4. Internal structure of chenier sand and mud flat materials. (Modified after LeBlanc, 1972; permission to publish by AAPG).

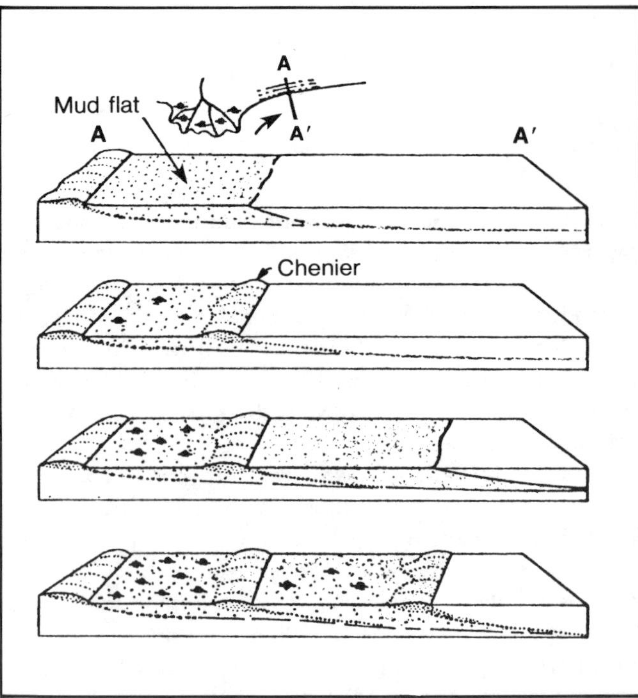

Fig. 9-5. Stages in development of a chenier plain. (Modified after LeBlanc, 1972; permission to publish by AAPG).

of the São Francisco River of northeastern Brazil have been reported (H. A. Chaves, Salvador, Brazil, personal commun., 1967) to be more than 40 ft (12 m) thick. Similar thicknesses are postulated for some of their ancient counterparts.

Cheniers as Reservoirs

Cheniers in the subsurface could be misinterpreted as either barrier bars or offshore bars; both are deposited parallel with the shoreline, may be plano-convex in profile, and have fairly smooth seaward margins. As pointed out, however, cheniers generally are present in multiples at the same stratigraphic position. Also, in comparison to barrier bars, they generally are narrower in plan view and thin more abruptly basinward. Cheniers are deposited at the shoreline, whereas barrier bars generally are deposited some distance offshore and are separated from the mainland by a lagoon.

Most petroleum exploration geologists have not been looking for cheniers in the subsurface because information about this type of sandstone body is incomplete, and to date they have not been positively documented from subsurface studies. None of the standard textbooks on petroleum geology make any reference to cheniers as potential sites of stratigraphic oil and gas accumulation. Fisk (1955) and Byrne et al. (1959) noted the similarity of chenier sands to certain subsurface lenticular sandstones. They referred to the Chanute pool (Fig. 9-6) of southeastern Kansas (Dillard et al., 1941) as a possible example of an ancient chenier-type reservoir. The sandstone body of this pool exceeds 40 ft (12 m) in thickness and diverges in a manner similar to that of modern cheniers.

Chenier Criteria

In identifying and tracing cheniers in the subsurface the following geometric and internal characteristics must be considered:

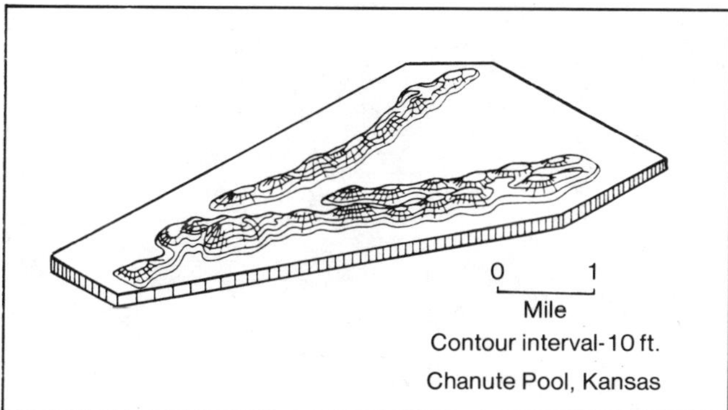

Fig. 9-6. Possible example of a chenier. (Modified after Dillard et al., 1941; permission to publish by AAPG).

A. Geometric Aspects
   1. Length—up to 30 mi (48 km) and without interruptions
   2. Orientation—parallel with basin contours of its GIS
   3. Width—100–1,500 ft (30–457 m)
   4. Thickness—2–15 ft (0.6–4.6 m) in Gulf Coast and up to 50 ft (15 m) elsewhere
   5. Cross section—plano-convex upward
   6. Plan view—slightly concave basinward
   7. Occur in multiples, merge, and curve upstream on both sides of a river

B. Internal Aspects
   1. Composition—predominantly quartz sand with up to 22% shell fragments in both thin layers and scattered
   2. Structure
      a. Upper one-third, cross stratified
      b. Middle one-third, horizontally bedded
      c. Lower one-third, bioturbated
   3. Sorting—to date there are no data available in regard to sorting, hence this criterion cannot be used
   4. E-log–SP Responses—subsurface cheniers remain to be positively documented in the subsurface—until they are, it is postulated that these logs will exhibit a cylindrical "signature" with an abrupt base and top
   5. Associated Strata
      a. Brackish-water silty shales on both sides and below

Not one of the above criteria is diagnostic of a chenier. They should be used in combination. As the explorationist "plugs" in multiple criteria, the degree of certainty increases almost exponentially as the number of criteria increase arithmetically.

BARRIER BARS

The terms "barrier bar", "barrier island" and "barrier beach" are synonymous. Other terms, such as "offshore bars" and "longshore bars" have been confused with the term barrier bar in the literature. The term barrier bar, as used in this text, is an elongate sandy island deposited parallel with the coastline and is separated from it by a lagoon that commonly is marshy. It may have beach ridges on the upper surface, which, in turn, may be partially modified by wind activity into dunes. An important aspect of barrier bars is tidal inlets with their associated subtidal-intertidal deltas. Barrier bars may be either regressive (prograding) or transgressive, and therefore a single model does not serve to illustrate the internal complexities that are known to occur within this type of sand body.

Numerous studies of modern barrier bars have been made within the past 15 years; e.g., Hayes (1975) Hayes and Kana (1976), Kraft (1971), Kraft and John (1979), and Reinson (1979). These studies serve as modern analogues for interpretation of their ancient counterparts. Most students of recent barrier bars devote little attention to the likelihood of preservation of these sand bodies in the subsurface and even less attention to the possibility of their serving as reservoirs for hydrocarbon accumulation. Furthermore, their detailed observations of such factors as bedding characteristics, mineralogical composition, sorting, and bioturbation, will apply in subsurface studies only when cores are available, which all too frequently is not the case.

Much has been written about the origin of modern barrier bars, but no one description will suffice inasmuch as the environmental variables are too numerous.

Reinson (1984) states

> Barrier islands are most prevalent in coastal settings which have the following characteristics: 1) a low gradient continental shelf adjacent to a low-relief coastal plain, 2) an abundant sediment supply, and 3) moderate to low tidal ranges (Glaeser, 1978). Both the shelf and coastal plain are composed of unconsolidated sediments, which are the material source for the building of barrier islands by nearshore processes. Glaeser noted that only 10 percent of the world's barrier-islands are present along coastlines where tidal ranges exceed three metres. However, it was Hayes (1975, 1976) who focused attention on the importance of tidal range in controlling the occurrence and morphology of barrier-island systems. Hayes observed not only that barrier islands were rare on macrotidal coastlines (greater than 4 m tidal range) but that there were geomorphological differences between barrier islands of microtidal regions (less than 2 m tidal range) and those of mesotidal regions (2 to 4 m tidal range). In general, microtidal barrier islands are long and linear with extensive storm washover features . . . mesotidal barrier islands are short and stunted, and characterized by large tidal inlets and deltas. Microtidal barriers are overwashed frequently by storm waves because of the lack of large enough tidal inlets to allow storm surges to flow past the barrier, rather than overtopping it . . . microtidal barrier islands can be considered to be wave dominated as opposed to mesotidal barriers which are affected by both wave and current processes.

Figure 9-7 is a block diagram of a barrier-island system illustrating the various subenvironments.

Barrier Beach Complex

The barrier-beach complex includes (1) the shoreface (or subtidal zone), (2) the foreshore (intertidal zone, or beach), (3) the backshore-dune (subaerial zone), and (4) the backshore-dune (supratidal to subaerial wave and wind-formed washover flats), as illustrated Figures 9-7 and 9-8. Reinson (1979) defines the shoreface environment "as the area seaward of the barrier from the low tide mark to a depth of about 10 to 20 m. The lower limits of the shoreface correspond to the position at which waves begin to affect the

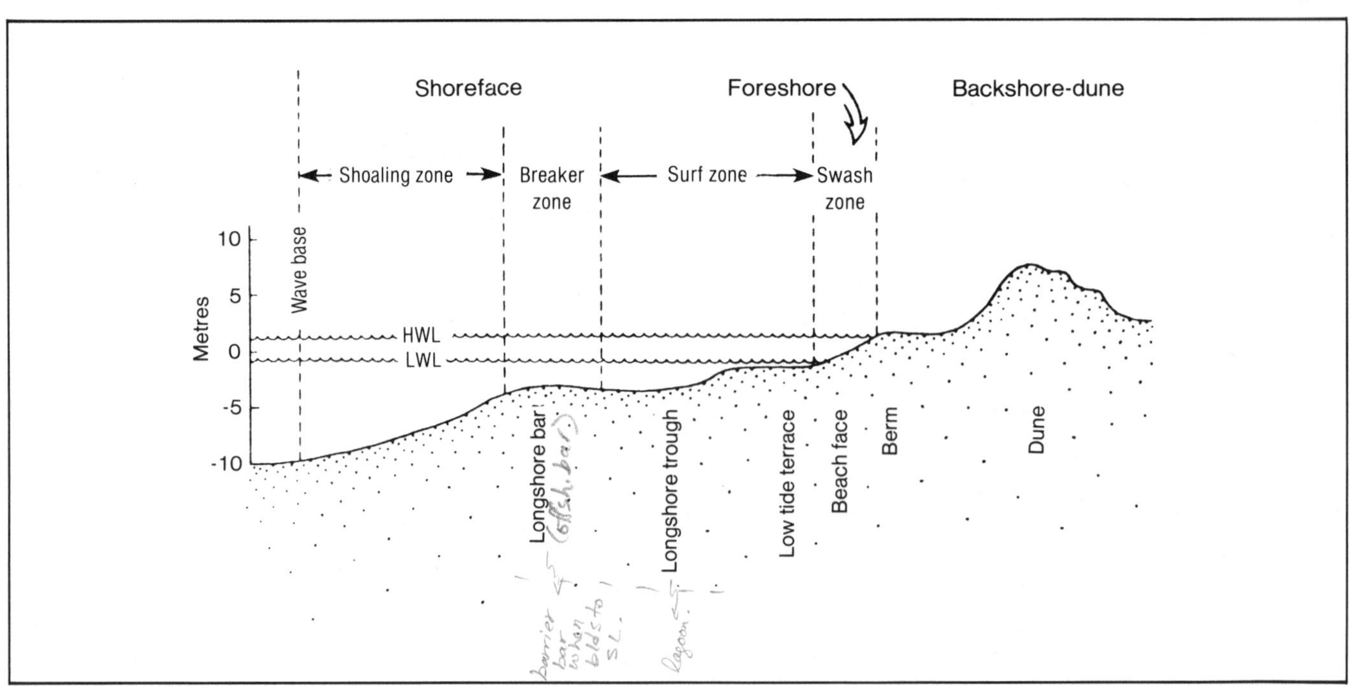

Fig. 9-7. Generalized profile of the barrier beach and shoreface environments. (From Reinson, 1984, as modified after Reinson, 1979; permission to publish by Geological Association of Canada).

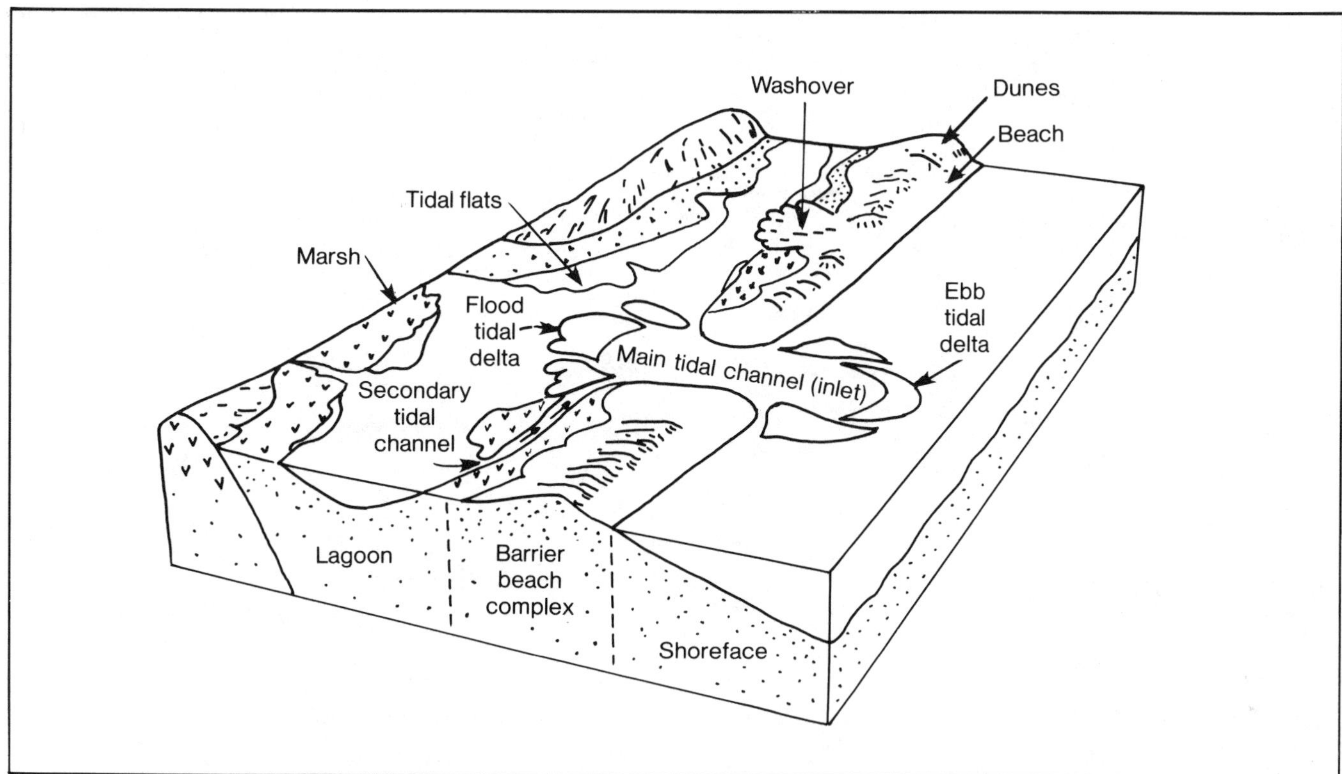

Fig. 9-8. Block diagram illustrating the various subenvironments in a barrier island system. (From Reinson, 1984, as modified after Reinson, 1979; permission to publish by Geological Association of Canada).

sea bed. Hence the shoreface is an environment in which depositional processes are governed by wave energy." It is within the middle and lower shoreface environments that multiple offshore bars will form in response to varying intensity of storm-generated waves.

It is primarily within the upper shoreface environment that barrier bars are formed

> "because they are situated in the high-energy surf zone just seaward of the beachface and landward of the breaker zone . . . The complex hydraulic environment of the surf zone (i.e., shore normal currents generated by plunging waves superimposed on shore-parallel wave-driven currents) gives rise to the complex sequence of multidirectional sedimentary structures and variable sediment textures characteristic of these deposits" (Reinson, 1979).

The intertidal zone is known as the foreshore environment, and it is here that plunging breakers sweep the surf zone. The dissipation of wave energy in this zone is followed immediately with backward runoff. Within this zone, the sand laminae are subparallel and have a gentle seaward slope. Along the landward margin of the foreshore environment there is a very narrow ridge called the "berm." Reinson notes that "sediment is transported to the berm crest by high spring tides or storms and is distributed over the backshore area by winds and washover." The backshore-dune environment is one of predominantly wind activity where sand dunes may occur in abundance.

Geomorphic Framework of Barrier Bars

A barrier bar system consists of three principal parts and Reinson (1979) states that "This tripartite geomorphic framework clearly demonstrates that barrier-island systems are composites of three major clastic depositional environments: 1) the subtidal to subaerial barrier-beach complex, 2) the back-barrier region or subtidal-intertidal lagoon, and 3) the subtidal-intertidal delta and inlet-channel complex," as illustrated in Figure 9-8. There is not just one barrier-bar stratigraphic sequence, but rather three are needed to properly interpret cores recovered from drilling operations. The stratigraphic sequences from barrier bars are even more complicated by the fact that they may be deposited under either regressive (prograding) or transgressive conditions.

Regressive (Prograding) Barrier Bars[1]

Regressive modern barrier bars present the most likely conditions for preservation in the subsurface. This is due primarily to their large dimensions. The geometric and in-

[1]Much of the discussion of regressive barrier bars is taken from Busch, 1974, Stratigraphic Traps in Sandstone—Exploration Techniques: AAPG Memoir 21.

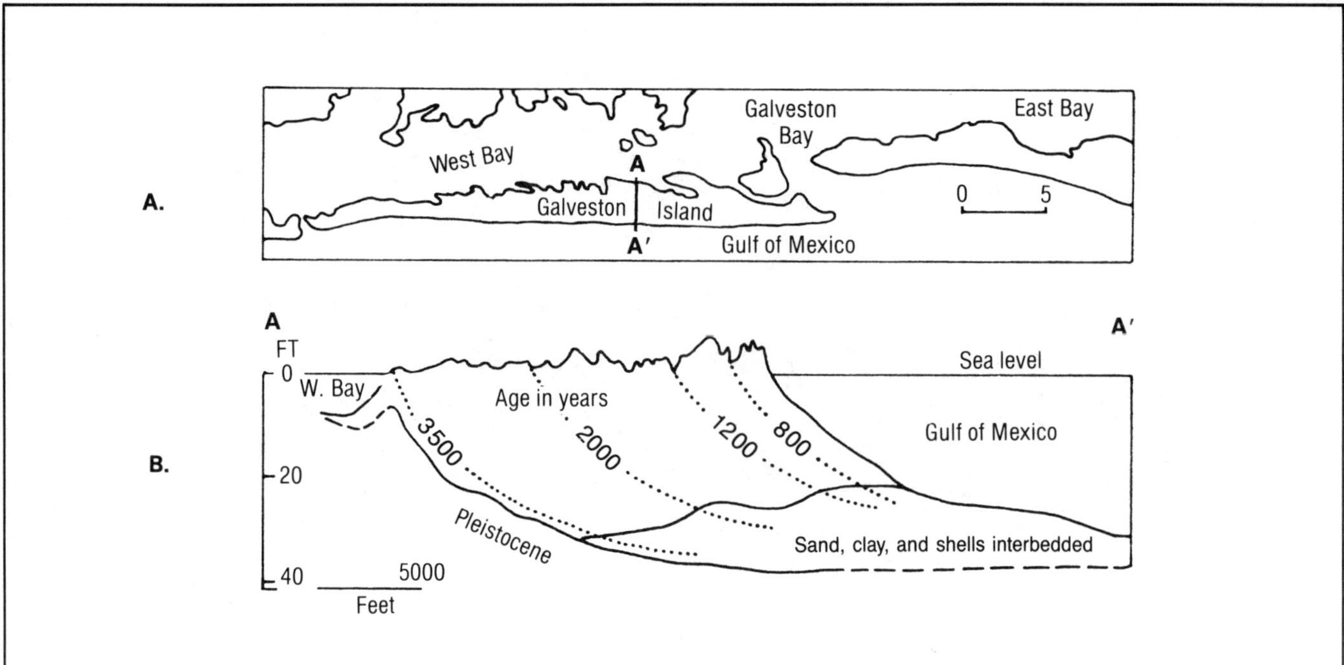

Fig. 9-9. **A.** *Galveston Island, Texas Gulf Coast.* **B.** *Profile A-A', Galveston Island, showing radiocarbon dates of shell beds.* (From Bernard et al., 1962; permission to publish by Houston Geological Society).

ternal characteristics of several classic modern examples are reviewed as a partial basis for the recognition of their ancient counterparts.

Galveston Barrier Island, Texas. Galveston Island, off the Texas Gulf Coast, is an excellent example of a barrier bar. It grew southwestward, in the direction of longshore drift, by beach-spit, tidal-channel, and tidal-delta accretion. Radiocarbon dating of shell beds within the sand body clearly indicates seaward accretionary growth of this barrier island. Abandoned beach ridges and intervening low swales, which are well preserved in the central part of the island, demonstrate the seaward growth of the island by continued addition of sand by longshore currents (Fig. 9-9). Sea level reached its present position a little over 3,500 years ago when Galveston Island was a small bar situated about 4 mi (6.4 km) offshore on the southwest side of the mouth of Galveston Bay. The combined effects of winds, surface currents, and waves refracting shoreward produced the westerly longshore currents necessary to add material to this bar, that grew into a barrier island. Most of the sand was derived from an easterly source, whereas much of the silt and clay probably were contributed by the nearby Trinity and San Jacinto Rivers. Bernard et al. (1962) point out that the depositional features ". . . consist principally of numerous narrow, parallel beach ridges that vary in elevation from a few inches to 12 ft (3.7 m), and intervening swales. These features trend almost parallel with the present shoreline, but are recurved along the backside and younger western ends of the islands." Four breaker bars normally may be seen from the beach. Their positions, height, and the number visible vary with wave height and direction, tides, and possibly longshore currents. Depending on prevailing conditions, these bars may move up the upper part of the shoreline and beach, transporting sand to the upper shoreface and beach environments.

The surface sands of Galveston Island are very well sorted, fine to very fine grained. Sand containing shells extends gulfward to the 30 ft (9 m) subaqueous contour; seaward from this the bottom consists of silt and clay with shells. Sedimentary structures are useful in distinguishing most of the Holocene facies of the barrier island. Most of the bay deposits are burrowed and churned by organisms. Bedding and laminations are common in the sand and clay in the tidal-delta and bay deposits near channels where the sediments frequently are reworked and deposited by currents. Lamination and very low-angle cross-laminae are the most common structures in the sand deposited in the upper foreshore of the beach environment. The middle-shoreface deposits consist of markedly burrowed, churned, laminated, bedded, and in places crossbedded shelly-sand and shell layers. Elongate sand grains in the beach deposits are preferentially aligned at right angles to the long dimension of the barrier island. This orientation is parallel with the direction of motion of the last waves to act upon the grains of sand.

The three components of this barrier island complex are illustrated in Figures 9-10 and 9-12. Under regressive conditions the seaward side of a barrier island is primarily an

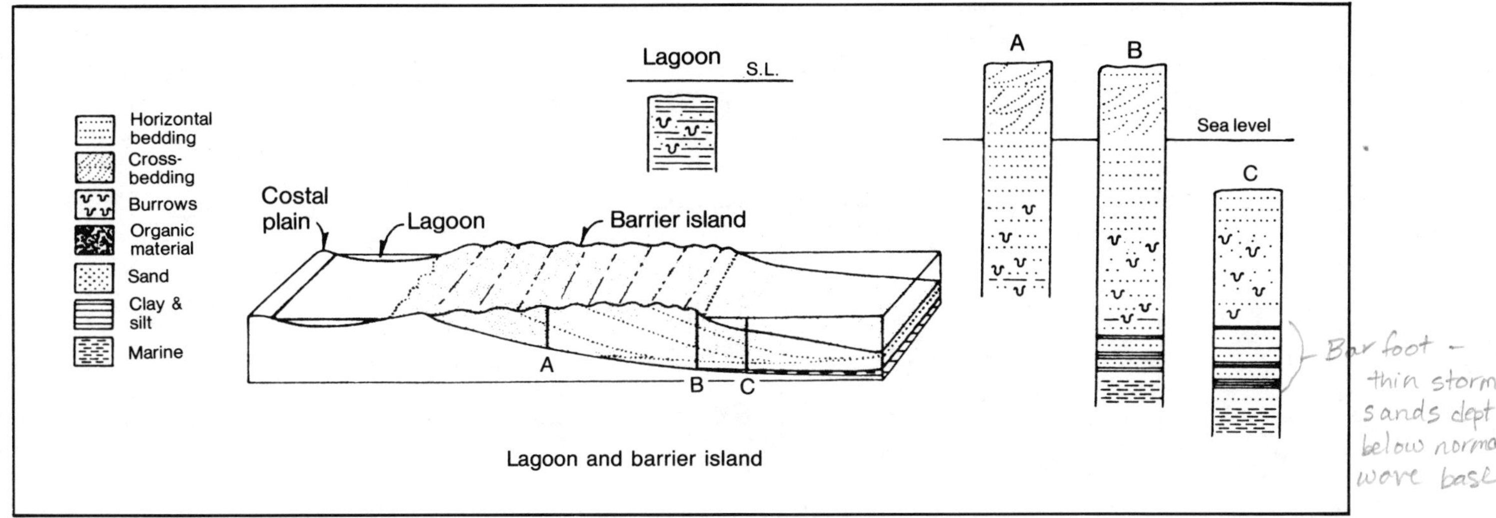

Fig. 9-10. Generalized sedimentary sequences of a barrier island and lagoon. (From LeBlanc, 1972; permission to publish by AAPG).

area of active sand accumulation. This sand is originally transported to the marginal marine environment by rivers and then is moved away from the river mouth (or delta) by longshore drift in a direction parallel with the coastline. LeBlanc states:

> Under ideal conditions, a barrier grows seaward by a beach shore-face accretion process to produce a typical barrier-island sequence of sediments which grades upward from fine to coarse. The various organisms which live in the beach, shoreface, and adjacent offshore areas usually have a significant influence on the character of sedimentary structures. Dry beach sand can be transported inland by the wind and redeposited as dune sand on the barrier, in the lagoon, or on the mainland across the lagoon.

The lithologies of three hypothetical cores, taken from positions A, B, and C on the barrier bar are generalized in Figure 9-10. Cores A and B are from the thicker portions of the bar and exhibit a tripartite zonation. The basal one-third consists of fine-grained, highly bioturbated sand. The middle one-third consists of horizontally bedded sand, and the uppermost one-third is cross-stratified eolian sand. Core B has a thinly interbedded zone of very fine-grained sand, silt, and shale at the base called the "bar foot" (Weber, 1971). Core C is situated in the upper foreshore environment, below sea level, and, therefore, is devoid of the uppermost dune sand. The bar-foot zone at the base, however, is even thicker than that of Core B. Fairly typical SP-GR log responses of these three cores are shown in Figure 9-11 A, B, and C.

Tidal inlets characteristically interrupt the otherwise linear trend of a barrier bar. They may be either fairly shallow or deep, depending on the magnitude of the tidal currents. Such channels have a length comparable to the width of the barrier bar and split up into a system of delta distributaries where the flood tide enters the lagoon behind the bar. Similar ebb-tidal deltas are formed on the seaward side of the tidal inlet. Tidal channels migrate laterally in response to, and in the direction of, longshore drift. Thus, erosion of the barrier bar takes place on the downdrift side of the tidal inlet, and accretionary deposition of spits and tidal flat sediment occurs on the opposite side, as shown in Figure 9-12. As a result of this simultaneous erosion and deposition on opposite sides, the cross section of the tidal inlet is asymmetric. The maximum velocity of the flood and ebb tides is in the deepest part of the channel and, therefore, the coarsest sand grains occur in the lower portion (with fining upward) of the tidal channel sequence. A single core taken from such a sequence (Fig. 9-12) exhibits cross-stratification in response to the flood- and ebb-tide directional flows. The SP-GR log response in a borehole is bell-shaped, as shown in Figure 9-11D. The core of a tidal delta sand is generalized in Figure 9-12. It is dominated by unidirectional dips of sand separated by thin sand or silt shoals.

Ojo de Liebre, Baja California, Mexico. It generally is thought that barrier bars are separated from the mainland by marshy lagoons containing comparatively stagnant water. In a study made by Phleger and Ewing (1962) in the Ojo de Liebre area of Baja California, it is clearly shown that a lagoon behind a barrier bar may be a high-energy environment in which the dominant sediment deposited is sand. The submarine topography of the lagoon behind the barrier bar is illustrated in Figure 9-13. Numerous, deep, flood tidal delta channels fan out landward and are bordered by large intertidal flats. The walls of these channels are quite steep. Deep channels are present on both sides of the inlet directly behind the barrier. The upper surface of the barrier bar has numerous high barchan dunes. The leeward faces of some of these dunes merge directly with the edges of

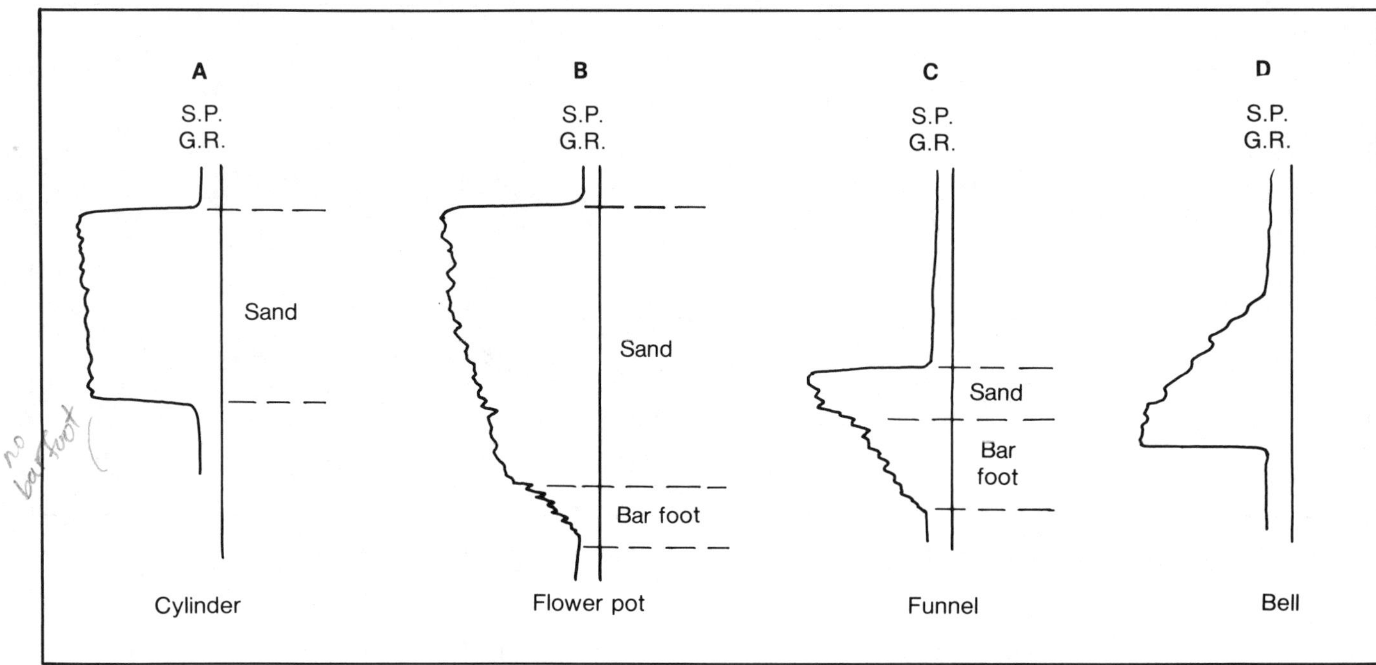

Fig. 9-11. SP-GR log responses to wells drilled at positions **A**, **B**, *and* **C** *of Figure 9-10, and position* **D** *of Figure 9-12.* (Profile A-A'–LeBlanc, 1972; permission to publish by AAPG).

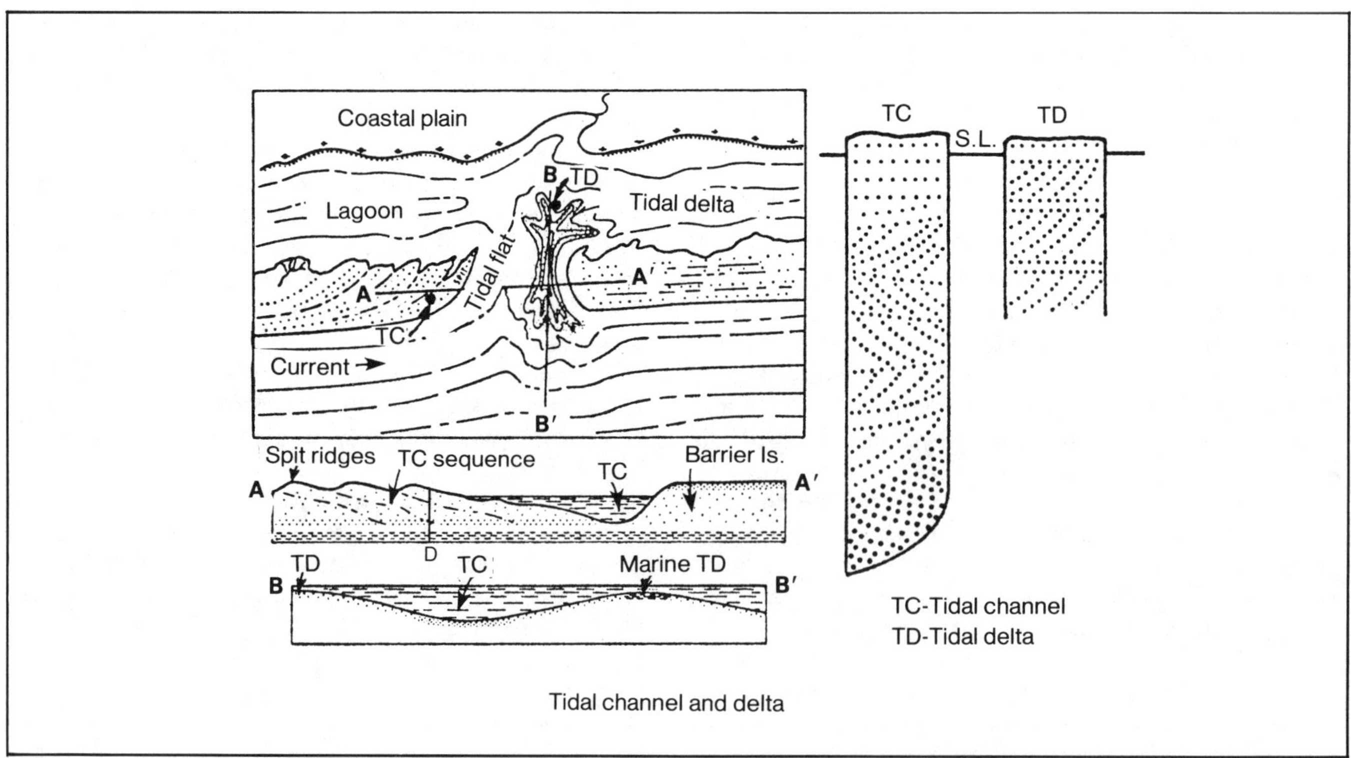

Fig. 9-12. Generalized sedimentary sequences of a tidal channel and delta. (From LeBlanc, 1972; permission to publish by AAPG).

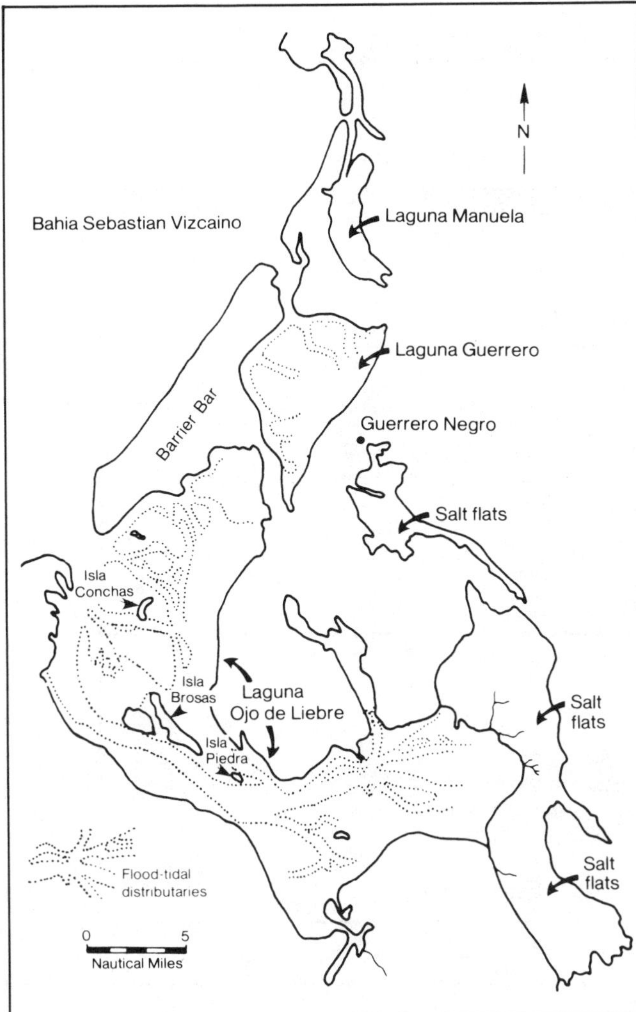

Fig. 9-13. Barrier bars and lagoons, Ojo de Liebre, Baja California. (From Phleger and Ewing, 1962; permission to publish by GSA).

the two major lagoonal channels. The depth of the tidal inlet ranges from 20 to 48 ft (6.0–14.6 m). The northern major channel behind the bar is 40–50 ft (12.2–15.2 m) deep, and the channel south of the inlet is approximately 40 ft (12.2 m) deep. These deeply scoured channels are attributed chiefly to a maximum tidal current of 2.5 knots, which in turn is the result of a tidal range of 4–9 ft (1.2–2.7 m). The water in the channels is quite turbulent.

A narrow strip of land connecting the barrier bar with the mainland separates an originally larger lagoon into two parts. The northern lagoon is Laguna Guerrero and the southern one is Laguna Ojo de Liebre. This lagoon "divider" consists of sand that was deposited where the flood tides, sweeping through the two tidal inlets, moved in opposite directions toward each other behind the barrier bar.

Phleger and Ewing (1962) write:

> Most sediment is fine to very fine, well-sorted and slightly skewed sand. The immediate source of the sand is the lagoon barrier, open ocean beach, and near shore zone . . . The average sand is appoximately 95 per cent quartz, 4 per cent dark minerals, and 1 per cent or less Foraminifera and shell . . . Processes of sedimentation include wind delivery of sand from the barrier and distribution by turbulent tidal currents. Most deposition occurs on intertidal flats and in the inner lagoon, due to loss of water turbulence. The channels next to the barrier tend to maintain themselves.

In explaining the barrier bar at Ojo de Liebre, Phleger and Ewing considered three conditions to be essential. They are (1) abundant supply of sand, (2) gently sloping foundation, and (3) wave action along an exposed coast. Most of the sand is distributed along the coast by longshore drift. Three stages are exhibited in Figure 9-14 showing the development of this regressive barrier bar. It is believed to have begun as a fringing beach on a gently sloping coastal plain when sea level was 40 ft (12.2 m) lower than at present. In Stage 1, sea level is shown at a position 30 ft (9.1 m) lower than at present. Phleger and Ewing (1962) state that:

> This depth is based on the presence of older sediment at about -40 feet in the inlets to lagunas Ojo de Liebre and Guerrero Negro . . . This older sediment is believed to be the foundation on which the present barrier was deposited. Sea level must have remained at 40 feet below the modern level long enough to establish an extensive beach . . .
>
> When the beach sand was above high tide, wind action began the formation of dunes, and the barrier grew upward. With the slow rise of sea level the barrier built up in the same location since a ridgelike foundation had been established against which sand could be piled. At Stage 2 . . . the sea was at its present level. A deposit of beach molluscs is found in the barrier approximately 1 mile inland from the present beach . . . The material in this beach has been dated by radiocarbon as 1800 ± 200 years B.P. At this time the barrier was 1 mile narrower than at present. Stage 3 is a diagrammatic representation of the present barrier.

Middle Sound Area, Onslow Bay, North Carolina. The Middle Sound area of Onslow Bay was studied by Miller (1962), who presents a barrier-bar-forming situation, that appears to have considerable application in the subsurface. The Atlantic coastal plain in this area, according to Miller, is a structurally stable depositional surface in which the shoreline is regressing as a result of deposition. Blanton (1963), however, disagrees regarding the stability of the depositional surface. He presents convincing evidence that the Middle Sound bars have persisted through as much as 15 ft (4.6 m) of subsidence. Blanton writes, "This creates a paradox of regressive sediments in a transgressive sea." The Middle Sound is an area of a 30 in. (760 mm) diurnal tide and a northeasterly flowing longshore current. A barrier bar parallels the shoreline and consists of a series of narrow, elongate islands with an average width of 1,050 ft (320 m) and a maximum height of 35-40 ft (11-12 m)

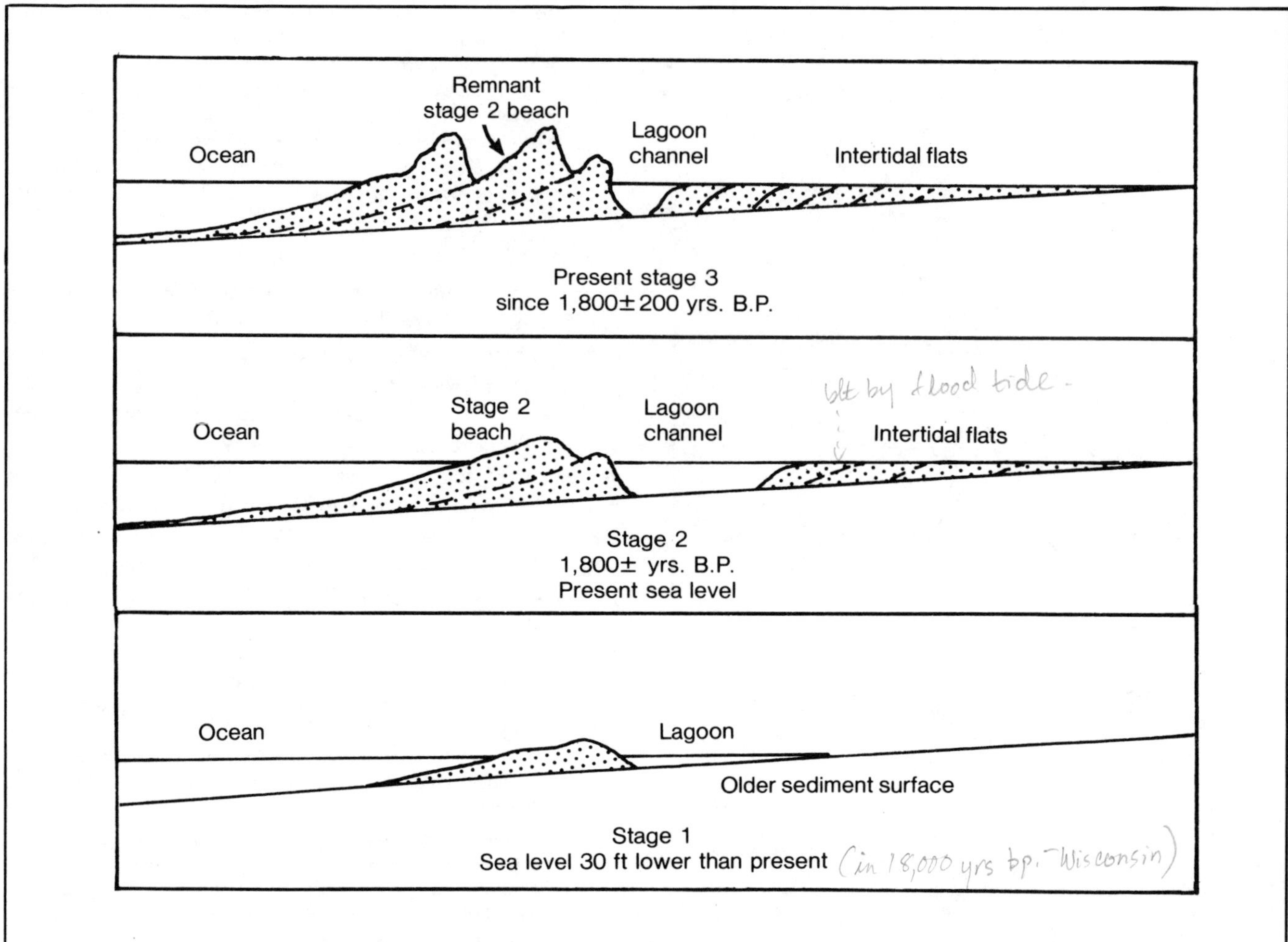

Fig. 9-14. Stages of barrier bar and lagoon development, Ojo de Liebre, Baja California, Mexico. (From Phleger and Ewing, 1962; permission to publish by GSA).

above sea level (Fig. 9-15). Miller (1962) pointed out that:

> With the exception of the strand-line deposits the sediments are extensively cross bedded in typical eolian fashion, well defined by the laminae of dark-colored heavy minerals . . . The steepest relief along the bar (from sediment surface to sediment surface) is on the shoreward side where it is backed up by a lagoonal channel. The seaward side of the bar slopes irregularly toward the beach through a series of low relief terraces. The thirty-inch tide has developed a beach approximately one hundred yards wide that is made up wholly of clean, fine and medium grained sand, shells, and shell fragments. With only minor exceptions, the dip (source direction) of the individual sand lentils of the beach is toward the sea. Localized drainage "guts" in the littoral zone create a gentle hill and valley relationship, with one or two feet of relief, at right angles to the shoreline . . .
>
> The inlet channels that separate the individual islands of the barrier bar are 500 to 1000 feet (152 to 305 m) wide and have a hard clay-packed substratum 10 to 15 feet (3–4.5 m) below mean sea level. There is little chance for detritus to be deposited in the inlet because of the nearly continuous currents that sweep through the opening during the flood and ebb of the tides. During flood tides, the surface water moves through the inlet at 10 to 15 miles (16 to 24 km) per hour (visual examination). Sand has accumulated in a deltaic pattern on both the marine and lagoon sides of the inlets in conjunction with this diminishing current flow.

Miller (1962) points out that the sands of the barrier bars thin and slope gradually in a seaward direction. The seaward gradient of the depositional surface ranges from 20 to 25 ft/mi (3.8–4.7 m/km) for a distance of 1.0–1.5 mi (1.6–2.4 km), which coincides with a water depth of approximately 40 ft (12 m). Beyond the 40 ft (12 m) depth the sand consists of thin veneer (measured in inches) lying on a hard, clayey base. The sand veneer has an average gradient of 2–3 ft/mi (0.38–0.57 m/km) and extends many miles out to sea to water depths of 50–60 ft (15–18 m). Miller postulates that this veneer, or apron, of sand is in

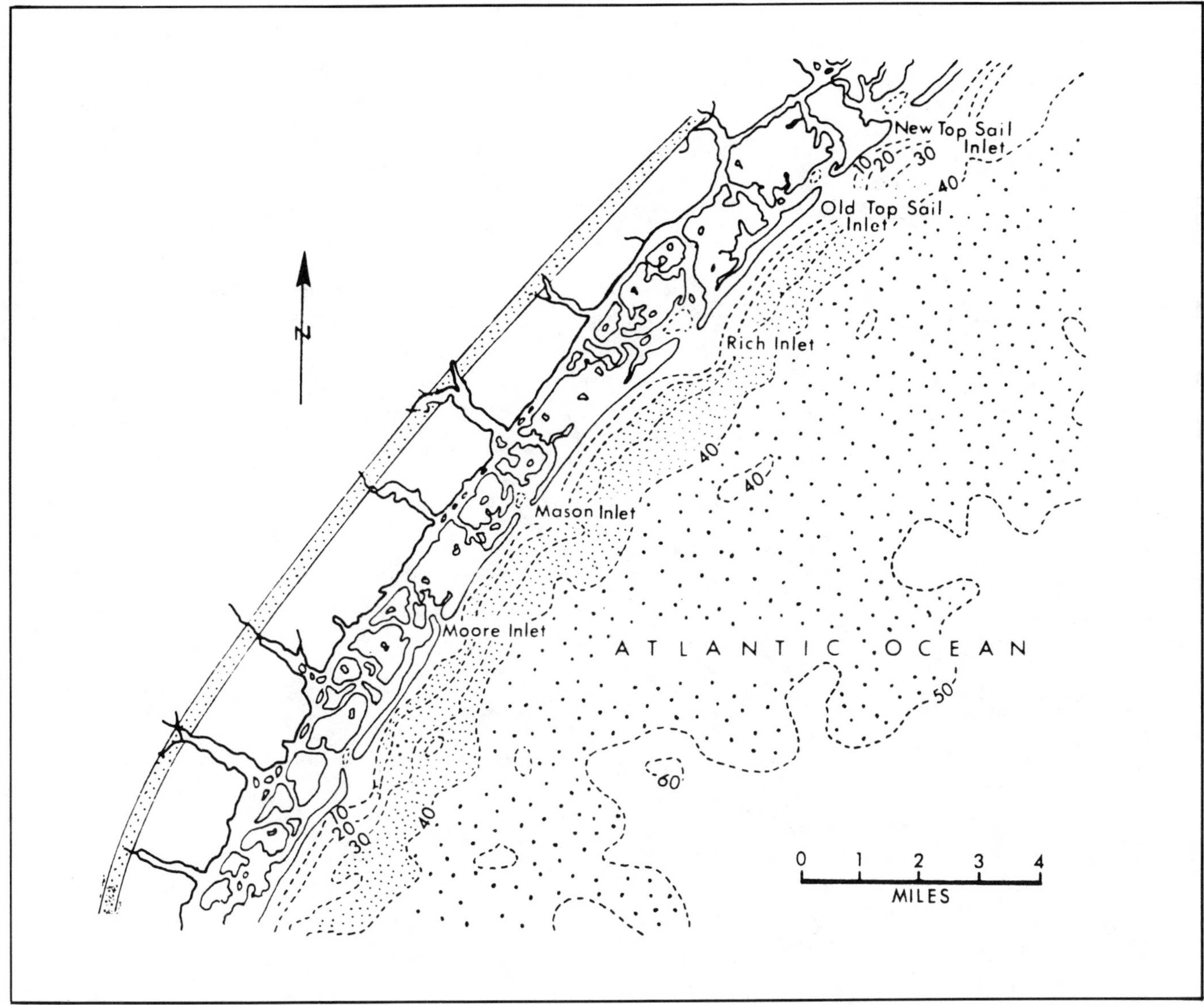

Fig. 9-15. Depositional pattern of shoreline features, Middle Sound, North Carolina. (Modified after Miller, 1962).

transit and ultimately will be deposited in water depths of less that 40 ft (12 m). He makes special note of ". . . the reoccurrence of the 40 ft thickness (and depth) value which seems to be a critical height (see Bernard et al., 1962, and LeBlanc et al., 1961) as well as critical depth for shore line and bar deposits just as it has been throughout geologic time."

The dimensions of the depositional features of Middle Sound are illustrated in Figure 9-16. From this figure it is apparent that the width of the lagoon ranges from 1.1 to 1.7 mi. (1.8–2.7 km). Except for the lagoonal channels, it consists principally of intertidal mud flats and low, flat-topped, heavily vegetated islands. The islands consist of oyster reefs covered with a thin veneer of black mud. Marsh grasses and reeds grow in this mud. The principal sediment of the lagoon is a black organic mud. Thin, discontinuous patches of sand are present along the sides of the distributary channels.

Miller postulates that the position and width of the part of the barrier bar above sea level are established and controlled by the profile of equilibrium. He writes:

> . . . in general (and within limits), the steeper the submarine gradient, the narrower the bar. If the gradient becomes too steep, no bar is developed because the sediments are distributed laterally. If the gradient becomes too shallow, the bar reverts to a broad flat beach that accretes seaward . . . The profile in turn is established in response to the amount and type of detritus available and the ability of the currents and

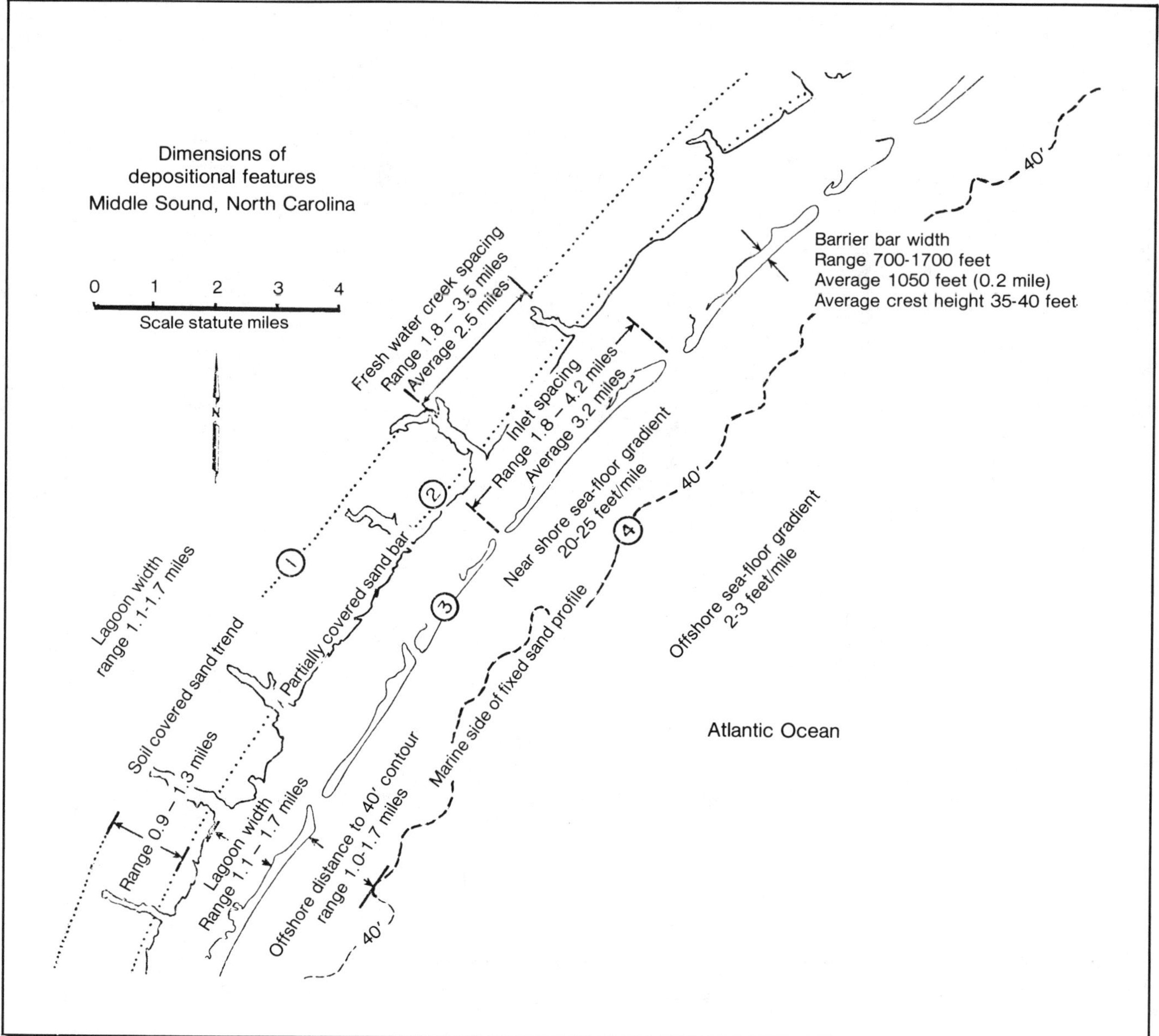

Fig. 9-16. Dimensions of depositional features shown in Figure 9-23, Middle Sound, North Carolina. (Modified after Miller, 1962).

> wave action to transport the detritus . . . hurricanes and other violent disturbances . . . alter the pattern periodically, but for the most part, the principal direction of current movement is from the deltas to the sea and then from the sea to the barrier bar. From a practical standpoint, the beginning and the end of each significant depositional pattern is directly related to changes in position of sea level with respect to river deltas.

The positions and trends of two soil-covered ridges on the shore can be seen in Figures 9-15 and 9-16. Miller considers these ridges to represent earlier stages of barrier-bar development. He infers that the barrier bar at Middle Sound represents, ". . . a single stage in an overall cyclic process of "shore line jump" in which each new mature bar-lagoon complex will add 1.5 miles (or progressively less near the deltas) of new shore to the coastal plain. In other words, the present barrier bar was preceded by similarly trending bars in the past and will be succeeded by more barrier bars in the future. . . ."

Figures 9-17 and 9-18 are his diagrams of the stage of development of a regressive barrier-bar sequence. In Figure 9-18 the plus signs over the deltas indicate an excess supply of sediment. The middle part of the embayment has not yet reached equilibrium and, therefore, is designated with a

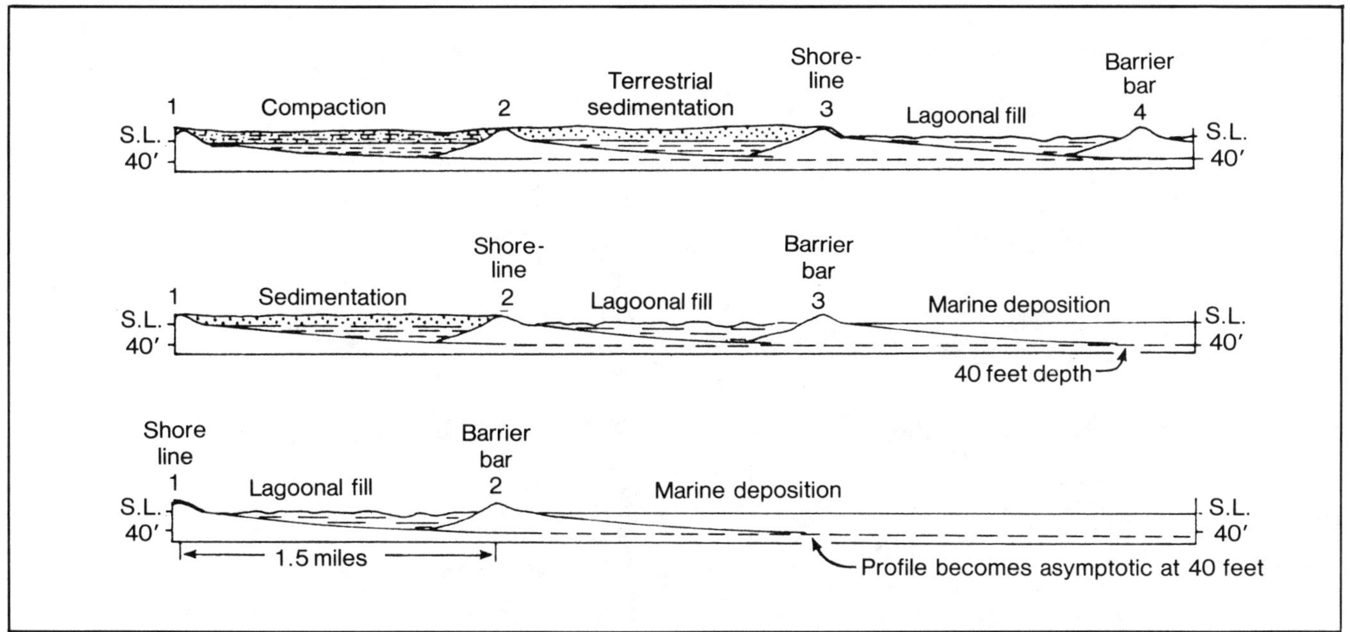

Fig. 9-17. Diagrammatic cross sections, showing stage development of regressive barrier bar sequence. (Modified after Miller, 1962).

minus sign. When equilibrium ultimately is reached in the middle of the embayment, ". . . a new and younger bar will begin to form at the delta and the process will be repeated as shown."

Miller was well aware that his multiple-bar hypothesis does not necessarily apply along all shorelines because of the numerous variables involved in the establishment of an equilibrium condition. The authors believe that Miller's ex-

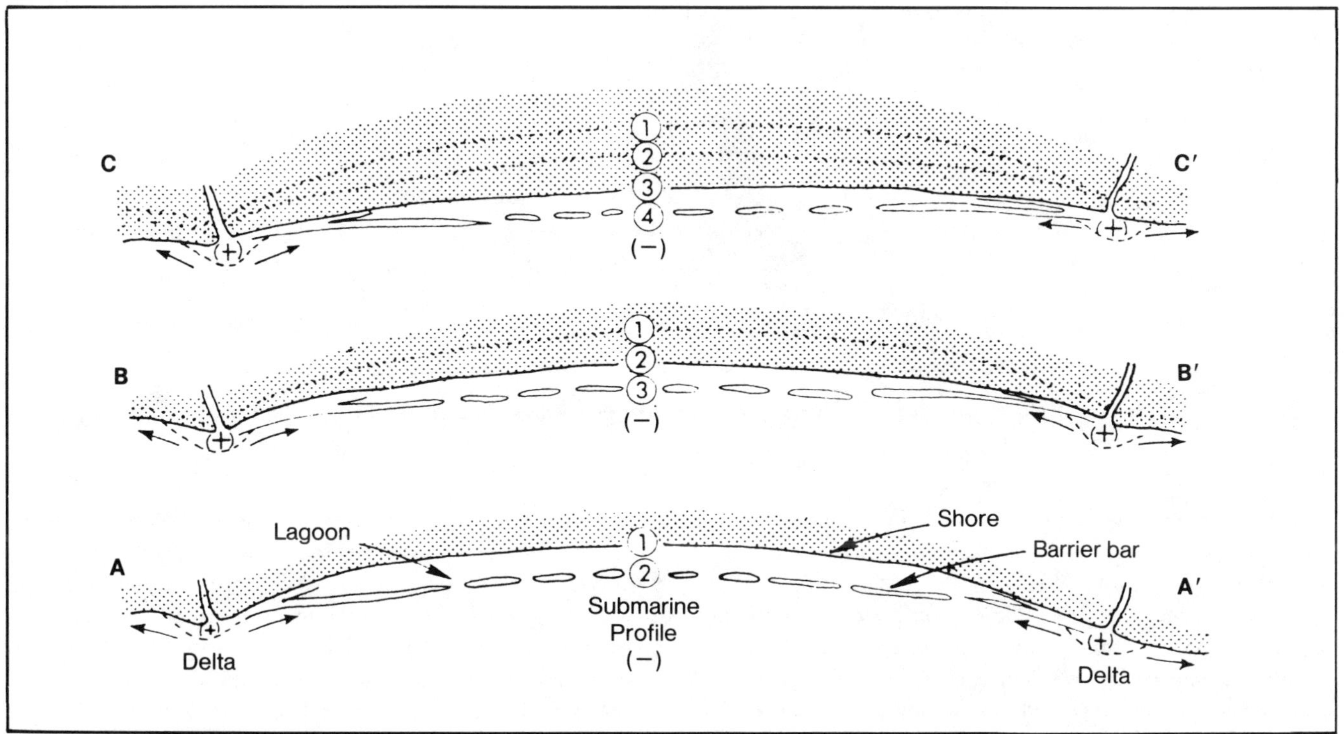

Fig. 9-18. Diagrammatic plan view of stage development of regressive shorelines. (Modified after Miller, 1962).

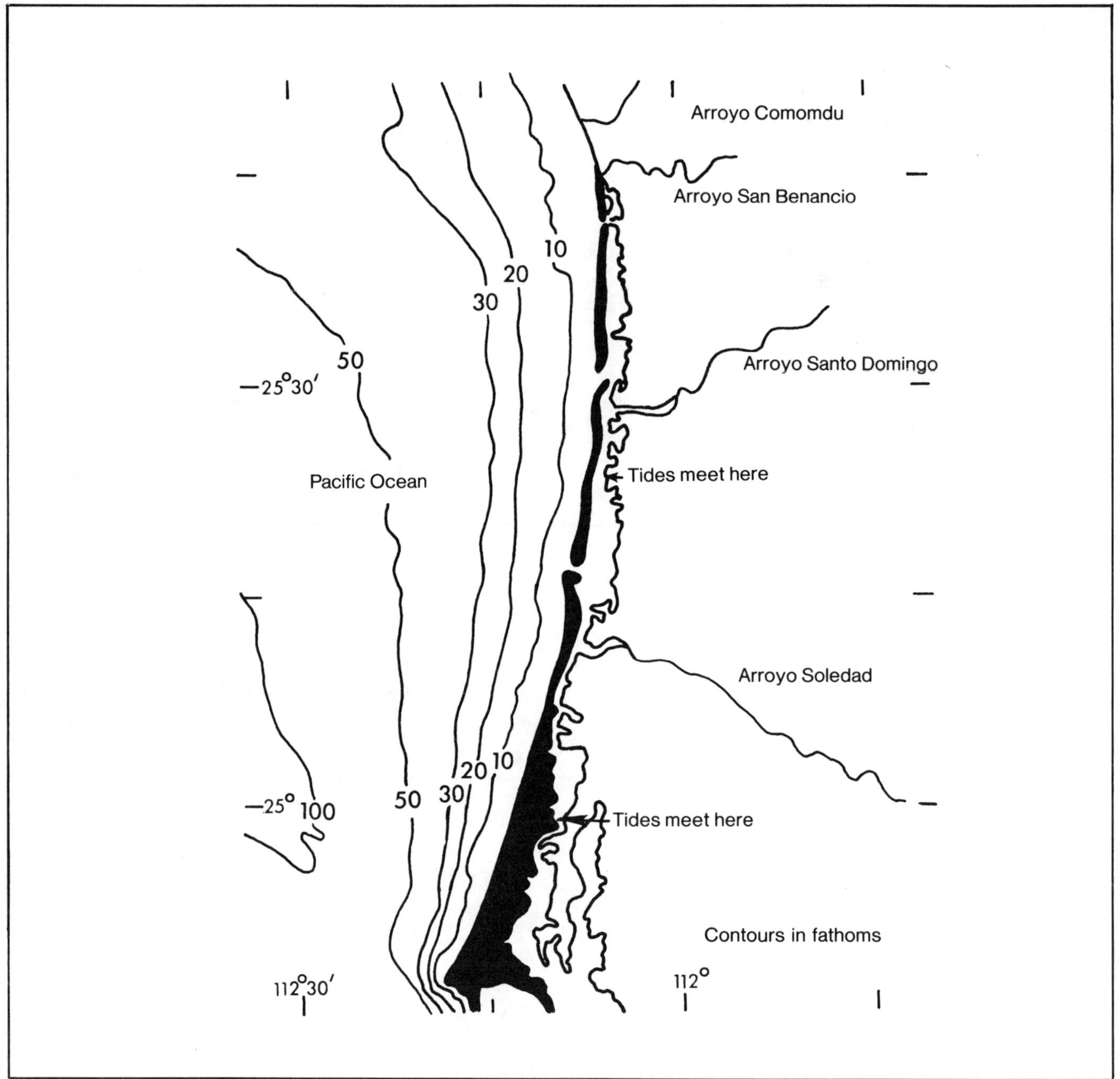

Fig. 9-19. Laguna Santo Domingo, Baja California, Mexico. (Modified after Phleger and Ewing, 1962; permission to publish by GSA).

planation for multiple lenses of marginal-marine sands has considerable application in subsurface stratigraphic analysis.

Laguna Santo Domingo, Baja California, Mexico. An extensive barrier beach is located approximately 200 mi (320 km) south of Ojo de Liebre in an area known as Laguna Santo Domingo (Fig. 9-19). This barrier-bar complex averages 0.5 mi (0.8 km) in width except near the southern end, where it is much wider. Barchans ranging from 40 to 50 ft (12.2–15.2 m) high are present on the upper surface. The barrier bar is bordered by relatively deep channels on the west side of the lagoon. The lagoon ranges from 0.5 to 2 mi (0.8–3.2 km) in width and, except for the deep channels, consists of tidal flats and marsh areas, bordered by mangrove. Phleger and Ewing (1962) report that "at most stations the sediment is fine sand with good sorting and little skewness . . . All samples contain significant quantities of apatite. An analysis of the lagoon barrier sand made by a mining company revealed an average of approximately 4 per cent apatite in the upper 20–40 feet of sediment." This barrier beach appears to be in an earlier stage of development than that of Ojo de Liebre. The absence of "divider" ridges of sand at the several positions where flood tides meet behind the barrier is evidence for this observation. This barrier bar is "tied" to the mainland to the south, where it exhibits its maximum width, and tapers northward in the direction of longshore drift. The southern one-third has an extremely irregular lagoonal margin, whereas the northern two-thirds has a much smoother margin. In several respects this barrier bar serves as the modern analogue for its ancient subsurface counterpart, namely, the "Second Berea" barrier bar of Early Mississippian age in central Ohio.

"Second Berea" Barrier Bar, Ohio. An excellent subsurface example of a barrier bar is present in east-central Ohio, extending from Gallia County north to Muskingum County (Fig. 9-20). This barrier bar is composed of "Second Berea" sandstone. It borders the eastern part of a large, south trending deltaic wedge of redbeds (Bedford), which also are Early Mississippian. Both this delta and the "Second Berea" sandstone are described and illustrated by Pepper et al. (1944) in a classic paper. The remarkable part

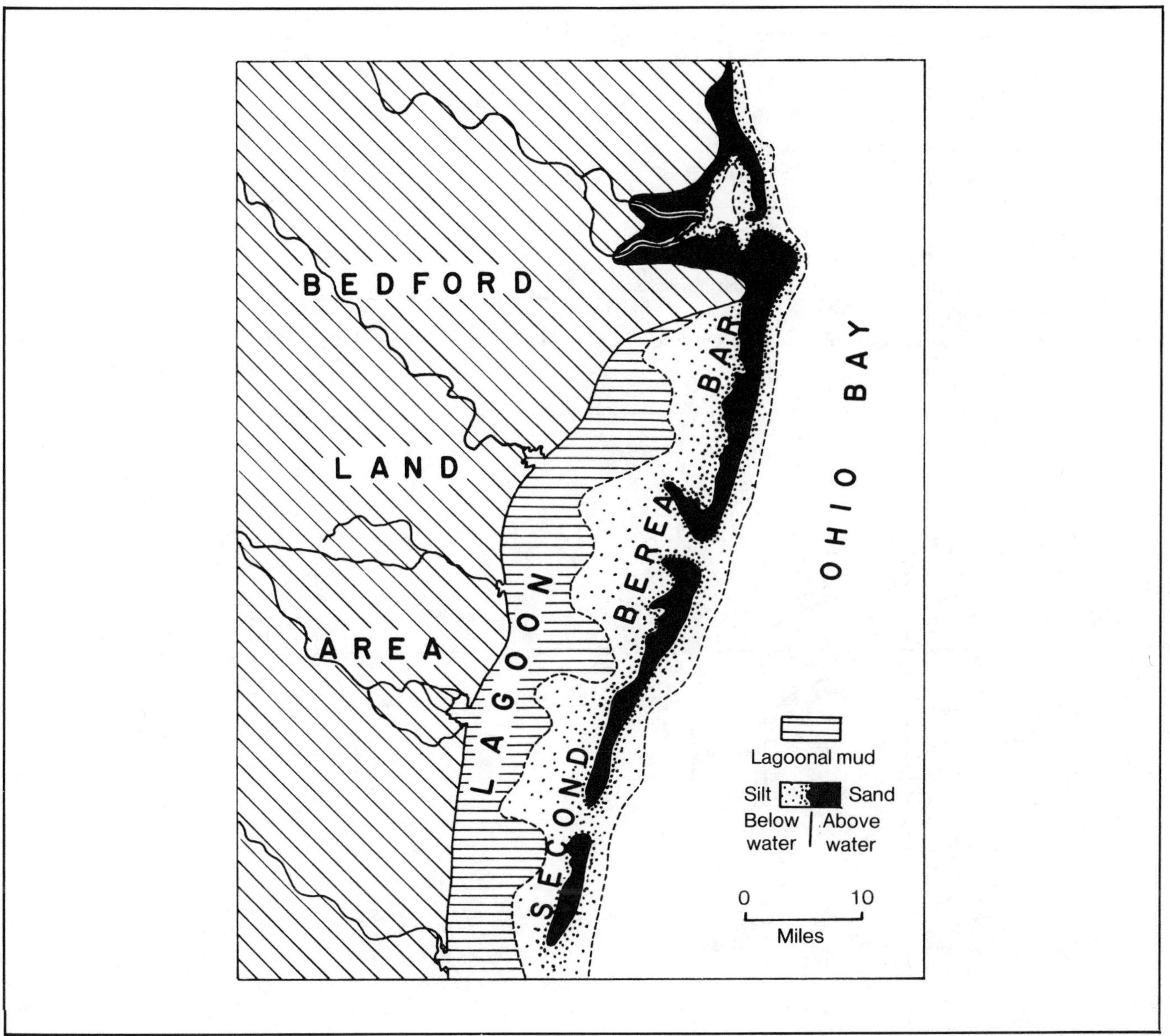

Fig. 9-20. "Second Berea" bar during time of Bedford deposition. (Modified after Pepper et al., 1944; permission courtesy of U. S. Geological Survey).

of this study is that it was based almost entirely on drillers' logs and outcrop data because almost no electric-log data were available for this area at the time of the investigation. They used more than 1,300 drillers' logs in outlining the shape, contouring the variations in thickness, and plotting cross sections of this barrier bar and described the "Second Berea sand" (Pepper et al., 1944) as:

> . . . an ancient shoreline deposit strongly resembling a modern barrier bar and lagoonal deposits, such as those along the present Gulf Coast of Texas. The Second Berea sand body is about 85 miles long, 3 to 15 miles wide. The eastern margin which is relatively straight represents the seaward side of an offshore bar and the irregular western margin, on the other hand is the result of unequal deposition in the lagoon on the landward side of the bar.

Over much of the area of its development, the "Second Berea" barrier bar is less than 20 ft (6.1 m) thick but locally exceeds 30 ft (9.1 m). The base is flat and the upper surface is convex. It is attached to the Bedford delta at its north end (Fig. 9-20) where one of the distributary systems of the delta discharged its sediment. To the north of this area of discharge, a south-trending spit approximates a baymouth bar in that it isolates a restricted lagoon environment. A tidal inlet is conspicuous in the central part of the barrier bar. On either side of this inlet the sandstone of the bar bends abruptly back into the lagoon. A tidal delta with diverging fingers presumably fanned out into the

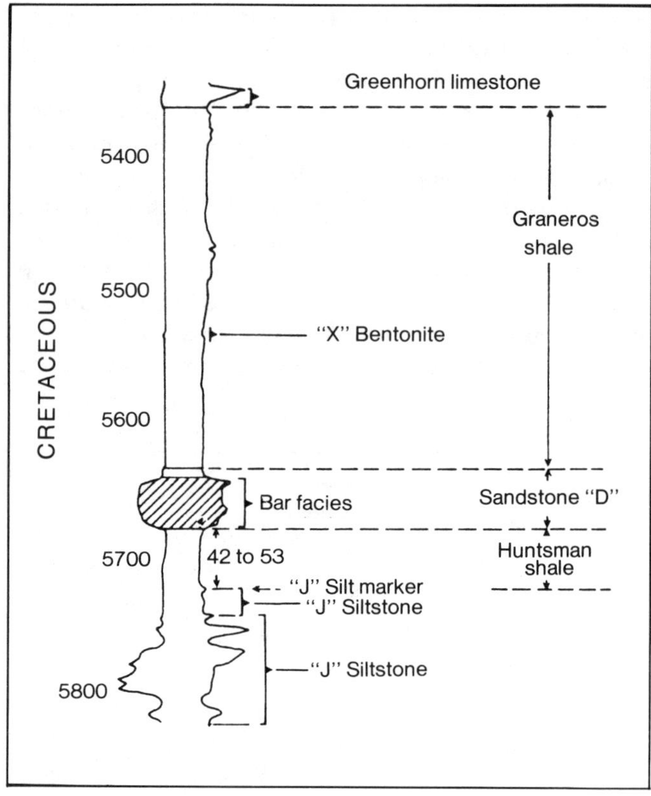

Fig. 9-21. Typical electric log in Saber field, Colorado. (Modified after Griffith, 1966; permission to publish by AAPG).

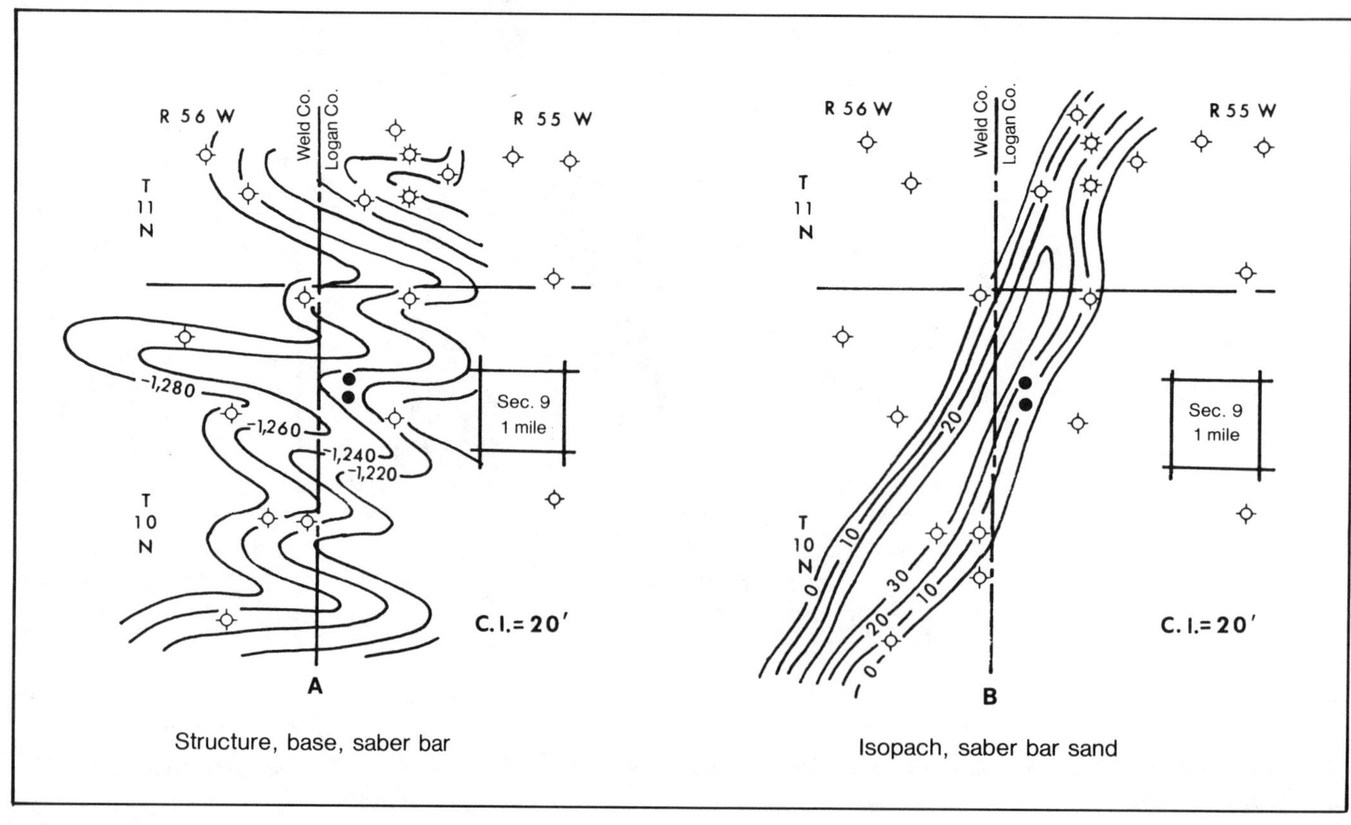

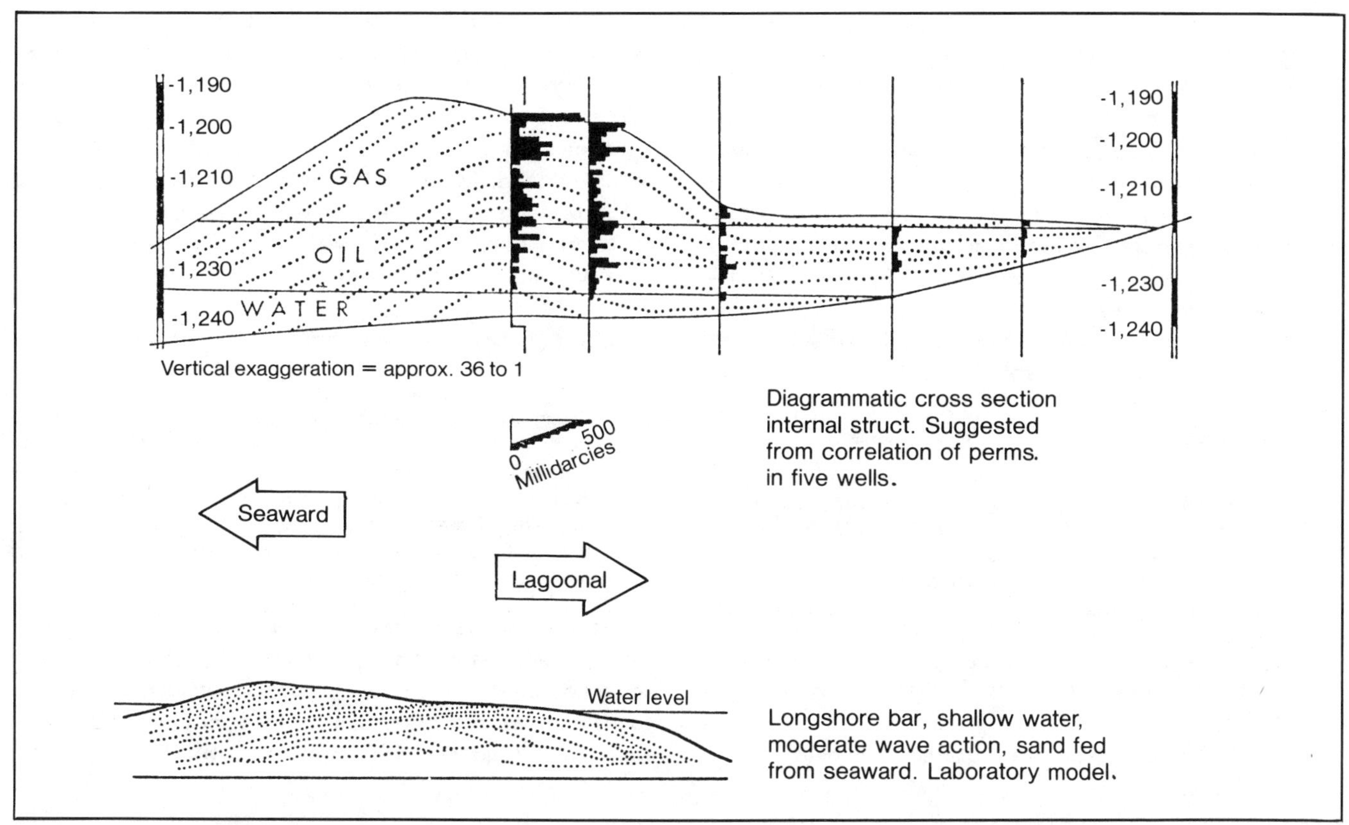

Fig. 9-23. West-east cross section across Saber bar (upper) showing internal structure and external shape. Compare with bar produced in laboratory (lower) by McKee and Sterret (1961). (Modified after Griffith, 1966; permission to publish by AAPG).

lagoon. Another tidal inlet may be noted approximately 30 mi (48 km) to the south.

The thicker parts of this barrier bar have produced considerable quantities of gas and a minor amount of oil. Pepper et al. (1944) pointed out,

> Production of natural gas in the Second Berea is controlled both by the thickness of the sand body and by the grain size of the sand. Normally, the best gas wells occur in the thicker sand. However, in the thick sand east of the crest of the bar a high percentage of wells have been dry, probably because the grain size is small and because the sand contains a high percentage of silt. On the west or lagoon side of the crest, however, production is sometimes obtained in as little as 10 feet of sand, but seldom in less than 10 feet. Northwest of Morgan Township, Morgan County, in the area described as a bay enclosed by a spit and along the northward extension of the oil mainland shore, the sand is rarely productive.

In this bay the sand is exceptionally silty and very low in permeability.

Saber Barrier Bar, Colorado. Griffith (1966) presents an instructive analysis of the Saber barrier bar, which is developed in the Upper Cretaceous "D" sandstone on the westward-sloping east flank of the Denver-Julesburg basin. The mapped part of this feature is about 10 mi (16.0 km) long and approximately 1 mi (1.6 km) wide. The maximum thickness is about 45 ft (13.7 m). The only oil-productive parts of this barrier bar are where west-plunging structural noses cross its depositional trend. The base ranges from 42 to 55 ft (12.8–16.8 m.) above the "J" siltstone marker, and the interval between is called the Huntsman shale (Fig. 9-21). This fairly uniform thickness of shale above a lithologic reference datum indicates that the base of the overlying barrier bar was essentially flat at the time of deposition. Griffith states that "A subsurface structural map was made on a datum described as the base of the sandstone bar or its correlative away from the bar," and that the illustration (Fig. 9-22A) ". . . shows the topography of the sea floor on which the bar was deposited." The essentially flat base of a barrier bar is controlled by wave base, and

Fig. 9-22. **A.** *Structural contour map of base of Saber bar; contours in ft.* **B.** *Isopach map of Saber bar; contours in ft.* (Modified after Griffith, 1966; permission to publish by AAPG).

if Figure 9-22A represents topography on the seafloor, the sandstone body should be roughly parallel with the sinuous trends of the structural contours of this figure. Rather than representing topography, however, Figure 9-22A actually represents the present structure of the base of the Saber bar; the structure was superimposed in post-Cretaceous time. Griffith further states, "The base of the sea floor on which the bar was deposited appears to be relatively smooth, gently sloping surface," an observation with which the authors agree completely.

The Saber bar is one of the few examples of subsurface barrier bars for which permeability studies of cores have been used to identify and correlate internal structural features, as shown in Figure 9-23. Griffith states,

> Within the Saber bar, the presence of several layers can be inferred by correlating intervals of similar permeability between wells (all of which were cored). The bar graphs next to each of the five wells on the cross section illustrate both the vertical and horizontal changes in permeability that occur. Permeability generally increases toward the upper surface and the seaward side of the bar. Whereas lateral changes are gradational, vertical changes are abrupt and emphasize the marked difference in energy levels at the time of deposition of various layers.

Bisti Barrier Bars, New Mexico. The Bisti field, of the San Juan basin of New Mexico, is an excellent example of an ancient barrier bar complex. This field consists of a series of three closely related bars, which Sabins (1963) referred to as the "Bisti bar complex." The oil field itself is approximately 3 mi (4.8 km) wide and more than 30 mi (48.3 km) long. Estimated recoverable primary reserves are 1,100 bbl/acre, and estimated secondary reserves are 900 bbl/acre. Accumulation is strictly stratigraphic; the strata have a regional northeastward homoclinal dip of approximately 75 ft/mi (14.2 m/km).

Sabins (1963) points out that,

> The Bisti stratigraphic trap is a sand bar complex developed during one of the regressive and transgressive sedimentary cycles that characterized Late Cretaceous sedimentation of the San Juan basin. Throughout most of Late Cretaceous time the San Juan basin was a marginal embayment of the sea that covered the Western Interior. It was not until the end of Cretaceous time that the San Juan structural basin formed as a result of marginal tectonic uplifts.

The trap itself is made up of three individual barrier bars, which Sabins designates as the Marye, Huerfano, and Carson sandstones. The principal producing member is the Marye sandstone, which ranges from 1 to 2 mi (1.6–3.2 km) in width, has a maximum thickness of 40 ft (12.2 m), and is

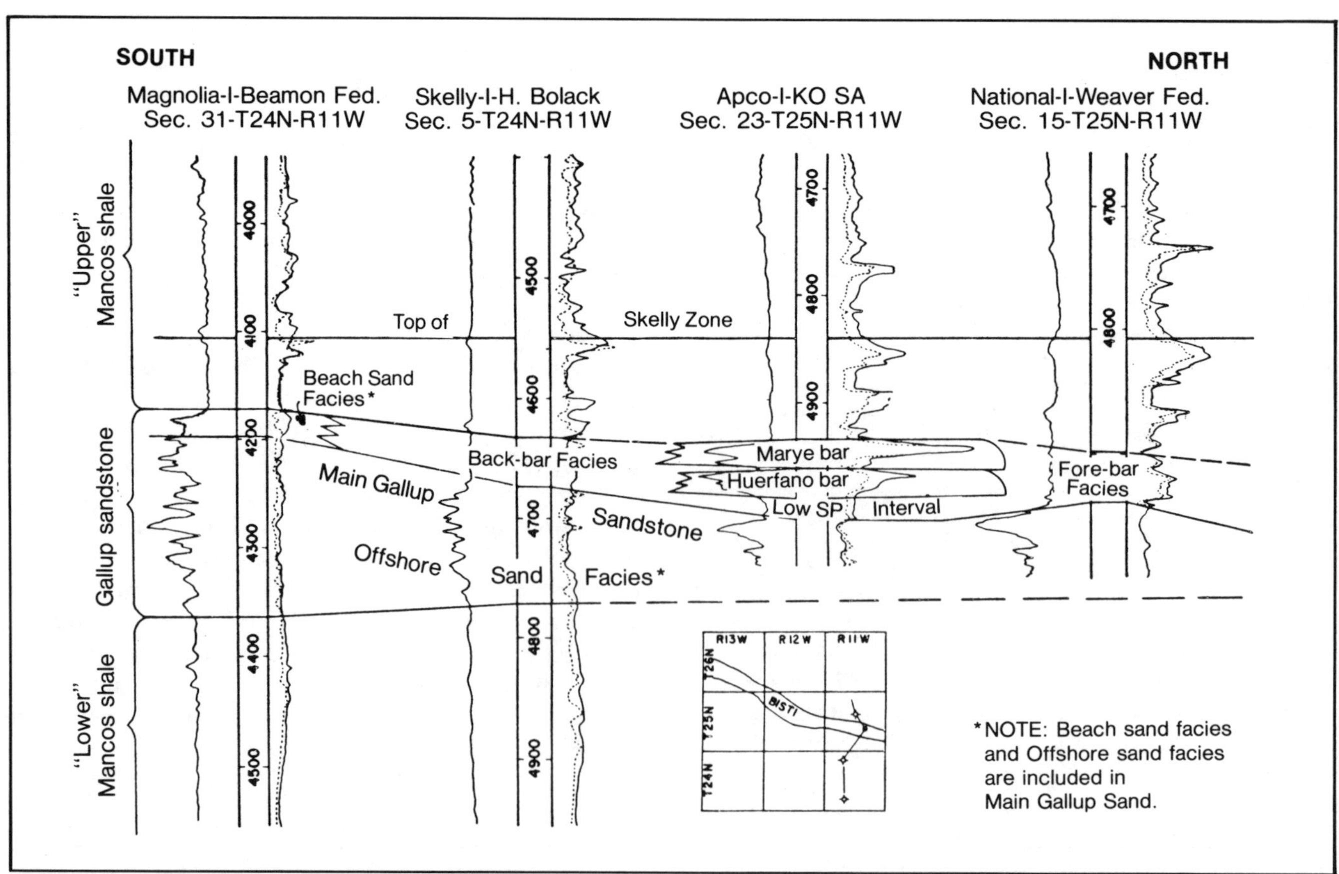

Fig. 9-24. Stratigraphic nomenclature in Bisti field. (Modified after Sabins, 1963; permission to publish by AAPG).

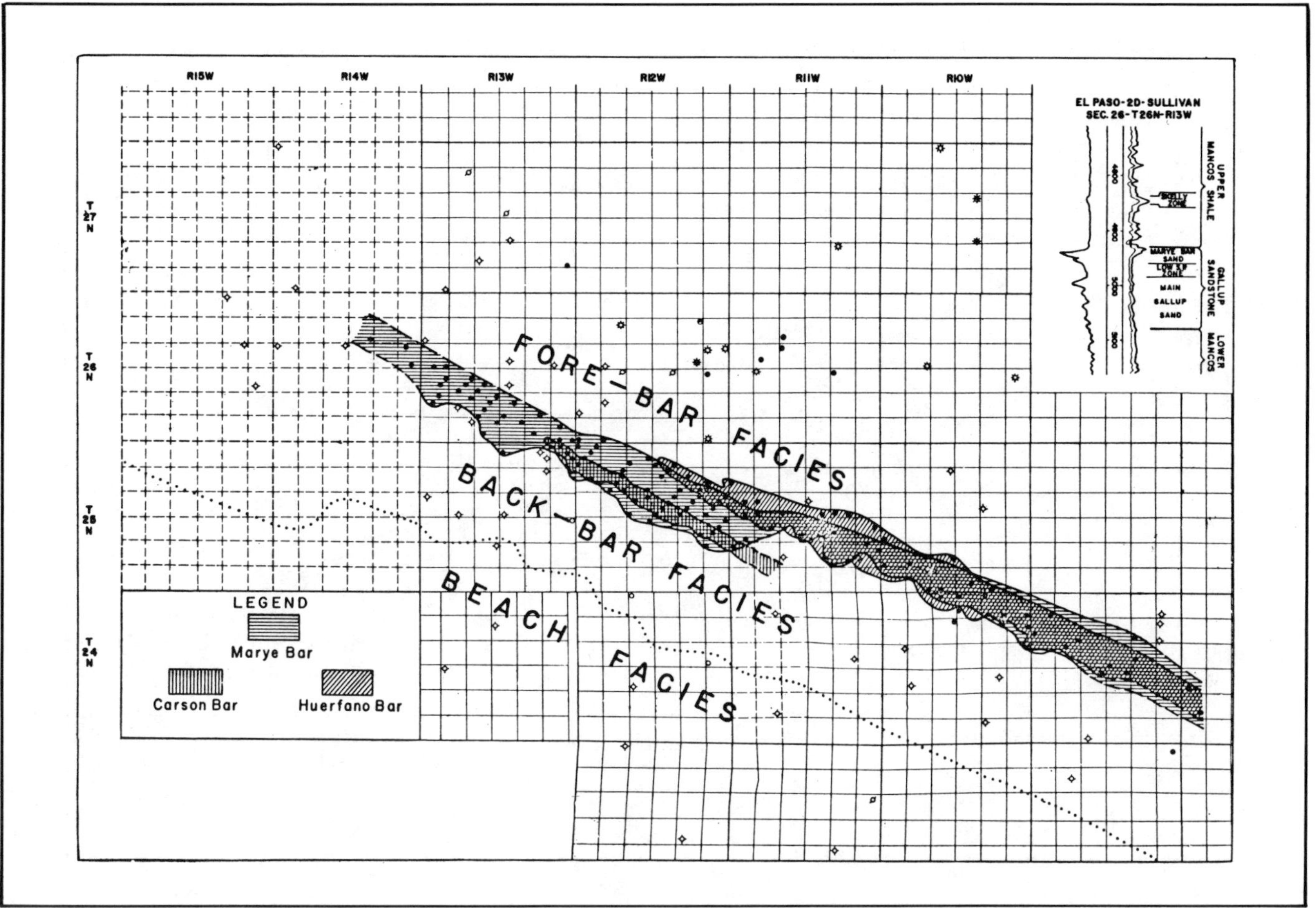

Fig. 9-25. Outlines of effective bar sandstones. (From Sabins, 1963; permission to publish by AAPG).

more than 30 mi (48.3 km) long. The Huerfano sandstone member underlies the Marye sandstone and is separated from it by a thin shale (Fig. 9-24). It is limited to the southeastern part of the field area, as shown in Figure 9-25. It is less than 30 ft (9.1 m) thick, approximately 2 mi (3.2 km) wide, and about 24 mi (38.6 km) long. The Carson sandstone is present only in the central part of the Bisti field. It is less than 30 ft (9.1 m) thick and less than 1 mi (1.6 km.) wide, and has a known length of 9 mi (14.5 km). It is situated southwest of, and trends parallel with, the Huerfano bar, and the two bars are at the same stratigraphic position relative to the underlying main Gallup sandstone. An isopach map of all bar sandstones of the Bisti field is shown in Figure 9-26.

Sabins' study is of particular value in subsurface studies of barrier bars because he presents core thin-section criteria for distinguishing bar facies, forebar, backbar, and beach sand facies. These thin-section criteria, which include texture, mineralogy, and paleontology, are summarized in Table 9-1 (See Sabins for more detailed tables of criteria.)

In summarizing his observations regarding the Bisti bar complex, Sabins states,

These petrographic characteristics constitute criteria for recognizing the individual facies. This makes it possible to discriminate facies that have identical megascopic and electric log characteristics. For example, at Bisti the high glauconite content of Bar sands distinguishes them from the similar appearing Beach sand facies which lacks glauconite. The open marine assemblage of Inoceramus, collophane, and calcite-filled planktonic Foraminifera in the Fore-Bar facies distinguishes it from the Back-Bar facies which contains a restricted marine fauna of pyrite-filled benthonic Foraminifera.

Vertical gradients of sandstone texture and primary dolomite grain distribution are valuable for reconstructing the depositional history. These gradients, plus paleotopographic mapping show that the Gallup sedimentary cycle in the Bisti area began with regressive deposition of the off-shore sand facies. Northwest-trending longshore bar and trough topography was formed on the upper surface of the off-shore sand unit A pulse of subsidence and possibly a minor disconformity account for the sharp contact between the off-shore sand facies and the overlying Low SP facies. Wave action concentrated at a seaward-facing break in slope then winnowed the clay and concentrated the sand of the Low SP facies to form the Bisti bar complex. On the protected shoreward side the restricted marine Back-Bar facies was deposited and on the seaward side the open Marine Fore-Bar facies accumulated. A basinward pulse of subsidence resulted in the burial of all these Gallup

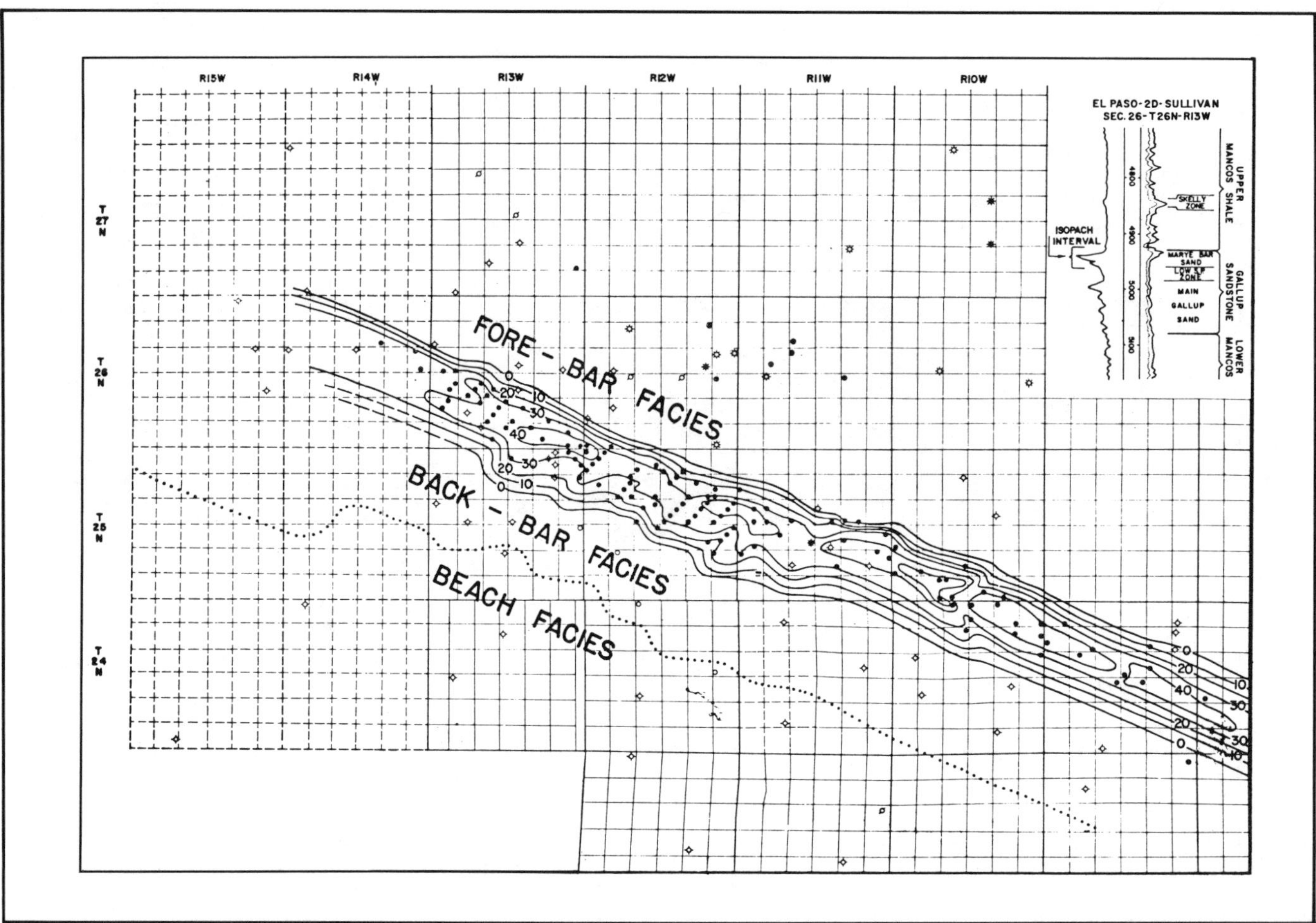

Fig. 9-26. Isopach map of all bar sandstones of Bisti field, based on selfpotential curve of electric log. Isopach interval 10 ft. (Modified after Sabins, 1963; permission to publish by AAPG).

facies beneath the transgressive "Upper" Mancos Shale.

Thus the Bisti stratigraphic trap was formed and preserved by marine depositional processes, which may be deciphered through detailed study of the rocks.

Transgressive Barrier Bars

In the past 15 years, numerous detailed studies have been conducted along the Atlantic seaboard of the eastern United States where a gradual, relative, sea-level rise is occurring. This results in a transgressive shoreline along which transgressive barrier bars may develop. In more local areas, shoreline erosion also occurs under relatively stable sea level conditions under intensive storm conditions. Klein (1974) suggests transgressive sequences have a low preservation potential relative to regressive sequences. This is particularly true along the New England coastline where narrow barrier bars develop during the summer months (under normal wave-energy conditions) and are then severely eroded by storm waves during the winter months. In fact, they may be completely destroyed as they migrate landward and blend with the sediments of the upper shoreface. The transgressive sequence resulting from the landward migration of such a narrow ridge and runnel during a storm is illustrated in Figure 9-27. Reinson (1979) states, "Such a sequence results when storm waves erode the beachface, removing sediment to the shoreface. The sediment is returned to the beach during the post-storm recovery, in the form of a ridge and runnel (bar and trough), which develops on the low tide terrace just seaward of the foreshore (Davis et al., 1972; Owens and Frobel, 1977). The ridge migrates shoreward eventually welding onto the beachface and creating a distinctive sequence of upper shoreface-foreshore deposits."

Kraft and John (1979) state that:

> Probably no precise set of parameters or sequences can be set to identify a transgressive or regressive barrier from a single point of control . . . Recognition of lagoon, estuarine, and nearshore shelf muds by microfauna or other means is the only way in which transgression or regression can be proved

TABLE 9-1. Summary of Petrographic Characteristics, Bisti Area[1]

	Median Grain Size	Maximum Grain Size	Clay Matrix Content	Dolomite Grain Percent	Glauconite Content	Benthonic Forams	Planktonic Forams Collophane Inoceramus
Main Gallup Sandstone							
Beach sand	Medium to fine	Very coarse	Low	2 to 22	Trace to present	Trace	Absent
Offshore sand	Very fine	Medium to fine	Moderate	15 to 30	Absent to trace	Present	Trace
Low-SP facies	Fine to very fine	Very coarse to medium	Very high	5 to 20	Trace to present	Trace	Absent
Bar sands	Medium to fine	Very coarse to coarse	Low, but increase toward base	<5 restricted to basal part	Very abundant through-out	Trace	Absent, except on seaward flank
Backbar facies	Fine to very fine	Coarse to medium	High	5 to 15	Trace to present	Very abundant	Absent, except for displaced fragments
Forebar facies	Fine to very fine	Medium to fine	High	8 to 25	Trace to present	Common	Very abundant
"Upper" Mancos Shale	Very fine	Medium to fine	High	5 to 30	Absent to trace	Common	Very abundant

[1]After Sabins (1963).

> from a single point of control or stratigraphic section. . . . transgressions and regessions must be identified from local and regional data used to contrast facies models that allow for extremes of variation in vertical and lateral facies sequences. Time units cross lithostratigraphic units in a transgression and regression (Fig. 9-28). Recognition of this factor is critical in stratigraphic correlation of coastal facies.

The principal difference in distinguishing a transgressive from a regressive barrier bar lies in the relationship of the lagoonal facies. Lagoonal facies underlie or occur within the lower or middle portions of the transgressive barrier bar, but overlie the sand facies of the regressive barrier bar. Idealized stratigraphic successions for regressive and transgressive barrier bars and the barrier inlet are illustrated in Figure 9-29. It is only when one or more cores are available that such "end-member" models can be applied.

Washover Fans. Washover fans occur abundantly on the lagoonal side of a barrier bar and are the principal factor contributing to the escaloped margin on the landward side. They are particularly significant on barrier bars developed in microtidal environments, and they present one of the main mechanisms in the landward migration of transgressive barrier bars. Such deposits generally are thin and range up to 2 m in thickness for each washover event. They are semi-circular, tabular bodies up to a few hundred meters in radius. Where they coalesce, they make extensive washover flats that furnish source material for eolian transport of sand. Schematic cross sections of two washover fans are illustrated in Figure 9-30. Two principal types of sedimentary structures are planar stratification and delta foreset strata. Andrews (1970) and Schwartz (1975) have noted the occurrence of textural and heavy mineral laminations and graded bedding, depending on the nature of the source material. The bulk of the washover deposits, however, consists of fine-grained to medium-grained sand.

Tidal Inlets and Deltas

Tidal inlet deposits and tidal delta deposits are extremely important features of the barrier bar complex. Modern analogues have been studied by Land (1972), Kumar and Sanders (1974), Hayes (1975, 1976), Hubbard and Barwis (1976), Hayes and Kana (1976), Reinson (1977, 1979), Hobday and Horne (1977), Horne and Ferm (1978), Barwis and Makurath (1978), Carter (1978), Armon (1979).

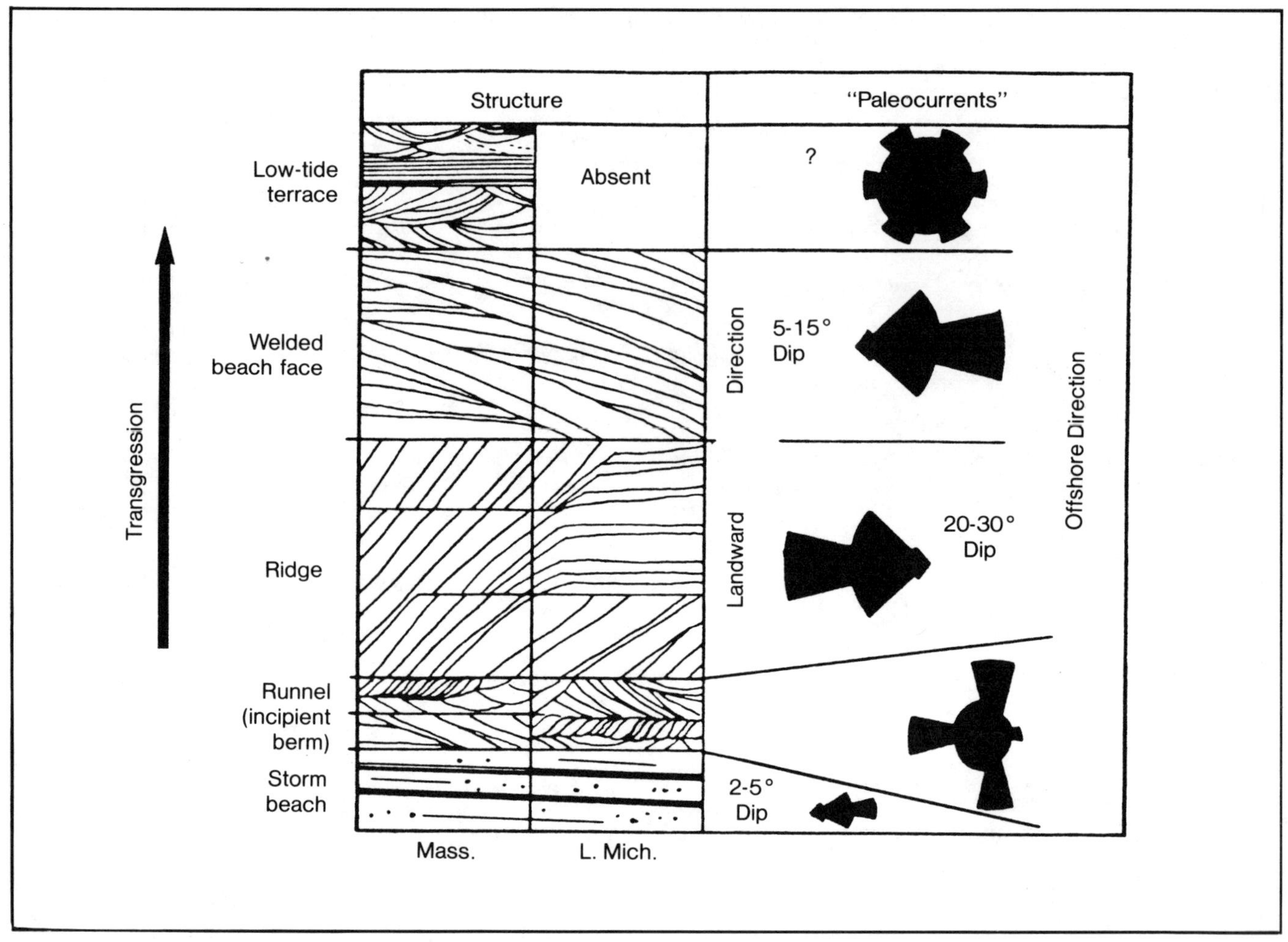

Fig. 9-27. Transgressive sequence formed by the landward migration of ridge-and-runnel during beach constructional phase. Vertical sequence would be about 1 m thick. (From Davis et al., 1972).

Their studies serve as a basis for a better understanding of subsurface counterparts within a barrier bar complex. One or more cores generally are essential to accomplish this.

Tidal Inlet Deposits. Tidal inlets seldom are stationary but rather migrate laterally in the prevailing direction of longshore drift. Simultaneous with this migration, there is erosion on the downdrift side of the inlet and accretionary spit deposition on the updrift side. The significance of the accretionary (channel) deposition is diagrammatically illustrated in Figure 9-31. From this figure, it may be seen that a significant "stretch" of the barrier bar (on the updrift side) has been completely reworked and therefore has a strikingly different internal structure than that of the orginally deposited material. As the inlet migrates laterally, flood- and ebb-tidal delta deposits shift in the same direction. The thickness of the inlet channel sand is the same as the depth of the inlet. Beach, dune, and washover deposits on top can increase the overall thickness. Reinson (1979) states that such sedimentary channel sequences . . . "have the following general characteristics: 1) an erosional base often marked by a coarse lag deposit, 2) a deep channel facies consisting of bidirectional large-scale planar and/or medium scale trough cross beds, 3) a shallow channel facies consisting of bidirectional small- to medium scale trough cross beds and/or plane-beds and "washed-out" ripple laminae; 4) a fining-upward textural trend and a thinning-upward of cross bed set thickness."

One should look for these features in a core. The fining-upward sequence frequently may be noted by a bell-shaped SP-GR curve on mechanical logs of wells drilled through such a sequence.

Tye and Moslow (1983) are of the opinion that 40% or more of a barrier bar sequence is tidal inlet related and therefore exhibits fining-upward sequences. Based on their

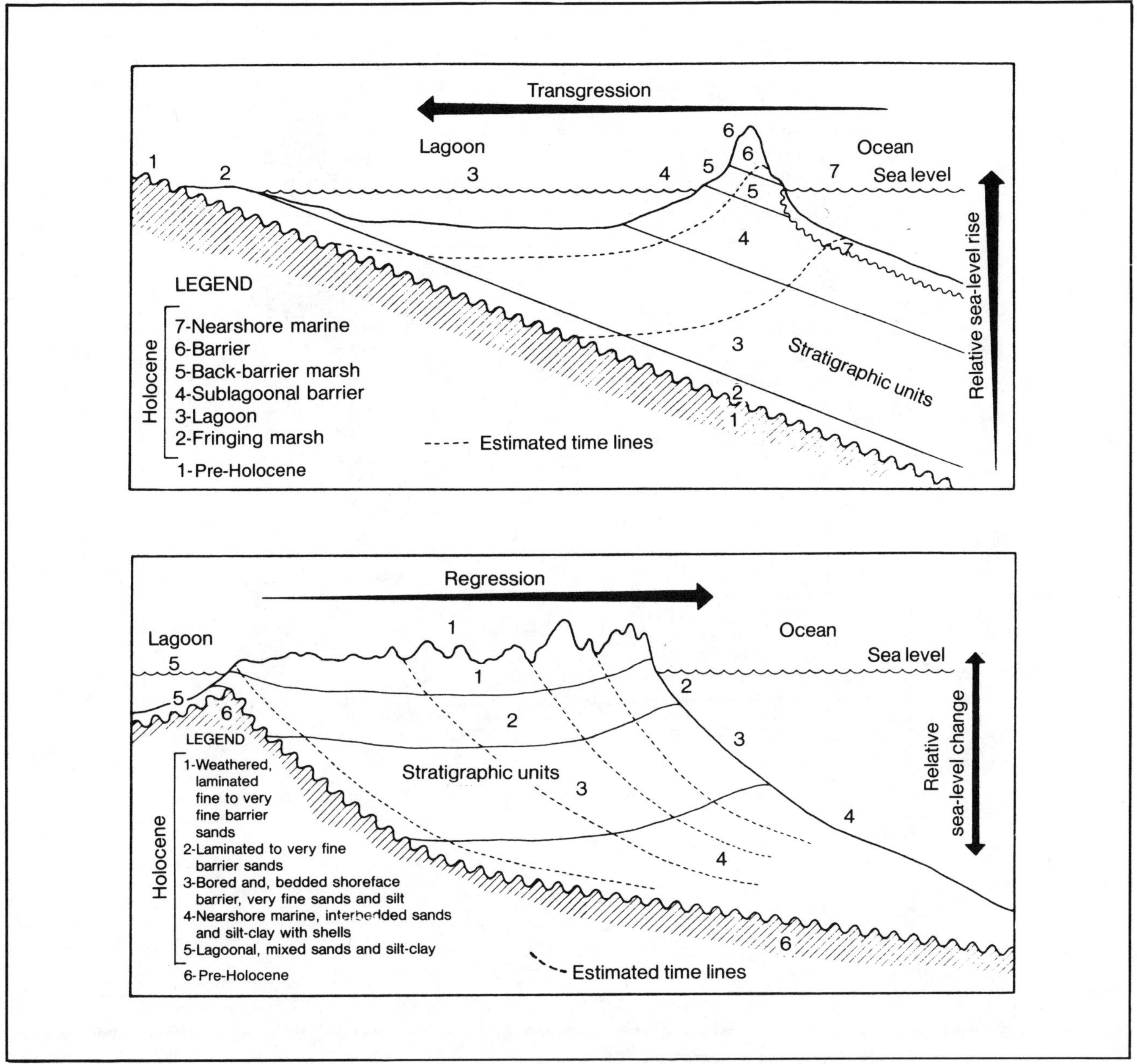

Fig. 9-28. Sequence models of transgressive and regressive barriers showing variants of depositional environmental sequences. Lithostratigraphic units cross time lines based on radiocarbon dates. Regressive model based on Bernard et al. (1970). (Modified after Kraft and John, 1979; permission to publish by AAPG).

studies of the North Carolina and South Carolina coasts, they place considerable importance on whether the coastal area is dominated by wave energy or tidal energy. Along wave-dominated inlets longshore transport results in . . . "a fining-upward sequence of: (1) inlet floor of coarse shell and pebble lag; (2) channel deposits of planar and trough cross-bedded sand and shell; and (3) a spit platform, composed of planar cross-bedded and horizontally laminated fine-grained sand. Channel migration and abandonment results in preservation of isolated shore-parallel (strike) wedge to lenticular-shaped inlet-fill sand bodies occurring randomly along the shoreline."

Tye and Moslow point out that migration rates of tide-dominated inlet mouths are less than those of the wave-dominated inlet mouths. "Abandoned inlet channels, exhibit symmetrical, U-shaped strike geometries and crescentic,

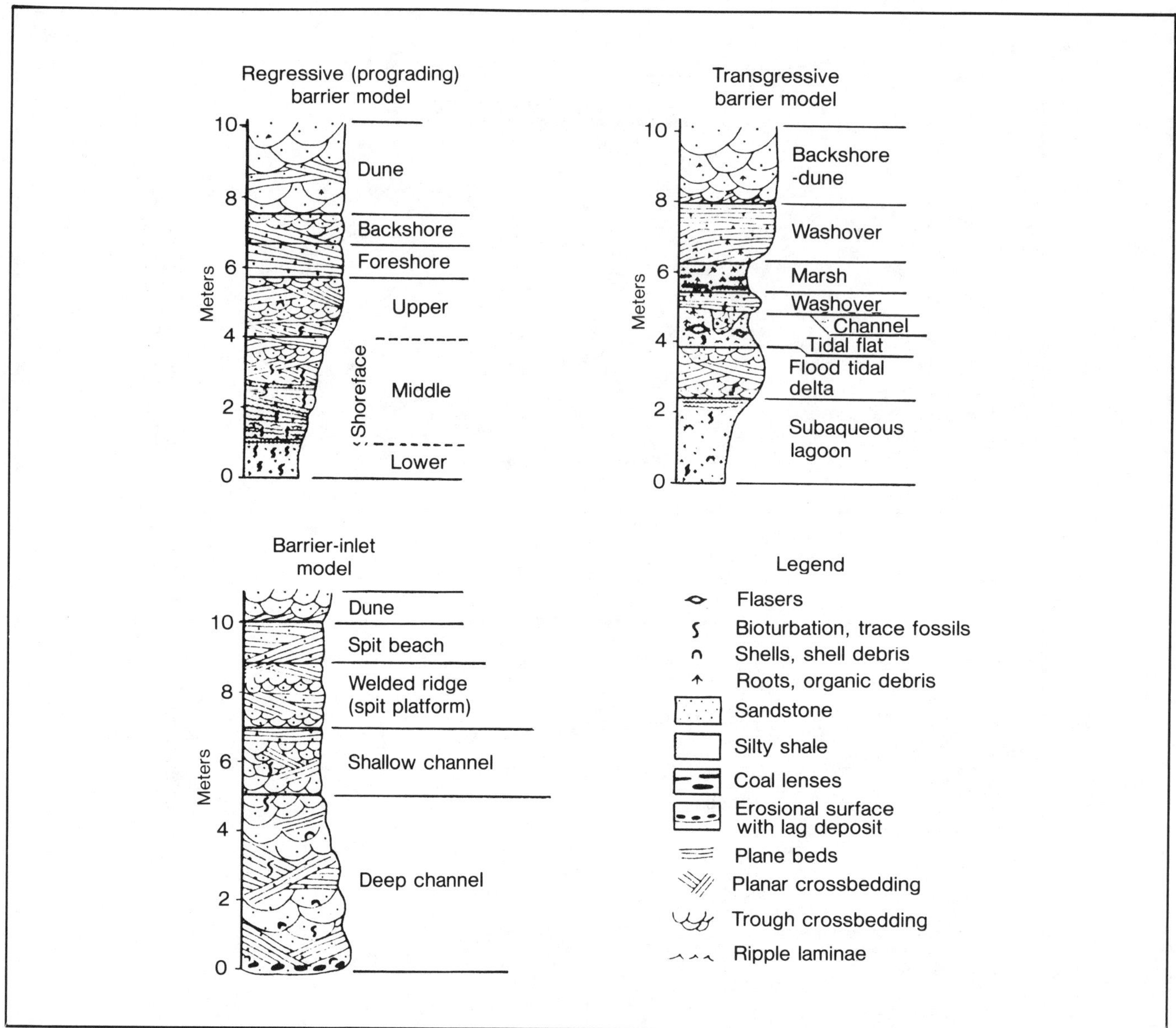

Fig. 9-29. The three "end members" models of barrier-bar stratigraphic sequences. (From Reinson, 1984, modified after Reinson, 1979; permission to publish by Geological Association of Canada).

concave-upward, dip geometries." They suggest tidal inlets make up only 5% of North Carolina's wave-dominated coast but the barrier bars contain 35% inlet fill. Along the South Carolina coast, tidal inlets make up 20% of the coast with only 15% of inlet fill. Thus, it appears that tidal inlets along tidal-dominated coasts are more stable than those of wave-dominated coastal areas. Tye and Moslow state, "With a higher preservation potential for inlet deposits over shoreface or foreshore deposits, fining upward sand and/or mud-rich inlet sequences will dominate ancient clastic shorelines. Porous, fining-upward quartz-rich inland sand bodies are the most preservable facies in barrier shoreline sequences and exhibit consistent thickness to width ratios and lateral geometries." For wave-dominated shorelines, this ratio is about 1:7.

Tidal Delta Deposits. The relative positions of the flood-tidal and ebb-tidal deltas are shown in Figure 9-8. The flood-tidal delta consists of a sand body deposited landward from the tidal inlet where flood-tidal energy is dissipated as it enters the quiet water of the lagoon. Reinson (1979) states that "tidal deltas can occur in a variety of forms (from linear shoals to complex channel-shoal systems) depending on tidal range, wave climate, and sediment supply, but the basic morphological pattern . . . is generally

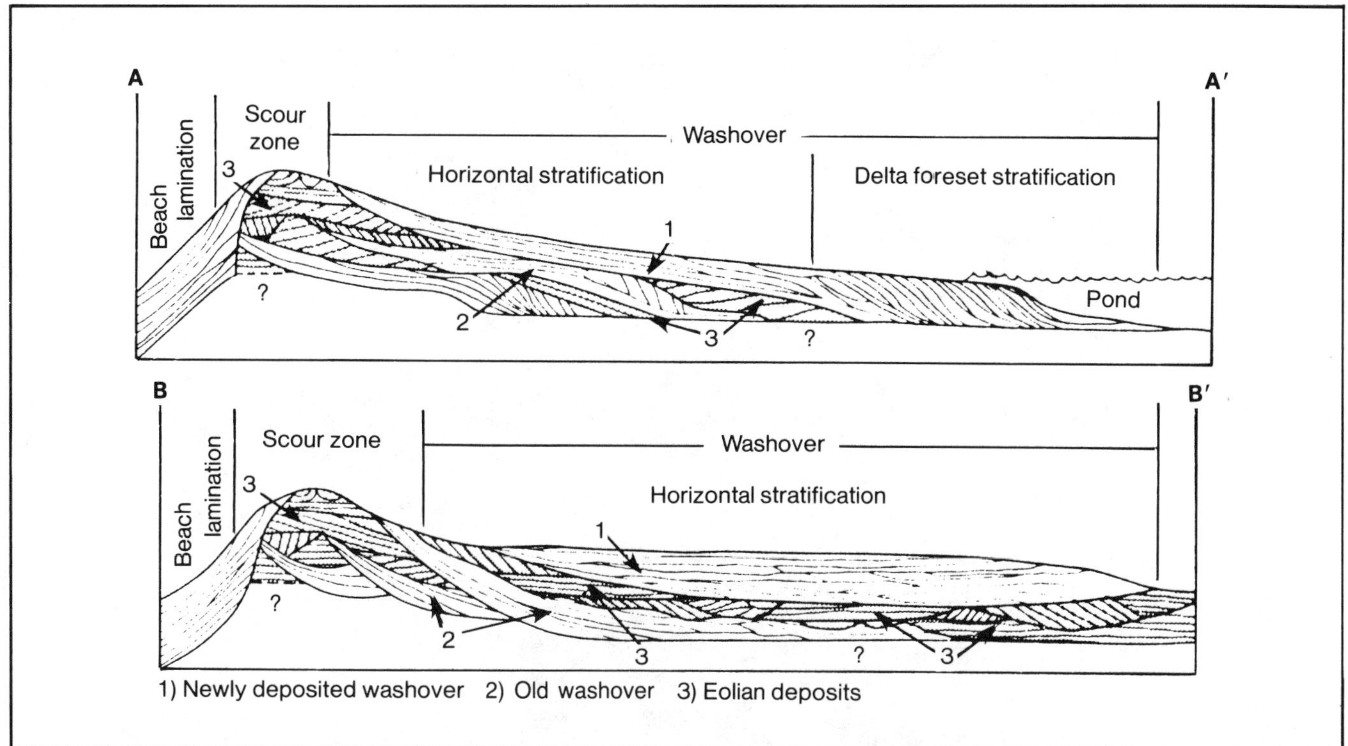

Fig. 9-30. Schematic cross sections of two washover fans showing sequences of sedimentary structures. The transgressive sedimentary structures represent the upper few meters, or less, of the sand body complex. Section A-A′ shows a horizontal stratification to delta-foreset stratigraphic sequence resulting from flow across a subaerial surface into a body of standing water. Section B-B′ shows horizontal stratification resulting from flow across a subaerial surface. Eolian processes may modify or bury washover deposits to various degrees. (Modified after Schwartz, 1975; permission courtesy of Army Corps of Engineers).

clearly evident." Figure 9-32 illustrates the morphology of tidal deltas and the inferred tidal- and longshore-current patterns at the mouth of the Miramichi estuary of New Brunswick. Reinson (1979) states:

> Hubbard and Barwis (1976) proposed a lithologic sequence for a flood-tidal delta as follows: 1) basal bidirectional cross-strata (megaripples)—represents early phases of deposition, 2) interbedded seaward-oriented trough cross-strata (megaripples) and landward-oriented planar cross beds (sand waves)—represents deposition prior to ebb-shield development, and 3) landward-oriented planar cross-strata with upward-decreasing set thickness (sand waves)—represents deposition on flood ramp. Deposits adjacent to this sequence would be characterized by bidirectional trough cross-strata (megarip-

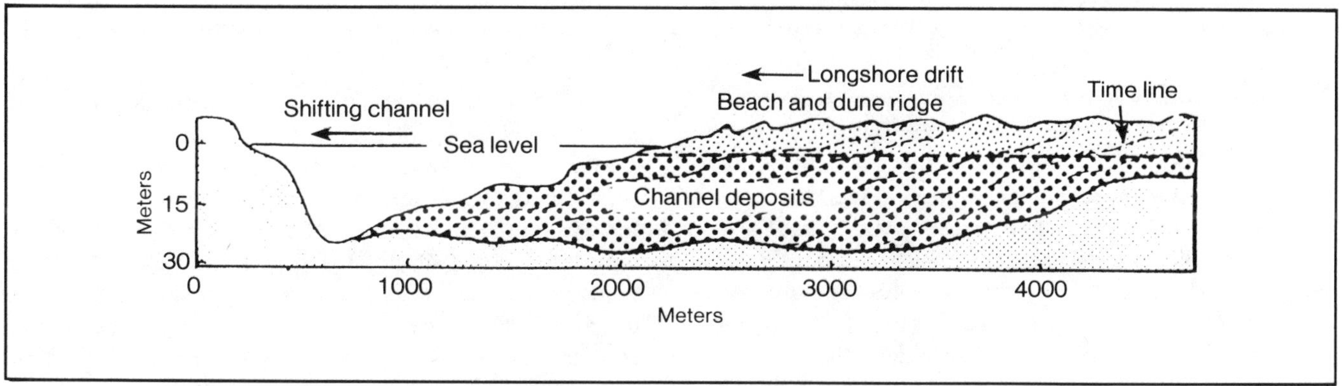

Fig. 9-31. Generalized cross section parallel with shoreline, illustrating the development of a barrier-inlet sand body by lateral inlet migration. (Modified after Hoyt and Henry, 1965).

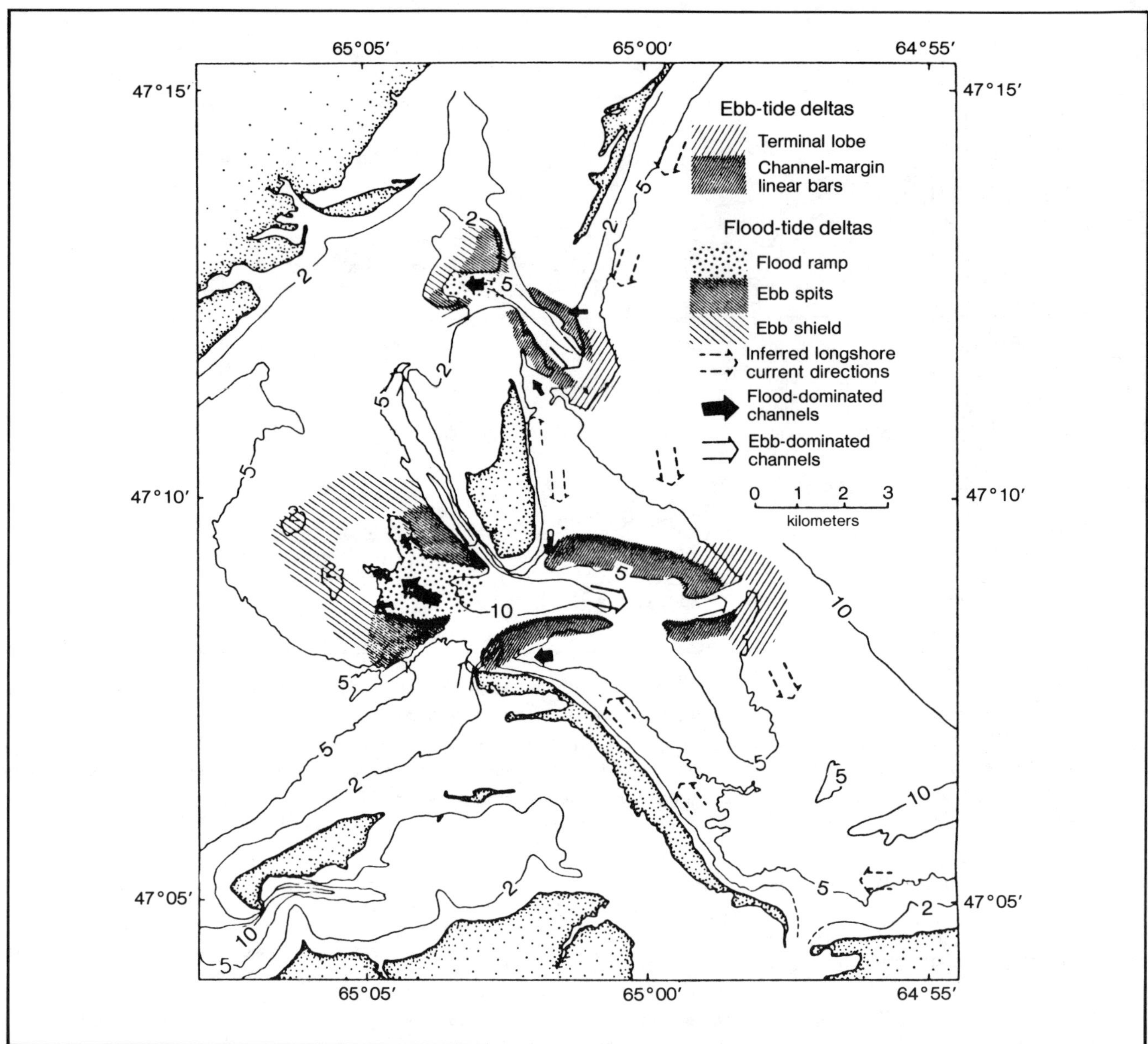

Fig. 9-32. Morphology of tidal deltas and inferred tidal-and longshore-current patterns at the mouth of the Miramichi estuary, New Brunswick. (From Reinson, 1984, modified after Reinson, 1977; permission to publish by Geological Association of Canada).

plies)—representing ebb-shield and ebb spit deposition. The total thickness of such a sequence would be in the order of 10 m.

Figure 9-33 is a hypothetical regressive sedimentary sequence for a flood-tidal delta as proposed by Hayes (1976). The basal section is dominated by ebb-oriented planar cross-beds of an ebb channel. Directly above this zone are flood-oriented planar cross-beds of a flood channel. This, in turn, is overlain by a thick section of dominantly flood-oriented planar cross-beds of the flood-tidal delta. Flood-tidal energy was dissipated in this latter section as it encountered the still-water environment of the lagoon.

The vertical sequence in ebb-delta deposits exhibits extreme variations in internal structure from one location to another. Within this environment there is an interaction of wave energy, longshore drift, and ebb-tidal flow and it is for this reason that there is extreme variability of the internal structure. Reinson (1979) points out that "the major difference between ebb-tidal delta deposits and flood-delta deposits is the occurrence of multidirectional cross beds in ebb delta sequences, as opposed to the predominantly flood-oriented or bidirectional crossbeds of flood-tidal delta sequences."

Texture and sedimentary structures of the tidal inlet fill

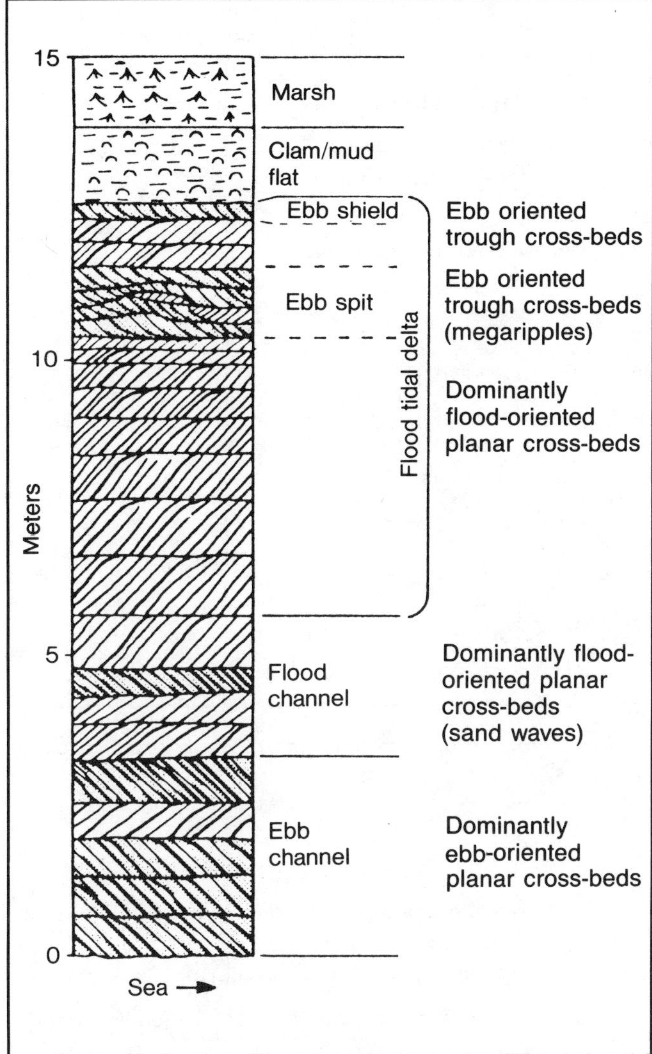

Fig. 9-33. Hypothetical regressive sedimentary sequence for a mesotidal flood-delta complex. (Modified after Hayes, 1976; Reinson, 1984).

and flood-and ebb-tidal deltas have sufficient similarities to render positive identification in the subsurface, based solely on these parameters, difficult. Occasional cores are the only tools for their identification. For this reason, the geometric and stratigraphic aspects of the barrier-bar complex are of extreme importance. Once these have been determined (without cores), the data available from an occasional core become much more meaningful.

Lagoonal (Back Barrier) Facies

Lagoonal deposits occur between the barrier bar and the mainland and generally are laid down in a low energy environment. The sediments are quite variable, consisting of interbedded shale, siltstone, coal, and thin sandstone. Sandstone may consist of washover fans, flood-delta channel fills, and tidal flats. Thin, lenticular coal beds reflect local marsh and swamp flatland conditions and occur on washover sand and mud flats along the margin of the lagoon. Shale and siltstone beds frequently contain brackish-water fauna. Fossilized plant remains also may occur in the shales. Tidal-flat deposits exhibit fine-grained to medium-grained ripple-laminated sand interbedded with shale. Although the barrier bar sand interfingers with the predominantly shale-siltstone sequence of the lagoon, the transitional zone is relatively narrow.

Figure 9-34 is a generalized stratigraphic section illustrating a lagoonal sequence of sediments in the Carboniferous of eastern Kentucky and southern West Virginia.

Barrier Bar Criteria

From the foregoing discussion, it should be apparent that the sediments of the barrier-bar complex are quite variable and that no single criterion may be used in either identification or tracing of its component parts in the subsurface. Most of the recent literature on the barrier-bar complex treats modern analogues. This information is valuable for an understanding of both the sedimentary types and their respective environments of deposition. However, to apply much of this information in subsurface studies, the explorationist must sift through that which is practical and that which has considerably less application. It is for this reason that the authors attempt to list those aspects of the barrier-bar complex that most readily apply to their subsurface identification and delineation:

A. Geometric Aspects
   1. Length; many miles, but interrupted by tidal inlets
   2. Width
      a. regressive barrier bars; 1 to 3 or 4 mi (1.6 to 4.8 or 6.4 km)
      b. transgressive barrier bars; several hundred feet (100+ m)
   3. Thickness; 45–65 feet (13.7–19.8 m)
   4. Cross section; asymmetrically plano-convex upward
      a. Thin abruptly on lagoonal margin
      b. Sand apron 2–3 mi (3.2–4.8 km) wide on seaward side
   5. Orientation; parallel with basin contours of GIS
   6. Seaward margin; smooth
   7. Lagoonal margin; escaloped
      a. Due to washover fans

B. Internal Aspects
   1. Composition; predominantly quartz sand with scattered carbonized plant remains and shell fragments, and occasional thin shell layers

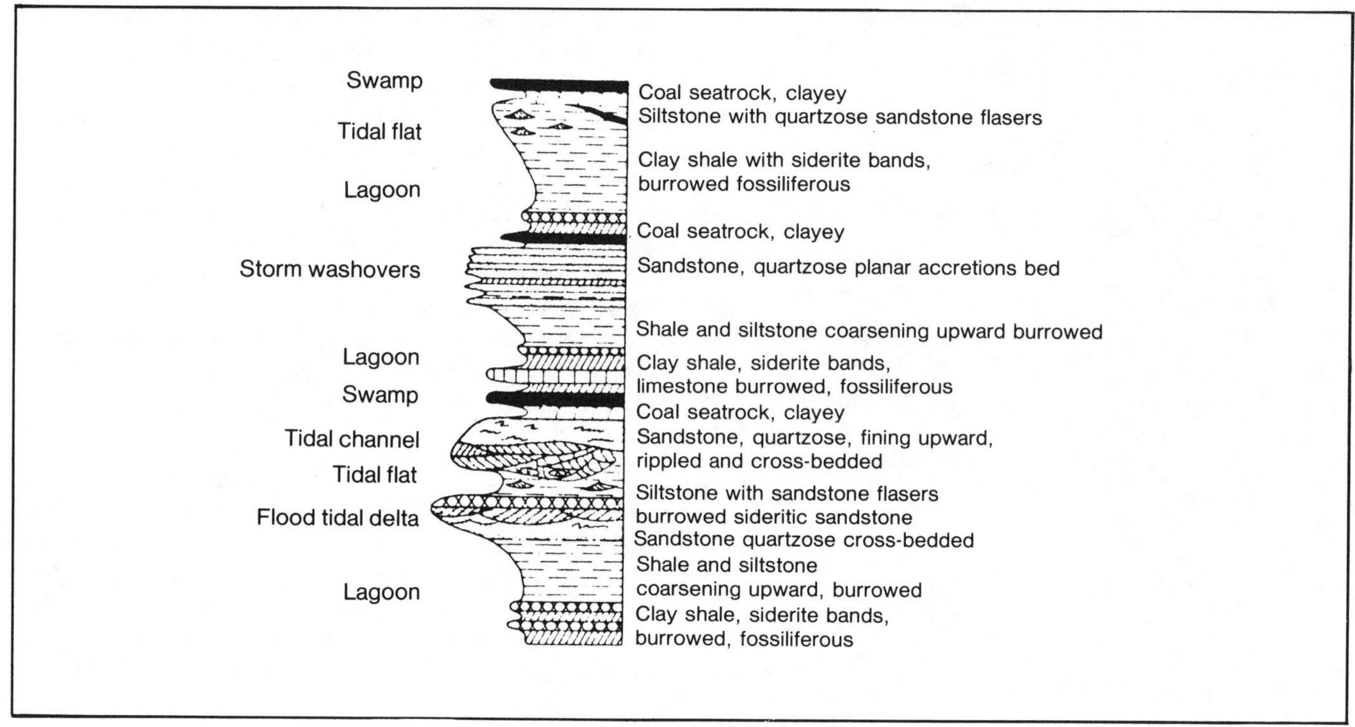

Fig. 9-34. Generalized lagoonal sequence in the Carboniferous of eastern Kentucky and southern West Virginia. Such sequences may range from 7.5 to 24 m thick. (Modified after Horne and Ferm, 1978).

2. Grain size; generally fine-grained to medium-grained
3. Sorting; generally well-sorted with:
   a. gradual increase in grain size upward in that portion not related to tidal inlet lateral migration
   b. gradual decrease in grain size upward in accretionary deposits on updrift side of tidal inlets
4. SP-GR "signatures"
   a. flower pot profile in thicker portions not related to tidal inlet lateral migration (abrupt top, transitional base)
   b. funnel-shaped profile on seaward margin where sand apron occurs (abrupt top, transitional base)
   c. bell-shaped profile in accretionary deposits on updrift side of tidal-inlets (abrupt base, transitional upper portion)
   d. cylinder-shaped profile along landward third where there frequently is no bar foot (abrupt top and base)
5. Internal structure
   a. upper one-third cross stratified (eolian)
   b. middle one-third horizontally bedded with some cross-stratification
   c. basal one-third bioturbated
6. Tidal inlet sandstone; cross stratified in two directions
7. Flood-tidal delta sandstone—cross stratified with preponderance of flood-tidal dip
8. Ebb-tidal delta sandstone—cross-stratified in multidirections

C. Associated strata
   1. Marine shale on seaward side
   2. Lagoonal (brackish-water shale) on landward side
   3. Brackish-water shale overlying regressive barrier bar
   4. Marine shale overlying transgressive barrier bar

There probably are exceptions to all of the above criteria, but when they are used in combination, it is not difficult to identify and delineate a barrier-bar complex in the subsurface. No one of these criteria is considered diagnostic but when used in combination, there is a synergetic relationship. In the above outline, there is limited emphasis placed on internal features due to the usual paucity of available cores.

SELECTED BIBLIOGRAPHY

Andrews, P. B., 1970, Facies and genesis of a hurricane-washover fan, St. Joseph Island, central Texas coast: Texas Univ., Austin, Bur. Econ. Geology, Rept. Inv. 67, 147 p.

Armon, J. W., 1979, Changeability in small flood tidal deltas and its effects, the Malpeque barrier system, Prince Edward Island: *in* S.B. McCann, ed., The Coastline of Canada: Geol. Survey Canada Paper, 1979.

Asquith, D. O., 1974, Sedimentary models, cycles, and deltas, Upper Cretaceous, Wyoming: AAPG Bull., v. 58, no. 11, p. 2274-2283.

Barwis, J. H., and J. H. Makurath, 1978, Recognition of ancient tidal inlet sequences: an example from the Upper Silurian Keyser limestone in Virginia: Sedimentology, v. 25, no. 1, p. 61-82.

Berg, R. R., and D. K. Davies, 1968, Origin of Lower Cretaceous Muddy sandstone at Bell Creek field, Montana: AAPG Bull., v. 52, no. 10, p. 1888-1898.

Bernard, H. A., C. F. Major, Jr., and B. S. Parrott, 1959, The Galveston barrier island and environs: a model for predicting reservoir occurrence and trend: Gulf Coast Assoc. Geol. Socs. Trans., v. 9, p. 221-224.

_______, R. J. LeBlanc, and C. F. Major, 1962, Recent and Pleistocene geology of southeast Texas: *in* Geology of the Gulf Coast and Central Texas and Guidebook of Excursions: Houston, Houston Geol. Soc.–Geol. Soc. Amer. Ann. Mtg., p. 175-205.

_______, and _______, 1965, Resume of the Quaternary geology of the northwestern Gulf of Mexico Province, *in* H. E. Wright, Jr., and D. G. Frey, eds., The Quaternary of the United States: Princeton, Princeton University Press, p. 137-185.

Blanton, S. L., 1963, Birth and death of an offshore bar: Gulf Coast Assoc. Geol. Socs. Trans., v. 13, p. 95-97.

Boyles, J. M., and A. J. Scott, 1982, A model for migrating shelf-bar sandstones in upper Mancos shale (Campanian), northwestern Colorado: AAPG Bull., v. 66, no. 5, p. 491-508.

Brenner, R. L., 1978, Sussex sandstone of Wyoming—example of Cretaceous offshore sedimentation: AAPG Bull., v. 62, no. 3, p. 181-200.

Busch, D. A., 1974, Stratigraphic traps in sandstones—exploration techniques: AAPG Mem. 21, 174 p.

Byrne, J. V., D. O. LeRoy, and C. M. Riley, 1959, The chenier plain and its stratigraphy, southwestern Louisiana: Gulf Coast Assoc. Geol. Socs. Trans, v. 9, p. 237-260.

Campbell, C. V., 1971, Depositional model—Upper Cretaceous Gallup beach shoreline, Ship Rock area, northwestern New Mexico: Jour. Sed. Pet., v. 41, no. 2, p. 395-409.

Carter, C. H., 1978, A regressive barrier and barrier-protected deposit: Depositional environment and geographic setting of the late Tertiary Cohansey sand: Jour. Sed. Pet., v. 48, no. 3, p. 933-949.

Chaves, H. A., 1967, personal communication, Salvador, Brazil.

Clifton, H. E., R. E. Hunter, and R. L. Phillips, 1971, Depositional structures and processes in the non-barred high energy nearshore: Jour. Sed. Pet., v. 41, no. 3, p. 651-670.

Davidson-Arnott, R. G. D., and B. Greenwood, 1976, Facies relationships on a barred coast, Kouchibouguac Bay, New Brunswick, Canada: *in* R. J. Davis, Jr., ed., Beach and Nearshore Sedimentation: SEPM Sp. Pub. no. 24, p. 149-168.

Davies, D. K., and F. G. Ethridge, 1971, The Clariborne Group of Central Texas: a record of Middle Eocene marine and coastal plain deposition: Trans. Gulf Coast Assoc. Geol. Soc. Trans. v. 21, p. 115-124.

_______, _______, and R. R. Berg, 1971, Recognition of barrier environments: AAPG Bull., v. 55, no. 4, p. 550-565.

Davis, R. A., ed., 1978, Coastal sedimentary environments: New York, Springer-Verlag, 420 p.

_______, W. T. Fox., M. O. Hayes, and J. C. Boothroyd, 1972, Comparison of ridge-and-runnel systems in tidal and non-tidal environments, Jour. Sed. Petrology, v. 32, no. 2, p. 413-421.

_______ and _______, 1972, Coastal processes and nearshore sand bars: Jour. Sed. Pet., v. 42, no. 2, p. 401-412.

Dickinson, K. A., H. L. Berryhill, Jr., and C. W. Holmes, 1972, Criteria for recognizing ancient barrier coast-lines, *in* J. K. Rigby and W. K. Hamblin, eds., Recognition of Ancient Sedimentary Environments: SEPM Sp. Pub. no. 16, p. 192-214.

Dillard, W. R., D. P. Oak, and N. W. Bass, 1941, Chanute oil pool, Neosho County, Kansas—a water-flooding operation, *in* A. I. Levorsen, ed., Stratigraphic type oil field, a symposium: Tulsa, AAPG, p. 57-77.

Doran, E., Jr., 1955, Land forms of the southeast Bahamas: Texas Univ. Pub. No. 5509, Dept. of Geography, p. 1-38.

Exum, F. A., and J. C. Harms, 1968, Comparison of marine-bar with valley-fill, stratigraphic traps, western Nebraska: AAPG Bull., v. 52, no. 10, p. 1851-1868.

Field, M. E., and D. B. Duane, 1976, Post-Pleistocene history of the United States inner continental shelf: significance of origin of barrier islands: GSA Bull., v. 87, no. 5, p. 691-702.

Fisher, J. J., 1968, Barrier island formation: discussion: GSA Bull., v. 79, no. 10, p. 1421-1426.

Fisk, H. N., 1948, Geological investigations of the lower Mermantau River basin and adjacent areas in coastal Louisiana: Vicksburg, Mississippi, U.S. Army Corps Engineers, Mississippi River Comm., 78 p.

———, 1955, Sand facies of Recent Mississippi delta deposits: 4th World Petroleum Cong. Proc., Sec. I/C, p. 1-21.

———, 1959, Padre Island and the Laguna Madre flats, coastal south Texas, *in* R. J. Russell, chm., 2d Coastal Geography Conf., April 6-9: Louisiana State Univ. Coastal Studies Inst., p. 103-151.

——— and E. McFarlin, Jr., 1955, Late Quaternary deltaic deposits of the Mississippi River, *in* A. Poldervaart, ed., The crust of the earth: GSA Spec. Paper 62, p. 279-302.

Galloway, W. E., D. K. Hobday, and K. Magara, 1982, Frio formation of Texas Gulf Coastal plain: depositional systems, structural framework, and hydrocarbon distribution: AAPG Bull., v. 66, no. 6, p. 649-688.

Glaeser, J. D., 1978, Global distribution of barrier islands in terms of tectonic setting: Jour. Geol., v. 86, no. 3, p. 283-297.

Gould, H. R., and E. McFarland, Jr., 1959, Geologic history of the chenier plain, southwestern Louisiana: Gulf Coast Assoc. Geol. Socs. Trans., v. 9, p. 261-270.

——— and J. P. Morgan, 1962, Coastal Louisiana swamps and marshlands, Field Trip No. 9 in Geology of the Gulf Coast and central Texas and guidebook of excursions, GSA 1962 Ann. Mtg.: Houston, Tex., Houston Geol. Soc., p. 287-341.

Griffith, E. G., 1966, Saber bar, Colorado: AAPG Bull., v. 50, no. 10, p. 2112-2118.

Hammond, E. H., 1954, A geomorphic study of the Cape region of Baja, California: California Univ. Pubs. Geography, v. 10, p. 45-112.

Hayes, M. O. 1975, Morphology of sand accumulations in estuaries, *in* L. E. Cronin, ed., Estuarine Research, v. 2, Geology and Engineering: New York, Academic Press, p. 3-22.

———, 1976, Transitional-coastal depositional environments, *in* M. O. Hayes and T. W. Kana, eds., Terrigenous clastic depositional environments, some modern examples: AAPG Field Course, Univ. South Carolina, Tech. Rept. No., 11-CRD, 315 p.

———, 1980, General morphology and sediment patterns in tidal inlets: *Sedimentary Geol.*, v. 26, p. 139–156.

Hobday, D. K., and J. C. Horne, 1977, Tidally influenced barrier-island and estuarine sedimentation in the Upper Carboniferous of southern West Virginia: Sedimentary Geol., v. 18, no. 1/3, p. 97-122.

Homewood, P., and P. Allen, 1981, Wave, tide, and current-controlled sand bodies of Miocene molasse, western Switzerland: AAPG Bull., v. 65, no. 12, p. 2534-2545.

Horne, J. C., and J. C. Ferm, 1978, Carboniferous depositional environments: Eastern Kentucky and southern West Virginia: Dept. of Geol., Univ. South Carolina, 151 p.

Howe, H. V., R. J. Russell and J. H. McGuirt, 1935, Physiography of coastal southwest Louisiana: reports on the geology of Cameron and Vermillion Parishes: Louisiana Geol. Survey Geol. Bull., no. 6, p. 1-72.

Hoyt, J. H., 1967, Barrier island formation: GSA Bull., v. 78, no. 9, p. 1125-1136.

———, 1969, Chenier versus barrier, genetic and stratigraphic distinction: AAPG Bull., v. 53, no. 2, p. 299-306.

———, 1970, Development and migration of barrier islands, north Gulf of Mexico: discussion: GSA Bull., v. 81, no. 12, p. 3779-3782.

——— and V. J. Henry, Jr., 1965, Significance of inlet sedimentation in the recognition of ancient barrier islands, *in* Sedimentation of Late Cretaceous and Tertiary outcrops, Rock Springs uplift: Casper, Wyoming, Wyoming Geol. Assoc. 19th Field Conf. Guidebook, p. 190-194.

——— and ———, 1967, Influence of island migration on barrier island sedimentation: GSA Bull., v. 78, no. 1, p. 77-86.

Hubbard, D. K. and J. H. Barwis, 1976, Discussion of tidal inlet sand deposits: examples from the South Carolina coast, in M. O. Hayes et al., eds., Terrigenous clastic depositional environments: S. C. Univ. Coastal Res. Div. tech. rep. no. 11, vol. II, p. 128-142.

Kaczorowski, R. T., 1980, The Louisiana chenier system—some preliminary reinterpretations and refinements: Gulf Coast Assoc. Geol. Socs. Trans., v. 30, p. 427-430.

Klein, G. de V., 1974, Estimating water depths from analysis of barrier island and deltaic sedimentary sequences: Geology, v. 2, p. 409-412.

Kolb, C. R., and J. R. van Loplic, 1958, Geology of the Mississippi River de Haire plain, southeastern Louisiana: Vicksburg, Miss., U.S. Army Engr. W.E.S. Tech. Report 3-483.

Kraft, J. C., 1971, Sedimentary facies patterns and geologic history of Holocene marine transgression: GSA Bull., v. 82, no. 8, p. 2131-2158.

———, and C. J. John, 1979, Lateral and vertical facies relations of transgressive barrier: AAPG Bull., v. 63, no. 12, p. 2145-2163.

Krumbein, W. C., L. L. Sloss, and E. C. Dapples, 1949, Sedimentary tectonics and sedimentary environments: AAPG Bull., v. 33, no. 11, p. 1859-1891.

Kumar, N., and J. E. Sanders, 1974, Inlet sequence: a vertical succession of sedimentary structures and textures created by the lateral migration of tidal inlets: Sedimentology, v. 21, no. 4, p. 491-532.

Land, C. B., Jr., 1972, Stratigraphy of Fox Hills sandstone and associated formations, Rock Springs uplift and Wamsutter arch area, Sweetwater County, Wyoming: a shoreline estuary sandstone model for the Late Cretaceous: Quart. Colorado School of Mines, v. 67, no. 2, p. 69.

LeBlanc, R. J., 1972, Geometry of sandstone bodies, *in* T. D. Cook ed., Underground waste management and environmental implications: AAPG Mem. 18, p. 133-189.

———, and W. D. Hodgson, 1959a, Origin and development of the Texas shoreline, *in* R. J. Russell, chm., 2d Coastal Geography Conf., April 6-9: Louisiana State Univ. Coastal Studies Inst., p. 57-101.

——— and ———, 1959b, Origin and development of the Texas shoreline, *in* Symposium on Late Cretaceous rocks, Wyoming and adjacent areas: Casper, Wyoming, Wyoming Geol. Assoc. 16th Ann. Field conf., p. 253-275.

Leckie, D. A., and R. G. Walker, 1982, Storm-and tide-dominated shoreline in Cretaceous Moosebar-Lower Gates interval-outcrop equivalents of Deep basin gas trap in western Canada: AAPG Bull., v. 66, no. 2, p. 138-157.

McGregor, A. A., and C. A. Briggs, 1968, Bell Creek field, Montana, a rich stratigraphic trap: AAPG Bull., v. 52, no. 10, p. 1869-1887.

McKee, E. D. and T. S. Sterret, 1961, Laboratory experiments on form and structure of longshore bars and beaches, *in* J. A. Peterson and J. C. Osmond, eds., Geometry of sandstone bodies, a symposium: Tulsa, AAPG, p. 13-28.

Miller, D. N., Jr., 1962, Patterns of barrier bar sedimentation and its similarity to Lower Cretaceous Fall River stratigraphy, *in* Symposium on Early Cretaceous rocks of Wyoming and adjacent areas: Casper, Wyoming, Wyoming Geol. Assoc. 17th. Ann. Field Conf., p. 232-247.

Ogunyomi, O. M. and L. C. Hills, 1977, Depositional environments, Formost Formation (Late Cretaceous) Milk River area, southern Alberta: Canadian Petrol. Geol. Bull., v. 25, p. 929-968.

Owens, E. J., and D. Frobel, 1977, Ridge and runnel systems in the Magdalen Islands, Quebec: Jour. Sed. Pet., v. 47, no. 1, p. 191-198.

Otvos, E. G., and W. A. Price, 1979, Problems of chenier genesis and terminology—an overview: Mar. Geol., v. 31, no. 3/4, p. 251-263.

Pepper, J. F., D. F. Demarest, R. D. Holt, W. de Witt, Jr., H. H. Gray, H. H. Mead, C. W. Merrels, II, and P. Averitt, 1944, Map of the Second Berea sand in Gallia, Meigs, Athens, Morgan, and Muskingum Countries, Ohio: U.S. Geol. Survey Oil and Gas Inv. Prelim. Map 5.

———, W. R. de Witt, Jr., and D. F. Demarest, 1955, Geology of the Bedford shale and the Berea sandstone in the Appalachian basin: U.S. Geol. Survey Prof. Paper 259, 111 p.

Phelger, F. B., and G. C. Ewing, 1962, Sedimentology and oceanography of coastal lagoons in Baja California, Mexico: GSA Bull., v. 73, no. 2, p. 145-181.

Price, W. A., 1955, Environment and formation of the chenier plain: Quaternaria, v. 2, p. 75-86.

Rautman, C. A., 1978, Sedimentology of Late Jurassic barrier-island complex-Lower Sundance formation of Black Hills: AAPG Bull., v. 62, no. 11, p. 2275-2289.

Reinson, G. E., 1977, Facies models 6, barrier island systems: Geoscience Canada, 1976-1979, Reprint 1, Geol. Assoc. Canada, p. 57-74.

———, 1979, Facies models 14, barrier island systems: Geoscience Canada, v. 6, no. 2, p. 51-68.

———, 1984, Barrier-island and associated strand plain systems, *in* R. G. Walker, ed., Facies Models, 2nd edition; Geoscience Canada, p. 119–140.

Russell, R. J., 1940, Quaternary history of Louisiana: GSA Bull., v. 51, no. 8, p. 1199-1234.

——— and H. V. Howe, 1935, Cheniers of southwestern Louisiana: Geog. Review, v. 25, p. 449-461.

Sabins, F. F., Jr., 1963, Anatomy of stratigraphic trap, Bisti field, New Mexico: AAPG Bull., v. 47, no. 2, p. 193-228.

Shelton, J. W., 1965, Trend and genesis of lowermost sandstone unit of Eagle sandstone at Billings, Montana: AAPG Bull., v. 49, no. 9, p. 1385-1397.

Swartz, R. K., 1975, Nature and genesis of some storm washover deposits: U.S. Army Corps of Engineers, Tech. Memo., No. 61, 69 p.

Tanner, W. F., 1973, West Louisiana chenier plain history: Gulf Coast Assoc. Geol. Soc. Trans., v. 23, p. 389-393.

Tercier, J., 1940, Depots marine actuel et series geologigues: Ecologae Geologicae Helvetiae, v. 32, p. 47-100.

Turvill, J. A., and G. W. Troy, 1983, AAPG Bull. Abs., v. 67, no. 3, p. 560.

Tye, R. S., and T. F. Moslow, 1983, Tidal inlet: dominant facies of clastic barrier shoreline (abs): AAPG Bull., v. 67, no. 3, p. 560.

Wilkinson, B. H., 1975, Matagorda Island, Texas: The evolution of a Gulf Coast barrier complex: GSA Bull., v. 86, no. 7, p. 959-967.

Williams, J. J., D. C. Conner, and K. E. Peterson, 1975, Piper oil field, North Sea: fault block structure with Upper Jurassic beach/bar reservoir sands: AAPG Bull., v. 59, no. 9, p. 1585-1601.

van Straaten, L. M. J. U., 1959, Littoral and submarine morphology of the Rhone delta, *in* R. J. Russell, chm., 2d Coastal Geography Conf., April 6-9: Louisiana State Univ. Coastal Studies Inst., p. 233-264.

———, 1965, Coastal barrier deposits in south-Holland and north-Holland, in particular in the areas around Scheveningen and Ijmuiden: Netherland Geol. Stichting, Med., n.s., no. 17, p. 41-75.

Zenkovich, V. P., 1962, Some new exploration results about sand shore development during the sea transgression: De Ingenieur, no. 17, Bouw en Waterbouwkunde, 9, p. 113-121.

10 CLASTIC DEPOSITION IN MARGINAL MARINE AREAS

Busch (1974) points out that the marginal-marine environment is one of comparatively high-energy conditions in contrast to terrestrial, lacustrine, and truly marine environments. Significant oscillations of the shoreline occur herè, which may be the result of tidal conditions, tectonic movements, or eustatic changes in sea level—or of all three. Regardless of cause, the lateral shifts of the shoreline in the marginal-marine depositional environment are accompanied by similar shifts in the sites of deposition of distinct sedimentary types. Inasmuch as a working knowledge of the facies relations within this environment is essential to meaningful stratigraphic analysis, it is desirable to review here the fundamental concepts and principles. In a subsequent section these concepts and principles are illustrated by specific examples.

TRANSGRESSION AND REGRESSION

The concepts of marine transgression and regression are fundamental even though the meaning of these terms varies among geologists. In the American Geological Institute *Glossary of Geology and Related Sciences* (Howell, 1960), instead of giving a definition of transgression, reference is made to the terms *onlap* and *overlap*. The meaning of the term *onlap* is considered to be the same as *transgressive overlap*. Howell also refers to the definition of *offlap* for the meaning of regression. He further notes that the term *offlap* is described by some as *regressive overlap*. Clearly, *processes* have been confused with *results*—the stratigraphic relations result from the processes of transgression and regression. More recently, the AGI *Glossary of Geology* (Gary et al., 1972) defines transgression and regression as processes, more consistent with the authors' views.

The following definitions of transgression and regression provide for a more compatible explanation. *Transgression* is the process of migration of a shoreline in a landward direction. *Regression* is the process of migration of a shoreline in a seaward direction. As defined here, these terms may be applied to either a marine or a lacustrine environment. Neither definition implies wedgeout of sedimentary rocks, either toward or away from the depositional basin. Only where the stratigrapher correlates time units and observes that certain sedimentary units have either a transgressive or a regressive relation to the time lines are such terms as *onlap, overlap, offlap,* and *regressive overlap* applicable. Moreover, only from the observation of these physical relations may ancient transgressions and regressions of the shoreline be inferred. The concept of transgression and regression, as determined from the overlapping relations of relatively shallow or deeper water facies, is extremely pertinent to the petroleum exploration geologist because of certain salient factors, among which are the following:

1. Paleogeomorphology of bordering land—was it a deeply dissected, rugged terrain, a gently undulating peneplain, or possibly a tidal flat?
2. Nature of source of sediment supply—was it igneous, sedimentary, meta-igneous, metasedimentary, or possibly a combination of these?
3. Paleoenvironments of deposition—which types existed, and to what extent did they control the trend, distribution, and thickness of sand bodies?
4. Mechanisms of deposition—what sedimentary processes existed at each locale, and to what extent did they predetermine the shape and internal structure of a body of sand?

5. Cause of migration of a shoreline—were shoreline movements the result of eustatic changes in sea level? If so, the facies would be shifted laterally, either seaward or landward. If, however, shoreline movements were the result of tectonism (e.g., Rocky Mountain uplift during Paleocene time), abundant new influxes of terrigenous clastic sediment and shoreline regression would have occurred.

Other currently less practical, perhaps academic, aspects of geology related to the transgression-regression concept are:

1. Time—how far back in geologic history was a particular sediment deposited?
2. Rates of sediment supply
3. Dispersal of sediments

These are some of the more difficult problems related to transgression and regression. The solution to any one of these problems may have an important application in petroleum exploration.

VARIABLES AFFECTING TERRIGENOUS DEPOSITION IN MARGINAL-MARINE AREAS

The overlapping and wedgeout relations of the sediments deposited in a marginal-marine embayment may be complex. However, the relations always give an indication of the behavior of the surface of deposition—subsiding, rising, or stationary. Also, the relations reflect the nature of the tectonic movement, for example, gradual or cyclic. The rate of sediment supply relative to the rate of submergence or emergence of the depositional surface also may be determined. The following outline is a summary of important variables affecting deposition in the marginal-marine environment. The comparisons between supply of sediment and the rate of subsidence, as shown in the following table (A1a, A1b, and A1c), are modified from Grabaü (1913) who first recognized the interrelations of the two variables. This outline also shows the relation of the supply of sediment to cyclic subsidence (A2a, A2b, A2c, and A2d). Similarly, the supply of sediment is related to gradual emergence (B1a and B1b) and cyclic emergence (B2a). Finally, a stationary surface of deposition (C) with a stationary sea level is considered.

Table 10.1. Variables Affecting Terrigenous Deposition in Marginal-Marine Areas

- A. Sinking Bottom
  - 1. Gradual Subsidence
    - a. Supply of Sediment Less Than Rate of Subsidence
    - b. Supply of Sediment More Than Rate of Subsidence
    - c. Supply of Sediment Equal to Rate of Subsidence
  - 2. Cyclic Subsidence
    - a. Limited Supply of Sand; Abundant Supply of Mud
    - b. Limited Sand, Moderate Mud, and Abundant Supply of Carbonate
    - c. Moderate Supply of Sand, Abundant Supply of Mud
    - d. Abundant Supply of Sand and Mud
- B. Rising Bottom
  - 1. Gradual Emergence
    - a. Limited Supply of Sand, Abundant Supply of Mud
    - b. Moderate to Abundant Supply of Sand and Mud
  - 2. Cyclic Emergence
    - a. Moderate, Steady Supply of Sand and Abundant Supply of Mud
- C. Stationary Bottom

The existence of predictable, cyclic sequences of strata was emphasized first by Wanless and Weller (1932) in coal-measure sections in the Eastern Interior basin. A review of the cyclic sequences of sediments deposited in marginal-marine environments reveals that the principle of cyclic sedimentation applies to deposits dating from earliest Paleozoic time to the Holocene. The extent to which cyclic sedimentation may be related to either tectonism (rising and lowering of sea bottom) or to cyclic changes in sea level is not known. As a mechanical convenience, the sedimentary patterns illustrated in Figures 10-1 to 10-33 are related to changes in sea level. Similar sedimentary patterns would result if sea level had remained constant and the surface of deposition had moved up or down in either a gradual or cyclic manner. There is considerable vertical exaggeration in all of these diagrammatic illustrations.

SINKING BOTTOM

Gradual Subsidence

Supply of Sediment Less than Rate of Subsidence. Figure 10-1 is a diagram of a sedimentary sequence deposited in a marginal-marine environment in which the surface of deposition is subsiding more rapidly than sediment is being supplied. As a result, the shoreline gradually transgresses from right to left. Within the area of net transgression, the marine sandstone lies unconformably on continental strata. The several sand-mud lines are "phantom" lines that represent the maximum depth of wave agitation for the several shown successive positions of sea level. A high-energy environment characterizes the site of marine sand accumulation above this line. Above this phantom line, wave and

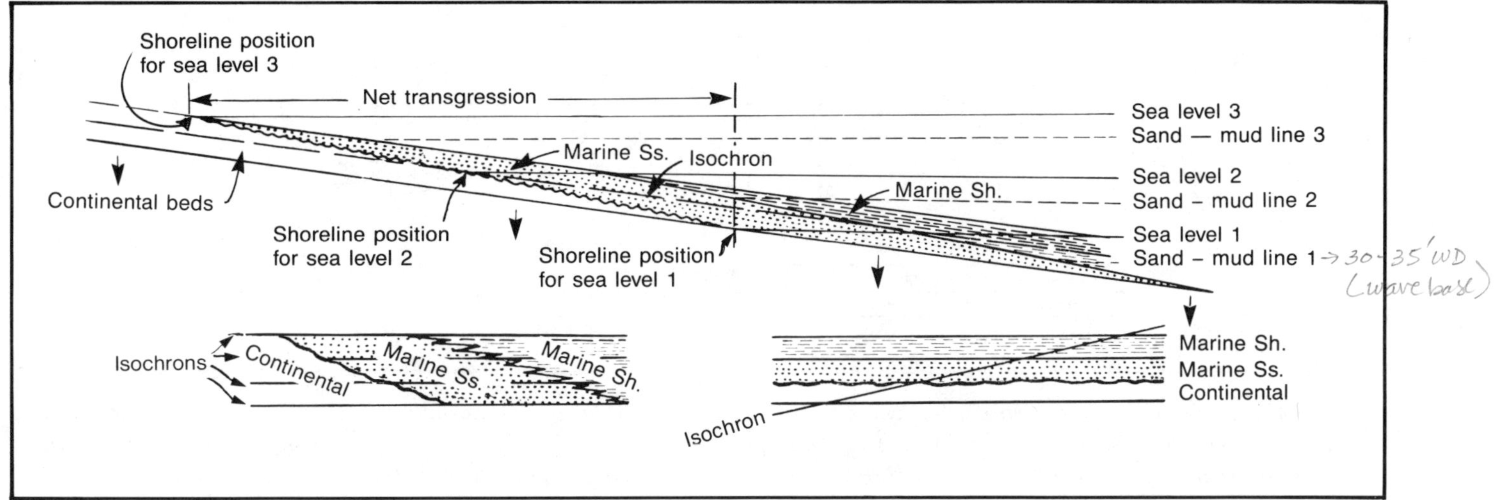

Fig. 10-1. Supply of sediments less than rate of subsidence in a marginal-marine environment. Low border lands with sluggish streams are inferred. The different lithologies transgress the time lines. This is the principal explanation for a widespread marine sandstone. (Modified after Grabau, 1913; permission to publish by Dover Publications, Inc.)

intertidal energy is sufficiently great to keep most of the silt-size and clay-size particles in suspension. These finer particles remain in suspension until they are transported into an environment of lower energy (e.g., protected bay or lagoon), where they settle and form either a silty shale or a clayey deposit. The water below the sand-mud line is essentially quiet. Thus, clay-size or silt-size particles in this area settle to form a marine shale that is the time equivalent of the more shoreward marine sandstone. In Figure 10-1 the supply of sediment is only half that necessary to keep the shoreline in a constant position. This condition suggests the presence of a low-lying borderland being drained by sluggish streams. The lithologic units (continental beds, marine sandstone, and shale) clearly transgress the time lines (isochrons). The illustration in the lower left of Figure 10-1 is drawn using isochrons as data of reference. This is a graphic means of illustrating that the three lithologic units are depositional time equivalents. In the lower right of Figure 10-1, lithologic boundaries are used arbitrarily as data of reference. In this portrayal, most of the

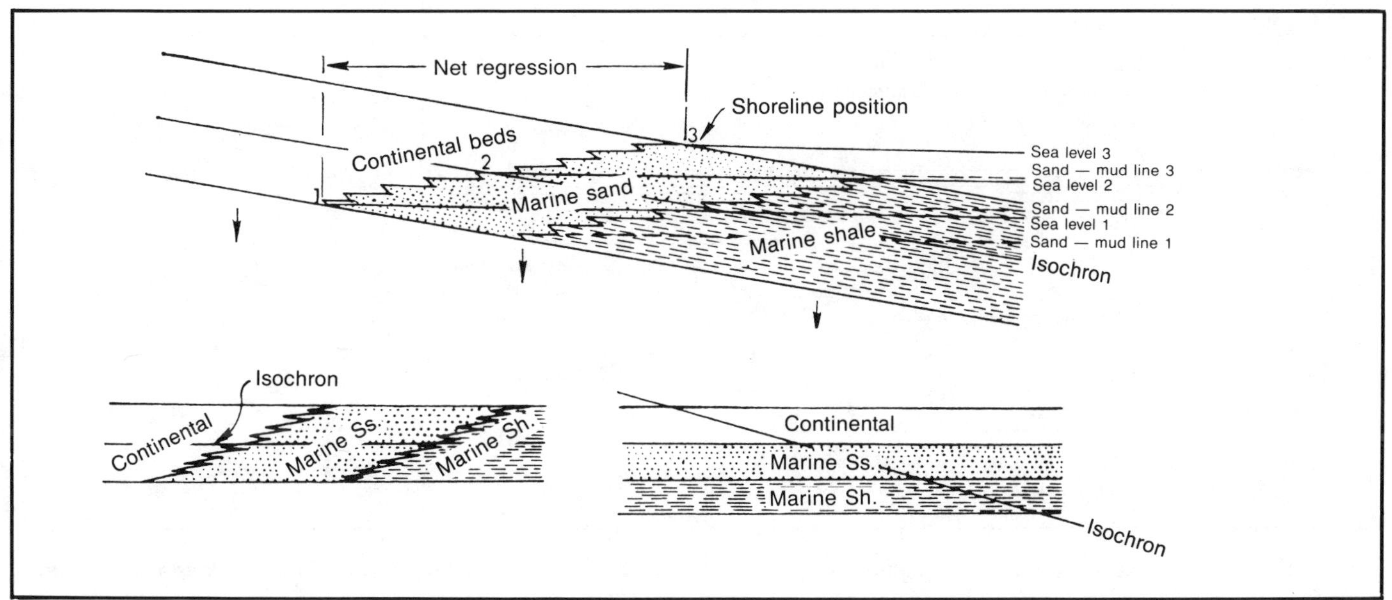

Fig. 10-2. Supply of sediments exceeds rate of subsidence in a marginal-marine environment. An active source area for sediments is inferred. The different lithologies regress the time lines. (Modified after Grabau, 1913; permission to publish by Dover Publications, Inc.)

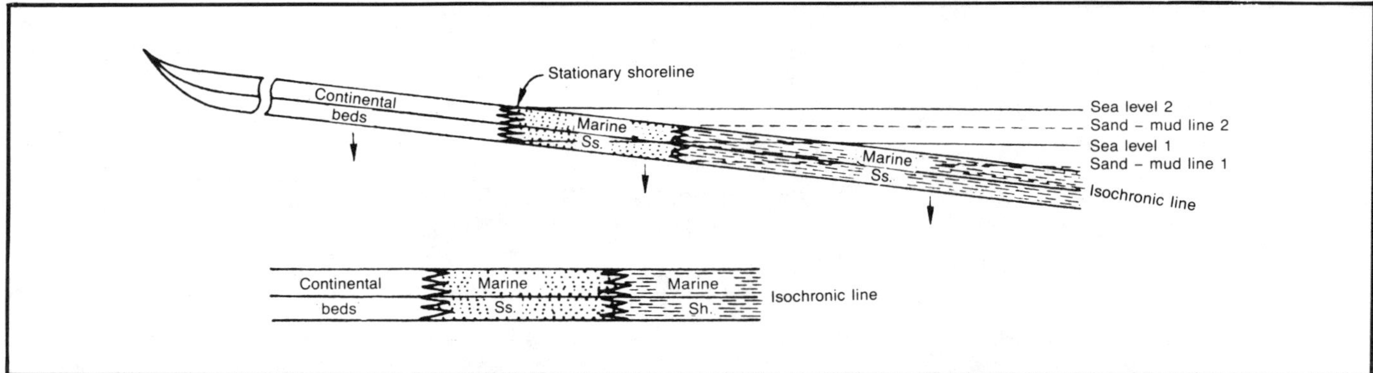

Fig. 10-3. Supply of sediments in equilibrium with rate of subsidence in a marginal-marine environment. Sandstone and shale units are of same age. Isochronic and bedding planes are the same. This is the only explanation for very thick (thousands of feet) marine sandstone. (Modified after Grabau, 1913; permission to publish by Dover Publications, Inc.)

time equivalence of the several lithologic units is obscured. The interpretation shown in Figure 10-1 accounts for the formation of a widespread, blanket-type marine sandstone.

Supply of Sediment More Than Rate of Subsidence. Figure 10-2 gives an interpretation of a marginal-marine environment in which the supply of sediment to the site of deposition is more than the rate of subsidence. In fact, the supply of sediment is twice that necessary to keep the position of the shoreline constant. As in Figure 10-1, the respective sites of marine sandstone and shale deposition are controlled largely by the vertical position of the sand-mud line. The three dominant lithologic types are clearly regressive to the time lines. This relation may be interpreted to indicate that an active neighboring source area at the left is being drained by streams having both moderate and steep gradients. In this situation, the continental (nonmarine) beds are as thick as their marine time equivalents. Isochrons are used as data of reference in the lower left of Figure 10-2. This method of portrayal emphasizes both the regressive nature and the time equivalence of the continental beds, marine sandstone, and marine shale. In the lower right of Figure 10-2, the lithologic boundaries are used arbitrarily as the reference data. The manner in which the isochrons are "crossed" by the three regressive lithologic units contrasts sharply with that shown in Figure 10-1. Here is a second explanation for a widespread marine sandstone.

A comparison of Figures 10-1 and 10-2 shows that the stratigraphic sequence in Figure 10-1 consists of continental beds unconformably overlain by sandstone, which, in turn,

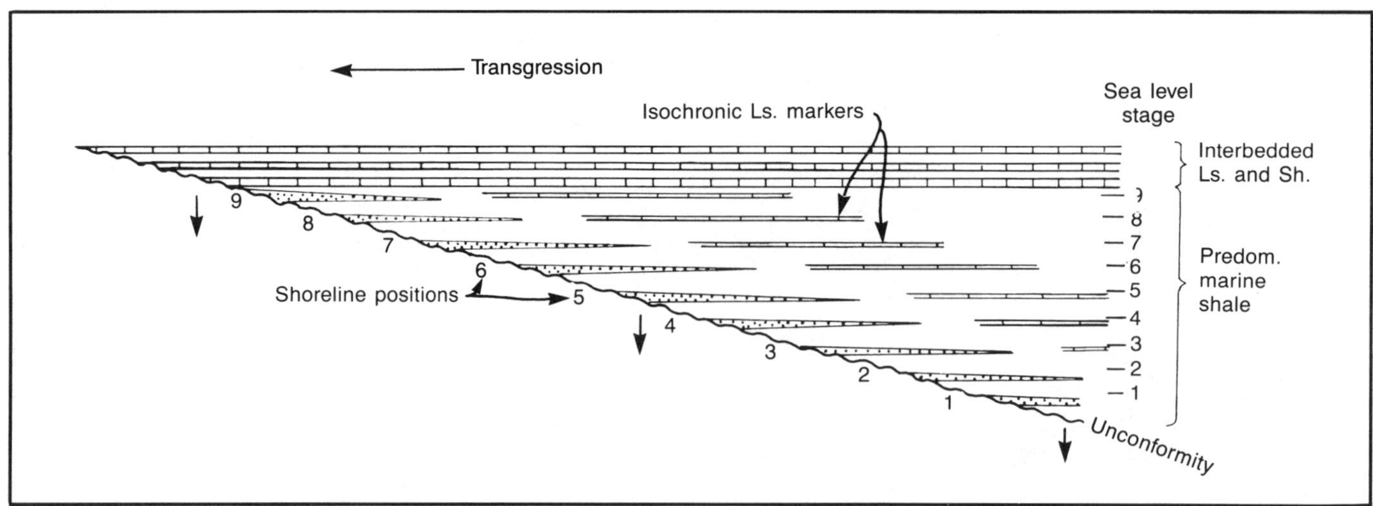

Fig. 10-4. Cyclic subsidence with very limited supply of sand and abundant mud and silt. Sand wedges may either impinge against unconformity and/or occur as isolated lenses basinward from shoreline (barrier bars and offshore bars) and are the result of a succession of stillstands of the shoreline in overall marine transgression. (Modified after Busch. 1974; permission to publish by AAPG).

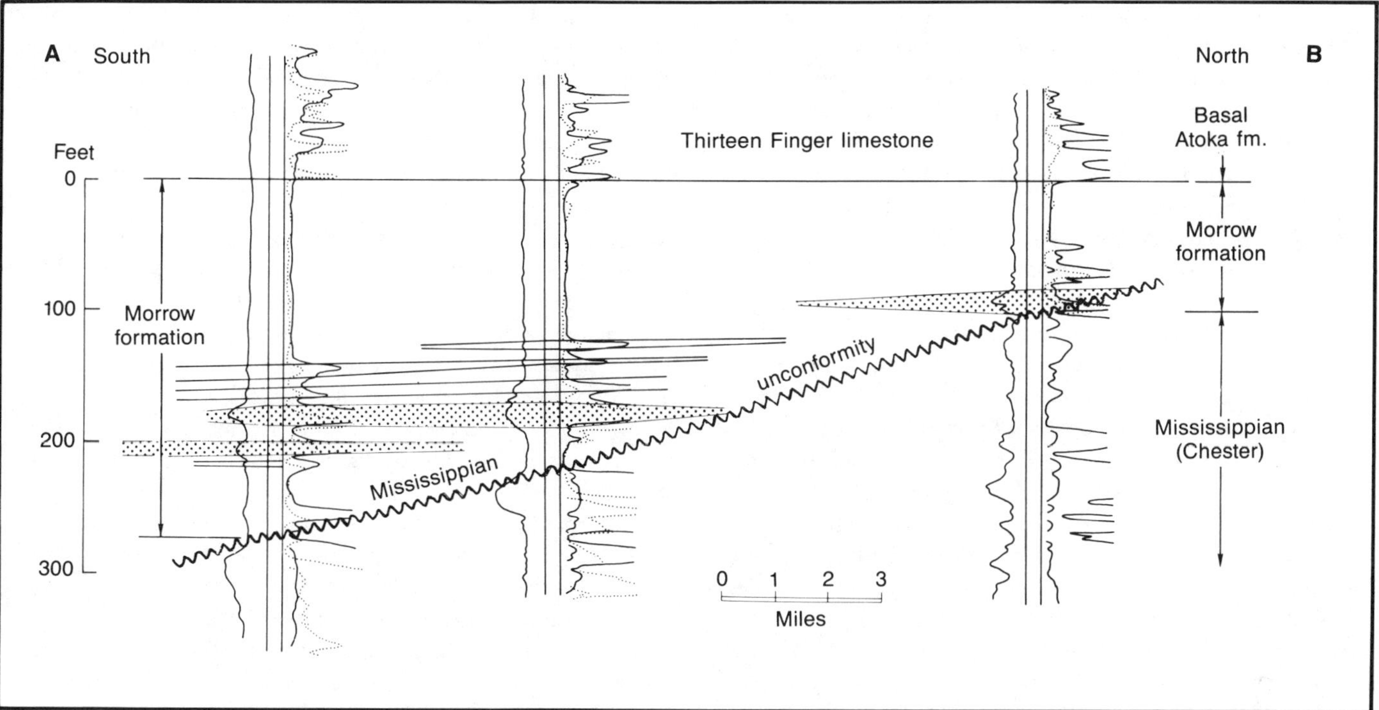

Fig. 10-5. Simplified stratigraphic profile of the Morrow series, northwestern Oklahoma, showing en échelon arrangement of three basal sandstone members. Location of profile A-B is shown on Figure 10-11. (Modified after Busch, 1959; permission to publish by AAPG).

is overlain by shale. In Figure 10-2 the stratigraphic sequence is completely reversed; the shale is overlain by sandstone, which is overlain by continental beds; unconformities[1] are absent.

In places where faunal assemblages are directly related to the depositional environment, a lateral shift (either transgressive or regressive) of the site of either sand or mud deposition is accompanied by a similar lateral shift of the fauna, especially the benthos. The net effect is a lithofacies-controlled fauna, which may either transgress or regress time lines, or both.

Supply of Sediment Equal to Rate of Subsidence. Figure 10-3 illustrates a situation in which the rate of sedimentation equals the rate of subsidence. As a result, the shoreline remains stationary. Because the shoreline neither transgresses nor regresses, great thicknesses of continental beds, sandstones (including reservoir types), and shale units accumulate. The sandstone and shale units are of the same age, and the transitional zones separating the marine sandstone from the continental beds and from the marine shale are nearly vertical. Bedding planes and time planes are, in effect, identical and cross the transitional zones at right angles. Very thick marine sandstone sequences must be explained by such a situation in which the rate of subsidence is in equilibrium with the rate of sedimentation.

Although several hundred feet of marginal-marine sandstone might be explained by the highly generalized Figure 10-3, it will not, however, explain thousands of feet of marginal-marine sandstone, i.e., the Frio (Oligocene) sandstones of the Louisiana-Texas-Mexican Gulf Coast. Such great thicknesses require a modification of Figure 10-3 with a growth fault on the landward side of the marine sandstone. These faults are common throughout the Gulf Coast Tertiary, Mid-Continent, Beaufort Sea, MacKenzie delta, Niger delta, offshore southwest Africa, Indonesia, etc. The influence of growth faulting on sedimentation is of such great importance that Chapter 11 is devoted to this subject.

Cyclic Subsidence

Limited Supply of Sand, Abundant Supply of Mud. A fairly common example of the facies relations that result from cyclic subsidence is illustrated in Figure 10-4. The supply of sediment is insufficient to offset the rate of subsidence; thus transgression occurs. In this example, there is a limited supply of sand-size material and an abundant sup-

[1]The types of unconformities referred to in this report are adopted from Dunbar and Rodgers (1957, p. 118-119) and are defined as follows: *nonconformity*—stratified rocks lie on nonstratified igneous or metamorphic rocks; *angular unconformity*—an angular discordance sepqrates two units of stratified rocks; *disconformity*—all the strata are parallel, but the contact between two units is an uneven erosion surface; *paraconformity*—the contact between two successive strata is a simple bedding plane, at which position a hiatus is inferred.

ply of clay-size and silt-size material. The available sand is insufficient to form a continuous blanket overlying the unconformity. This condition results in a series of beach sands that impinge on the unconformity and wedge out basinward. They bear a vertical en echelon relation to each other. The position and shape of each sandstone body shown on Figure 10-4 are extremely generalized, and each position represents a stillstand of the shoreline in a period of overall marine transgression. The mud and clay sequences separating the sand wedges form as a result of cyclic subsidence (under conditions of limited sand supply) in which transgression is too rapid to permit continuous sand accumulation (the nature of the beach environment of deposition is discussed in a subsequent section.) Each of the sand wedges in Figure 10-4 is a composite of the sediments deposited in the upper and lower foreshore environment. Thus, one wedge of sand, as generalized here, may consist in part of beach sand and in part of one or more offshore bars. A vertical series of en echelon, thin, discontinuous marine carbonate beds may be present basinward from each of the wedges of sand. In most places these carbonate beds consist of a diagenetically altered mixture of clastic material and fossil excrement. The mud and clay sequences separating any two of them generally have a constant thickness, any deviation from which is a slight increase in thickness basinward. The individual carbonate beds are deposited contemporaneously with shoreward sand bodies at the same level. The carbonate beds are very extensive, parallel with the shoreline, and are a few miles to 10–15 mi (16–24 km) wide perpendicular to the shoreline. Therefore, they are excellent lithologic time markers for detailed stratigraphic analysis. The uppermost part of Figure 10-4 indicates depletion of the sand source and shows the sediments consist entirely of an alternating sequence of thin carbonate and shale beds.

An isopach map of the stratigraphic interval from the unconformity to any one of the several thin carbonate beds shown at the top of the diagram would be a simulated reconstruction of the topography prior to the cyclic marine transgression. The irregularities on this unconformity surface directly control the irregularities of the shoreline trend at any stage of cyclic subsidence. This trend, in turn, partly controls the sinuosity of that part of the marginal-marine depositional environment in which sands are deposited. The seaward gradient of this unconformity surface and the rate of sediment supply also are factors in determining not only the irregularities of the depositional trend, but also the variations in width of a particular sand body. Such an isopach map is an excellent tool for tracing and projecting each of the en echelon sands in the direction of paleodepositional strike.

The Morrowan (basal Pennsylvanian) series of sandstones in the Anadarko basin of the mid-continent area afford an excellent example of the situation illustrated in Figure 10-4. They consist of a repetitious sequence of gas-bearing beach sandstones (interrupted by cross-trending channel sandstones) with all hydrocarbon accumulations occurring in stratigraphic traps. The essentials of prospecting in this area were presented by Busch (1959) in the early stages of prospecting, but subsequent development drilling now affords an opportunity for more detailed analysis. Figure 10-5 is a generalized stratigraphic cross section of the Morrowan strata in northwestern Oklahoma. These strata form a genetic sequence of rocks lying unconformably on the Mississippian; this GSS is defined at the top by the base of the "Thirteen-Finger" limestone. The Morrowan represents a south-to-north transgressive unit with southward thickening that reflects the inclusion in that direction of progressively older beds at the base. Time lines are essentially parallel with the base of the "Thirteen-Finger" limestone

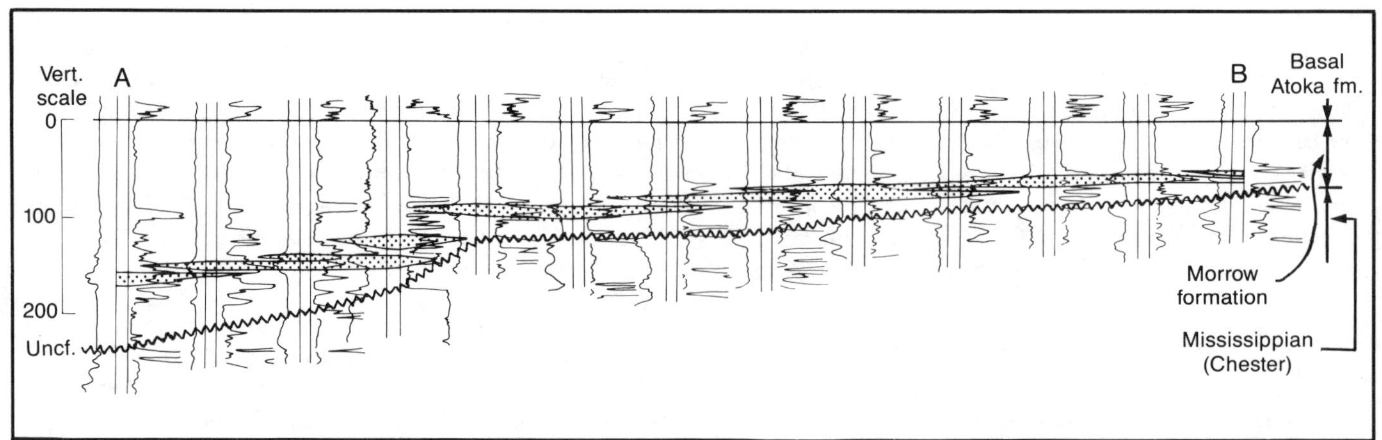

Fig. 10-6. Stratigraphic profile of the Morrow (basal Pennsylvanian) series, northwestern Oklahoma, showing en échelon onlap relationship of reservoir sandstones. Location of profile A-B is shown on Figure 10-11. (Modified after Busch, 1963).

and impinge against the truncated surface of the underlying Mississippian.

Three sandstones may be identified from the cross section in Figure 10-5. They bear an en echelon relation to each other, and each wedges out basinward to the south. The uppermost sandstone impinges against the old Mississippian land surface, whereas the lowest of the three sandstones wedges out before it reaches the unconformity. It cannot be determined from this profile whether the middle sandstone grades into shale before it reaches the unconformity. A more detailed cross section is shown in Figure 10-6. This cross section was drawn at approximately the same position as that of Figure 10-5, but many more logs were used in its construction. Individual en echelon sandstone bodies are stippled. Most of the sandstones grade into either shale or siltstone in the direction of the unconformity. Several of the sandstones grade basinward (south) into siltstone. Each of these sandstone bodies was deposited during a stillstand of the shoreline under conditions of cyclic marine transgression of the Early Pennsylvanian sea. They consist of either an offshore bar or fringing beach sands or both, and they trend parallel with the ancient shoreline positions. Each of the sandstones is a separate reservoir, which are abundantly gas productive.

An isopach map of the genetic sequence (Morrowan series), containing the gas-producing sandstone members, is shown in Figure 10-7. The Morrowan sequence represents more or less continuous sedimentation, although the shoreline was transgressing in a cyclic manner. In essence, this isopach map represents the simulated paleotopographic surface of the eroded Mississippian over which the Early Pennsylvanian sea transgressed. The known trends of individual sandstone members of the Morrowan are indicated by sinuous stippled bands. Each of these sandstones follows

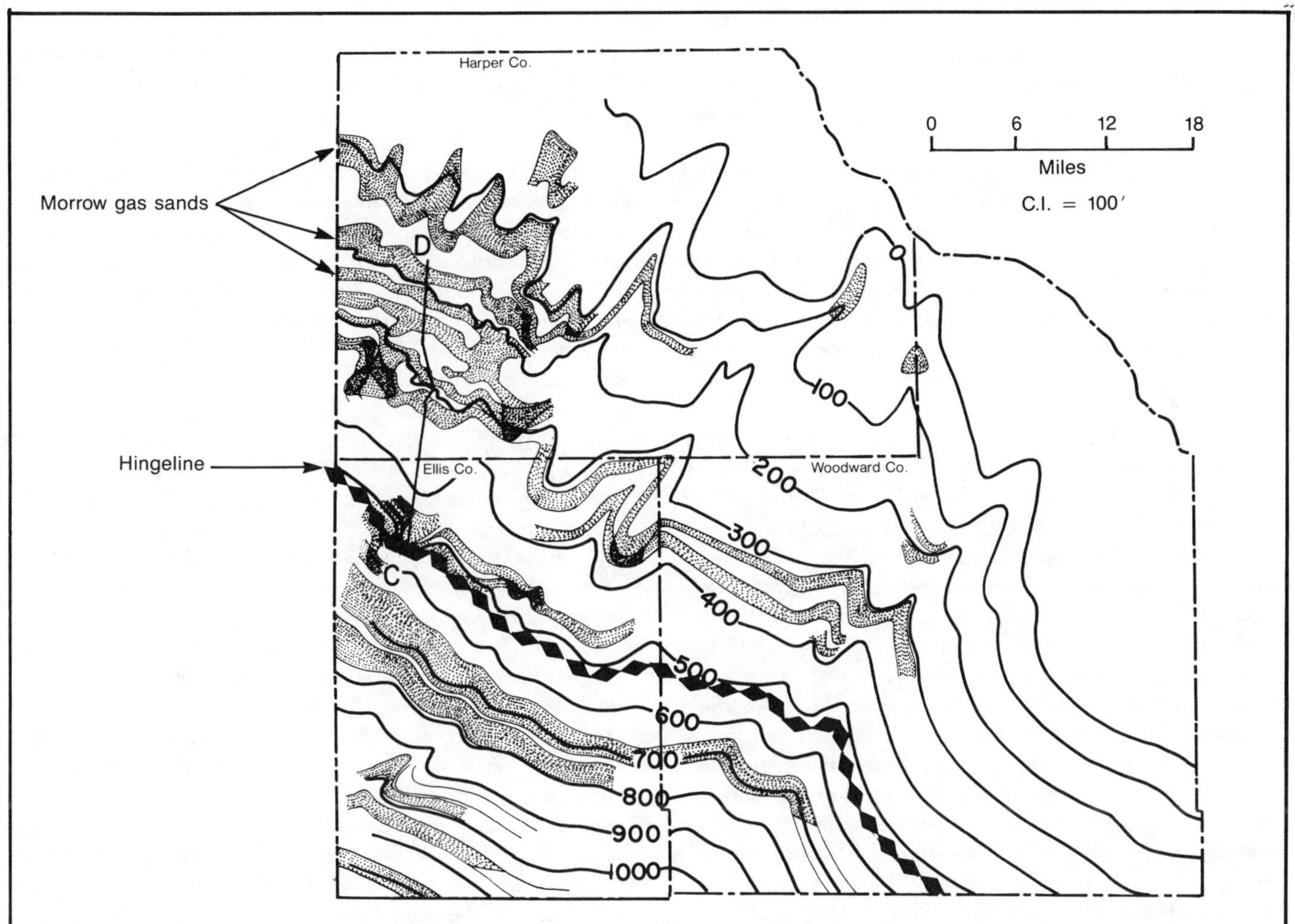

Fig. 10-7. Early isopach interpretation of Morrow Series, northwestern Oklahoma, showing generalized trends of several gas-bearing sandstones (stippled). Note manner in which individual sandstones follow restricted thickness intervals of the Morrow. (Modified after Busch, 1963).

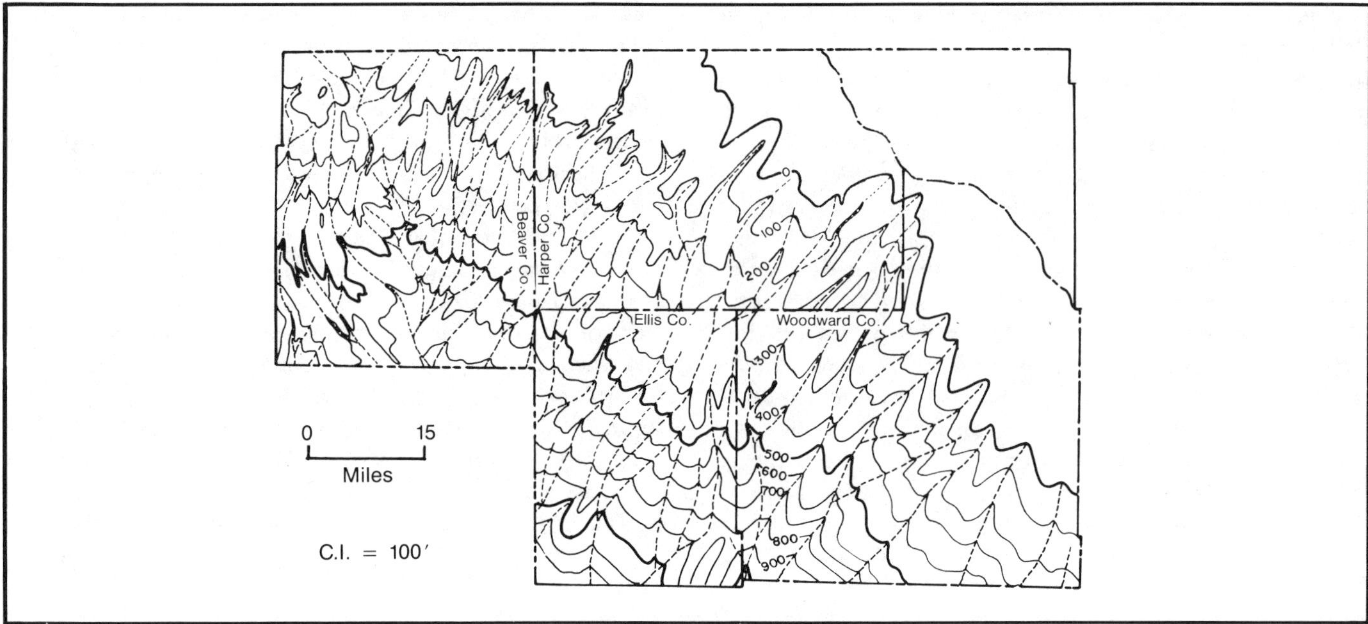

Fig. 10-8. Updated isopach map of the Morrowan series, northwestern Oklahoma, which simulates the pre-Pennsylvanian paleotopographic surface; stream channels are indicated by dashed lines. (Modified after Khaiwka, 1968; permission to publish by University of Oklahoma).

a different critical thickness interval of the genetic increment to which it is related. In all of this analysis the base of the "Thirteen-Finger" limestone is considered as an isochron essentially parallel with sea level. The basinward thickening of the Morrowan series simulates the basinward slope of the unconformable Mississippian surface. Because this was an irregular surface, the cyclic shoreline trends were equally irregular, causing the sinuosity of the en échelon, "stair-step" sandstones.

Figure 10-8 (Khaiwka, 1968) was drawn approximately five years after Figure 10-7 when more than twice as many log data were available. This isopach map (Fig. 10-8) of the Morrowan clearly shows the southwesterly-flowing pre-Pennsylvanian drainage courses (dotted lines) that were eroded into the Mississippian surface. As an adjunct to this study, Khaiwka has identified every sandstone and siltstone unit within the Morrowan by number. The trends and positions of their respective northeastward wedge-outs are shown in Figure 10-9. In most of the area there is considerable overlap of several sandstones and siltstones; therefore, it is impossible to portray graphically the geographic extent of each sandstone on one map. Most of the shoreline positions are represented by siltstone that terminates against the Mississippian unconformity surface. In several places, however, sandstone impinges against the unconformity. The shoreline trends of Figure 10-9 show striking parallelism with the simulated paleotopographic configuration of the Mississippian (Fig. 10-8).

In Figure 10-10 the trends, thicknesses, and widths of three of the numbered Morrowan sandstone and siltstone members are shown. The landward (northern) margins of these sandstones parallel the shoreline. The irregularities of the seaward margins probably resulted from such variable factors as irregularities on the seafloor (caused partly by differential compaction), direction of wave motion, longshore currents, and delta positions. Sandstone 29 not only exhibits a very irregular, fringing beach sand, but also several subparallel offshore bars. This type of map may be constructed for each of the Morrowan sandstone members.

Figure 10-11 is a structure map of the Mississippian surface in northwestern Oklahoma. With two local exceptions there is almost complete absence of structural closure. This surface has been tilted basinward (south) in post-Morrowan time, as indicated by steeper basinward dip than that of the paleotopographic surface shown in Figure 10-8. Also, the structural contour lines are less sinuous than the contours of the paleotopographic map.

In all the studies of the Morrowan sandstones, it has been possible to trace individual sandstone members systematically, as outlined previously. These maps (Figures 10-8, 10-9, 10-10, and 10-11), however, constitute only a geologic framework for systematic sampling, detailed thin-section studies, and other petrographic investigations. There should be diagnostic petrographic differences for the offshore-bar, fringing-beach, channel, and delta sandstones—all of which are present in the Morrowan series. The maximum density of well control for these studies is one well per square mile. In the southeastern part of the study area, the density of well control is considerably less.

Limited Sand, Moderate Mud, and Abundant Supply of

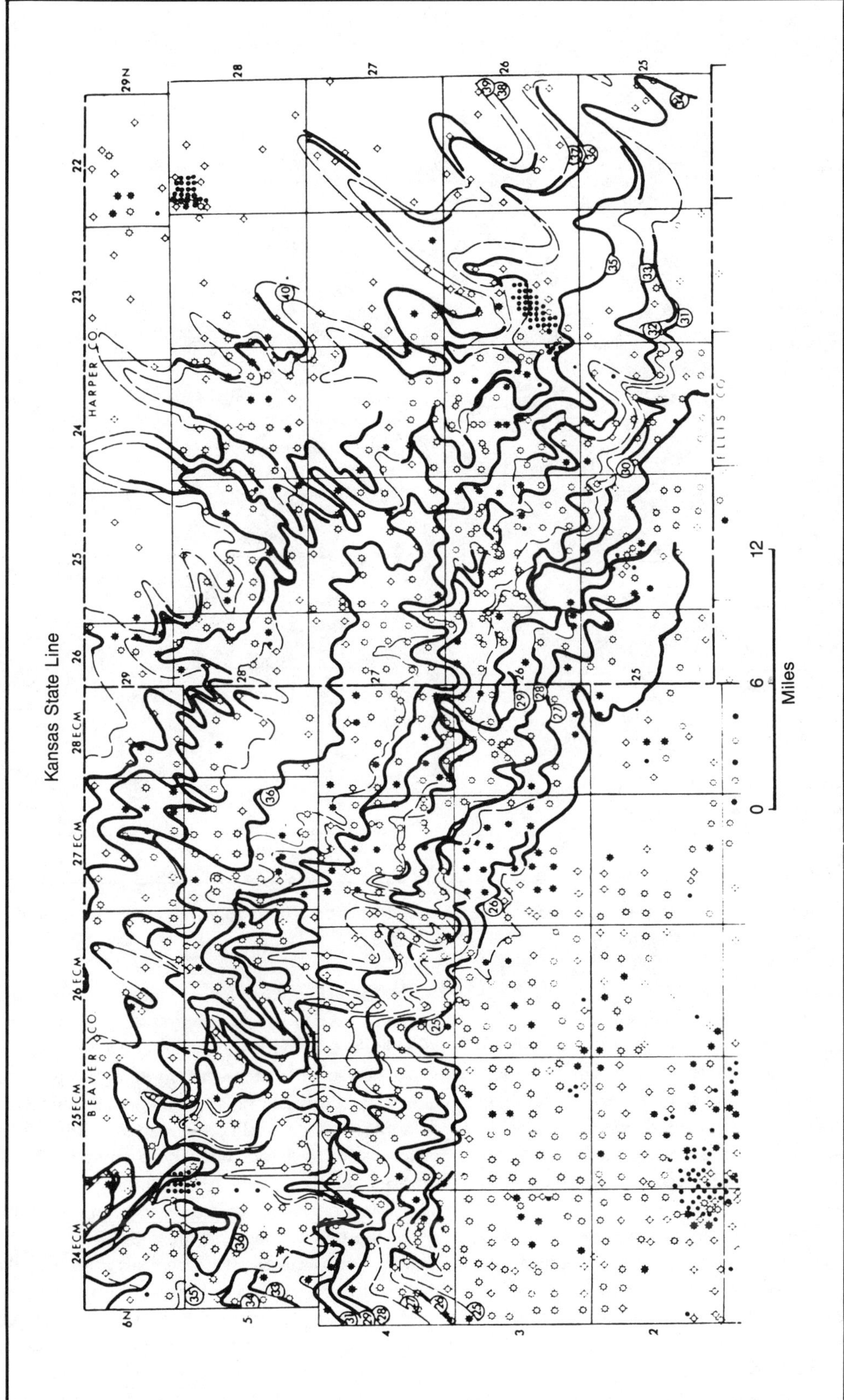

Fig. 10-9. Several successive stages of shoreline trends in Morrowan series, northwestern Oklahoma. Heavy lines are shorelines with sand development; dashed lines are shoreline without sand development. (Modified after Khaiwka, 1968; permission to publish by The University of Oklahoma).

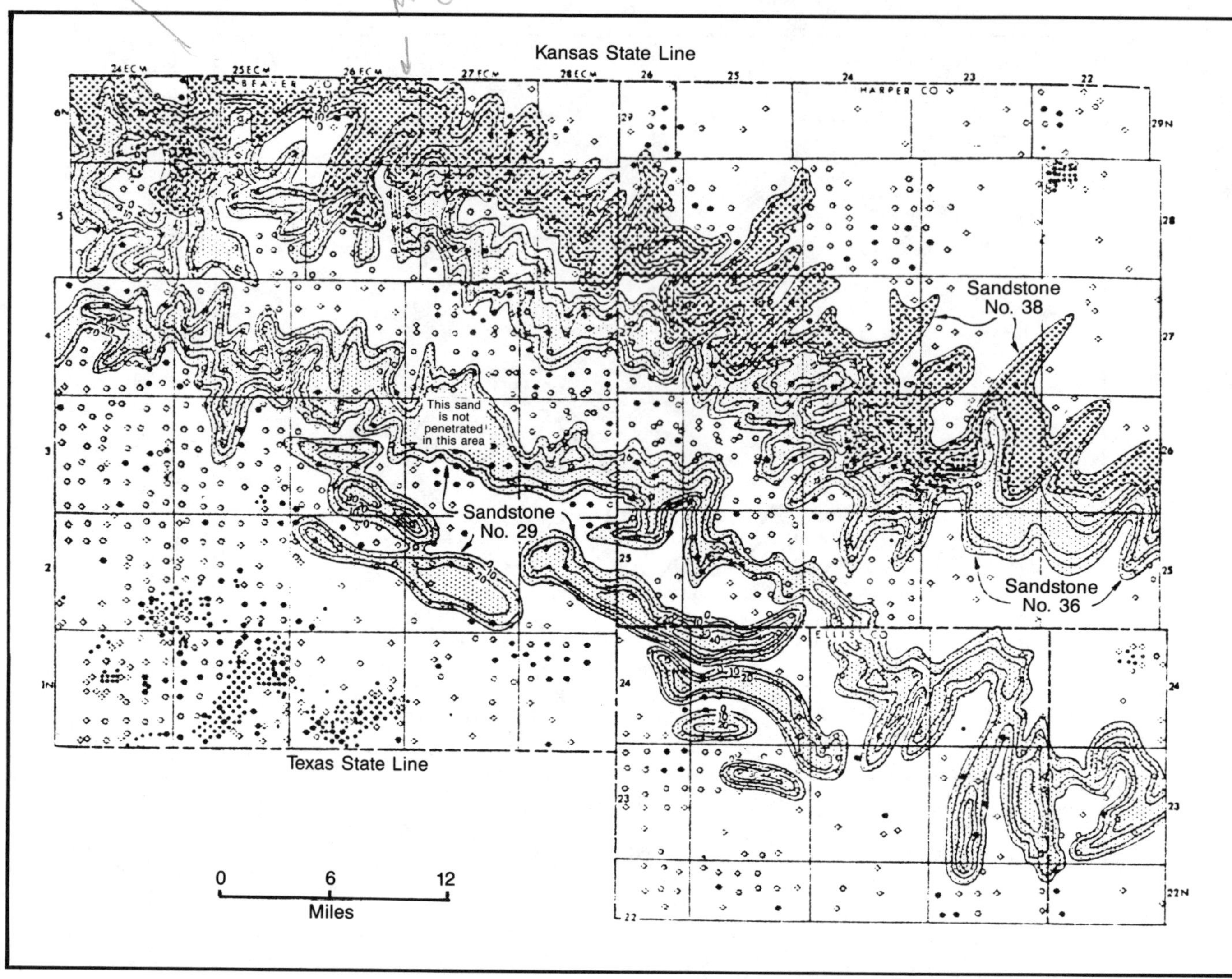

Fig. 10-10. Isopach map of three Morrowan sandstone members, northwestern Oklahoma. (Modified after Khaiwka, 1968; permission to publish by The University of Oklahoma).

Carbonate. An example of cyclic subsidence in which there is a limited supply of sand, a moderate supply of mud, and an abundant supply of carbonate material, is illustrated in Figure 10-12. As in Figure 10-4, separate bodies of beach sand are deposited en échelon (as seen in vertical section) along the unconformable slope. Each sand body has a time-equivalent, thin, carbonate unit present basinward. The beach sands are deposited parallel with successive shoreline positions. The topography of the unconformity surface can be reconstructed by drawing an isopach map of the interval from the top of the carbonate wedge to the unconformity. The relatively persistent thickness of the predominantly mud-clay interval between the unconformity and the carbonate wedge is the most striking feature of this illustration. This mud interval may extend as a continuous lithologic unit for hundreds of miles both normal to and parallel with the depositional trend.

The Geological Survey of Canada defines a sedimentary formation as ". . . a lithologically distinctive product of essentially continuous sedimentation selected from a local succession of strata as a convenient unit for purposes of mapping, description, and reference" (American Commission on Stratigraphic Nomenclature Note 1948). This definition refers to a rock unit whose differentiation is not based on time relations. The predominantly shale unit of Figure 10-12 is clearly transgressive in relation to the time lines; all of the unit is older on the right side of the profile than its lithologic equivalent on the left. Most of the contained marine faunas have the same pattern of distribution. The interpretation shown in Figure 10-12 is derived from the Lower Silurian stratigraphic sequence of the northwestern-most part of the Appalachian geosyncline. Although the dominant source of sediments in this basin was to the southeast, the northeast-trending western margin of this narrow

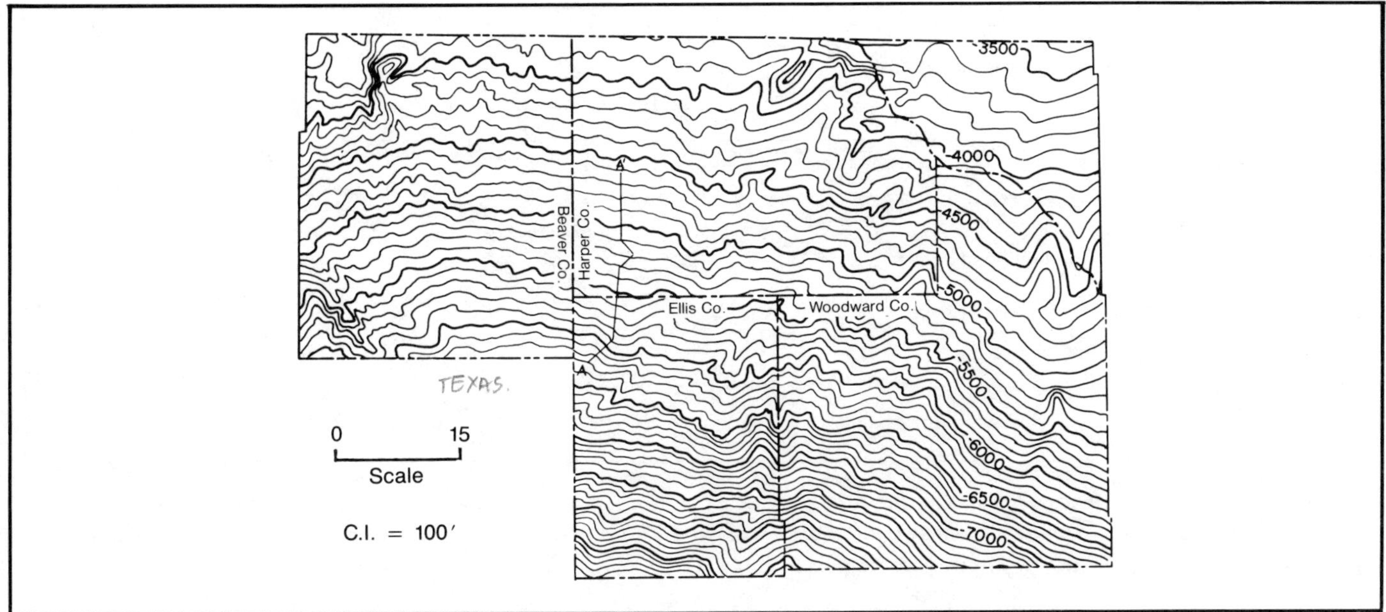

Fig. 10-11. Structure, top of Mississippian, northwestern Oklahoma. Note that axial trends of structural lows are similar to paleodrainage courses shown in Figure 10-8. Line A-A′ shows location of stratigraphic profiles of Figures 10-5 and 10-6. (Modified after Khaiwka, 1968; permission to publish by The University of Oklahoma).

embayment transgressed generally northwestward. The successive "shingles" of beach sand deposited along this northwestern shore clearly indicate the cyclic nature of the subsidence of the depositional surface, the cyclic rise of sea level, or the cyclic influx of large quantities of sediments from the southeast. The shale and sandstone illustrated in Figure 10-12 belong to the Clinton formation; the limestone wedge is the drillers' "Big Lime" (Siluro-Devonian). The thin limestone members, arranged vertically en échelon in the Clinton Shale, are referred to by drillers as "packer lime."

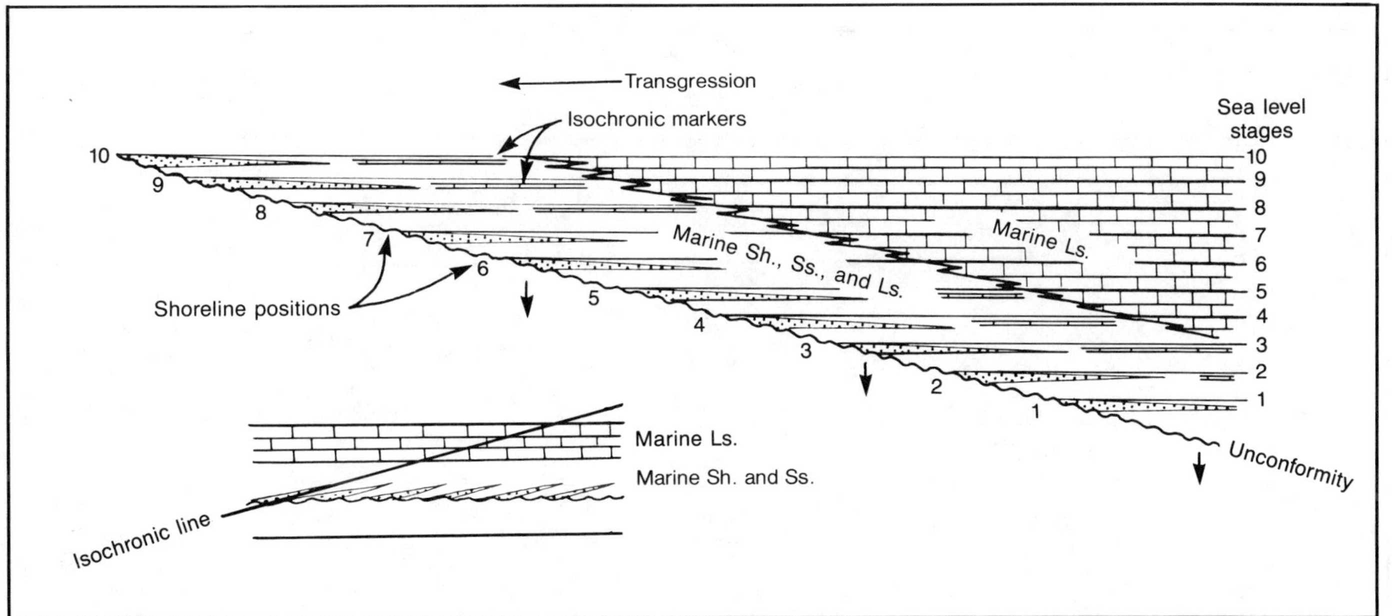

Fig. 10-12. Cyclic subsidence with limited supply of sand, moderate supply of mud, and abundant supply of limestone. Each wedge of sandstone is deposited parallel with, and adjacent to, successive stillstand positions of the shoreline. (Modified after Busch, 1974; permission to publish by AAPG).

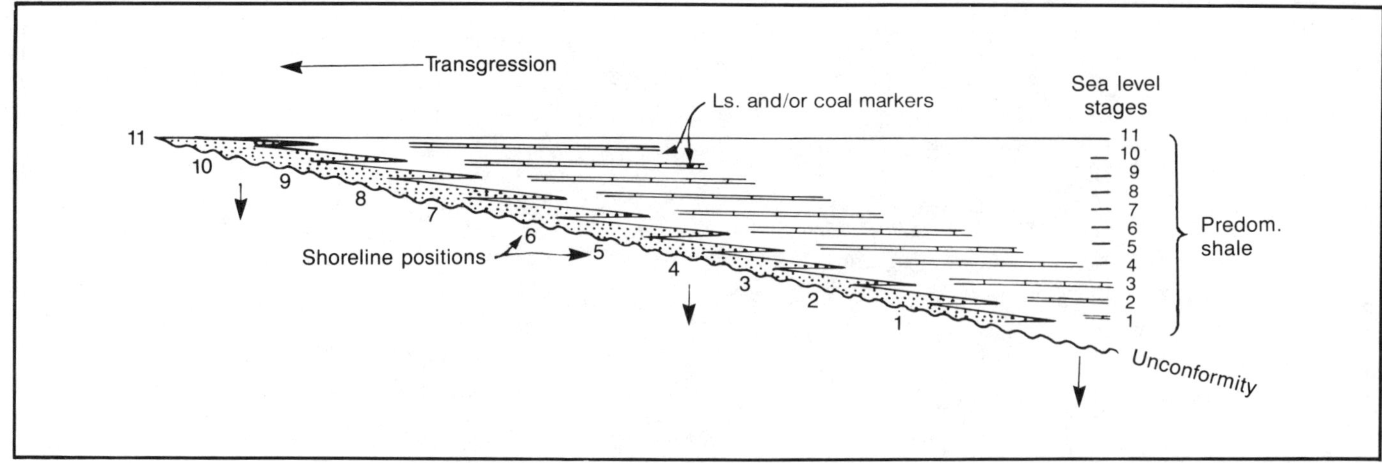

Fig. 10-13. Cyclic subsidence with moderate supply of sand and abundant supply of mud. Sandstone blankets unconformity. Upper surface of sandstone consists of a series of en echelon wedges, each of which represents a stillstand of shoreline. (Modified after Busch, 1974; permission to publish by AAPG).

They are lithologic time markers and are useful as reference data in selective isopach mapping of parts of the Clinton Shale that underlie these data.

Another example of a marine shale unit of fairly uniform thickness that is transgressive of time lines is the Chattanooga shale. This formation contains Late Devonian fossils at its type locality in Chattanooga, Tennessee. It has been traced as a continuous lithologic unit northward 400 mi (645 km) along the east side of the Nashville dome and across Kentucky into central Ohio along the east side of the Cincinnati arch. In east-central Ohio this shale is Early Mississippian in age and is underlain by the Bedford shale and Berea sandstone (earliest Mississippian). Thus, the Chattanooga shale is a transgressive unit of Devonian age to the southwest and Mississippian age to the northeast.

From the two cited examples, it appears that any marine shale that maintains a uniform thickness for a considerable distance in a direction normal to shoreline trends is likely to have either a transgressive or regressive relation to the time lines.

Moderate Supply of Sand, Abundant Supply of Mud. The sedimentary and stratigraphic situation illustrated in Figure

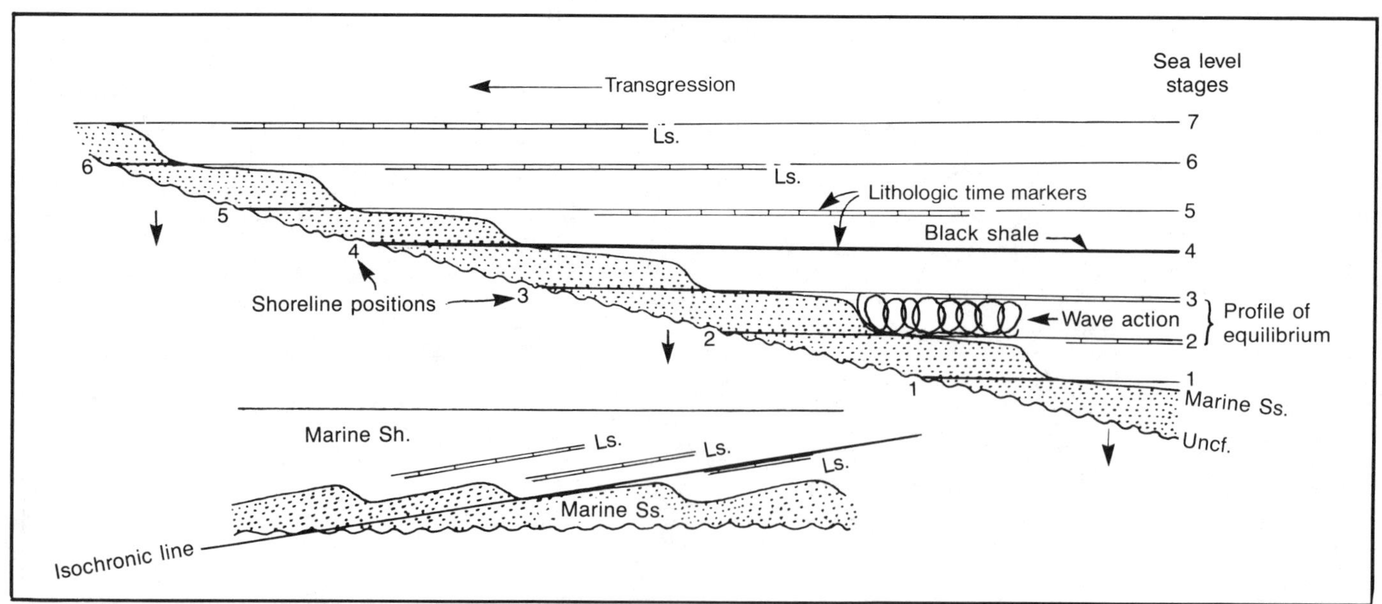

Fig. 10-14. Cyclic subsidence with abundant supply of sand and mud. Sandstone is a continuous, thick blanket lying on unconformity. Terraced upper surface is interpreted as result of wave erosion and deposition during successive stillstands of sea level. (Modified after Busch, 1974; permission to publish by AAPG).

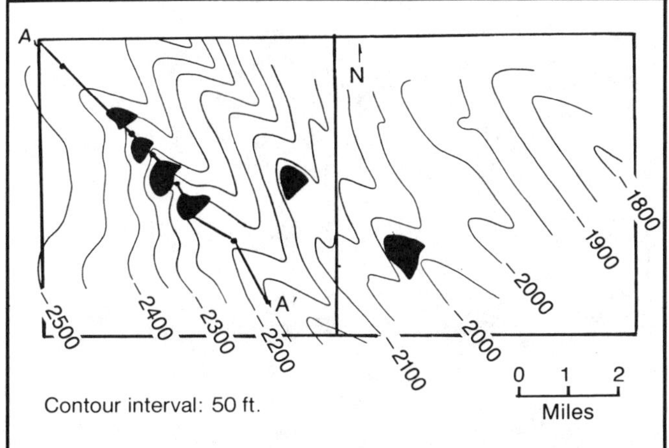

Fig. 10-15. Structure map of a black shale marker bed, showing six oil pools all occurring on the axes of several northwest-plunging structural noses. Note three sides of each pool are parallel with nearest structural contour line, whereas, the southeast margin of each pool is normal to the structural axis. (Modified after Busch, 1959; permission to publish by AAPG).

10-13 is similar to that in Figure 10-12 except that in Figure 10-13 the source area (at the left) furnished a moderate supply of sand and an abundance of mud. The shale is interrupted by numerous thin beds of limestone. There was sufficient sand available to blanket the unconformity. The principal evidence for cyclic subsidence is the "shingled" appearance of the upper part of the sand and the en'echelon arrangement of the thin marine carbonate units that are deposited basinward from the sand wedges. Although there may be effective porosity and permeability within such a sand body, there may be, nevertheless, separate and distinct reservoirs for hydrocarbon accumulation. These reservoirs may develop as a result of either structural or stratigraphic causes. A postdepositional tilt in a counterclockwise direction could cause each of the "shingles" of sand to become a separate reservoir, each having its own oil-water or gas-water contact. In such a situation the main body of the sand, where it lies on the unconformity, could be entirely water-bearing.

The paleotopography at the unconformity may be restored by selecting any of the several thin carbonate beds as a reference datum and constructing an isopach map of the predominantly shale interval from this datum to the underlying unconformity. Because each carbonate datum has only a limited geographic distribution, the use of a "phantom" horizon in areas away from the carbonate bed may be necessary.

Abundant Supply of Sand and Mud. Figure 10-14 illustrates cyclic subsidence under conditions of abundant sand and mud supply. The sand forms a continuous transgressive blanket lying unconformably on either marine or continental beds. The thickness of the sand varies considerably. Time lines are transgressed conspicuously by the sand blanket. Two distinct types of lithologic time markers—thin limestone and thin black shale—are shown. The latter is the better marker to use for a reference datum in this instance because it is more extensive. Black shale markers may extend into nearshore sandstones (Fig. 10-14). Such thin units of black shale produce a characteristic "pip" on the resistivity side of the electric log.

The irregular profile of the upper surface of the sand-

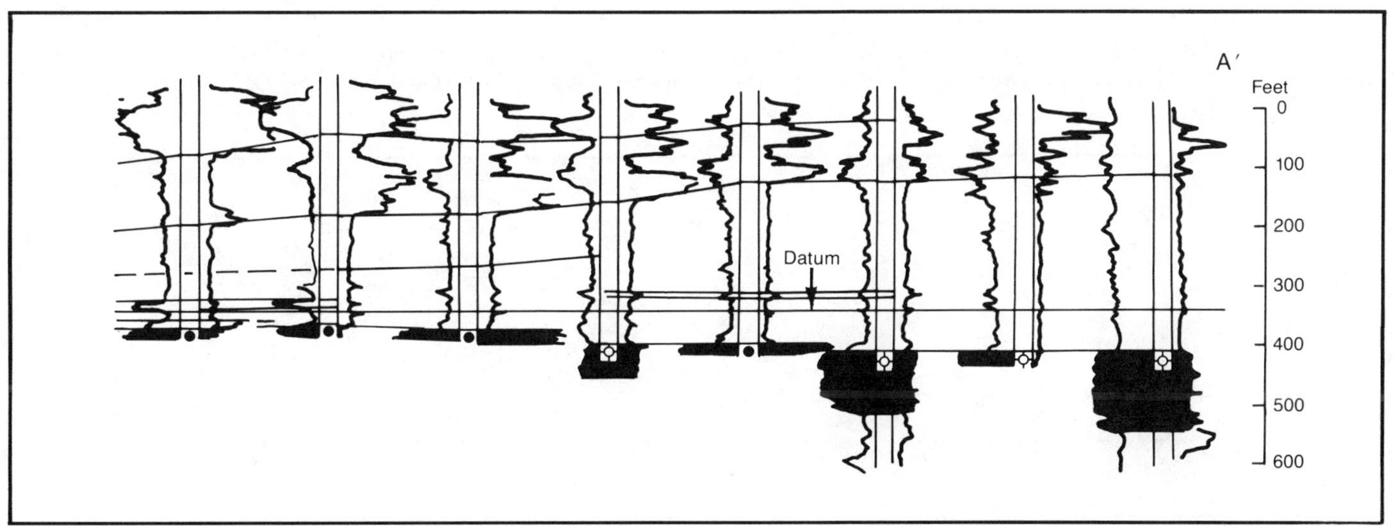

Fig. 10-16. Stratigraphic profile along southeast half of southwestern-most structural axis of Figure 10-15. The reference datum is a thin black shale resistivity "pip" on the E-logs and occurs above the reservoir sandstone. (Modified after Busch, 1959; permission to publish AAPG).

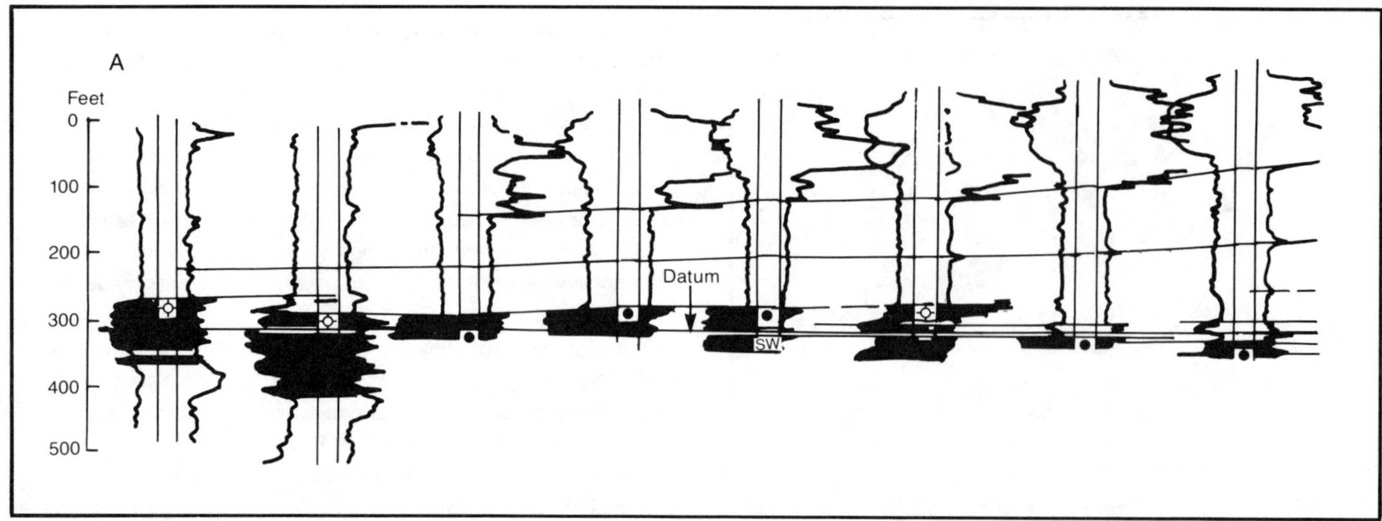

Fig. 10-17. Northwestward extension of stratigraphic profile shown in Figure 10-16. Note that the black shale reference datum extends into the sandstone (below the top) to the northwest, as diagrammatically shown in Figure 10-15. (Modified after Busch, 1959; permission to publish by AAPG).

stone is interpreted as the result of wave erosion during successive stillstands of sea level. This erosion produces a series of sand ridges that are steeper on the seaward side. A postdepositional counterclockwise rotation of this section would produce ideal traps for accumulation of oil and gas along the highest parts of the individual ridges. To map the structure of the upper surface of such a sandstone body, abundant well control is required. Thus, this type of map would have little exploratory value.

Figure 10-15 is a structure map (Busch, 1959) of a black shale marker bed occurring above (and extending into) such a "ridged" sandstone as shown in Figure 10-14. Six oil pools are situated on the northwest-plunging axes of several structural noses. Three sides of each oil pool are parallel with the nearest structural contour line, suggesting an oil-water contact. This is not, however, the case on the southeast side of each pool where the margin is a line normal to the trends of the structural axes. An understanding of this anomalous southeastern margin is essential to an explanation of these six oil pools and equally important if one is to define drillable prospects.

Figures 10-16 and 10-17 are a stratigraphic profile along the axis of the southwesternmost structural nose of Figure 10-15. The reference datum is a thin black shale resistivity "pip" on the E-logs. It occurs above the sandstone in Figure 10-16, and in much of Figure 10-17 it occurs below the top of this sandstone. The composite of these two profiles indicates that the depositional basin to the southeast, which is the direction of thickening of the shale overlying the sandstone. The present tilt of these sediments to the northwest is in Figure 10-18, which is a structural cross section of combined figures 10-16 and 10-17. The profile passes

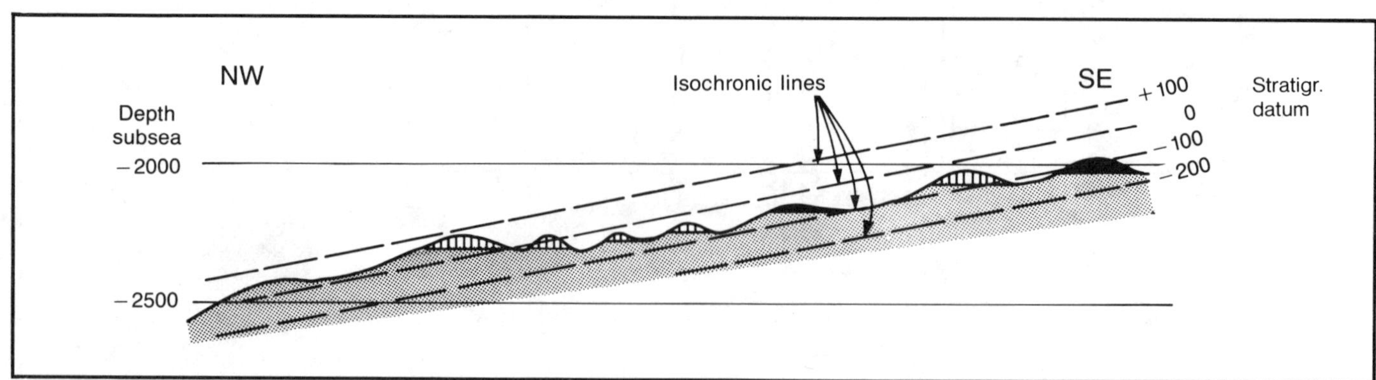

Fig. 10-18. Structural profile of combined Figures 10-16 and 10-17 showing a series of ridges. The profile passes through oil pools where ridges are shaded black and through prospect areas where ridges are vertically shaded. (Modified after Busch, 1959; permission to publish by AAPG).

through oil pools where ridges are shaded black and through prospect areas where the ridges are vertically shaded.

Figure 10-19 is an isopach map of the interval between the black shale marker bed and the top of the reservoir sandstone. This interval consists of shale in the southeastern half of the area and of sandstone in the northwestern half. The isopach map thus presents a simulated topographic map of the top of the sandstone. The upper surface is not smooth but, rather, consists of a series of "stairsteps" trending N 45° E. When tilted structurally northwest they appear as a series of ridges. The southeastern margins of all six oil pools may now be seen to occur along the steep sides of the "stairsteps" of the upper surface of the sandstone. In other words, there is an abrupt thinning of the sandstone along the southeastern margin of each pool.

From this analysis it is apparent that all six oil pools occur in restricted areas where depositional ridges are intersected by northwest-trending structural noses. Thus, prospect areas are shaded where the depositional ridges and structural noses intersect.

RISING BOTTOM

Gradual Emergence

Limited Supply of Sand, Abundant Supply of Mud. Figure 10-20 shows the situation in which the surface of deposition is rising gradually and sand supply is limited. Under these circumstances, shore deposits are destroyed progressively, and the sandstone and shale facies are regressive. In Stage 1, a marginal-marine sand is deposited in the area from the shoreline to the line of intersection of the sand-mud line and the submarine depositional surface. The result of weathering and erosion of the sand and continental beds under conditions of gradual emergence and limited sand supply is shown in Stage 2. The upper surface

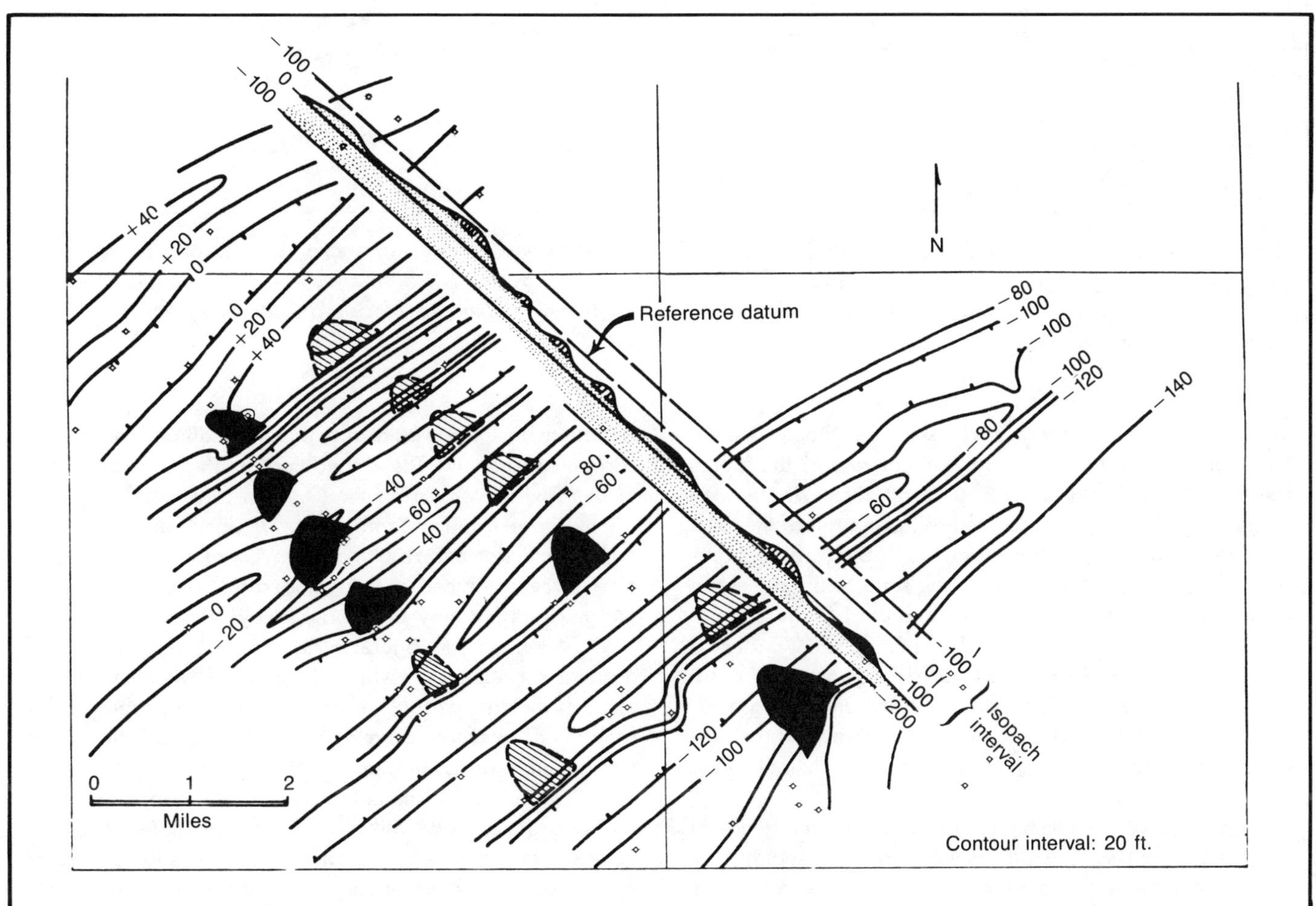

Fig. 10-19. Isopach map of the interval between the black shale marker and the top of the reservoir sandstone. Note that this interval consists of shale in the southeastern half of the area and sandstone in the northwestern half. The depositional trend of the sandstone ridges is clearly north 45° east. (Modified after Busch, 1959; permission to publish by AAPG).

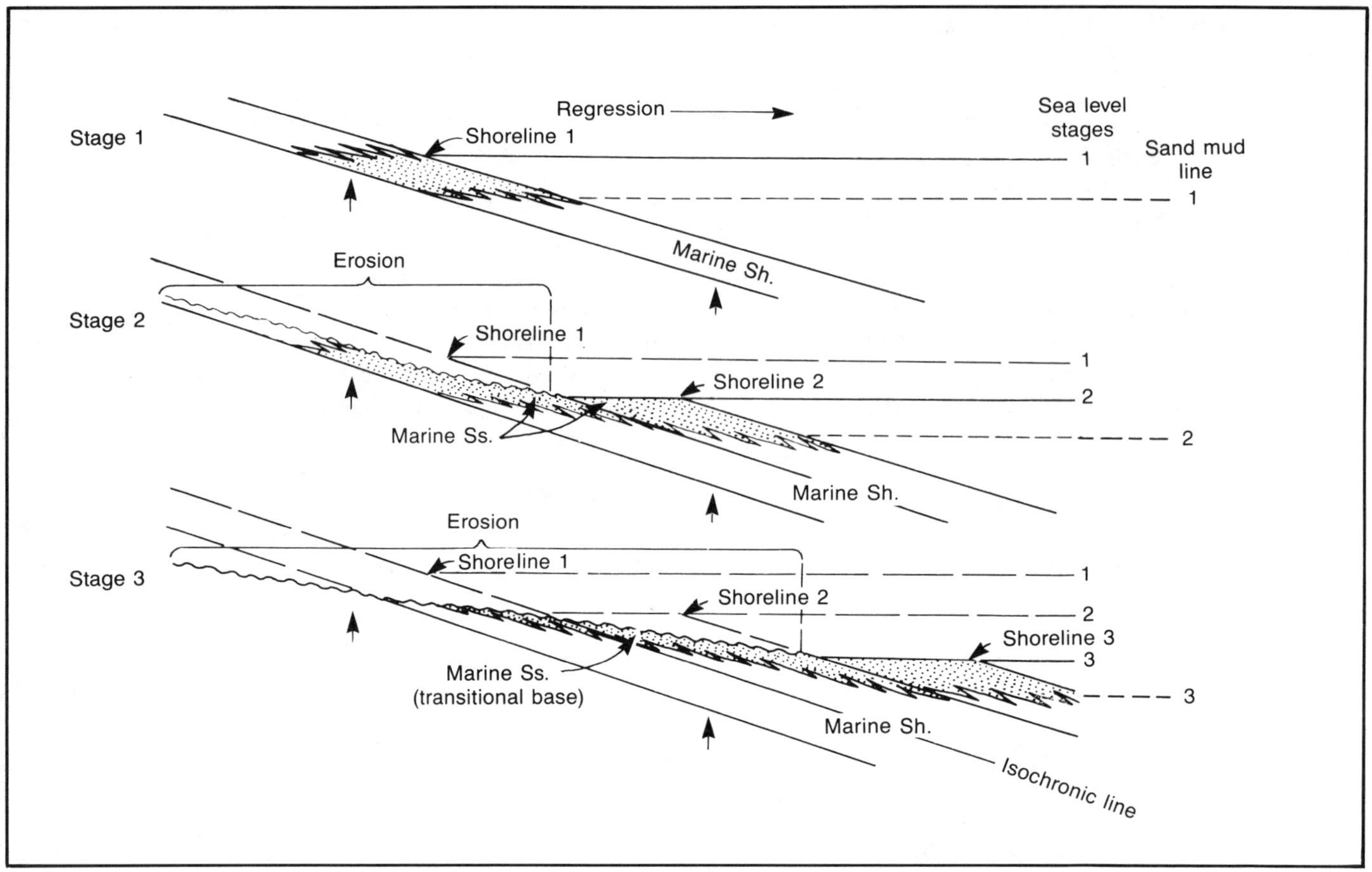

Fig. 10-20. Gradual uplift with limited supply of sand. There is continuous erosion of shore deposits and recycling of marginal-marine sands deposited in regressive sea. Basal contact of sandstone is transitional with underlying shale, and upper surface is partly unconformable and partly conformable with any sediments deposited later. (Modified after Busch, 1974; permission to publish by AAPG).

of the sand and continental beds (shown in Stage 1) is an unconformable surface in Stage 2. Furthermore, the shoreline has retreated (regressed) to a new position. The eroded material of Stage 2 is transported to the margin of the beds of Stage 1 are recycled in Stage 2. In Stage 3, all continental beds and most of the marine sand of Stage 1 have been eroded. Furthermore, the new sand of Stage 2 is actively weathered and eroded. The site of sand and mud deposition gradually has shifted farther seaward. The basal contact of the sand is transitional with the underlying mud. A part of the upper surface is disconformable, and the seaward part is conformable. In all three stages, lithologic types regress the isochrons.

As a result of this combination of gradual emergence of the depositional surface and limited sediment supply, the marine sand body has a very limited width but may have a linear trend of many miles. If the surface of deposition should begin to subside, the lenticular sand body shown in Stage 3 would be completely isolated because of mud deposition above it. This complete isolation of such a sandstone lens in shale would offer ideal conditions for hydrocarbon entrapment. These conditions existed in many areas of the Gulf Coast during the Cenozoic.

Moderate to Abundant Supply of Sand and Mud. The facies arrangement resulting from conditions of gradual emergence combined with a moderate to abundant sand and mud supply may be visualized by reference to Figure 10-2. Although this figure is discussed as an example of gradual subsidence in which the rate of sediment supply exceeds the rate of subsidence, it is also applicable to this situation. The source area of the sediments and the site of their ultimate deposition are uplifted concurrently at similar rates. Figure 10-2 is the only example known to the authors for which it might be difficult, if not impossible, to determine whether the depositional surface was submerging or emerging. In such situations, a correct analysis generally may be made by determing whether the stratigraphic sequences just above and below were deposited during submergence or emergence, because the intermediate stratigraphic unit is likely to represent the opposite condition.

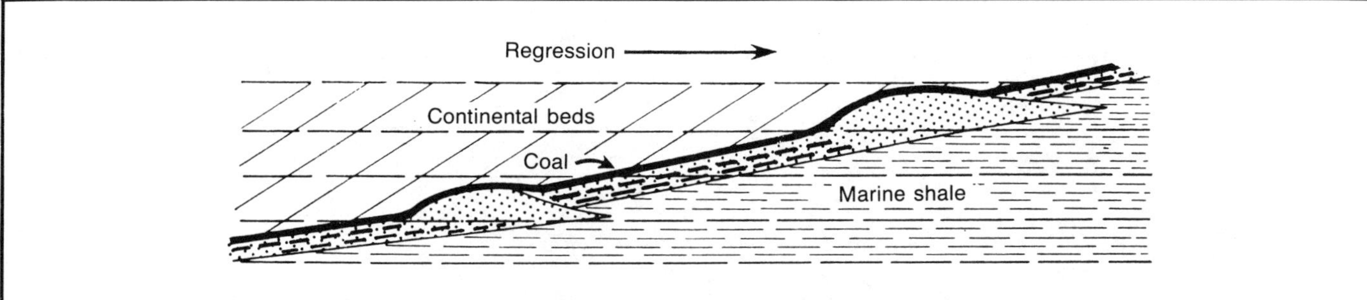

Fig. 10-21. Cyclic uplift with moderate, steady supply of sand. Each of the thicker sandstones is a beach sand and is the result of a stillstand of the shoreline. The sheet sands between are deposited under conditions of relatively rapid regression. Such sand is generally poorly sorted, shaly, thin bedded, and even interbedded with shale and, as such, is devoid of reservoir characteristics. (Modified after Busch, 1974; permission to publish by AAPG).

Cyclic Emergence

Moderate, Steady Supply of Sand and Abundant Supply of Mud. In some places, cyclic emergence of a depositional surface occurs. Figure 10-21 illustrates the facies relations resulting from cyclic emergence where the supply of sediment is steady and moderate in quantity. The overall picture is a regressive sheet sandstone bordered on the seaward side (right) by marine shale and on the landward side (left) by coal and continental beds. This sandstone differs, however, from the type shown in Figure 10-2 in that there are localized areas where it thickens. Such areas represent a stillstand of the beach environment as the shoreline regressed in a cyclic fashion. The areas of thin sandstone between the locally thick sandstones are due to fairly rapid regression. The thickness of this sheet sandstone is related directly to the rate of regression and the amount of sand available. The dashed horizontal lines in Figure 10-21 are isochrons. Each crosses, and thus illustrates, concurrent deposition of four lithologic types from left to right: continental beds, a coal-forming swamp, marine sand, and ma-

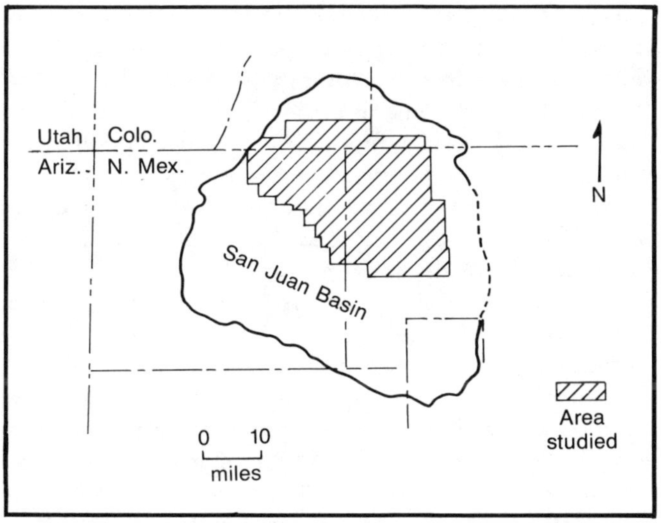

Fig. 10-22. Location of San Juan Basin and Four Corners area. (Modified after Hollenshead and Pritchard, 1961; permission to publish by AAPG).

System	Group	Lithologic Units San Juan Basin
Cretaceous		McDermott (Cgl.)
		Kirtland Sh.
		Fruitland Ss., Sh., Coal
		Pictured Cliffs Ss.
		Lewis Sh.
	Mesa-verde	Cliff House Ss.
		Menefee Ss., Sh., Coal
		Point Lookout Ss.
		Mancos Sh.
	Dak.	Dakota Ss., Sh., Coal

Fig. 10-23. Named Upper Cretaceous lithologic units in San Juan Basin area. (Modified after Hollenshead and Pritchard, 1961; permission to publish by AAPG).

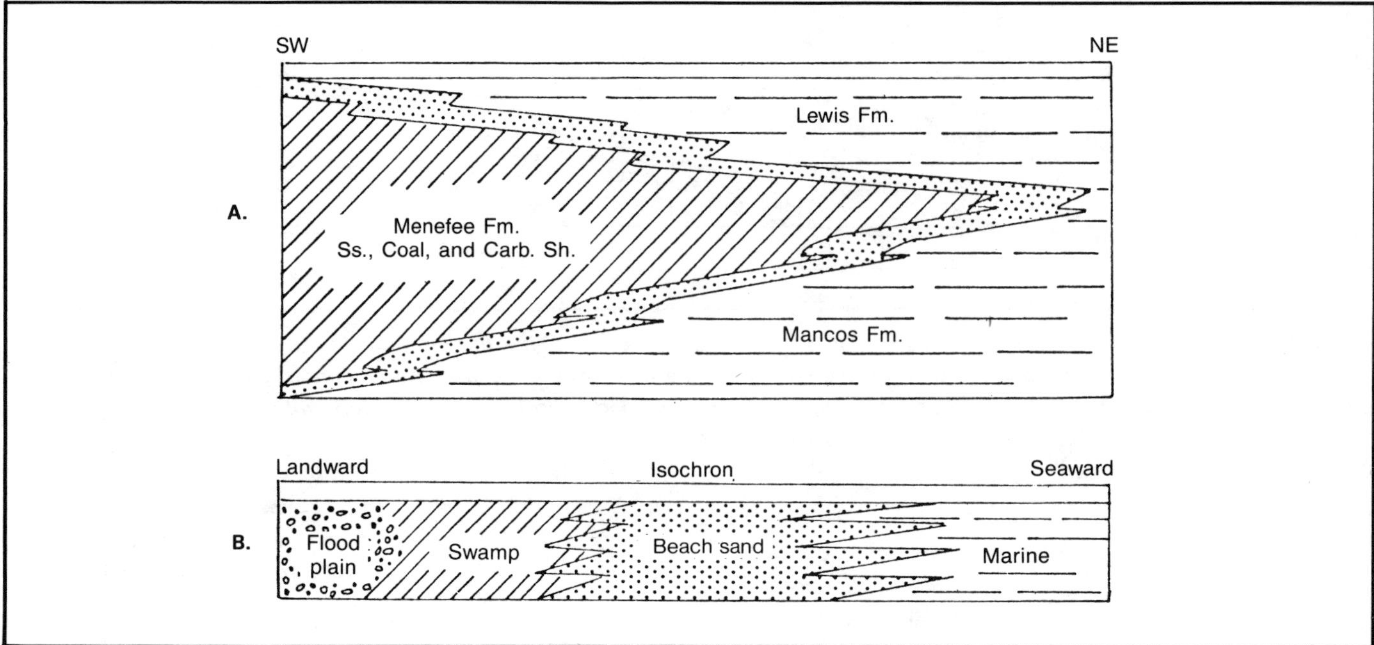

Fig. 10-24. **A.** *Southwest-northeast cross section of Mesa Verde in San Juan Basin, showing time relation.* **B.** *Mesa Verde depositional environments.* (Modified after Hollenshead and Pritchard, 1961).

rine mud. In this situation, where the coal bed overlies a regressive marine sandstone, it should not be used as a datum for subsurface stratigraphic analysis. This is because it was deposited in a narrow fringing swamp that regressed with the continental and marginal-marine environments. Under the conditions illustrated in Figure 10-21, the locally thick sandstone is likely to have well-sorted sand grains, better porosity, and better permeability than the thin sheet-sandstone beds. The sheet sandstone is not well-sorted, is thinner-bedded, and may even be interbedded with shale and siltstone.

An excellent example of the significance of both cyclic emergence and cyclic submergence is presented by Hollenshead and Pritchard (1961) relative to the Mesaverde sandstone of the San Juan basin of New Mexico. They present a concrete application of fundamental principles of stratigraphy. Large quantities of gas have been produced from sandstones at the bottom and top of the Mesaverde Group. The accumulations are stratigraphic and bear no relation to structure.

The San Juan Basin and the area studied by Hollenshead and Pritchard, relative to the Four Corners area, are shown in Figure 10-22. The gas-producing area is approximately 75 mi (120 km) long and 30 mi (48 km) wide and trends northwest-southeast. Approximately 1,900 electric and gamma-ray-neutron logs, well cuttings, cores, and thin sections were used.

The Mesaverde Group was deposited in Middle Cretaceous time in a broad, shallow sea. The embayment extended for several thousand miles north-south and was 500 to 1,000 mi (800–1600 km) east-west. The western shore lines in the Four Corners area, locally had a northwest-southeast trend. These shorelines fluctuated widely, first regressing and later transgressing. The San Juan Basin is a closed structural basin that lies within a small part of the total area of middle Cretaceous strata. The configuration of this structural basin is reproduced from Hollenshead and Pritchard in Figure 10-28.

Figure 10-23 is a chart that shows the named lithologic units deposited in Late Cretaceous time in the San Juan Basin area. The lowest formation of the Mesa Verde Group is the Point Lookout sandstone. It consists of fine-grained to very fine-grained marine sandstone, which contains considerable gas. The Menefee formation is a northeastward-thinning wedge of continental strata consisting of nonmarine shale, sandstone, and coal; these beds are erratic and have no oil or gas. The Menefee formation is overlain by the Cliff House sandstone, which is fine-grained to very fine-grained sandstone and gas-bearing.

The stratigraphic and facies relations of these three for mations are shown in the upper part of Figure 10-24. The Point Lookout sandstone bears a regressive relation to time lines through approximately 350 ft (107 m) of section. The Cliff House sandstone exhibits a transgressive relation through about 210 ft (64 m) of section (right to left, Fig. 10-24). Between the Point Lookout and Cliff House sandstones, the Menefee thins from 860 ft (262 m) in the southwest to 160 ft (49 m) in the northeast. The marine Mancos shale was

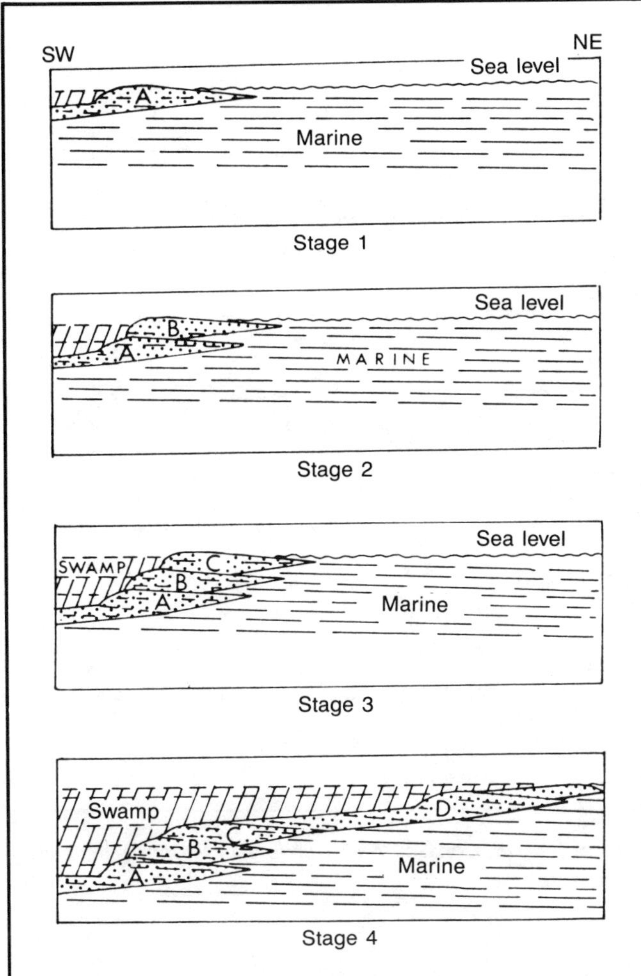

Fig. 10-25. Southwest-northeast cross sections, showing irregular transgressive Point Lookout shoreline. (Modified after Hollenshead and Pritchard, 1961; permission to publish by AAPG).

deposited concurrently with the lower Menefee and the Point Lookout sandstone, whereas the marine Lewis shale was deposited during the transgressive phase of Cliff House deposition. Thus, during the time of Mesaverde deposition, four depositional environments existed concurrently. They are diagramed in the lower part of Figure 10-24. They include a continental-floodplain environment that was separated from the shore- and nearshore-sand environment by a narrow, coal-forming swamp. Basinward from the marginal-marine environment, mud was being deposited. All four environments first regressed and then transgressed during the time of Mesa Verde deposition. The different sediment types of each environment migrated and were deposited laterally in response to the shoreline movements. A thin coal bed directly overlies the Point Lookout sandstone and another underlies the Cliff House sandstone. Neither coal is a meaningful time reference datum. The landward margins of the two sandstones are abrupt where they are in contact with coal, whereas they grade transitionally into the marine shale and siltstone on the seaward side.

Three sets of conditions are illustrated in the lower part of Figure 10-24:

1. The rate of sedimentation exceeded the rate of subsidence, resulting in a regression of the shoreline
2. The rate of subsidence predominated over the rate of sedimentation, resulting in a transgression of the shoreline
3. The rates of sedimentation and subsidence were in equilibrium, resulting in a stillstand of the shoreline and a relatively thick, local accumulation of shoreline and nearshore sands

Four stages of cyclic regression of the shoreline (Fig. 10-25) are recognized for the Point Lookout sandstone by Hollenshead and Pritchard (1961). The Point Lookout is not a uniform blanket sandstone. It contains four local areas of thicker sandstone, each of which is the result of a stillstand of the overall regressive shoreline. These thicker areas consist of well-sorted sandstone, whereas the thinner, platform-like sandstone is poorly sorted and interstratified with siltstone and shale. The latter was deposited in areas of

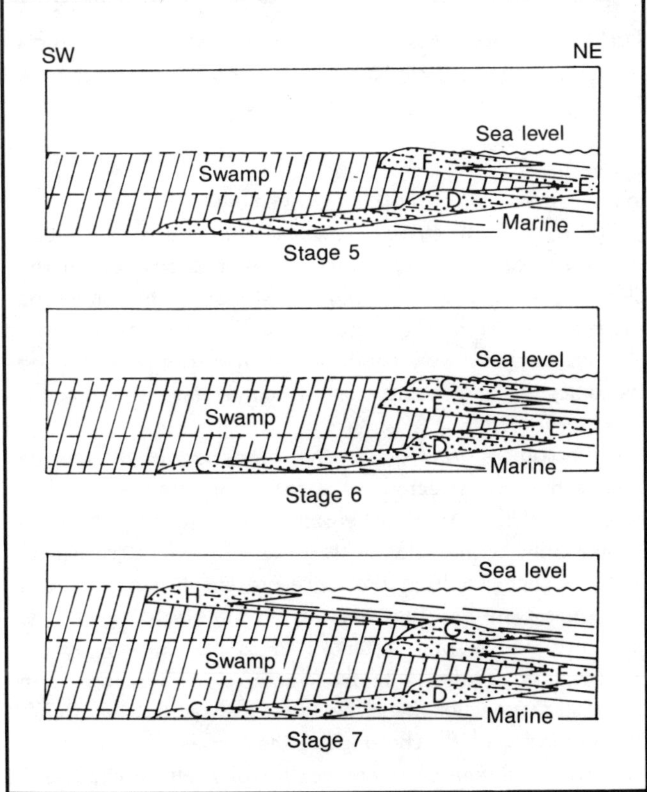

Fig. 10-26. Southwest-northeast cross sections showing irregular transgressive Cliff House sandstone. (Modified after Hollenshead and Pritchard, 1961; permission to publish by AAPG).

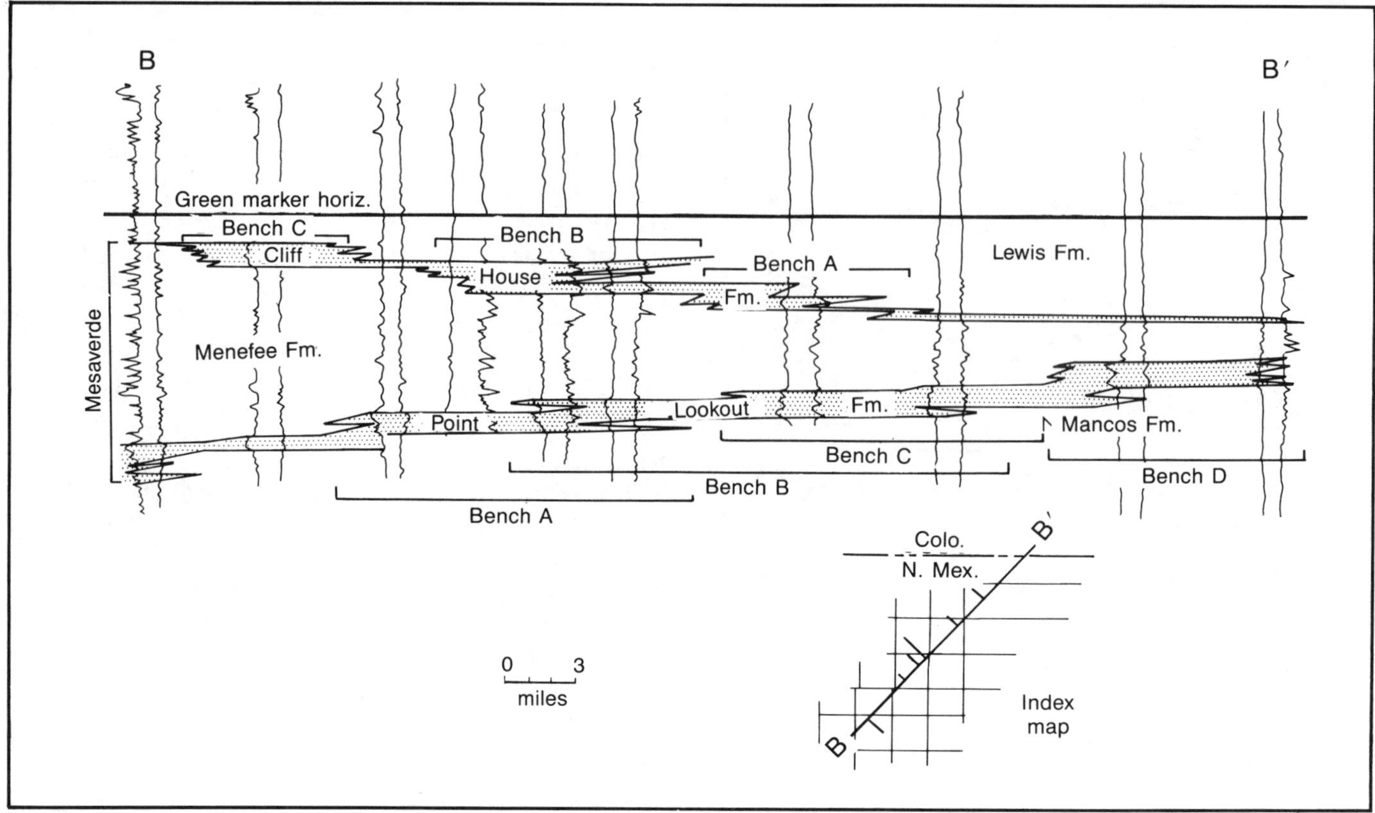

Fig. 10-27. Southwest-northeast cross section B-B′, showing stratigraphic relations of Mesaverde sandstones. (Modified after Hollenshead and Pritchard, 1961; permission to publish by AAPG).

comparatively rapid shoreline movement.

Figure 10-26 illustrates the three stages of cyclic subsidence that occurred during the time of deposition of the Cliff House sandstone. Again, stillstands of the shoreline resulted in locally thicker accumulations of well-sorted sand, whereas comparatively rapid marine transgressions caused deposition of the poorly-sorted, connecting sheet sandstones.

A lithologic marker bed in the lower part of the marine Lewis shale was selected as a datum of reference. Lithologically, it consists of a bentonite zone that appears on the majority of the electric logs as a subdued resistivity "kick." Other usable markers are present in the few areas where this zone becomes obscure on the electric logs. The bentonite marker bed, referred to as the "green marker horizon," is parallel with the other resistivity "lows." In plotting cross sections, such as that shown in Figure 10-27, Hollenshead and Pritchard used the "green marker horizon" as the datum of reference; thereby, the vertical distance down to the upper and lower boundaries of the Menefee could be established readily. By isopaching these intervals they reconstructed the depositional history of the Mesaverde in the restricted area of the San Juan structural

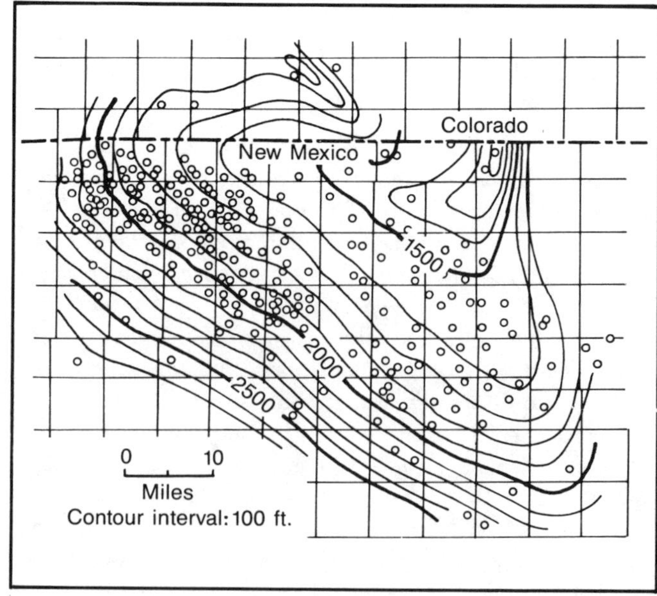

Fig. 10-28. Structure map, "green marker horizon," San Juan Basin, New Mexico. (Modified after Hollenshead and Pritchard, 1961; permission to publish by AAPG).

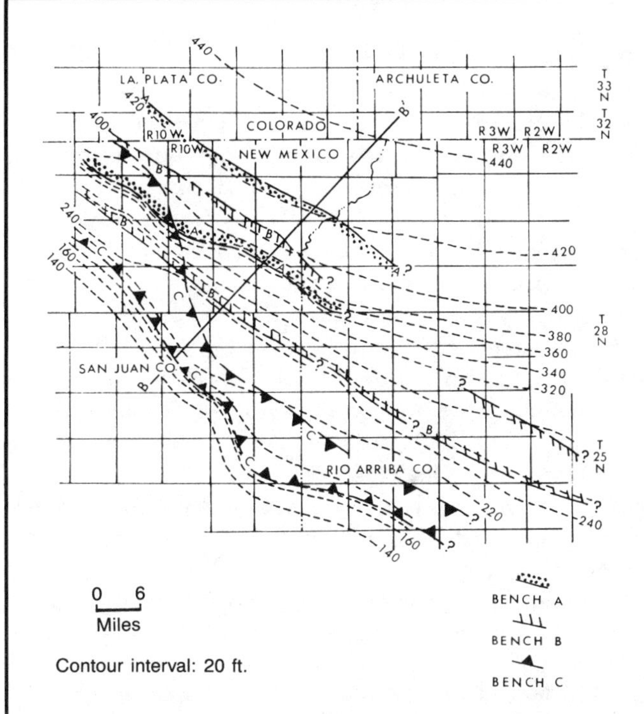

Fig. 10-29. Isopach map, "green marker horizon" to top of Menefee, showing principal Cliff House benches. (Modified after Hollenshead and Pritchard, 1961; permission to publish by AAPG).

basin. The Point Lookout and Cliff House sandstone benches are directly related to steps in the strandline. The sandstone benches terminate abruptly in a landward (southwest) direction against the swamp deposits of the Menefee Formation. The seaward margins (northeast) of these benches are transitional and cannot be defined as easily as the landward margins.

Figure 10-28 is a structural map of the San Juan Basin drawn on the "green marker horizon." A regional dip of approximately 1,500 ft (457 m) is mostly to the northeast. The overall picture is one of a very asymmetric syncline. Gas is trapped principally where sandstone grades into shale updip along the landward margins of the successive Point Lookout and Cliff House sandstone benches. Hollenshead and Pritchard constructed two isopach maps for the purpose of determining the depositional trends and widths of these benches. The first isopach map, shown in Figure 10-29, is of the stratigraphic interval from the "green marker horizon" down to the base of the Cliff House sandstone. This is a genetic increment of strata similar to that illustrated in Figure 2-1D. More stable shoreline areas and trends are identified by closely spaced contour lines, whereas the areas of transgression are indicated by wide spacing of the thickness contours. Thus, the paleodepositional trends of the shoreline have been reconstructed by a technique that removes any effects of postdepositional tectonism. These shoreline trends are completely independent of the present-day structural configuration of the San Juan Basin.

Figure 10-30 is an isopach map of a genetic sequence of strata with the "green marker horizon" at the top; the base of the Menefee was used as the bottom of the GIS because of the sharp contact between the coal bed and the underlying Point Lookout sandstone. Four distinct benches are identified: they represent different stages of regression that occurred during the time of Point Lookout deposition. This technique of selective isopach mapping of GIS's and GSS's serves as a means of projecting shorelines into unexplored areas in which there is a minimum of subsurface well control. Lease plays, as well as the selection of specific drillsites, can be based on this type of subsurface stratigraphic analysis. In this instance, a knowledge of the structure of the area is of little value from the standpoint of leasing and drilling.

Wanek (1954) mapped the trends of two benches (tongues) of Mesaverde sandstone in the outcrop area northwest of the San Juan Basin. The trends and positions of these two benches are remarkably similar to those of the subsurface Cliff House benches B and C (Fig. 10-31). Similar benches, exposed at the surface, have been reported southeast of the San Juan Basin. A surface stratigraphic study of the

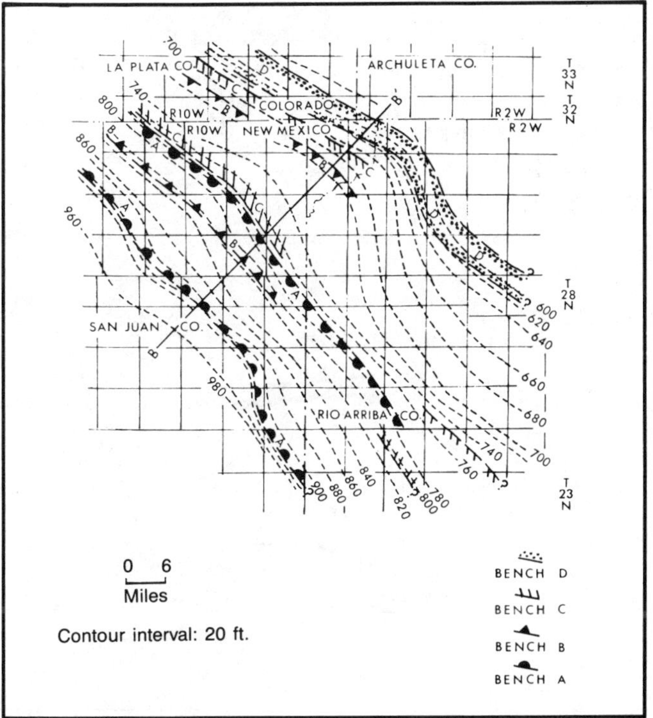

Fig. 10-30. Isopach map, "green marker horizon" to base of Menefee, showing principal Point Lookout benches. (Modified after Hollenshead and Pritchard, 1961; permission to publish by AAPG).

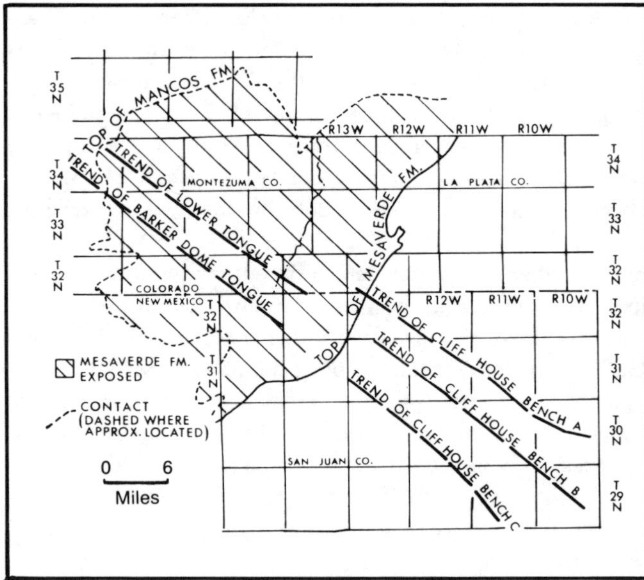

Fig. 10-31. Outcropping Mesaverde sandstone benches (Lower Tongue and Barker Dome Tongue) equivalents of subsurface Cliff House Bench B and Cliff House Bench C, respectively. (Modified after Hollenshead and Pritchard, 1961; permission to publish by AAPG).

Mesaverde in the areas to the southeast and northwest of the San Juan Basin, prior to the development of this extensive gas field, enabled one major company to project and correlate benches of sandstone across the area of the then-unknown subsurface equivalents. By "tying" together the outcrops and predicting the subsurface occurrences of these benches, this company was able to lease and drill considerably ahead of competition.

STATIONARY BOTTOM

Figure 10-32 taken from Grabaü (1913) illustrates the facies relations resulting from a stationary depositional surface. The arrangement of the facies is independent of the rate of sediment supply. The successive positions of the regressive shoreline are numbered to emphasize the direction of shoreline movement. Because there is progressive basinward thinning of the shale beds, the isochronic lines are curved. The coal bed separates the continental beds from the underlying marine sandstone and should not utilized widely as a reference datum because of its regressive nature. The conditions shown in Figure 10-32 are fairly typical of deltaic deposits but are greatly oversimplified. Topset, foreset, and bottomset beds are normally developed. Under marginal-marine conditions, however, none of these beds are likely to be deposited with such steep dips as those shown in Figure 10-32; the attitudes of the foreset beds are especially exaggerated.

The lower part of Figure 10-33 is a duplication of Figure 10-32. It illustrates some of the normal complexities that may confront the subsurface stratigrapher when he compares the logs of various wells. In this example, three wells were drilled through sedimentary sequences that are in sharp contrast but, nevertheless, are genetically related. Well A encountered a preponderance of continental beds underlain by a much thicker section of coal and, below that, a thin marine sandstone. In well B, a more variable stratigraphic section with a partially repetitious sequence of strata was drilled. Well C penetrated a preponderance of marine shale interrupted by a thin marine sandstone. Reconstruction of the depositional history, as seen in this profile, shows the following sequence: (a) stationary bottom with regressive

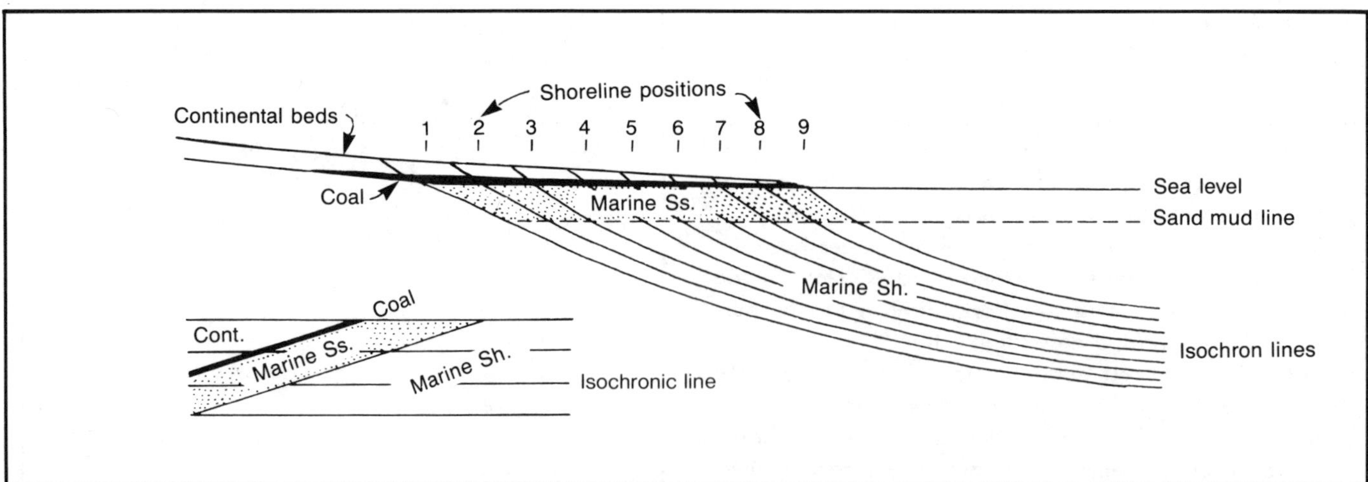

Fig. 10-32. Stationary surface of deposition with steady supply of sediments resulting in regression of shoreline. This sedimentary sequence is fairly typical of a deltaic environment. Tilt of foreset beds is greatly exaggerated for a marginal-marine delta. Isochrons are concave and upward and have a regressive relation to all four lithologic types illustrated. (Modified after Grabaü, 1913; permission to publish by Dover Publications, Inc.).

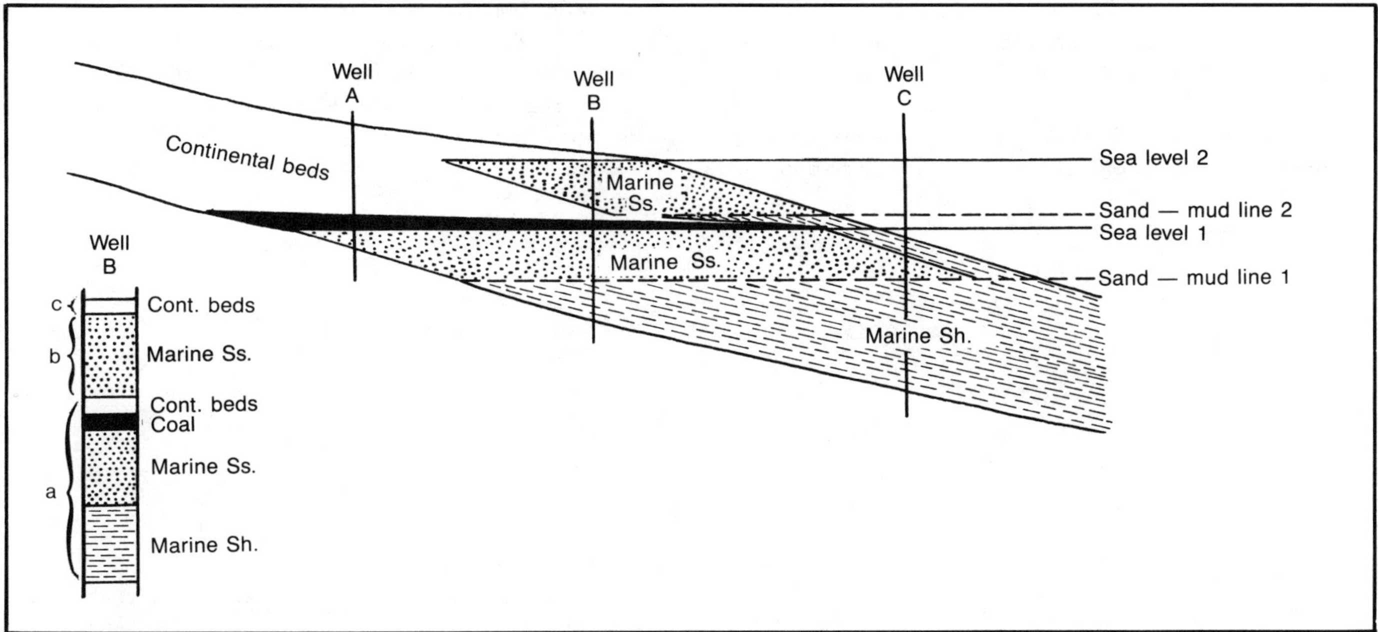

Fig. 10-33. From base upward: **(a)** *stationary surface of deposition with regression of shoreline;* **(b)** *rapid subsidence that exceeds rate of sediment supply, with transgression of shoreline;* **(c)** *stationary surface of deposition with regression of shoreline.* (Modified after Busch, 1974; permission to publish by AAPG).

sediments; (b) subsidence greater than the rate of sediment supply; and (c) stationary bottom with prograding sediments.

A single well may penetrate a combination of any of the cross sections illustrated in this chapter. Other wells in the same area may penetrate other combinations of these cross sections. Thus, it is paramount to identify marker beds and use these as a basis for both time correlations and the unraveling of sedimentary history.

SELECTED BIBLIOGRAPHY

American Commission on Stratigraphic Nomenclature, 1948, Note 3—Rules of geological nomenclature of the Geological Survey of Canada: AAPG Bull., v. 32, no. 3, p. 366-367.

Busch, D. A., 1959, Prospecting for stratigraphic traps: AAPG Bull., v. 43, no. 12, p. 2829-2843.

———, 1963, Methods of prospecting for stratigraphic oil and gas traps: Assoc. Francaise Tech. P'etrole Bull. 160, pt. 1, p. 459-464; Bull. 161, pt. 2, p. 633-643.

———, 1974, Stratigraphic traps in sandstones—exploration techniques: AAPG Mem. 21, 174 p.

Dunbar, C. O., and J. Rodgers, 1957, Principles of stratigraphy: New York, John Wiley & Sons, 356 p.

Emery, K. O., 1965, Characteristics of continental shelves and slopes: AAPG Bull., v. 49, no. 9, p. 1379-1384.

———, 1980, Continental margins—classification and petroleum prospects: AAPG Bull., v. 64, no. 3, p. 297-315.

Gary, M., R. McAfee, Jr., and C. L. Wolf, eds., 1972, Glossary of Geology: Washington, D.C., Amer. Geol. Inst., 805 p.

Grabaü, A. W., 1913, Principles of stratigraphy: New York, 1185 p., revised, 1924: New York, A. G. Seiler, 1185 p., reprinted, 1960: New York, Dover Publications, 2 v., 1185 p.

Hollenshead, C. T., and R. L. Pritchard, 1961, Geometry of producing Mesa Verde sandstones, San Juan Basin, *in* J. A. Peterson and J. C. Osmond, eds., Geometry of sandstone bodies–a symposium: Tulsa, AAPG, p. 98-118.

Howell, J. V., chm., 1960, Glossary of Geology and Related Sciences: Washington, D. C., Amer. Geol. Inst., 325 p.; Supplement, 72 p.

Khaiwka, M. H., 1968, Geometry and depositional environments of Pennsylvanian reservoir sandstones, northwestern Oklahoma: Ph.D. Diss., Univ. of Oklahoma, 126 p.

———, 1973, Geometry and depositional environments of Morrow reservoir sandstones, northwestern Oklahoma: Shale Shaker, v. 23, no. 9, p. 196-214; v. 23, no. 10, p. 228-232.

Milliman, J. D., 1978, Morphology and structure of the upper continental margin off southern Brazil: AAPG Bull.,

v. 62, no. 6, p. 1029-1048.

———, 1979, Morphology and structure of Amazon upper continental margin: AAPG Bull., v. 63, no. 6, p. 934-950.

Wanek, A. A., 1954, Geologic map of the Mesaverde area, Montezuma Co., Colo. U.S. Geol. Survey Oil and Gas Inv. Map OM152, scale 1:63,360.

Wanless, H. R., 1931, Pennsylvanian cycles in western Illinois: Illinois Geol. Survey Bull., v. 60, p. 179-193.

——— and Weller, J. M., 1932, Regional persistence of Pennsylvanian cycles (abs.): GSA Bull., v. 43, no. 1, p. 139.

11 GROWTH FAULTING AND ITS INFLUENCE ON SEDIMENTATION AND PROSPECT ANALYSIS

Growth faulting occurs along continental margins as well as at intermediate positions between the shoreline and the outer edge of the submerged shelf slope. Such faults develop on the landward side of the major depocenters simultaneously with deposition. As a result, sedimentary accumulations are considerably thicker on the downthrown side than on the upthrown. The shallower beds generally dip basinward, whereas there is a tendency for the development of progressively increasing counter-regional rollover with depth. The beds of the rollover generally exhibit thinning along structural axes. Thus, local thickening frequently occurs toward the fault in a landward direction.

Growth faults are not flat surfaces but rather describe a hyperbolic curve in a vertical profile. The dip of the faults in shallow beds may be 45°–60° from the horizontal. There is a gradual decrease in dip with depth. At depths of 15,000–20,000 ft (4572–6096 m) in the Gulf Coast, the growth fault flattens off and becomes a bedding plane fault where the deep strata consist of undercompacted shales. Carver (1968) attributes growth (contemporaneous) faults to basement tectonics, deep salt or shale movement, slump across flexures, slump at the shelf edge, differential compaction, response to crustal loading, or combinations of these factors.

MODEL STUDIES AND EXAMPLES

In studies of the Tertiary strata of south Texas, Bruce (1973) concluded that low-density, high-pressured shale is a "dominant factor in the formation of these (growth) fault systems." A diagrammatic cross section (Fig. 11-1) illustrates how local depocenters of Tertiary strata are related to underlying pre-Tertiary sediments in south Texas.

Thin, interstratified sandstones and shales to the west thicken abruptly in local depocenters and then grade into shale in an eastward (basinward) direction. The growth faults occur along the western borders of the succession of regressive depocenters. Bruce points out that "The heavy (wavy) black line extended across this section to separate Tertiary sandstone above from the underlying shale illustrates a 'ridge and valley' effect that becomes more pronounced in a seaward direction. These 'ridges' represent residual shale masses of undercompacted shale between sandstone-shale depo-axes along which greater compaction has occurred."

Figure 11-2 is a stage diagram illustrating the development of a residual shale mass bordered on the west by a transitional sand-to-shale zone and on the east by a growth fault. The initiation of the depocenters is due to "the instability of newly deposited clay materials—which results in compaction and downwarping of the unconsolidated clay under the heavier sand section" (Bruce, 1973, Fig. 11-2B). Once the slight tensile strength of the underlying shales on the landward side of the downwarp is exceeded, a growth fault develops and remains "active as long as the depositional axis is maintained along the same line" (Fig. 11-2C and 11-2D).

Figure 11-3 illustrates an interpreted variable area seismic profile that crosses a major growth fault system. The residual shale mass is devoid of traceable seismic events, has a vertical dimension of over 10,000 ft (3050 m) and is approximately 10 mi (16 km) wide at the base. Differential compaction, due to a local loading factor, and shale diagenesis offers a prime explanation for this residual shale mass. In early stages of shale compaction, the interstitial water escapes from both the shale and the adjacent interstratified sands and shale (Figs. 11-2A and 11-2B). Bruce

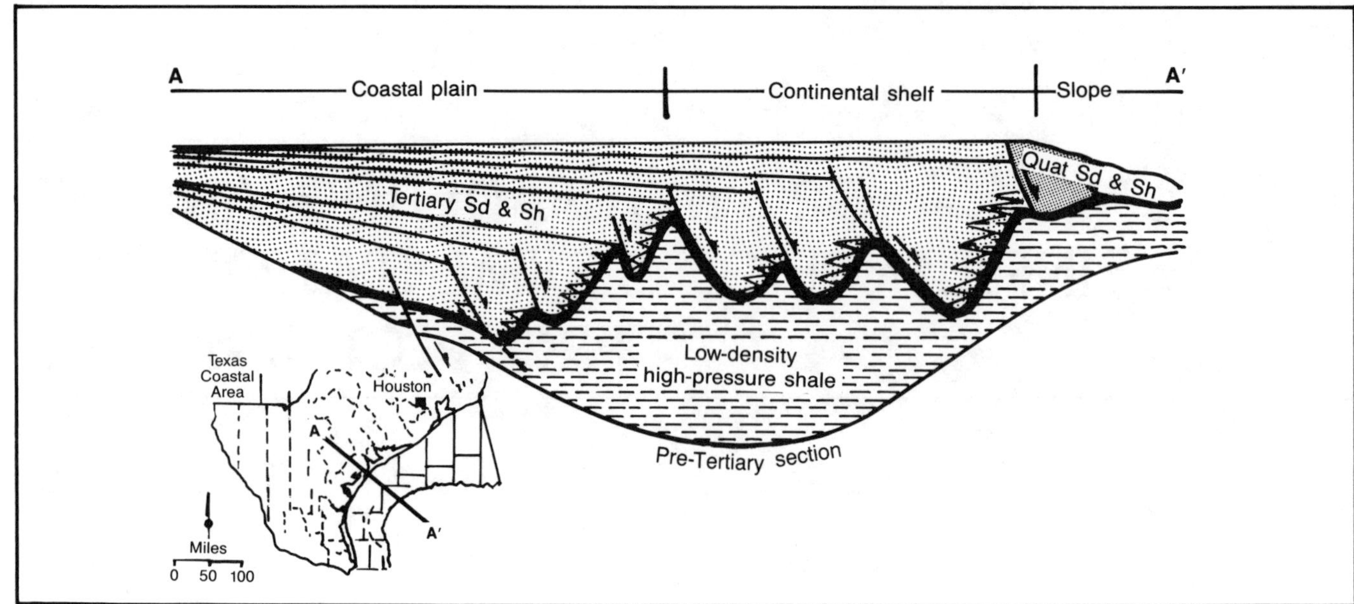

Fig. 11-1. Diagrammatic cross section across Texas portion of northern Gulf of Mexico basin. (Modified after Bruce, 1972; permission to publish by Gulf Coast Association of Geological Societies).

regressive depocenters overlap

states, "At first the water loss is higher from the shales, because of the relatively high water content. However, as subsidence continues, water loss from the shale decreases progressively as a result of decreasing permeability, until a critical depth is reached where orderly expulsion of water from the shale section is restricted and abnormally high pore pressure is developed within the shale mass. If subsidence continues to depths where fluid pressure within the shale approaches total overburden pressure, compaction ceases."

To the left (west) of the residual shale mass the strata consist of interbedded sands and shales. Within this area montmorillonite alters to illite when a critical temperature-pressure level is reached. The amount of water moved by this diagenetic mechanism may be as much as 10–15% of the compacted bulk volume of the shale involved. Thus, conspicuous downwarping of the sediments may occur in this area. This secondary water loss does not occur within the residual shale mass because of the absence of permeable strata.

The fault pattern of Figure 11-3 is the result of gravity slide and produces three principal types of gravity faults: growth (g), crestal (c), and antithetic (a) faults. Antithetic faults always dip in a landward direction, whereas the other two dip basinward.

A second type of faulting is shown on the left half of the variable area seismic profile of Figure 11-4. The faulting in this area is due primarily to differential compaction of deep-seated shales. In this case there is one growth fault and a series of crestal faults. Antithetic faults are rare to absent. Differential compaction faults probably are the result of differential water loss from adjacent blocks during the process of compaction within a depocenter.

Hubbert and Rubey (1959) postulated that high fluid pressure in the deep-seated shales produces a reduction in internal friction, which favors the development of low-angle gravitational faults. Figure 11-5 is a diagram of the relationship between fluid pressure and fault angle.

The normal vertical pressure gradient in the Gulf Coast Tertiary sediments is approximately 1.0 psi/ft of depth. This represents the combined pressures of 0.465 psi/ft for fluid and 0.535 psi/ft of sediment. Normal combined pressures occur down to an approximate depth of 5,000 ft (1524 m). Below this depth, there is a progressive increase in fluid pressure/overburden ratio. For example, this ratio is 0.702 at an approximate depth of 10,300 ft (3140 m) and 0.90 psi/ft at 15,000 ft (4572 m). This progressive change is accompanied by a decrease in the angle of faulting with depth. Down to about 5,000 ft (1524 m), the fault angle is 60°, whereas at 15,000 ft (4572 m) it is 15°. Other fault systems in the Gulf Coast exhibit a similar decrease with depth but may vary with the physical and chemical properties of the shale and the number and thickness of interbedded sandstone layers.

It is reasonable to postulate that there is a horizontal component of movement in the deeper zones that will cause a certain amount of shale diapirism far to the right of Figure 11-5. Numerous such diapirs are known to occur in the prodelta clays along the periphery of the Niger delta. A small amount of vertical movement of residual masses also may be noted occasionally from variable area seismic profiles.

Figure 11-6 diagrammatically illustrates three possible

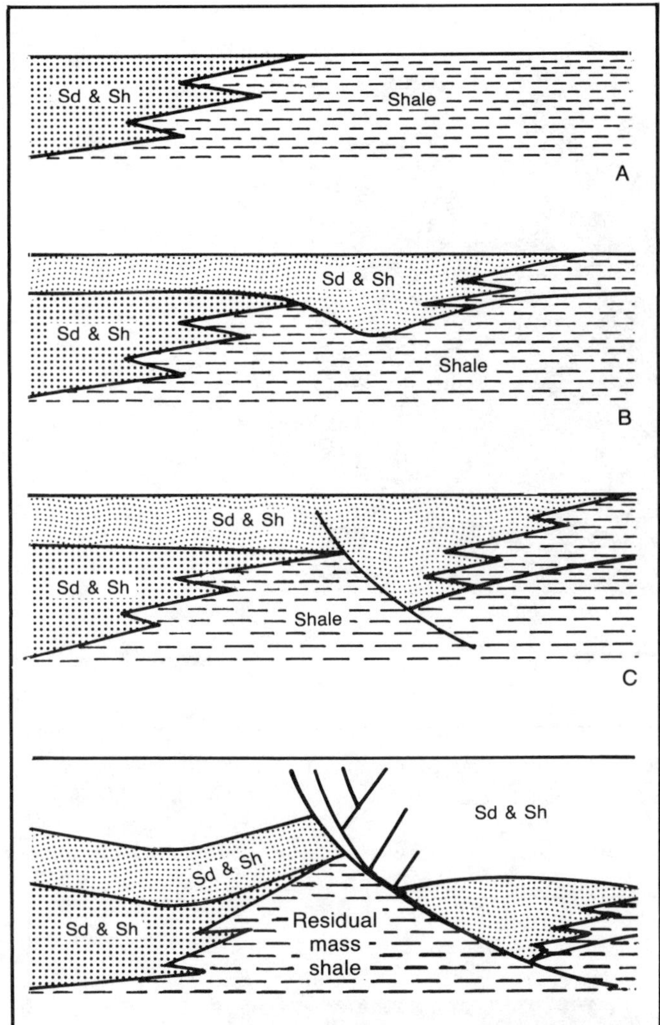

Fig. 11-2. Diagrammatic illustration showing four stages in the development of a residual shale mass. (Modified after Bruce, 1972; permission to publish by Gulf Coast Association of Geological Societies).

types of growth fault systems. In Figure 11-6A the rate of sediment influx exceeds the available space of a single depocenter. Thus, there is shown a succession of three depocenters that have developed in a regressive fashion. Each one is bordered on the landward side by a growth fault, and rollover into the fault occurs at depth. In Figure 11-6B the rate of influx of sediment is in equilibrium with the rate of subsidence of the depocenter, and as a result, sediment accumulation and fault displacement are maximal. There also is a maximum amount of rollover into the fault. In Figure 11-6C the rate of sediment influx is less than the rate of subsidence, and therefore a succession of depocenters is developed that become younger in a landward direction. In all three of these situations, there is formational thinning on structure and local thickening in a landward direction into the fault. Fault systems shown in Figures 11-6A and 11-6B are common in the Tertiary of the Gulf Coast of Louisiana, Texas, and Mexico. The transgressive system shown in Figure 11-6C is considerably less common than the other two, although it has been observed in south Texas.

Figure 11-7 is a stage diagram illustrating how a normal down-to-basin fault in shallow beds becomes a bedding-plane fault at depth. At depth this fault assumes the same low angle of dip into the basin as the strata themselves. Thus, there is a horizontal component of movement that produces a hypothetical gap, as shown in Figure 11-7C. The sediments, being unconsolidated, roll over into this gap and thus fill it. In collapsing into this gap, tensional faults, both crestal and antithetic, may develop, as shown in Figure 11-7D.

Figure 11-8 is an excellent example of a combination differential compaction and bedding plane fault system. The shallow portions of the growth and the crestal faults have an approximate tilt of 60°. The growth fault exhibits a greater amount of decrease in dip with depth than the crestal faults. Most of the crestal faults merge at depth with the growth fault and become secondary (or subsidiary) faults. All of them occur on the crestal portion of the rollover structure. Most of the antithetic faults maintain an approximate 60° dip and disappear where they "dead end" at depth against crestal faults. The combined effect of all of these faults is a collapsed structure. Mapping of a single, reservoir sandstone over such a structure requires detailed correlation and close attention to missing sections, the latter of which are indicative of normal faulting.

GULF COAST EXAMPLE

Boyd and Dyer (1964), in an earlier study in southeast Texas than that of Bruce (1972), recognized the influence of the Frio-Vicksburg flexure (growth fault) on the tremendous thickness of the Frio sandstones. They also related the series of asymmetric closures on the downthrown side of this fault to repeated movements along the fault simultaneous with deposition (see Fig. 11-9). Multiple Frio sandstone reservoirs in southeast Texas have produced in excess of 5 billion barrels of oil from such closures.

Figure 11-10 is one of several cross sections they published. The maximum thickness of the predominantly sandstone section is 5,000 ft (1524 m). This section, of course, contains numerous shale interbeds that generally serve as vertical permeability barriers. At the top of the middle Frio, a single, thin, shale bed may be noted that contains the foraminifera *Cibicides hazzardi*. In detailed studies of the Frio of the Mexican Gulf Coast, the author has traced this fossil for approximately 500 mi (800 km) south of the Rio Grande. An abrupt facies change, from predominantly sandstone on the west to 100% marine shale to the east,

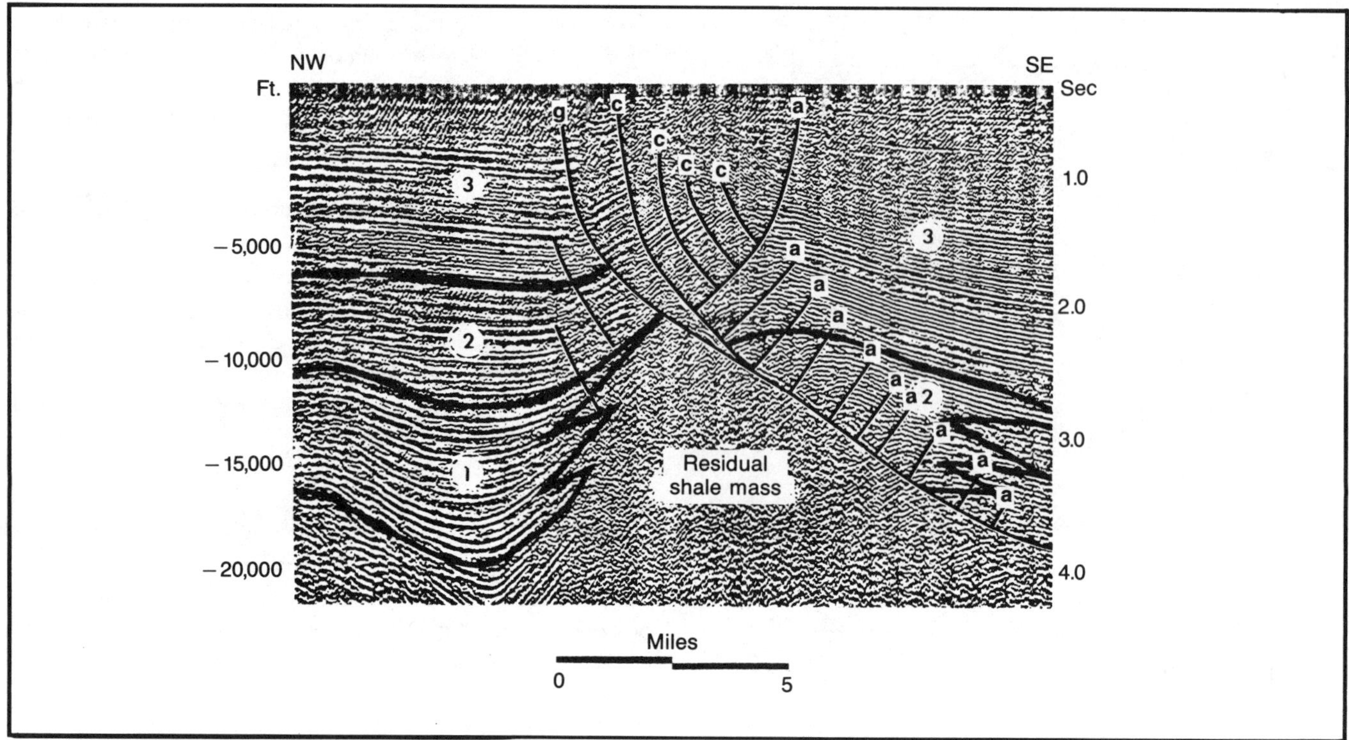

Fig. 11-3. Seismic illustration of a residual shale mass. Numbers on either side of fault system indicate sedimentary sequence of sand and shale units. (Modified after Bruce, 1972; permission to publish by Gulf Coast Association of Geological Societies).

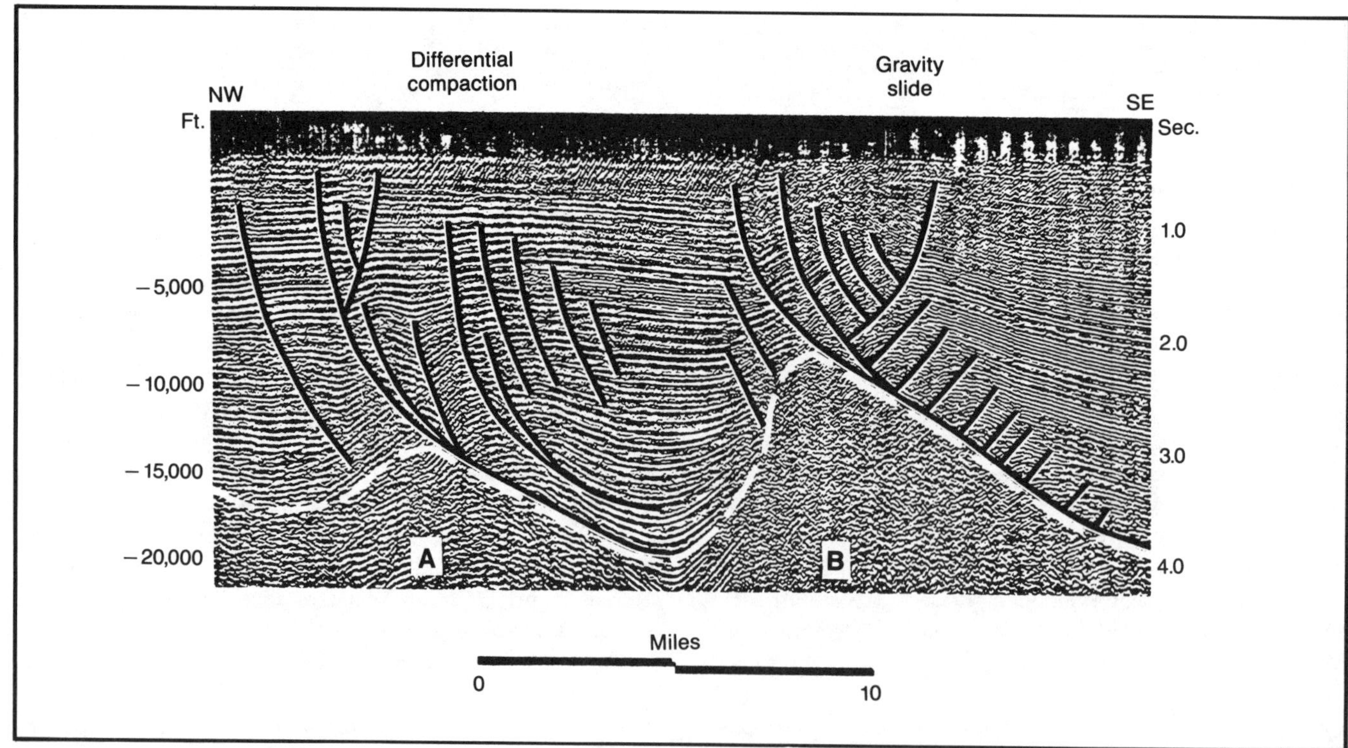

Fig. 11-4. Seismic illustration showing differences between fault systems formed by differential compaction and gravity slide (dashed white line). (Modified after Bruce, 1972; permission to publish by Gulf Association of Geological Societies).

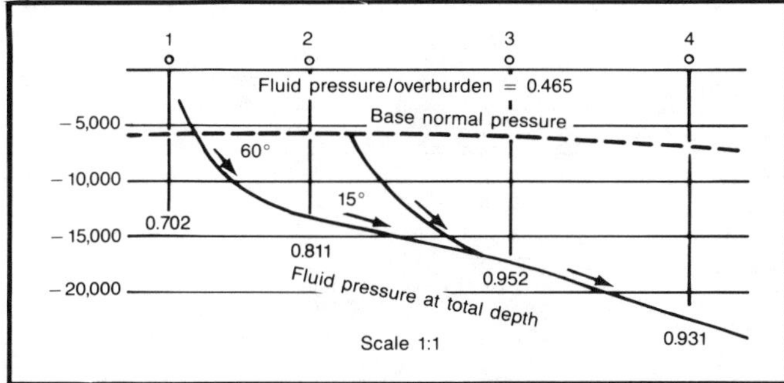

Fig. 11-5. Diagrammatic illustration showing fluid pressure-fault angle relationship. (Modified after Bruce, 1972; permission to publish by Gulf Coast Association of Geological Societies).

may be noted when comparing the two easternmost wells of Figure 11-10.

Although Boyd and Dyer refer to the 5,000 ft (1524 m) section of Frio as a barrier bar, it undoubtedly consists of many dozens of stacked barrier bars, all interrupted by offshore bars, deltaic distributary sandstones, and distributary mouth bars. Boyd and Dyer point out that the shale between the barrier-bar complex and the Frio-Vicksburg growth fault contains a brackish-water fauna and, therefore, interpret these sediments as being lagoonal.

The fault trace is shown as a very steep, straight line. The steepness is due to the vertical exaggeration of the profile. At the depths shown, it should have about a 60° tilt. At greater depth it should decrease in dip (basinward) and at a depth of 15,000–20,000 ft (4572–6096 m) become a bedding plane fault.

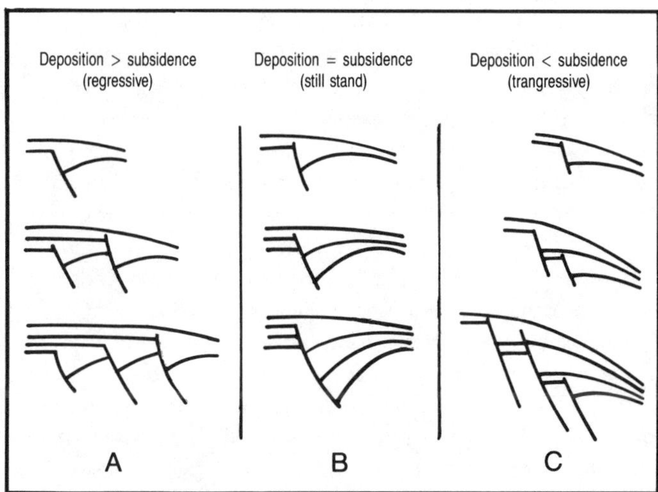

Fig. 11-6. Diagrammatic illustration showing three types of growth (contemporaneous) fault systems. (Modified after Bruce, 1972; permission to publish by Gulf Coast Association of Geological Societies).

GROWTH AND ASSOCIATED FAULTS OF NORTHEAST MEXICO

The Oligocene sediments of the Burgos basin of northeastern Mexico are broken abundantly by both growth faults and postdepositional faults. The named fault blocks of Figure 11-11 have a sinuous north-south trend and are many kilometers in length. Individually they consist of a series of macrostructures, and collectively these subparallel fault blocks are referred to as megastructures. Although there is a general basinward thickening of the sediments, there is also a local counter-regional thickening into the fault planes. Vertical profiles of these growth faults describe hyperbolic curves that flatten with depth.

Postdepositional faults are more numerous than the growth faults, more closely spaced, and generally have less throw. Their most distinguishing characteristic is that the sedimentary section on either side of a postdepositional fault is of the same thickness, whereas the sediments on the downthrown side of a growth fault are thicker everywhere than their counterparts on the upthrown side.

Five fault blocks have been identified in the Oligocene sediments of the Burgos basin in northeastern Mexico, as shown in Figures 11-11 and 11-12. From west to east, they are the Becerro, McAllen, Altamirano, Brasil, and Marzo blocks. In each instance, the blocks are named after the fault that borders the block on the west. These blocks vary in width from about 8 to 20 mi (12.9–32 km). Each of the

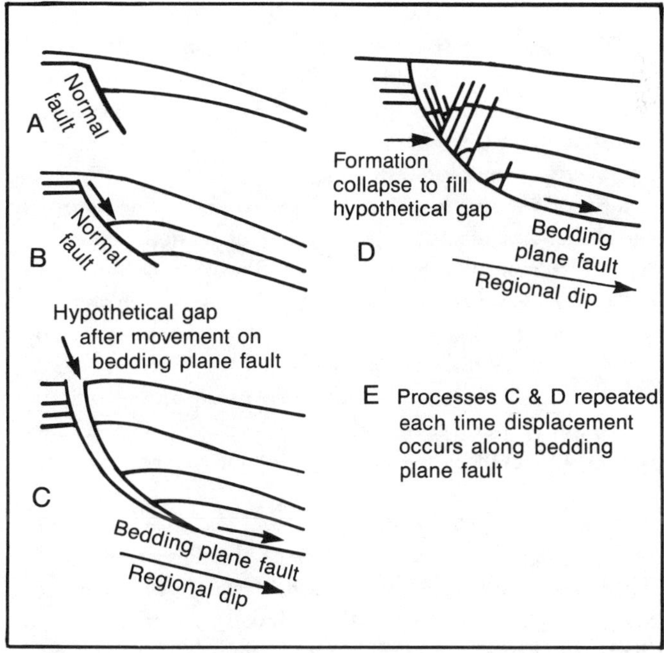

Fig. 11-7. Diagrammatic illustration showing development of a bedding plane fault. (Modified after Bruce, 1972; permission to publish by Gulf Coast Association of Geological Societies).

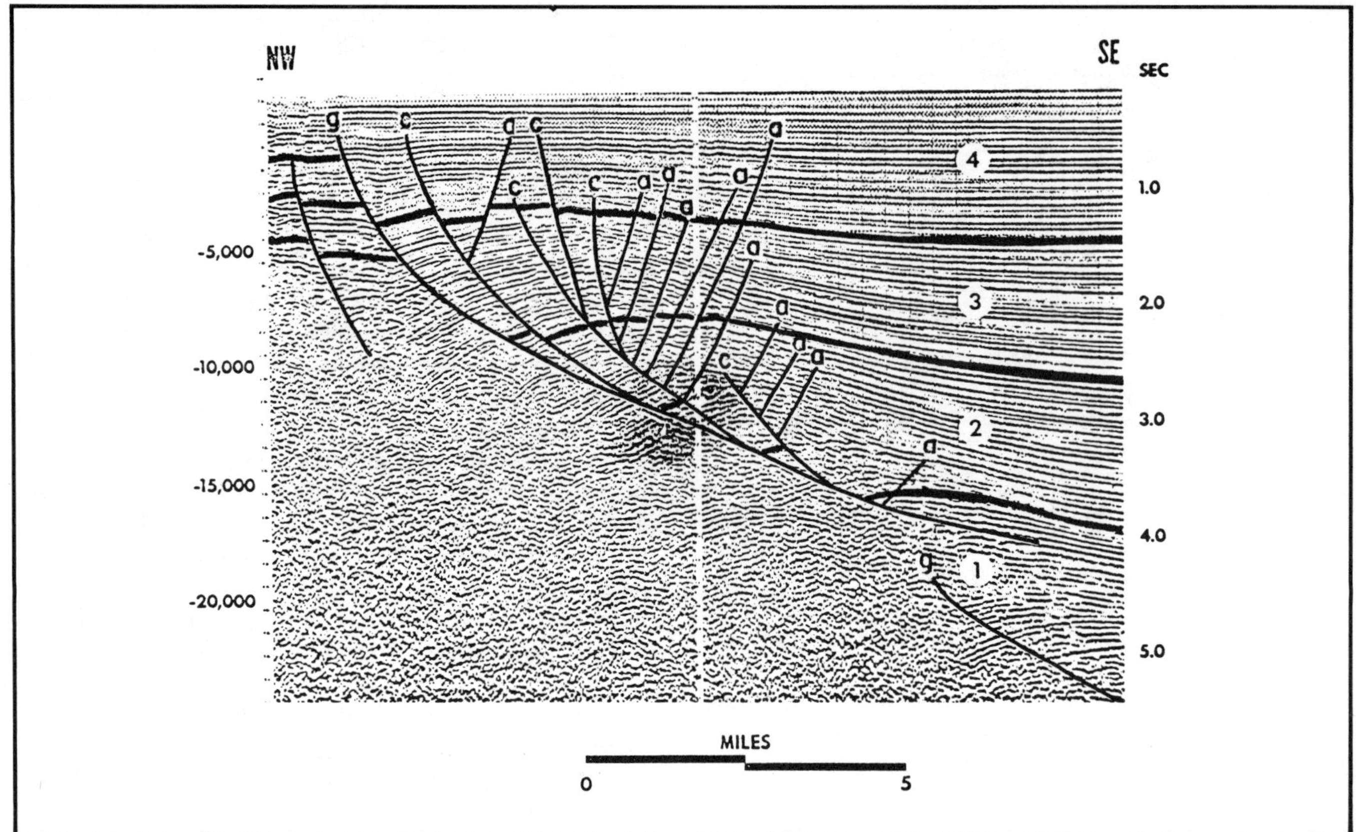

Fig. 11-8. Seismic illustration of combination differential compaction and bedding plane fault system. (Modified after Bruce, 1972; permission to publish by Gulf Coast Association of Geological Societies). (AQUAPULSE—courtesy Western Geophysical Company).

named blocks is, in turn, abundantly broken by a series of discontinuous postdepositional faults. All of the faults have a down-to-basin displacement. The basis for this fault interpretation is a combination of seismic, subsurface geologic, and micropaleontologic studies.

Most of the postdepositional faults are based on detailed correlations of electric logs in areas of moderately dense subsurface control. Approximately 75% of these faults are not easily identified from variable-area seismic profiles. If it were not for the fact that the faults have been cut by closely spaced wells, their presence, in all probability, would not be known. In the areas of few to no postdepositional faults, many additional such faults probably are present and their "absence" is due to lack of subsurface data control.

Figure 11-12 is a highly generalized west-east cross section of the Oligocene sediments of the Burgos Basin of northeastern Mexico. The minimum thickness of these sediments is approximately 450 m (1476 ft) on the west, and the maximum penetrated thickness is about 3,000 m (9843 ft) in the central part of the profile. Considerably greater thicknesses are present progressively toward the east, but drilling depths are inadequate for precise determination. The Becerro, McAllen, Altamirano, and Brasil growth faults of Figure 11-12 define the western boundary of each of the four named fault blocks of Figure 11-11.

Figures 11-13 and 11-14 are lithofacies maps of the middle and upper parts of the Vicksburg formation, respectively. A typical electric log of the middle Vicksburg is shown in the middle of Figure 11-13. The entire section consists of a series of alternating sandstones (black) and shales (white). Only those sandstones exhibiting a microlog porosity thickness of 2 m (7 ft), or more are plotted graphically on the log. The thickness of the middle Vicksburg in this well is 245 m (804 ft); number of sandstone beds, 58; and the composite sandstone thickness, 184 m (604 ft). Similar plots were made for all the wells drilled into the Becerro macrostructure. In contouring the number of middle Vicksburg sandstones (having a microlog porosity thickness of 2 m (7 ft), or more) it is immediately apparent that the sandstones are concentrated in a narrow belt close to and parallel with the Becerro fault. Likewise, the maximum composite sandstone thickness is parallel with and close to this fault. Only the western one-third of the fault block offers any possibility of reservoir sandstone development in the middle Vicksburg.

Figure 11-14 is a similar quantitative study of the upper Vicksburg. The thickness of the upper Vicksburg in the type log is 1,205 m (3944 ft), the number of sandstone beds is 75, and the composite sandstone thickness is 324 m (1063 ft). A contour map showing the number of upper Vicksburg sandstones in the Becerro block is shown in the left half of Figure 11-14. A north-south depositional alignment of sandstones, parallel with and in close proximity to the Becerro fault, again is shown clearly. Equally significant is the total absence of sandstones in the eastern quarter of the block. An isopach map of composite sandstone thickness is shown on the right half of Figure 11-14. Again, the area for maximum Upper Vicksburg sandstone is in close proximity to the Becerro fault. The principal value of such maps is to high-grade and low-grade structural anomalies that have been delineated on this block by a combination of subsurface geologic and seismic methods.

The middle and upper Vicksburg sandstones of the Becerro fault block collectively have a vertical V-shaped distribution. As such the stratigraphically lowest sandstones are the narrowest, and their respective east-west widths become progressively greater in an upward direction.

Sandstone distribution in the middle and upper Vicksburg of the Becerro block clearly is controlled by the trend and position of the Becerro fault (on the west). A depocenter developed in the western part of the block, and the sands

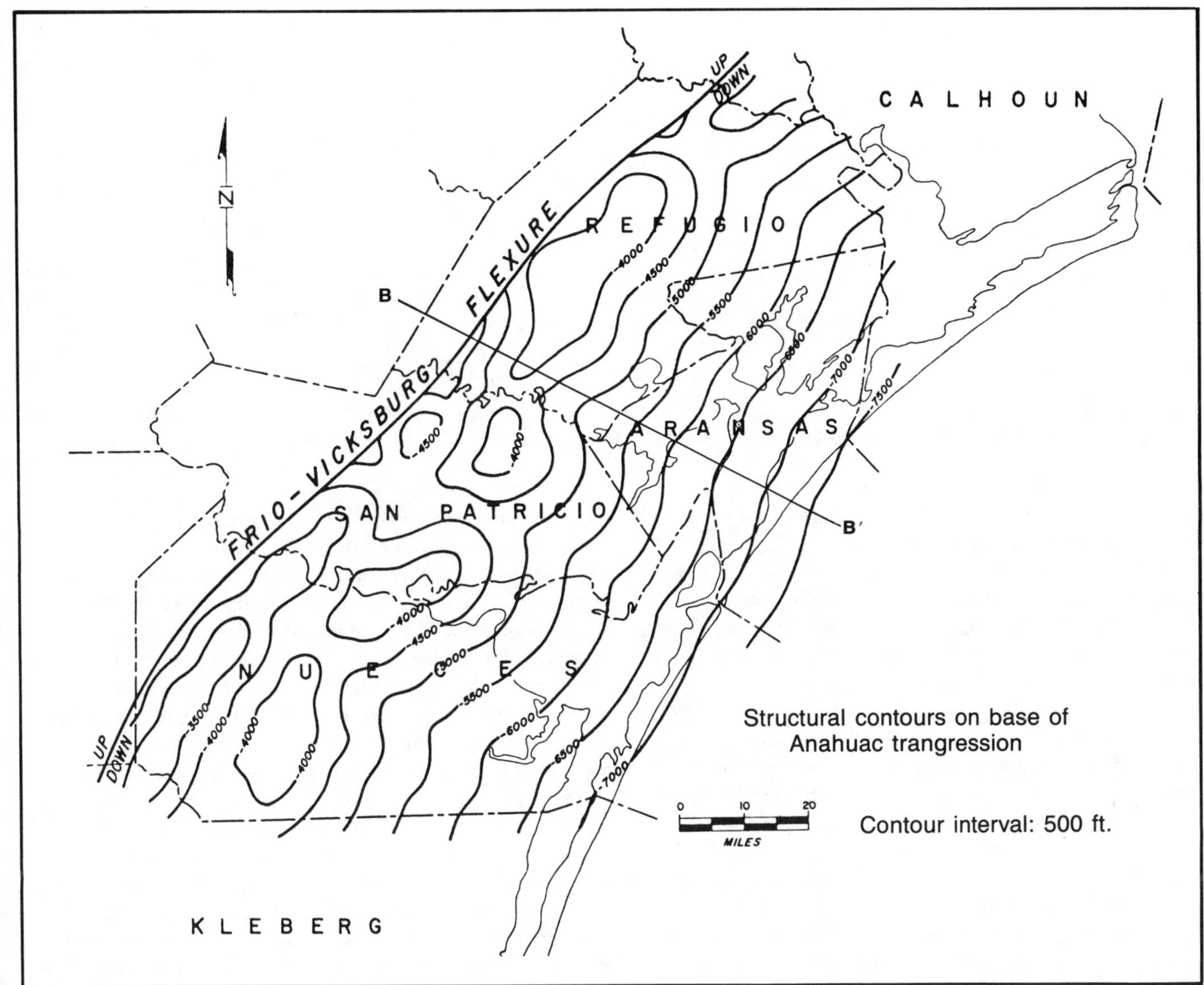

Fig. 11-9. Structure map of the base of the Anahuac (top of the Oligocene Frio) showing a series of asymmetric structural closures on the basinward side of the Frio-Vicksburg flexure (fault). Profile B-B′ is shown in Figure 11-10. (Modified after Boyd and Dyer, 1964; permission to publish by Gulf Coast Association of Geological Societies).

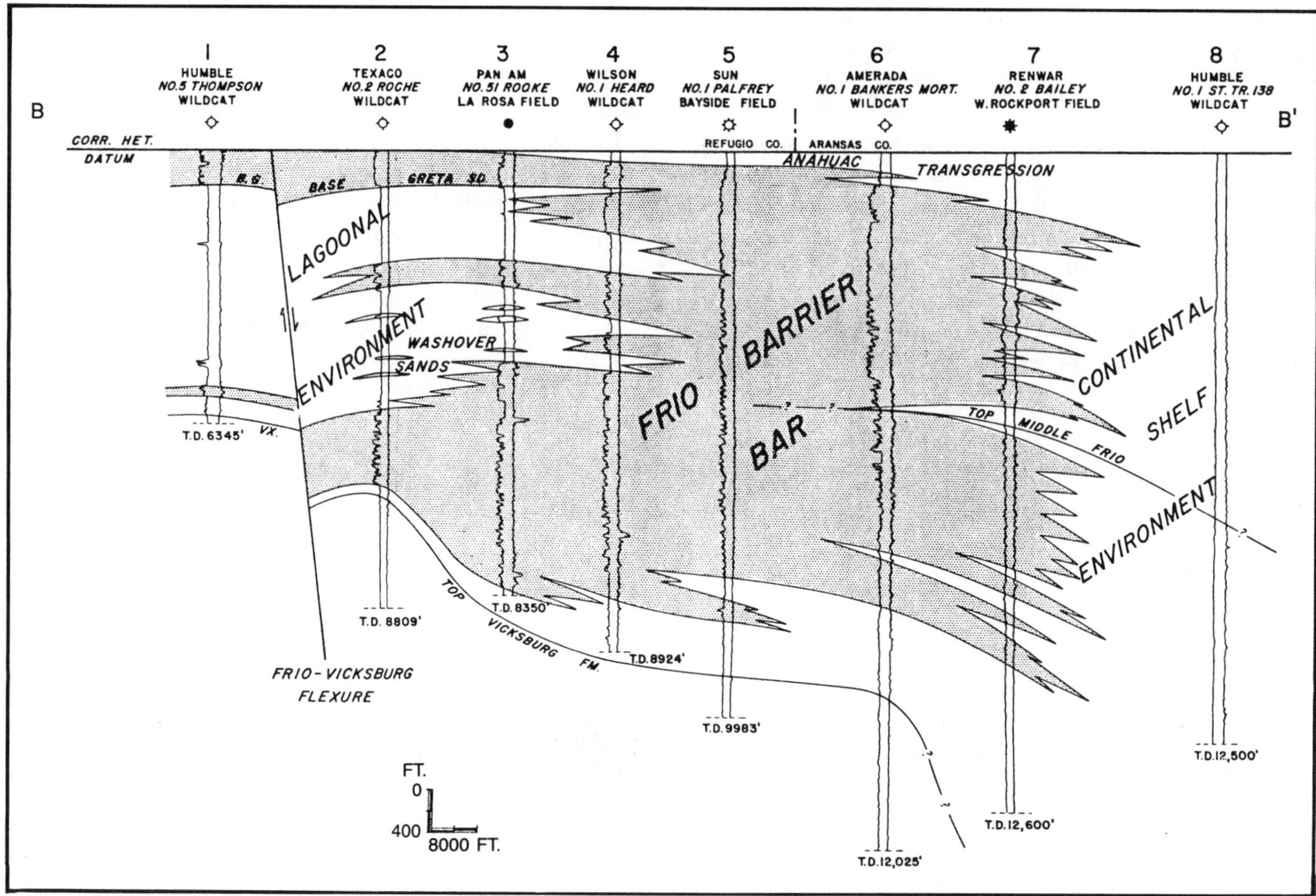

Fig. 11-10. Frio stratigraphic dip section BB', in Refugio and Aransas counties of southeast Texas. For location of profile see Figure 11-9. (Modified after Boyd and Dyer, 1964; permission to publish by Gulf Coast Association of Geological Societies).

were redistributed, probably by longshore currents, in a direction parallel with the Becerro fault.

The McAllen fault block (Figs. 11-15, 16, and 17) is defined on the west by the sinuous McAllen growth fault and on the east by the Altamirano growth fault. These faults have been identified from both subsurface well data and variable-area seismic profiles.

A type E-log for the upper Frio of this block is shown on Figure 11-15. The upper and lower limits of this stratigraphic interval were used for purposes of seismic structural mapping because (1) they are correlated readily on all electric logs of the wells, and (2) they are at the top and near the base, respectively, of a zone of abundant sandstones. There is an additional 250 m (820 ft) zone of sandstone development below the middle Frio seismic horizon of the type log shown on Figure 11-15. No potentially attractive sandstone is known below this lower sandstone zone.

The right half of Figure 11-15 shows an isopach map of the upper Frio of the McAllen fault block. The interval-thickness data from which this map was made could have as many as five different interpretations, but, obviously, only one could be correct. This interpretation is considered to be the most realistic. The stratigraphic interval exhibits an axial thickness trend subparallel with the McAllen growth fault with several bifurcating, divergent trends splitting off to the southeast. The southern part of the Reynosa field is shown in the northernmost part of this isopach map. There are several dozen productive lenticular sandstones in this field that fan out toward the north, east, and south. They were deposited in a north-south-trending depocenter just east of the McAllen growth fault. In the type log of this fault block, the stratigraphic interval of thickness is 790 m (2792 ft); the number of sandstone beds, 49; and the composite sandstone thickness, 310 m (1017 ft).

Figure 11-16 shows lithofacies maps of the number of sandstones and a composite sandstone isopach map. The data might be contoured variously, but the interpretations shown on Figure 11-16 are deemed the most likely. The map showing the number of sandstones illustrates the number of sandstone beds that might reasonably be anticipated within this stratigraphic interval at any drillsite on the McAllen fault block. The south and southeastward trends of the

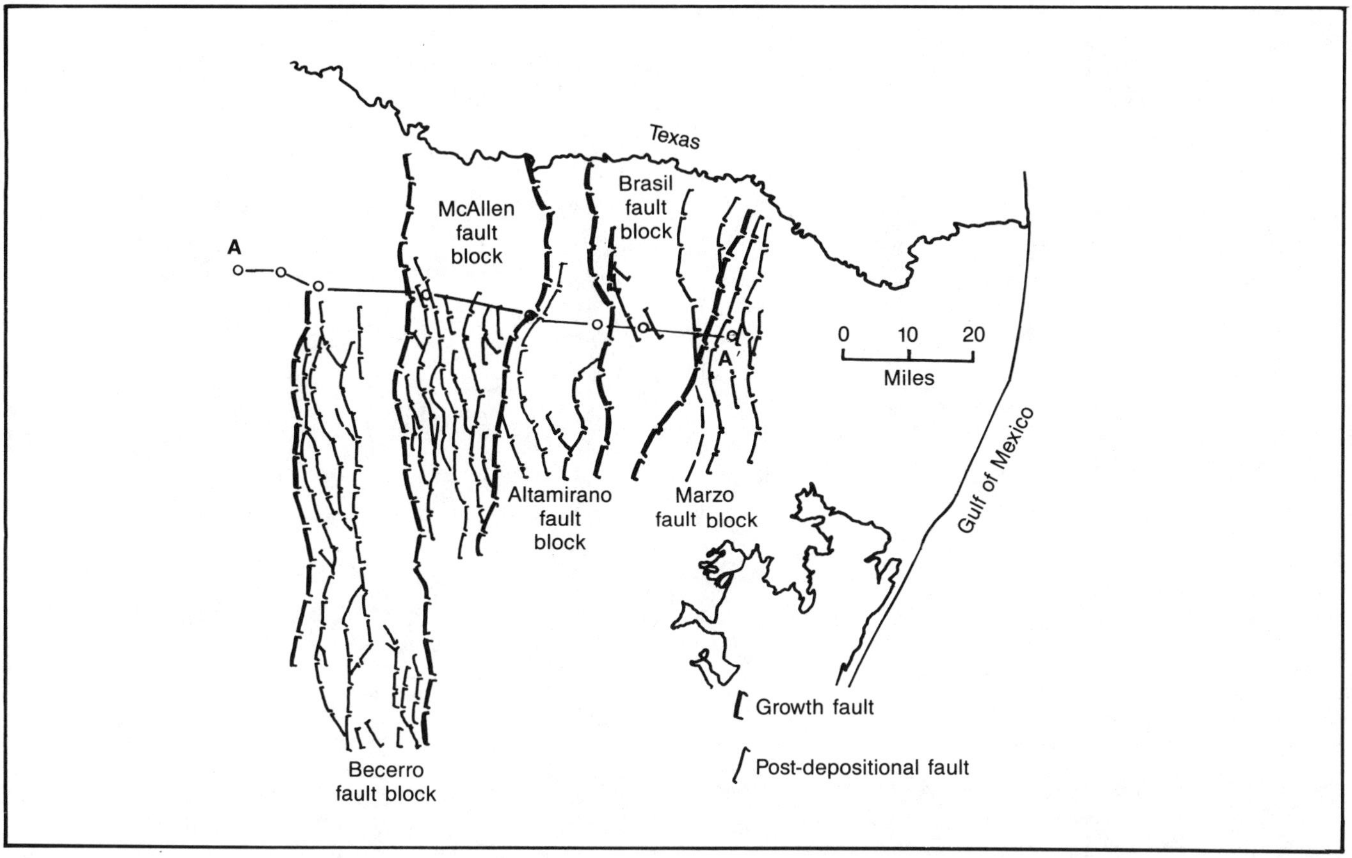

Fig. 11-11. Fault pattern in Burgos Basin of northeastern Mexico. (Modified after Busch 1975; permission to publish by AAPG).

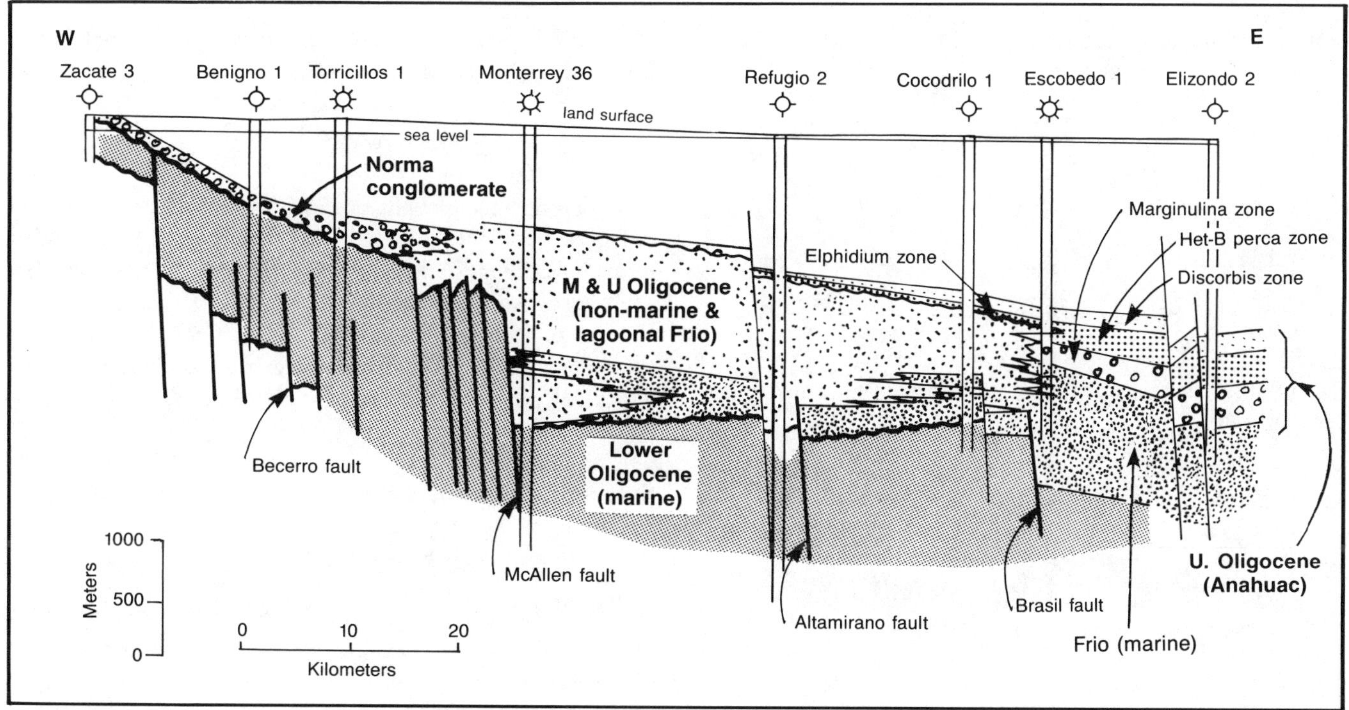

Fig. 11-12. Regional cross section A-A′ of Oligocene sediments of the Burgos Basin, northeast Mexico. For location of cross section see Figure 11-11. (Modified after Busch, 1975; permission to publish by AAPG).

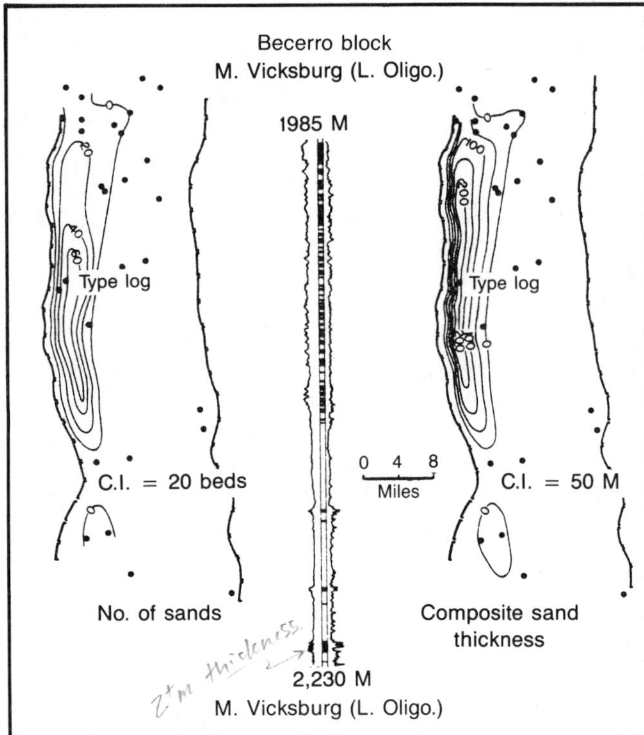

Fig. 11-13. Lithofacies maps of the middle Vicksburg (Lower Oligocene) showing number of sandstone beds and composite sandstone thickness, Becerro fault block. For location of this block see Figure 11-11. (From Busch, 1975; permission to publish by AAPG).

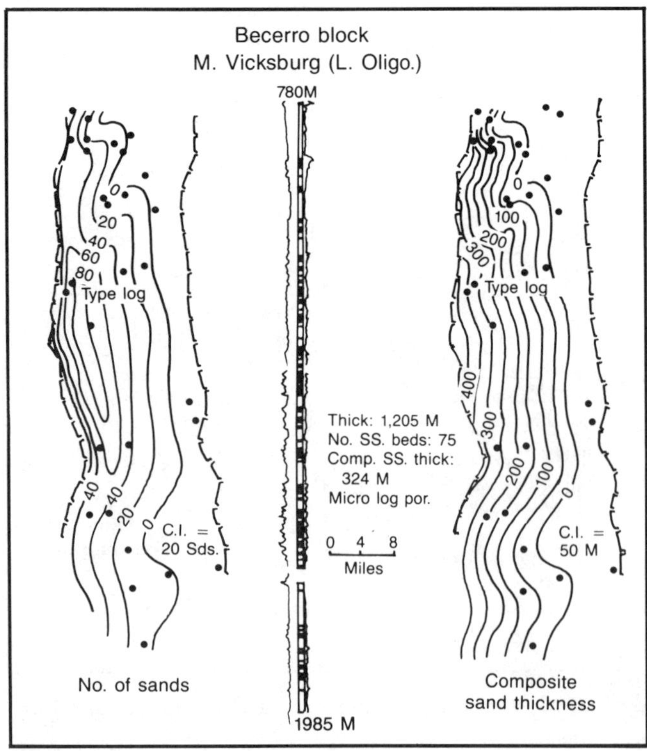

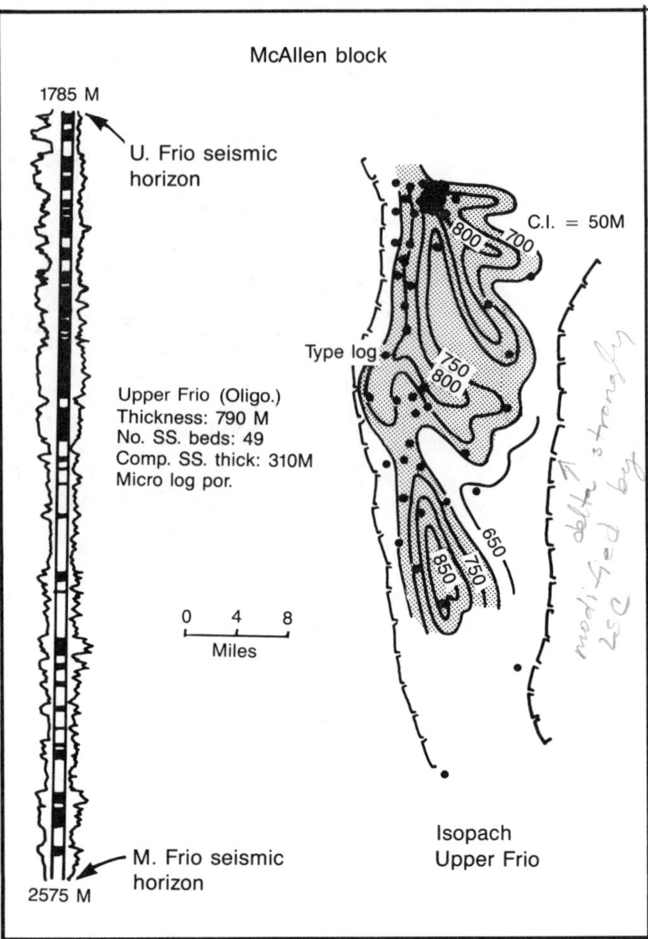

Fig. 11-15. McAllen fault block showing type electric log and isopach of upper Frio (Upper Oligocene) formation. For location of this block see Figure 11-11. (From Busch. 1975; permission to publish by AAPG).

maximum number of beds are consistent with similar trends in Figure 11-15. These sandstones are present in the Monterrey and Reynosa fields in the north and appear to be southward extensions of the deltaic-distributary systems of sandstones in these fields. With the number of sandstones ranging between 20 and 65, there are more than enough individual potential reservoirs for hydrocarbon accumulation at any proposed drillsite on this fault block. The right half of Figure 11-16 shows the composite thickness of all porous sandstone (as determined from the micrologs) within the upper Frio of the McAllen fault block. Again, the data on which this map is based lend themselves to multiple inter-

Fig. 11-14. Lithofacies maps of upper Vicksburg (Lower Oligocene) showing number of sandstone beds and composite sandstone thickness, Becerro fault block. For location of this block see figure 11-11. (From Busch, 1975; permission to publish by AAPG).

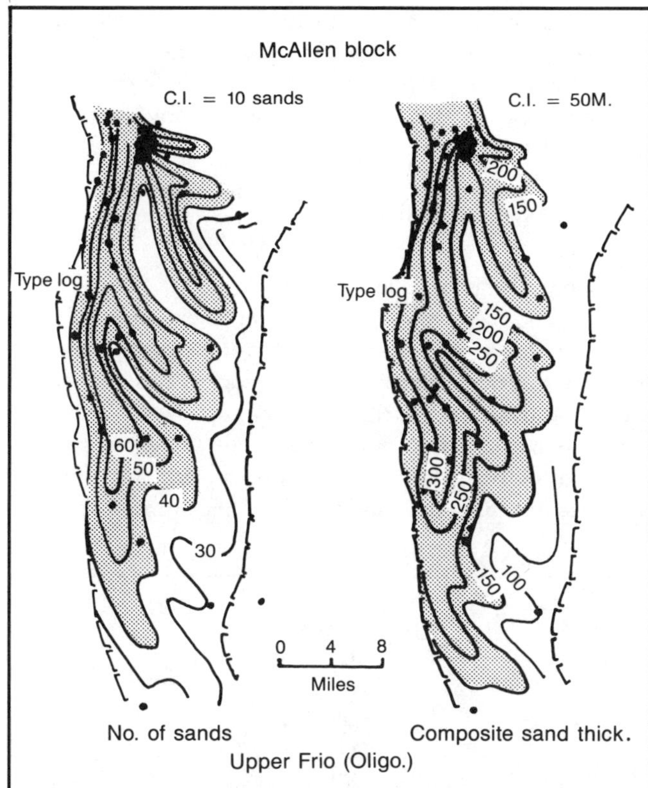

Fig. 11-16. McAllen fault block showing number of sand stones and composite sandstone thickness of the upper Frio (Upper Oligocene) formation. For location of block see Figure 11-11. (From Busch, 1975; permission to publish by AAPG).

pretations, but this one is considered to be the most realistic and meaningful. The south and southeastward trends of maximum reservoir porosity reflect a source on the north and northwest. The principal value of this map is to demonstrate that anywhere within the McAllen fault block, where structural conditions are favorable, there is an abundance of sandstone porosity for possible hydrocarbon accumulation. Certain linear trends, however, are more favorable than others.

Figure 11-17 shows two seismic, structural interpretations, one of the top of the Frio (at the base of the *Heterostegina* zone) and the other of the middle Frio (at the top of the *Cibicides hazzardi* zone). Both interpretations are integrated with subsurface well data and micropaleontological data.

A series of postdepositional faults breaks up the McAllen fault block into a group of smaller fault blocks. In all probability, velocity gradients exist within the area of the McAllen fault block, but a velocity survey was available for only one well. One may not be certain which reflection on one side of a fault corresponds to the same reflection on the other side. Consequently, reflections corresponding to the known depth positions (from E-logs) of the upper and middle Frio were selected for each of the smaller blocks. Subsea depths of these two horizons on the E-logs were converted to travel time using the calculated velocities of the single well for which a velocity survey was available. Thus, in a single, smaller fault block, the seismic time reflection is picked by reference to one or more wells within it. Any errors caused by velocity variations are thereby minimized, and this also serves to integrate the geology and geophysics.

The reason for preparing two structural maps rather than one in Figure 11-17 is that structures on the downthrown side of a growth fault tend to roll over and increase in amplitude with depth. Also, structural axes migrate basin-

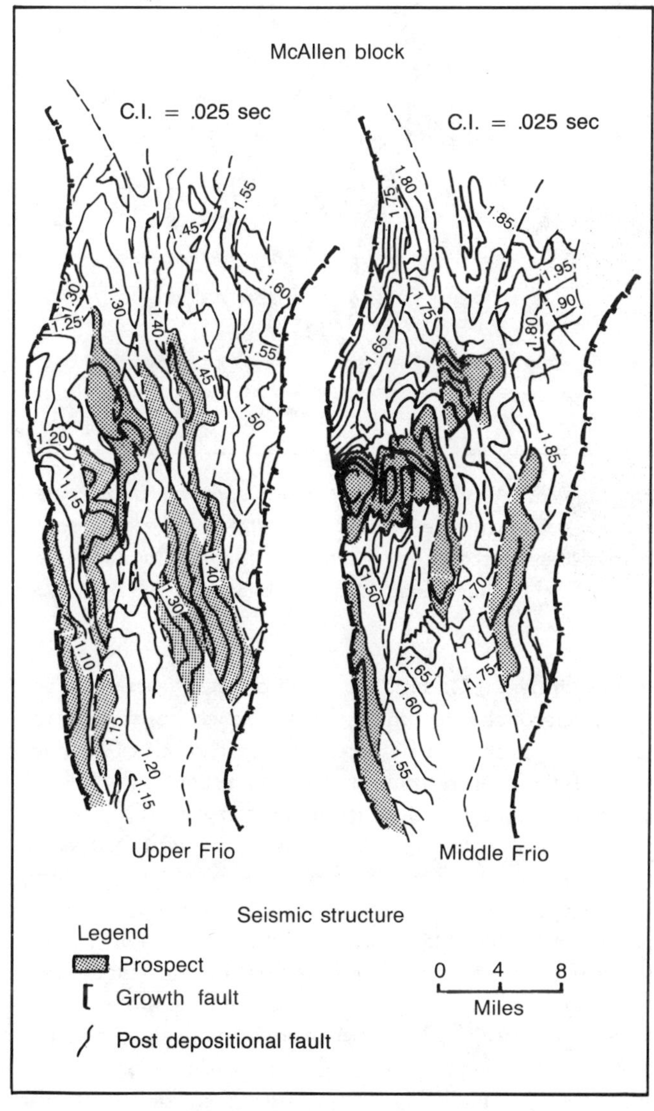

Fig. 11-17. McAllen fault block showing seismic-structural interpretation of upper and middle Frio (Upper Oligocene) formation. For location of block see Figure 11-11. (From Busch, 1975; permission to publish by AAPG).

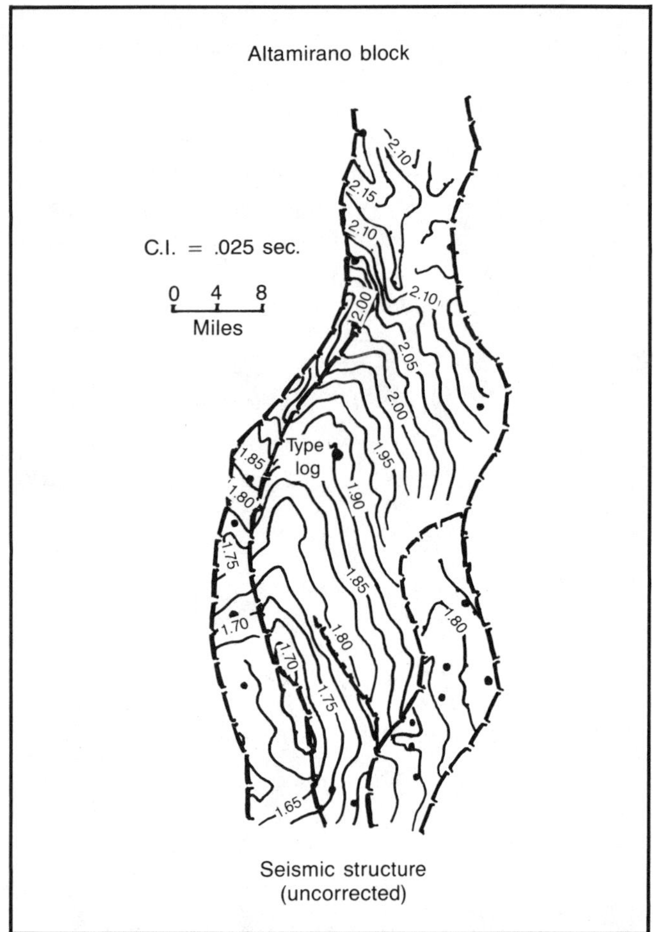

*Fig. 11-18. Altamirano fault block showing uncorrected seismic-structural interpretation of the Middle Oligocene (***Cibicides hazzardi*** horizon). For location of block see Figure 11-11.* (From Busch, 1975; permission to publish by AAPG).

ward with depth. On both maps, all prospect (shaded) areas are structural closures against faults. Several of these prospect areas proved to be productive as a result of drilling.

A seismic interpretation of the Altamirano fault block is shown in Figure 11-18. A large structural closure, cut off by a sinuous postdepositional fault on the west, is present in the central and southern parts of the block. This picture, however, is more apparent than real, for it has not been corrected for possible velocity variations. As a velocity survey is available from only one well on this fault block, (see type log location on Fig. 11-18), it is not possible to construct a velocity-gradient map from velocity-survey data.

The base of the Middle Oligocene, as shown on the type log of Figure 11-20, is correlated readily on all electric logs of wells drilled into the Altamirano block. Similarly, a seismic reflection at this horizon can be correlated readily within the entire fault block. It is 1,315 m (4429 ft) below the top of the Oligocene. The uncorrected seismic-structure map of Figure 11-18 is predicated on the assumption that velocities within the entire fault block are the same as those determined from the type log. By using the following equation and applying it at each drillsite, it may be quickly determined if the velocity is uniform or variable:

$$\text{Aver. Vel. (in m/sec)} = \frac{\text{Depth (m)}}{\text{Time (millisecs)}}$$

In applying this equation at every drillsite, the average velocity is nowhere the same. The variations in numerical values are, however, systematic. Thus, it is possible to construct an average velocity contour map and by superimposing the average velocity contour map over the uncorrected seismic-structure map, a new set of data can be derived at all points of intersection of contour lines for the two maps. At these new depth points, the average velocity

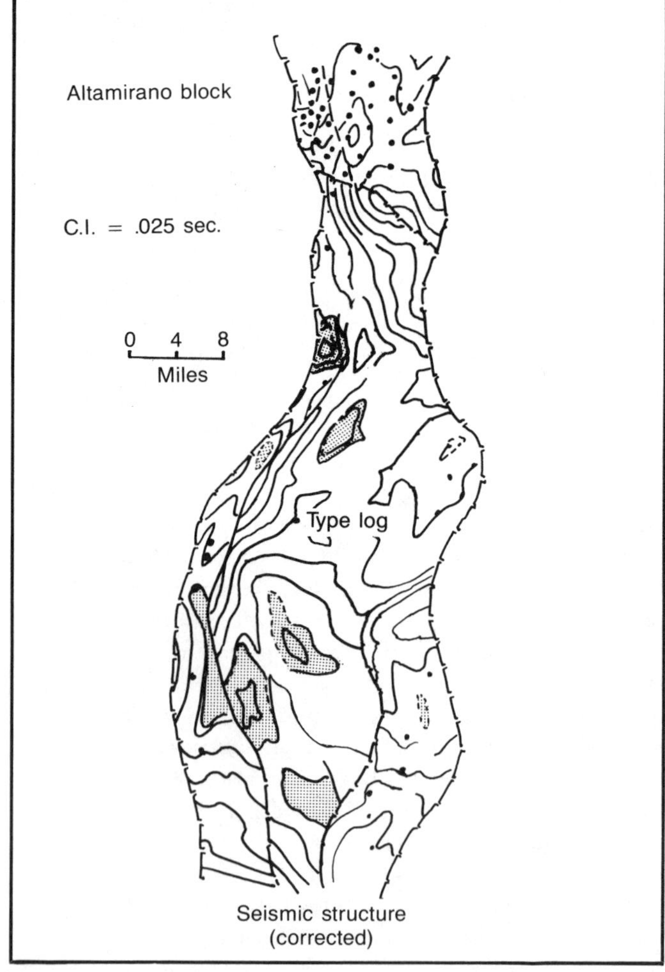

*Fig. 11-19. Altamirano fault block showing seismic-structural interpretation of the Middle Oligocene (***Cibicides hazzardi*** horizon) corrected for velocity variations. For location of block see Figure 11-11.* (From Busch, 1975; permission to publish by AAPG).

value is multiplied by the time value to obtain an approximate subsea-depth value in meters. These new values, after contouring, show an approximation of a true structural configuration of the Altamirano fault block shown in Figure 11-19.

Figure 11-19 offers a logical explanation for every dry hole on the fault block (they are off-structure). Furthermore, it serves to delineate drillable structural closures that are either obscure or not present on the uncorrected seismic interpretation of Figure 11-18.

It should be recognized that, if velocity surveys were available from all of the wells drilled into this block, a more refined interpretation of Figure 11-19 would be feasible. Another limitation of this technique is that calculated average-velocity values have been used, and this generalized approach is bound to introduce some errors in size and shape of the anomalies. Seven significant structural anomalies are shaded in Figure 11-19, all of which are considered to be drillable prospects.

The isopach map of Figure 11-20 is of the combined Middle and Upper Oligocene and exhibits a thickness range from 1,150–1,450 m (3773–4757 ft). This interval includes most of the potentially productive sandstones of the Oligocene.

In detailed studies of the Francisco Cano field (to the north), most of the sandstones become thinner on the highest parts of the structure. The same is true of the composite Upper and Middle Oligocene sedimentary section. Similar observations have been made of many other sandstones and stratigraphic intervals present over structural anomalies on the downthrown side of growth faults in the Burgos basin. Thinning on the crests and thickening on the flanks are normal consequences of these structures, having formed simultaneously with deposition.

South of the F. Cano field on the Altamirano block, there are insufficient well data to show details of thickening and thinning of the sediments. There is every reason, however, to believe that sedimentary thinning occurs over all the structures, in a manner similar to that over the F. Cano and other growth structures of the Burgos basin. For this reason the very limited subsurface well data have been contoured subjectively in Figure 11-20. Objective contouring of these data has no meaning and produces a picture completely inconsistent with the observed facts in field areas of abundant well control.

The F. Cano field produces gas from 26 different Frio sandstone reservoirs, all of which were deposited on the downthrown (eastern) side of the Altamirano growth fault. Two gas-productive and one water-bearing sandstones, all in the F. Cano field, illustrate sharply contrasting sandstone configurations. These, together with their respective structural configurations, are shown in Figures 11-21, 11-22, and 11-23.

Figure 11-21 illustrates the structure and sandstone dis-

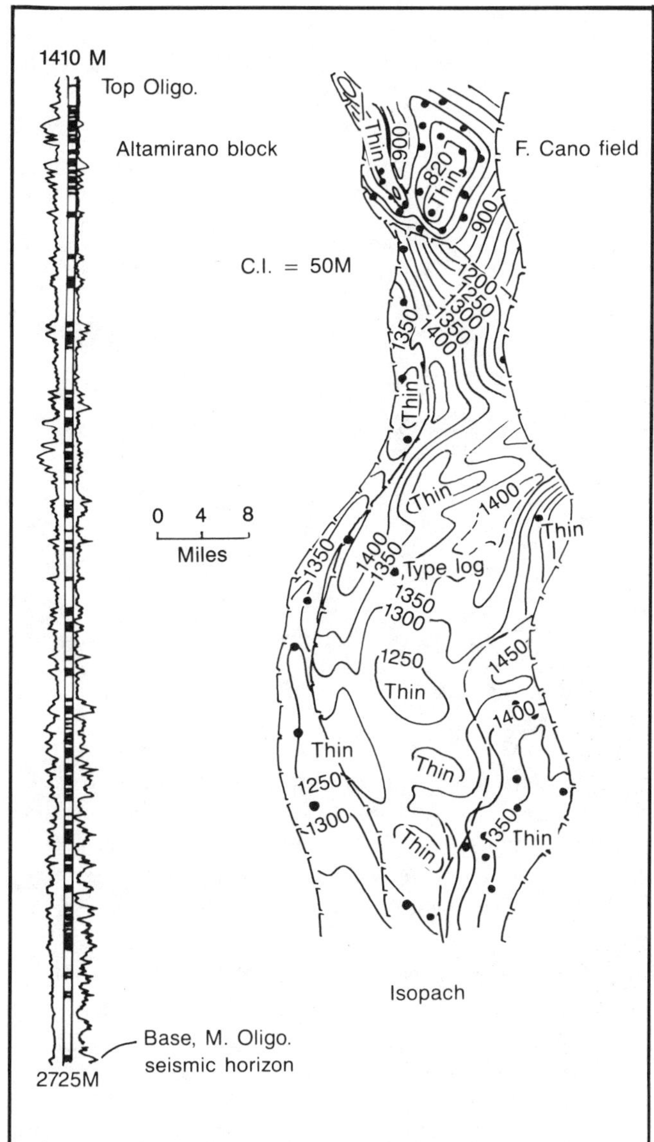

Fig. 11-20. Altamirano fault block showing type electric log and isopach of Upper and Middle Oligocene. For location of block see Figure 11-11. (From Busch, 1975; permission to publish by AAPG).

tribution of a one-well pool. The well was drilled on the highest structural position of the eastern fault block. The original gas-water contact is estimated at -2,062 m (-6765 ft). The thickest sand accumulation is on the structurally highest parts of the structural closures. Also, there are two bar-like lenticular sandstones that were deposited on the structural flank of the eastern fault block of Figure 11-21.

The structure and sandstone configuration of another gas-bearing reservoir in the F. Cano field are shown in Figure 11-22. Gas accumulated on the highest part of the eastern structural closure. The original gas-water contact was -2,169 m (-7116 ft), whereas the 1972 contact was at -2,155 m

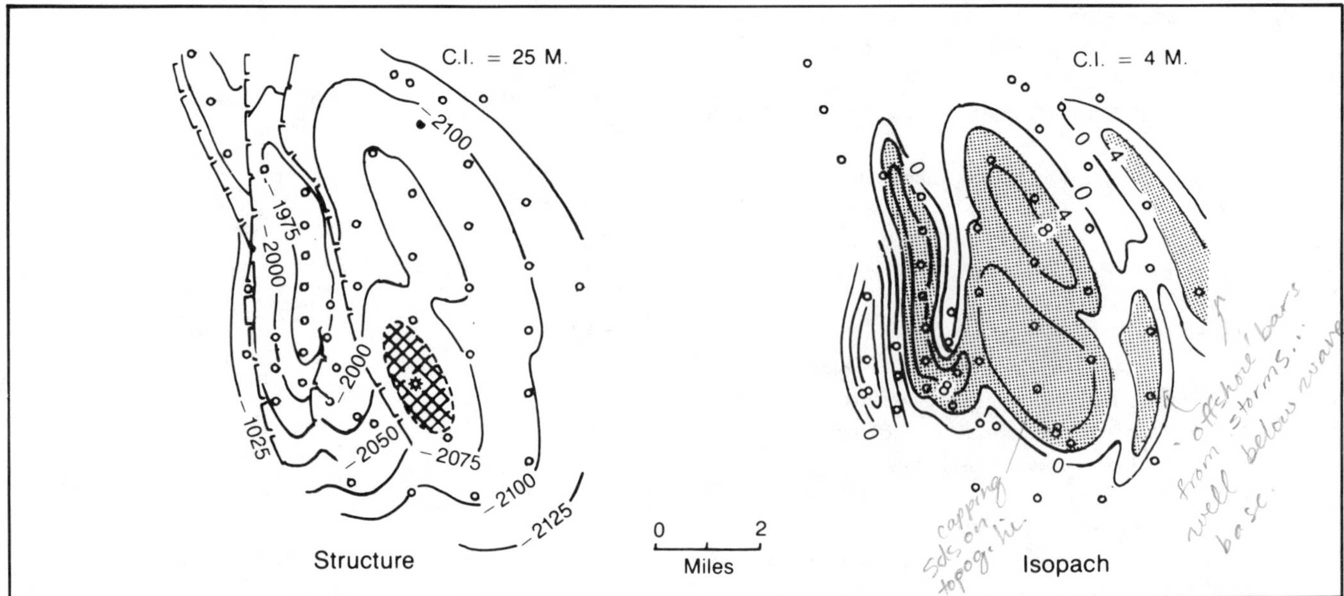

Fig. 11-21. *Structure and isopach maps of a single well gas pool in the upper Frio of the Cano field, northeast Mexico. For location of Cano field see Figure 11-20.* (From Busch, 1975; permission to publish by AAPG).

(-7070 ft). The main body of the sandstone is confined largely to the eastern fault block and is thinnest over the crest of the anticline. Its thickest development is a crescent-shaped lens draped around the northern, eastern, and southern flanks of the structure—leaving no doubt that its shape and thickness were controlled by this growth structure at the time of deposition.

Figure 11-23 is a structure and isopach map of still another sand occurring at the top of the upper Frio of the F. Cano field, it is not productive. The structure is very similar to that of the sands below. It is apparent from the isopach map that this water-bearing reservoir consists of a deltaic-distributary system. It fails to exhibit any genetic relation to the growth structure over which it was deposited. This being the uppermost sand of the upper Frio, it is reasonable to assume that a slightly tilted flat surface of shale

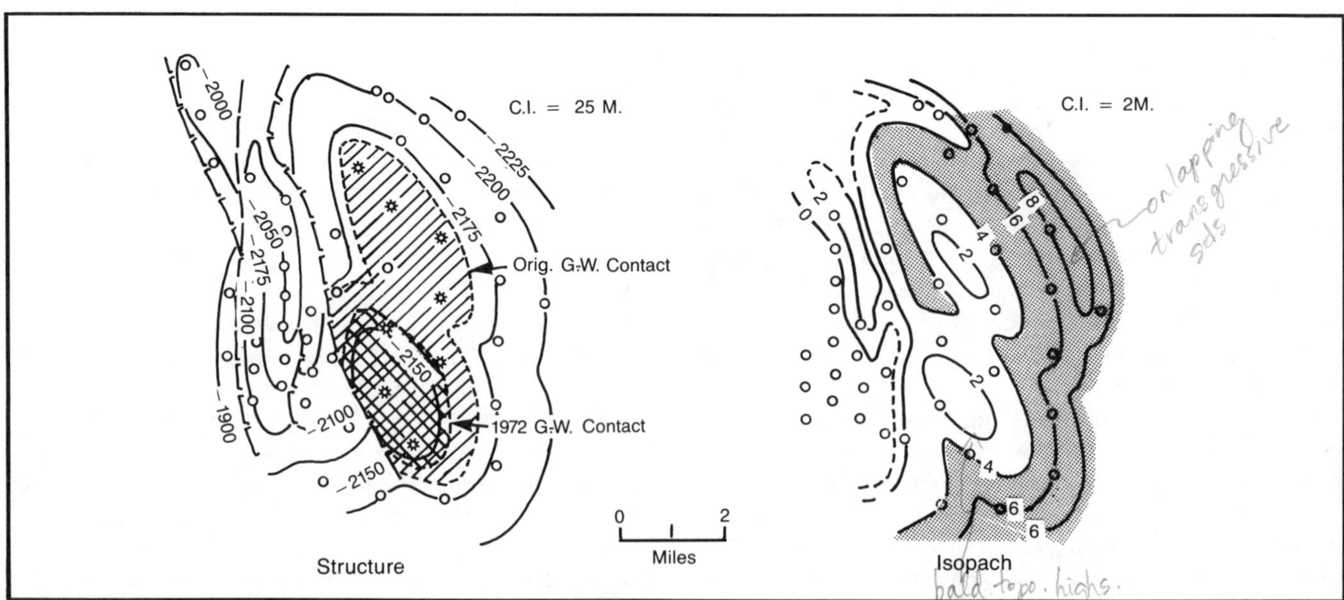

Fig. 11-22. *Structure and isopach maps of gas-productive pool in the upper Frio of the Cano field, northeast Mexico. For location of Cano field see figure 11-20.* (From Busch, 1975; permission to publish by AAPG).

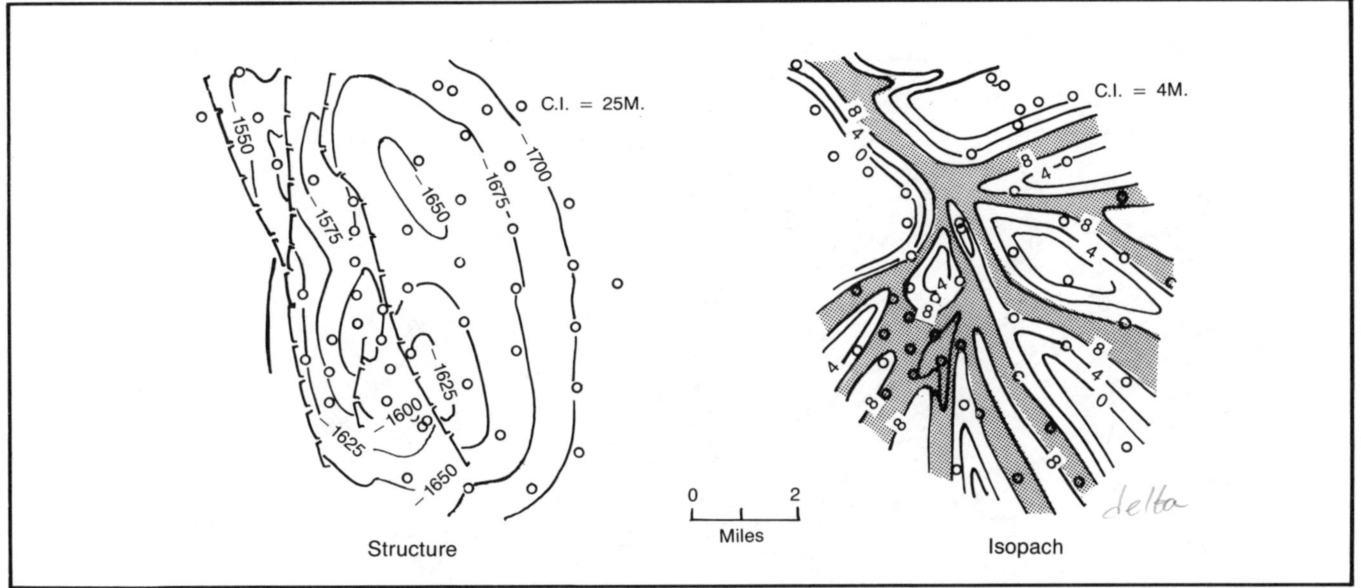

Fig. 11-23. Structure and isopach maps of a water-bearing sandstone near the top of the Frio of the Cano field, northeast Mexico. For location of Cano field see Figure 11-20. (From Busch, 1975; permission to publish by AAPG).

was selectively eroded and then filled by a series of distributary channel sands. The distributary sandstones thicken downward at the expense of the subjacent shale.

Most of the 26 gas-bearing reservoirs of the F. Cano field consist of one to three bar-like "halo" sandstones. Their shape and distribution are related directly to such environmental factors as growth structures, water depth, and relative intensity of wave energy.

GROWTH AND ASSOCIATED FAULTS OF NIGER DELTA

The Niger delta of west-central Africa presents an excellent example of the influence of growth and associated faults on sedimentation (Weber, 1971; Weber and Daukora, 1975). The great abundance of growth faults, which characterize the Niger delta, is shown in Figure 11-24. The great majority are concave basinward in plan view and of limited linear extent. They form the landward margin of individual depocenters, generally directly opposite the point of sediment discharge of individual deltaic distributaries. The faults are generally younger in a seaward (southwestward) direction because of the overall progradation of the sediments.

A generalized stratigraphic profile of the Niger delta is shown in Figure 11-25. The uppermost zone of continental (fresh-water) sands has a maximum thickness of 5,000–6,000 ft (1524–1829 m) and is devoid of hydrocarbons. The middle (white) portion of the delta is also thousands of feet thick and consists of multiple paralic sequences of sandstones and shale. The entire zone is progradational in nature and is broken by the growth faults shown in Figure 11-24. The lowest zone (stippled) consists of thousands of feet of marine, mobile, pro-delta clay-shale. The broad S-shaped area is a vertically exaggerated time-stratigraphic unit. Each of the areas between successive growth faults was a local depocenter opposite a deltaic distributary. It is within these prograding, multiple depocenters that a multiplicity of reservoir sands was deposited.

Figure 11-26 is a block diagram of a restricted coastal area of the Niger delta showing sites of deposition of genetically related sediments. These include a barrier bar, point bar, distributary channel, tidal channel, and river-mouth bar. A growth fault generally occurs on the landward (lagoonal) side of a barrier bar. The barrier bar, barrier foot, marine clay, and thin, transgressive marine sandstone comprise one paralic sequence, and all tend to thicken and "rollover" into the growth fault (see vertical face of block diagram, Fig. 11-26). Each time there is movement along the growth fault, deposition of one of these paralic sequences is abruptly terminated. The shoreline transgresses, and a thin transgressive, highly calcareous marine sand is deposited to initiate deposition of the basal member of the next paralic sequence. Such a sandstone has a minor unconformity at its base.

Growth faults generally exhibit a crescent shape (concave) in plan view toward the embayment. This crescent shape causes the layered sediments on the downthrown side to bend parallel with the fault and to assume asymmetric egg shapes. Several such faults, when along the same strike line, may merge at their distal ends to produce an undulating fault trend subparallel with the shoreline. These fault

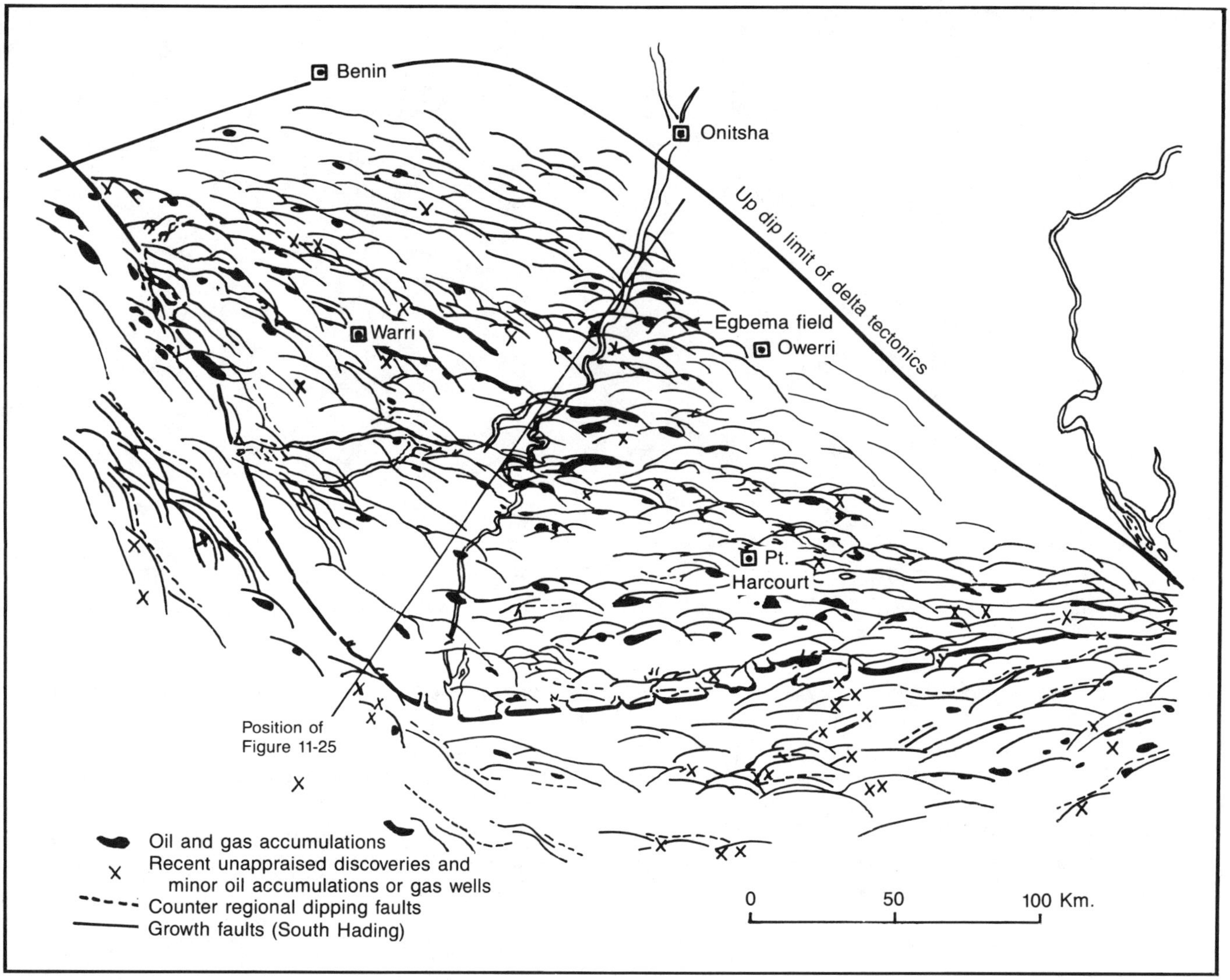

Benin
Onitsha
Up dip limit of delta tectonics
Egbema field
Warri
Owerri
Pt. Harcourt
Position of Figure 11-25
Oil and gas accumulations
Recent unappraised discoveries and minor oil accumulations or gas wells
Counter regional dipping faults
Growth faults (South Hading)
0
50
100 Km.

Fig. 11-24. Growth faults and known hydrocarbon accumulations of the Niger delta. (Modified after Evamy et al., 1978; permission to publish by AAPG).

Coast
Continental sands
Time stratigraphic unit
Paralic
SW
NE
Sea
Mobile clay substratum
Tertiary delta
Cret.
5
10
Km
0
50
Km.
Basement

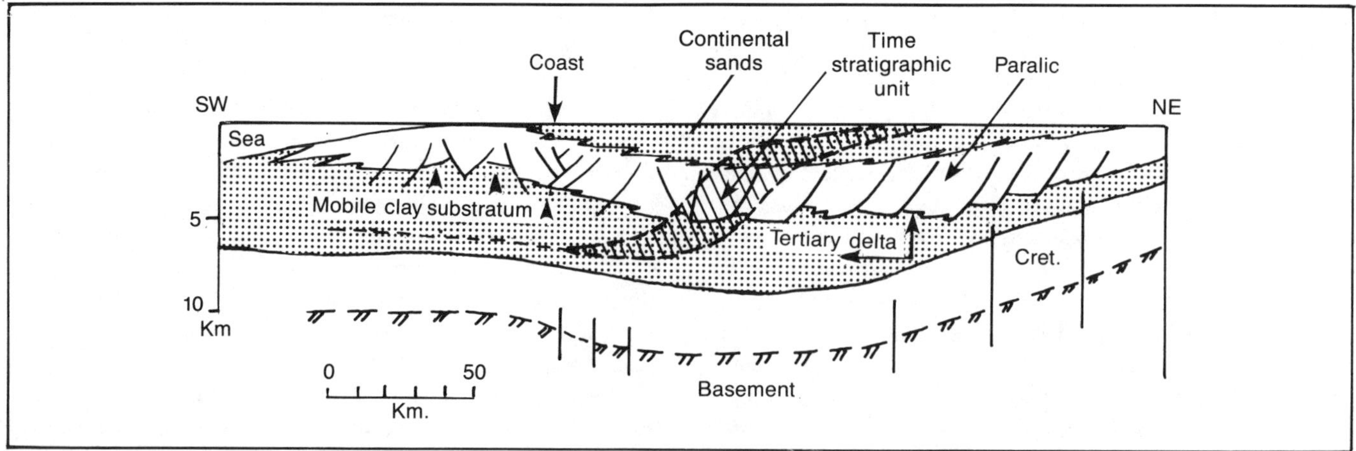

Fig. 11-25. Schematic cross section of the Niger delta perpendicular to the coastline. See Figure 11-24 for location. (Modified after Weber, 1971; permission to publish by Geologie en Mijnbouw).

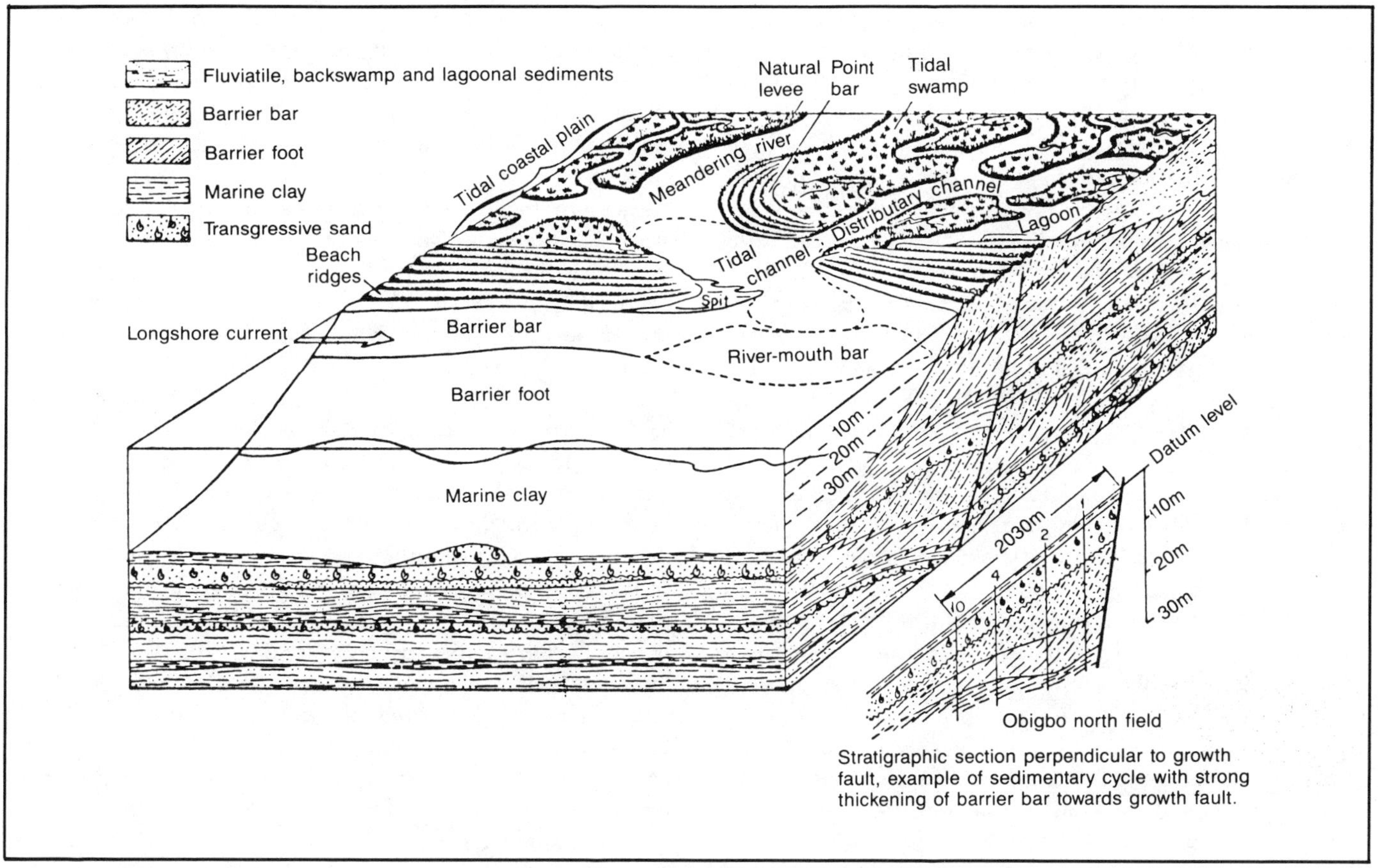

Fig. 11-26. Block diagram of the Niger delta coastal zone showing geomorphology, cyclic sediments and an active growth fault. (Modified after Weber, 1971; permission to publish by Geologie en Mijnbouw).

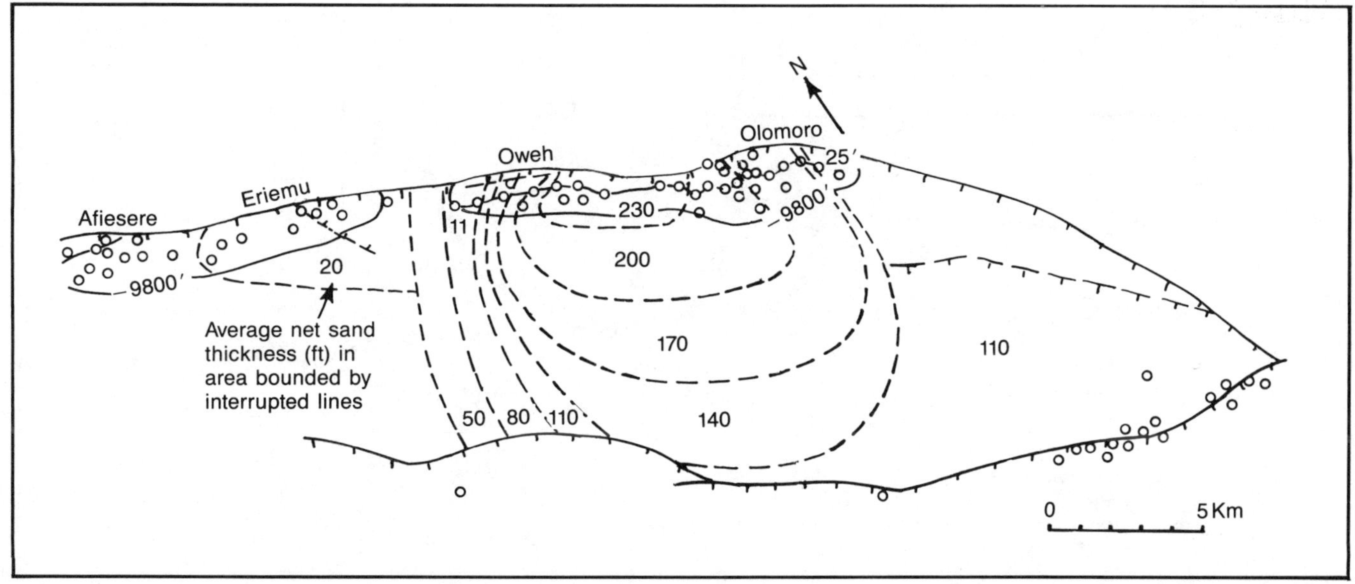

Fig. 11-27. Growth-fault block showing net sandstone below the Gumbelina-4 marker bed, together with the estimated and measured sand thickness in the underlying sedimentary cycle. (Modified after Weber, 1971; permission to publish by Geologie en Mijnbouw).

trends occur along the southwestern margin of Figure 11-27. Weber (1971) states, "Growth faults are probably initiated by a locally very fast rate of sedimentation, the focal point of the fault crescent coinciding with the area of maximum sediment accumulation . . . opposite both ends of the growth faults, the sedimentation is less rapid and thus the sedimentary cycles tend to be somewhat lens-shaped. This lens shape is emphasized by the high sand/shale ratio opposite the centre of the growth faults (Fig. 11-27 Short and Stäuble, 1967)." The locally thicker area of net sand in the middle part of Figure 11-27 probably is the "dump" site of one of the delta distributaries. The skewing of the thickness contours in a general southerly direction probably is due to the partial redistribution effects of a long shore drift in this direction.

Figure 11-28 is a diagram of the macrostructural concept. A macrostructure consists of a structural unity genetically related to one "structure-building" (growth) fault. It has one deep axis, but may consist of several fault blocks resulting from crestal, flank, and antithetic faults. In many instances the growth fault along the landward margin of the depo-center is thought to consist of a narrow zone of slip planes. Thus, the thicker the fault trace, the greater the amount of growth with depth. It may be noted from Figure 11-28 that even restricted portions of the crestal and flank faults exhibit some movement concurrent with deposition.

The massive, mobile, clay-shales (Akata formation) that underlie the prograding Niger delta are thought to be the principal source rocks for generation of hydrocarbons. These shales were "deposited under anoxic conditions on the continental slope in front of the delta where the nutrient supply for planktonic organisms must have been plentiful" (Weber and Daukoru, 1975). Shales associated with the overlying paralic sequences are generally immature, as contrasted to the mature shales of the underlying prodelta Akata formation. Weber and Daukoru have translated outcrop observations of fault zones in Germany to those of the Niger delta. They point out, "If the throw of the fault reaches several hundred feet, the fault zones develop a laminated character. This consists of streaks of sand, silt and clay, smeared into the fault zones. The thickness of the fault zones in the outcrops ranges from 20 to 60 cm (8-24 in.), which is similar to the thickness that can be deduced for Nigerian faults of 1:20 scale dip-meter log recordings.

The clay streaks in the fault zone can hamper or prevent flow across a fault whilst the sand streaks can give the fault zones a certain permeability along the fault. In Nigeria the best evidence for the vertical conductivity of the major boundary faults is the fact that in most cases the fault intersection with the upper bedding plane of the reservoir functions as the spill point of the accumulation (Fig. 11-29). It is thought likely that these spill points are also the entry points of the hydrocarbons from the fault zone into the reservoir.

At the level of the Akata formation, the major growth faults offset a thickness of up to several thousand feet of overpressured shale against paralic sediments in the down-

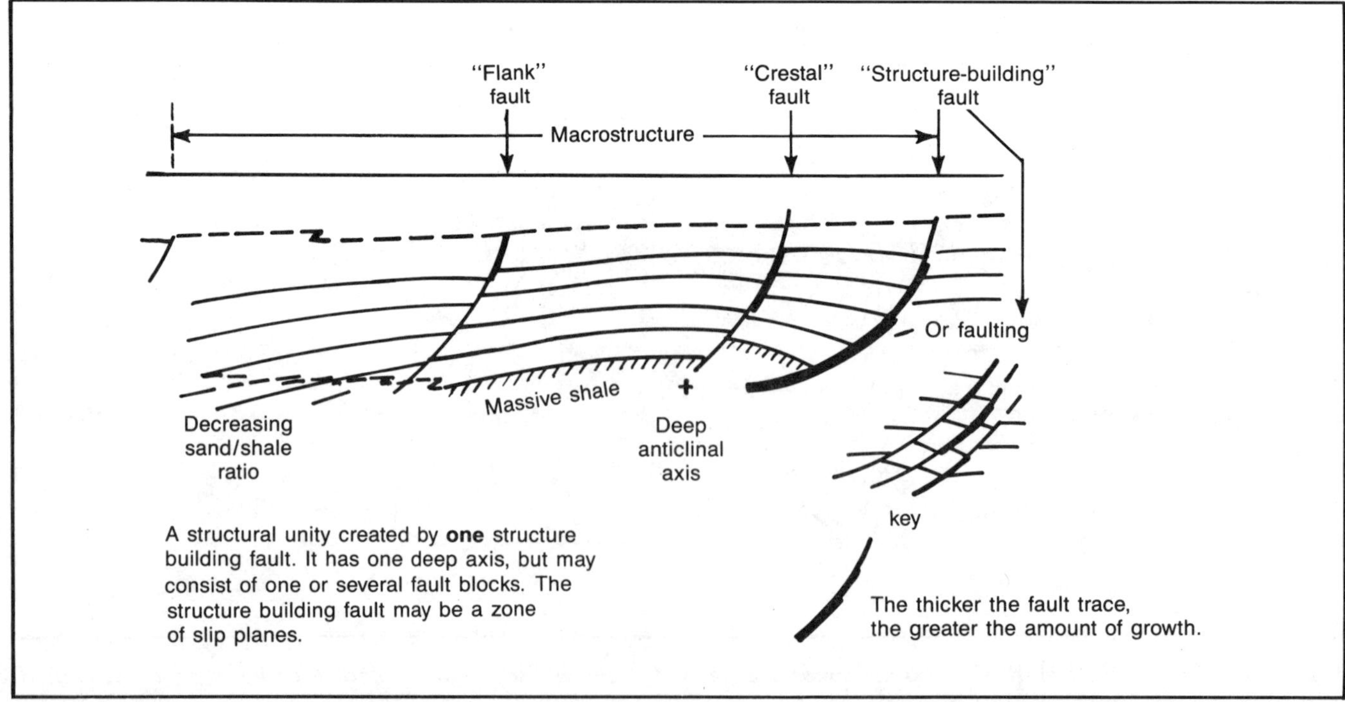

Fig. 11-28. Macrostructural concept, Niger delta. (Modified after Weber, 1971).

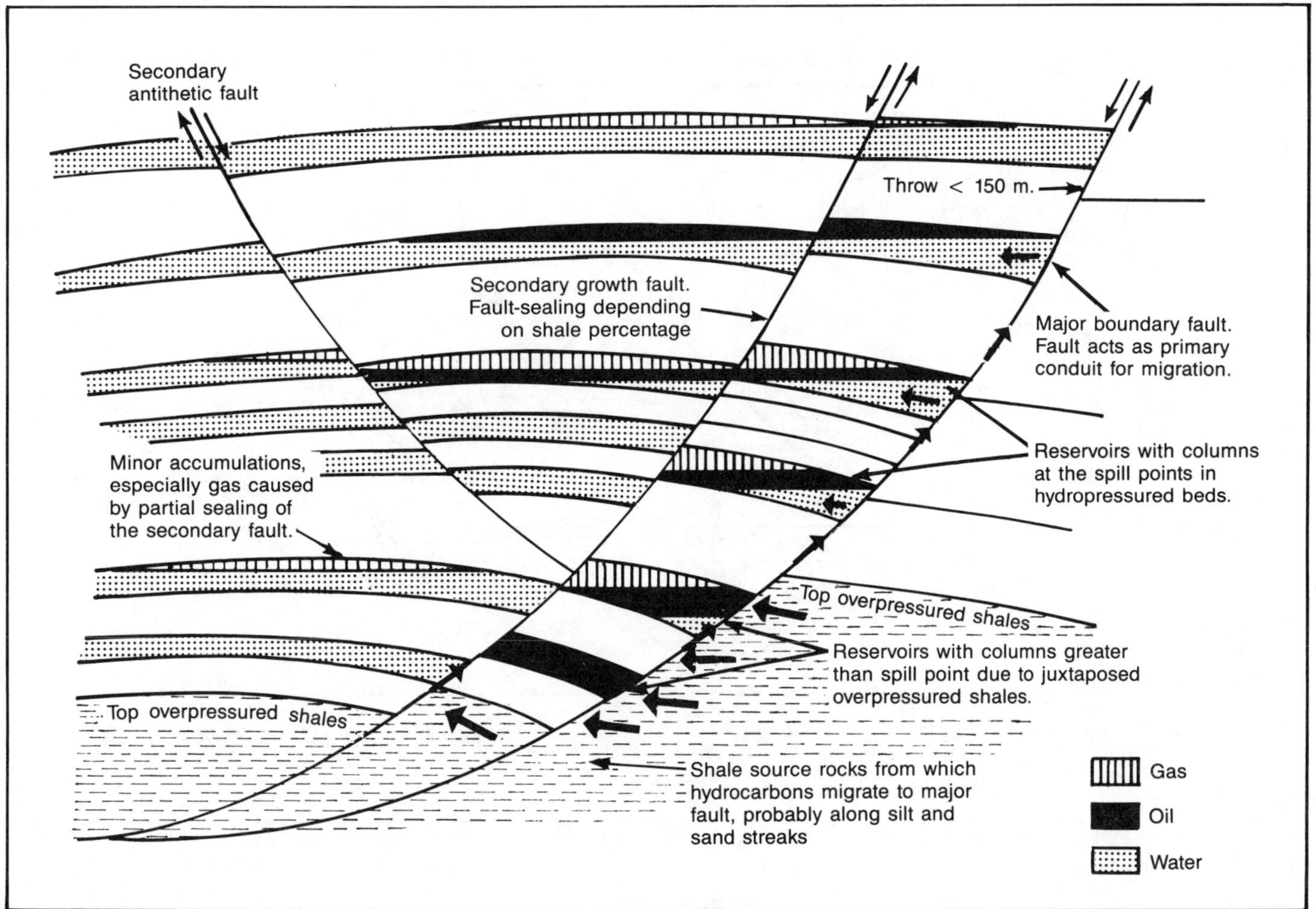

Fig. 11-29. Schematic section of a Nigerian oil field showing the principal features of the proposed accumulation model. (Modified after Weber et al., 1978).

thrown block. A plausible migration path may be from the overpressured shale into and through the fault zone (Fig. 11-29). The sands juxtaposed against overpressured formations are the only downthrown block reservoirs that occasionally have hydrocarbon accumulations trapped against growth faults. From the conductive fault zones the hydrocarbons appear to flow into the downthrown blocks only. This may be related to the effect of the specific gravity of the hydrocarbons that tends to bring them into that part of the fault zone which is adjacent to the downthrown blocks. A study of the relationship between the level where hydrocarbons are found and the throw of the growth faults indicates that the vertical conductivity of the fault zones for oil probably ceases when the throw drops below 150 m (492 ft).

Apart from the along-fault migration, other migration routes from the Akata formation shales must not be ruled out. In this respect the most likely migration path is along regional flanks, i.e., from seaward facies change updip into the south flank of a rollover structure.

The great majority of Nigerian fields, however, have at least one fault besides the major growth fault that influences the accumulations. When the same sands are juxtaposed across a fault, the faults are non-sealing. When the fault throw is larger, the sealing capacity of a fault appears to depend on the amount of shale smeared into the fault zone. This shale smearing is a function of the number and thickness of the clay layers over the throw of the fault.

It has been established statistically that a fault zone is usually sealing at a given level when it has been passed on the downthrown side by an interval consisting of more than 25% of shale. The larger the shale percentage, the larger the trapping capacity of the fault appears to be. Besides trapping by shale smearing, trapping can also occur when a reservoir is juxtaposed against shale.

As outlined above, faults with a throw of 150 m (492 ft) or more must be considered as potential vertical leaks. This fact together with the small dip closure of the structural traps is probably responsible for the small average hydrocarbon column height in the Niger delta. Only about 5%

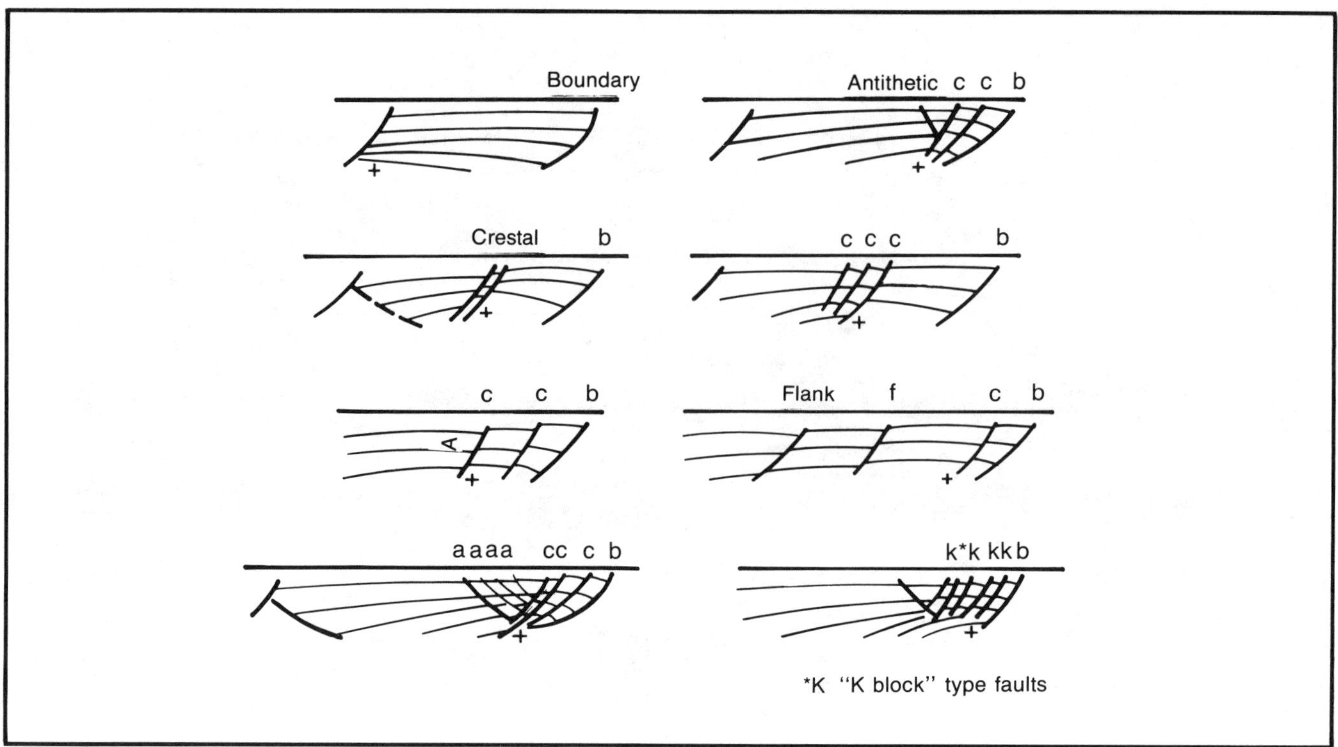

Fig. 11-30. Examples of macrostructures, Niger delta. (Modified after unpublished illustration by Weber, 1975).

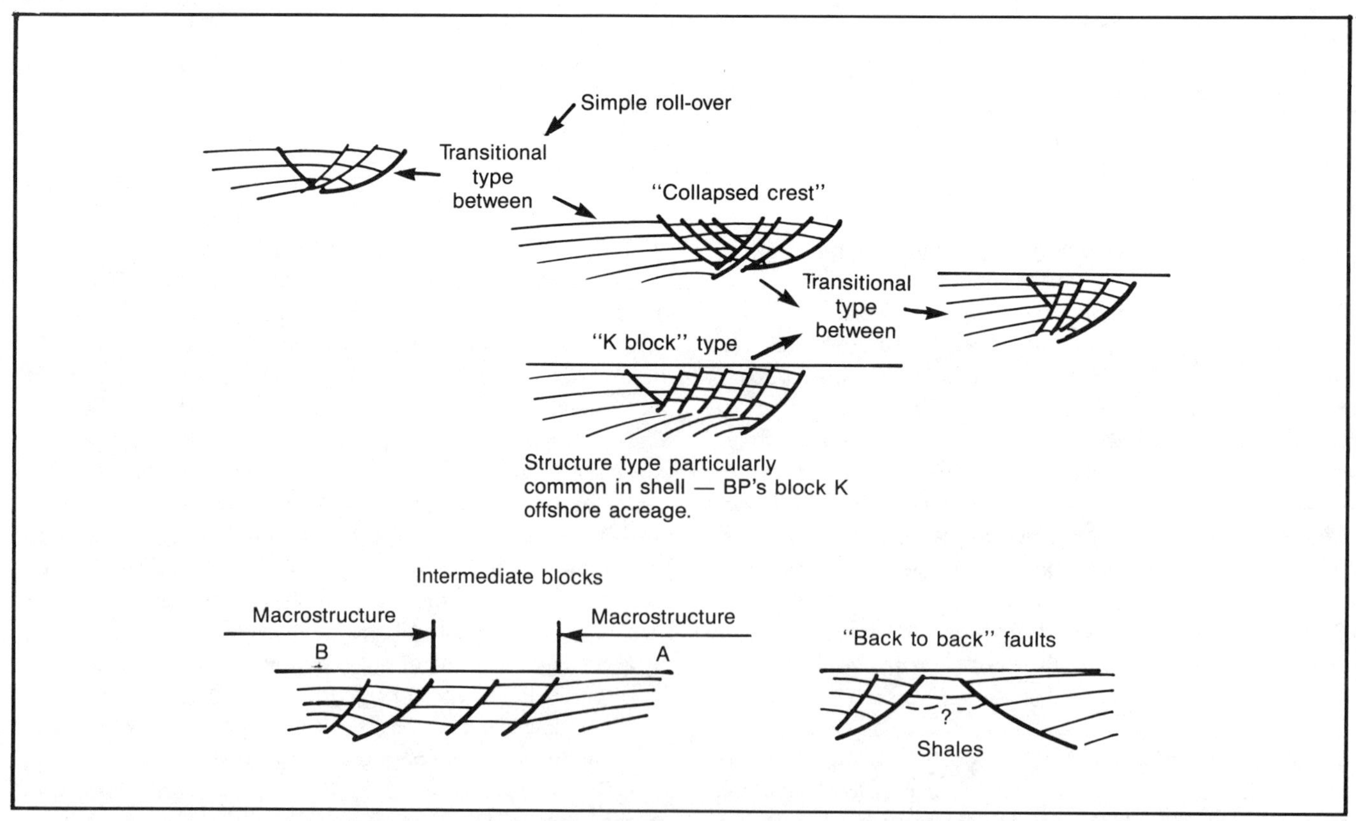

Fig. 11-31. Complex structural types, Niger delta. (Modified after unpublished illustration by Weber, 1975).

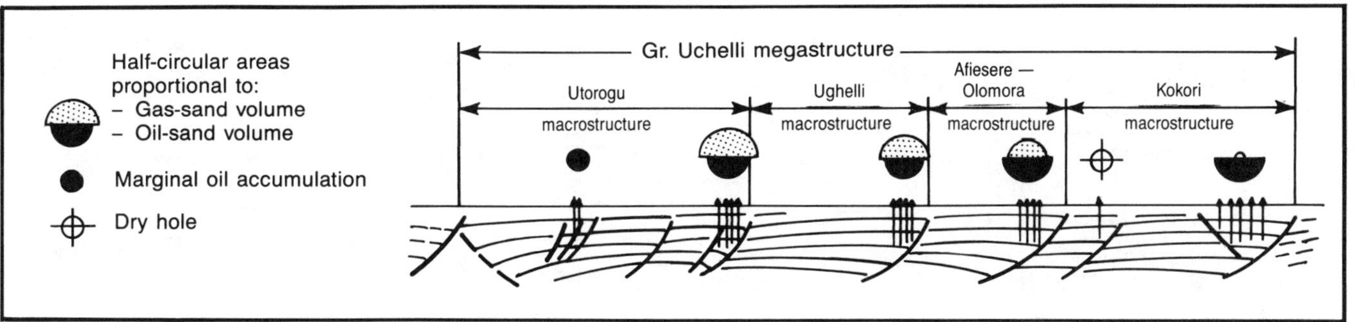

Fig. 11-32. A megastructure of the Niger delta. (Modified after Evamy et al., 1978; permission to publish by AAPG).

of the oil columns exceed 50 m (164 ft) and hydrocarbon columns in excess of 150 m (492 ft) are rare.

Examples of macrostructures are shown on Figure 11-30. They may range from quite simple to very complex, depending on the number and types of faults. In each instance the major (or boundary) growth fault occurs on the landward side. The seaward side may be defined by either another boundary (growth) fault or a counter-regional growth fault. In between, there may be either no faults, crestal faults, or combinations of crestal, flank, and antithetic faults.

Various combinations of faults related to a macrostructure are illustrated in Figure 11-31. Perhaps the most complicated structural type is that in which the structural crest is collapsed. This is the result of a combination of crestal and antithetic faults. Areas of "back-to-back" faulting are defined by two growth faults dipping in opposite directions. It is in such areas that the underlying mobile clay-shales are distorted and sometimes "squeezed" upwards. This zone of back-to-back faulting (when present) marks the seaward margin of a macrostructure.

Multiple macrostructures collectively make up a megastructure, and their recognition is based on structure, time stratigraphy, hydrocarbon distribution, and possibly hydrocarbon generation and migration. These relationships are illustrated in Figure 11-32. In this example, four macrostructures make up a megastructure. The megastructure is defined on the right (landward) side by a major growth fault and on the left (seaward) side by a counter-regional growth fault. The macrostructure of the Kokori field principally produces oil on the crest. The other three macrostructures (to the southwest) also produce oil but increasing quantities of gas reserves may be noted.

Postdepositional faulting is a significant aspect of most

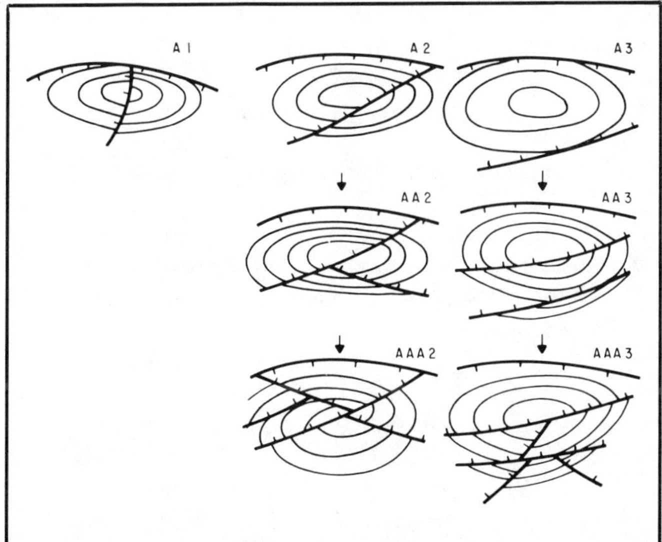

Fig. 11-33. Structural types of Niger delta fields. (Modified after unpublished illustration by Weber, 1975).

A= antithetics

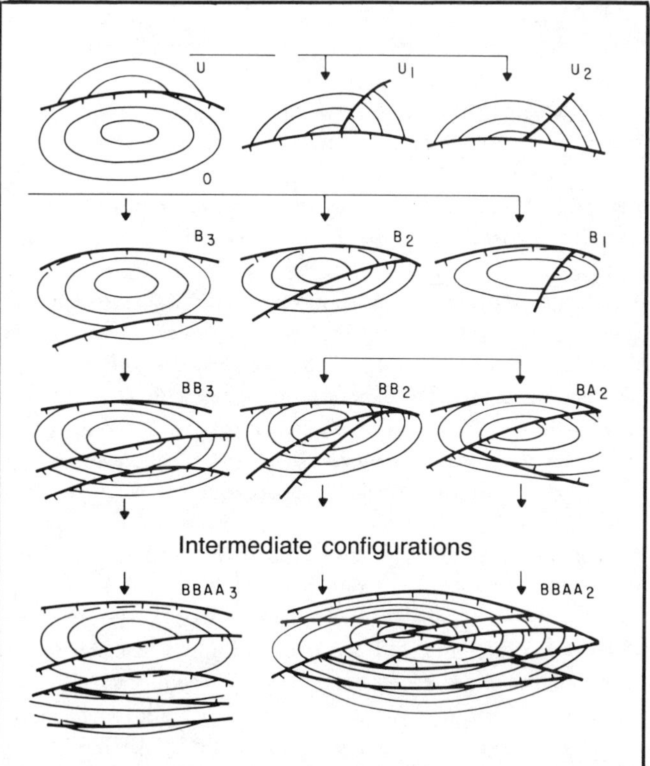

Fig. 11-34. Structural types of Niger delta fields. (Modified after unpublished illustration by Weber, 1975).

B= synth. crestals, flank fault.

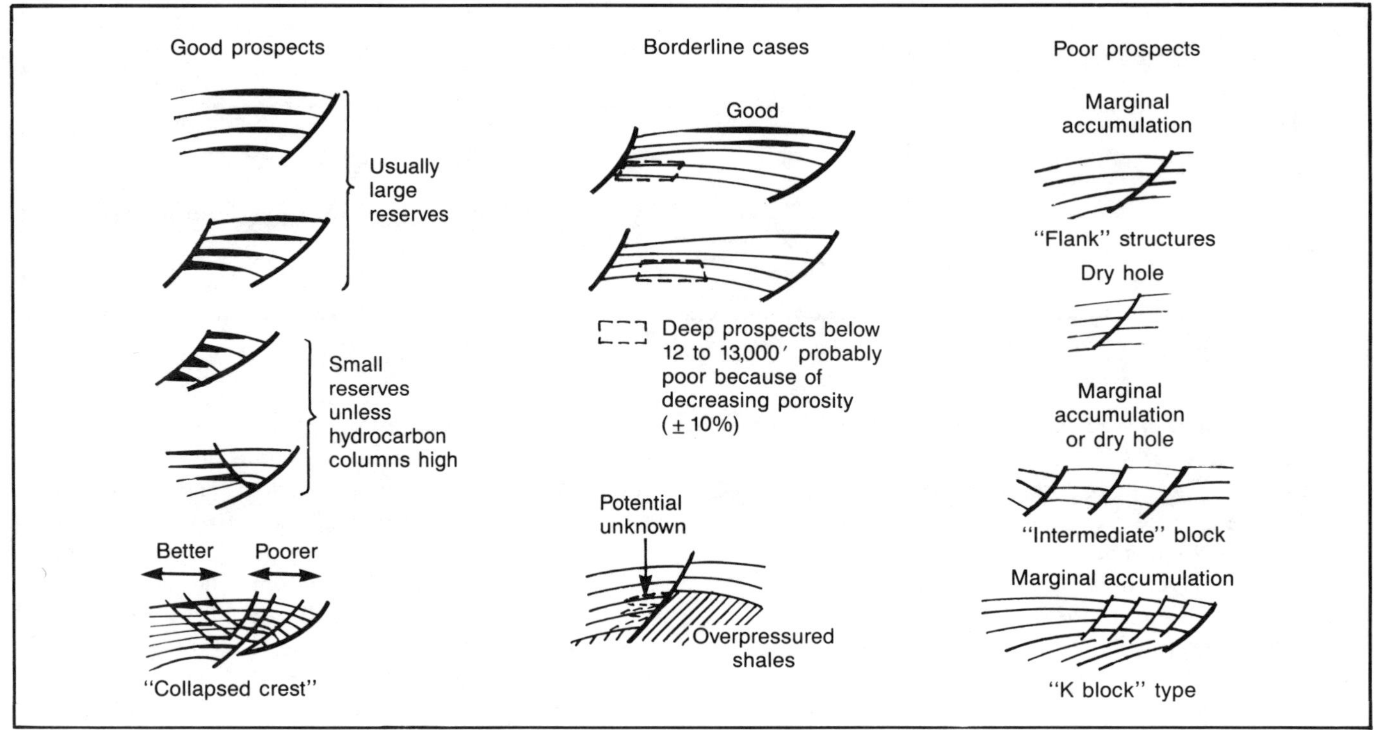

Fig. 11-35. Rating of structures, Niger delta. (Modified after unpublished illustration by Weber, 1975).

of the anticlinal closures that occur on the downthrown side of the growth faults of the Niger delta. These structures have been classified according to the number and nature of antithetic faults and flank faults.

In Figure 11-33, all of the anticlines are bordered at the top by a growth fault. The top horizontal row of anticlines each have one antithetic fault. This fault may be nearly perpendicular to (A1), diagonal to (A2), or parallel with (A3) the growth fault. The two anticlines in the middle horizontal row each have two antithetic faults. They may be either diagonal to (AA2) or subparallel with (AA3) the growth fault. In the bottom horizontal row of anticlines there are three or more antithetic faults that may be diagonal to (AAA2) or diagonal to and subparallel with (AAA3) the growth fault. The large open structure of A3 would generally be the most commercially attractive because of the large accumulation area for hydrocarbons.

In Figure 11-34 the top row of anticlines all have structural closure on the upthrown side of the growth fault. In the second row down of anticlines, there is a single flank fault. It may be nearly perpendicular to (B1), diagonal to (B2), or parallel with (B3) the growth fault. In the third row down of anticlines there are two flank faults that may be parallel (BB3), diagonal (BB2), or one diagonal flank fault in combination with a diagonal antithetic (BA2) fault. The two intensely faulted anticlines in the bottom row contain combinations of subparallel and diagonal antithetic and flank faults (BBAA3 and BBAA2). In general, the more intensely faulted an anticline is the smaller the area for hydrocarbon accumulation within individual fault blocks.

The oil and gas reserves recoverable from the previously cited structures vary widely. The height of the oil column above the spill-point and the geographic spread of an oil pool are directly related to the type of closure in which the hydrocarbons are trapped. This is graphically illustrated in Figure 11-35. Individual prospects may be good, borderline, or poor, as shown on this figure.

SELECTED BIBLIOGRAPHY

Bornhauser, M., 1958, Gulf Coast tectonics: AAPG Bull., v. 42, no. 2, p. 339-370.

Boyd, D. R., and B. F. Dyer, 1964, Frio barrier bar system of south Texas: Gulf Coast Assoc. Geol. Socs. Trans., v. 14, p. 309-322 1965; South Texas Geol. Soc. Bull., v. 5, no. 4, p. 3-16.

Bruce, C. H., 1972, Pressured shale and related sediment deformation: a mechanism for development of regional contemporaneous faults: Gulf Coast Assoc. Geol. Socs., Trans., v. 22, p. 23-31.

———, 1973, Pressured shale and related sediment deformation: a mechanism for development of regional contemporaneous faults: AAPG Bull., v. 57, no. 5, p. 878-886.

Burke, K. C. A. and J. T. Wilson, 1976, Hot spots on the earth's surface: Scientific American, August, p. 46-57.

Burst, J. F., 1969, Diagenesis of Gulf Coast clayey sediments and its possible relation to petroleum migration: AAPG Bull., v. 53, no. 1, p. 73-93.

Busch, D. A., 1973, Oligocene studies, northeast Mexico: Gulf Coast Assoc. Geol. Socs. Trans., v. 23, p. 136-145.

_______, 1975, Influence of growth faulting on sedimentation and prospect evaluation: AAPG Bull., v. 59, no. 2, p. 217-230.

Carver, R. E., 1968, Differential compaction as a cause of regional contemporaneous faults: AAPG Bull., v. 52, no. 3, p. 414-419.

Cloos, H., 1968, Experimental analysis of Gulf Coast fracture patterns: AAPG Bull., v. 52, no. 3, p. 420-444.

Dailly, G. C., 1976, A possible mechanism relating progradation, growth faulting, clay diapirism and overthrusting in a regressive sequence of sediments: Bull. Can. Petrol. Geol., v. 24, no. 1, p. 92-116.

Dickinson, K. A., 1968, Upper Jurassic stratigraphy of some adjacent parts of Texas, Louisiana, and Arkansas: U. S. Geol. Survey Prof. Paper 594 E, 25 p.

_______, 1969, Upper Jurassic carbonate rocks in northeastern Texas and adjoining parts of Arkansas and Louisiana: Gulf Coast Assoc. Geol. Socs. Trans., v. 19, p. 175-187.

Edwards, M. B., 1976, Growth faults in Upper Triassic deltaic sediments: AAPG Bull., v. 60, no. 3, p. 341-355.

Evamy, D. D., J. Haremboure, P. Kamerling, W. A. Knapp, F. A. Molloy, and P. H. Rowlands, 1978, Hydrocarbon habitat of Tertiary Niger delta: AAPG Bull., v. 62, no. 1, p. 1-39.

Hager, D. C., and C. M. Burnett, 1960, Mexia-Talco fault line in Hopkins and Delta Counties, Texas: AAPG Bull., v. 44, no. 3, p. 316-356.

Hamblin, W. K., 1965, Origin of "reverse drag" on the downthrown side of normal faults: GSA Bull., v. 76, no. 10, p. 1145-1164.

Hardin, F. R., and G. C. Hardin, Jr., 1961, Contemporaneous normal faults of Gulf Coast and their relation to flexures: AAPG Bull., v. 45, no. 2, p. 238-248.

Hazzard, R. T., W. C. Spooner, and B. W. Blanpied, 1947, Notes on the stratigraphy of the formations which underlie the Smackover limestone in south Arkansas, northeast Texas, and north Louisiana: Shreveport Geol. Soc. Reference Rept., v. 2, p. 483-503.

Hubbert, M. K., and W. W. Rubey, 1959, Role of fluid pressure in mechanics of overthrust faulting. 1. Mechanics of fluid-filled porous solids and its application to overthrust faulting: GSA Bull., v. 70, no. 2, p. 115-166.

Hughes, D. J., 1968, Salt tectonics as related to several Smackover fields along the northeast rim of the Gulf of Mexico basin: Gulf Coast Assoc. Geol. Socs. Trans., v. 18, p. 320-330.

Jux, U., 1961, The palynologic age of diapiric and bedded salt in the Gulf Coastal province: Louisiana Geol. Survey Bull. 38, 46 p.

Kirkland, D. W., and J. E. Gerhard, 1971, Jurassic salt, central Gulf of Mexico, and its temporal relation to circum-Gulf evaporites: AAPG Bull., v. 55, no. 5, p. 680-686.

Koinm, D. N., and P. A. Dickey, 1967, Growth faulting in McAlester Basin of Oklahoma: AAPG Bull., v. 51, no. 5, p. 710-718.

Magara, K., 1975, Reevaluation of montmorillonite dehydration as cause of abnormal pressure and hydrocarbon migration: AAPG Bull., v. 59, no. 2, p. 292-302.

Murray, G. E., 1961, Geology of the Atlantic and Gulf coastal province of North America: New York, Harper and Bros., 692 p.

Musgrave, A. W., and W. G. Hicks, 1968, Outlining shale masses by geophysical methods, *in* Diapirism and diapirs: AAPG Mem. 8, p. 122-136.

Ocamb, R. D., 1961, Growth faults of south Louisiana: Gulf Coast Assoc. Geol. Socs. Trans., v. 11, p. 265-278.

Powers, M. C., 1967, Fluid-release mechanisms in compacting marine mudrocks and their importance in oil exploration: AAPG Bull., v. 51, no. 7, p. 1240-1254.

Quarles, M. W., Jr., 1953, Salt ridge hypothesis on origin of Texas Gulf Coast type of faulting: AAPG Bull., v. 37, no. 3, p. 489-508.

Rider, M. H., 1978, Growth faults in Carboniferous of western Ireland: AAPG Bull., v. 62, no. 11, p. 2191-2213.

Rosenkrans, R. R., and J. D. Marr, 1967, Modern seismic exploration of the Smackover trend: Jour. Geophys., v. 32, no. 2, p. 184-206.

Seglund, J. A., 1974, Collapse-fault systems of Louisiana Gulf Coast: AAPG Bull., v. 58, no. 12, p. 2389-2397.

Shelton, J. W., 1968, Role of contemporaneous faulting during basinal subsidence: AAPG Bull., v. 52, no. 3, p. 399-413.

Short, K. C., and A. J. Stäuble, 1967, Outline of geology of Niger delta: AAPG Bull., v. 51, no. 5, p. 761-779.

Thomas, W. A., 1968, Contemporaneous normal faults on flanks of Birmingham anticlinorium, central Alabama: AAPG Bull., v. 52, no. 11, p. 2123-2136.

Wagner, W. R., 1976, Growth faults in Cambrian and Lower Ordovician rocks of western Pennsylvania: AAPG Bull., v. 60, no. 3, p. 414-427.

Walthall, B. H., and J. L. Walper, 1967, Peripheral Gulf rifting in northeast Texas: AAPG Bull., v. 51, no. 1, p. 102-110.

Weber, K. J., 1971, Sedimentological aspects of oil fields

in the Niger delta: Geologie en Mijnbouw, v. 50, no. 3, p. 559-576.

———, 1975, Unpub. personal comm.

——— and Daukoru, E., 1975, Petroleum geology of the Niger delta: Proc., 9th World Petroleum Congr., Panel Disc. 4 (1), p. 1-14.

———, G. Mandl, W. F. Pilaar, F. Lehner, and R. G. Precious, 1978, The role of faults in hydrocarbon migration and trapping in Nigerian growth fault structures: Houston, Texas, Shell Int. Pet. 10th Annual Offshore Technology Conference, Proc. 10, Vol 4, p. 2643-2653.

12 TURBIDITES

INTRODUCTION

Submarine fans are now recognized as important hydrocarbon reservoirs. Exploration of continental margins and failed rift basins has discovered important accumulations of oil and gas in fan deposits. Producing fan reservoirs occur in the Los Angeles and Ventura basins of California, the Forties, Montrose, and Frigg fields in the North Sea, and areas of the U.S. Gulf coast. As exploration efforts move farther offshore, the importance of submarine fans as sources and reservoirs of hydrocarbons will increase.

It is imperative that exploration and development geologists, geophysicists, engineers, and managers understand the nature of the fan environment in order to better exploit it. Reasonable and expectable horizontal and vertical variations in reservoir lithology and geometry must be understood to successfully and most efficiently develop fan resources. Within the last 25 years, a number of papers have appeared in the literature that have greatly expanded our knowledge of fan morphology, sediments, facies, and depositional processes. Some of these papers provided the information necessary for the exploitation of the Forties and Frigg fields. A partial listing of the references cited by authors writing about the fields (Parker, 1975; Heritier, et al., 1979; and Hill and Wood, 1980) include the writings of Bouma (1962), Normark (1970a, 1978), Haner (1971), Mutti and Ricci Lucchi (1972), and Walker (1978).

In the consideration of submarine fans, two things must be continuously borne in mind: although fans are most commonly thought of as occurring in the deep marine environment, the processes that produce fans can be equally effective in shallow marine water below wave base and in lakes (Normark and Dickson, 1976); and a submarine fan is a complex consisting of a submarine canyon or sediment feeder channel(s), the fan proper with a system of superimposed valleys, and the proximal part of the surrounding basin plain.

FAN MORPHOLOGY

A submarine fan consists of a mass of sediment derived from the shelf or upper slope and redeposited at the base of a submarine slope. A submarine canyon cut into the slope usually acts as a conduit for the sediment. At its mouth, the canyon acts as a point source for the sediment deposited on the fan. Barring restrictions related to basin morphology or bottom currents, a fan will radiate out from the canyon mouth in much the same way an alluvial fan radiates out from its source canyon. Beyond the fan, deep-sea channels may conduct sediment farther out on the abyssal plain. If the fan deposits from a number of adjacent point sources (canyons) coalesce, they form a sedimentary apron at the base of a submarine slope or escarpment (Fig. 12-1).

Submarine Valleys

Several types of valleys on the ocean floor associated with submarine fans act as conduits for sediments. They are submarine canyons, fan valleys, and deep-sea channels. Location within the shelf-fan-basin system and mode of formation distinguish between the valleys.

Submarine canyons. A submarine canyon is a straight or sinuous valley cut into a slope. It feeds sediment from the shelf or upper slope to a fan on the lower slope or in the basin. Canyons often have steep to vertical walls and are V-shape or U-shape in cross section (Shepard, 1963; Stanley, 1975; Ayers and Cleary, 1980; McGregor et al., 1982)

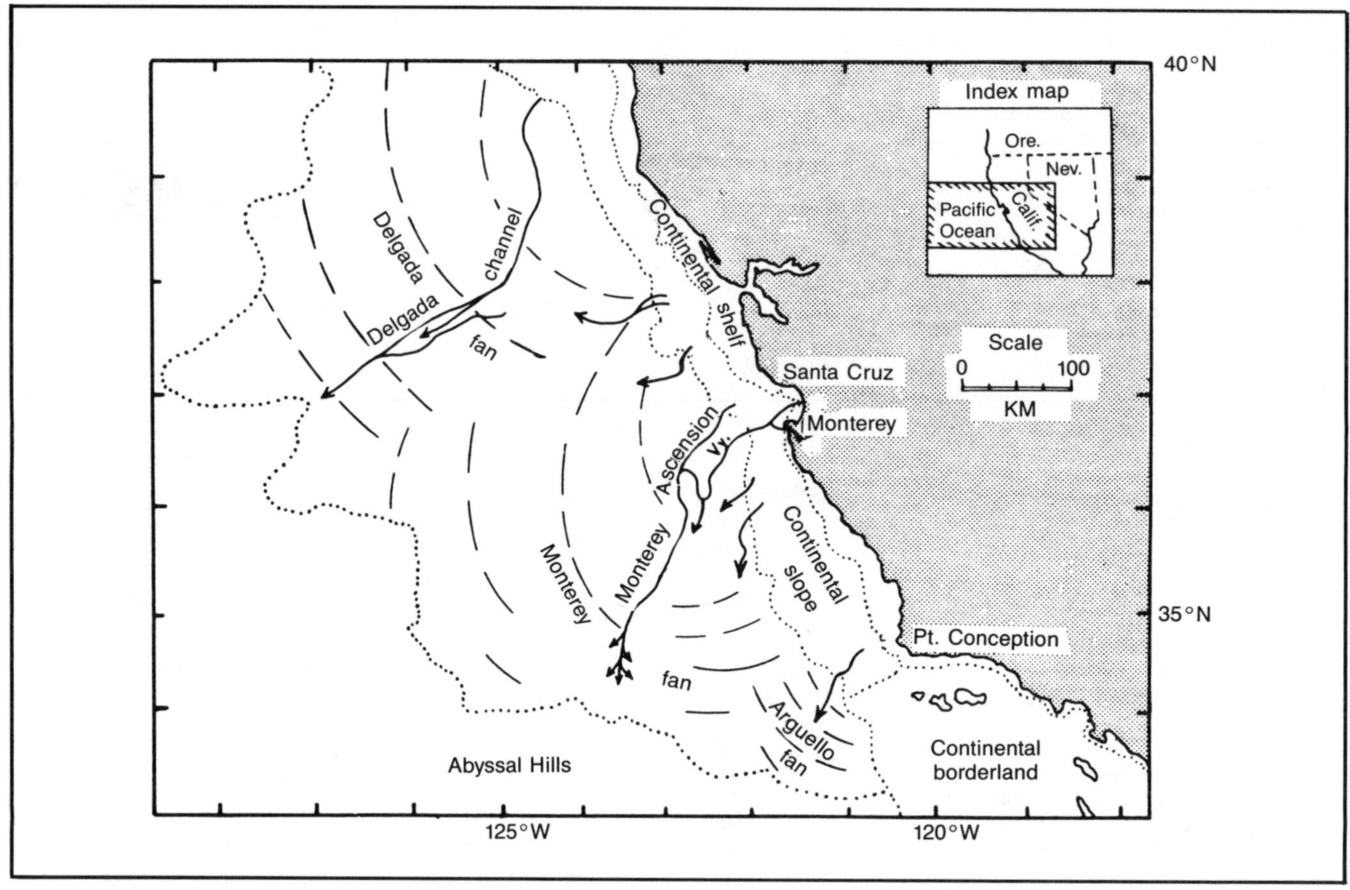

Fig. 12-1. Schematic diagram of the sediment apron off central California coast caused by coalescing submarine fans. (From Wilde et al., 1978; permission to publish by AAPG).

Some canyons head on the slopes into which they are cut (South Wilmington and North Heyes Canyons off the mid-Atlantic coast of the United States; McGregor et al., 1982), and others head on the shelf or in estuaries (Congo Canyon, Shepard, 1963). At its lower end a submarine canyon opens onto a submarine fan.

Submarine canyons appear to be produced by one or more processes. Sea level changes, subaerial erosion, faulting, slumping, turbidity and debris flows, and plate tectonics are all involved in canyon initiation and development (Shepard, 1981). Additional erosion of canyon walls is effected by abrasion from sand falls cascading down steep walls, gravity creep of loose sand moving slowly down steeply inclined canyon floors, and boring organisms burrowing into and undermining canyon walls (Shepard, 1964, 1981).

Some modern submarine canyons are obvious extensions of subaerial canyons and probably were eroded by the same processes that produced the subaerial canyons. Today, these submarine canyons lie at depths below those that can reasonably be expected from eustatic fluctuations. Submarine canyons on the island of Corsica extend into estuaries and may have been cut during the Messinian (Miocene) dessication of the Mediterranean (Hsu et al., 1977; Shepard, 1981). Off western Africa, the Congo Canyon, which extends into the Zaire River estuary, may have been initiated subaerially during Cretaceous dessication of a basin related to the opening of the Atlantic (Berger and Winterer, 1974; Leyden et al., 1978; Shepard, 1981). Turbidity currents and possibly slumping maintain the canyons today (Shepard and Emery, 1973).

Submarine fault valleys constitute at least a part of some submarine canyons. On the southwest side of Baja California, a subaerial fault valley is on trend and apparently continuous with a submarine fault valley. In northern California, a portion of the Monterey submarine canyon is a fault valley with granite and sedimentary rocks forming opposite canyon walls (Shepard, 1981).

Slumps, slides, and sediment gravity flows are instrumental in the initiation and development of submarine canyons. A mass failure on a slope may leave a steeply inclined scar making the slope above the scar unstable. This leads to more slumping and upslope extension of the scar (Belderson and Kenyon, 1976; Kenyon et al., 1978; Lewis, 1982; McGregor et al., 1982). In addition, the failures erode the substrate as they move downslope. Turbidity currents may be triggered by the slumps, which also erode the slope. Once topographic lows are established by erosional activity, subsequent mass failures and sediment gravity flows will preferentially follow the lows and further develop a canyon.

Eustatic changes are involved in development of submarine canyons. During low stands of sea level, rivers extend across the exposed continental shelf. Valleys are incised into the shelf and sediments are deposited at the shelf edge or in canyon heads at the shelf break (Shanmugan and Moiola, 1982). Slumps and turbidity currents derived from unstable sediment build-ups at the shelf break erode the substrate extending the canyons downslope (Stanley et al., 1972; Pitman, 1978).

Rising sea level generally inhibits canyon development. Coarse fluvial sediment is trapped in estuaries or in other ways retained on the shelf (Curray, 1965; Kulm and Byrne, 1966; Swift, 1970; Walker, 1978; Bouma, 1979). This prevents coarse sediment from reaching a canyon and may result in canyon abandonment. Fine-grained clastics may then fill the canyon creating a shale plug (Sabate, 1968; Walker, 1978; Lewis, 1982). Subsequent lowering of sea level may lead to renewed erosion and cleaning out of preexisting canyons allowing them to become active again (Burke, 1972). Several erosional cycles related to eustatic fluctuations may cause extensive canyon development (Burke, 1972; Shepard, 1981).

Fan valleys. A distributary network of valleys on the fan surface radiates out from a single valley that is an extension of the submarine canyon (Fig. 12-2). Three types of valleys occur on modern fans (Fig. 12-3). Depositional and ero-

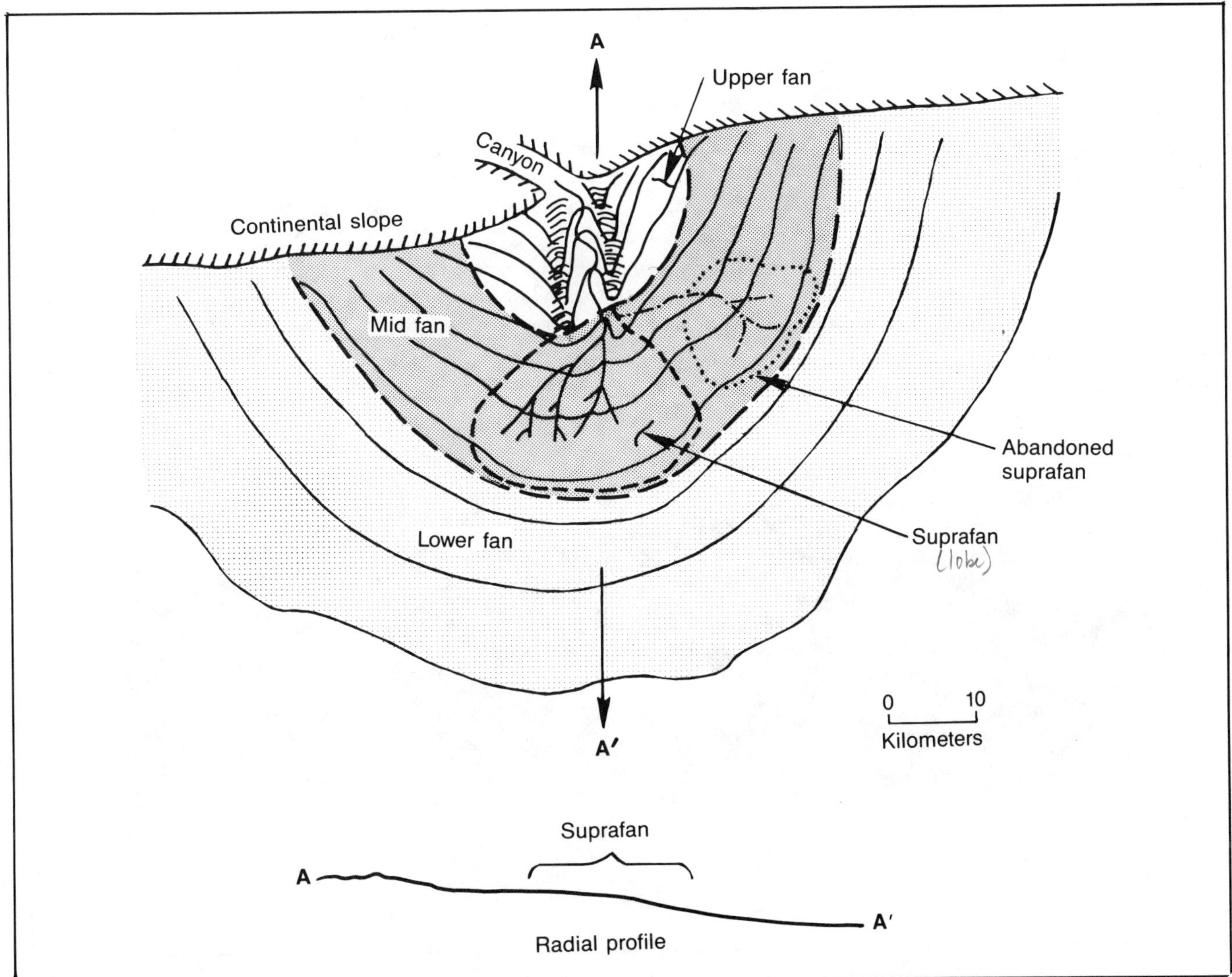

Fig. 12-2. Morphological divisions of a submarine fan as used by Normark. (From Normark, 1978; permission to publish by AAPG).

sional valleys are two ends of a series with depositional-erosional valleys between the end members (Nelson and Kulm, 1973; Normark, 1978; Walker, 1978; Wilde et al., 1978).

Depositional valleys form by sedimentation from turbidity currents. If a turbidity current flows over the banks of its confining channel, deposition occurs adjacent to the channel due to decreased turbulence and a drop in velocity. The levees formed are analagous to the natural levees in a fluvial system. As the levees build in height, the fan valley floor also aggrades and the whole system builds up above the surrounding fan surface. With simultaneous deposition in the channel and on the levees, sediments of both features may grade into or interfinger with each other (Normark, 1978) forming continuous beds between the channel and the levees (Fig. 12-3A).

Many depositional valleys have characteristics that strongly resemble fluvial features. Valley floors have braided channels, meandering thalwegs, and terraces (Normark, 1978; Walker, 1978; Damuth et al., 1983). Fan valleys themselves also meander (Shepard, 1966; Normark, 1970b; Stow, 1981; Damuth et al., 1983). Lateral migration of channels occurs by cut-and-fill and channel avulsion if fan valley deposition reduces the gradient too much (Curray and Moore, 1971; Wilde et al., 1978; Damuth et al., 1983).

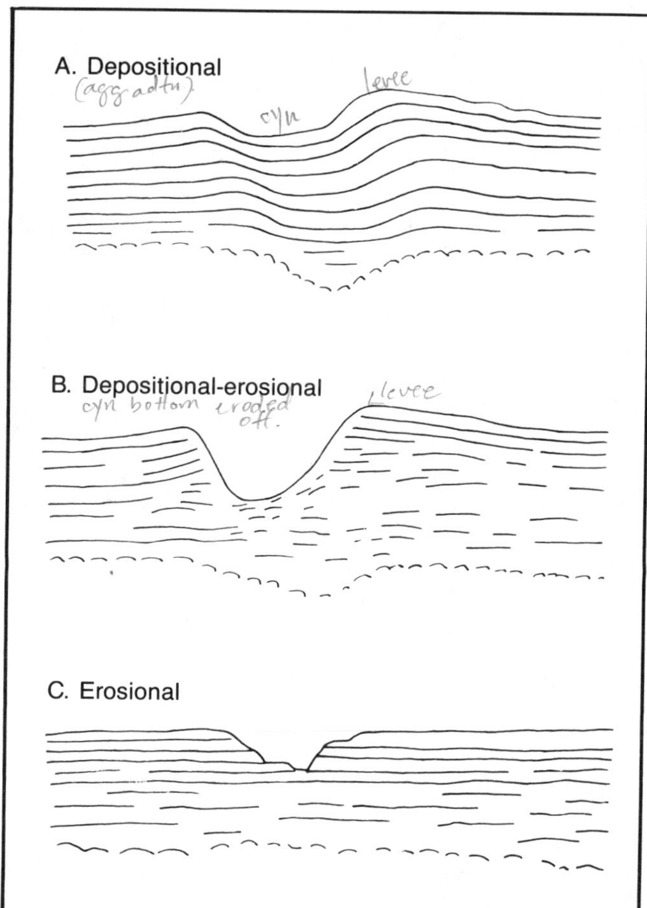

Fig. 12-3. Three types of fan valleys. (Modified after Nelson and Kulm, 1973; permission to publish by SEPM).

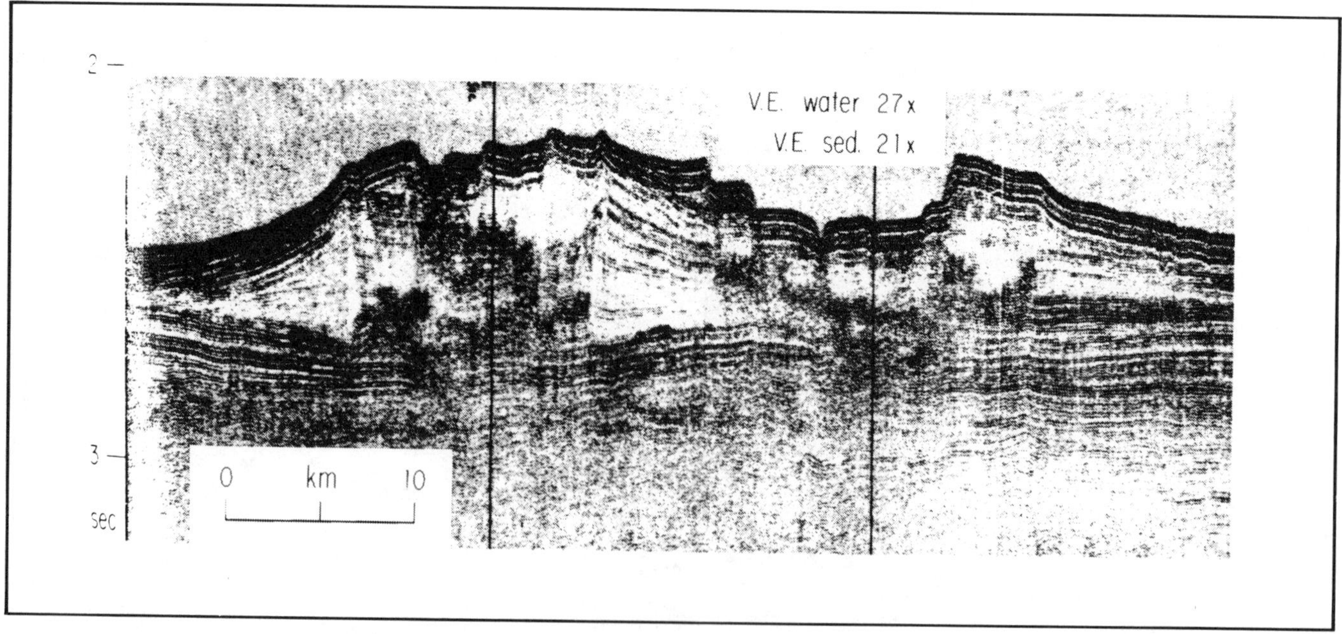

Fig. 12-4. Reflection profile of elevated channels and natural levees on the Bengal Fan. Abandoned, nearly filled, leveed channel occurs on the left. Note depression of fan sediments under depositional valley system. (From Curray and Moore, 1971; permission to publish by GSA).

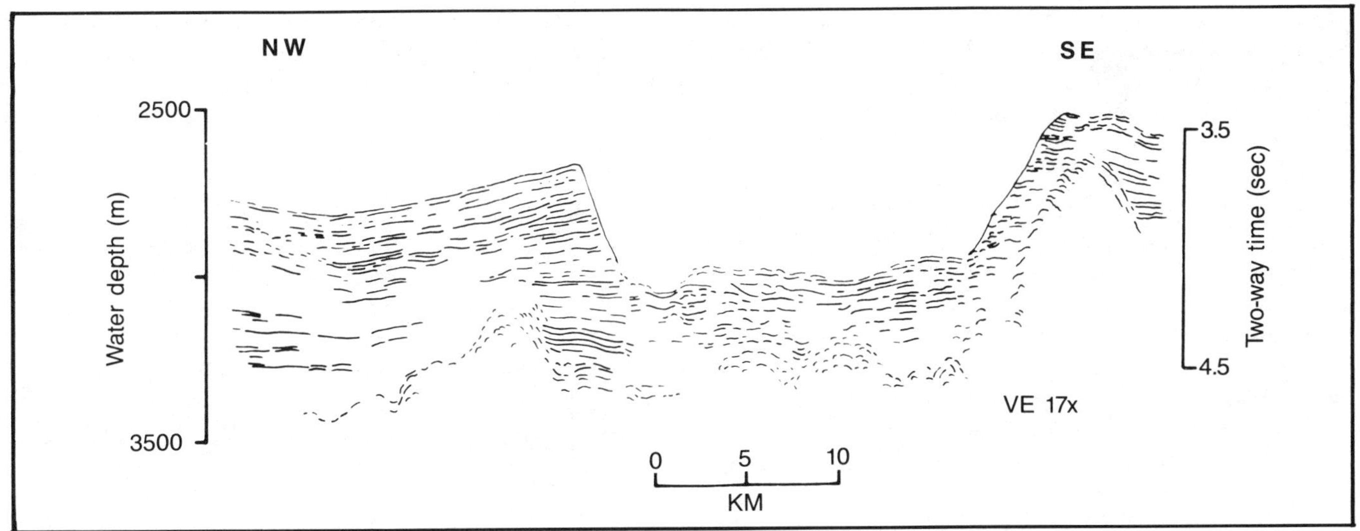

Fig. 12-5. Line tracing of depositional-erosional fan valley. (From Nelson and Kulm, 1973; permission to publish by SEPM).

Depositional fan valley dimensions vary considerably. Levees range from several to hundreds of meters above the fan surface and distance between levee crests varies from hundreds of meters to 50 km (Fig. 12-4). The valley floor may be as much as 200 m below the levee crests yet still be several hundred meters above the general fan surface. Valley size is roughly proportional to fan size (Normark, 1978), and the relief of the valley decreases downfan until it is imperceptibly lost on the lower part of the fan (Curray and Moore, 197l; Nelson and Kulm, 1973; Damuth et al., 1983).

While depositional fan valleys represent sites of sediment accumulation on the fan, sediment bypass and loss are indicated by erosional fan valleys (Normark and Piper, 1969; Normark, 1970b; Piper, 1970; Normark, 1978). These valleys may cut several hundred meters into pre-existing fan and depositional valley floor sediments. Levees are absent or poorly formed. Erosional valley floors may be terraced (Fig. 12-3C) and coarse channel sediments juxtaposed with finer-grained, valley-wall deposits (Nelson and Kulm, 1973; Normark, 1978; Wilde et al. 1978).

Depositional-erosional valleys are the most common type of fan valley (Fig. 12-3B and 12-5). Some depositional valleys are modified by later erosion (Nelson and Kulm, 1973), while other valleys are erosional in their upper reaches and depositional in their lower reaches (Normark, 1970a). These composite valleys show truncated valley wall and levee sediments and a channel floor that may be cut below the surrounding fan surface (Fig. 12-3B). Channel relief may be several hundred meters (Nelson and Kulm, 1973).

Valley type seems to be a function of the materials carried by the flows moving through the valley. Fan valleys acting as pathways for turbidity currents carrying fine-grained sediments tend to be depositional valleys. When currents are sufficiently strong to transport gravel, the valleys are depositional-erosional (Nelson and Kulm, 1973).

Deep-sea channels. A deep-sea channel is a low gradient valley continuous with a submarine canyon or outer fan valley that extends onto the abyssal plain. As with other submarine valleys, comparisons between deep-sea channels and fluvial systems are readily made. Some channels have thalwegs, point bars, natural levees, and tend to meander (Ness and Kulm, 1973; Chough and Hesse, 1976).

Deep-sea channel morphology indicates depositional and depositional-erosional modes of formation. Changes in channel morphology are more gradual than those on fan valleys, occurring over hundreds of kilometers of channel length. Channel width in its upper reaches varies from a few to 10 km and channel depth may reach 200 m. Down channel, width increases and depth, after initially increasing, ultimately decreases until the channel becomes lost on the abyssal plain. Local tectonism or sea floor irregularities may alter these generalities (Nelson and Kulm, 1973; Ness and Kulm, 1973; Chough and Hesse, 1976). Levees flank depositional deep-sea channels and are formed in the same manner as levees bounding fan valleys (Menard, 1955). Levees may build up over 250 m above the surrounding plain and be 30 km between crests.

Submarine Fan

A submarine fan is divisible into three parts based on morphology and sedimentary characteristics. Normark (1970a), using studies of modern fans, differentiates between upper, mid, and lower fan (Fig. 12-2). Mutti and Ricci-Lucchi (1972), using facies analysis of Late Tertiary fans in the Apennines, differentiate between inner, middle, and outer fan. Normark's notation will be used as it seems to have received acceptance in, at least, the North American literature (Nelson and Kulm, 1973; Walker, 1978; Stow, 1981; Damuth et al., 1983).

Fan morphology and the proportions of the three divisions of a fan are a function of type and rate of sediment supply, and basin shape (Piper, 1975; Normark, 1978). Small fans found in continental borderlands and restricted basins tend to have coarser sediments derived from rivers and littoral drift sands brought to the fan by submarine canyons heading on the shelf. Such fans have relatively steep gradients and limited lateral extent. Large fans found in the open ocean have finer-grained sediments derived from major rivers, relatively low gradients, and great lateral extent (Nelson and Kulm, 1973).

Effects of basin morphology on fan morphology are illustrated by the Laurentian and Amazon fans. The lower Laurentian fan is well developed except for some restriction to the northeast by a local topographic rise (Fig. 12-6). Due to interference from and ponding of sediment by the Mid-Atlantic Ridge, the Amazon fan has a narrow, restricted lower fan (Fig. 12-7).

Upper fan. Just below the submarine canyon is the upper fan (Figs. 12-2 and 12-8). A leveed fan valley that is continuous with the submarine canyon is the principal morphological feature of the upper fan (Curray and Moore, 1971; Normark, 1978; Damuth et al., 1983). Potentially high levees flanking the active fan valley and recently abandoned fan valleys can make the topographic surface of the upper fan relatively rough (Curray and Moore, 1971; Nelson and Kulm, 1973).

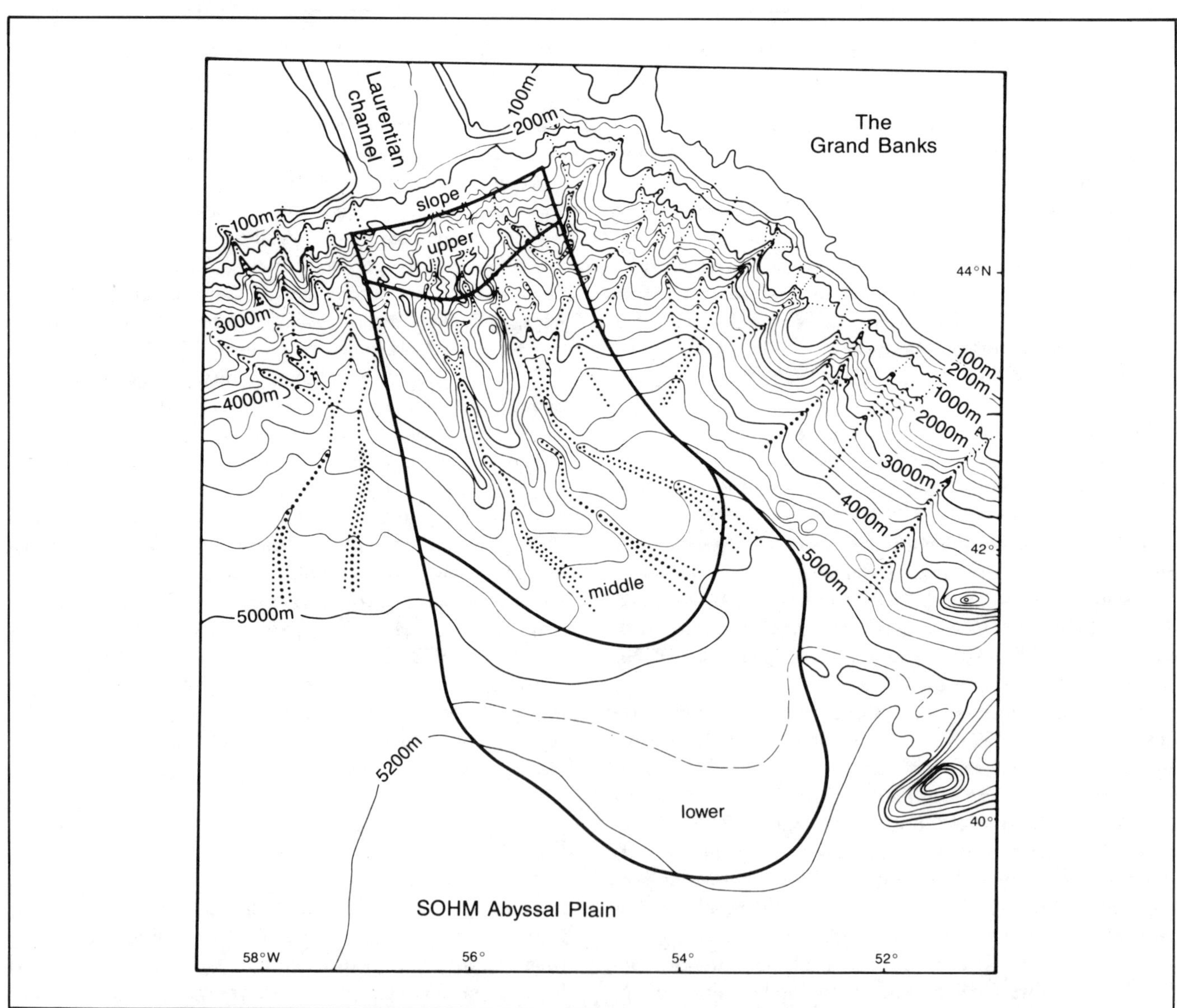

Fig. 12-6. Morphology of the Laurentian submarine fan. (From Stow, 1981; permission to publish by AAPG).

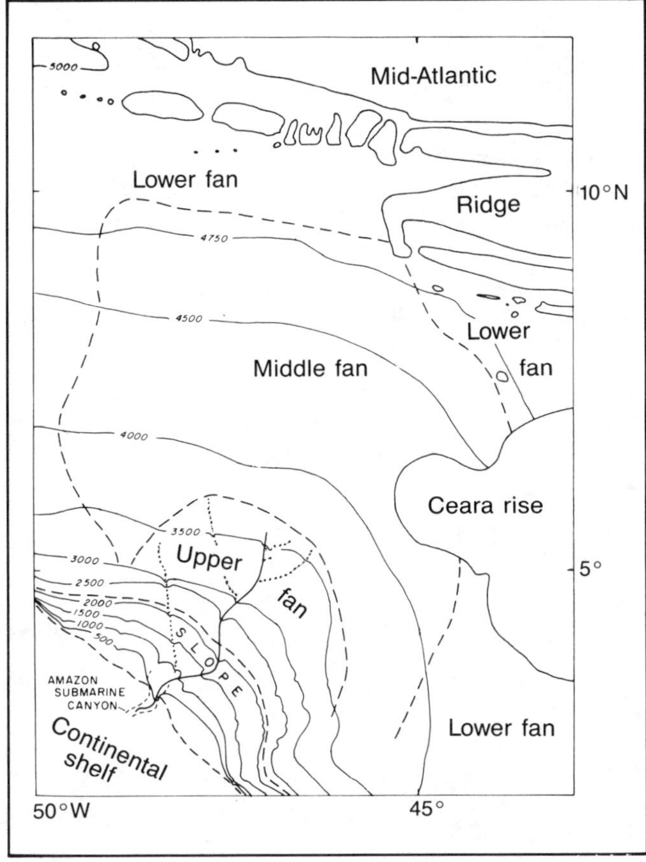

Fig. 12-7. Morphology of the Amazon submarine fan. (From Normark, 1978; permission to publish by AAPG).

Middle fan. The middle fan starts at the terminus of the active upper fan valley where the valley divides into a series of unleveed distributary channels (Figs. 12-2 and 12-8). Channel migration is rapid and deposition builds a bulge or suprafan above the general fan surface. Braided channels extend part way across the middle fan, dying out downfan and merging with the suprafan surface (Normark, 1978; Walker, 1978).

If the upper fan valley is abandoned and a new one forms elsewhere on the upper fan, a new suprafan lobe will form at the terminus of the new, upper fan valley. An abandoned suprafan will eventually become covered with fine-grained sediment and its topographic expression lost (Normark, 1978). Continual switching of the upper fan valley and the suprafan lobes broadly builds the fan surface.

Lower fan and basin plain. The lower fan is morphologically undistinctive. There are no channels and the gradient is very low. Gradation of the lower fan into the basin plain is imperceptible (Normark, 1978).

SUBMARINE FAN SEDIMENTATION

A wide variety of clastic sediments may occur on a submarine fan (Fig. 12-8). Grain size and type of deposit are functions of where on the fan the deposit occurs and the sediment available at the source. General sedimentation patterns show coarser sediments accumulating in the fan valleys while associated fine sediments accumulate in the valleys, on the levees, and on the fan surface in the inter-channel areas. Horizontal distribution of sediments is accomplished by lateral migration of fan valleys and overflow of the valleys by turbidity currents. Deposition is intermittent as shown by repetitive sequences of thick, coarse-grained sediments grading upward into thinner, fine-grained sediment (Nelson and Kulm, 1973).

Fan sediments are usually derived from shallow water. Fragments of large shells, displaced shallow water biota, glauconite, and wood and plant debris (Normark and Piper, 1972; Selley, 1979) are redeposited with the sediment of the fan. If a wide range of sediment sizes is available at the source, conglomerate through claystone may be found on the fan. If only fine-grained sediments occur at the source, only siltstones and claystones will be deposited on the fan.

Depositional Processes

A sediment gravity flow is the mechanism for redeposition of shallow water sediment in deeper water. In this kind of flow, the sediment moves parallel to the bed under the pull of gravity, and interstitial or entrained fluids move with the sediment. This differs from a fluid gravity flow in which the fluid moves parallel to the bed under the pull of gravity, and entrained sediment moves with the fluid, e.g., a river (Middleton and Hampton, 1976).

Four kinds of sediment gravity flow are distinguished, theoretically, based on the way the sediment is maintained above the bed (Fig. 12-9). *Turbidity currents* support sediment by the upward component of fluid turbulence; *fluidized sediment flows* support grains by upward flowing fluid escaping between grains settling due to gravity; *grain flows* support sediment by grain-to-grain collisions and close approaches; and *debris flows* support large clasts in a matrix of fine sediment and interstitial fluid of sufficient concentration to have a finite yield strength. Each type is a theoretical end member of a spectrum of flows and in a sedimentation event, one or more types may occur or evolve from the other types (Middleton and Hampton, 1976).

Turbidity current. Among sediment gravity flows, only the turbidity current has been demonstrated to move coarse sediment into deep water. Turbidity currents are surges initiated by events such as earthquakes, storms, and waves that cause suspension of sediment in water. The suspended material flows downslope in a channel, such as a submarine canyon, out onto the fan. Once the surge goes beyond the confines of the channel by spreading either over the channel banks or far enough downslope to where the channel no longer exists, it then spreads laterally as well as continuing to move farther downslope (Nelson and Kulm, 1973; Middleton and Hampton, 1976).

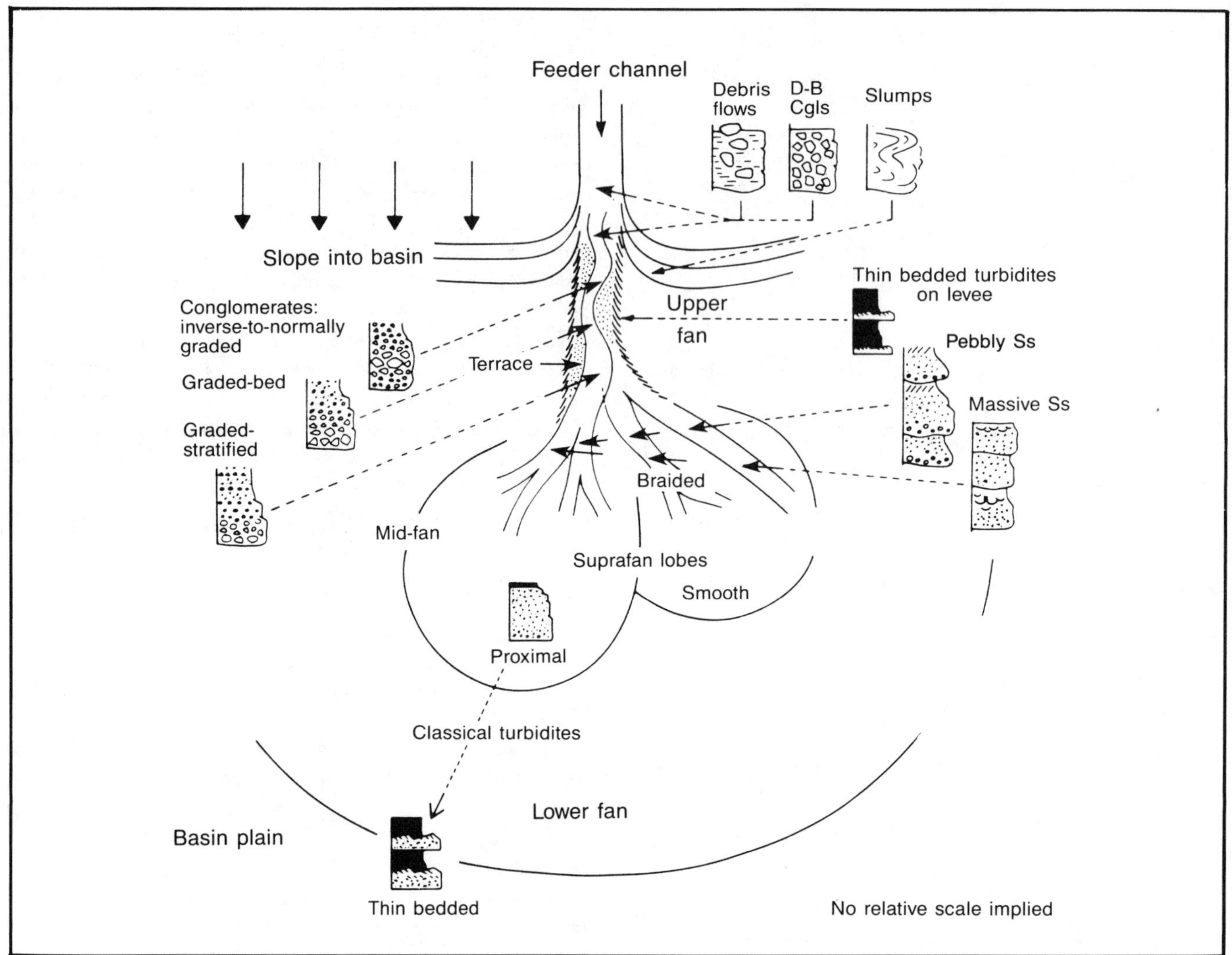

Fig. 12-8. Distribution of sediment facies on a submarine fan. (From Walker, 1978; permission to publish by AAPG).

A turbidity current is divisible into three segments: head, body, and tail (Fig. 12-10A). The head is the thickest part of the flow and moves more slowly than the body. Behind the head and somewhat thinner is the body, while the tail is last, thinner and more dilute than the body. Above the body and moving more slowly is a dilute suspension of turbidity current sediment and entrained water caused by mixing between the current and the overlying water (Middleton and Hampton, 1976).

Flow vectors in a turbidity current (Fig. 12-10B) show that sediment and water move forward and upward through the head and then around toward the back of the head. Eddies from the back of the head may contribute fine-grained sediment to the entrained layer above the surge while the coarse sediment settles back into the flow and is recirculated. This results in an increased concentration of coarse material in the head. Circulation of the coarse material abrades the bed producing the variety of sole marks associated with turbidity currents, e.g., flutes and grooves. The body continually supplies sediment to the slower moving head and may, at the same time, deposit sediment on the surface scoured by the head (Middleton and Hampton, 1976).

Deposition from a turbidity current may be rapid or slow. Rapid deposition occurs when turbulence in the current is quickly reduced. Losses in turbulence are related to energy dissipation caused by waning of the mechanism originally producing the surge, a rapid decrease in slope as at the mouth of a submarine canyon, and overflow of channel banks. Resulting deposits are massive to graded beds with sharp, erosional bases corresponding to the T_a division of the Bouma sequence (Fig. 12-11A). Grading in this division is often produced by deposition from flows with high sediment concentration and involves only the coarser grain sizes (Middleton and Hampton, 1976; Walker, 1978).

Slightly less rapid deposition from a turbidity current

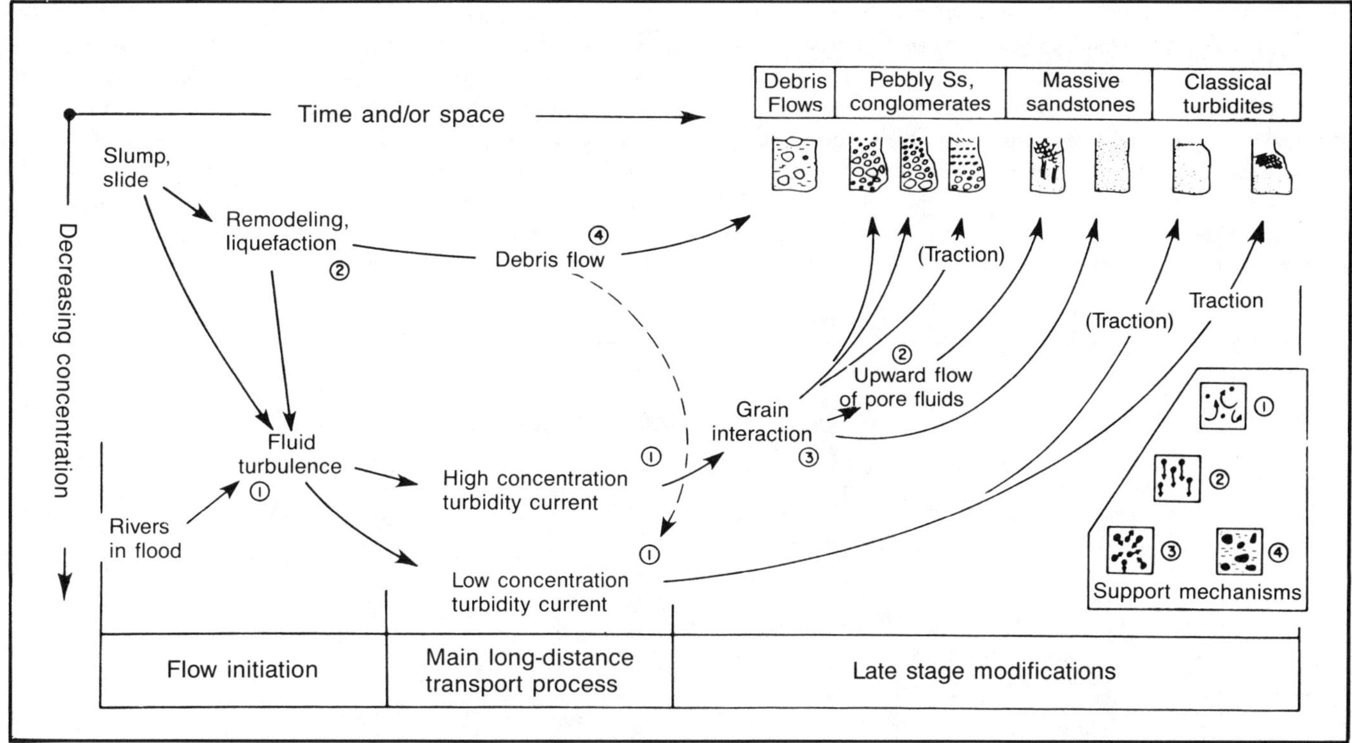

Fig. 12-9. Relationships between transportation and sediment support mechanisms and submarine fan deposits. (From Walker, 1978; permission to publish by AAPG).

produces the laminated Bouma T_b division. Lamination may be due, in part, to plane bed deposition in the upper flow regime (Fig. 12-11A), or it may be caused by deposition that is sufficiently rapid to prevent the formation of bed forms yet slow enough to allow some grain size segregation. Middleton and Hampton (1976) indicate lamination is not due to traction, whereas Walker considers T_b division laminations to be caused by upper-flow regime traction (Walker, 1978, Fig. 3).

Ripple cross-bedding of the T_c division (Fig. 12-11A) results from traction reworking of turbidity current deposits by either the tail of the current or other ambient currents. Climbing ripple sets indicate traction reworking during aggradation of the bed (Middleton and Hampton, 1976; Walker, 1978).

T_d and T_e intervals are combined by Walker (1978). They represent slow deposition of fine-grained sediment by the tail of a turbidity current and the usual "background" or pelagic deposition between currents (Fig. 12-11A).

A Bouma sequence represents deposition from a turbu-

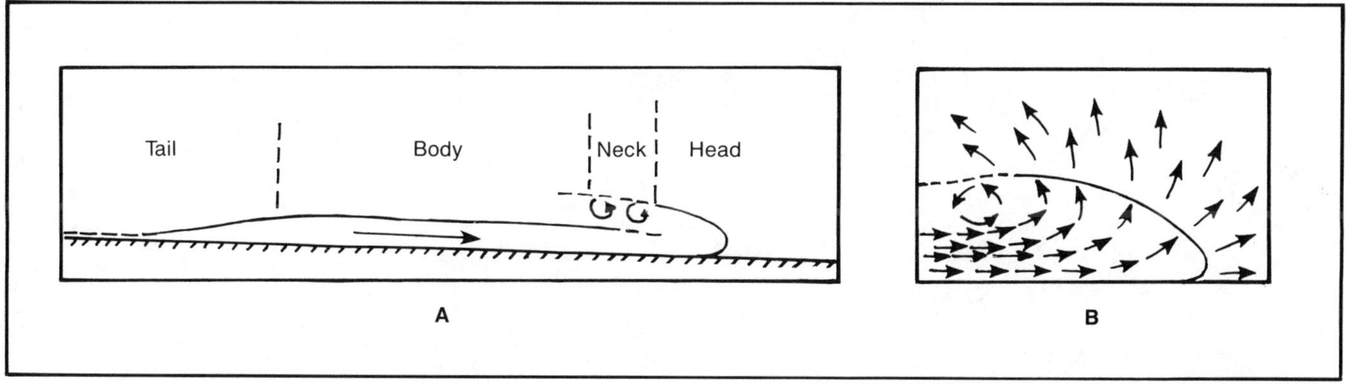

Fig. 12-10. **A.** *Divisions of a turbidity current.* **B.** *Flow in and near the head.* (From Middleton and Hampton, 1973; permission to publish by SEPM).

lent flow at continuously decreasing energy levels and may be developed both vertically and laterally. In vertical development, rapid deposition produces a T_a bed. With decreasing energy, the T_b division is deposited on the T_a. At still lower energy levels, deposition from the body and tail of the flow lays down T_c and T_d divisions on the T_b. Hemipelagic sedimentation deposits the T_e interval after the current has passed.

In the case of lateral development, T_a deposition occurs under high-energy conditions near the source or at the base of a relatively steep slope, e.g., the mouth of a canyon. On less steep slopes, T_b deposition occurs downslope from the T_a. Continued energy loss from decreasing slopes farther out on the fan and/or overflowing of the fan valleys causes deposition of the T_{cde} intervals (Fig. 12-11B). Levee and overbank deposits are thinly bedded, low-energy deposits commonly of the T_{cde} (Mutti and Ricci-Lucchi, 1975; Walker, 1978).

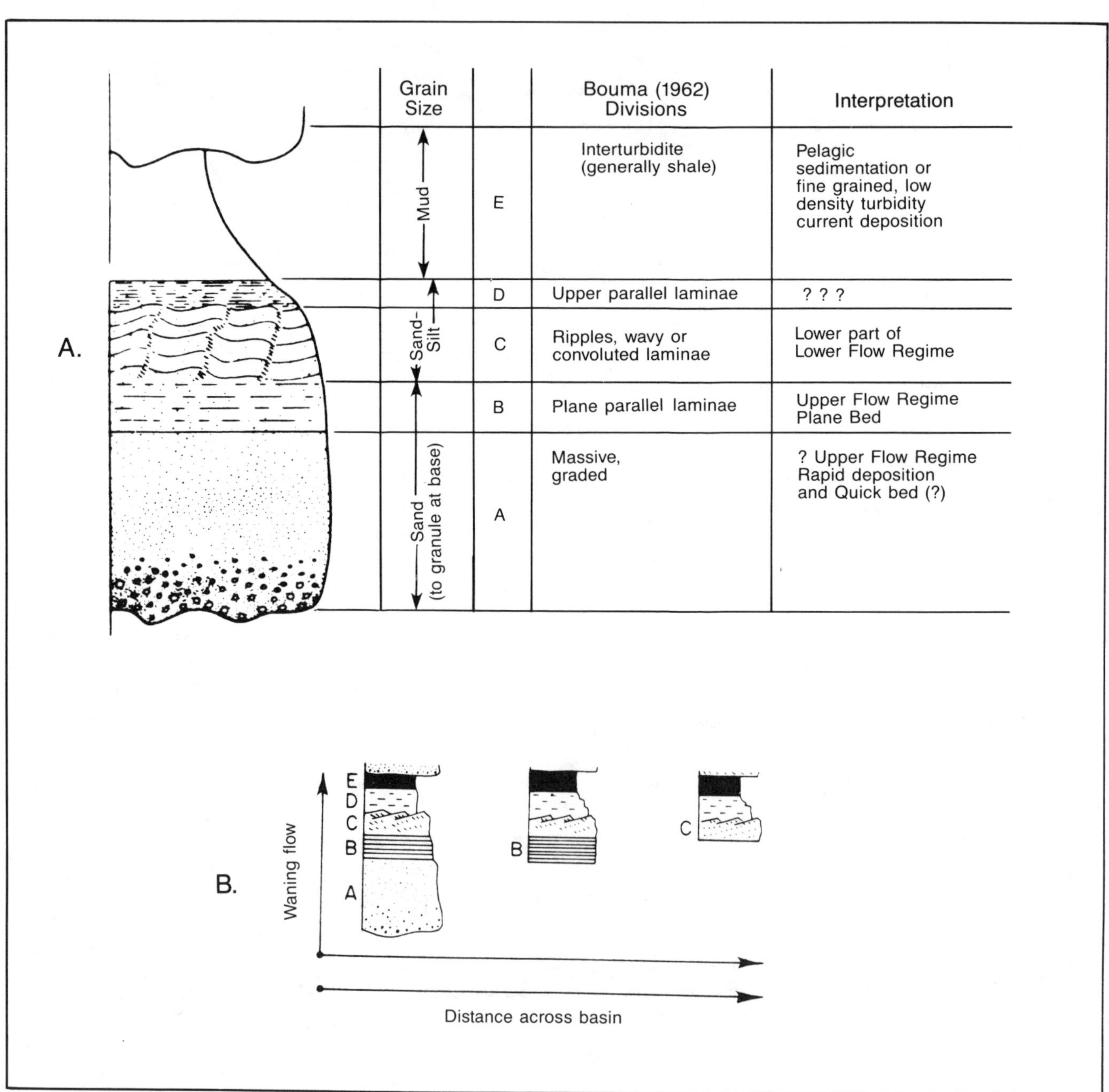

Fig. 12-11. **A.** *Complete Bouma sequence.* (From Middleton and Hampton, 1976; permission to publish by Wiley and Sons, Inc.). **B.** *Base missing Bouma sequences.* (From Walker, 1978; permission to publish from AAPG).

Fluidized sediment flow. Fluidized sediment flows differ from turbidity currents in the method for supporting the grains. In a fluidized flow, sediment is dispersed by upward-moving fluids, and the mixture behaves as a viscous fluid, moving rapidly down slopes of three to ten degrees (Middleton, 1969). As the particles in a flow settle and displace the water upward, the flow loses fluidity and congeals from the bottom up. Structures in such a deposit are produced by the escaping fluids and include dish structure and injection features such as sandstone dikes and sills. Somewhat finer-grained sediment may occur in the injection features due to transportation by the escaping fluid (Middleton and Hampton, 1976).

Walker (1978) indicates the massive sandstone facies found on submarine fans is deposited by fluidized sediment flow and resembles a group of stacked or amalgamated T_a deposits. This facies differs from a simple T_a by having thicker and coarser sandstones, more channeling, occasional fluid escape structures, and no interbedded shales. Sand transportation is, for the most part, by turbidity current. As transport ceases and deposition begins, the sediment becomes very concentrated toward the base of the flow and the water is displaced upward, producing the fluid escape structures. Massive sandstones with no fluid escape features probably experienced less forceful expulsion of pore water and greater interaction between sediment particles such as grain-to-grain collisions.

Grain flow. Grain flows support sediment particles by interaction between grains. In a concentrated, cohesionless suspension of sediment moving downslope, dispersive pressure caused by grain-to-grain collisions prevents the sediment from settling out. In the presence of a viscous fluid, close approaches of the grains to each other also maintain the dispersion. With increasing energy in a grain flow, fluid turbulence intensifies and at some point aids in the support of the sediment. Thus, there is a gradation between grain flows and turbidity currents and some sediment gravity flows may be combinations of both (Middleton and Hampton, 1976).

Deposition from grain flow occurs when the force moving the flow becomes less than that necessary to maintain the flow. A loss in force occurs principally when the angle of the slope down which the flow is moving is reduced, causing simultaneous deposition of several layers of grains. Such deposition is in contrast to grain-by-grain deposition from ocean currents and rivers (Middletown and Hampton, 1976).

Walker (1978) suggests that deposition of pebbly sandstones on submarine fans is from grain flows and similar to the deposition of the massive sandstone facies from fluidized sediment flows. He envisions transportation of the sediment into the basin by suspension due to fluid turbulence. During deposition, late-stage particle support comes from intergranular collisions. Stratification and cross-stratification in the pebbly sandstone facies indicate that traction of the bed occurs.

Pebbly sandstones are important reservoirs in the Los Angeles and Ventura basins in southern California. On the outcrop, the beds vary from one-half to five meters thick,

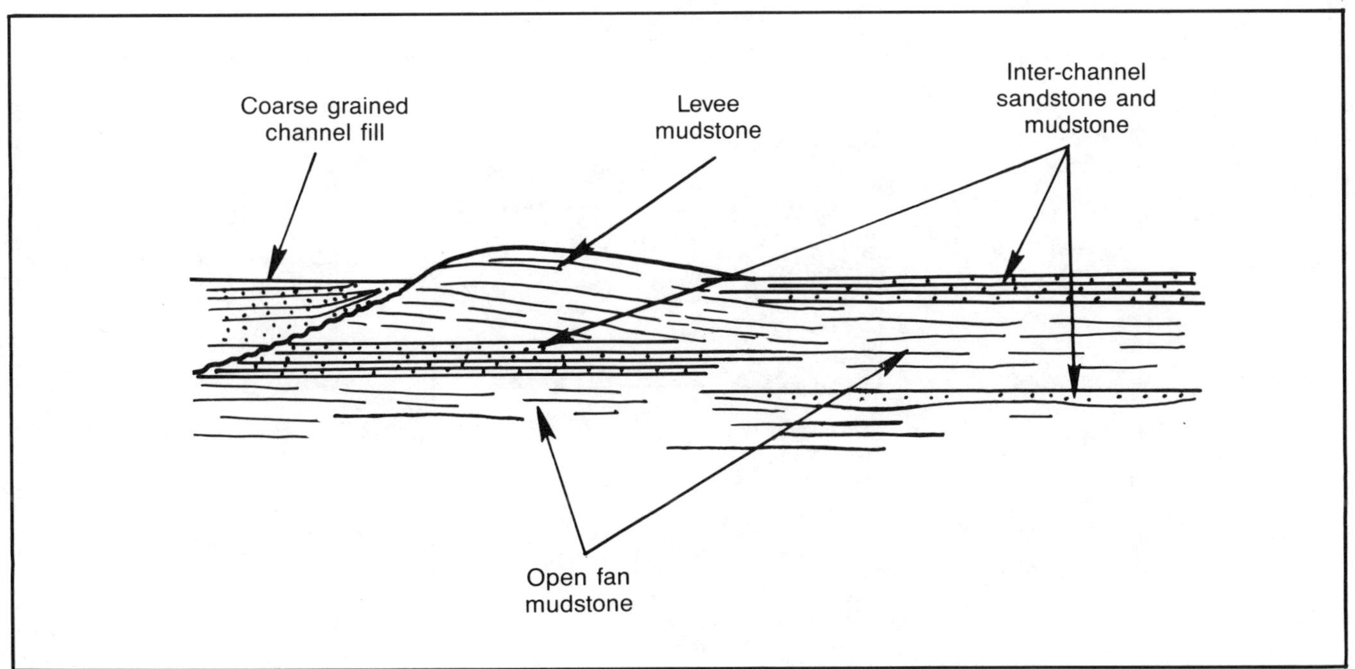

Fig. 12-12. Lateral relationship between channel, levee, and overbank turbidity current deposits. (Modified after Mutti and Ricci-Lucchi, 1975; permission to publish by International Congress of Sedimentology).

have sharp bases, and no shale interbeds. Stratification is shown by layers with varying amounts of pebbles. Cross-stratification on the 10–20 cm scale also occurs (Walker, 1978).

Individual pebbly sandstones are lenticular and have irregular, scoured bases. Coalescing of individual graded beds may create laterally extensive units ten to hundreds of meters thick. Downdip interfingering with turbidities creates an ideal juxtaposition of possible source and reservoir rocks, e.g., in the Los Angeles basin (Walker, 1978).

Clast-supported conglomerates are implied by Walker (1978) to be deposited from grain flows (Fig. 12-9). He also states that finer-grained and stratified conglomerates may be gradational with coarser-grained pebbly sandstones. In outcrop, clast-supported conglomerates range in thickness from 1 to 50 m, have sharp, channeled bases, and are discontinuous laterally. Structures include reverse grading on a 20–30 cm scale, normal grading, cross-stratification, and stratification from alternating coarse and fine layers. Three types of bedded conglomerate occupy theoretical positions downcurrent relative to one another (Fig. 12-13). Gradation between types is *not* suggested but proximal versus distal position is implied (Walker, 1978).

Debris flow. Debris flows support large clasts in a cohesive matrix of water and relatively fine-grained material such as clay or sand. Matrix strength is sufficient to support and carry relatively coarse materials (Middleton and Hampton, 1976). Slumps, by virtue of their matrix strength, may be considered debris flows (Walker, 1978).

Features of debris flows include grooves in underlying beds from large clasts or blocks moving in contact with the bed, internally chaotic beds displaying varying degrees of disruption, and large clasts rafted up and protruding above the top of the flow (Walker, 1978).

Hemipelagic deposition. Near the continental margin and over the adjacent abyssal plain, deposition of very fine-grained detritus occurs. Fine-grained material derived from the land by volcanism, wind, or mild erosion, coupled with biogenic debris from the sea, settles on the ocean floor. This slow, steady "rain" of clastics with no real point source occurs regardless of input of other sediments and may be considered background sedimentation. Hemipelagic sediments are masked by rapid influxes of other types of sediments. If enough time elapses between sediment gravity flows on a fan, sufficient hemipelagic sediment may accumulate to be preserved and recognized.

Fan Sediments

In considering sedimentation on a submarine fan, it should always be kept in mind that the sediment found in the fan depends on the detritus available at the source. Regardless of the competence of a sediment gravity flow, if pebbles and cobbles are not available, no conglomerate will be deposited. In discussing fan sediments, all sizes of clastics are assumed to be available.

Upper fan. The coarsest, most poorly sorted, and least structured sediments occur in the upper fan valley (Nelson and Kulm, 1973). Conglomerates and coarse sandstones constituting the valley fill have limited lateral continuity, and

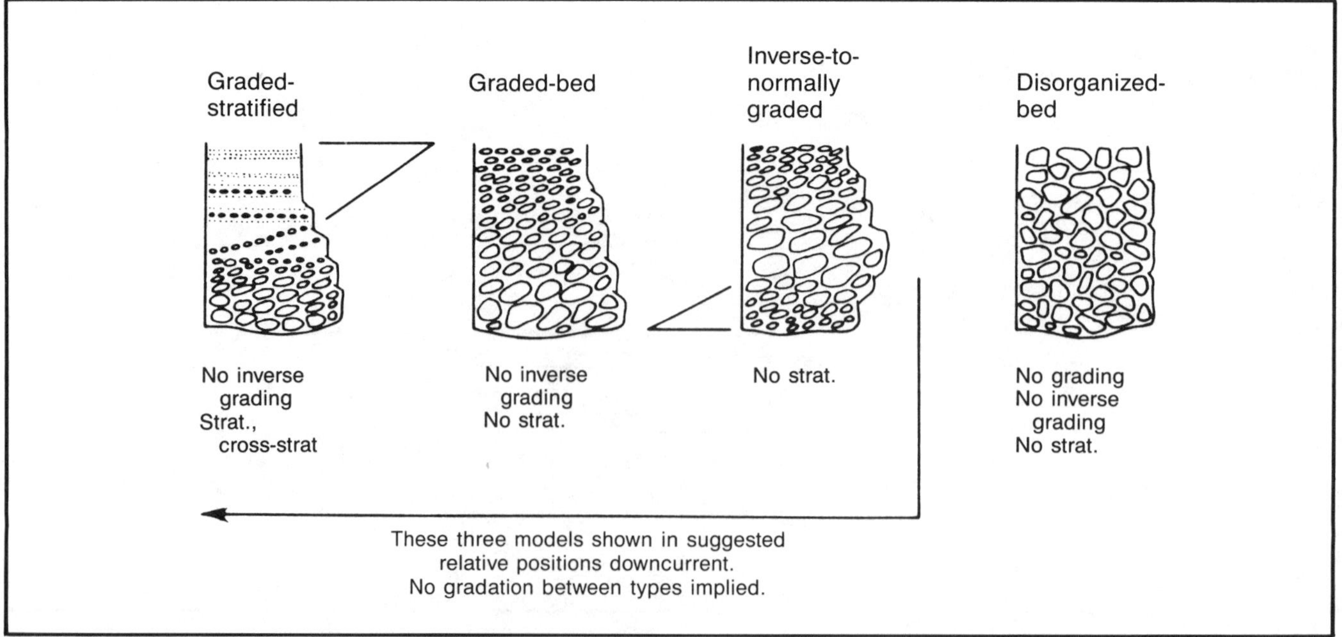

Fig. 12-13. Types of clast-supported conglomerates. (Modified after Walker, 1978; permission to publish by AAPG).

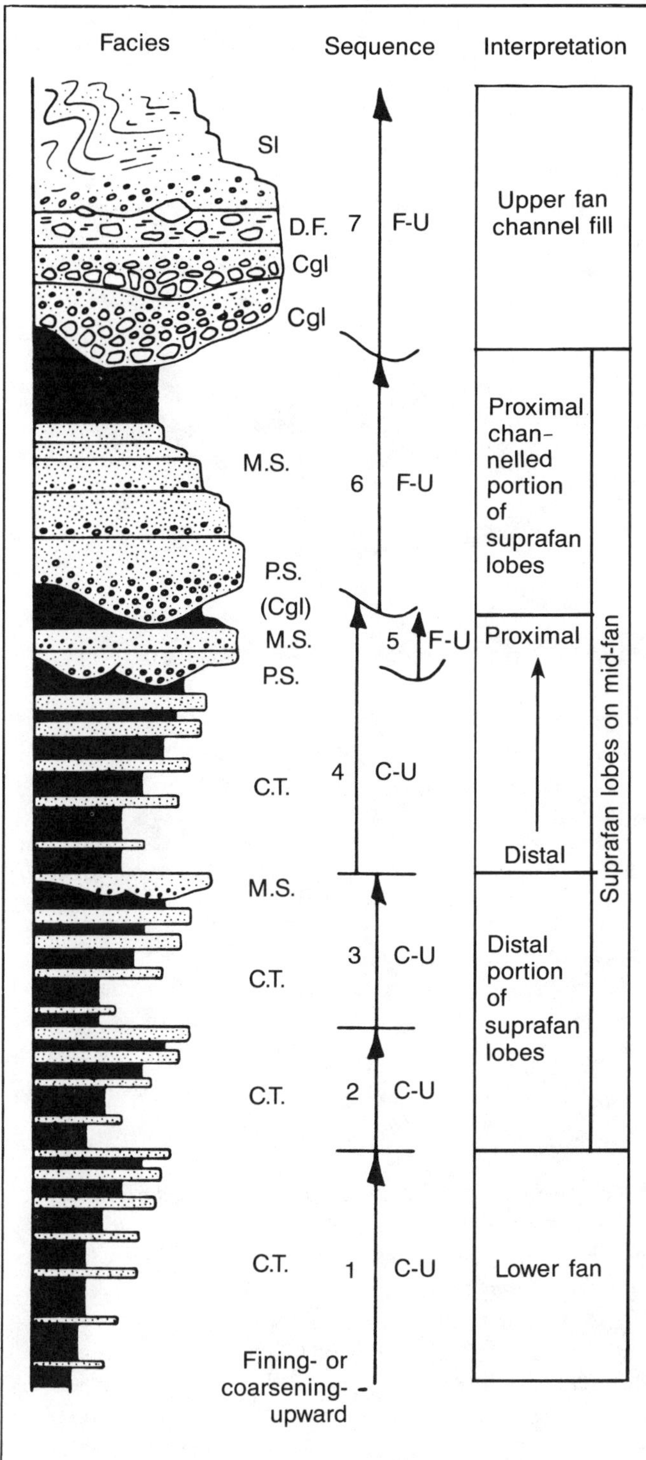

Fig. 12-14. Hypothetical stratigraphic sequence that could be developed during fan progradation. C-U represents thickening- and coarsening-upward sequence; F-U represents thinning- and fining-upward sequence; C.T., classic turbidites; M.S., massive sandstones; P.S., pebbly sandstones; CGL, conglomerate; D.F., debris flows; SL, slumps. (Modified after Walker, 1978; permission to publish by AAPG).

thickness may be from tens to hundreds of meters (Fig. 12-8; Mutti and Ricci-Lucchi, 1972; Walker, 1978). Abandonment of the fan valley is indicated by a fining upward of the fill (Fig. 12-14, sequence 7).

If the upper fan valley is a depositional valley, the coarse valley fill grades laterally into fine-grained levee deposits (Fig. 12-3A; Normark, 1978). Levee deposits consist of laminated siltstones and medium- to fine-grained sandstones a few centimeters thick. They are lenticular and resemble thin-bedded turbidites found on the basin plain (Fig. 12-8; Nelson and Kulm, 1973; Walker, 1978). Levee deposits grade laterally into thin, fine-grained sandstones and mudstones of base-absent Bouma sequences, hemipelagic mudstones, and deposits from dilute suspensions, such as nepheloid layers and dilute turbidity currents (Fig. 12-12; Walker and Mutti, 1973).

In an erosional upper fan valley, the coarse valley fill terminates abruptly against the channel margin deposits. Erosional valley fills also become finer upwards indicating channel abandonment (Normark, 1978; Walker, 1978).

Middle fan. Sediments in the proximal, braided channel portion of the middle fan are a downfan continuation of the clastics found in the distal part of the upper fan valley. Conglomerates deposited here and pebbly to massive sandstones deposited just downfan from the conglomerate form extensive coarse-grained units due to the lateral and vertical coalescing of the braided channel deposits (Fig. 12-15). These channel deposits, like those of the upper fan, may decrease in grain size and bed thickness upward, indicating channel abandonment (Fig. 12-14, sequence 6). Fining-upward and thinning-upward intervals a few meters to tens of meters thick should be discernible on electric logs (Fig. 12-15; Walker, 1978).

Sediments on the more distal, unchanneled part of the suprafan are deposited as classical turbidites (Fig. 12-8). Complete Bouma sequences and varying degrees of base-absent sequences occur (Fig. 12-11B; Mutti and Ricci-Lucchi, 1972; Walker, 1978).

As a part of a delta may be abandoned by upstream channel switching, so may a suprafan lobe be abandoned by shifting of the upper fan valley. As a new suprafan lobe forms elsewhere on the middle fan, the abandoned lobe will be covered by fine-grained hemipelagic sediments and possibly overbank materials from the new suprafan lobe. A few meters of fine-grained sediment will effectively seal the abandoned lobe that now becomes a good reservoir (Fig. 12-15). Considering that suprafan lobes have lateral dimensions of several to tens of kilometers, thicknesses of tens to more than 100 m and are encased in shale, the lobes make excellent exploration targets.

Outer fan and basin plain. Hemipelagic sediments and turbidites characterize distal deposits of the outer fan and basin plain. Beds are thin and laterally continuous on a scale of several kilometers (Mutti and Ricci-Lucchi, 1972).

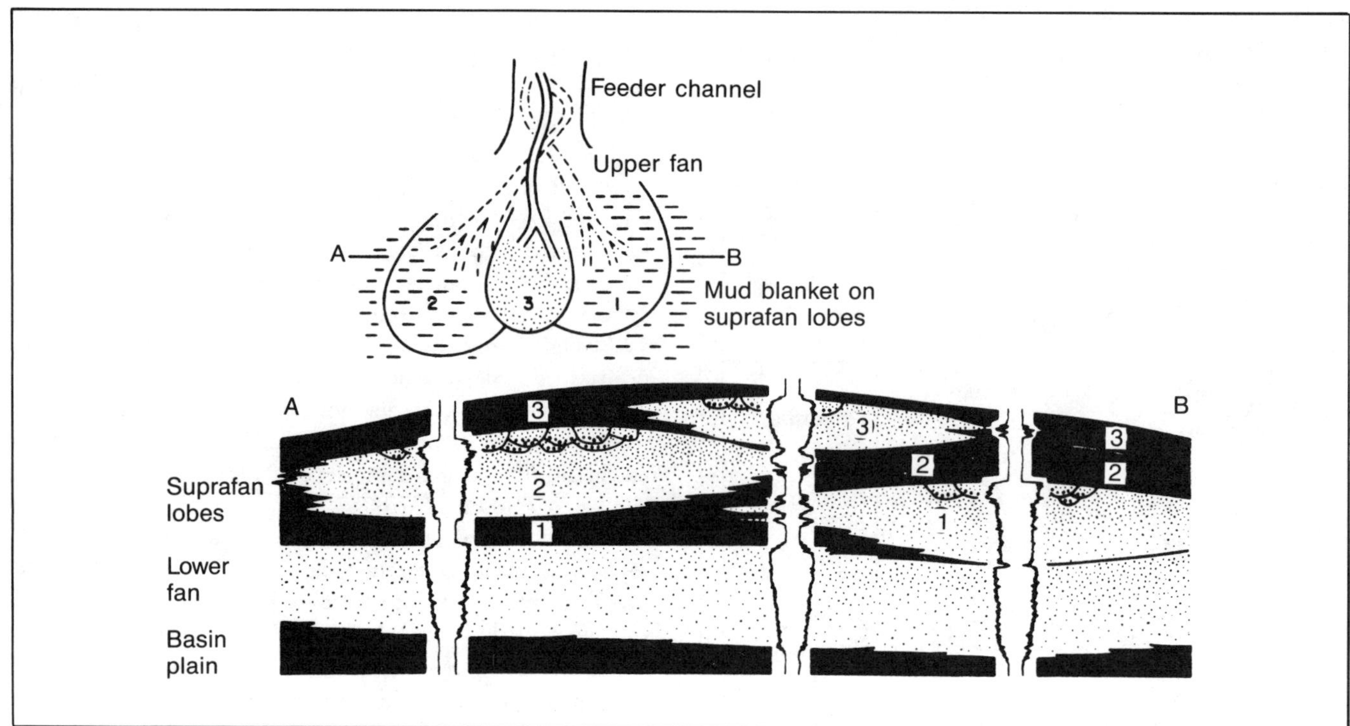

Fig. 12-15. Theoretical cross section across prograding lower and middle fan system. Hypothetical electric logs show progradation by coarsening-upward SP/GR. Channel abandonment indicated by fining upward SP/GR at top of some sandstone units. (Modified after Walker, 1978; permission to publish by AAPG).

Distinction between the lower fan and basin plain is difficult and probably academic. Suffice it to say that bed thickness and grain size tend to increase with proximity to the fan (Normark and Piper, 1972; Walker, 1978).

With the sediment and bedding characteristics of the different fan divisions in mind, consider an hypothetical section through a prograding fan (Fig. 12-14). Grain size and bed thickness show an overall increase upward but reversals do occur within channel deposits. Even in the absence of very coarse clastics, there is some size segregation in the sands, and variation in bed thickness is evident. All of this should be discernible on electric logs.

When working with electric logs in the fan environment or a suspected fan environment, it is appropriate to remember the vertical scale of the depositional units. Many individual turbidites are only a few centimeters thick and cannot be resolved on logs. Shale breaks between some units are thin and discontinuous and may not be seen by the logging tool. If they are, correlation of such breaks is useless. It is better to consider depositional packages or units when using logs rather than trying to work with individual beds.

EXPLORATION FOR HYDROCARBONS IN SUBMARINE FANS

A number of different methods used to identify the submarine fan environment in the subsurface are treated in the literature. Some of these include: the study of well logs, drilling cuttings, whole cores, and seismic lines; paleontological analysis; and mapping techniques such as isopach maps of genetic units and panel diagrams. No one method is definitive in identifying any environment of deposition, but a combination of methods should indicate the correct environment.

North Sea

Forties field. The Forties field, located in the British sector of the North Sea (Fig. 12-16), was discovered in 1970 in a previously unexplored area. By 1972, four appraisal wells had been drilled (Walmsley, 1975). From 1974 to 1981, a number of articles were published that show the evolution of geological thought with respect to the field. The following general discussion of the field's exploration history and stratigraphy is based on a paper by Thomas et al. (1974), which outlines what was publicly known about the Forties field up to mid-1973.

A reconnaissance seismic survey conducted prior to 1965 identified the Forties structure, showing a structural nose plunging southeast into the North Sea Tertiary basin. A five km by five km seismic survey shot in 1967 further defined the structure and indicated about 40 sq km of low amplitude closure in Block 21/10. The discovery well and three of four appraisal wells encountered oil in Paleocene sands. In the adjacent block to the east, Block 22/6, (Fig. 12-16, inset), a well also produced oil in Paleocene sands with the same oil-water contact. Sand quality in the eastern block was considerably less than that in Block 21/10.

In the early 1970's, no formal Paleocene stratigraphy for the area had been established. For convenience, the Paleocene section was divided into units based on lithology and Roman numerals applied in ascending order (Fig. 12-17). Unit I is divided into two parts. The lower part is sandstone, calcareous sandstone, limestone, and mudstone. Reworked Danian and Cretaceous faunas and detrital limestones indicate erosion of Danian and Cretaceous carbonates after uplift at the end of the Mesozoic. This part of Unit I thickens to the north-west across the field. The upper part of Unit I is interbedded sandstone, siltstone, and shale. Sandstones are generally fine-grained and argillaceous. A sonic log through this interval fluctuates rapidly due to the interbedding of thin strata of differing lithology (Fig. 12-18, well 21/10-I, Unit I).

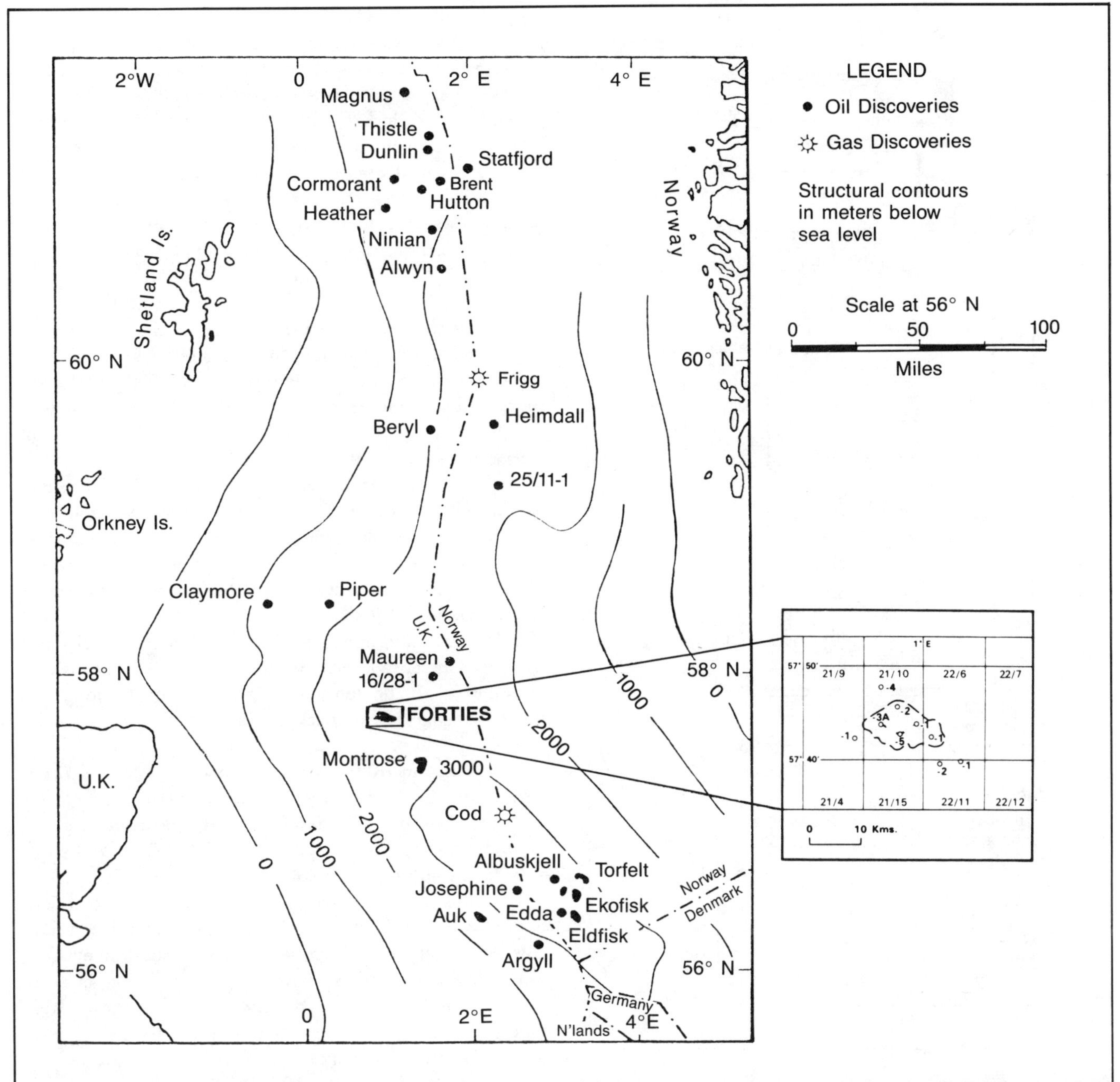

Fig. 12-16. Northern North Sea map and inset of region around Forties field. (Modified after Walmsley, 1975; permission to publish by Elsevier Applied Science Publishers Ltd.).

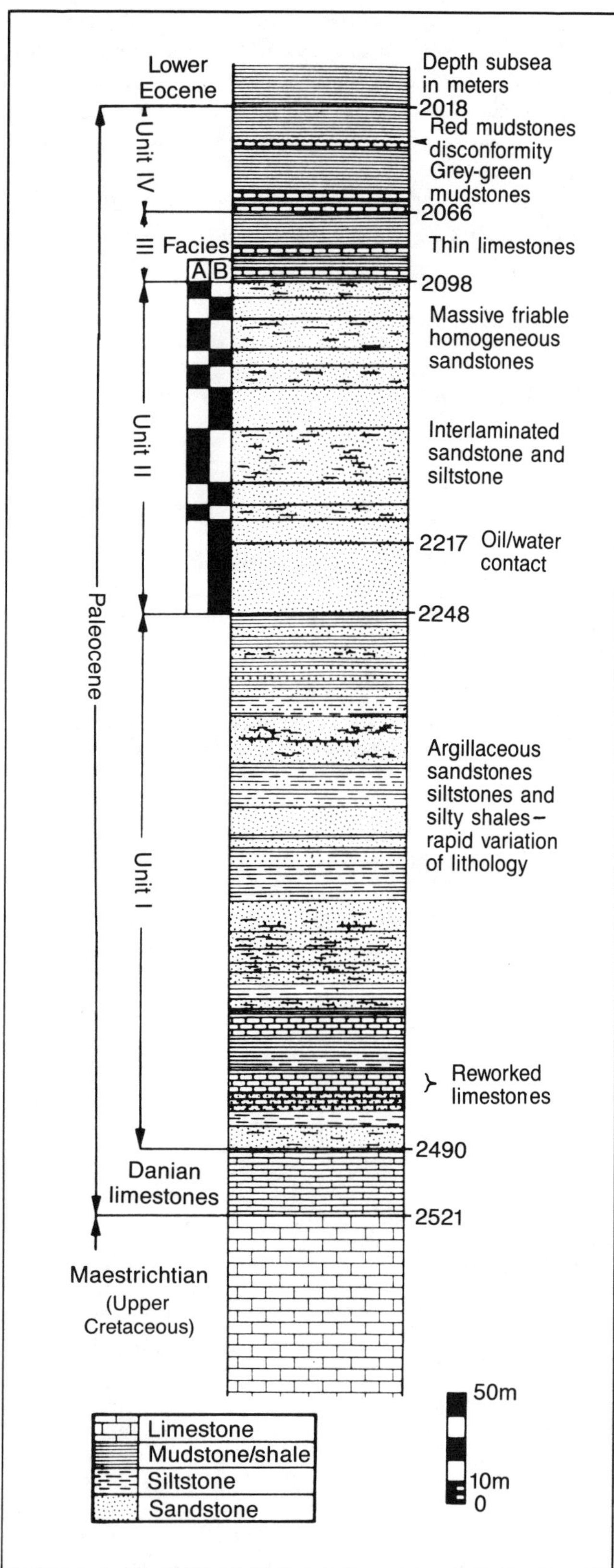

Fig. 12-17. Paleocene lithostratigraphic column. (From Thomas et al., 1974; permission to publish by AAPG).

Unit II is the productive unit in the Forties field. Sandstone and mudstone are the principal lithologies. Core studies in Unit II permit the recognition of four facies, and these are given letter designations. Facies A is interbedded fine-grained sandstones, laminated siltstones, and sparsely fossiliferous shales that form fining-upward, graded units from two cm to 1.7 m thick. Sandstones are locally silty and often have detrital mica and lignite. Intraformational conglomerates with shale clasts occur at a few horizons. Sedimentary structures include load casts, deformed cross-bedding, ripple-drift lamination, contorted bedding, slump folding, and soft sediment folding.

Facies B is clean, homogeneous, fine to coarse, poorly to moderately sorted sandstone. Clay laminae and pebbly layers with up to 50% igneous and metamorphic rock fragments occur in thick sandstones. Numerous fining-upward sections occur with convolute and laminar bedding near the top. Individual sandstones are tens of meters thick and some reach 80 m. These thick sandstones appear to be amalgamations of sandstone layers one-half to one meter thick.

Facies C consists of shales and graded siltstone-shale pairs. Occasional thin, fine-grained sandstones occur that are sometimes graded. Beds have sharp, erosional bases, and some have sole marks such as flute casts. Slump structures are common. Siltstones show lenticular bedding, wavy lamination, and microcross-lamination. Facies C shales have a sparse fauna and are rarely bioturbated in contrast with the shales constituting Facies D, which have an abundant marine fauna and are bioturbated.

Unit II consists mostly of the more sandy Facies A and B (Fig. 12-17). Facies C occurs mostly to the south and east in the field while Facies D is confined to the east. Distribution of the various facies is a function of the thickness of the unit. Facies B occurs where Unit II is thick and Facies C occurs where it is thin.

Unit III is a silty, lignitic mudstone with thin beds of sandstone that increase in proportion to the west. Siliceous diatoms and microplankton are the principal fossils along with some fish remains. Unit thickness and lithology are constant over the field and in adjacent areas.

Unit IV is a mudstone with occasional limestones, a reddish zone near the top, and distinctive, degraded volcanic ash horizons. Constant thickness and a characteristic sonic-log pattern (Fig. 12-18, well 21/10-1, Unit IV) make the unit easy to identify and correlate.

This, then, is what was known, or at least published, about the Forties field in 1974. Core descriptions and samples were sufficient to suggest a submarine-fan environment. Paleontological evidence for water depth, if available, was not given. No environments of deposition were suggested and it was stated for Unit II that the environment was under study and that results were not expected for some time (Thomas et al., 1974).

One year later, the informal Paleocene stratigraphy was

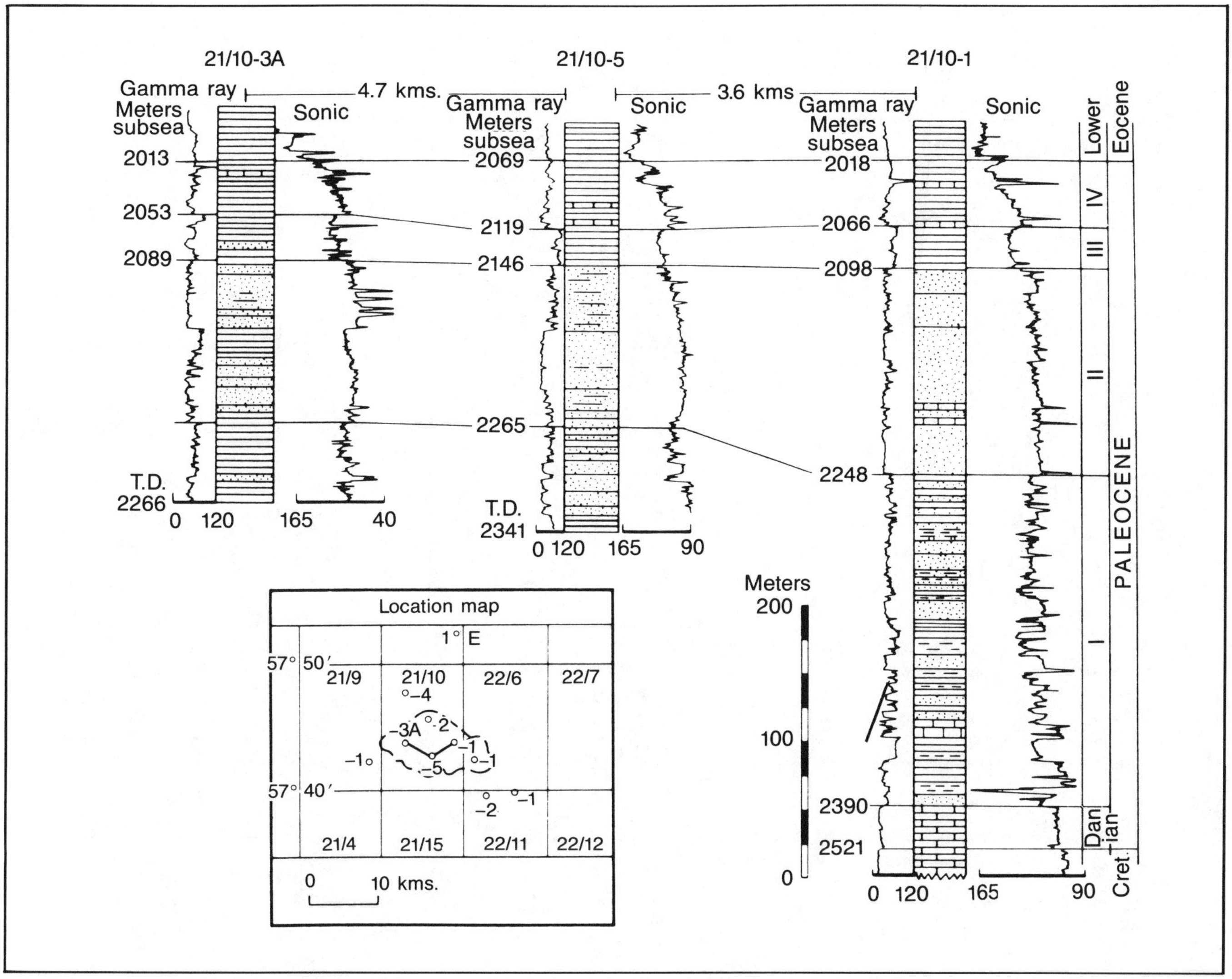

Fig. 12-18. Log characteristics and well correlation E-W across Forties field. (From Thomas et al., 1974; permission to publish by AAPG).

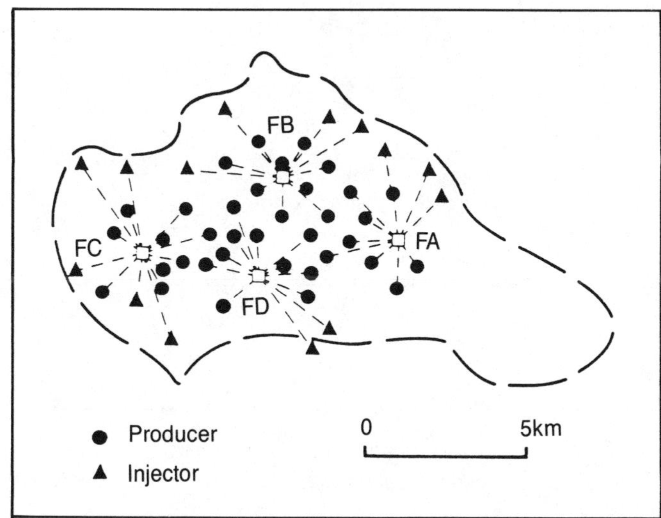

revised slightly (Walmsley, 1975). The alternating sandstones, siltstones, and shales originally in the upper part of Unit I were added to the bottom of Unit II, and Unit IV was divided into three parts.

In a related paper, Parker (1975) identifies the reservoir sands in Forties field and Montrose field to the south as turbidite sandstones. He recognizes a basinal sequence of alternating shales and fine- to very coarse-grained sandstones. Sandstone thickness varies from very thin up to 60 m. Thinner sandstones are graded and have load casts, contorted bedding, and groove marks. Some of the thick

Fig. 12-19. Distribution of development wells, Forties field, in early 1979. FA, FB, etc., are platform locations. (From Hill and Wood, 1980; permission to publish by AAPG).

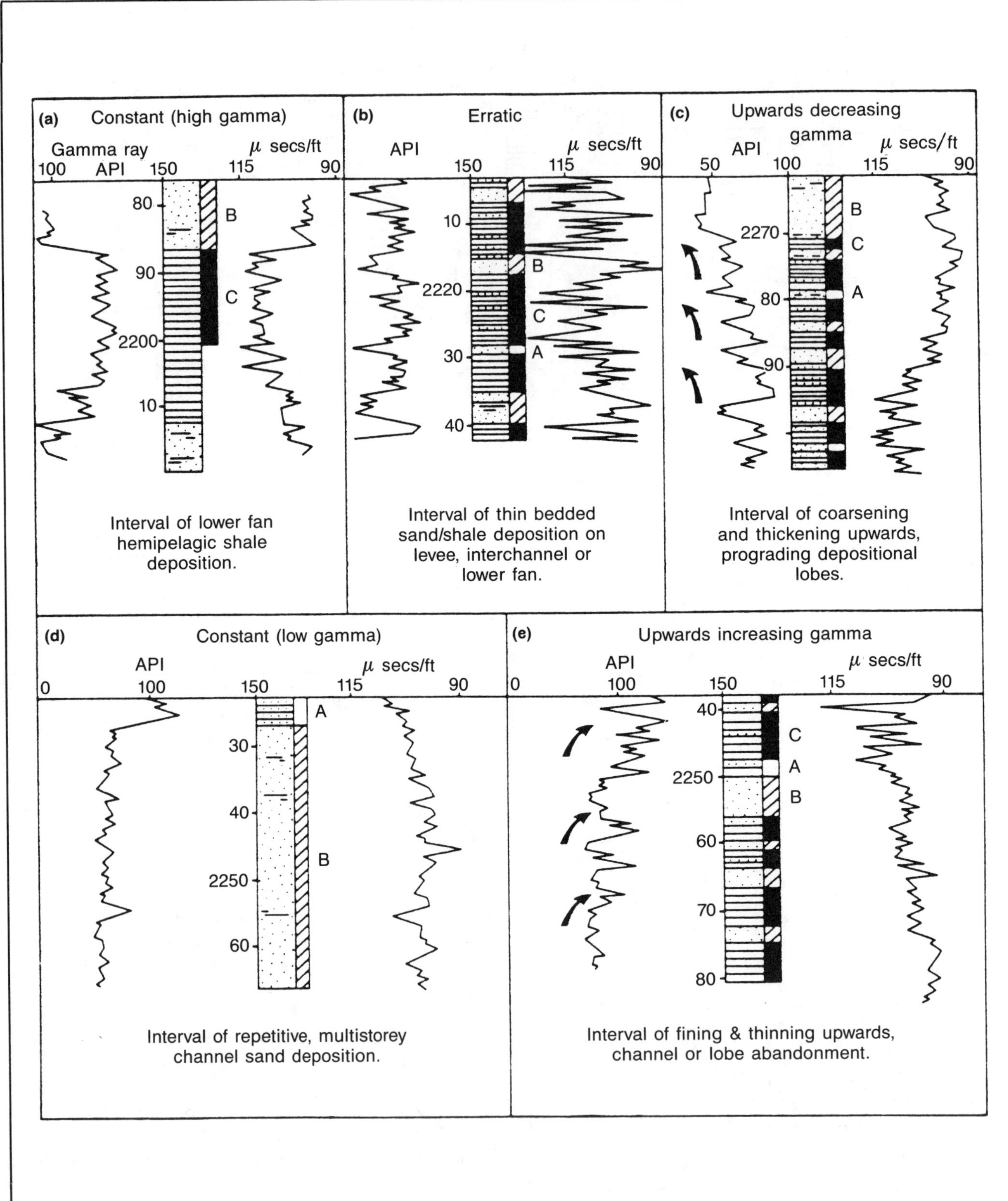

Fig. 12-20. Characteristic log patterns of submarine fan strata of Forties field. Interpreted fan environment given below each pattern. Facies types identified in cores shown to right of lithology column. True vertical depths in meters below rotary table. (From Hill and Wood, 1980; permission to publish by AAPG).

sands have dish structure. Faunas in the interbedded shales indicate deposition in water depths of 180 to 900 m. This confirms the relief between shallow and deep water units as seen on seismic lines (Parker, 1975).

By 1980, core and log data from 50 development wells were available (Fig. 12-19). Cores and logs show rapid facies variations across the field with some sand bodies having widths less than the 700 m prime well spacing (Hill and Wood, 1980). Correlation between whole cores and logs over the same intervals shows a correspondence between individual fan environments and certain log shapes and patterns. Environmental interpretations were then made on logs from uncored wells and intervals using the known log shapes as reference (Hill and Wood, 1980).

Five log patterns are recognized by Hill and Wood (1980). The constant high gamma-ray (or SP) pattern is stable over a 5 to 25 m interval (Fig. 12-20A). Sonic and density logs are also stable and resistivity is low. A constant, high, GR pattern coincides with Facies C and D in cored intervals and represents fine-grained deposition on the lower fan and basin plain.

An erratic log pattern (Fig. 12-20B) varies over two meter intervals and coincides with thinly interbedded sandstones and shales. The pattern represents thin-bedded turbidites of Facies C and, occasionally, thicker proximal turbidites of Facies A. These kinds of deposits occur on levees, in interchannel areas throughout the fan, along channel margins, and in abandoned channel fills. Due to the number of environments associated with the erratic log pattern, vertical and nearby horizontal associations may help in determining the correct environment of deposition. Association with thick Facies B channels suggests an interchannel or levee environment. In proximity to Facies C or D, lower fan deposition is indicated.

An upward-decreasing gamma-ray pattern occurs over a five to ten meter interval (Fig. 12-20C). Density and sonic logs show decreasing shale and rising porosity and permeability. Cores indicate this pattern starts in Facies A turbidites that vertically become thicker-bedded and more proximal. This sequence represents prograding lobes on the middle fan.

A constant low gamma-ray pattern over several to many meters (Fig. 12-20D) indicates a thick, uniform, coarse-grained deposit such as coarse, laminated sandstone, pebbly sandstone, and/or conglomerate. Amalgamation of individual units forms sequences up to 20 m thick. Individual sequences may be separated by thin layers of Facies C shale or siltstone, or Facies A graded sandstone. Stacked amalgamated units with thin shaly interbeds achieve thicknesses of 50 m in the Forties field. These units are coarse-grained channel fills that are overlain by units indicating abrupt or gradual channel abandonment.

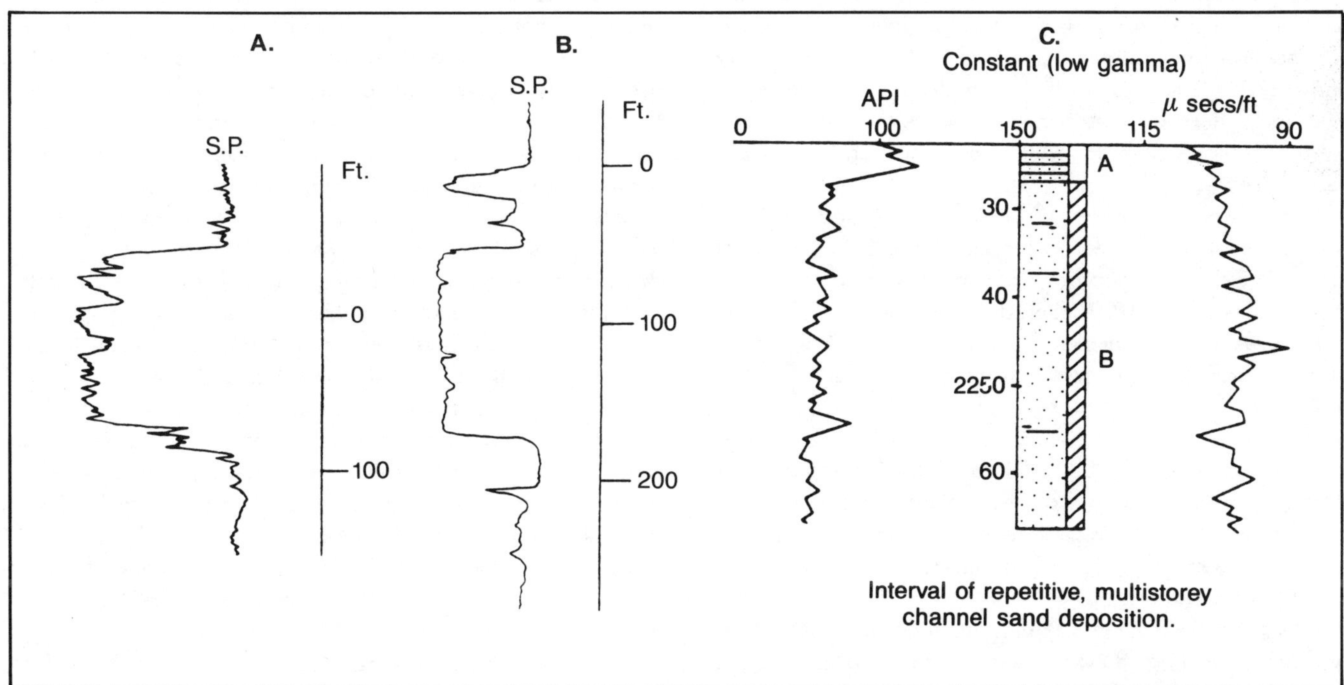

Fig. 12-21. Comparison of electric log patterns of submarine fan channels. A.*-SP, Miocene, Vermilion Parish, Louisiana.* B.*-SP, Miocene, Terrebonne Parish, Louisiana. Depth scale 100 ft (30 m) per interval.* C.*-GR, Paleocene, Forties field, North Sea, UK. Depth scale in meters. (*A *and* B *from Lock, 1982; permission to publish by Gulf Coast Association of Geological Societies;* C *from Hill and Wood, 1980; permission to publish by AAPG).*

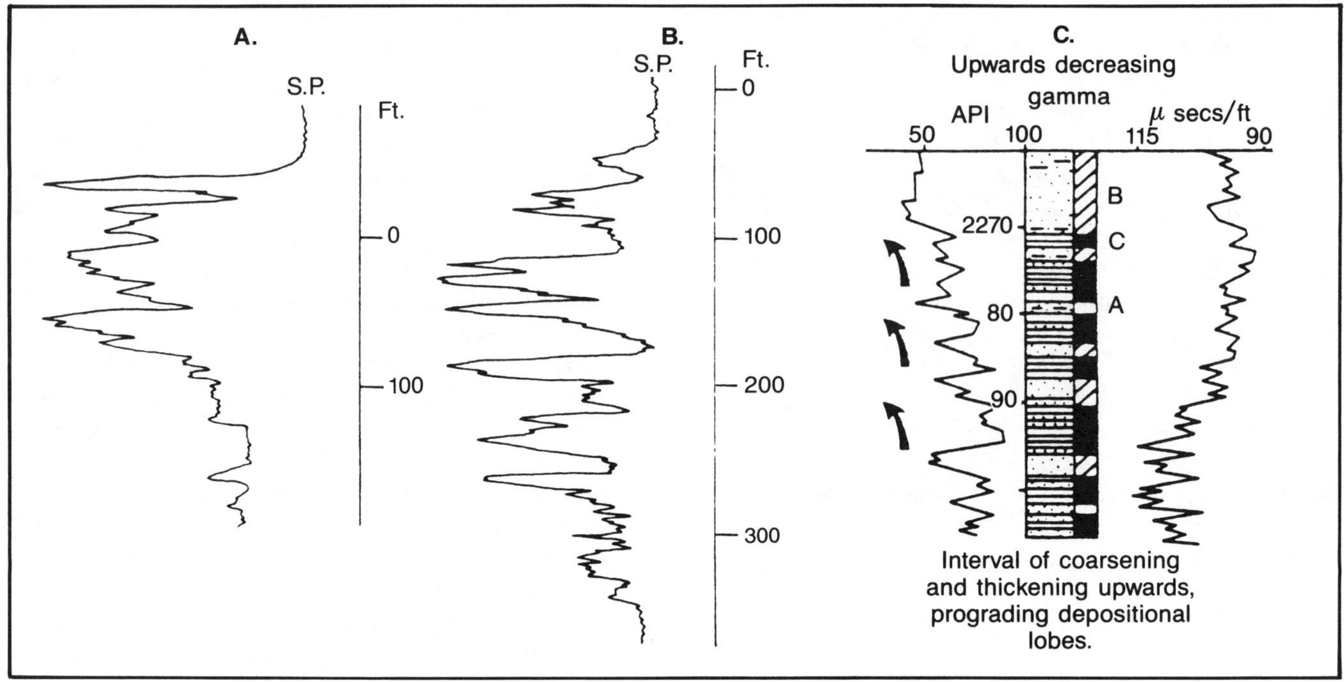

Fig. 12-22. Comparison of electric log patterns of depositional lobes on submarine fans. A.*—Miocene, East Vermilion Parish, Louisiana.* B.*—Miocene, St. Mary Parish, Louisiana.* C.*—GR, Paleocene, Forties field, North Sea, U.K.* (A and B from Lock, 1982; permission to publish by Gulf Coast Association of Geological Societies; C from Hill and Wood, 1980; permission to publish by AAPG).

An upward increasing GR pattern occurs over intervals varying from 5 to 40 m (Fig. 12-20E). Sonic logs show a corresponding increase in interval transit time, while density and resistivity decrease with increasing shaliness upward. This log pattern overlies Facies B channel sandstones and indicates channel abandonment as grain size and bed thickness decrease.

Comparison of the Forties submarine fan log patterns with patterns from Miocene fans in the U.S. Gulf Coast (Lock, 1982) shows good agreement between patterns for channel sands (Fig. 12-21) and depositional lobes (Fig. 12-22). Agreement is expected since similar depositional processes produce similar rocks and log response is a function of the rocks. However, it is re-emphasized that log patterns alone are not sufficient to identify a submarine fan. Log patterns in Figures 12-20, 12-21, and 12-22 could be produced in a delta, mature alluvial valley, or a submarine fan. Seismic data, basin geology and morphology, and especially paleontology can help in differentiating among the possible environments.

Other North Sea fields. Fields, such as Magnus, Frigg, and Montrose (Fig. 12-16), among others, also produce from submarine fans. The exploration histories of these fields and the use of cores to determine environments of deposition are similar to those already discussed for the Forties field. Some of the different information obtained from cores to identify the fan environment is worthy of mention, however.

Production in the Magnus field is from the Magnus sandstone member of the Kimmeridge Clay formation (Upper Jurassic). It is a poorly sorted, fine- to medium-grained, occasionally coarse-grained sandstone with local zones of granules. Sorting and grain size vary strongly in a vertical direction, and member thickness varies sharply over the field. Sedimentary features and structures include ripped-up mudstone clasts, injection sandstone pipes, load features, graded units, parallel lamination, ripple crosslamination, thin-bedded sandstones with sharp bases, and dish structures. Trends in sandstone thickness both thicken upward and thin upward (Fig. 12-23). Individual sandstones within an overall thickening-upward trend may thin upward and are interpreted as progessive channel abandonment or migration on prograding, middle fan lobes. Upward thickening of indivdual sandstones within a thinning-upward trend represent depositional lobes on a generally receding fan (Fig. 12-23). A westerly source for the Magnus submarine fan is indicated by a west to east fining of mean grain size and areal distributions of distal turbidites, channel sandstones, and lobe deposits (De'Ath and Schuyleman, 1981).

Sedimentary features and structures, along with the trends in grain size and distribution of types of deposits, strongly indicate a submarine fan. Paleontological evidence of deep water deposition should be obtained to be certain of the environment.

In the Frigg field, gas is produced from the Frigg sandstone (Lower Eocene). The Frigg structure is in the deepest

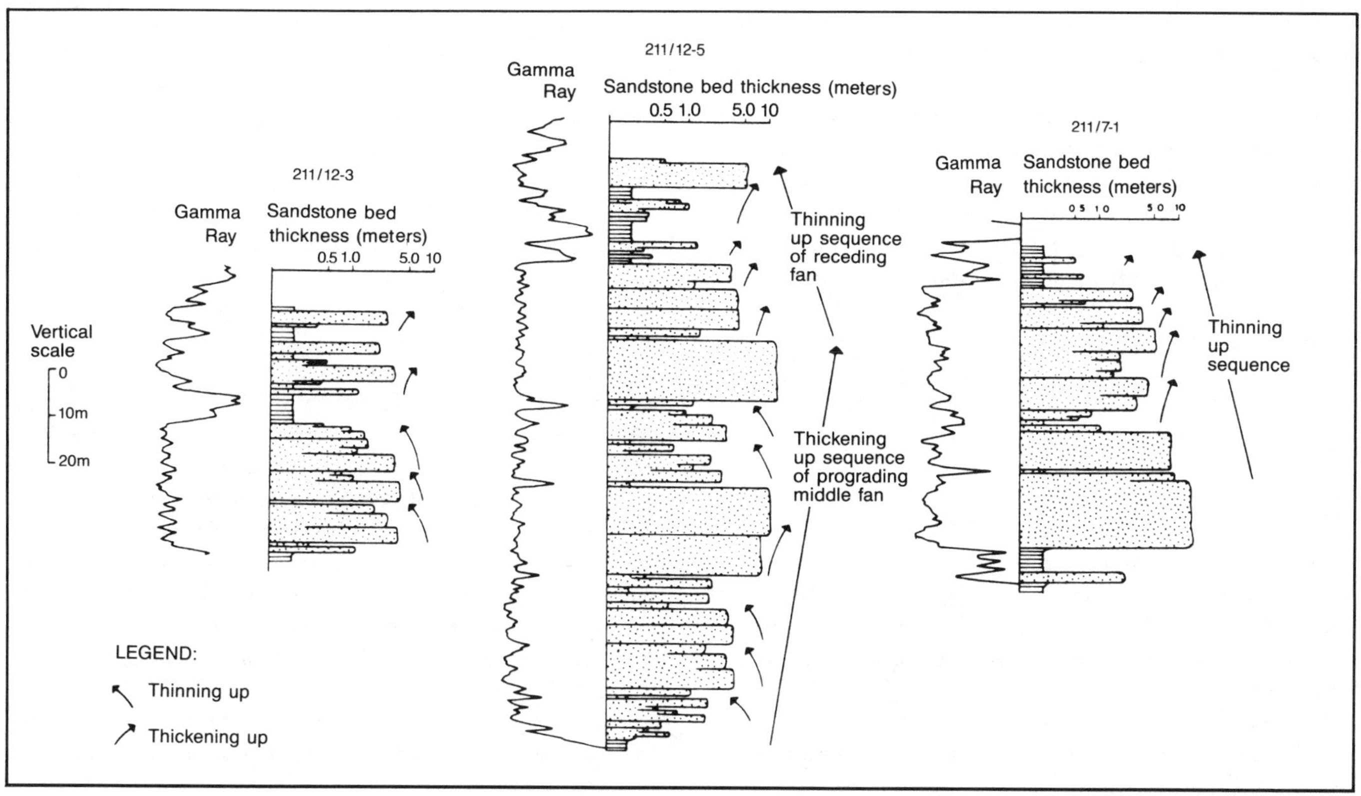

Fig. 12-23. Sandstone bed thickness variation, Magnus field. (From De'Ath and Schuyleman, 1981; permission to publish by Institute of Petroleum, London).

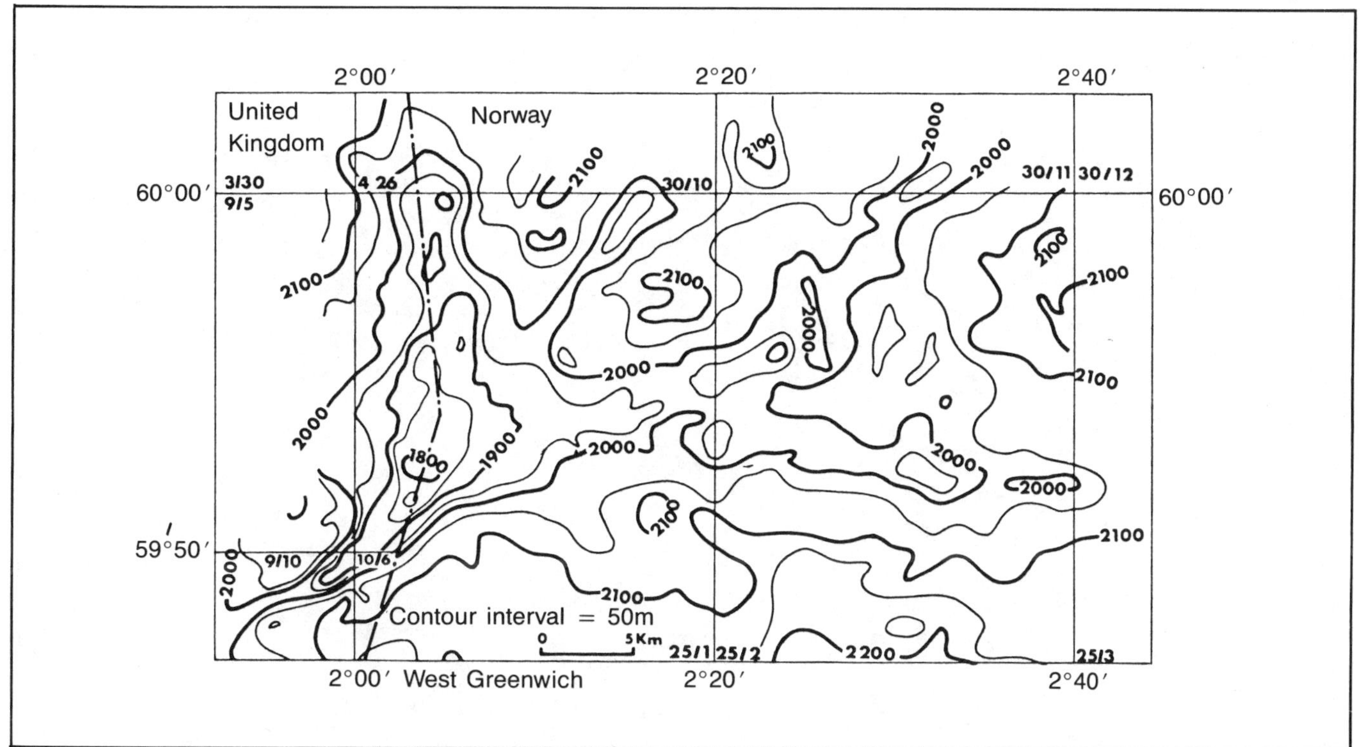

Fig. 12-24. Seismic structure map of Frigg field at top of Frigg sandstone. (From Heritier et al., 1979; permission to publish by AAPG).

part of the North Sea Tertiary basin, east of the Shetland platform escarpment, and its morphology suggests a fan with lobes (Fig. 12-24). Sandstones are fine- to medium-grained, thinly to massively bedded, and contain shale clasts. Grain size decreases downfan. Glauconite and carbonaceous detritus are commonly associated with the sandstones (Heritier et al., 1979), which, according to Selley (1976), is strongly suggestive of the submarine fan environment. Glauconite is indicative of the marine environment, and carbonaceous matter indicates poor winnowing as it has not been subjected to abrasion or oxidation in shallow water, suggesting rapid burial. As with the Magnus field, paleobathymetric data derived from paleontology would be useful.

South-central U.S.

Miocene, offshore Louisiana. A useful method of mapping submarine fans and other environments is the panel diagram. Reduced-scale SP or gamma-ray curves of the stratum of interest are mounted on a large-scale map in their corresponding well locations. Depositional environments indicated by SP shape for the different wells may be seen, and avenues of similar environments connected. In this way, the geometry of the environments is seen, and an interpretation of the distribution of environments can be made. Once the overall environment is known, sand isopach maps or other kinds of maps may be subjectively made to conform to the environment.

Lock (1982) shows an hypothetical sandstone isopach map in the Miocene, offshore Louisiana. The example is hypothetical because of the proprietary nature of the data, but it is based on actual data from several fields (Lock, 1983, personal communication). Objective contouring of the data results in a nondescript map (Fig. 12-25). A panel diagram for the same area permits an interpretation of the environments, and the construction of a sandstone isopach map subjectively contoured based on the environments (Fig. 12-26). Log shapes and their distribution outline channel and interchannel areas, but paleontological data are needed to indicate deep-water deposition (Lock, 1982).

Midland basin, Texas. With abundant well control in mature petroleum provinces, the geometry of mapped sandstone bodies may be sufficient to identify the environments of deposition. Bloomer (1977) examined several thousand logs from wells drilled in west-central Texas and was able to map two fluvial-deltaic-basin slope depositional systems from the logs. The following discussion of the slope deposits

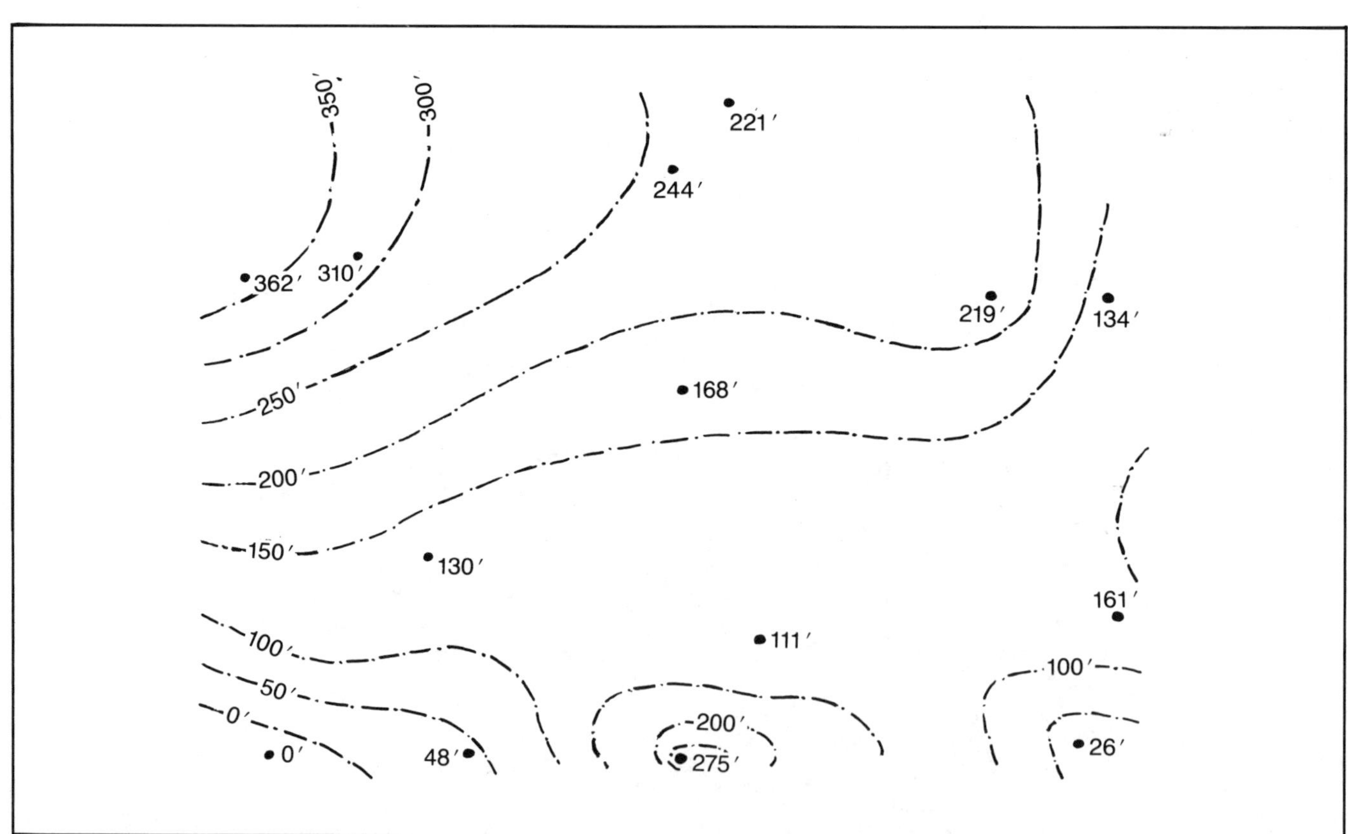

Fig. 12-25. Sandstone isopach map, objectively contoured, of hypothetical Miocene submarine fan, U.S. Gulf coast. (From Lock, 1982; permission to publish by Gulf Coast Association of Geological Societies).

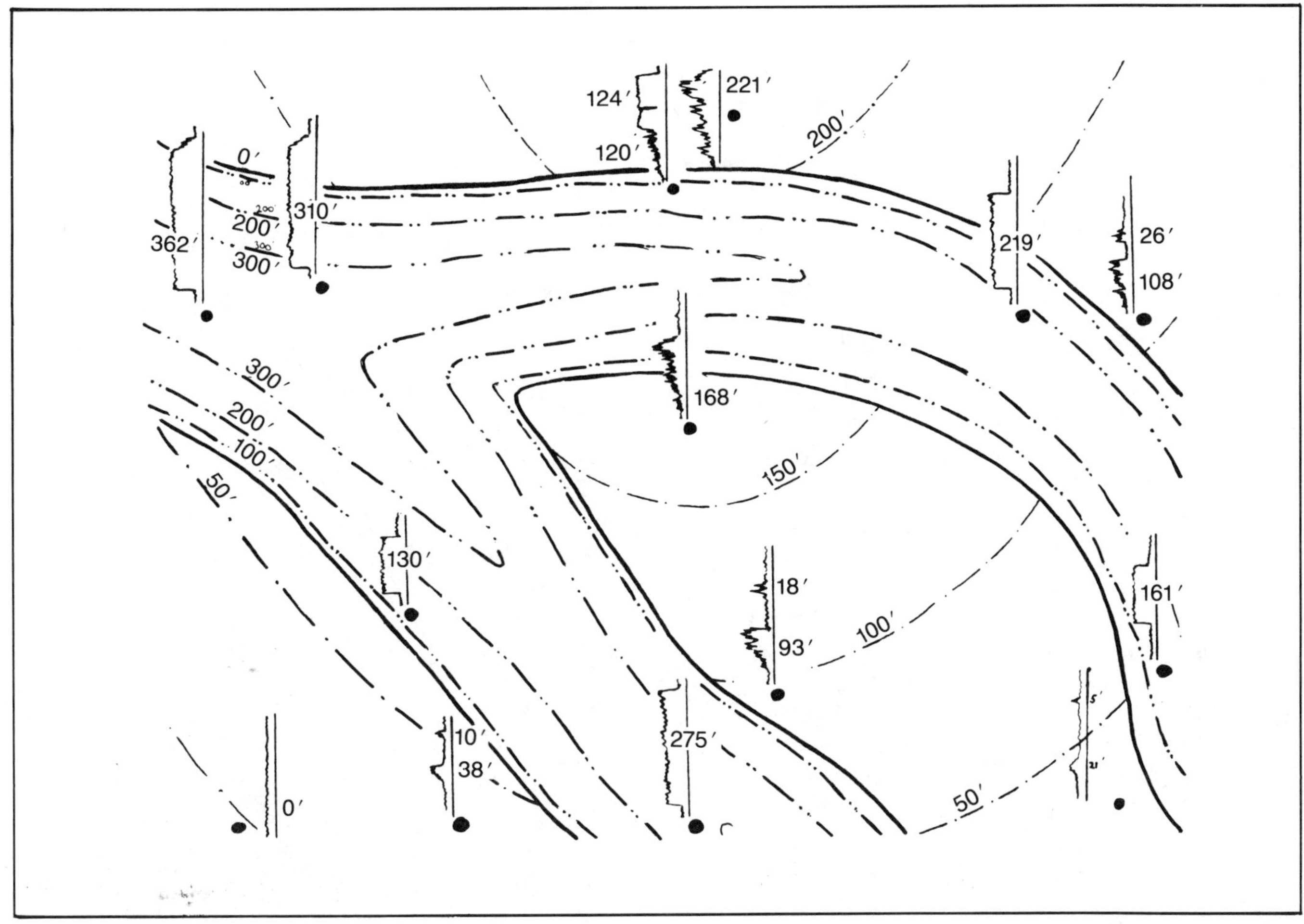

Fig. 12-26. Panel diagram, interpretation of environments of deposition, and subjectively contoured sandstone isopach map of same data in Fig. 12-24. (From Lock, 1982; permission to publish by Gulf Association of Geological Societies).

is based entirely on Bloomer's (1977) work.

In Early Permian (Wolfcamp) time, the Midland basin in west Texas was bounded on the east by the Eastern shelf (Fig. 12-27). Meandering streams crossed the shelf from east to west and formed deltas that prograded into the Midland basin (Fig. 12-28). Basinward from the deltas, slope deposits were laid down.

Within the Wolfcampian stratigraphic section, the interval of interest is the Waldrip shale above the Crystal Falls limestone (Fig. 12-29). Three thin limestones occur in the shale, the uppermost of which is called the Flippen limestone. Any sandstone between the Crystal Falls and Flippen limestones is referred to informally as Cook sandstone.

A cross section from the Eastern shelf into the Midland basin shows the change in slope of the base of the Flippen limestone that marks the prograding shelf edge (Fig. 12-30). Sandstones in wells to the west of well 7 are clearly basin-slope deposits. Well 4 is in the Northeast Bloodworth field, a stratigraphic trap that produces from the Cook sandstone. An isopach map of the GIS from the base of the Flippen limestone to the base of the Cook sandstone (Fig. 12-31, left) and a cross section of the same interval (Fig. 12-32) show the convex downward or channel-like nature of the sandstone in the field. Near coincidence of the channel thalweg and the axis of maximum sandstone thickness (Fig. 12-31) is a further indication that the Cook sandstone is a channel fill.

Cook sandstone morphology and its position on the basin slope indicate the sandstone is a part of a submarine canyon fill. Four other lines of evidence obtained from the logs support the canyon fill hypothesis: the apparent erosional contact between the Cook sandstone and subjacent shale; a fining upward of the channel fill; alternating thin sands and shales suggesting interbedded turbidite sandstones and pelagic shales; and a lack of correlation between wells in the lower sandstones.

Still farther basinward, the Cook sandstone is found in the Jameson field (Fig. 12-30, well 1; Fig. 12-33), which is also a stratigraphic trap. Northeast of the field is North Jameson field which is long and narrow, with sands that

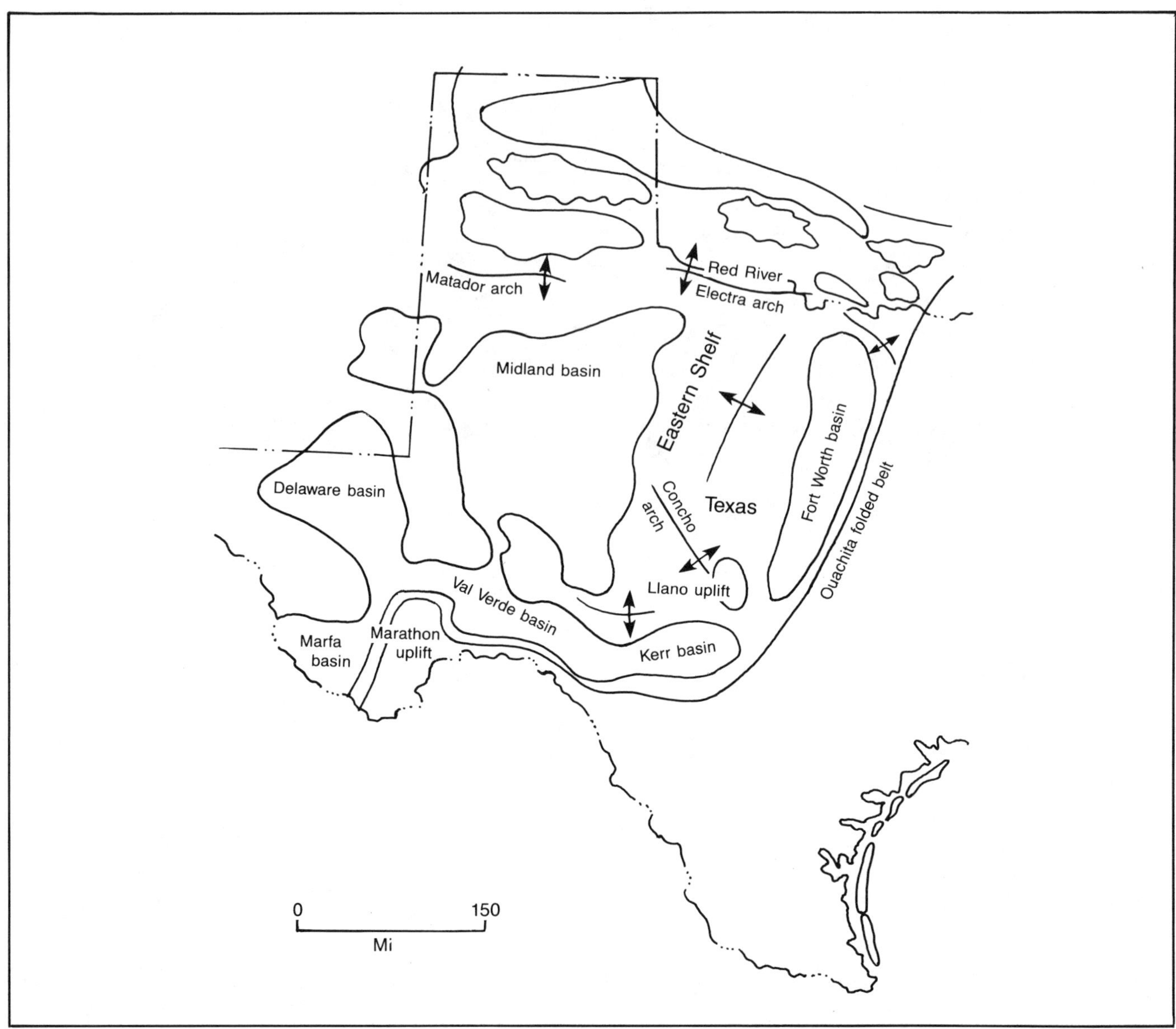

Fig. 12-27. Tectonic features of west Texas.

are convex downward in cross section indicating a submarine canyon fill. The southwestern part of this field is seen in the northeast corner of the Jameson field map (Fig. 12-33). Jameson field spreads fan-like from the south end of the channel suggesting a submarine fan with a feeder channel to the northeast. Cook sandstones in Jameson field are fine-grained, thin, lenticular, and uncorrelatable between wells. Scattered zones of carbonaceous and shale clasts indicate transport of shelf sediments into deep water.

This example is an effective demonstration of the use of well logs and maps of genetic units to identify not only a submarine fan but a fluvial-deltaic system as well. A block diagram illustrates how Bloomer envisions the entire system (Fig. 12-34).

Yoakum channel. In the examples given so far, submarine canyon and fan sediments have been the reservoirs. Canyon fills, however, may also act as seals and the Yoakum channel of the central Texas Gulf Coast is the classic example. Hoyt's (1959) description of the canyon is the source of information for the following discussion.

Fig. 12-29. Diagrammatic composite electric logs of Lower Permian section in Taylor County, Texas, showing stratigraphic sections with and without fluvial sandstones (right and left logs, respectively). (From Bloomer, 1977; permission to publish by AAPG).

Shelf edge
Stephens
Outcrop
Nolan
Taylor
Callahan
Dip
A-A′
N
Cook Ss
Shale
Cook Ss Oil Field
Coleman
Brown
10 Miles
Outcrop

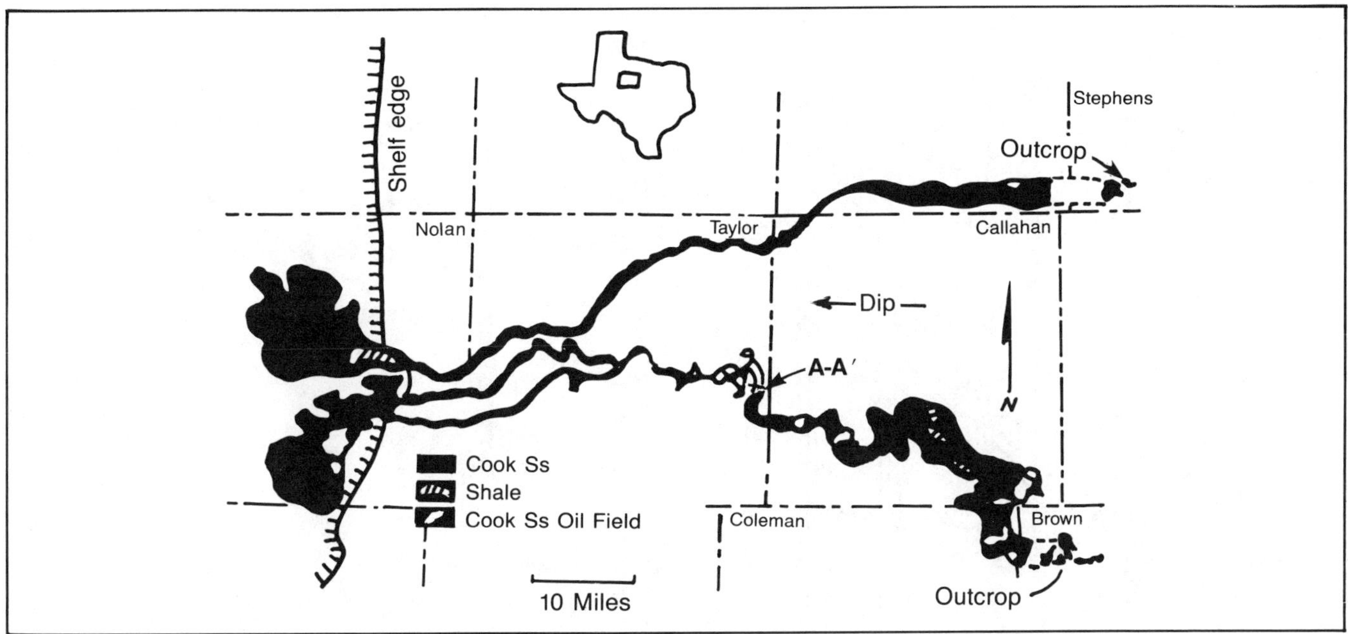

Fig. 12-28. Two paleodrainage systems on the Eastern shelf. Deltas are to the west of the Permian shelf edge. (Modified after Bloomer, 1977; permission to publish by AAPG).

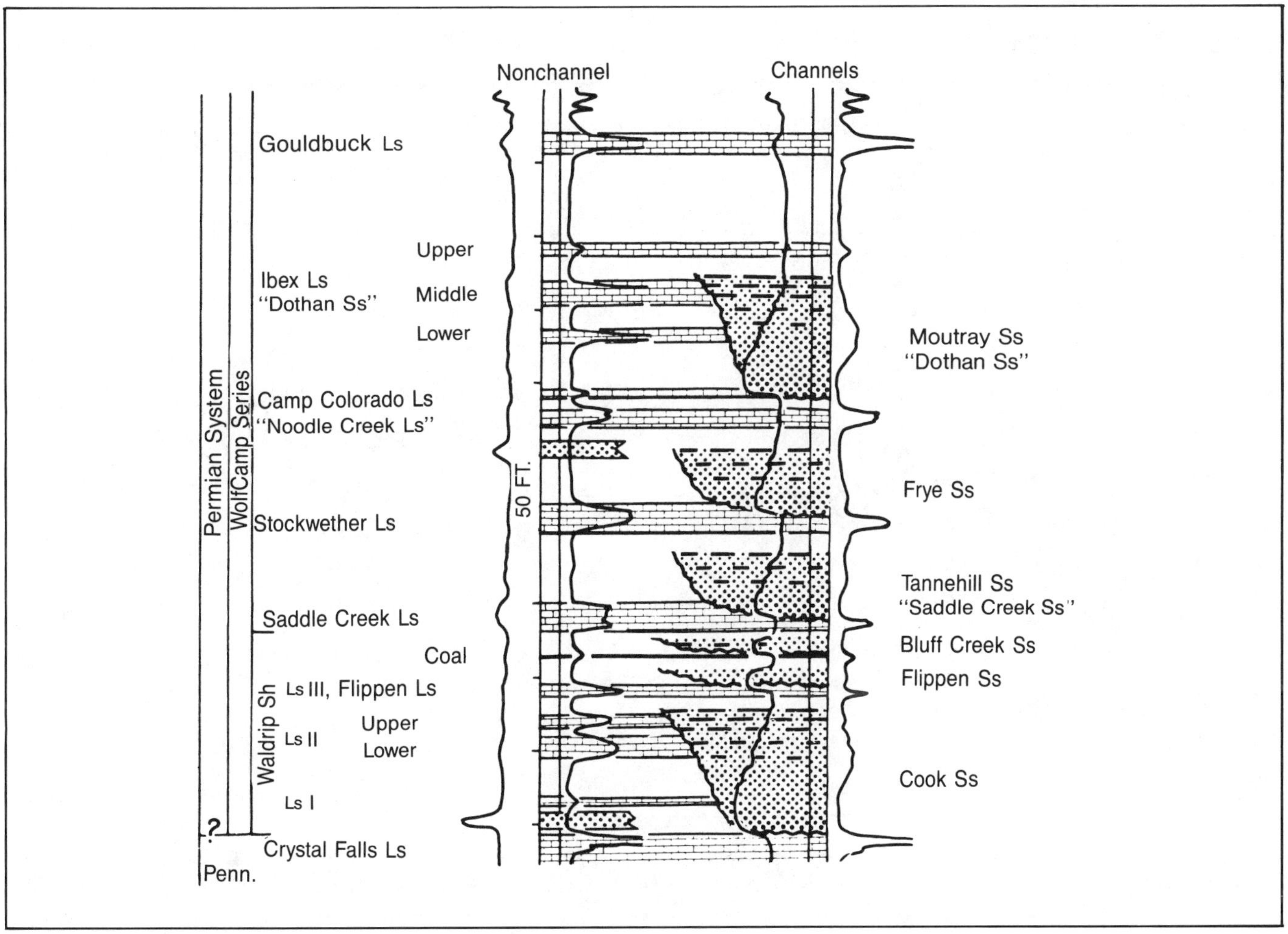

Nonchannel
Channels
Permian System
WolfCamp Series
Gouldbuck Ls
Upper
Ibex Ls
"Dothan Ss"
Middle
Lower
Moutray Ss
"Dothan Ss"
Camp Colorado Ls
"Noodle Creek Ls"
50 FT.
Frye Ss
Stockwether Ls
Tannehill Ss
"Saddle Creek Ss"
Saddle Creek Ls
Bluff Creek Ss
Coal
Flippen Ss
Waldrip Sh
Ls III, Flippen Ls
Ls II
Upper
Lower
Cook Ss
Ls I
?
Crystal Falls Ls
Penn.

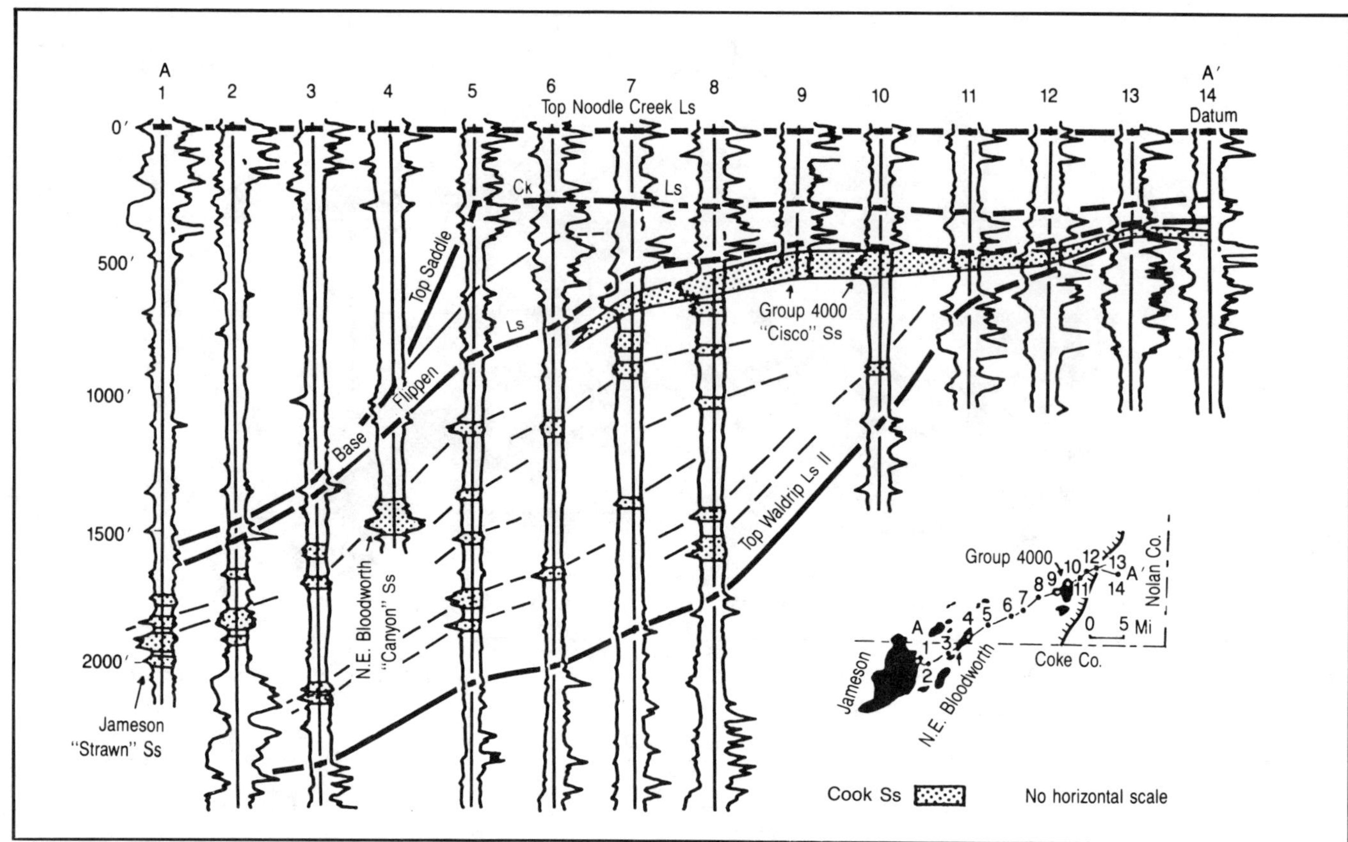

Fig. 12-30. Cross section of basal Permian clastic wedge from Eastern shelf into Midland basin. No horizontal scale between wells. (From Bloomer, 1977; permission to publish by AAPG).

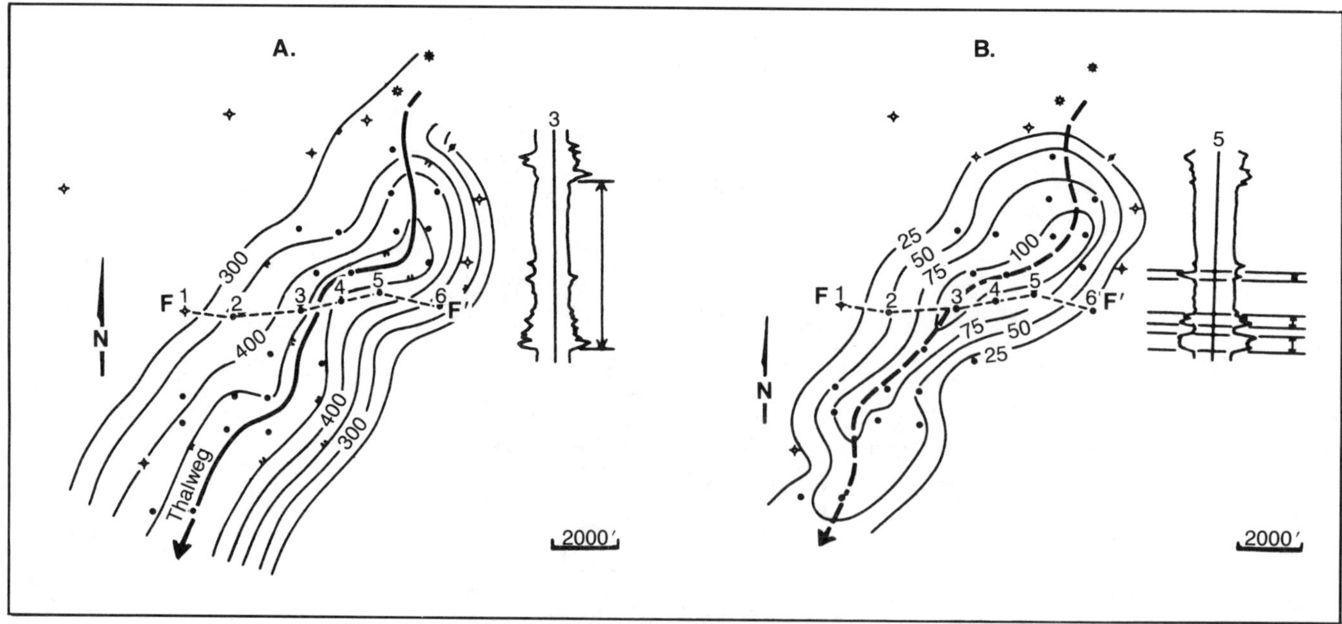

Fig. 12-31. **A.**–*Isopach map of the GIS from the base of the Flippen limestone to the base of the Cook sandstone and* **B.**–*Isopach of the Cook sandstone in the Northeast Bloodworth field. Note similarity of axial trends on both maps.* (From Bloomer, 1977; permission to publish by AAPG).

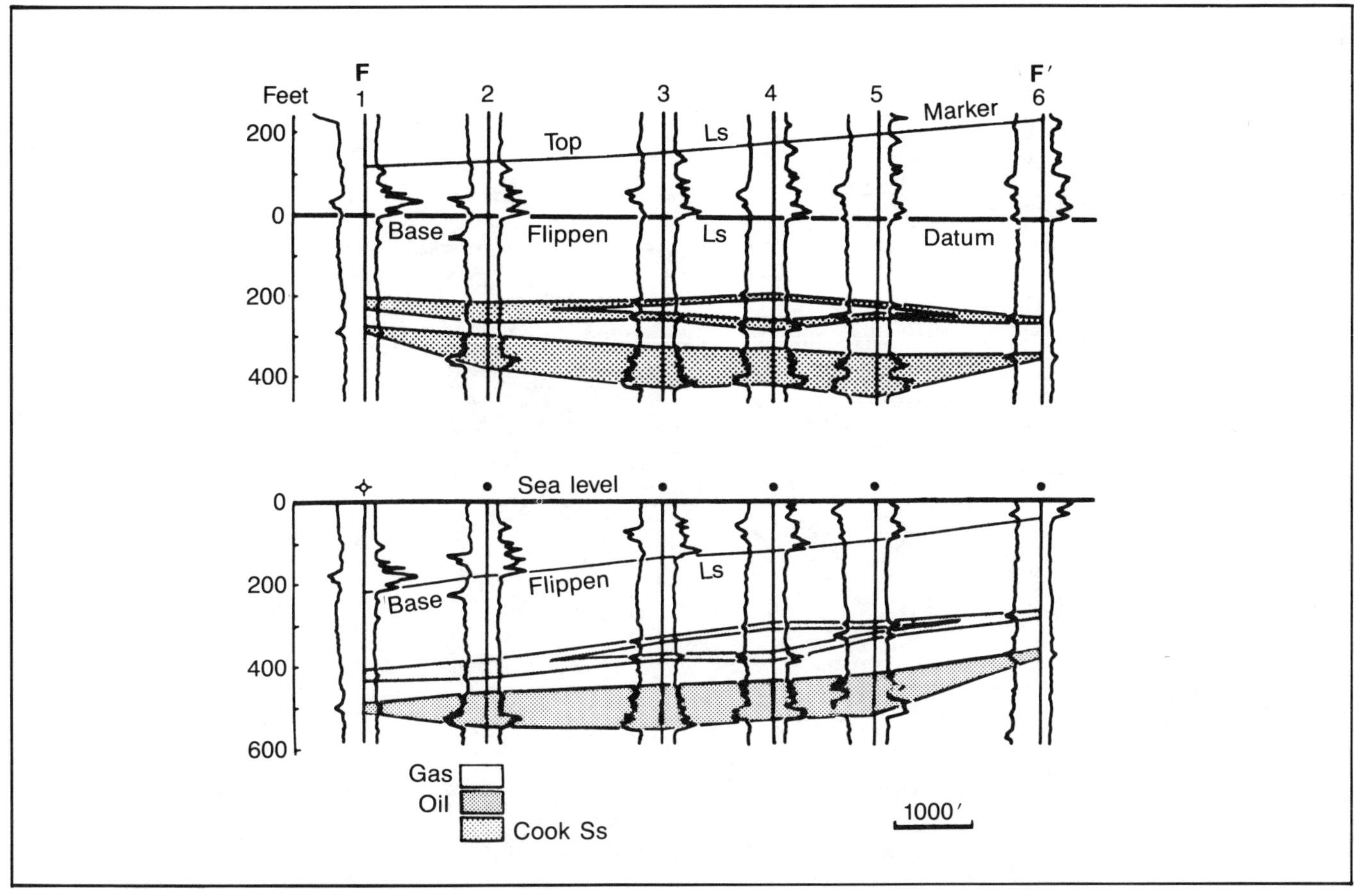

Fig. 12-32. Stratigraphic and structural cross sections F-F′ of Northeast Bloodworth field. Distribution of gas and oil in Cook sandstone shown in structure section. See Fig. 12-31 for location of cross sections. (From Bloomer, 1977; permission to publish by AAPG).

The Yoakum gas field was discovered in 1945 in lower Wilcox (Eocene) sandstones. In an attempt to extend the field to the east, a step-out unexpectedly encountered 1585 ft (483 m) of shale in what should have been a section of sandstones and shales. Later, another step-out to the northwest encountered 1750 ft (534 m) of shale in the same section. Lithological change and faulting were among the reasons cited for the shale but no satisfactory explanation was offered.

Drilling subsequent to the discovery of the Yoakum gas field showed a shale mass of decreasing dimensions extending to the northwest. Correlation of more than 50 wells provided data on the depth and thickness of the shale mass. An isopach map of the shale shows a thin, widespread, normal Wilcox shale thickening dramatically to over 2500 ft (765 m) along a narrow, generally northwest-southeast trending belt (Figs. 12-35 and 12-36). The belt is over 55 mi (89 km) long and up to 10 mi (16 km) wide.

Cross sections across the shale belt referred to the top of the Wilcox show thickening of the shale at the expense of the underlying Wilcox beds (Figs. 12-37 and 12-38). Morphology dictates an erosional canyon. Differential compaction of the channel fill is accompanied by an increase in thickness of upper Wilcox sandstones over the canyon fill.

Truncation of Wilcox reservoirs is the trapping mechanism for the Yoakum field (Figs. 12-39 and 12-40). Lower Wilcox sandstones and shales form a southwest-plunging structural nose that is truncated on the northeast by the Middle Wilcox canyon. Shale filling the canyon acts as the updip seal for the gas.

Origin of the canyon is speculative. Subaerial erosion is unlikely as there is no evidence of a disconformity under the thin shale on either side of the canyon. Hoyt envisions an unstable mass of unconsolidated Wilcox sediments adjacent to deep water in a subsiding basin. Slumps and slides on the face of the sediment mass at the outer shelf set up turbidity currents to cut the canyon. Continued slumping and turbidity current activity eroded the canyon up-channel by a stoping process. Hoyt's process does not differ appreciably from submarine erosion processes used to explain some modern canyons (see text on formation of submarine

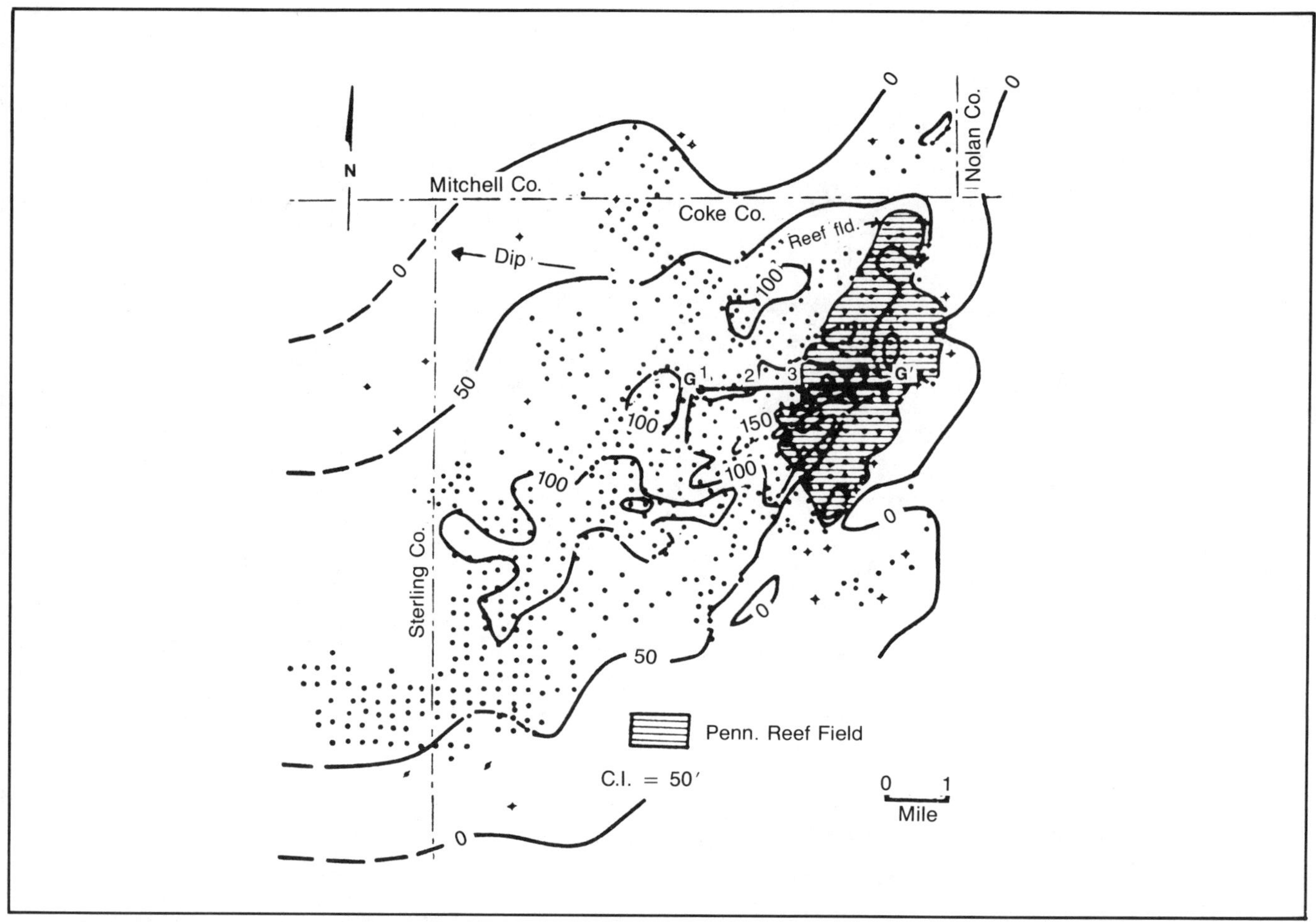

Fig. 12-33. Isopach map of Cook sandstone, Jameson field. (From Bloomer, 1977; permission to publish by AAPG).

canyons). Referring to global sea level curves (Fig. 12-41), there is a lowstand of sea level in the Early Eocene (end of Ypresian) followed by a Middle Eocene (end of Bartonian) highstand. The Yoakum canyon could have been cut by submarine erosion initiated during the lowstand and then filled with mud during the subsequent highstand of sea level.

It is reasonable to postulate a submarine fan beyond the mouth of the Yoakum canyon. Deep drilling near the southeast end of the canyon (as mapped by Hoyt) has failed to encounter submarine fan sediments (Vormelker, 1980). Possible fan sediments do occur just southwest of the canyon mouth. (Fig. 12-42, well 1; Vormelker, 1980). Vormelker suggests that turbidity currents flowing southeast down the canyon were deflected to the southwest by deep contour currents flowing in that direction parallel to the coast and slope. Fan sediments, then, should be found southwest of the canyon rather than along the channel trend to the southeast (Vormelker, 1980). This innovative approach is an example of how an understanding of the many processes operating in a specific environment of deposition may be used to generate new ideas that suggest where to look for petroleum.

The Yoakum canyon is not the only fine-grained channel fill that has acted as a trap for hydrocarbons. In the Sacramento basin of central California, the Paleocene Meganos canyon is filled with fine clastics and is the updip seal for four Paleocene fields (Dickas and Payne, 1967). Its dimensions are similar to those of the Yoakum canyon. Variations in sea level (Fig. 12-41) show a Paleocene lowstand followed by a Paleocene highstand, which may have been instrumental in the formation and filling of the Meganos canyon.

Mexico

Tampico-Misantla Basin. During the Early Tertiary, the Tampico-Misantla basin was a deep-water embayment lying immediately southwest of the Golden Lane fields in east-central Mexico (Fig. 12-43). Many hundreds of wells have been drilled in this area with Jurassic reservoirs being the prime targets. As a consequence of this drilling, oil and gas in commercial quantities were found in the Chicontepec formation (Lower Tertiary) with production coming from a scattering of fields and isolated wells (Fig. 12-44). Detailed structural mapping around and over these fields and wells reveals that none of the accumulations is structurally controlled (Busch and Govela, 1978).

A study of the Chicontepec formation was undertaken to: establish the environment of deposition; determine its distribution and geometry; map the subcrops of underlying formations; map the structure and determine the mechanism(s) for trapping hydrocarbons; and delineate prospective area(s) for hydrocarbons. Toward this end, a team of stratigraphers, micropaleontologists, and a geophysicist was assembled. Busch and Govela (1978) describe the multidisciplinary approach to the study and their paper is the reference for this example.

Chicontepec sediments fill a major submarine canyon fed by multiple tributaries. Canyon sediments thicken downward at the expense of underlying formations, which range in age from Early Eocene to Late Jurassic (Fig. 12-45). Based on micropaleontological data, most of the Chicontepec is Early Eocene in age with only the lowermost portion possibly being Late Paleocene.

Surface observations of the Chicontepec west of Poza Rica together with well data from the Tampico-Misantla basin show that the formation consists of a repetitious sequence of graded beds typical of many turbidites. Approximately one-half of the section is multiple, thin sandstone beds and zones of sandstone beds with the remainder of the formation made up of shales and silty shales. The sandstones are fine- to very fine-grained, frequently argillaceous, and of less than ideal reservoir quality. Benthonic microfaunal assemblages in the subsurface Chicontepec indicate deposition in deep water.

Cross section II-II′ (Fig. 12-46), using the top of the Chicontepec for a datum, trends in a general northeast-southwest direction. For all practical purposes this datum was deposited parallel to sea level and serves to illustrate the southwestward component of dip of the Jurassic-Cretaceous strata during Chicontepec time. A pronounced erosional unconformity separates the Chicontepec from the underlying Velasco (Tertiary) and Cretaceous. Maximum thickness of the Chicontepec along this cross section is approximately 650 m.

Development of sandy zones in the Chicontepec is best in the upper half of well 2. This development does not persist, however, to the northeast, and in well 3, sandy zones occur only in the basal third of the Chicontepec. Still farther to the northeast, sandy zones are either absent or sparsely developed. What little sandstone there is in wells 6 and 7 occurs only in the uppermost portion of the Chicontepec.

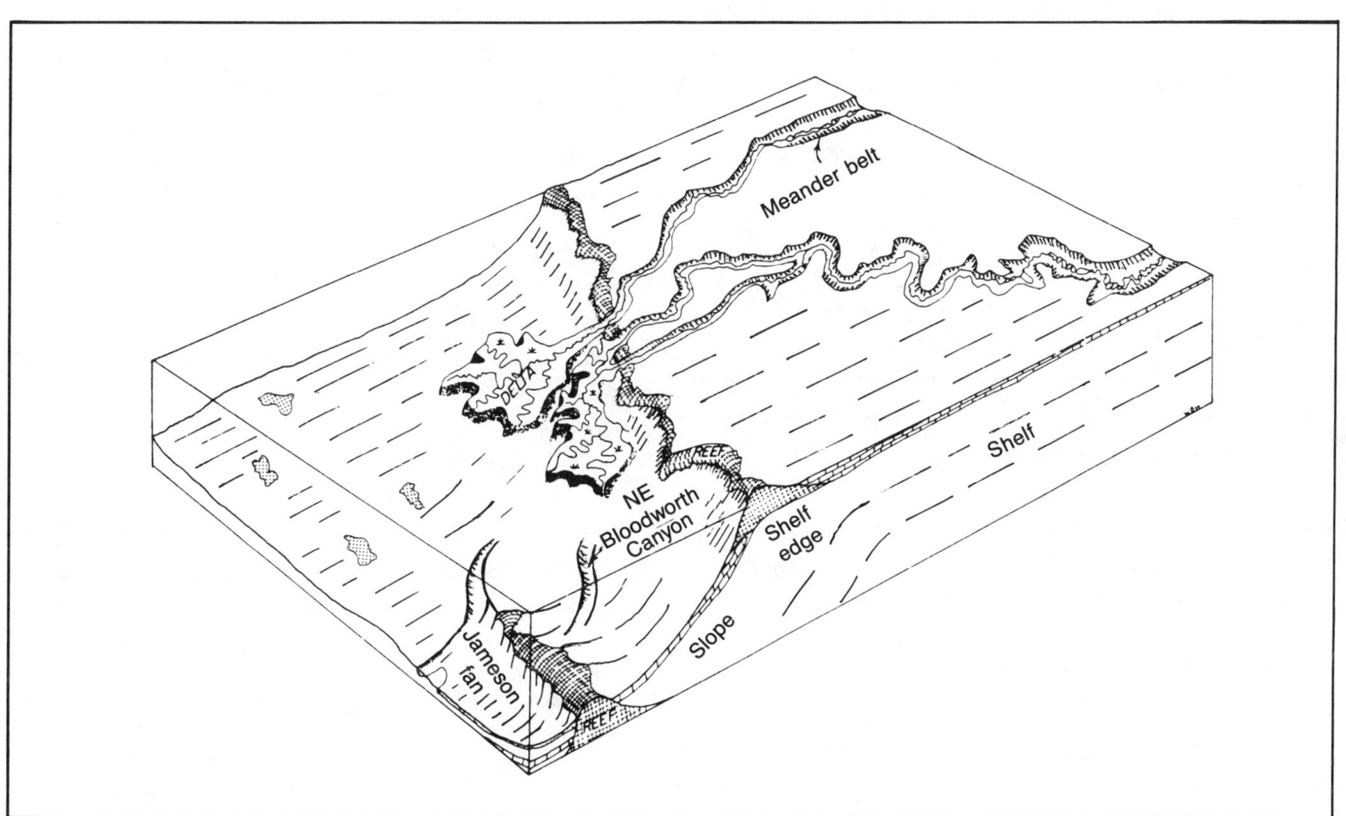

Fig. 12-34. Block diagram of Eastern shelf and Midland basin slope illustrating depositional environments of Cook sandstone. (From Bloomer, 1977; permission to publish by AAPG).

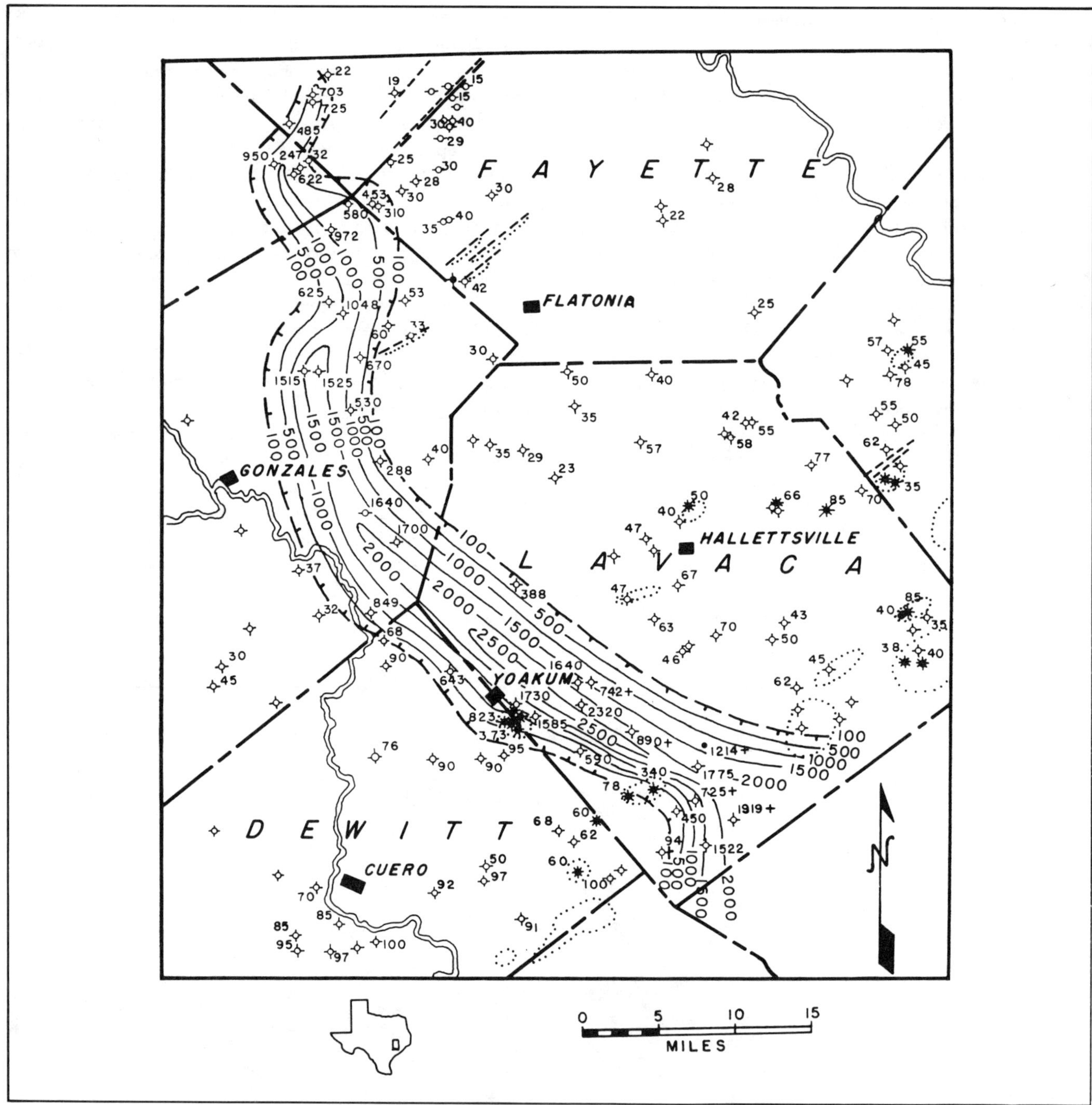

Fig. 12-35. Isopach map of shale filling Yoakum canyon. (From Hoyt, 1959; permission to publish by Gulf Coast Association of Geological Societies).

Cross section IV-IV′, east of section II-II′ and referred to the same datum, illustrates the southwestern dip component of the Jurassic, Cretaceous, and Velasco beds (Fig. 12-47). Asymmetry of the channel in both cross sections is caused by the northeastern half of the channel being the dip slope of the underlying beds and the steeper, southwestern half being a scarp slope.

Well 11 (Fig. 12-47) exhibits the thickest channel fill of

Fig. 12-37. Log section A-A′. See Figure 12-36 for location. (From Hoyt, 1959; permission to publish by Gulf Coast Association of Geological Societies).

560 m along this cross section. In all of the wells of this cross section, the channel fill consists primarily of a series of sandy zones with thin interbeds of shale. Sandstone constitutes a higher percentage of the Chicontepec in this section because it is located closer to the primary source of the sediments than section II-II′. Extreme lenticularity of the sandstone zones of the Chicontepec is the most conspicuous feature of this cross section.

Cross section V-V′ trends west-northwest and coincides generally with the axis of the main Chicontepec canyon. Undulations in the unconformity occur where the trend of the cross section fails to coincide with the axis of the main channel. Local downward thickening of the Chicontepec occurs where a well was drilled in a tributary valley, and local thinning occurs on the divides separating the tributaries. Dip of the subjacent Jurassic, Cretaceous, and Velasco formations is generally south-southwest. A structural cross section using the same wells as cross section V-V′ shows all of

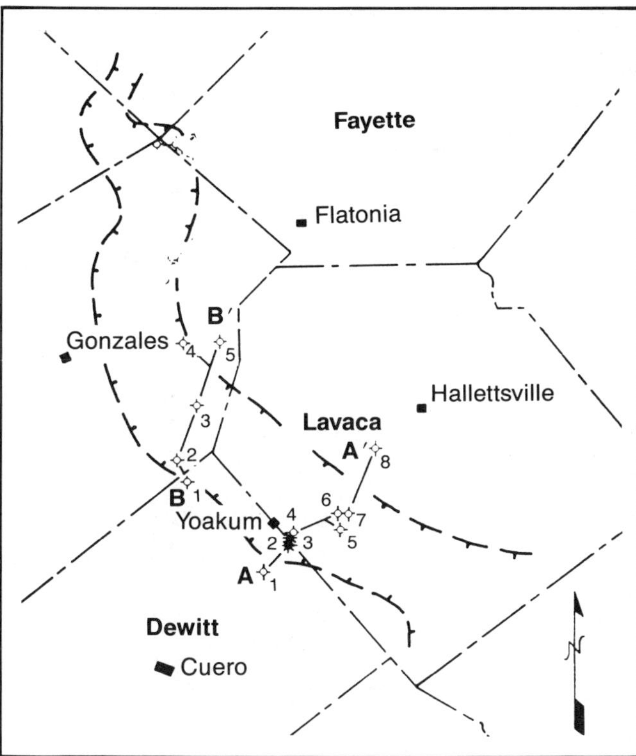

Fig. 12-36. Outline of the Yoakum canyon showing location of cross sections A-A′ and B-B′ of Figures 12-36 and 12-37. (From Hoyt, 1959; permission to publish by Gulf Coast Association of Geological Societies).

submarine erosn thru poorly consol. sd/sh sectn

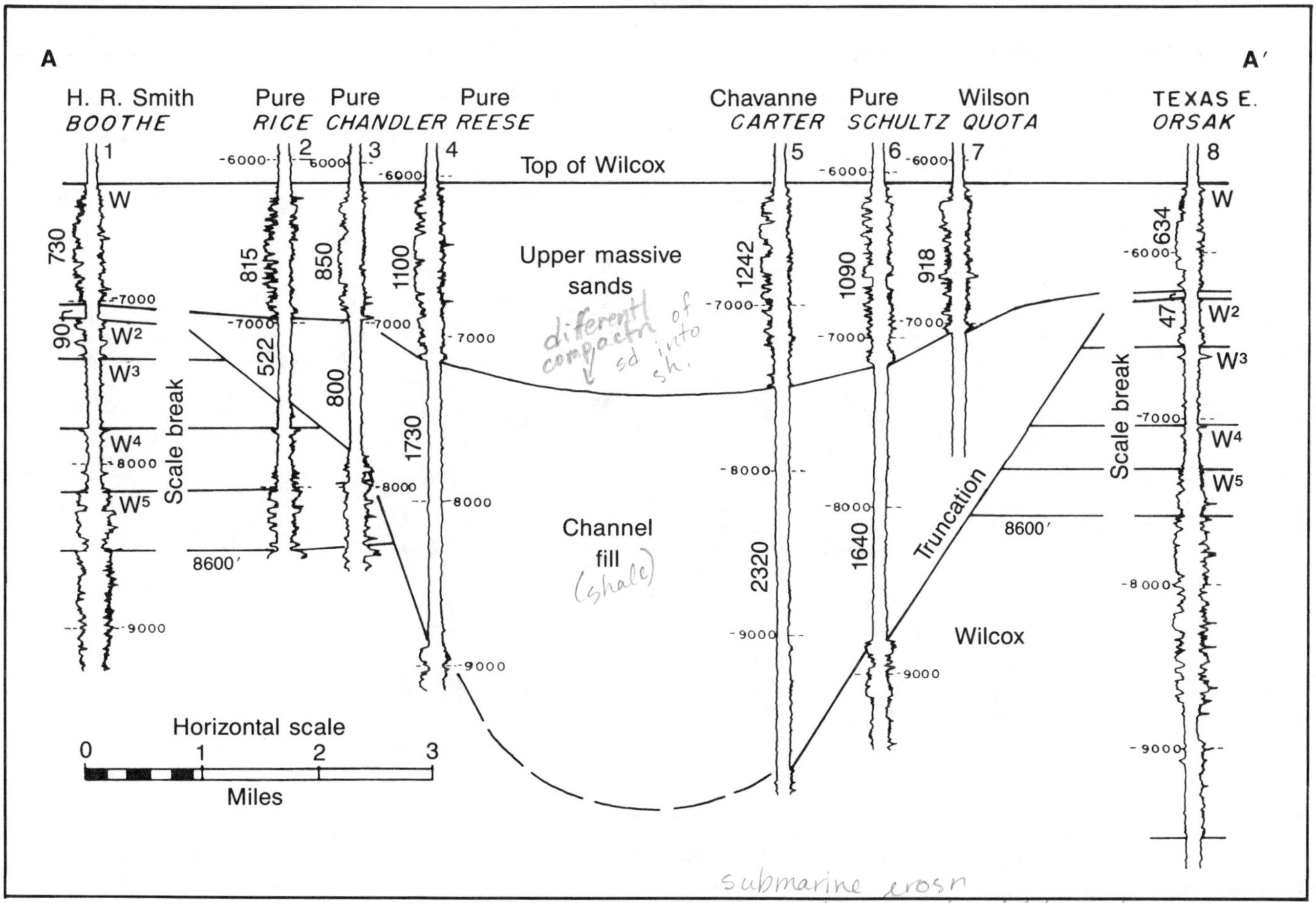

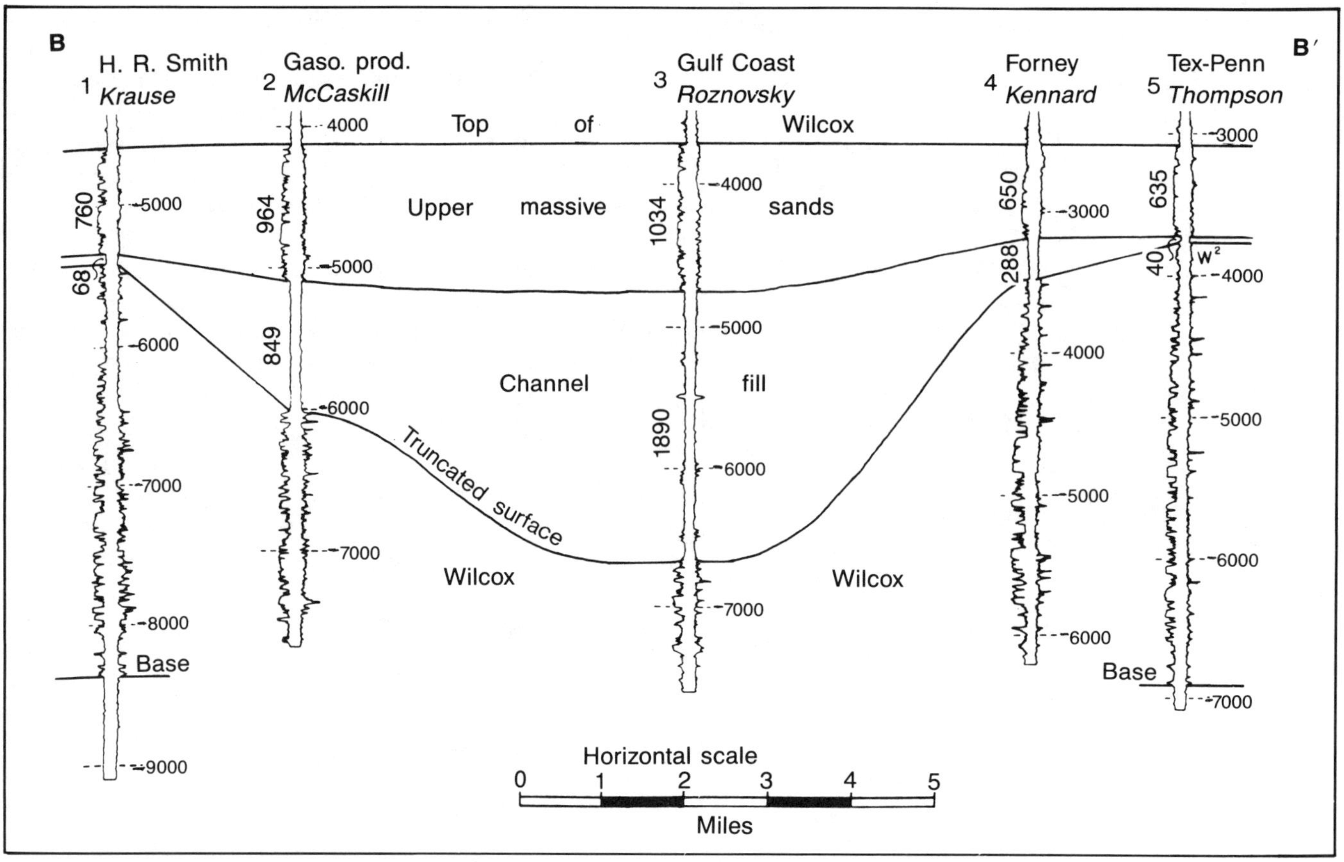

B
B′
1 H. R. Smith Krause
2 Gaso. prod. McCaskill
3 Gulf Coast Roznovsky
4 Forney Kennard
5 Tex-Penn Thompson
Top of Wilcox
Upper massive sands
Channel fill
Truncated surface
Wilcox
Wilcox
Base
Base
760
68
964
849
1034
1890
650
288
635
40
W²
Horizontal scale
0 1 2 3 4 5
Miles

Fig. 12-38. Log section B-B′. See Figure 12-36 for location. (From Hoyt, 1959; permission to publish by Gulf Coast Association of Geological Societies).

Structure 8600′ Sand
Dewitt
Lavaca
Truncated surface
-8351
-8360
-8299
-8349
-8675
-8324
-8320
-8299
-8300
-8500
-9000
-8726
-8355
-8350
-8400
-8470
1 Mile

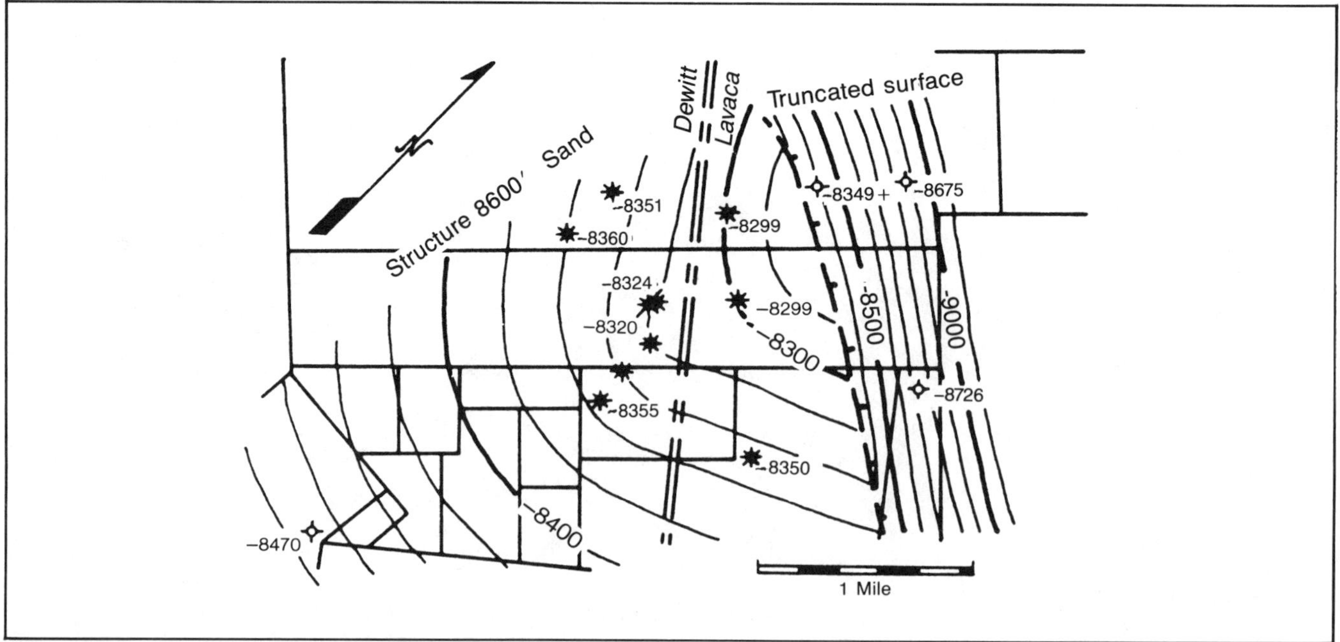

Fig. 12-39. Structure contour map of 8600′ sandstone, Yoakum gas field. Truncating edge of the canyon to the northeast. (From Hoyt, 1959; permission to publish by Gulf Coast Association of Geological Societies).

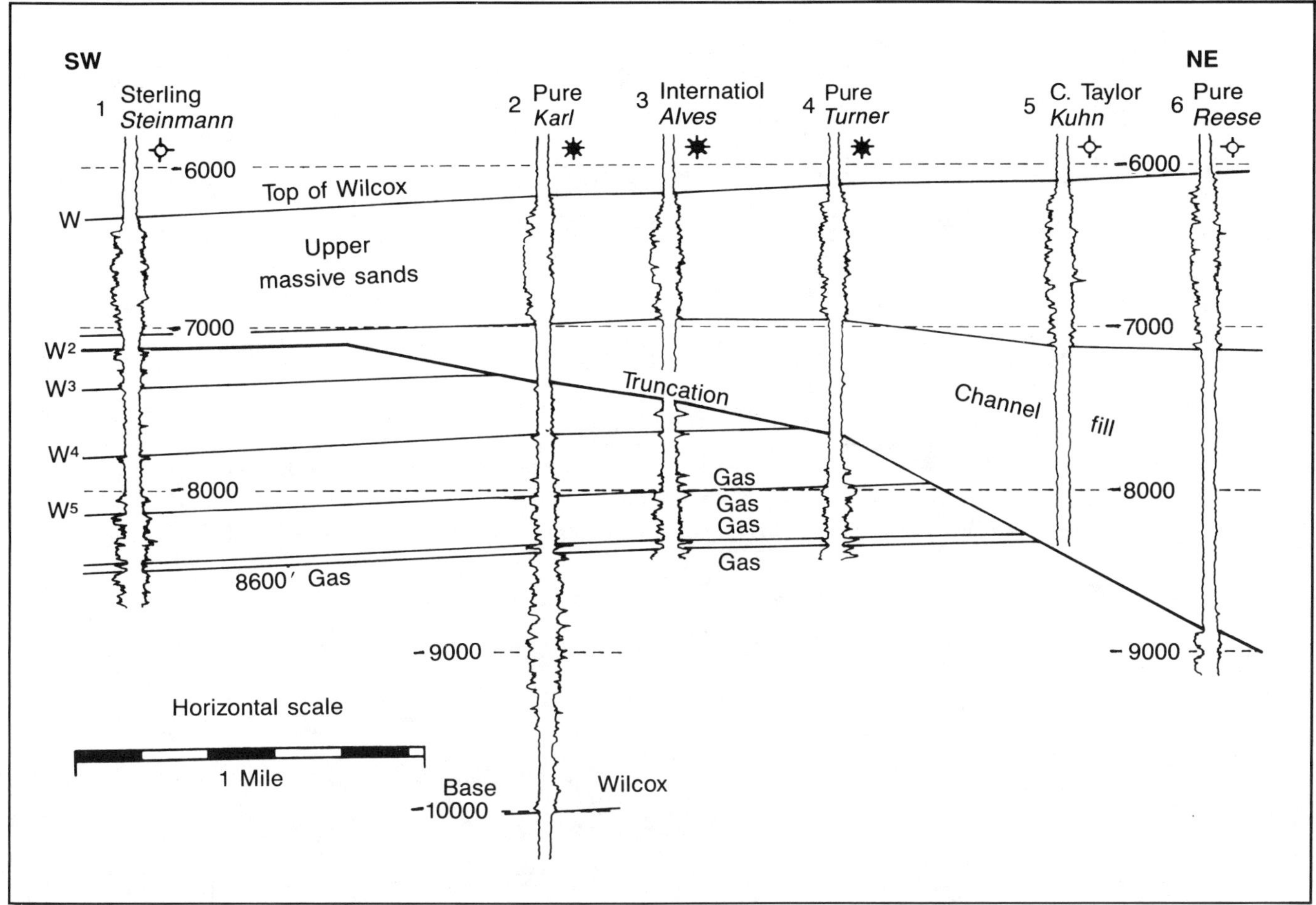

Fig. 12-40. Northeast-southwest cross section through the Yoakum gas field, illustrating trapping of gas against shale fill of Yoakum canyon. (From Hoyt, 1959; permission to publish by Gulf Coast Association of Geological Societies).

the formations, including the Chicontepec, were later tilted to the east (inset, Fig. 12-48).

A vast majority of the wells drilled within the study area passed through the Cretaceous and penetrated the underlying Jurassic. All pre-canyon formations were identified and correlated by a combination of lithological and electric-log characteristics supplemented by paleontological studies. With a substantial data base from which to work, a subcrop map of Paleocene, Cretaceous, and Jurassic formations in contact with the Chicontepec was readily made (Fig. 12-49). Owing to the south-southwest dip of pre-Chicontepec sediments (Fig. 12-48), cutting of the Chicontepec canyon by submarine erosion resulted in the greatest loss of stratigraphic section in the east where the Chicontepec rests on the Jurassic (Figs. 12-48 and 12-49). Westward, the channel fill overlies progressively younger formations as subcrop bands of Cretaceous and Paleocene formations on either side of the canyon converge and disappear under the canyon axis.

Defined at the top by a stratigraphic time marker (Horizon "C") and at the base by a erosional unconformity, the Chicontepec formation is a Genetic Sequence of Strata (GSS). An isopach map of this GSS shows a submarine canyon fill with numerous tributary channel fills along the northern margin (Fig. 12-50). Tributaries along the north side of the canyon flowed down a dip slope of the strata underlying the Chicontepec and are more numerous and longer than those along the southern margin. Being a scarp slope, the southern margin of the main channel is steeper than the north slope. Relatively few tributaries are seen along the south margin, due in part to the paucity of subsurface data in this area.

Thickness of the Chicontepec along the main axis of the channel ranges from approximately 500 m in the east to an estimated 900 m in the west (Fig. 12-50). This gradual east-to-west increase in thickness indicates a deepening of the canyon to the west. Consequently, downcanyon submarine current movement and sediment transport was to the west in Chicontepec time.

Chicontepec formation sediments consist of individual beds

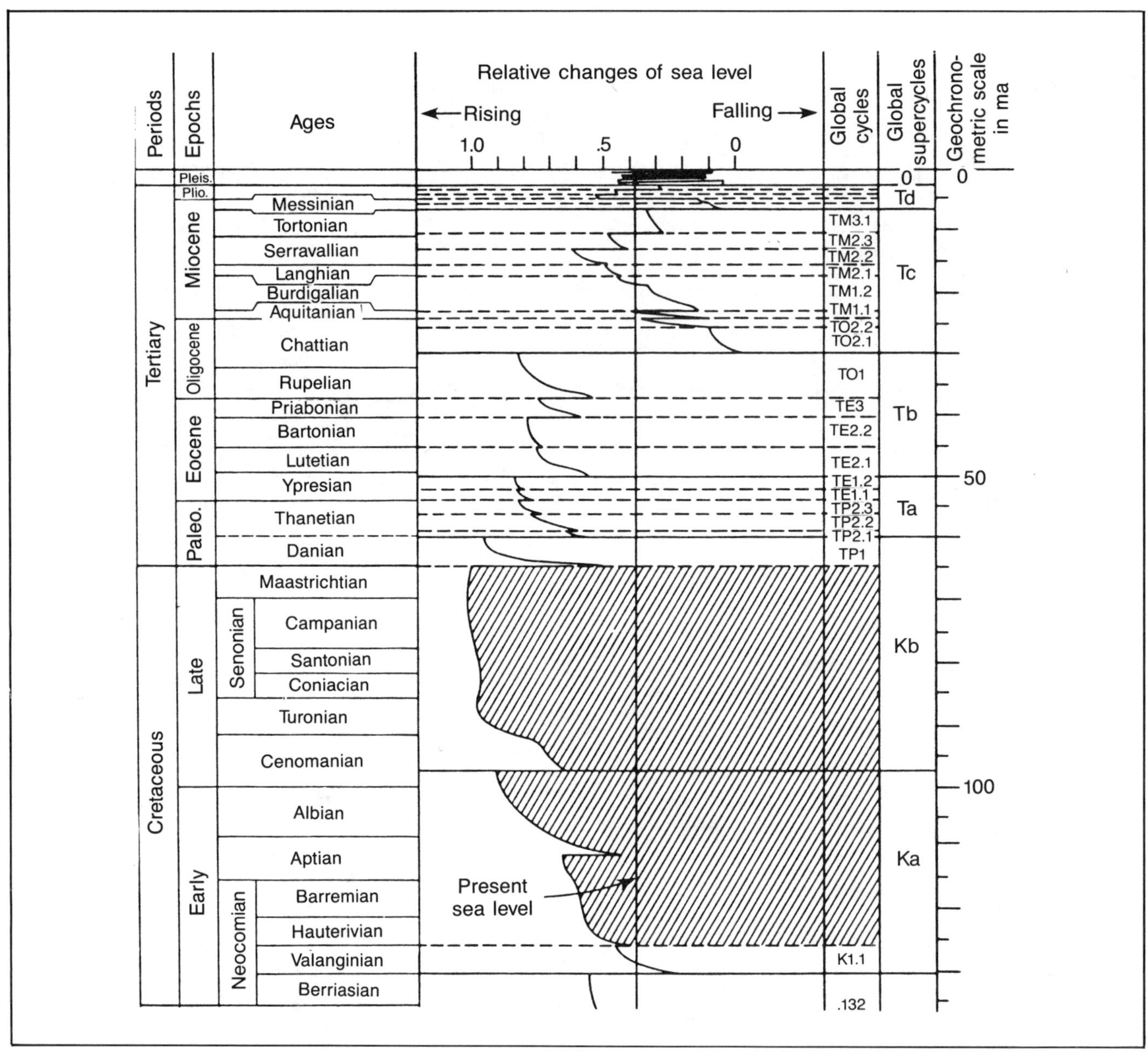

Fig. 12-41. Relative sea-level curve for Tertiary and Cretaceous. (From Vail et al., 1977; permission to publish by AAPG).

and zones of argillaceous, fine-grained sandstone alternating with shale and siltstone. To construct a net sandstone map for the Chicontepec, it was necessary to differentiate between sandstone and finer-grained sedimentary rocks using electric logs. The distinction between siltstone and sandstone is an arbitrary one involving a judgmental factor applied to log interpretation. In order to eliminate differences in electric-log evaluation among individuals, a single investigator was assigned the task of distinguishing the sands from the shales and siltsones. Net Chicontepec sandstone distribution is more restricted than that of the Chicontepec formation with its shales and siltstones (Fig. 12-51). As would be expected, the greatest thicknesses of net sandstone occur in the deepest portion of the channel fill, and the estimated maximum thickness within the study area exceeds 600 m.

Fig. 12-44. Locations of Chicontepec oil pools and wells supplying paleontological data, southeastern Tampico-Misantla basin. (From Busch and Govela, 1978; permission to publish from AAPG).

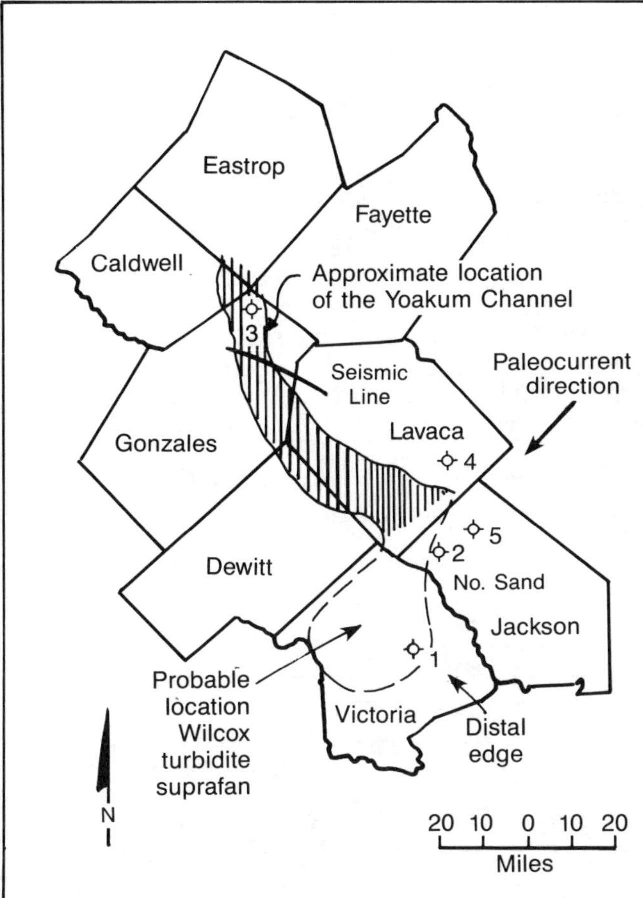

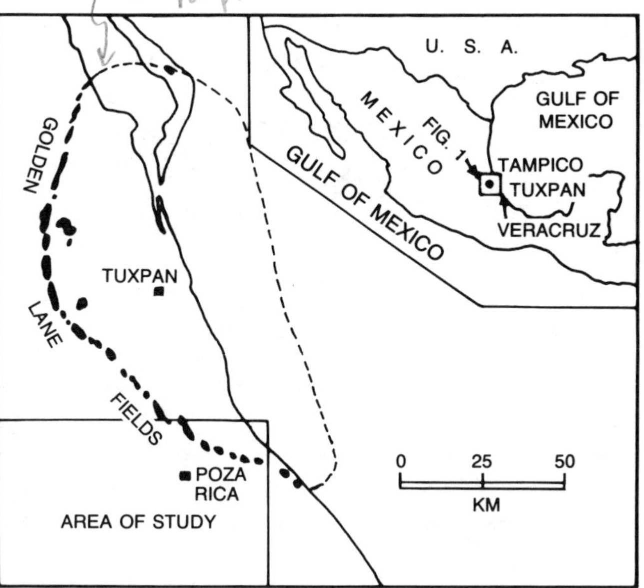

Fig. 12-43. Location of Chicontepec formation study area in Tampico-Misantla basin. (From Busch and Govela, 1978; permission to publish by AAPG).

Fig. 12-42. Deep well locations beyond mouth of Yoakum canyon. (From Vormelker, 1980; permission to publish by Oil and Gas Journal).

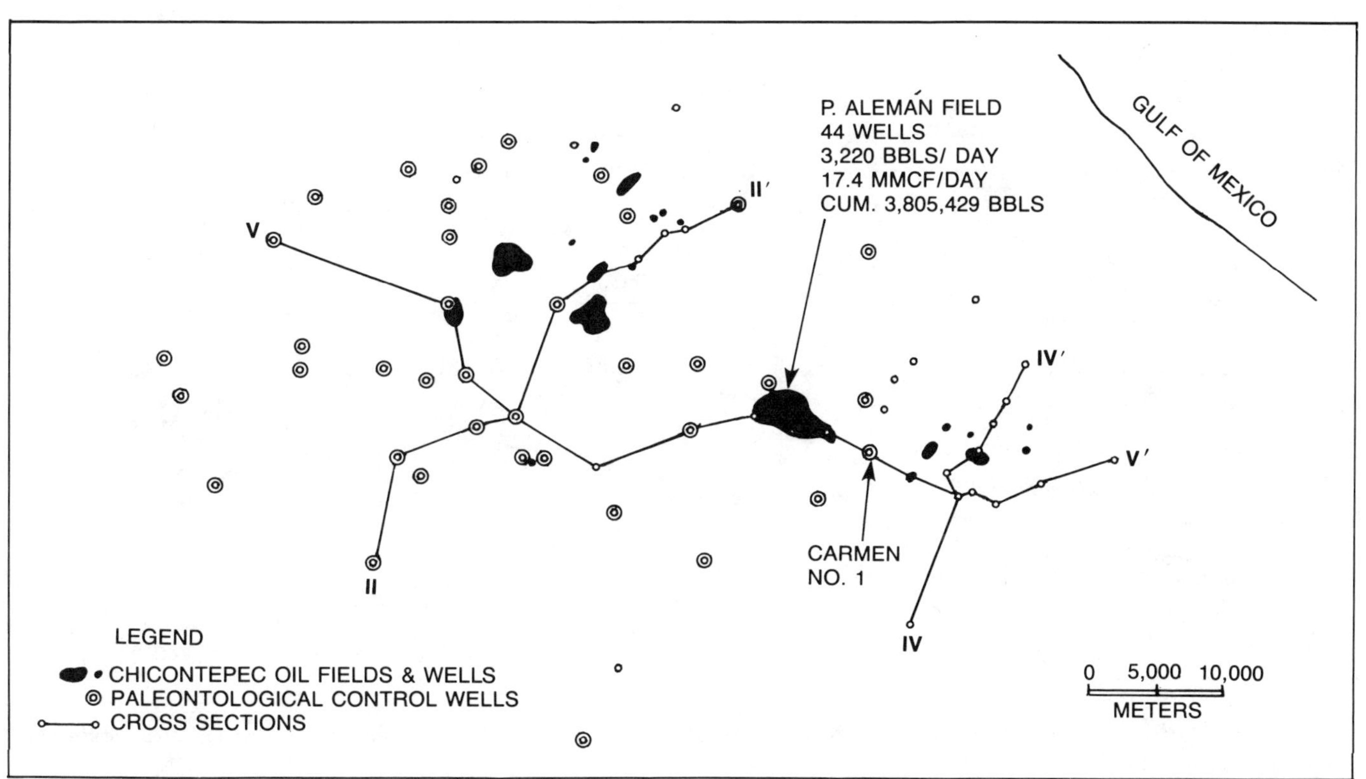

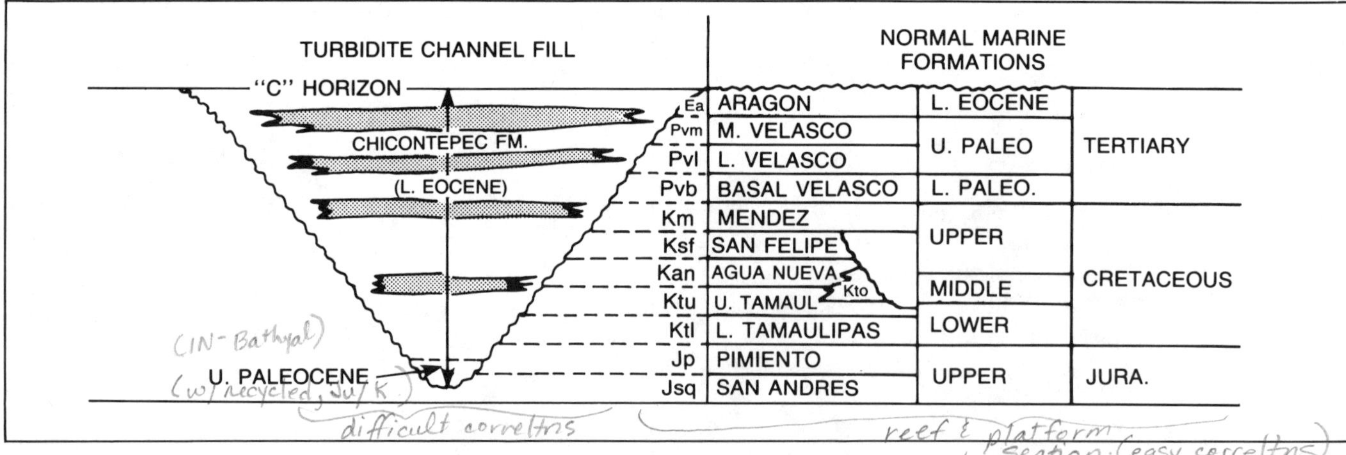

Fig. 12-45. Stratigraphy of Upper Jurassic, Cretaceous, and Lower Tertiary and of Tampico-Misantla basin, showing unconformable Chicontepec channel fill. (From Busch and Govela, 1978; permission to publish from AAPG).

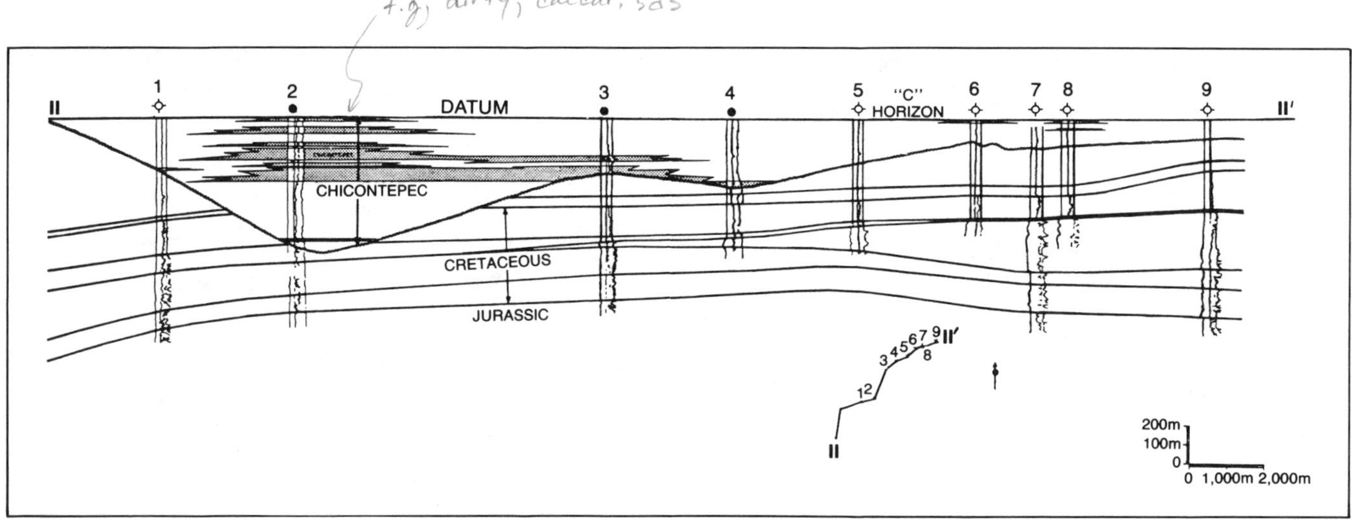

Fig. 12-46. Stratigraphic profile II-II'. See Figures 12-44 and 12-49 for location. (From Busch and Govela, 1978; permission to publish from AAPG).

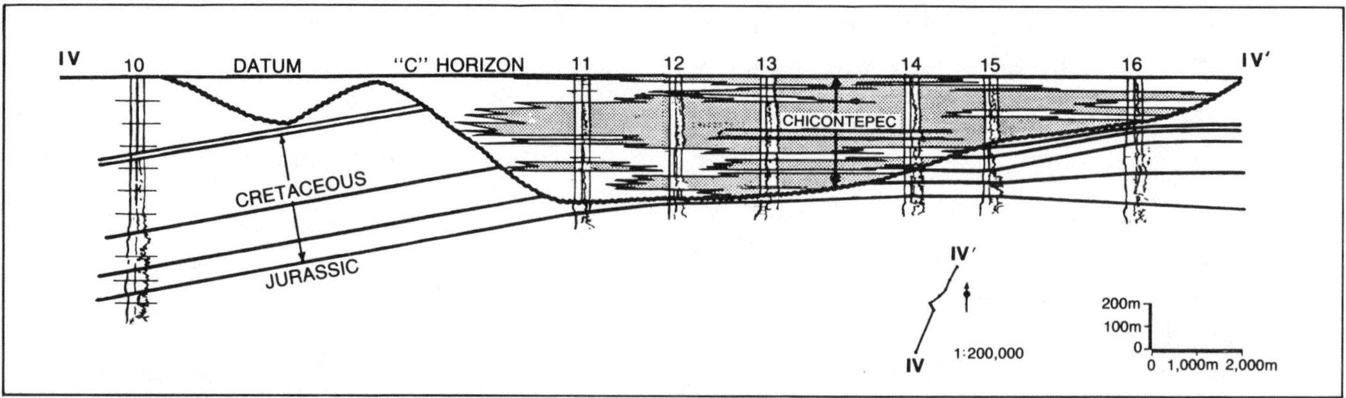

Fig. 12-47. Stratigraphic profile IV-IV'. See Figures 12-44 and 12-49 for location. (From Busch and Govela, 1978; permission to publish from AAPG).

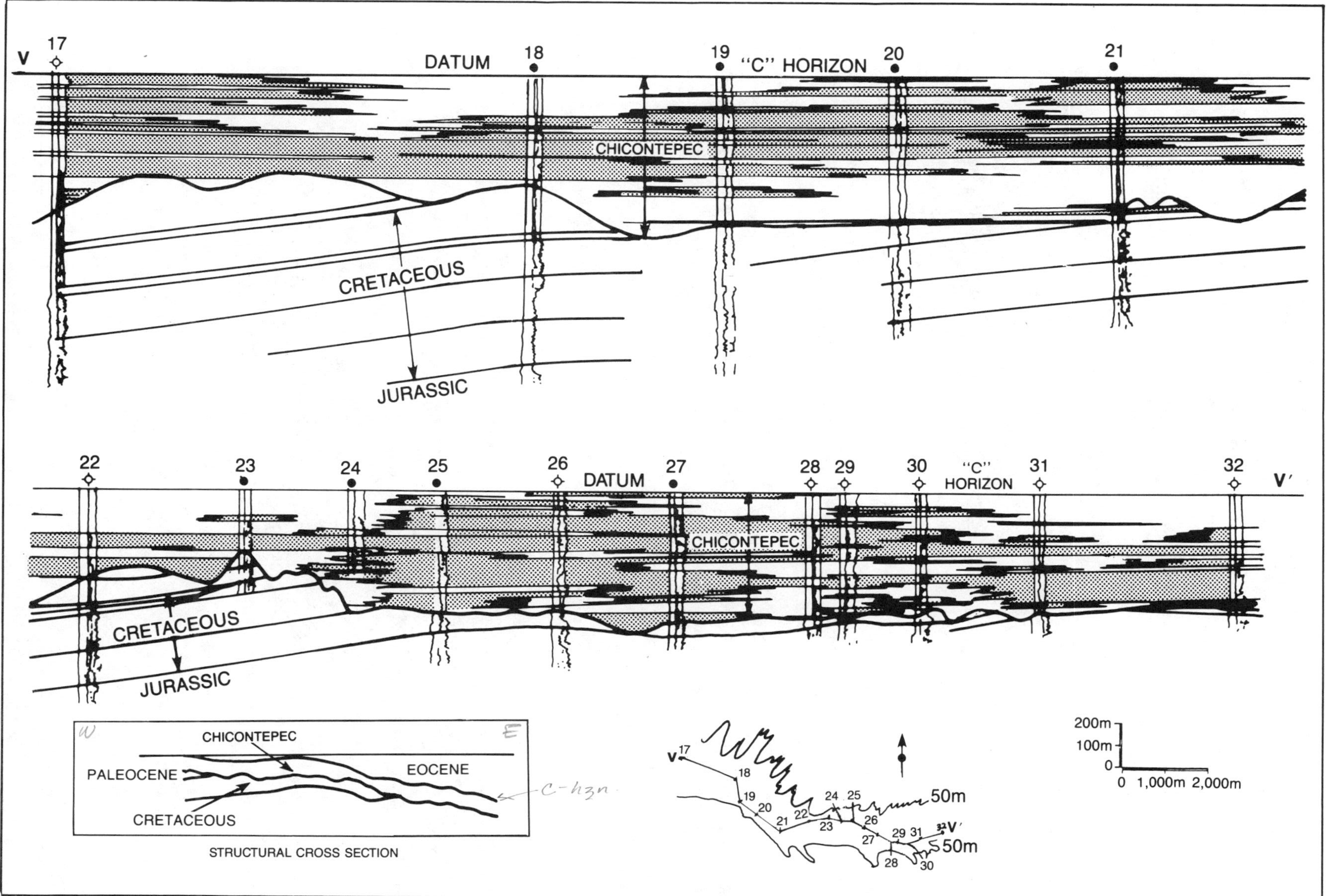

Fig. 12-48. Stratigraphic profile V-V'. See Figures 12-44 and 12-49 for for location. (From Busch and Govela, 1978; permission to publish from AAPG).

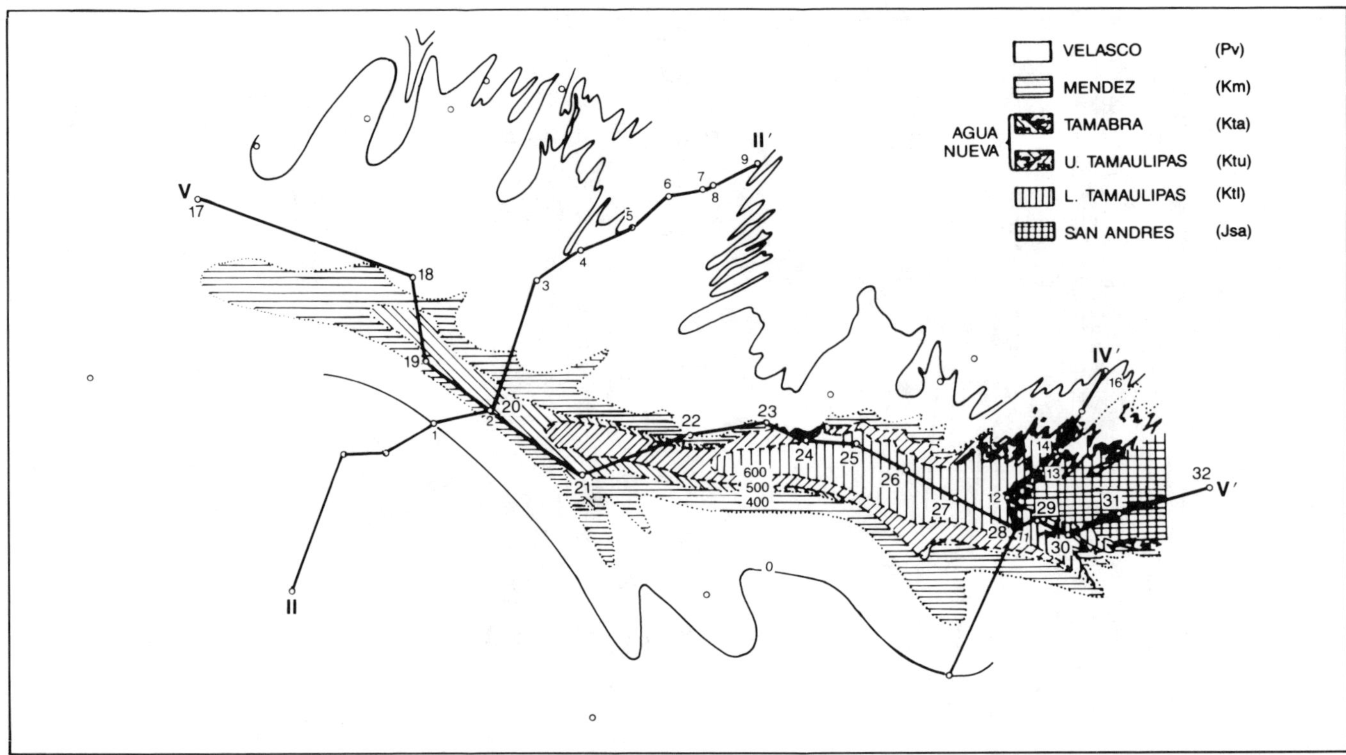

Fig. 12-49. Subcrop map of Jurassic, Cretaceous, and Paleocene formations. (see Fig. 12-44 for location). (Modified after Busch and Govela, 1978; permission to publish by AAPG).

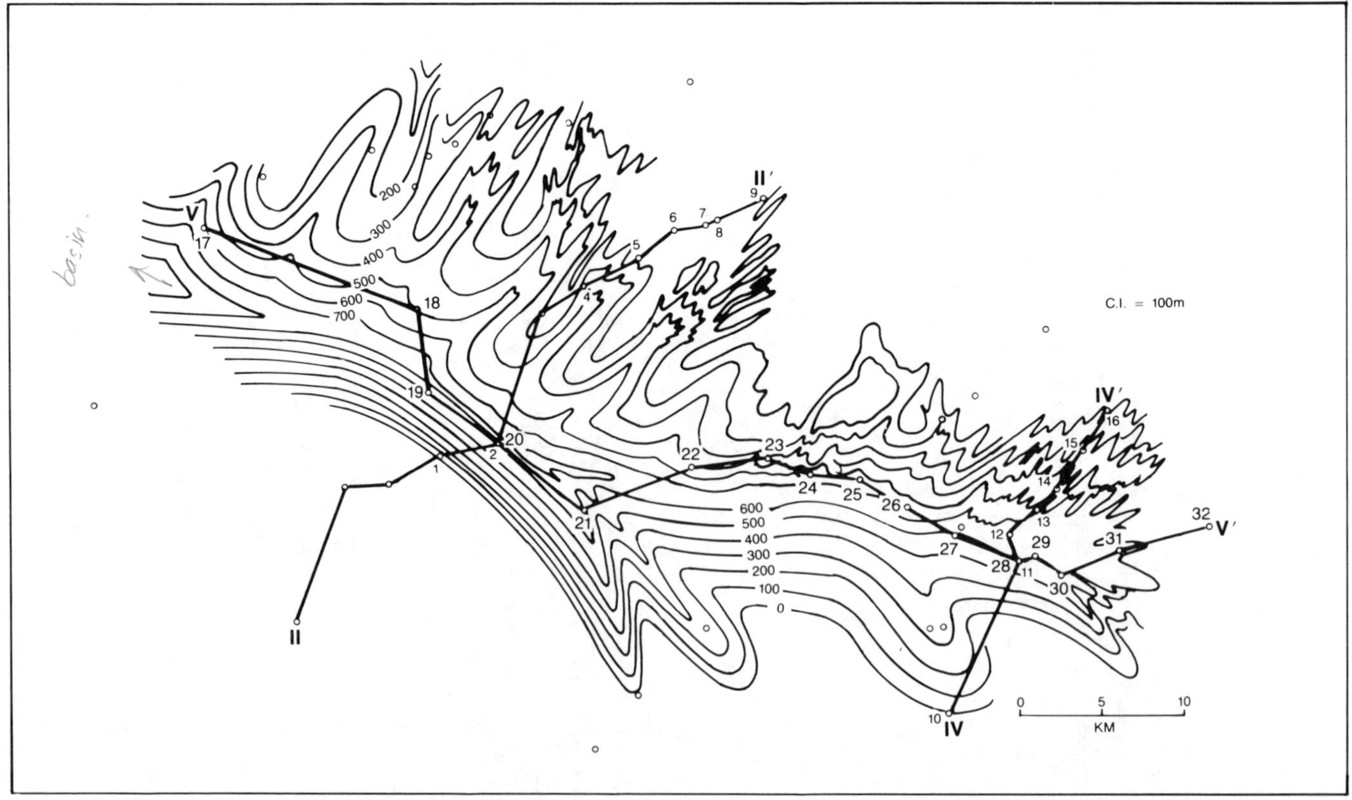

Fig. 12-50. Isopach map of Chicontepec formation. (Modified after Busch and Govela, 1978; permission to publish by AAPG).

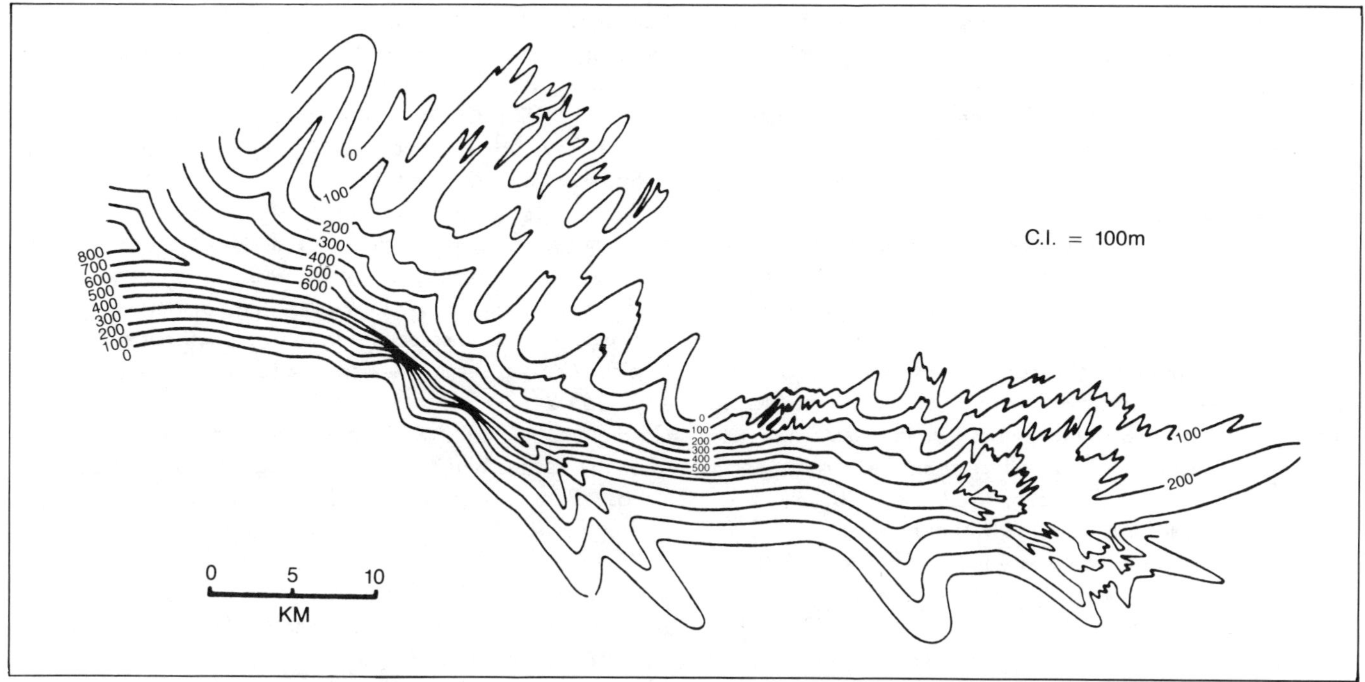

Fig. 12-51. Net sandstone isopach map of Chicontepec formation. (From Busch and Govela, 1978; permission to publish by AAPG).

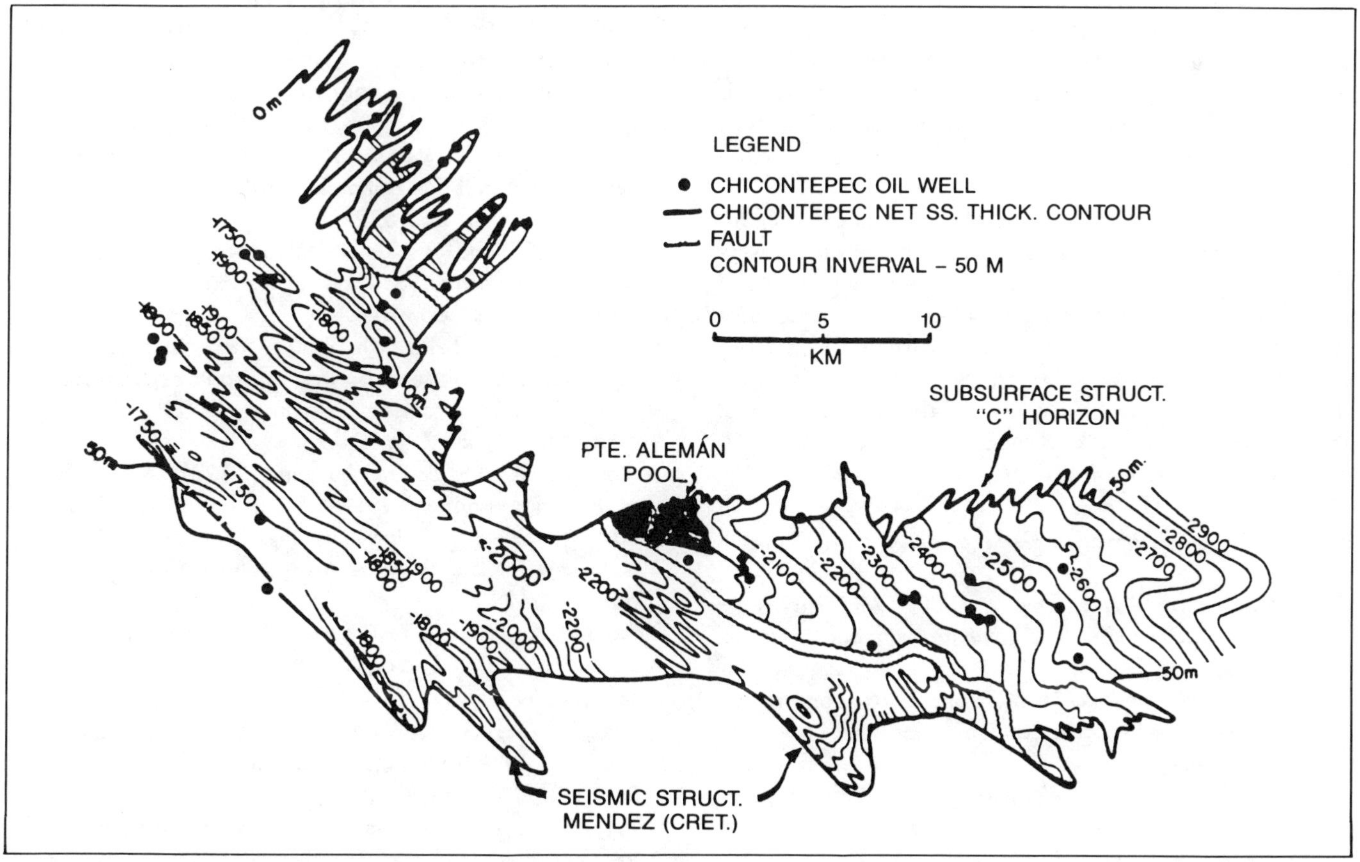

Fig. 12-52. Structure map of top of Chicontepec (C horizon) to the east and the Upper Cretaceous Mendez formation to the west within the area of channel fill. (From Busch and Govela, 1978; permission to publish from AAPG).

Production from the Presidente Aleman field (Fig. 12-44) is from a restricted area along the north flank of the channel fill where the net sandstone ranges in thickness from 0–250 m. Impingement of the Chicontepec sandstone against the northern side of the canyon controls the northern margin of the field. Drilling has yet to define the limits of the field to the east, south, and west.

In post-Early Eocene time, the sediments of the Tampico-Misantla basin were uplifted and tilted to the northeast. To the east there is sufficient, subsurface well control to construct a meaningful structure map of the "C" horizon at the top of the Chicontepec. Well control, however, is too scattered for accurate portrayal of this horizon to the west. Although seismic coverage does exist for this western area, there are no reliable seismic reflectors in the Tertiary section. For this reason, the top of the Mendez (Upper Cretaceous), which is approximately 500 m below the "C" horizon, was mapped in the western part of the area (Fig. 12-52). In spite of mapping on two different horizons, structural contours intersecting Chicontepec oil wells to the west are 700 m higher than those intersecting Chicontepec wells to the east. When the approximately 500 m of stratigraphic interval between the two mapped horizons is added, there is about 1200 m of elevation difference between Chicontepec producing wells from west to east. None of the production can be related to structural closure. It is apparent that the sandstones of the Chicontepec constitute a gigantic stratigraphic trap within a submarine canyon fill. Trapping of the oil and gas is by lenticular Chicontepec sandstones that shale out laterally and are separated by impermeable shales.

Chemically, Chicontepec oil is almost identical with Jurassic oil. Undoubtedly the subcrop of the Jurassic to the east furnished the oil, which saturates the more porous and permeable sandstones of the Chicontepec. It is reasonable to expect hydrocarbon accumulation anywhere Chicontepec sandstones have reservoir capabilities within the mapped area of net sandstones (Fig. 12-51). The axial portion of this channel fill should afford an optimum number of sandstones and best reservoir quality.

SELECTED BIBLIOGRAPHY

Ayers, M. W., and W. J. Cleary, 1980, Wilmington fan: mid-Atlantic lower rise development: Jour. Sed. Pet., v. 50, no. 1, p. 235-246.

Belderson, R. H., and N. H. Kenyon, 1976, Long-range sonar views of submarine canyons: Mar. Geol., v. 22, p. M69-M74.

Berger, W. H., and E. L. Winterer, 1974, Plate stratigraphy and the fluctuative carbonate line: Internat. Assoc. Sedimentol. Spec. Pub. 1, p. 11-48.

Bloomer, R. R., 1977, Depositional environments of a reservoir sand stone *in* west-central Texas: AAPG Bull., v. 61, no. 3, p. 344-359.

Bouma, A. H., 1962, Sedimentology of some flysch deposits, a graphic approach to facies interpretation: Amsterdam, Elsevier, 168p.

———, 1979, Continental slopes, *in* L. J. Doyle and O. H. Pilkey, eds., Geology of Continental Slopes: SEPM Sp. Pub. no. 27, p. 1-15.

Burke, K., 1972, Longshore drift, submarine canyons and submarine fans in development of Niger Delta: AAPG Bull., v. 56, no. 10, p. 1975-1983.

Busch, D. A., and S. A. Govela, 1978, Stratigraphy and structure of Chicontepec turbidites, southeastern Tampico-Misantla basin, Mexico: AAPG Bull., v. 62, no. 2, p. 235-246.

Carman, G. J., and R. Young, 1981, Reservoir geology of the Forties oilfield: *in* L. V. Illing and G. D. Hobson, eds., Petroleum geology of the continental shelf of northwest Europe: London, Institute of Petroleum Geology, p. 371-379.

Casnedi, R., 1983, Hydrocarbon-bearing submarine fan system of Cellino formation, central Italy: AAPG Bull., v. 67, no. 3, p. 359-370.

Chough, S., and R. Hesse, 1976, Submarine meandering thalweg and turbidity currents flowing for 4,000 km in the Northwest Atlantic Mid-Ocean Channel, Labrador Sea: Geology, v. 4, no. 9, p. 529-533.

Curray, J. R., 1965, Late Quaternary history, continental shelves of the United States, *in* H. E. Wright and D. G. Frey, eds., The Quaternary of the United States: New Jersey, Princeton Univ. Press, p. 723-735.

——— and D. G. Moore, 1971, Growth of the Bengal deep-sea fan and denudation in the Himalayas: GSA Bull., v. 82, no. 3, p. 563-572.

Damuth, J. E., and N. Kumar, 1975, Amazon cone: morphology, sediments, age, and growth pattern: GSA Bull., v. 86, no. 6, p. 863-878.

———, et al., 1983, Distributary channel meandering and bifurcation patterns on the Amazon deep-sea fan as revealed by long-range side-scan sonar (GLORIA): Geology, v. 11, no. 2, p. 94-98.

De'Ath, N. G., and S. F. Schuyleman, 1981, The geology of the Magnus oilfield, *in* L. V. Illing, and G. D. Hobson, eds., Petroleum geology of the continental shelf of northwest Europe, Institute of Petroleum Geology, London, p. 342-351.

Dickas, A. B., and J. L. Payne, 1967, Upper Paleocene buried channel in Sacramento Valley, California: AAPG Bull., v. 51, no. 6, p. 873-882.

Egloff, J., and G. L. Johnson, 1975, Morphology and structure of the southern Labrador Sea: Can. Jour. Earth Sci., v. 12, . 2111-2133.

Hamilton, E. L., 1967, Marine geology abyssal plains in the Gulf of Alaska: Jour. Geophys. Res., v. 72, p. 4189-4213.

Haner, B. E., 1971, Morphology and sediments of Redondo submarine fan, southern California: GSA Bull., v.

82, no. 9, p. 2413-2432.

Heritier, F. E., P. Lossel, and E. Wathne, 1979, Frigg field—a large submarine-fan trap in lower Eocene rocks of North Sea Viking Graben: AAPG Bull., v. 63, no. 11, p. 1999-2020.

Hess, G. R., 1974, Submarine fan fare; a comparison of modern and Miocene deep-sea fans: Master's Thesis, University of Minnesota, 118p.

Hill, P. J., and G. V. Wood, 1980, Geology of the Forties Field, U.K. continental shelf, North Sea, *in* M. T. Halbouty, ed., Giant Oil and Gas Fields of the Decade 1968-1978: AAPG Mem. 30, p. 81-93.

Hoyt, W. V., 1959, Erosional channel in the Middle Wilcox near Yoakum, Lavaca County, Texas: Gulf Coast Assoc. of Geol. Soc. Trans., v. 9, p. 41-50.

Hsu, K. J., 1977, Studies of Ventura Field, California, I: Facies geometry and genesis of Lower Pliocene turbidities: AAPG Bull., v. 61, no. 2, p. 137-168.

_______, et al., 1977, History of the Mediterranean salinity crisis: Nature, v. 267, p. 399-403.

Jacka, A. D., R. H. Beck, L. C. St. Germain, and S. C. Harrison, 1968, Permian deep-sea fans of the Delaware Mountain Group, Delaware Basin, *in* Field trip guidebook for 1968 symposium, Guadalupian facies, Apache Mountain area, West Texas: SEPM, Permian Basin section, Sp. Pub. 68-11, p. 49-90.

Judice, P. C., and S. J. Mazzullo, 1983, Gray sandstones (Jurassic) in Terryville field—basinal depositional and exploration model: Oil and Gas Jour., v. 81, no. 16, p. 191-206.

Kenyon, N. H., R. H. Belderson, and A. H. Stride, 1978, Channels, canyons and slump folds on the continental slope between southwest Ireland and Spain: Oceanologica Acta, v. 1, p. 369-380.

Kulm, L. D., and J. V. Byrne, 1966, Sedimentary response to hydrography in an Oregon estuary: Marine Geology, v. 4, p. 85-118.

Lewis, D. W., 1982, Channels across continental shelves: corequisites of canyon-fan systems and potential petroleum conduits: New Zealand Jour. Geol. and Geophys., v. 25, p. 209-225.

Leyden, R., J. E. Damuth, L. K. Ongley, J. Kostecki, and W. van Stevenick, 1978, Salt diapirs on Sao Paulo plateau, south-eastern Brazilian continental margin: AAPG Bull., v. 62, no. 4, p. 657-666.

Link, M. H., and J. E. Welton, 1982, Sedimentology and reservoir potential of Matilija sandstone: an Eocene sand-rich deep-sea fan and shallow-marine complex, California: AAPG Bull., v. 66, no. 10, p. 1514-1534.

Lock, B. E., 1982, Towards a better understanding of Gulf Coast Miocene deep water sediments: Gulf Coast Assoc. Geol. Soc., v. 32, p. 283-288.

_______, 1983, Personal comm.

MacPherson, B. A., 1978, Sedimentation and trapping mechanism in Upper Miocene Stevens and older turbidite fans of southeastern San Joaquin Valley, California: AAPG Bull., v. 62, no. 11, p. 2243-2274.

McGregor, B., W. L. Stubblefield, W. B. F. Ryan, and D. C. Twichell, 1982, Wilmington submarine canyon: a marine fluvial-like system: Geology, v. 10, no. 1, p. 27-30.

Menard, H. W., 1955, Deep-sea channels, topography and sedimentation: AAPG Bull., v. 39, no. 2, p. 236-255.

Middleton, G. V., 1969, Turbidity currents and grain flows and other mass movements down slopes, *in* D. J. Stanley, ed., The new concepts of continental margin sedimentation, Amer. Geol. Inst. Short Course Notes, p. GM-A-1 to GM-B-14.

_______, and M. A. Hampton, 1973, Sediment gravity flows: mechanics of flow and deposition, *in* G. V. Middleton and A. H. Bouma, eds., Turbidites and deep water sedimentation: Anaheim, California, SEPM Pacific Sec. Short Course, p. 1-38.

_______ and _______, 1976, Subaqueous sediment transport and deposition by sediment gravity flows, *in* D. J. Stanley, and D. J. P. Swift, eds., Marine sediment transport and environmental management: New York, John Wiley and Sons, Inc., p. 197-218.

Moore, G. T., and T. J. Fullam, 1975, Submarine channel systems and their potential for petroleum localization, *in* M. L. Broussard, ed., Deltas: Houston Geological Society, p. 165-189.

Mutti, E., 1977, Distinctive thin-bedded turbidite facies and related depositional environments in the Eocene Hecho Group (south-central Pyrenees, Spain): Sedimentology, v. 24, no. 1, p. 107-131.

_______, and F. Ricci-Lucchi, 1972, Le torbiditi dell'E Apennino settentrionale: introduzione all' analisi di facies: Memorie della Societa Geologica Italiana, p. 161-199. English translation in International Geology Review, 1978, v. 20, no. 2, p. 125-166.

_______ and _______, 1975, Turbidite facies and facies associations, *in* Examples of turbidite facies and facies associations from selected formations of the northern Apennines: Nice, IXth International Congress of Sedimentology, Field Trip A-11, p. 21-36.

Nelson, C. H., 1976, Late Pleistocene and Holocene depositional trends, processes, and history of Astoria deep-sea fan, northeast Pacific: Mar. Geol., v. 20, no. 2, p. 129-173.

_______ and L. D. Kulm, 1973, Submarine fans and deep-sea channels, in G. V. Middleton and A. H. Bouma, eds., Turbidites and deep water sedimentation: Anaheim, California, SEPM Pacific Section Short Course, p. 39-70.

Ness, G. E., and L. D. Kulm, 1973, Origin and development of Surveyor deep sea channel: GSA Bull., v. 84, no. 10, p. 3339-3354.

Normark, W. R., 1970a, Growth patterns of deep-sea fans: AAPG Bull., v. 54, no. 11, p. 2170-2195.

_______, 1970b, Channel piracy on Monterey deep-sea fan: Deep Sea Res., v. 17, p. 837-846.

_______, 1974, Submarine canyons and fan valleys: factors affecting growth pattern of deep-sea fans, *in* R. H. Dott, Jr., and R. H. Shaver, eds., Modern and ancient geosynclinal sedimentation: SEPM Sp. Pub. no. 19, p. 56-68.

_______, 1978, Fan valleys, channels, and depositional lobes on modern submarine fans: characters for recognition of sandy turbidite environments: AAPG Bull., v. 62, no. 6, p. 912-931.

_______ and F. H. Dickson, 1976, Sublacustrine fan morphology in Lake Superior: AAPG Bull., v. 60, no. 7, p. 1021-1036.

_______ and D. J. W. Piper, 1969, Deep-sea fan-valleys, past and present: GSA Bull., v. 80, no. 9, p. 1859-1866.

_______ and _______, 1972, Sediments and growth pattern of Navy deep-sea fan, San Clemente basin, California borderland: Jour. of Geol. v. 80, no. 2, p. 198-223.

Parker, J. R., 1975, Lower Tertiary sand development in the central North Sea, *in* A. W. Woodland, ed., Petroleum and the continental shelf of north-west Europe, 1. Geology: London, Applied Sci. Pubs., p. 447-452.

Piper, D. J. W., 1970, Transport and deposition of Holocene sediment on La Jolla deep-sea fan, California: Mar. Geol., v. 8, p. 211-227.

_______, 1975, Late Quaternary deep-water sedimentation off Nova Scotia and the western Grand Banks: Canadian Soc. Petrol. Geol. Mem. 4, p. 195-204.

Pitman, W. C., III, 1978, Relationship between eustacy and stratigraphic sequences of passive margins: GSA Bull., v. 89, no. 9, p. 1389-1403.

Sabate, R. W., 1968, Pleistocene oil and gas in coastal Louisiana: Gulf Coast Assoc. of Geol. Soc. Trans. v. 18, p. 373-386.

Selley, R. C., 1976, Subsurface environmental analysis of North Sea sediments: AAPG Bull., v. 60, no. 2, p. 184-195.

_______, 1979, Dipmeter and log motiffs in North Sea submarine fan sands: AAPG Bull., v. 63, no. 6, p. 905-917.

Shanmugan, G., and R. J. Moiola, 1982, Eustatic control of turbidites and winnowed turbidites: Geology, v. 10, no. 5, p. 231-235.

Shepard, F. P., 1963, Submarine Geology, 2nd ed.: New York, Harper & Row, 557p.

_______, 1964, Sea-floor valleys of Gulf of California, *in* T. H. van Andel and G. G. Shor, Jr., eds., Marine geology of the Gulf of California: AAPG Mem. 3, p. 157-192.

_______, 1966, Meander in valley crossing a deep-ocean fan: Science, v. 154, no. 3747, p. 385-386.

_______, 1981, Submarine canyons: multiple causes and long-time persistence: AAPG Bull., v. 65, no. 6, p. 1062-1077.

_______, R. F. Dill, and U. von Rad, 1969, Physiography and sedimentary processes of La Jolla submarine fan and fan-valley California: AAPG Bull., v. 53, no. 2, p. 390-420.

_______ and K. O. Emery, 1973, Congo submarine canyon and fan valley: AAPG Bull., v. 57, no. 9, p. 1679-1691.

Siemers, C. T., 1978, Submarine fan deposition of the Woodbine-Eagle Ford interval (Upper Cretaceous), Tyler County, Texas: Gulf Coast Assoc. Geol. Soc. Trans. v. 28, p. 493-533.

Stanley, D. J., 1975, Submarine canyon and slope sedimentation (Gres d'Annot) in the French Maritime Alps: Nice, Ninth International Sedimentological Congress, 1975, 131p.

_______, P. Fenner, and G. Kelling, 1972, Currents and sediment transport at the Wilmington Canyon shelf break as observed by underwater television, *in* D. J. P. Swift, D. B. Duane, and O. H. Pilkey, eds., Shelf sediment transport: process and pattern: Stroudsberg, Dowden, Hutchinson and Ross, Inc., p. 621-644.

Stow, D. A. V., 1981, Laurentian fan: morphology, sediments, processes, and growth pattern: AAPG Bull., v. 65, no. 3, p. 375-393.

Stubblefield, W. L., et al., 1982, Reconnaissance in DSRV *Alvin* of a "fluvial-like" meander system in Wilmington Canyon and slump features in South Wilmington Canyon: Geology, v. 10, no. 1, p. 31-36.

Sullwold, H. H., Jr., 1960, Tarzana fan, deep submarine fan of Late Miocene age, Los Angeles County, California: AAPG Bull., v. 44, no. 4, p. 433-457.

Swift, D. J. P., 1970, Quaternary shelves and the return to grade: Mar. Geol., v. 8, no. 1, p. 5-30.

Thomas, A. N., P. J. Walmsley, and D. A. L. Jenkins, 1974, Forties field, North Sea: AAPG Bull., v. 58, no. 3, p. 396-406.

Vail, P. R., R. M. Mitchum, Jr., and S. Thompson III, 1977, Seismic stratigraphy and global changes of sea level, Part 4: Global cycles of relative changes of sea level, *in* C. E. Payton, ed., Seismic stratigraphy— applications to hydrocarbon exploration, AAPG Mem. 26, p. 83-97.

Vormelker, R. S., 1980, Texas' Middle Wilcox channel: deep exploration potential: Oil and Gas Jour., v. 78, no. 10, p. 136-154.

Walker, R. G., 1978, Deep-water sandstone facies and ancient submarine fans: models for exploration for stratigraphic traps: AAPG Bull., v. 62, no. 6, p. 932-966.

_______ and E. Mutti, 1973, Turbidite facies and facies associations, *in* G. V. Middleton and A. H. Bouma, eds.,

Turbidites and deep-water sedimentation: Anaheim, California, SEPM Pacific Section Short Course, p. 119-157.

Walmsley, P. J., 1975, The Forties field, *in* A. W. Woodland, ed., Petroleum and the continental shelf of northwest Europe, 1. Geology: London, Applied Sci. Pubs., p. 477-484.

Webb, G. W., 1981, Stevens and earlier Miocene turbidite sandstones, southern San Joaquin valley, California: AAPG Bull., v. 65, no. 3, p. 438-467.

Wilde, P., 1965, Recent sediments of the Monterey deep-sea fan: Univ. California, Berkeley, Hydrol. Eng. Lab. Report HEL 2-13, 153p.

———, W. R. Normark, and T. E. Chase, 1978, Channel sands and petroleum potential of Monterey deep-sea fan, California: AAPG Bull., v. 62, no. 6, p. 967-983.

Winn, R. D., Jr., and R. H. Dott, Jr., 1977, Large-scale traction-produced structures in deep-water fan-channel conglomerates in southern Chile: Geology, v. 6, no. 1, p. 41-44.